Vaccines and Autoimmunity

Vaccines and Autoimmunity

EDITED BY

Yehuda Shoenfeld
Zabludowicz Center for Autoimmune Diseases
Sheba Medical Center
Tel Hashomer, Israel

Sackler Faculty of Medicine
Tel Aviv University
Tel Aviv, Israel

Nancy Agmon-Levin
Zabludowicz Center for Autoimmune Diseases
Sheba Medical Center
Tel Hashomer, Israel

Sackler Faculty of Medicine
Tel Aviv University
Tel Aviv, Israel

Lucija Tomljenovic
Neural Dynamics Research Group
University of British Columbia
Vancouver, BC, Canada

WILEY Blackwell

Published by John Wiley & Sons, Inc., Hoboken, New Jersey
Published simultaneously in Canada

Library of Congress Cataloging-in-Publication Data:

Vaccines and autoimmunity / edited by Yehuda Shoenfeld, Nancy Agmon-Levin and Lucija Tomljenovic.
p. ; cm.
Includes bibliographical references and index.
ISBN 978-1-118-66343-1 (cloth)
I. Shoenfeld, Yehuda, editor. II. Agmon-Levin, Nancy, editor.
III. Tomljenovic, Lucija, editor.
[DNLM: 1. Vaccines–immunology. 2. Adjuvants, Immunologic–adverse effects. 3. Autoimmunity. 4. Drug Discovery. 5. Vaccines–adverse effects. QW 805]
RA638
615.3′72–dc23

2015006774

Typeset in 8.75/11pt MeridienLTStd by SPi Global, Chennai, India

M WEP278171 160424

Contents

Contributors

Jacob N. Ablin
Department of Rheumatology
Tel Aviv Sourasky Medical Center and Sackler Faculty of Medicine
Tel Aviv University
Tel Aviv, Israel

Nancy Agmon-Levin
Zabludowicz Center for Autoimmune Diseases
Sheba Medical Center
Tel Hashomer, Israel

Sackler Faculty of Medicine
Tel Aviv University
Tel Aviv, Israel

Howard Amital
Department of Medicine B
Sheba Medical Center
Tel Hashomer, Israel

Sackler Faculty of Medicine
Tel Aviv University
Tel Aviv, Israel

Juan-Manuel Anaya
Center for Autoimmune Diseases Research (CREA)
School of Medicine and Health Sciences
Del Rosario University
Bogotá, Colombia

Alessandro Antonelli
Department of Clinical and Experimental Medicine
University of Pisa
Pisa, Italy

María-Teresa Arango
Zabludowicz Center for Autoimmune Diseases
Sheba Medical Center
Tel Hashomer, Israel

Doctoral Program in Biomedical Sciences
Del Rosario University
Bogotá, Colombia

François-Jérôme Authier
Faculty of Medicine
University of Paris East
Paris France

Neuromuscular Center
H. Mondor Hospital
Paris, France

Tadej Avčin
Department of Allergology
Rheumatology and Clinical Immunology
University Children's Hospital
University Medical Centre Ljubljana
Ljubljana, Slovenia

Nicola Bassi
Division of Rheumatology
Department of Medicine
University of Padua
Padua, Italy

Sharon Baum
Department of Dermatology
Sheba Medical Center
Tel Hashomer, Israel

Rotem Baytner-Zamir
Department of Medicine E, Meir Medical Center
Kfar Saba, Israel

Sackler Faculty of Medicine
Tel Aviv University
Tel Aviv, Israel

Luigi Bernini
Rheumatology Unit
Department of Internal Medicine

University of Modena and Reggio Emilia
Medical School
Modena, Italy

Miri Blank
Zabludowicz Center for Autoimmune Diseases
Sheba Medical Center
Tel Hashomer, Israel

Dimitrios P. Bogdanos
Institute of Liver Studies
King's College London School of Medicine
King's College Hospital
London, UK

Department of Medicine
School of Health Sciences
University of Thessaly
Larissa, Greece

Eloisa Bonfá
Division of Rheumatology
Children's Institute Faculty of Medicine
University of São Paulo
São Paulo, Brazil

Elisabetta Borella
Division of Rheumatology
Department of Medicine
University of Padua, Padua
Italy

Zabludowicz Center for
Autoimmune Diseases
Sheba Medical Center
Tel Hashomer, Israel

Dan Buskila
Rheumatic Disease Unit
Department of Medicine
Soroka Medical Center
Beersheba, Israel

Josette Cadusseau
Faculty of Medicine
University of Paris East
Paris, France

John Castiblanco
Center for Autoimmune Diseases Research
(CREA)
School of Medicine and Health Sciences
Del Rosario University
Bogotá, Colombia

Joab Chapman
Zabludowicz Center for Autoimmune Diseases
and Department of Neurology
Sheba Medical Center
Tel Hashomer, Israel

Paola Cruz-Tapias
Doctoral Program in Biomedical Sciences
Del Rosario University
Bogotá, Colombia

Andrea Di Domenicantonio
Department of Clinical and Experimental
Medicine
University of Pisa
Pisa, Italy

Pilar Cruz Dominguez
Research Division
Hospital de Especialidades
"Dr Antonio Fraga Mouret,"
Mexican Social Security Institute
National Autonomous University of Mexico
Mexico City, Mexico

Andrea Doria
Division of Rheumatology
Department of Medicine
University of Padua
Padua, Italy

Poupak Fallahi
Department of Clinical and Experimental
Medicine
University of Pisa
Pisa, Italy

Ele Ferrannini
Department of Clinical and Experimental
Medicine
University of Pisa
Pisa, Italy

Silvia Martina Ferrari
Department of Clinical and Experimental
Medicine
University of Pisa
Pisa, Italy

Clodoveo Ferri
Rheumatology Unit
Department of Internal Medicine
University of Modena and Reggio Emilia
Medical School
Modena, Italy

Mariele Gatto
Division of Rheumatology
Department of Medicine
University of Padua
Padua, Italy

Romain K. Gherardi
Faculty of Medicine
University of Paris East
Paris, France

Neuromuscular Center H. Mondor Hospital
Paris, France

Anna Ghirardello
Division of Rheumatology
Department of Medicine
University of Padua
Padua, Italy

Eitan Giat
Rheumatology Unit
Sheba Medical Center
Tel Hashomer, Israel

Gili Givaty
Zabludowicz Center for Autoimmune Diseases
Department of Neurology and Sagol
Neuroscience Center
Sheba Medical Center
Tel Hashomer, Israel

Carla Gonçalves
Division of Rheumatology
Children's Institute, Faculty of Medicine
University of São Paulo
São Paulo, Brazil

Rotem Inbar
Zabludowicz Center for Autoimmune Diseases
Sheba Medical Center
Tel Hashomer, Israel

Eitan Israeli
Zabludowicz Center for Autoimmune Diseases
Sheba Medical Center
Tel Hashomer, Israel

Luis J. Jara
Direction of Education and Research
Hospital de Especialidades "Dr Antonio Fraga
Mouret," Mexican Social Security Institute
National Autonomous University of Mexico
Mexico City, Mexico

Dimitrios Karussis
Department of Neurology
Multiple Sclerosis Center and Laboratory of
Neuroimmunology
The Agnes-Ginges Center for Neurogenetics
Hadassah University Hospital
Jerusalem, Ein Karem, Israel

Nurit Katz-Agranov
Department of Medicine
Wolfson Medical Center
Tel Aviv, Israel

Shaye Kivity
Zabludowicz Center for Autoimmune Diseases
Rheumatic Disease Unit
and The Dr Pinchas Borenstein Talpiot Medical
Leadership Program 2013
Sheba Medical Center
Tel Hashomer, Israel

Aaron Lerner
Pediatric Gastroenterology and Nutrition Unit
Carmel Medical Center
B. Rappaport School of Medicine
Technion – Israel Institute of Technology
Haifa, Israel

Roger A. Levy
Faculty of Medical Sciences
Rio de Janeiro State University
Rio de Janeiro, Brazil

Yair Levy
Department of Medicine E
Meir Medical Center
Kfar Saba, Israel

Sackler Faculty of Medicine
Tel Aviv University, Tel Aviv, Israel

Merav Lidar
Rheumatology Unit
Sheba Medical Center
Tel Hashomer, Israel

Sackler Faculty of Medicine
Tel Aviv University
Tel Aviv, Israel

Hussein Mahagna
Department of Medicine B
Sheba Medical Center
Tel Hashomer, Israel

Sackler Faculty of Medicine
Tel Aviv University, Tel Aviv, Israel

Naim Mahroum
Department of Medicine B
Sheba Medical Center
Tel Hashomer, Israel

Sackler Faculty of Medicine
Tel Aviv University, Tel Aviv, Israel

Raffaele Manna
Periodic Fevers Research Center
Department of Internal Medicine
Catholic University of the Sacred Heart
Rome, Italy

Carlo Umberto Manzini
Rheumatology Unit
Department of Internal Medicine
University of Modena and Reggio Emilia
Medical School
Modena, Italy

Maria Martinelli
Zabludowicz Center for Autoimmune Diseases
Sheba Medical Center
Tel Hashomer, Israel

Rheumatology Division, Department of Medicine
University of Brescia
Brescia, Italy

Gabriela Medina
Clinical Epidemiological Research Unit
Hospital de Especialidades "Dr Antonio Fraga Mouret,"
Mexican Social Security Institute
National Autonomous University of Mcxico
Mexico City, Mexico

Quan M. Nhu
The W. Harry Feinstone Department of Molecular Microbiology and Immunology
Center for Autoimmune Disease Research, and Department of Pathology
The Johns Hopkins Medical Institutions
Baltimore, MD, USA

Giovanna Passaro
Periodic Fevers Research Center
Department of Internal Medicine
Catholic University of the Sacred Heart
Rome, Italy

Carlo Perricone
Rheumatology, Department of Internal and Specialized Medicine
Sapienza University of Rome
Rome, Italy

Roberto Perricone
Rheumatology, Allergology, and Clinical Immunology
Department of Internal Medicine
University of Rome Tor Vergata
Rome, Italy

Panayiota Petrou
Department of Neurology, Multiple Sclerosis Center, and Laboratory of Neuroimmunology
The Agnes-Ginges Center for Neurogenetics
Hadassah University Hospital
Jerusalem, Israel

Rodrigo Poubel V. Rezende
Faculty of Medical Sciences
Rio de Janeiro State University
Rio de Janeiro, Brazil
Brazilian Society of Rheumatology
Rio de Janeiro, Brazil

Maurizio Rinaldi
Rheumatology, Allergology, and Clinical Immunology
Department of Internal Medicine
University of Rome Tor Vergata
Rome, Italy

Ignasi Rodriguez-Pintó
Department of Autoimmune Disease
Hospital Clínic de Barcelona
Barcelona, Spain

Noel R. Rose
The W. Harry Feinstone Department of Molecular Microbiology and Immunology
Center for Autoimmune Disease Research, and Department of Pathology
The Johns Hopkins Medical Institutions
Baltimore, MD, USA

Schahin Saad
Division of Rheumatology
Children's Institute
Faculty of Medicine
University of São Paulo
São Paulo, Brazil

Miguel A. Saavedra
Department of Rheumatology
Hospital de Especialidades "Dr Antonio Fraga Mouret" Mexican Social Security Institute
National Autonomous University of Mexico
Mexico City, Mexico

Lazaros I. Sakkas
Department of Medicine
School of Health Sciences
University of Thessaly
Larissa, Greece

Minoru Satoh
School of Health Sciences
University of Occupational and Environmental Health
Kitakyushu, Japan

Christopher A. Shaw
Department of Ophthalmology and Visual Sciences
Programs in Experimental Medicine and Neuroscience
University of British Columbia
Vancouver, BC, Canada

Yehuda Shoenfeld
Zabludowicz Center for Autoimmune Diseases
Sheba Medical Center
Tel Hashomer, Israel
Sackler Faculty of Medicine
Tel Aviv University
Tel Aviv, Israel

Clóvis A. Silva
Pediatric Rheumatology Unit
Children's Institute, Faculty of Medicine
University of São Paulo
São Paulo, Brazil

Daniel S. Smyk
Institute of Liver Studies
King's College London School of Medicine
King's College Hospital
London, UK

Alessandra Soriano
Zabludowicz Center for Autoimmune Diseases
Sheba Medical Center
Tel Hashomer, Israel

Department of Clinical Medicine and Rheumatology
Campus Bio-Medico University
Rome, Italy

Vera Stejskal
Department of Immunology
University of Stockholm
Stockholm, Sweden

Lucija Tomljenovic
Neural Dynamics Research Group
University of British Columbia
Vancouver, BC, Canada

Nataša Toplak
Department of Allergology
Rheumatology and Clinical Immunology
University Children's Hospital
University Medical Centre Ljubljana
Ljubljana, Slovenia

Guido Valesini
Rheumatology, Department of Internal and Specialized Medicine
Sapienza University of Rome
Rome, Italy

Mónica Vázquez del Mercado
Institute of Research in Rheumatology and Musculoloeskeletal System
Hospital Civil JIM
University of Guadalajara
Jalisco, Mexico

Olga Vera-Lastra
Department of Internal Medicine
Hospital de Especialidades "Dr Antonio Fraga Mouret," Mexican Social Security Institute
National Autonomous University of Mexico
Mexico City, Mexico

Abdulla Watad
Zabludowicz Center for Autoimmune Diseases and Department of Internal Medicine B
Sheba Medical Center
Tel Hashomer, Israel

Yaron Zafrir
Department of Dermatology and Zabludowicz Center for Autoimmune Diseases
Sheba Medical Center
Tel Hashomer, Israel

Gisele Zandman-Goddard
Department of Medicine
Wolfson Medical Center
Tel Aviv, Israel

Sackler Faculty of Medicine
Tel Aviv University
Tel Aviv, Israel

Introduction

Yehuda Shoenfeld,[1,2] Nancy Agmon-Levin,[1,4] and Lucija Tomljenovic[3]

[1]Zabludowicz Center for Autoimmune Diseases, Sheba Medical Center, Tel Hashomer, Israel
[2]Incumbent of the Laura Schwarz-Kipp Chair for Research of Autoimmune Diseases, Sackler Faculty of Medicine, Tel Aviv University, Tel Aviv, Israel
[3]Neural Dynamics Research Group, University of British Columbia, Vancouver, BC, Canada
[4]Sackler Faculty of Medicine, Tel Aviv University, Tel Aviv, Israel

Vaccines and Autoimmunity is a result of decades of experience in vaccinology, immunology, and autoimmunity, and of a review of the vast literature in this field. The book has three parts. Part I deals with general mechanisms of vaccine- and adjuvant-induced autoimmunity. In Parts II and III, we have asked the different authors to summarize, on one hand, individual vaccines and which common autoimmune diseases they may trigger in susceptible individuals (Part III), and on the other, the common autoimmune diseases and identified vaccines which may trigger their emergence (Part III).

The editors of this book are quite confident that vaccinations represent one of the most remarkable revolutions in medicine. Indeed, vaccines have been used for over 300 years and are probably one of the most effective strategies for preventing the morbidity and mortality associated with infections. Like other drugs, vaccines can cause adverse events, but unlike conventional drugs, which are prescribed to people who are ill, vaccines are administered to healthy individuals, which increases the concern over adverse reactions. Most side effects attributed to vaccines are mild, acute, and transient. Nonetheless, rare reactions, such as hypersensitivity and induction of autoimmunity, do occur, and can be severe and even fatal. In this regard, the fact that vaccines are delivered to billions of people without preliminary screening for underlying susceptibilities is thus of concern (Bijl *et al.*, 2012; Tomljenovic and Shaw, 2012; Soriano *et al.*, 2014).

Indeed, it is naive to believe that all humans are alike. Notably, autoimmune diseases have been increasingly recognized as having a genetic basis, mediated by HLA subtypes. For instance, celiac disease has been strongly associated with HLA haplotype DR3-DQ2 or DR4-DQ8 (Liu *et al.*, 2014), multiple sclerosis with HLA-DRB1 (Yates *et al.*, 2014), rheumatoid arthritis with HLA-DR4 and HLA-DQ8 (Vassallo *et al.*, 2014), and type I diabetes with HLA-DR3/4 (Steck *et al.*, 2014). Thus, certain HLA genes create a genetic predisposition toward development of autoimmune disease, typically requiring some environmental trigger to evolve into a full-blown disease state (Luckey *et al.*, 2011). One such environmental trigger which is commonly associated with development of autoimmunity is viral (Epstein Barr virus, cytomegalovirus, and hepatitis C virus) or bacterial (*Heliobacter pylori*) challenge (Rose, 2010; Magen and Delgado, 2014).

The multifacet associations between infectious agents and subsequent development of autoimmune or autoinflammatory conditions have been well established, and a number of mechanisms by which infectious agents can bring about such responses have been identified (molecular mimicry, epitope spreading, polyclonal activation, and others) (Molina and Shoenfeld, 2005; Kivity *et al.*, 2009; Shoenfeld, 2009; Rose, 2010).

Vaccines and Autoimmunity, First Edition. Edited by Yehuda Shoenfeld, Nancy Agmon-Levin, and Lucija Tomljenovic.

Recently, we and others have suggested another mechanism, namely the adjuvant effect, by which infections may relate to autoimmunity in a broader sense (Rose, 2010; Rosenblum *et al.*, 2011; Shoenfeld and Agmon-Levin, 2011; Zivkovic *et al.*, 2012; Perricone *et al.*, 2013). Adjuvants are substances which enhance the immune response. For this purpose, they are routinely included in vaccine formulations, the most common of which are aluminum compounds (alum hydroxide and phosphate). Although the mechanisms of adjuvancy are not fully elucidated, adjuvants seem to modulate a common set of genes, promote antigen-presenting cell recruitment, and mimic specific sets of conserved molecules, such as bacteria components, thus increasing the innate and adaptive immune responses to the injected antigen (Agmon-Levin *et al.*, 2009; Israeli *et al.*, 2009; McKee *et al.*, 2009; Exley *et al.*, 2010; Perricone *et al.*, 2013).

Although the activation of autoimmune mechanisms by both infectious agents and substances with adjuvant properties (such as those found in vaccines) is common, the appearance of an autoimmune disease is not as widespread and apparently not always agent-specific. The adjuvant effect of microbial particles, namely the nonantigenic activation of the innate and regulatory immunity, as well as the expression of various regulatory cytokines, may determine if an autoimmune response remains limited and harmless or evolves into a full-blown disease. Additionally, as already mentioned, the genetic background of an individual may determine the magnitude of adverse manifestations. For example, it has been shown that the vaccine for Lyme disease is capable of triggering arthritis in genetically susceptible hamsters and that, when the adjuvant aluminum hydroxide is added to the vaccine, 100% of the hamsters develop arthritis (Croke *et al.*, 2000). Other studies have shown that the development of inflammatory joint disease and rheumatoid arthritis in adults in response to the HepA and HepB vaccines, respectively, is correlated to the HLA subtype of the vaccinated individual (Ferrazzi *et al.*, 1997; Pope *et al.*, 1998). Given that aluminum works as an adjuvant by increasing expression of MHC (Ulanova *et al.*, 2001), it perhaps should not be surprising that in individuals susceptible to autoimmune disease on the basis of the MHC, HLA subtype might be adversely affected by the use of aluminum hydroxide in vaccines. In addition to aluminum, the vaccine preservative thimerosal has also been demonstrated to induce a systematic autoimmune syndrome in transgenic HLA-DR4 mice (Havarinasab *et al.*, 2004), while mice with a genetic susceptibility for autoimmune disease show profound behavioral and neuropathological disturbances. These results are not observed in strains of mice without autoimmune sensitivity.

We have recently reported a new syndrome: "autoimmune/inflammatory syndrome induced by adjuvants" (ASIA), which encompasses a spectrum of immune-mediated diseases triggered by an adjuvant stimulus such as chronic exposure to silicone, tetramethylpentadecane, pristane, aluminum, and other adjuvants, as well as infectious components, which may also have an adjuvant effect. All these environmental factors have been found to induce autoimmunity and inflammatory manifestations by themselves, both in animal models and in humans (Israeli *et al.*, 2009; Shaw and Petrik, 2009; Shoenfeld and Agmon-Levin, 2011; Gherardi and Authier, 2012; Israeli, 2012; Cruz-Tapias *et al.*, 2013; Lujan *et al.*, 2013; Perricone *et al.*, 2013).

The definition of the ASIA syndrome thus helps to detect those subjects who have developed autoimmune phenomena upon exposure to adjuvants from different sources. For example, the use of medical adjuvants has become common practice, and substances such as aluminum adjuvant are added to most human and animal vaccines, while the adjuvant silicone is extensively used for breast implants and cosmetic procedures (Kaiser *et al.*, 1990; Molina and Shoenfeld, 2005; Israeli *et al.*, 2009; Shoenfeld and Agmon-Levin, 2011; Cohen Tervaert and Kappel, 2013). Furthermore, "hidden adjuvants" such as infectious material and house molds have also been associated with different immune-mediated conditions associated with the so-called "sick-building syndrome" (Israeli and Pardo, 2010; Perricone *et al.*, 2013).

Although ASIA may be labeled a "new syndrome," in reality it reflects old truths given a formal label (Meroni, 2010). Notably, in 1982, compelling evidence from epidemiological, clinical, and animal research emerged to show that Guillain-Barre syndrome and other demyelinating autoimmune neuropathies (i.e., acute disseminated encephalomyelitis and multiple sclerosis) could occur up to 10 months following vaccination (Poser and Behan, 1982). In such cases, the disease would first manifest with vague symptoms (arthralgia, myalgia, paraesthesia, weakness; all of which are typical ASIA symptoms), which were frequently deemed insignificant and thus ignored by the treating physicians. However, these symptoms would progress slowly and insidiously until the patient was exposed to a secondary immune stimulus (in the form of either infection or vaccination). This would then trigger the rapid and acute clinical manifestation of the disease (Poser and Behan, 1982). In other words, it was the secondary anamnestic response that would bring about the acute overt manifestation of an already present subclinical long-term persisting disease.

Thus, it was already recognized in the early 1980s that vaccine-related manifestations often presented themselves as unspecific, yet clinically relevant symptoms (termed "bridging symptoms" Poser and Behan (1982) or "nonspecific ASIA symptoms" by us (Shoenfeld and Agmon-Levin, 2011)). These manifestations pointed to a sub-clinical, slowly evolving disease. Whether this disease would eventually progress to its full-blown clinically apparent form depended on whether the individual was further exposed to noxious immune stimuli, including subsequent vaccinations. As a case in point, we recently described six cases of systemic lupus following HPV vaccination (Gatto *et al.*, 2013). In all six cases, several common features were observed; namely, a personal or familial susceptibility to autoimmunity and an adverse response to a prior dose of the vaccine, both of which were associated with a higher risk of post-vaccination full-blown autoimmunity. Similarly, in an analysis of 93 cases of autoimmunity following hepatitis B vaccination (Zafrir *et al.*, 2012), we identified two major susceptibility factors: (i) exacerbation of adverse symptoms following additional doses of the vaccine (47% of patients); and (ii) personal and familial history of autoimmunity (21%).

It should further be noted that some individuals who are adversely afflicted through exposure to adjuvants do not satisfy all of the criteria that are necessary to diagnose a full-blown and clinically apparent autoimmune disease (Perricone *et al.*, 2013). Nonetheless, these individuals are at higher risk of developing full-blown autoimmunity following subsequent adjuvant exposure, whether that be via infections or vaccinations (Poser and Behan, 1982; Zafrir *et al.*, 2012; Gatto *et al.*, 2013).

A casual glance at the US Centers for Disease Control and Prevention (CDC, 2013)_immunization schedule for infants shows that according to the US prescribed guidelines, children receive up to 19 vaccinations during infancy, many of which are multivalent in the first 6 months of their life (Table I.1).

The various vaccines given to children, as well as adults, may contain either whole weakened infectious agents or synthetic peptides and genetically engineered antigens of infectious agents and adjuvants (typically aluminum). In addition, they also contain diluents, preservatives (thimerosal, formaldehyde), detergents (polysorbate), and residuals of culture growth media (*Saccharomyces cerevisiae*, gelatin, bovine extract, monkey kidney tissue, etc.; Table I.2). The safety of these residuals has not been thoroughly investigated, primarily because they are presumed to be present only in trace amounts following the vaccine manufacture purification process. However, some studies

Table I.1 Typical pediatric vaccine schedule for preschool children currently recommended by the US Centers for Disease Control and Prevention (2013a). Shaded boxes indicate the age range in which the vaccine can be given. Asterisks denote Al-adjuvanted vaccines. Hep A is given in 2 doses spaced at least 6 months apart. According to this schedule, by the time a child is 2 years of age, they would have received 27 vaccinations (3 × HepB, 3 × Rota, 4 × DTaP, 4 × Hib, 4 × PCV, 3 × IPV, 2 × Influenza, 1 × MMR, 1 × Varicella, and 2 × HepA)

Birth	1 month	2 months	4 months	6 months	12 months	15 months	18 months	19–23 months	2–3 years	4–6 years
HepB*	HepB*				HepB*					
		Rota	Rota	Rota						
		DTaP*	DTaP*	DTaP*		DTaP*				DTaP*
		Hib*	Hib*	Hib*	Hib*					
		PCV*	PCV*	PCV*	PCV*					
		IPV	IPV		IPV					IPV
							Influenza (yearly)			
					MMR					MMR
					Varicella					Varicella
							HepA*			

Hep A, hepatitis A; Hep B, hepatitis B; Rota, rotavirus; DTaP, diphtheria-pertussis-tetanus; Hib, *Haemophilus influenzae* type b; PCV, pneumococcal; IPV, inactivated polio; MMR, measles-mumps-rubella

Table I.2 Complete list of vaccine ingredients (i.e., adjuvants and preservatives) and substances used during the manufacture of commonly used vaccines. Adapted from US Centers for Disease Control and Prevention (2013b)

Vaccine	Vaccine excipient and media summary
DT (Sanofi)	aluminum potassium sulfate, peptone, bovine extract, formaldehyde, thimerosal (trace), modified Mueller and Miller medium
DTaP (Daptacel)	aluminum phosphate, formaldehyde, glutaraldehyde, 2-phenoxyethanol, Stainer–Scholte medium, modified Mueller's growth medium, modified Mueller–Miller casamino acid medium (without beef heart infusion)
DTaP (Infanrix)	formaldehyde, glutaraldehyde, aluminum hydroxide, polysorbate 80, Fenton medium (containing bovine extract), modified Latham medium (derived from bovine casein), modified Stainer–Scholte liquid medium
DTaP (Tripedia)	sodium phosphate, peptone, bovine extract (US sourced), formaldehyde, ammonium sulfate, aluminum potassium sulfate, thimerosal (trace), gelatin, polysorbate 80 (Tween 80), modified Mueller and Miller medium, modified Stainer–Scholte medium
DTaP-HepB-IPV (Pediarix)	formaldehyde, gluteraldehyde, aluminum hydroxide, aluminum phosphate, lactalbumin hydrolysate, polysorbate 80, neomycin sulfate, polymyxin B, yeast protein, calf serum, Fenton medium (containing bovine extract), modified Latham medium (derived from bovine casein), modified Stainer-Scholte liquid medium, Vero (monkey kidney) cells
DTaP-IPV/Hib (Pentacel)	aluminum phosphate, polysorbate 80, formaldehyde, gutaraldehyde, bovine serum albumin, 2-phenoxethanol, neomycin, polymyxin B sulfate, Mueller's Growth Medium, Mueller–Miller casamino acid medium (without beef heart infusion), Stainer–Scholte medium (modified by the addition of casamino acids and dimethyl-beta-cyclodextrin), MRC-5 (human diploid) cells, CMRL 1969 medium (supplemented with calf serum)
Hib (ActHIB)	ammonium sulfate, formalin, sucrose, Modified Mueller and Miller medium
Hib (Hiberix)	formaldehyde, lactose
Hib (PedvaxHIB)	aluminum hydroxphosphate sulfate
Hib/Hep B (Comvax)	yeast (vaccine contains no detectable yeast DNA), nicotinamide adenine dinucleotide, hemin chloride, soy peptone, dextrose, mineral salts, amino acids, formaldehyde, potassium aluminum sulfate, amorphous aluminum hydroxyphosphate sulfate, sodium borate
Hep A (Havrix)	aluminum hydroxide, amino acid supplement, polysorbate 20, formalin, neomycin sulfate, MRC-5 cellular proteins
Hep A (Vaqta)	amorphous aluminum hydroxyphosphate sulfate, bovine albumin, formaldehyde, neomycin, sodium borate, MRC-5 (human diploid) cells
Hep B (Engerix-B)	aluminum hydroxide, yeast protein, phosphate buffers
Hep B (Recombivax)	yeast protein, soy peptone, dextrose, amino acids, mineral salts, potassium aluminum sulfate, amorphous aluminum hydroxyphosphate sulfate, formaldehyde
Hep A/Hep B (Twinrix)	formalin, yeast protein, aluminum phosphate, aluminum hydroxide, amino acids, phosphate buffer, polysorbate 20, neomycin sulfate, MRC-5 human diploid cells
Human Papillomavirus (HPV) (Cerverix)	vitamins, amino acids, lipids, mineral salts, aluminum hydroxide, sodium dihydrogen phosphate dehydrate, insect cell and viral protein
Human Papillomavirus (HPV) (Gardasil)	yeast protein, vitamins, amino acids, mineral salts, carbohydrates, amorphous aluminum hydroxyphosphate sulfate, L-histidine, polysorbate 80, sodium borate
Influenza (Afluria)	beta-propiolactone, thimerosol (multi-dose vials only), monobasic sodium phosphate, dibasic sodium phosphate, monobasic potassium phosphate, potassium chloride, calcium chloride, sodium taurodeoxycholate, neomycin sulfate, polymyxin B, egg protein
Influenza (Fluarix)	sodium deoxycholate, formaldehyde, octoxynol-10 (Triton X-100), α-tocopheryl hydrogen succinate, polysorbate 80 (Tween 80), hydrocortisone, gentamicin sulfate, ovalbumin
Influenza (Fluvirin)	nonylphenol ethoxylate, thimerosal (multidose vial–trace only in prefilled syringe), polymyxin, neomycin, beta-propiolactone, egg proteins
Influenza (Flulaval)	thimerosal, α-tocopheryl hydrogen succinate, polysorbate 80, formaldehyde, sodium deoxycholate, ovalbumin
Influenza (Fluzone: standard, high-dose, & intradermal)	formaldehyde, octylphenol ethoxylate (Triton X-100), sodium phosphate, gelatin (standard formulation only), thimerosal (multidose vial only), egg protein
Influenza (FluMist)	ethylene diamine tetraacetic acid (EDTA), monosodium glutamate, hydrolyzed porcine gelatin, arginine, sucrose, dibasic potassium phosphate, monobasic potassium phosphate, gentamicin sulfate, egg protein

Table I.2 *(Continued)*

Vaccine	Vaccine excipient and media summary
Meningococcal (MCV4Menactra)	formaldehyde, phosphate buffers, Mueller Hinton agar, Watson Scherp media, Modified Mueller and Miller medium
Meningococcal (MCV4Menveo)	formaldehyde, amino acids, yeast extract, Franz complete medium
Meningococcal (MPSV4Menomune)	thimerosal (multidose vial only), lactose, Mueller Hinton agar, Watson Scherp media
MMR (MMR-II)	vitamins, amino acids, fetal bovine serum, sucrose, sodium phosphate, glutamate, recombinant human albumin, neomycin, sorbitol, hydrolyzed gelatin, chick embryo cell culture, WI-38 human diploid lung fibroblasts
MMRV (ProQuad)	sucrose, hydrolyzed gelatin, sorbitol, monosodium L-glutamate, sodium phosphate dibasic, human albumin, sodium bicarbonate, potassium phosphate monobasic, potassium chloride, potassium phosphate dibasic, neomycin, bovine calf serum, chick embryo cell culture, WI-38 human diploid lung fibroblasts, MRC-5 cells
Pneumococcal (PCV13 – Prevnar 13)	casamino acids, yeast, ammonium sulfate, Polysorbate 80, succinate buffer, aluminum phosphate
Polio (IPV – Ipol)	2-phenoxyethanol, formaldehyde, neomycin, streptomycin, polymyxin B, monkey kidney cells, Eagle MEM modified medium, calf serum protein
Rabies (Imovax)	albumin, neomycin sulfate, phenol, MRC-5 human diploid cells
Rabies (RabAvert)	β-propiolactone, potassium glutamate, chicken protein, ovalbumin, neomycin, chlortetracycline, amphotericin B, human serum albumin, polygeline (processed bovine 14 gelatin)
Rotavirus (RotaTeq)	sucrose, sodium citrate, sodium phosphate monobasic monohydrate, sodium hydroxide, polysorbate 80, cell culture media, fetal bovine serum, vero cells (DNA from porcine circoviruses (PCV) 1 and 2 has been detected in RotaTeq; PCV-1 and PCV-2 are not known to cause disease in humans)
Rotavirus (Rotarix)	amino acids, dextran, sorbitol, sucrose, calcium carbonate, xanthan, Dulbecco's Modified Eagle Medium (DMEM) (Porcine circovirus type 1 (PCV-1) is present in Rotarix; PCV-1 is not known to cause disease in humans)
Td (Decavac)	aluminum potassium sulfate, peptone, formaldehyde, thimerosal, bovine muscle tissue (US sourced), Mueller and Miller medium
Td (Tenivac)	aluminum phosphate, formaldehyde, modified Mueller–Miller casamino acid medium without beef heart infusion
Td (Mass Biologics)	aluminum phosphate, formaldehyde, thimerosal (trace), ammonium phosphate, modified Mueller's media (containing bovine extracts)
Tdap (Adacel)	aluminum phosphate, formaldehyde, glutaraldehyde, 2-phenoxyethanol, ammonium sulfate, Mueller's growth medium, Mueller–Miller casamino acid medium (without beef heart infusion)
Tdap (Boostrix)	formaldehyde, glutaraldehyde, aluminum hydroxide, polysorbate 80 (Tween 80), Latham medium derived from bovine casein, Fenton medium containing a bovine extract, Stainer–Scholte liquid medium
Typhoid (inactivated – Typhim Vi)	hexadecyltrimethylammonium bromide, phenol, polydimethylsiloxane, disodium phosphate, monosodium phosphate
Typhoid (oral – Ty21a)	yeast extract, casein, dextrose, galactose, sucrose, ascorbic acid, amino acids
Varicella (Varivax)	sucrose, phosphate, glutamate, gelatin, monosodium L-glutamate, sodium phosphate dibasic, potassium phosphate monobasic, potassium chloride, sodium phosphate monobasic, EDTA, residual components of MRC-5 cells including DNA and protein, neomycin, fetal bovine serum, human diploid cell cultures
Yellow Fever (YF-Vax)	sorbitol, gelatin, egg protein
Zoster (Shingles – Zostavax)	sucrose, hydrolyzed porcine gelatin, monosodium L-glutamate, sodium phosphate dibasic, potassium phosphate monobasic, neomycin, potassium chloride, residual components of MRC-5 cells including DNA and protein, bovine calf serum

suggest that even these trace amounts may not be inherently safe, as was previously assumed (Moghaddam *et al.*, 2006; Rinaldi *et al.*, 2013).

What is obvious, nonetheless, is that a typical vaccine formulation contains all the necessary biochemical components to induce autoimmune manifestations. With that in mind, our major aim is to inform the medical community regarding the various autoimmune risks associated with different vaccines. Physicians need to be aware that in certain individuals, vaccinations can trigger serious and potentially disabling and even fatal autoimmune manifestations. This is not to say that we oppose vaccination, as it is indeed an important tool of preventative medicine. However, given the fact that vaccines are predominantly administered to previously healthy individuals, efforts should be made to identify those subjects who may be at more risk of developing adverse autoimmune events following vaccine exposure. In addition, careful assessment should be made regarding further vaccine administration in individuals with previous histories of adverse reactions to vaccinations. The necessity of multiple vaccinations over a short period of time should also be considered, as the enhanced adjuvant-like effect of multiple vaccinations heightens the risk of post-vaccine-associated adverse autoimmune and inflammatory manifestations (Tsumiyama *et al.*, 2009; Lujan *et al.*, 2013). Finally, we wish to encourage efforts toward developing safer vaccines, which should be pursued by the vaccine manufacturing industry.

References

Agmon-Levin, N., Paz, Z., Israeli, E., and Shoenfeld, Y. (2009). Vaccines and autoimmunity. *Nat Rev Rheumatol*, **5**: 648–2.

Bijl, M., Agmon-Levin, N., Dayer, J.M., *et al.* (2012). Vaccination of patients with auto-immune inflammatory rheumatic diseases requires careful benefit-risk assessment. *Autoimmun Rev*, **11**: 572–6.

CDC (Centers for Disease Control and Prevention). (2013). Recommended immunizations for children from birth through 6 years old. Available from: http://www.cdc.gov/vaccines/parents/downloads/parent-ver-sch-0-6yrs.pdf [last accessed 11 December 2014].

Cohen Tervaert, J.W. and Kappel, R.M. (2013). Silicone implant incompatibility syndrome (SIIS): a frequent cause of ASIA (Shoenfeld's syndrome). *Immunol Res*, **56**: 293–8.

Croke, C.L., Munson, E.L., Lovrich, S.D., *et al.* (2000). Occurrence of severe destructive lyme arthritis in hamsters vaccinated with outer surface protein A and challenged with Borrelia burgdorferi. *Infect Immun*, **68**: 658–3.

Cruz-Tapias, P., Agmon-Levin, N., Israeli, E., *et al.* (2013). Autoimmune/inflammatory syndrome induced by adjuvants (ASIA) – animal models as proof of concept. *Current Med Chemistry* **20**: 4030–4.

Exley, C., Siesjo, P., and Eriksson, H. (2010). The immunobiology of aluminium adjuvants: how do they really work? *Trends Immunol*, **31**: 103–9.

Ferrazzi, V., Jorgensen, C., and Sany, J. (1997). Inflammatory joint disease after immunizations. A report of two cases. *Rev Rhum Engl Ed*, **64**: 227–232.

Gatto, M., Agmon-Levin, N., Soriano, A., *et al.* (2013). Human papillomavirus vaccine and systemic lupus erythematosus. *Clin Rheumatol*, **32**: 1301–7.

Gherardi, R. and Authier, F. (2012). Macrophagic myofasciitis: characterization and pathophysiology. *Lupus*, **21**: 184–9.

Havarinasab, S., Lambertsson, L., Qvarnstrom, J., and Hultman, P. (2004). Dose–response study of thimerosal-induced murine systemic autoimmunity. *Toxicol Appl Pharmacol*, **194**: 169–79.

Israeli, E. (2012). Gulf War syndrome as a part of the autoimmune (autoinflammatory) syndrome induced by adjuvant (ASIA). *Lupus*, **21**: 190–4.

Israeli, E. and Pardo, A. (2010). The sick building syndrome as a part of the autoimmune (auto-inflammatory) syndrome induced by adjuvants. *Mod Rheumatol*, **21**: 235–9.

Israeli, E., Agmon-Levin, N., Blank, M., and Shoenfeld, Y. (2009). Adjuvants and autoimmunity. *Lupus*, **18**: 1217–25.

Kaiser, W., Biesenbach, G., Stuby, U., *et al.* (1990). Human adjuvant disease: remission of silicone induced autoimmune disease after explanation of breast augmentation. *Ann Rheum Dis*, **49**: 937–8.

Kivity, S., Agmon-Levin, N., Blank, M., and Shoenfeld, Y. (2009). Infections and autoimmunity – friends or foes? *Trends Immunol*, **30**: 409–14.

Liu, E., Lee, H.S., Aronsson, C.A., *et al.* (2014). Risk of pediatric celiac disease according to HLA haplotype and country. *N Engl J Med*, **371**: 42–9.

Luckey, D., Bastakoty, D., and Mangalam, A.K. (2011). Role of HLA class II genes in susceptibility and resistance to multiple sclerosis: studies using HLA transgenic mice. *J Autoimmun*, **37**: 122–8.

Lujan, L., Perez, M., Salazar, E., *et al.* (2013). Autoimmune/autoinflammatory syndrome induced by adjuvants (ASIA syndrome) in commercial sheep. *Immunol Res*, **56**: 317–24.

Magen, E. and Delgado, J.S. (2014). Helicobacter pylori and skin autoimmune diseases. *World J Gastroenterol*, **20**: 1510–16.

McKee, A.S., Munks, M.W., MacLeod, M.K., *et al.* (2009). Alum induces innate immune responses through macrophage and mast cell sensors, but these sensors are not required for alum to act as an adjuvant for specific immunity. *J Immunol*, **183**: 4403–14.

Meroni, P.L. (2010). Autoimmune or auto-inflammatory syndrome induced by adjuvants (ASIA): old truths and a new syndrome? *J Autoimmun*, **36**: 1–3.

Moghaddam, A., Olszewska, W., Wang, B., *et al.* (2006). A potential molecular mechanism for hypersensitivity caused by formalin-inactivated vaccines. *Nat Med*, **12**: 905–7.

Molina, V. and Shoenfeld, Y. (2005). Infection, vaccines and other environmental triggers of autoimmunity. *Autoimmunity*, **38**: 235–45.

Perricone, C., Colafrancesco, S., Mazor, R.D., *et al.* (2013). Autoimmune/inflammatory syndrome induced by adjuvants (ASIA) 2013; unveiling the pathogenic, clinical ad diagnostic aspects. *J Autoimmun*, **47**: 1–16.

Pope, J.E., Stevens, A., Howson, W., and Bell, D.A. (1998). The development of rheumatoid arthritis after recombinant hepatitis B vaccination. *J Rheumatol*, **25**: 1687–93.

Poser, C.M. and Behan, P.O. (1982). Late onset of Guillain-Barre syndrome. *J Neuroimmunol*, **3**: 27–41.

Rinaldi, M., Perricone, R., Blank, M., *et al.* 2013. Anti-Saccharomyces cerevisiae autoantibodies in autoimmune diseases: from bread baking to autoimmunity. *Clin Rev Allergy Immunol*, **45**(2): 152–61.

Rose, N.R. (2010). Autoimmunity, infection and adjuvants. *Lupus*, **19**: 354–8.

Rosenblum, H., Shoenfeld, Y., and Amital, H. (2011). The common immunogenic etiology of chronic fatigue syndrome: from infections to vaccines via adjuvants to the ASIA syndrome. *Infect Dis Clin N Am*, **25**: 851–63.

Shaw, C.A. and Petrik, M.S. (2009). Aluminum hydroxide injections lead to motor deficits and motor neuron degeneration. *J Inorg Biochem*, **103**: 1555–62.

Shoenfeld, Y. (2009). Infections, vaccines and autoimmunity. *Lupus*, **18**: 1127–8.

Shoenfeld, Y. and Agmon-Levin, N. (2011). "ASIA" – autoimmune/inflammatory syndrome induced by adjuvants. *J Autoimmun*, **36**: 4–8.

Soriano, A., Nesher, G., and Shoenfeld, Y. (2014). Predicting post-vaccination autoimmunity: who might be at risk? *Pharmacol Res*, pii: S1043-6618(15)00139-X. doi: 10.1016/j.phrs.2014.08.002. [Epub ahead of print].

Steck, A.K., Dong, F., Wong, R., *et al.* (2014). Improving prediction of type 1 diabetes by testing non-HLA genetic variants in addition to HLA markers. *Pediatr Diabetes*, **15**: 355–62.

Tomljenovic, L. and Shaw, C.A. (2012). One-size fits all? *Vaccine*, **30**: 2040.

Tsumiyama, K., Miyazaki, Y., and Shiozawa, S. (2009). Self-organized criticality theory of autoimmunity. *PLoS One*, **4**: e8382.

Ulanova, M., Tarkowski, A., Hahn-Zoric, M., and Hanson, L.A. (2001). The common vaccine adjuvant aluminum hydroxide up-regulates accessory properties of human monocytes via an interleukin-4-dependent mechanism. *Infect Immun*, **69**: 1151–9.

Vassallo, R., Luckey, D., Behrens, M., *et al.* (2014). Cellular and humoral immunity in arthritis are profoundly influenced by the interaction between cigarette smoke effects and host HLA-DR and DQ genes. *Clin Immunol*, **152**: 25–35.

Yates, R.L., Esiri, M.M., Palace, J., *et al.* (2014). The influence of HLA-DRB1*15 on motor cortical pathology in multiple sclerosis. *Neuropathol Appl Neurobiol*, doi: 10.1111/nan.12165. [Epub ahead of print].

Zafrir, Y., Agmon-Levin, N., Paz, Z., *et al.* (2012). Autoimmunity following hepatitis B vaccine as part of the spectrum of "autoimmune (auto-inflammatory) syndrome induced by adjuvants" (ASIA): analysis of 93 cases. *Lupus*, **21**: 146–52.

Zivkovic, I., Petrusic, V., Stojanovic, M., *et al.* (2012). Induction of decreased fecundity by tetanus toxoid hyper-immunization in C57BL/6 mice depends on the applied adjuvant. *Innate Immun*, **18**: 333–42.

I Mosaic of Autoimmunity

1 Role of Adjuvants in Infection and Autoimmunity

Eitan Israeli,[1] Miri Blank,[1] and Yehuda Shoenfeld[1,2]

[1]Zabludowicz Center for Autoimmune Diseases, Sheba Medical Center, Tel Hashomer, Israel

[2]Incumbent of the Laura Schwarz-Kipp Chair for Research of Autoimmune Diseases, Sackler Faculty of Medicine, Tel Aviv University, Tel Aviv, Israel

Introduction

Commonly used vaccines are a cost-effective and preventive way of promoting health, compared to the treatment of acute or chronic disease. However, not all vaccines are as efficient and easy to administer as the vaccine against smallpox (Vaccinia). Usually, upon injection of a pure antigen, the antigen is not taken up at the injection site, and an immunological reaction fails. In order to help the immune system to recognize the antigen, adjuvants are added to the antigens during the process of developing and producing a vaccine. For the last few years, researchers have been striving to elucidate the mechanisms by which adjuvants exert their immunological effects. By deciphering these mechanisms, scientists hope to design more efficient and less harmful adjuvants. As of 2013, the action mechanisms of the most used and "veteran" of adjuvants, alum, are being revealed. It seems that alum acts on multiple pathways, each of which can enhance immunological reactions to antigens independently.

Parts of this manuscript were published previously by our group (Israeli *et al.*, 2009). Permission to reuse them was granted by Sage Publications.

The different types of adjuvants

Old and novel adjuvants are currently used in human and animal vaccination programs, as well as in experimental models, some of which are listed in this section.

Aluminum salts

Aluminum salt (alum) is an inorganic reagent that carries the potential to augment immunogenicity. Alum salts include alum phosphate and alum hydroxide, which are the most common adjuvants in human vaccines. The organic compound squalene (originally obtained from shark liver oil and a biochemical precursor to steroids) is sometimes added to the preparation.

Oil-based adjuvants

Oil-based adjuvants (e.g., Freund's adjuvant, pristine, etc.) are commonly found in some formulations of veterinary vaccines. Incomplete Freund's adjuvant (IFA) contains water-in-oil emulsion, while complete Freund's adjuvant (CFA) additionally contains killed mycobacteria. The mycobacteria added to the adjuvant attract macrophages and other cells to the injection site, which enhances the immune response. Thus, CFA is usually used for the primary vaccination, while the incomplete version is applied for boosting. Some novel oil-in-water emulsions are being developed by pharmaceutical companies, such as MF59 (Novartis), AS03 (GalxoSmithKline), Advax (Vaxine Pty), and Qs-21/ISCOMs (see further on).

Virosomes

During the last 2 decades, a variety of technologies have been investigated for their ability to

Vaccines and Autoimmunity, First Edition. Edited by Yehuda Shoenfeld, Nancy Agmon-Levin, and Lucija Tomljenovic.

improve the widely used alum adjuvants (Holzeret *et al.*, 1996), which may induce local inflammation. Thus, other novel adjuvants that can also be used as antigen-carrier systems, the virosomes, have been developed. Virosomes contain a membrane-bound hemagglutinin and neuraminidase derived from the influenza virus, both of which facilitate uptake into antigen-presenting cells (APCs) and mimic the natural immune response (Gluck, 1999).

Novel and experimental adjuvants

In the search for new and safer adjuvants, several new ones have been developed by pharmaceutical companies utilizing new immunological and chemical innovations.

Toll-like receptor-related adjuvants

IC31 is a two-component synthetic adjuvant that signals through toll-like receptor (TLR)-9. This novel adjuvant is tested as of 2008 in influenza vaccine combinations (Riedlet *et al.*, 2008). Four others, ASO4, ASO2A, CPG 7907, and GM-CSF, are investigated for highly relevant vaccines, such as those against papilloma virus, hepatitis B, and malaria (Pichichero, 2008). Other TLR-dependent adjuvant candidates are as yet only in clinical development, such as RC-529 and ISS, Flagellin and TLR-agonists. AS02 and AS04 are proprietary adjuvants of GlaxoSmithKline (GSK). AS02 contains MPL and QS-21 in an oil-in-water emulsion. AS04 combines MPL with alum. MPL is a series of 4′monophosphoryl lipid A that varies in the extent and position of fatty acid substitution. It is prepared from lipopolysaccharide (LPS) of *Salmonella minnesota* R595 by treating the LPS with mild acid and base hydrolysis, followed by purification of the modified LPS. Unmethylated CpG dinucleotides are the reason why bacterial DNA, but not vertebrate DNA, is immunostimulatory. Vertebrate DNA has relatively low amounts of unmethylated CpG compared to bacterial DNA. The adjuvant effect of CpG is enhanced when conjugated to protein antigens. CPG7909, an adjuvant developed by Coley Pharmaceuticals, has been tested in a few vaccines directed at infectious agents (such as Hepatitis B allergen: Creticos *et al.*, 2006) and tumor cells (Alexeevet *et al.*, 2008; Kirkwood *et al.*, 2009).

New formulated adjuvants

MF59 is a submicron oil-in-water emulsion of a squalene, polyoxyethylene sorbitan monooleate (Tween 80), and sorbitan trioleate. MF59 was approved in Europe and is found in several vaccines, including influenza. It has also been licensed to other companies and is being actively tested in vaccine trials. Other oil-in-water emulsions include Montanide (Seppic), adjuvant 65 (in use since the 1960s), and Lipovant. QS-21, a natural product of the bark of the *Quillaja saponaria* tree, which is native to Chile and Argentina, is currently under investigation (Ghochikyan, 2006). Immune-stimulating complexes (ISCOMs) are honeycomb-like structures composed mainly of *Quillaja saponins*, cholesterol, phospholipid, and antigen. Some ISCOMs are formed without antigen and then mixed with antigen, so that the antigen is absorbed on to or conjugated with the ISCOM. Specific isoforms of ADVAX, an adjuvant developed in Australia based on inulin (a natural plant-derived polysaccharide consisting of a chain of fructose molecules ending in a single glucose), are prepared and formulated into compositions suitable for use as adjuvants. A synergistic effect is obtained by combining gamma inulin with an antigen-binding material such as inulin; the product is called Algammulin.

Xenobiotic adjuvants (the natural adjuvants)

Some of the adjuvant properties of the bacterial walls of Gram-negative bacteria have been clearly attributed to the lipid A fraction of LPSs (Ulrich, 1995). Similarly, the xenobioitic muramyl dipeptide, shown to be the smallest peptidic moiety of bacteria cell walls, can replace mycobacteria in CFA (Bahr, 1986).

More recently, interest has been focused on another well-defined natural structure endowed with adjuvanticity: the bacterial DNA. Studies on bacterial DNA have shown that unmethylated CpG motifs displaying 5′ Pu-Pu-CpG-Pyr-Pyr 3′ (Pu: purine, A or G; Pyr: pyrimidine, C or T) nucleotide sequences are recognized by, and can activate, cells of the immune system (Krieget *et al.*, 1995). Such motifs allow the immune system to discriminate pathogen-derived foreign DNA from self-DNA. CpG motifs have been found to activate antigen-presenting cells, leading to upregulation of major histocompatibility complex (MHC) and costimulatory molecules, the secretion of proinflammatory cytokines (TNFα, IFNγ, IL1, IL6, IL12, and IL18), and the switching on of T helper 1 (Th1) immunity (Lipfordet *et al.*, 1997; Millan, 1998; Zimmerman, 1998).

Tuftsin autoadjuvant

Tuftsin is a physiological natural immunostimulating tetrapeptide (Thr-Lys-Pro-Arg), a fraction of the IgG heavy-chain molecule produced by enzymatic cleavage in the spleen. Tuftsin deficiency, either hereditary or following splenectomy, results in increased susceptibility to certain infections caused by capsulated organisms, such as *H. influenza, pneumococci,* and *meningococci* and *Salmonella*. Tuftsin, being a self-immunostimulating molecule, can be termed an "autoadjuvant" on the basis of its biological functions, which encompass the following:

1. Binding to receptors on neutrophils and macrophages, to stimulate their phagocytic activity. Tuftsin is able to increase the efficacy of antimicrobial agents. Tuftsin-based therapy was proven successful, by activity of a Gentamicin combined with tuftsin conjugate, in treating experimental keratitis caused by *Pseudomonas aeruginosa* and *Candida peritonis* infections in a murine model. Murine peritoneal macrophages activated by tuftsin killed the intracellular protozoan *Leishmania major*, as well. Moreover, the tuftsin derivative Thr-Lys-Pro-Arg-NH-(CH2)2-NHCOC15H31 protected mice against *Plasmodium berghei* infection. In human studies, tuftsin showed stimulation of the antimicrobial activity of blood monocyte macrophages in leprosy patients.

2. Increasing tumor necrosis factor alpha (TNFα) release from human Kupffer cells.

3. Enhancing secretion of IL1 by activating macrophages (Phillips *et al.*, 1981; Dagan *et al.*, 1987).

4. Interaction with macrophages, resulting in expression of nitric oxide (NO) synthase to produce NO (Dagan *et al.*, 1987).

5. Enhancement of murine natural cell-mediated cytotoxicity (Phillips *et al.*, 1981). Being a natural autoadjuvant small molecule, its implementation may include, in addition to antimicrobial and antifungal activities, the restoration of the innate immune system in immunocompromised hosts, such as AIDS (Fridkin *et al.*, 2005) and cancer (Khan *et al.*, 2007; Yuan *et al.*, 2012) patients. In addition, tuftsin may serve as a good adjuvant for a new generation of vaccines, with minimal or no side effects (Pawan *et al.*, 1994; Gokulan *et al.*, 1999; Wardowska *et al.*, 2009; Liu *et al.*, 2012).

Liu *et al.* (2012) introduced a novel vaccine against influenza A virus, based on a multimer of tuftsin with the extracellular domain of influenza A matrix protein 2 (M2e). Following animal studies, the tuftsin-M2e construct has been proposed as a promising candidate for a universal vaccine against influenza A virus. Assessing malaria vaccine, tuftsin was chemically linked to EENVEHDA and DDEHVEEPTVA repeat sequences of ring-infected erythrocyte surface-antigen protein (an asexual blood-stage antigen) of *Plasmodium falciparum*. Mice immunized with these synthetic constructs had higher antibody titers and better secondary immune responses and antigen-induced T cell proliferation than the peptide dimers alone. Thus, tuftsin-based synthetic conjugates were proposed to be useful for the development of malaria vaccines. In an additional trial, a fusion protein composed of antiidiotypic scFv antibodies mimicking CA125 and tuftsin manifested a number of biological activities, including activation of macrophages and stimulation of the T cell response against cancer (Yuan *et al.*, 2012). Another trial using a chimeric molecule composed of multimeric tuftsin and synthetic peptides of HIV gp41 and gp120 proteins was successful (Gokulan *et al.*, 1999). A significantly stronger immune response was observed in mice immunized with the peptide polytuftsin conjugates than in mice receiving the peptide dimers (peptide–peptide); therefore, this chimeric molecule was proposed as a future candidate for the treatment of AIDS patients.

Tuftsin autoadjuvant is an immunomodulator small molecule in some autoimmune diseases (Lukács *et al.*, 1984; Bhasin *et al.*, 2007; Wu *et al.*, 2012). Tufsin improved the clinical score of naive mice with experimental autoimmune encephalomyelitis (EAE) induced by myelin oligodendrocyte glycoprotein (MOG), a model commonly used for multiple sclerosis. During the progression of EAE, microglia, the immunocompetent cells of the brain, were activated; these accumulated around demyelinated lesions. Microglial activation is mediated by the extracellular protease tissue plasminogen activator (tPA). Successful treatment with tuftsin, a macrophage/microglial activator, revealed that the disease progression could be manipulated favorably in its early stages by altering the timing of microglial activation, which upregulates T helper 2 cells and inhibits disease progression. In systemic lupus erythematosus patients, an impairment in monocyte macrophage chemotaxis can be demonstrated *in vitro* and *in vivo*, in concert with defective phagocytic activity. Exposing defective, lupus-originated monocytes and macrophages *in vitro* to tuftsin resulted in improved chemotaxis similar to that of healthy individuals (Lukács *et al.*, 1984).

Mechanisms of adjuvanticity

Adjuvants accomplish their task by mimicking specific sets of evolutionarily conserved molecules, including liposomes, LPS, molecular cages for antigen, components of bacterial cell walls, and endocytosed nucleic acids, such as double-stranded RNA (dsRNA), single,stranded DNA (ssDNA), and unmethylated CpG dinucleotide-containing DNA. Because immune systems have evolved to recognize these specific antigenic moieties, the presence of adjuvant in conjunction with the vaccine can greatly increase the innate immune response to the antigen by augmenting the activities of dendritic cells (DCs), lymphocytes, and macrophages by mimicking a natural infection. Furthermore, because adjuvants are attenuated beyond any function of virulence, they have been thought to pose little or no independent threat to a host organism. But is this really true? Adjuvants may exert their immune-enhancing effects according to five immune functional activities, summarized in Table 1.1 (Schijns, 2000).

Adjuvants and the adaptive and innate immune response

In order to understand the links between the innate immune response and the adaptive immune response, in order to help substantiate an adjuvant function in enhancing adaptive immune responses to the specific antigen of a vaccine, the following points should be considered: innate immune-response cells such as DCs engulf pathogens through phagocytosis. DCs then migrate to the lymph nodes, where T cells (adaptive immune cells) wait for signals to trigger their activation (Bousso and Robey, 2003). In the lymph nodes, DCs process the engulfed pathogen and then express the pathogen clippings as antigen on their cell surface by coupling them to the MHC. T cells can then recognize these clippings and undergo a cellular transformation, resulting in their own activation (Mempelet *et al.*, 2004). Macrophages can also activate T cells, in a similar manner. This process, carried out by both DCs and macrophages, is termed "antigen presentation" and represents a physical link between the innate and adaptive immune responses. Upon activation, mast cells release heparin and histamine to effectively increase trafficking and seal off the site of infection, allowing immune cells of both systems to clear the area of pathogens. In addition, mast cells also release chemokines, resulting in a positive chemotaxis of other immune cells of both the innate and adaptive immune responses to the infected area (Kashiwakura *et al.*, 2004). Due to the variety of mechanisms and links between the innate and adaptive immune responses, an adjuvant enhanced innate immune response results in an enhanced adaptive immune response.

Adjuvants and TLRs

The ability of the immune system to recognize molecules that are broadly shared by pathogens

Table 1.1 Adjuvants exert their immunological effect by different modes of action. Schijns, V. E. Immunological concepts of vaccine adjuvant activity. *Curr Opin Immunol* 12(4): 456–63. Copyright © 2000, Elsevier

No.	Mode of action	Immunological effect
1	Translocation of antigens to the lymph nodes, where they can be recognized by T cells	Greater T cell activity, heightened clearance of pathogen throughout the organism
2	Protection to antigens, granting a prolonged delivery and longer exposure	Upregulation of the production of the B and T cells necessary for greater immunological memory in the adaptive immune response
3	Increased capacity to cause local reactions at the injection site	Greater release of danger signals by chemokine-releasing cells such as helper T cells and mast cells
4	Induction of the release of inflammatory cytokines	Recruitment of B and T cells at sites of infection and increasing transcriptional events, leading to a net increase of immune cells as a whole
5	Interaction with pattern-recognition receptors (PRRs) (specifically, Toll-like receptors, TLRs) on accessory cells	Increased innate immune response to antigen

is due, in part, to the presence of special immune receptors called TLRs that are expressed on leukocyte membranes. TLRs were first discovered in Drosophila and are membrane-bound pattern-recognition receptors (PRRs) responsible for detecting most (although certainly not all) antigen-mediated infections (Beutler, 2004). In fact, some studies have shown that in the absence of TLRs, leukocytes become unresponsive to some microbial components, such as LPS (Poltoraket *et al.*, 1998). There are at least 13 different forms of TLR, each with its own characteristic ligand. Prevailing TLR ligands described to date (all of which elicit adjuvant effects) include many evolutionarily conserved molecules, such as LPSs, lipoproteins, lipopeptides, flagellin, double-stranded RNA, unmethylated CpG islands, and various other forms of DNA and RNA classically released by bacteria and viruses. The binding of ligand, either in the form of adjuvant used in vaccinations or in the form of invasive moieties during times of natural infection, to the TLR marks the key molecular event that ultimately leads to innate immune responses and the development of antigen-specific acquired immunity (Takeda and Akira, 2005). The very fact that TLR activation leads to adaptive immune responses to foreign entities explains why so many adjuvants used today in vaccinations are developed to mimic TLR ligands.

It is believed that upon activation, TLRs recruit adapter proteins within the cytosol of the immune cell in order to propagate the antigen-induced signal-transduction pathway. To date, four adapter proteins have been well characterized: MyD88, Trif, Tram, and Tirap (also called "Mal") (Shizuo, 2003). These recruited proteins are responsible for the subsequent activation of other downstream proteins, including protein kinases (IKKi, IRAK1, IRAK4, and TBK1), which further amplify the signal and ultimately lead to the upregulation or suppression of genes that orchestrate inflammatory responses and other transcriptional events. Some of these events lead to cytokine production, proliferation, and survival, while others lead to greater adaptive immunity. MyD88 is essential for inflammatory cytokine production in response to all TLR ligands, except the TLR3 ligand. TIRAP/Mal is essential for TLR2- and TLR4-dependent inflammatory cytokine production but is not involved in the MyD88-independent TLR4 signaling pathway. TRIF is essential for TLR3 signaling, as well as the MyD88-independent TLR4 signaling pathway.

Mechanisms of adjuvant adverse effects

The mechanisms underlying adjuvant adverse effects are under renewed scrutiny because of their enormous implications for vaccine development. Additionally, new, low-toxicity adjuvants are being sought, to enhance vaccine formulations. Muramyl dipeptide (MDP) is a component of the peptidoglycan polymer and has been shown to be an active but low-toxicity component of CFA, a powerful adjuvant composed of mycobacteria lysates in an oil emulsion. MDP activates cells primarily via the cytosolic nucleotide binding domain and Leucine-rich repeat-containing family (NLR) member Nod2 (nucleotide binding oligomerization domain containing 2), and is therefore linked to the ability of adjuvants to enhance antibody production. Moreira *et al.* (2008) tested the adjuvant properties of the MDP-Nod2 pathway and found that MDP, compared to the TLR agonist LPS, has minimal adjuvant properties for antibody production under a variety of immunization conditions. They also observed that the oil emulsion IFA supplemented the requirements for the TLR pathway, independent of the antigen. Nod2 was required for an optimal IgG1 and IgG2c response in the absence of exogenous TLR or NLR agonists. By combining microarray and immunofluorescence analysis, Mosca *et al.* (2008) monitored the effects of the adjuvants MF59 oil-in-water emulsion, CpG, and alum in the mouse muscle. MF59 induced a time-dependent change in the expression of 891 genes, whereas CpG and alum regulated 387 and 312 genes, respectively. All adjuvants modulated a common set of 168 genes and promoted antigen-presenting cell recruitment. MF59 was the stronger inducer of cytokines, cytokine receptors, adhesion molecules involved in leukocyte migration, and antigen-presentation genes. In addition, MF59 triggered a more rapid influx of $CD11b^+$ blood cells compared with other adjuvants. The authors proposed that oil-in-water emulsions are the most efficient human vaccine adjuvants, because they induce an early and strong immunocompetent environment at the injection site by targeting muscle cells. Emerging data suggest that alum phosphate and alum hydroxide adjuvants do not promote a strong commitment to the helper T cell type 2 (Th2) pathway when they are coadministered with some Th1 adjuvants. Iglesias *et al.* (2006) have shown that subcutaneous immunization, in alum phosphate, of a mixture comprising three antigens (the surface and core antigens of hepatitis B virus (HBV) and the multi-epitopic protein CR3 of human immunodeficiency virus type 1) elicits a CR3-specific Th1 immune response. Although alum is known to induce the production of proinflammatory cytokines *in vitro*, it has been repeatedly demonstrated that it does not require intact TLR signaling to activate the immune system. This was suggested by Gavin *et al.* (2006), who reported that mice deficient in the critical signaling components for TLR mount robust antibody responses to T cell-dependent antigen given in four typical adjuvants: alum, CFA, IFA, and monophosphoryl lipid A/trehalose dicorynomycolate adjuvant. They concluded that TLR signaling does not account for the action of classical adjuvants and does not fully explain the action of a strong adjuvant containing a TLR ligand. Eisenbarth *et al.* (2008) showed that alum adjuvants activated the intracellular innate immune response system, the Nalp3 (also known as cryopyrin, CIAS1, or NLRP3) inflammasome. Production of the proinflammatory cytokines IL-1 and IL-18 by macrophages in response to alum *in vitro* required intact inflammasome signaling. Furthermore, *in vivo*, mice deficient in Nalp3, ASC (apoptosis-associated speck-like protein containing a caspase recruitment domain), or caspase1 failed to mount a significant antibody response to an antigen administered with alum adjuvants, whereas the response to CFA remained intact. The authors identified the Nalp3 inflammasome as a crucial element in the adjuvant effect of alum adjuvants; in addition, they showed that the innate inflammasome pathway can direct a humoral adaptive immune response. Recently, Kool *et al.* (2008) succeeded in exposing an angle of its mysterious mechanism: they found that alum activates DCs *in vivo* by provoking the secretion of uric acid, a molecule that is triggered by tissue and cell trauma. The injection of alum induced an influx of neutrophils and inflammatory cytokines and chemokines, a combination that had previously been seen in response to the injection of uric acid into mice. In mice injected with a mixture of antigens, ovalbumin peptide, and alum, uric acid levels increased within hours. The uric acid may have been released by the cells lining the body's cavities, which turn necrotic after contacting the alum. In response to the uric acid, inflammatory monocytes flocked to the injection site, took up the antigens, and broke them down into T cell-stimulating epitopes. The monocytes then migrated to lymph nodes, where they matured into DCs and activated $CD4^+$ T cells. Without alum, the antigens were not taken up at the injection site. Still, they eventually reached the lymph nodes via the flowing lymph. The resident node DCs, however, did not efficiently process the alum-free antigens or express T cell co-stimulating receptors. The resulting subdued immunity was similar to that seen in mice that were depleted of inflammatory monocytes or injected with enzymes that degrade uric acid. These findings suggest that alum is immunogenic through exploitation of "nature's adjuvant," via induction of the endogenous danger signal: **uric acid**. In another study, Kool *et al.* (2008) showed that alum adjuvant induced the release of IL1β from macrophages and DCs, and that this is abrogated in cells lacking various NALP3 inflammasome components. The NALP3 inflammasome is also required *in vivo* for the innate immune response to ovalbumin in alum. The early production of IL1β and the influx of inflammatory cells into the peritoneal cavity is strongly reduced in NALP3-deficient mice. The activation of adaptive cellular immunity to ovalbumin-alum is initiated by monocytic DC precursors, which induce the expansion of antigen-specific T cells in a NALP3-dependent way. The authors proposed that, in addition to TLR stimulators, agonists of the NALP3 inflammasome should also be considered vaccine adjuvants. Flach *et al.* (2011) reported that, independent of inflammasome and membrane proteins, alum binds DC plasma membrane lipids with substantial force. Subsequent lipid sorting activates an abortive phagocytic response, which leads to antigen uptake. Such activated DCs, without further association with alum, show high affinity and stable binding with $CD4^+$ T cells via the adhesion molecules intercellular adhesion molecule 1 (ICAM1) and lymphocyte function-associated antigen 1 (LFA1). The authors proposed that alum triggers DC responses by altering membrane lipid structures. This study therefore suggests an unexpected mechanism for how this crystalline structure interacts with the immune system and how the DC plasma membrane may behave as a general sensor for solid structures. Marichal *et al.* (2011) reported that, in mice, alum caused cell death and the subsequent release of host-cell DNA, which acted as a potent endogenous immunostimulatory signal, mediating alum adjuvant activity. Furthermore, the authors proposed that host DNA signaling differentially regulated IgE and IgG1 production following alum-adjuvanted immunization. They suggested that, on the one hand, host DNA induces primary B cell responses, including IgG1 production, through interferon response factor 3 (Irf3)-independent mechanisms, but that, on the other, host DNA may also stimulate "canonical"

T helper type 2 (Th2) responses, associated with IgE isotype switching and peripheral effector responses, through Irf3-dependent mechanisms. The finding that host DNA released from dying cells acts as a damage-associated molecular pattern that mediates alum adjuvant activity may increase our understanding of the mechanisms of action of current vaccines and help in the design of new adjuvants.

Compiling all the evidence concerning alum's mechanism of action, it seems that alum may play a role in a few parallel and alternative pathways: through the inflammasome, by causing inflammation either directly or by uric acid; by binding DC plasma membrane lipids with substantial force and activating an abortive phagocytic response that leads to antigen uptake; or by causing cell death and the subsequent release of host-cell DNA, which acts as a potent endogenous immunostimulatory signal.

Autoimmunity and environmental/natural adjuvants

Genetic, immunological, hormonal, and environmental factors (i.e., infections, vaccines, xenobiotics, etc.) are considered to be important in the etiology of autoimmunity. Overt autoimmune disease is usually triggered following exposure to such environmental factors, among which infectious agents are considered of great importance (Molina and Shoenfeld, 2005). Some researchers consider adjuvants to be environmental factors involved in autoimmune diseases. Several laboratories are pursuing the molecular identification of endogenous adjuvants. Among those identified so far, sodium monourate and the high-mobility group B1 protein (HMGB1) are well known to rheumatologists. However, even the complementation of apoptotic cells with potent adjuvant signals fails to cause clinical autoimmunity in most strains: autoantibodies generated are transient, do not undergo epitope/spreading, and do not cause disease. Lastly, as vaccines may protect or cure autoimmune diseases, adjuvants may also play a double role in the mechanisms of these diseases. Myasthenia gravis (MG) and its animal model, experimental autoimmune gravis (EAMG), are caused by interference with neuromuscular transmission by autoantibodies against the nicotinic acetylcholine receptor (AChR) on muscle. Two peptides, denoted RhCA 67-16 and RhCA 611-001 and designed to be complementary in structure to the main immunogenic region and the dominant Lewis rat T cell epitope (α-chain residues 100–116) of the AChR, respectively, are effective vaccines that prevent EAMG in rats by inducing antiidiotypic/clonotypic antibodies (Ab) and lowering levels of AChR Ab. Their study employed keyhole limpet hemocyanin (KLH) as a carrier and the CFA. In advance of a clinical trial, McAnally *et al.* (2001) tested the efficacy of RhCA 611-001 when combined with different adjuvants approved for use in humans: IFA and alum hydroxide. As a second goal, the authors evaluated diphtheria toxin (DT) as an alternative carrier protein to KLH. Alum was found to be an effective adjuvant, particularly when used with the peptide conjugated to DT. This combination of carrier and adjuvant provided protection against EAMG comparable with that observed with CFA and KLH. It was found that disease protection is qualitatively, but not quantitatively, related to the antipeptide antibody response. This work demonstrated a vaccine formulation that should be useful in the first soon-to-be-conducted clinical trials of peptide vaccines to specifically correct aberrant T and B cell responses in an autoimmune disease.

Adjuvant-related diseases

Alongside their supportive role, adjuvants have themselves been found to inflict illnesses of autoimmune nature, termed "the adjuvant diseases" (Agmon-Levin, 2008).

Mineral oils as a cause of autoimmunity

Mineral oils are generally considered "nontoxic" and have been used extensively in food, cosmetics, medicines, and other products. Subcutaneous injection of mineral oil induces sclerosing lipogranulomas, a chronic local inflammatory reaction (Di Benedetto *et al.*, 2002). The oil is absorbed through the intestine and distributes throughout the body, causing lipogranulomas in the lymph nodes, liver, and spleen of healthy individuals. Oral or intraperitoneal administration of mineral oil induces similar lesions in laboratory animals. Pristane (2,6,10,14-tetramethylpentadecane) and mineral oil induce plasmacytomas in susceptible strains of mice (Anderson and Potter, 1969). Pristane, IFA, and squalene (2,6,10,15,19,23-hexamethyl-2,6,10,14,18,22-tetracosahexaene) induce chronic arthritis in mice and rats (Cannon *et al.*, 1993; Carlson *et al.*, 2000). Reeves and colleagues reported that, in addition to pristane (Satoh and Reeves 1994; Satoh *et al.*, 1995), IFA and squalene, but not medicinal mineral oils, can induce lupus-related anti-nRNP/Sm and Su autoantibodies in nonautoimmune-prone strains of mice. These data suggest that hydrocarbons can have a variety of immune effects. Kuroda *et al.* (2004) investigated whether medicinal mineral oils can induce other types of autoantibodies and whether structural features of hydrocarbons influence autoantibody specificity. Induction of autoantibodies by mineral oils considered nontoxic also may have pathogenetic implications in human autoimmune diseases. Moreover, Kuroda *et al.* (2004) have reported that a single intraperitoneal injection of the adjuvant oils pristane, IFA, or squalene induces lupus-related autoantibodies to nRNP/Sm and Su in nonautoimmune BALB/c mice. Induction of these autoantibodies appears to be associated with the hydrocarbon's ability to induce IL-12, IL-6, and TNFα, suggesting a relationship with hydrocarbon's adjuvanticity. Whether this is relevant in human vaccination is a difficult question, due to the complex effects of vaccines and the fact that immunotoxicological effects vary depending on species, route, dose, and duration of administration. Nevertheless, the potential of adjuvant hydrocarbon oils to induce autoimmunity has implications in the use of oil adjuvants in human and veterinary vaccines, as well as in basic research (Table 1.2).

Human adjuvant disease, silicone as an adjuvant, and connective tissue diseases

Spiera *et al.* (1994) reviewed the literature examining the association of silicone gel-filled implants with connective tissue disease. They stated that silicones are not biologically inert. Injectable and implantable silicones have proven capable of eliciting inflammatory and fibroproliferative responses. The physical and biological properties of silicone gel-filled implants and their behavior *in vivo* are compatible with the hypothesis that they may contribute to the development of connective-tissue disease. The association seems most likely with scleroderma; however, there are currently inadequate epidemiological data to definitively establish causality. Janowsky *et al.* (2000) performed a meta-analysis of the relation between silicone breast implants and the risk of connective tissue diseases. There was no evidence that breast implants were associated with a significant increase in the adjusted relative risk of individual connective tissue diseases. Nor was there evidence of significantly increased risk in the unadjusted analyses or in the analysis restricted to silicone gel-filled implants. Vasey *et al.* (2003) proposed a definition for a silicone-related disorder, by major and minor criteria: tenderness, capsule formation, change in shape or position, and/or rapture of envelope; chronic fatigue lasting

Table 1.2 Adjuvant involvement in autoimmune manifestation

Adjuvant	Manifestations/disease/Ab	Species	References
MDP, LPS, Gram + CoxackieB3,IL1β,TNF	Experimental thyroiditis; Myocarditis	Mice	Rose (2008)
Mineral oils	Sclerosing lipogranulomas	Mice human?	Di Benedetto *et al.* (2002)
Pristane, mineral oils	Plasmacytomas	Mice	Anderson and Potter (1969)
Pristane, squalene, IFA	Chronic arthritis	Mice, rats	Cannon *et al.* (1993), Carlson *et al.* (2000)
Pristane, squalene, IFA	Lupus-related anti-nRNP/Sm /Su antibodies	Mice	Satoh and Reeves (1994), Satoh *et al.* (1995)
Pristane, squalene, IFA, mineral oils	Anticytoplasmic Ab, anti-ssDNA/chromatin Ab	Mice	Kuroda *et al.* (2004)
Pristane, squalene, IFA	Lupus-related anti-nRNP/Sm /Su antibodies	Mice	Kuroda *et al.* (2004)
Silicone	Human adjuvant disease, connective tissue diseases	Human	Hennekens *et al.* (1996)
Silicone	Scleroderma, SLE, RA	Human	Spiera *et al.* (1994)
Alum in vaccines (HBV,HAV, tetanus)	MS, chronic fatigue syndrome, polymyalgia rheumatica	Human	Gherardi (2003)
Aluminum hydroxide, squalene	Gulf War syndrome, antibodies to squalene	Human	Asa *et al.* (2000)

6 months, myalgias with tender muscles; bladder dysfunction, dry eyes or mouth, impaired cognition, and a few more symptoms.

Macrophagic myofasciitis and Gulf War syndrome

Macrophagic myofasciitis was first reported in 1998 but its cause remained obscure until 2001 (Gherardi, 2003). The condition manifests by diffuse myalgias and chronic fatigue, forming a syndrome that meets both Centers for Disease Control and Prevention (CDC) and Oxford criteria for the so-called "chronic fatigue syndrome" in about half of patients. One-third of patients develop an autoimmune disease, such as multiple sclerosis. Electron microscopy, microanalytical studies, experimental procedures, and an epidemiological study recently demonstrated that the lesion results from persistence for years at the site of injection of an alum adjuvant used in vaccines against hepatitis B virus, hepatitis A virus, and tetanus toxoid. Alum hydroxide is known to potently stimulate the immune system and to shift immune responses toward a Th2 profile. Interestingly, special emphasis has been put on Th2-biased immune responses as a possible explanation of chronic fatigue and associated manifestations known as the "Gulf War syndrome" (GWS). Results concerning macrophagic myofasciitis may well open new avenues for etiologic investigation of this syndrome. Indeed, both the type and the structure of symptoms are strikingly similar in Gulf War veterans and patients with macrophagic myofasciitis. Multiple vaccinations performed over a short period of time in the Persian Gulf area have been recognized as the main risk factor for GWS. Moreover, the war vaccine against anthrax, which is administered in a six-shot regimen and seems to be crucially involved, is adjuvanted by alum hydroxide and, possibly, squalene, another Th2 adjuvant. Asa *et al.* (2000) sought to determine whether the presence of antibodies to squalene correlates with the presence of signs and symptoms of GWS. All (100%) GWS patients immunized for service in Desert Shield/Desert Storm who did not deploy but had the same signs and symptoms as those who did had antibodies to squalene. In contrast, no (0%) deployed Persian Gulf veterans not showing signs and symptoms of GWS had antibodies to squalene. Neither patients with idiopathic autoimmune disease nor healthy controls had detectable serum antibodies to squalene. If safety concerns about the long-term effects of alum hydroxide are confirmed, it will become mandatory to propose novel and alternative vaccine adjuvants in order to rescue vaccine-based strategies and the enormous benefit for public health they provide worldwide.

Autoimmune (autoinflammatory) syndrome induced by adjuvants – ASIA

Siliconosis, GWS, macrophagic myofasciitis (MMF) syndrome, and post-vaccination phenomena have all been linked with previous exposure to an adjuvant. Furthermore, these four diseases share a similar complex of signs and symptoms, which further support a common denominator. Shoenfeld and Agmon–Levin (2011) recently suggested that these four somehow enigmatic conditions should be included under a common syndrome entitled the "autoimmune (auto-inflammatory) syndrome induced by adjuvants" (ASIA). The authors further proposed several major and minor criteria, which, although they require further validation, may aid in the diagnosis of this newly defined syndrome. Recently, the sick building syndrome was also suggested as part of ASIA (Israeli and Pardo, 2011). Comparison of the clinical manifestations, symptoms, and signs of the four conditions described by Shoenfeld and Agmon-Levin (2011) with those described for SBS shows that nine out of ten main symptoms are in correlation in all five conditions: namely, myalgia, arthralgias, chronic fatigue, neurological cognitive impairment, fever, gastrointestinal symptoms, respiratory symptoms, skin manifestations, and appearance of autoantibodies. Thus, ASIA may be a common syndrome for all five conditions mentioned. The amassed data regarding each condition may enable a new view of the immune responses to environmental adjuvants, as well as a better definition and diagnosis of these conditions. Moreover, unraveling the pathogenesis of this newly defined syndrome may facilitate the search for preventive and therapeutic interventions.

Conclusions

Due to the adverse effects exerted by adjuvants, there is no controversy over the need for safer adjuvants for incorporation into future vaccines.

The problem with the pure recombinant or synthetic antigens used in modern-day vaccines is that they are generally far less immunogenic than older-style live or killed whole-organism vaccines. This has created a major need for improved and more powerful adjuvants for use

Table 1.3 Adjuvants in human vaccines. Reed, S. G., S. Bertholet, et al. New horizons in adjuvants for vaccine development. *Trends Immunol* 30(1): 23–32. Copyright © 2009, Elsevier

Adjuvants	Human vaccines
Alum	DPT, DT, HBV, HAV, H. influenza B, inactivated polio, strep. pneumonia, HPV, meningococcal
Oil and water-MF59	Herpes simplex, HBV, HIV
MPL AS04/AS01B/AS02A	HBV,HAV,HPV, malaria, tuberculosis, leishmania, HIV, vesicular stomatitis
Virosomes-VLP / IRIV	HBV,HPV/HAV
Cholera toxin B subunit	Cholera

DPT, diphtheria–pertussis–tetanus; DT, diphtheria–tetanus; HAV, hepatitis A; HBV, hepatitis B; HIV, human immunodeficiency virus; HPV, human papilloma virus

Table 1.4 Adjuvants in development

Adjuvants	Formulation	Preclinical or clinical trials
Montanides	Water-in-oil emulsions	Malaria (Phase I), HIV, cancer (Phase I/II)
Saponins (QS-21)	Aqueous	Cancer (Phase II), herpes (Phase I), HIV (Phase I)
SAF	Oil-in-water emulsion containing squalene, Tween 80, Pluronic L121	HIV (Phase I – Chiron)
AS03	Oil-in-water emulsion containing α-tocopherol, squalene, Tween 80	Pandemic flu (GSK)
MTP-PtdEtn	Oil-in-water emulsion	HSV
Exotoxins	*P. aeruginosa*	*P. aeruginosa*, cystic fibrosis (AERUGEN – Crucell/Berna)
	E. coli heat-labile enterotoxin LT	ETEC (Phase II – Iomai Corp.)
ISCOMs	Phospholipids, cholesterol, QS-21	Influenza, HSV, HIV, HBV, malaria, cancer
TLR ligands		
MPL®-SE	Oil-in-water emulsion	Leishmania (Phase I/II – IDRI)
Synthetic Lipid A	Oil-in-water emulsion	Various indications (Avanti/IDRI)
MPL®-AF	Aqueous	Allergy (ATL), cancer (Biomira)
AS01	Liposomal	HIV (Phase I), malaria (AS01, Phase III, GSK), cancer (Phase II/III, Biomira/MerckKGaA)
AS02	Oil-in-water emulsion containing MPL and QS-21	HPV (Cervarix), HIV, tuberculosis, malaria (Phase III), herpes (GSK)
AS04	Alum + aqueous MPL	HPV, HAV (GSK)
AS15	AS01 + CpG	Cancer therapy (GSK)
RC529	Aqueous	HBV, pneumovax

in these vaccines (Petrovsky and Aguilar, 2004). With few exceptions, alum remains the major adjuvant approved for human use in the majority of countries worldwide. Although alum is able to induce a good antibody (Th2) response, it has little capacity to stimulate cellular (Th1) immune responses, which are so important for protection against many pathogens. In addition, alum has the potential to cause severe local and systemic side effects, including sterile abscesses, eosinophilia, and myofasciitis, although, fortunately, most of the more serious side effects are relatively rare. Consequently, there is a major unmet need for safer and more effective adjuvants suitable for human use. In particular, there is demand for safe and nontoxic adjuvants capable of stimulating cellular (Th1) immunity. Several other adjuvants besides alum have been approved to date for use in human vaccines, among them MF59 in some viral vaccines, MPL, AS04, AsO1B and AS02A against viral and parasitic infections, virosomes for HBV, HPV, and HAV, and cholera toxin for cholera (Table 1.3) (Reed *et al.*, 2009).

Other needs in light of new vaccine technologies are adjuvants suitable for use with mucosally delivered vaccines, DNA vaccines, cancer, and autoimmunity vaccines. Each of these areas is highly specialized, with its own unique needs with respect to suitable adjuvant technology.

Although controversial, the high sensitivity of TLR for microbial ligands is what makes adjuvants that mimic TLR ligands such a prime candidate for enhancing the overall effects of antigen-specific vaccinations on immunological memory. The expression of TLRs is vast, as they are found on the cell membranes of innate immune cells (DCs, macrophages, natural killer cells), cells of the adaptive immunity (T and B lymphocytes), and nonimmune cells (epithelial cells). This further substantiates the importance of administering vaccines with adjuvants in the form of TLR ligands, as they will be capable of eliciting their positive effects across the entire spectrum of innate and adaptive immunity. Nevertheless, there are certainly adjuvants whose immune stimulatory function completely bypasses the putative requisite for TLR signaling (Table 1.4). In short, all TLR ligands are adjuvants but not all adjuvants are TLR ligands. We can conclude that there are, in all likelihood, other receptors besides TLRs that have not yet been characterized, opening a field of future research. Perhaps future adjuvants occupying these putative receptors will be employed to bypass the TLR signaling pathway completely, in order to circumvent common side effects of adjuvant-activated TLRs, such as local inflammation and the general malaise felt because of the costly whole-body immune response to antigen. Surely, such issues will be the subject of much debate for future researchers.

References

Agmon-Levin Nancy, Y.S. (2008). Chronic fatigue syndrome with autoantibodies – the result of an augmented adjuvant effect of hepatitis-B vaccine and silicone implant. *Autoimmun Rev*, **8**(1): 52–5.

Alexeev, V., Mucci, T., Igoucheva, O. (2008). Immunotherapeutic strategies for the treatment of malignant melanoma. *G Ital Dermatol Venereol*, **143**(2): 139–49.

Anderson, P.N. and Potter, M. (1969). Induction of plasma cell tumours in BALB-c mice with 2,6,10,14-tetramethylpentadecane (pristane). *Nature* **222**(5197): 994–5.

Asa, P.B., Cao, Y., Garry, R.F. (2000). Antibodies to squalene in Gulf War syndrome. *Exp Mol Pathol*, **68**(1): 55–64.

Bahr, G.M. and Chedid, L. (1986). Immunological activities of muramyl peptides. *Fed Proc*, **45**: 2541–4.

Beutler, B. (2004). Inferences, questions and possibilities in Toll-like receptor signalling. *Nature*, **430**(6996): 257–63.

Bhasin, M., Wu, M., and Tsirka, S.E. (2007). Modulation of microglial/macrophage activation by macrophage inhibitory factor (TKP) or tuftsin (TKPR) attenuates the disease course of experimental autoimmune encephalomyelitis. *BMC Immunol*, **8**: 10.

Bousso, P. and Robey, E. (2003). Dynamics of CD8+ T cell priming by dendritic cells in intact lymph nodes. *Nat Immunol*, **4**(6): 579–85.

Cannon, G.W., Woods, M.L., Clayton, F., and Griffiths, M.M. (1993). Induction of arthritis in DA rats by incomplete Freund's adjuvant. *J Rheumatol*, **20**(1): 7–11.

Carlson, B.C., Jansson, A.M., Larsson, A., *et al.* (2000). The endogenous adjuvant squalene can induce a chronic T-cell-mediated arthritis in rats. *Am J Pathol*, **156**(6): 2057–65.

Creticos, P.S., Schroeder, J.T., Hamilton, R.G., *et al.* (2006). Immunotherapy with a ragweed-toll-like receptor 9 agonist vaccine for allergic rhinitis. *N Engl J Med*, **355**(14): 1445–55.

Di Benedetto, G., Pierangeli, M., Scalise, A., and Bertani, A. (2002). Paraffin oil injection in the body: an obsolete and destructive procedure. *Ann Plast Surg*, **49**(4): 391–6.

Dagan, S., Tzehoval, E., Fridkin, M., and Feldman, M. (1987). Tuftsin and tuftsin conjugates potentiate immunogenic processes: effects and possible mechanisms. *J Biol Response Mod*, **6**(6): 625–36.

Eisenbarth, S.C., Colegio, O.R., O'Connor, W., *et al.* (2008). Crucial role for the Nalp3 inflammasome in the immunostimulatory properties of aluminum adjuvants. *Nature*, **453**(7198): 1122–6.

Flach, T.L., Ng, G., Hari, A., *et al.* (2011). Alum interaction with dendritic cell membrane lipids is essential for its adjuvanticity. *Nat Med*, **17**(4): 479–87.

Fridkin, M., Tsubery, H., Tzehoval, E., *et al.* (2005). Tuftsin-AZT conjugate: potential macrophage targeting for AIDS therapy. *J Pept Sci*, **11**(1): 37–44.

Gavin, A.L., Hoebe, K., Duong, B., *et al.* (2006). Adjuvant-enhanced antibody responses in the absence of toll-like receptor signaling. *Science*, **314**(5807): 1936–8.

Gherardi, R.K. (2003). [Lessons from macrophagic myofasciitis: towards definition of a vaccine adjuvant-related syndrome]. *Rev Neurol (Paris)*, **159**(2): 162–4.

Ghochikyan, A., Mkrtichyan, M., Petrushina, I., *et al.* (2006). Prototype Alzheimer's disease epitope vaccine induced strong Th2-type anti-Abeta antibody response with Alum to Quil A adjuvant switch. *Vaccine*, **24**(13): 2275–82.

Gluck, R. (1999). Adjuvant activity of immunopotentiating reconstituted influenza virosomes (IRIVs). *Vaccine*, **17**(13–14): 1782–7.

Gokulan, K., Khare, S., and Rao, D.N. (1999). Increase in the immunogenicity of HIV peptide antigens by chemical linkage to polytuftsin (TKPR40). *DNA Cell Biol*, **18**(8): 623–30.

Holzer, B.R., Hatz, C., Schmidt-Sissolak, D., *et al.* (1996). Immunogenicity and adverse effects of inactivated virosome versus alum-adsorbed hepatitis A vaccine: a randomized controlled trial. *Vaccine*, **14**(10): 982–6.

Iglesias, E., Thompson, R., Carrazana, Y., *et al.* (2006). Coinoculation with hepatitis B surface and core antigen promotes a Th1 immune response to a multiepitopic protein of HIV-1. *Immunol Cell Biol*, **84**(2): 174–83.

Israeli, E., Agmon-Levin, N., Blank, M., and Schoenfeld, Y. (2009). Adjuvants and autoimmunity. *Lupus*, **18**(13): 1217–25.

Israeli, E. and Pardo, A. (2011). The sick building syndrome as a part of the autoimmune (auto-inflammatory) syndrome induced by adjuvants. *Mod Rheumatol*, **21**(3): 235–9.

Janowsky, E.C., Kupper, L.L., and Hulka, B.S. (2000). Meta-analyses of the relation between silicone breast implants and the risk of connective-tissue diseases. *N Engl J Med*, **342**(11): 781–90.

Kashiwakura, J., Yokoi, H., Saito, H., and Okayama, Y. (2004). T cell proliferation by direct cross-talk between OX40 ligand on human mast cells and OX40 on human T cells: comparison of gene expression profiles between human tonsillar and lung-cultured mast cells. *J Immunol*, **173**(8): 5247–57.

Khan, A., Khan, A.A., Dwivedi, V., *et al.* (2007). Tuftsin augments antitumor efficacy of liposomized etoposide against fibrosarcoma in Swiss albino mice. *Mol Med*, **13**(5–6): 266–76.

Kirkwood, J.M., Tawbi, H.A., Tarhini, A.A., and Moschos, S.J. (2009). Does pegylated interferon alpha-2b confer additional benefit in the adjuvant treatment of high-risk melanoma? *Nat Clin Pract Oncol*, **6**(2): 70–1.

Kool, M., Petrilli, V., De Smedt, T., *et al.* (2008a). Cutting edge: alum adjuvant stimulates inflammatory dendritic cells through activation of the NALP3 inflammasome. *J Immunol*, **181**(6): 3755–9.

Kool, M., Soullié, T., van Nimwegen, M., *et al.* (2008b). Alum adjuvant boosts adaptive immunity by inducing uric acid and activating inflammatory dendritic cells. *J Exp Med*, **205**(4): 869–82.

Krieg, A.M., Yi, A.K., Matson, S., *et al.* (1995). CpG motifs in bacterial DNA trigger direct B-cell activation. *Nature*, **374**(6522): 546–9.

Kuroda, Y., Akaogi, J., Nacionales, D.C., *et al.* (2004a). Distinctive patterns of autoimmune response induced by different types of mineral oil. *Toxicol Sci*, **78**(2): 222–8.

Kuroda, Y., Nacionales, D.C., Akaogi, J., *et al.* (2004b). Autoimmunity induced by adjuvant hydrocarbon oil components of vaccine. *Biomed Pharmacother*, **58**(5): 325–37.

Lipford, G.B., Bauer, M., Blank, C., *et al.* (1997). CpG-containing synthetic oligonucleotides promote B and cytotoxic T cell responses to protein antigen: a new class of vaccine adjuvants. *Eur J Immunol*, **27**(9): 2340–4.

Liu, X., Guo, J., Han, S., *et al.* (2012). Enhanced immune response induced by a potential influenza A vaccine based on branched M2e polypeptides linked to tuftsin. *Vaccine*, **30**(46): 6527–33.

Lukács, K., Szabó, G., Sonkoly, I., *et al.* (1984). Stimulating effect of tuftsin and its analogues on the defective monocyte chemotaxis in systemic lupus erythematosus. *Immunopharmacology*, **7**(3–4): 171–8.

Marichal, T., Ohata, K., Bedoret, D., *et al.* (2011). DNA released from dying host cells mediates aluminum adjuvant activity. *Nat Med*, **17**(8): 996–1002.

McAnally, J.L., Xu, L., Villain, M., and Blalock, J.E. (2001). The role of adjuvants in the efficacy of a peptide vaccine for myasthenia gravis. *Exp Biol Med (Maywood)*, **226**(4): 307–11.

Mempel, T.R., Henrickson, S.E., and Von Andrian, U.H. (2004). T-cell priming by dendritic cells in lymph nodes occurs in three distinct phases. *Nature*, **427**(6970): 154–9.

Millan C.L.B., Risini, W., Krieg, A.M., *et al.* (1998). CpG DNA can induce strong Th1 humoral and cell-mediated immune responses against hepatitis B surface antigen in young mice. *Proc Natl Acad Sci USA*, **95**: 15 553–8.

Molina, V. and Shoenfeld, Y. (2005). Infection, vaccines and other environmental triggers of autoimmunity. *Autoimmunity*, **38**(3): 235–45.

Moreira, L.O., Smith, A.M., DeFreitas, A.A., *et al.* (2008). Modulation of adaptive immunity by different adjuvant-antigen combinations in mice lacking Nod2. *Vaccine*, **26**(46): 5808–13.

Mosca, F., Tritto, E., Muzzi, A., *et al.* (2008). Molecular and cellular signatures of human vaccine adjuvants. *Proc Natl Acad Sci USA*, **105**(30): 10 501–6.

Pawan, K., Ivanov, B.B., Kabilan, L., and Rao, D.N. (1994). Construction of a synthetic immunogen: use of the natural immunomodulator polytuftsin in malaria vaccines against RESA antigen of *Plasmodium falciparum*. *Vaccine*, **12**(9): 819–24.

Petrovsky, N. and Aguilar, J.C. (2004). Vaccine adjuvants: current state and future trends. *Immunol Cell Biol*, **82**(5): 488–96.

Phillips, J.H., Babcock, G.F., and Nishioka, K. (1981). Tuftsin: a naturally occurring immunopotentiating factor. I. In vitro enhancement of murine natural cell-mediated cytotoxicity. *J Immunol*, **126**(3): 915–21.

Pichichero, M.E. (2008). Improving vaccine delivery using novel adjuvant systems. *Hum Vaccin*, **4**(4): 262–70.

Poltorak, A., He, X., Smirnova, I., *et al.* (1998). Defective LPS signaling in C3H/HeJ and C57BL/10ScCr mice: mutations in Tlr4 gene. *Science*, **282**(5396): 2085–8.

Reed, S. G., Bertholet, S., Coler, R.N., and Fried, M. (2009). New horizons in adjuvants for vaccine development. *Trends Immunol* **30**(1): 23–32.

Riedl, K., Riedl, R., von Gabain, A., *et al.* (2008). The novel adjuvant IC31 strongly improves influenza vaccine-specific cellular and humoral immune responses in young adult and aged mice. *Vaccine,* **26**(27–28): 3461–8.

Rose, N.R. (2008). The adjuvant effect in infection and autoimmunity. *Clin Rev Allergy Immunol,* **34**(3): 279–82.

Satoh, M. and Reeves, W.H. (1994). Induction of lupus-associated autoantibodies in BALB/c mice by intraperitoneal injection of pristane. *J Exp Med,* **180**(6): 2341–6.

Satoh, M., Kumar, A., Kanwar, Y.S., and Reeves, W.H. (1995). Anti-nuclear antibody production and immune-complex glomerulonephritis in BALB/c mice treated with pristane. *Proc Natl Acad Sci USA,* **92**(24): 10934–8.

Schijns, V.E. (2000). Immunological concepts of vaccine adjuvant activity. *Curr Opin Immunol,* **12**(4): 456–63.

Shizuo, A. (2003). Toll-like receptor signaling. *J Biol Chem,* **278**(40): 38105–8.

Shoenfeld, Y. and Agmon-Levin, N. (2011). "ASIA" – autoimmune/inflammatory syndrome induced by adjuvants. *J Autoimmun,* **36**(1): 4–8.

Spiera, R.F., Gibofsky, A., and Spiera, H. (1994). Silicone gel filled breast implants and connective tissue disease: an overview. *J Rheumatol,* **21**(2): 239–45.

Takeda, K. and Akira, S. (2005). Toll-like receptors in innate immunity. *Int Immunol,* **17**(1): 1–14.

Ulrich, J.T. and Myers, K.R. (1995). Monophosphoryl lipid A as an adjuvant. Past experiences and new directions. *Pharm Biotechnol,* **6**: 495–524.

Vasey, F.B., Zarabadi, S.A., Seleznick, M., and Ricca, L. (2003). Where there's smoke there's fire: the silicone breast implant controversy continues to flicker: a new disease that needs to be defined. *J Rheumatol,* **30**(10): 2092–4.

Wardowska, A., Dzierzbicka, K., Szaryńska, M., *et al.* (2009). Analogues of muramyl dipeptide (MDP) and tuftsin limit infection and inflammation in murine model of sepsis. *Vaccine,* **27**(3): 369–74.

Wu, M., Nissen, J.C., Chen, E.I., and Tsirka, S.E. (2012). Tuftsin promotes an anti-inflammatory switch and attenuates symptoms in experimental autoimmune encephalomyelitis. *PLoS One,* **7**(4): e34933.

Yuan, W., Xia, G., Zhao, C., *et al.* (2012). Anti-idiotypic single chain mimicking CA125 linked with tuftsin provides protective immunity against ovarian cancer in mice. *Mol Med Rep,* **5**(2): 388–94.

Zimmermann, S., Egeter, O., Hausmann, S., *et al.* (1998). CpG oligodeoxynucleotides trigger protective and curative Th1 responses in lethal murine leishmaniasis. *J Immunol,* **160**: 3627–30.

2 Infections as Adjuvants for Autoimmunity: The Adjuvant Effect

Quan M. Nhu and Noel R. Rose

The Johns Hopkins Medical Institutions, The W. Harry Feinstone Department of Molecular Microbiology and Immunology, Center for Autoimmune Disease Research, and Department of Pathology, Baltimore, MD, USA

Introduction

The concept of host immunity dates back to the 5th century BCE, when Thucydides observed that survivors of the plague did not develop the disease when subsequently re-exposed. In China circa the 10th century CE, survivors of smallpox were found to be protected from disease following reinfection. The practice of variolation, in which pulverized smallpox (variola) scabs or pustules were introduced into the skin or the nose, probably began in China around the 10th century CE to render protection against smallpox infection. In 1796, Edward Jenner (1749–1823) introduced vaccination, in which prior inoculation with cowpox (vaccinia) protected against a subsequent infection with smallpox in humans (Behbehani, 1983; Morgan and Parker, 2007).

Exuberant immune responses to infection, while protective, can be detrimental to the host. More than a century ago, the first human autoimmune disease, paroxysmal cold hemoglobulinuria (Donath–Landsteiner syndrome), was thought to be a late consequence of syphilis. Rheumatic fever and rheumatic heart disease are well-studied sequelae of streptococcal infection, while myocarditis and type 1 diabetes mellitus have been associated with Coxsackievirus infection. As noted by Sir William Osler (1849–1919), some patients seem to die not of their infection but of the body's response to it (Brem, 1914; Rose and Afanasyeva, 2003; Ercolini and Miller, 2009).

Autoimmune diseases in humans arise from the unfavorable inheritance of a group of genes that together dictate increased predilection for autoimmune responses. The most prominent susceptibility genes are members of the major histocompatibility complex (MHC), encoded by the human leukocyte antigen genes. In addition to genetic predisposition, an environmental trigger (e.g. infection) is required for the onset of autoimmune disease in humans. Based on the two-signal hypothesis, viruses and bacteria can provide the autoantigen-specific signal required for activation of the T cell receptor (TCR). Uncontrolled immune reactivity to endogenous self-antigens results in autoimmune diseases. Several mechanisms have been described to account for autoantigen-specific responses during or after infection, including molecular mimicry, T cell degeneracy, epitope spreading, release of normally inaccessible or cryptic antigens, and alteration of autologous antigen (Rose and Griffin, 1991; Davidson and Diamond, 2001; Fujinami *et al.*, 2006; Kuchroo *et al.*, 2012).

Until recent years, less attention had been directed to the requirement for the nonclonal, non-antigen-specific second signal required to initiate or exacerbate an autoimmune response. From a historical perspective, for most of the 20th century, antigen-specific recognition of pathogens

Vaccines and Autoimmunity, First Edition. Edited by Yehuda Shoenfeld, Nancy Agmon-Levin, and Lucija Tomljenovic.

by immunoglobulins and TCRs and the associated mechanisms of clonal selection dominated immunological thinking and research (Germain, 2004). Nevertheless, it was widely recognized that the survival of an infected host depended on the integrated cooperativity between both innate and adaptive immune limbs (Kaufmann, 2008). For example, activation of adaptive immunity required both innate immune antigen-presenting cells (APCs) (e.g. macrophages and dendritic cells) for antigen presentation (signal 1) and expression of co-stimulatory molecules (signal 2) (Palm and Medzhitov, 2009). Importantly, antigen presentation was known to be greatly enhanced *in vivo* by "adjuvants," stimuli contained in empirically developed concoctions (e.g. complete Freund's adjuvant, CFA) that augmented antigen-specific lymphocytic responses (White, 1976). In fact, adjuvants in the form of aluminum salts were widely used in vaccines, although their mechanism of action is still not fully understood (McKee *et al.*, 2007). Charles Janeway, Jr (1943–2003) referred to adjuvants as the immunologist's "dirty little secret" (Janeway, 1989). Recent understanding of innate immune pattern recognition of pathogens has helped explain the "adjuvant effect" (Medzhitov and Janeway, 2002; Ishii *et al.*, 2008; Ronald and Beutler, 2010; Hoffmann and Akira, 2013).

The adjuvant effect

The term "adjuvant effect" was first applied to the critical role of mycobacteria in the Freund's adjuvant used to induce experimental thyroiditis (N. Rose, discussed in Grabar and Miescher, 1959; Rose and Afanasyeva, 2003; Rose, 2008a). It refers to the coadministration of an exogenous microbial factor with an antigen to enhance an antigen-specific immune response *in vivo* (Freund, 1951; White, 1976; O'Hagan and Valiante, 2003). The microbial components of adjuvants activate APCs, resulting in the upregulation of molecules essential for antigen presentation, such as MHC class II (antigen-specific signal 1) and B7-1/2, and the production of proinflammatory cytokines ("nonspecific" signal 2). These innate immune events allow the coadministered antigen to be processed and presented to the adaptive immune system more effectively, resulting in the augmented activation and greater clonal expansion of T cells.

In a similar manner, when coadministered with self-antigens, adjuvants are also responsible for driving the development of autoimmune diseases by activating the innate immune response. Many animal models of organ-specific autoimmune disease (e.g. experimental autoimmune encephalomyelitis, EAE; experimental autoimmune uveitis, EAU; experimental autoimmune myocarditis, EAM) rely on the use of the powerful adjuvant CFA to generate autoantigen-specific T cells (Neu *et al.*, 1987; Caspi, 2008; Cihakova and Rose, 2008; Rangachari and Kuchroo, 2013). In other instances, autoreactive T cells that escape central and peripheral immune tolerance mechanisms are triggered by exogenous adjuvants to become autoaggressive. Adjuvants can also activate APCs to overcome the suppressive effects of $CD4^+$ $CD25^+$ regulatory T cells, leading to the activation of autoreactive T cells (Pasare and Medzhitov, 2003). While adjuvants are often empirical microbial components, possessing potent bioactivities, infectious agents naturally generate their own adjuvant effect and can induce autoimmunity (Fairweather *et al.*, 2001, 2005; Rose and Afanasyeva, 2003; Rose, 2008a).

Innate immune pattern recognition of pathogens and adjuvants

Innate immune cells recognize a wide range of pathogens, despite the relatively limited number of signal-transducing receptors expressed on their surface. To explain this phenomenon, Janeway proposed in 1989 that evolutionarily conserved molecular components of infectious microorganisms – pathogen-associated molecular patterns (PAMPs) – were detected by the innate immune system through a restricted set of nonclonal, germline-encoded, specialized receptors termed "pattern-recognition receptors" (PRRs). PRR–PAMP interactions activate the APCs to promote antigen-specific lymphocytic responses. Without such activation of innate immune cells via the recognition of pathogens through PRRs, the adaptive immune system would ignore or become tolerant of the antigens presented by the "quiescent" APCs (Janeway, 1989; Palm and Medzhitov, 2009).

In support of Janeway's hypothesis, Hoffmann and colleagues reported in 1996 that an antifungal immune response in *Drosophila* was mediated by Toll, a type I transmembrane protein found nearly a decade earlier by Nüsslein-Volhard and associates to be involved in *Drosophila* embryonic dorsal–ventral polarity development (Anderson *et al.*, 1985; Lemaitre *et al.*, 1996). In 1991, Gay and Keith had proposed that Toll had a cytosolic domain homologous to the human IL-1 receptor

(IL-1R), raising the possibility that Toll might contribute to host immunity in *Drosophila* (Gay and Keith, 1991). In 1994, the tobacco N protein that mediates resistance to the tobacco mosaic virus had been found to contain a signaling domain homologous to the human IL-1R and the *Drosophila* Toll cytoplasmic domains (Whitham *et al.*, 1994). These findings suggested the potential existence of an evolutionarily conserved Toll/IL-1R/resistance (TIR) domain involved in promoting immunity across a broad phylogenetic expanse.

In 1997, Janeway and colleagues reported an evolutionarily conserved human homolog of *Drosophila* Toll, a human Toll-like receptor (TLR) now known as TLR4 (Medzhitov *et al.*, 1997; Rock *et al.*, 1998). TLR4 was found to contain an extracellular leucine-rich repeat (LRR) domain and a cytoplasmic TIR domain. In 1998, Beutler and colleagues, and shortly thereafter, in 1999, Qureshi *et al.*, identified the transducing receptor for lipopolysaccharide (LPS) as TLR4, which, when mutated or deleted, was responsible for the LPS-resistant phenotype of the C3H/HeJ and C57BL/10ScCr mice, respectively (Poltorak *et al.*, 1998; Qureshi *et al.*, 1999).

It is now recognized that approximately 10 human TLRs, acting as PRRs, bind PAMPs (or co-receptor-bound PAMPs) through their LRR-containing extracellular domains. The signature cytoplasmic TIR domains of TLRs transmit intracellular signals to induce potent antimicrobial and proinflammatory responses (Akira *et al.*, 2006; Medzhitov, 2009; Hoffmann and Akira, 2013). The definition of PAMPs has now broadened, with the appreciation that the recognized structure need not be derived from a pathogen, leading to the concept of "microbe-associated molecular patterns" (MAMPs). Subsequent findings that endogenous host molecules released during an inflammatory response could also activate TLRs have led to the use of "danger/damage-associated molecular patterns" (DAMPs), based largely on the concept that the endogenous host molecules signal danger or damage to the immune system (Matzinger, 1994; Kono and Rock, 2008).

Innate immune response mediates the adjuvant effect

The PRRs sense PAMPs/MAMPs and DAMPs and alert the immune system of the presence of infection and tissue damage, respectively. In the context of an infection, innate immune PRRs recognize PAMPs and MAMPs such as LPS, peptidoglycan, flagella, and microbial nucleic acids. In the context of tissue damage and necrosis, PRRs sense DAMPs such as hyaluronic acid, heparan sulfate, heat-shock proteins, endogenous DNA and RNA, adenosine triphosphate, and uric crystals. PRR engagement on APCs provides the necessary co-stimulatory signal 2 and proinflammatory cytokines to generate the adjuvant effect that drives antigen-specific adaptive immune responses (Marshak-Rothstein, 2006; Sansonetti, 2006; Kono and Rock, 2008; Kawai and Akira, 2009; Medzhitov, 2009; Palm and Medzhitov, 2009).

The TLRs represent a family of single-transmembrane PRRs that are localized to the cell surface and the endosomal membranes. The cytosolic PRRs include retinoic acid-inducible gene-I (RIG-I)-like receptors (RLRs), inflammasome/nucleotide-binding oligomerization domain (NOD)-like receptors (NLRs), and the recently identified cytosolic DNA sensors. Working together, these transmembrane and cytosolic PRRs provide surveillance of all cellular compartments (Akira *et al.*, 2006; Ishii *et al.*, 2008; Hoffmann and Akira, 2013; Holm *et al.*, 2013).

The TLRs are the best studied PRRs to date and are covered more extensively in other chapters. Briefly, TLRs play crucial roles in the innate immune responses against potentially harmful microorganisms, in addition to being biosensors of tissue damage. The combinatorial utilization of the four known TLR adapters – MyD88, TIRAP/Mal, TRAM, TRIF – in TLR signaling tailors the innate immune responses to infection with different classes of microbes (Takeuchi *et al.*, 1999; Vogel *et al.*, 2003; Yamamoto *et al.*, 2004; Dunne and O'Neill, 2005). For example, TLR4 induces inflammatory responses to Gram-negative bacteria by recruiting all four TLR adapters, whereas TLR2 mediates inflammation elicited by Gram-positive bacteria by recruiting only TIRAP/Mal and MyD88. Infection with viruses with dsRNA genome or intermediates triggers TLR3 to elicit an antiviral response solely through TRIF. Interestingly, although the adjuvant monophosphoryl lipid A activates TLR4, only the TRIF pathway is engaged (Mata-Haro *et al.*, 2007). The ability of the TLRs to modify cellular responses to different pathogenic agents determines the appropriate immune responses to infections.

The binding of specific PAMPs to TLRs recruits distinct combinations of TLR adapters to activate a unique repertoire of *trans*-activating nuclear transcription factors. Activation of NF-κB induces expression of proinflammatory cytokines, such as IL-1β, IL-6, and TNF-α, which further potentiate the inflammatory response. Activation of the interferon regulatory factor-3 (IRF-3) induces transcription of IRF-3-responsive genes, such as RANTES and IFN-β. Once induced, IFN-β can act in an autocrine or paracrine fashion via the IFN-α/β receptor to activate the JAK-STAT pathway, which, in turn, generates a second wave of transcription factors that further tailors the immune response (Toshchakov *et al.*, 2002; Barton and Medzhitov, 2003; Vogel *et al.*, 2003; Dunne and O'Neill, 2005).

The TLRs, and other innate immune PRRs, play critical roles in regulating adaptive immunity. TLR-stimulated APCs upregulate expression of MHC class II and co-stimulatory molecules, such as CD80 and CD86, required for the activation of T cells (Palm and Medzhitov, 2009). TLR ligation on immature dendritic cells, which are highly phagocytic and weakly immunogenic, elicits maturation of the dendritic cells, which then become weakly phagocytic but highly immunogenic. These mature dendritic cells leave the tissue and migrate to the draining lymph node, where they interact with the T cells to tailor the appropriate T cell responses. TLR engagement on APCs also induces the production of immunomodulatory cytokines, such as IL-10, IL-12, IL-23, IL-6, and IFN-γ. These inflammatory responses are essential for creating the appropriate milieu for induction of adaptive immunity against pathogens (Iwasaki and Medzhitov, 2004). In addition, TLRs are also potent mediators of sterile, noninfectious inflammatory diseases, including atherosclerosis, ischemia-reperfusion injury, transplant rejection, and many autoimmune diseases (Toubi and Shoenfeld, 2004; Goldstein, 2006; Marshak-Rothstein and Rifkin, 2007; Arumugam *et al.*, 2009; Curtiss and Tobias, 2009).

The innate immune system also senses proteolytic enzymes derived from the host and pathogens that are generated during infection (Shpacovitch *et al.*, 2007, 2008). Host cells often encounter multiple PRR agonists concurrently when exposed to pathogens. Signaling pathways coordinately triggered by distinct innate immune PRRs upon activation by various microbial components, such as PAMPs and proteinases, can synergize with or antagonize one another to modulate the inflammatory response (Trinchieri and Sher, 2007; O'Neill, 2008; Nhu *et al.*, 2010). Thus, signaling integration from multiple PRRs might drive customized inflammatory responses to combinatorial stimuli from the environment.

Viral infection as adjuvant for autoimmunity: evidence from Coxsackievirus-induced myocarditis

Approximately half the myocarditis cases among humans are preceded by an acute viral infection. Myocarditis can be a serious heart disease, especially in active young adults. It can progress to dilated cardiomyopathy and heart failure requiring cardiac transplantation. Commonly caused by Coxsackievirus B3 (CB3), a positive ssRNA enterovirus of the *picornaviridae* family, infectious myocarditis in humans can be reproduced in experimental murine models of myocarditis (Rose *et al.*, 1987; Fairweather and Rose, 2007; Cihakova and Rose, 2008).

Cardiotropic CB3 induces acute cardiac inflammation with a mixed immune cellular infiltrate and cardiomyocyte damage that peaks by days 7–9 after infection and resolves by day 21 in most strains of mice. The few susceptible strains (e.g. BALB/c, A/J, and SJL/J mice) progress to chronic myocarditis characterized by generalized monocytic and lymphocytic infiltration, cardiomyocyte death, and the formation of IgG autoantibodies to cardiac myosin (Rose *et al.*, 1986; Wolfgram *et al.*, 1986; Alvarez *et al.*, 1987; Neumann *et al.*, 1994). During the chronic phase, viral RNA, but not infectious viral particles, may be detectable. Nevertheless, the chronic phase of viral myocarditis is an autoimmune process that mirrors experimental autoimmune myocarditis induced by immunization with cardiac myosin or its peptide derivatives (Neu *et al.*, 1987; Donermeyer *et al.*, 1995; Pummerer *et al.*, 1996; Rose, 2008b). It must be noted that cardiac myosin immunization requires the powerful CFA emulsion to consistently induce disease. Cardiac inflammatory lesions are absent if cardiac myosin is emulsified, instead, in incomplete Freund's adjuvant (IFA). Thus, the inflammatory cytokine milieu generated by the viral infection or provided by the exogenous administration of potent adjuvants likely determines the level of inflammation that dictates the progression to chronic myocarditis.

During infection, genetically determined viral components are sensed as PAMPs by the TLRs and other PRRs (Kawai and Akira, 2006). Viral RNA is recognized by TLRs 3, 7, and 8, as well as the RNA helicase cytosolic sensors (e.g. RIG-I and melanoma differentiation-associated gene 5, MDA5). As TLR3 recognizes CB3 dsRNA intermediate, host defense against CB3 *in vivo* requires intact TLR3 and its adapter TRIF. Mice deficient in TLR3 or TRIF develop significant myocarditis that progresses to chronic inflammatory cardiomyopathy (Negishi *et al.*, 2008; Riad *et al.*, 2011; Abston *et al.*, 2012, 2013). TLR3 mutants encoding mutations identified in patients with enteroviral myocarditis and dilated cardiomyopathy show a blunted CB3-mediated IFN-β antiviral response that fails to control CB3 replication *in vitro* (Gorbea *et al.*, 2010). The structurally unrelated proteinase-activated receptor 2 (PAR_2) inhibits TLR3-induced IFN-β antiviral response (Nhu *et al.*, 2010), resulting in the exacerbation of CB3-induced myocarditis (Weithauser *et al.*, 2013). In contrast, the thrombin receptor PAR_1 enhances TLR3-induced IFN-β antiviral response and thus protects against CB3-induced myocarditis (Antoniak *et al.*, 2013). While the potent antiviral immune responses are generally protective, they can contribute to the development of several autoimmune conditions in human (Baccala *et al.*, 2005; Munz *et al.*, 2009; Sozzani *et al.*, 2010).

CB3 infection also activates other innate PRRs to induce expression of proinflammatory cytokines (e.g. IL-1 and TNF-α) critical to the development of postviral autoimmune myocarditis. TLR4, TLR7, and TLR8 have been reported to mediate CB3-induced cytokine production *in vitro* (Triantafilou *et al.*, 2005), while TLR4 and TLR9 contribute to acute CB3-induced myocarditis *in vivo* (Fairweather *et al.*, 2003; Riad *et al.*, 2010), and MDA5 and its adapter protein, mitochondrial antiviral signaling (MAVS), are critical for survival following CB3 infection (Wang *et al.*, 2010). The absence of MyD88, the adapter protein used by most TLRs and IL-1R, reduces myocarditis induced by either CB3 infection or myosin-CFA immunization (Fuse *et al.*, 2005; Marty *et al.*, 2006; Blyszczuk *et al.*, 2009).

In CB3-infected susceptible mice, exogenous IL-1 administration exacerbates myocarditis, whereas blocking IL-1 significantly reduces disease severity (Neumann *et al.*, 1993; Huber *et al.*, 1994; Lim *et al.*, 2002). The antiinflammatory cytokine IL-10 confers a protective effect that limits myocarditis induced by either CB3 infection or myosin-CFA immunization (Kaya *et al.*, 2002; Fousteri *et al.*, 2011). In mice typically resistant to chronic myocarditis (e.g. B10), resistance can be overcome and the disease generated if IL-1, TNF-α, or their potent inducer LPS is administered during the innate phase of CB3 infection (Lane *et al.*, 1991, 1992, 1993). IL-1 and TNF-α may augment the local inflammatory response, leading to tissue damage. These cytokines may also induce APCs to provide the non-antigen-specific T cell-activating co-stimulatory signals that culminate in the adjuvant effect (Rose, 2008a).

The LPS-transducing receptor TLR4 contributes to CB3 myocarditis, secondary in part to TLR4-driven cardiac expression of IL-1β and IL-18, which augments inflammation (Fairweather *et al.*, 2003; Richer *et al.*, 2006). Similarly, the severity of myocarditis induced by immunization with cardiac myosin is significantly reduced in TLR4-defective C3H/H3J mice (Q. Nhu, unpublished observation). The co-adjuvants used in the EAM model, mycobacterial extracts and *Bordetella pertussis* toxin, have been shown to signal, in part, through TLR4 (Kerfoot *et al.*, 2004; Wang *et al.*, 2006; Nishida *et al.*, 2010; van de Veerdonk *et al.*, 2010). Additionally, the muramyl dipeptide component of mycobacterial extracts also activates the inflammasome, which requires LPS/TLR4 priming (Netea *et al.*, 2008). Therefore, the similarities observed in the development of CB3-induced myocarditis and the EAM model may be due, in part, to the ability of PAMPs present on CB3 and the mycobacterial component of CFA to engage PRRs during the innate immune response.

Early in infection, CB3 replication is de-repressed in TLR4-deficient heart, consistent with the antiviral function of the TLR4-IRF-3 pathway. However, TLR4 activation later, during the adaptive immune phase, promotes CB3-induced autoimmune disease. This observation accords with a recent report that the absence of TLR4 in $CD4^+$ T cells virtually abrogated autoimmune inflammation in EAE, secondary mostly to a blunted T helper 17 (Th17) response (Reynolds *et al.*, 2012). Thus, during the innate immune response, TLR4 activation alerts the immune system to protect the host against viral replication, but may subsequently promote autoimmune disease. In addition, TLR4 also senses endogenous damage signals to further augment the local inflammatory response, which, again, contributes to the adjuvant effect.

"Adjuvant disease"

Given that adjuvants are powerful enhancers of immune responses, the question has been raised of whether adjuvants alone can induce autoimmune disease in humans. Precedent for this possibility was raised by the report of Pearson in 1956 that administration of CFA, but not IFA, elicited a form of inflammatory mono- or polyarthritis in Wistar and Long–Evans rats (Pearson, 1956; Pearson *et al.*, 1961; Waksman, 2002; Whitehouse, 2007). Other systemic involvement includes the eyes, skin, and the gastrointestinal and genitourinary mucosa. The mechanism of the disease is not understood, but one possibility is that the adjuvant augments an autoimmune process already underway in these particular animals. As yet, similar results have not been reported in humans. For example, vaccine containing antigenic peptide of the H1N1 influenza A virus was given to large numbers of individuals in Europe and Canada using a "Freund-like" adjuvant. Similar vaccine used in the United States contained no added adjuvant. The reported adverse reaction rates to the two different formulations were comparable (Pellegrini *et al.*, 2009).

Harnessing the adjuvant effect and autoimmunity against cancer

The flipside of autoimmunity is the development of cancer when immune surveillance is diminished. The nonclonal augmentation of autoimmunity by the adjuvant effect can be harnessed to boost anticancer immune responses. Historically, severe infections in tumor-bearing patients have resulted in the spontaneous regression of the tumors (Hoption Cann *et al.*, 2003). William Coley (1862–1936) pioneered the use of Coley's toxins, comprising extracts of killed Gram-positive *Streptococcus pyogenes* and Gram-negative *Serratia marcescens*. This preparation has been reported to successfully treat a variety of tumors when injected directly into the primary and metastatic tumor sites (Thotathil and Jameson, 2007). Intravesical administration of bacillus Calmette–Guérin (BCG) is effective against superficial bladder cancer (Morales *et al.*, 1976; Lamm *et al.*, 1980). The agonist of the ssRNA-binding TLR7, imiquimod, has been used successfully as a topical cream for the treatment of several skin tumors (Gaspari *et al.*, 2009), while tumor cells genetically manipulated to express bacterial flagellin, which activates TLR5, have been reported recently to induce effective tumor-specific immune responses *in vivo* (Garaude *et al.*, 2012). While there are several mechanisms that may be involved in the antitumoral activities of these immunomodulators, activation of the innate immune system through specific PRR activation may signal infection and/or tissue damage to drive inflammatory responses and autoimmunity, central to the adjuvant effect.

Conclusions

Empirical formulations of adjuvants and breakthrough immunological theories in the 20th century have played important roles in advancing our understanding of the immune system. Adjuvants and natural infections can exert potent immunostimulatory activities through the adjuvant effect, which can lead to autoimmune disease. The powerful adjuvant effect can also be utilized to develop novel antitumor therapies and effective vaccines. The goal of a vaccine is to reproduce the protection afforded by an infection, while minimizing the risks. Recent advances in the understanding of the molecular mechanisms involved in immune cell activation have provided an opportunity to fine-tune vaccine development and the adjuvant effect through the strategic activation of specific immune pattern recognition receptors.

Acknowledgments

Q.M.N. is supported by NIH T32 HL-7227 and AI-7417. N.R.R. is supported by NIH RO1 HL113008.

References

Abston, E.D., Coronado, M.J., Bucek, A., *et al.* (2012). Th2 regulation of viral myocarditis in mice: different roles for TLR3 versus TRIF in progression to chronic disease. *Clin Dev Immunol*, **2012**: 1–12.

Abston, E.D., Coronado, M.J. Bucek, A., *et al.* (2013). TLR3 deficiency induces chronic inflammatory cardiomyopathy in resistant mice following coxsackievirus B3 infection: role for IL-4. *Am J Physiol Regul Integr Comp Physiol*, **304**: R267–77.

Akira, S., Uematsu, S., and Takeuchi, O. (2006). Pathogen recognition and innate immunity. *Cell*, **124**: 783–801.

Alvarez, F.L., Neu, N., Rose, N.R., *et al.* (1987). Heart-specific autoantibodies induced by Coxsackievirus B3: identification of heart autoantigens. *Clin Immunol Immunopathol*, **43**: 129–39.

Anderson, K.V., Bokla, L., and Nusslein-Volhard, C. (1985). Establishment of dorsal-ventral polarity in the Drosophila embryo: the induction of polarity by the *Toll* gene product. *Cell*, **42**: 791–8.

Antoniak, S., Owens, A.P. 3rd,, Baunacke, M., *et al.* (2013). PAR-1 contributes to the innate immune response during viral infection. *J Clin Invest*, **123**: 1310–22.

Arumugam, T.V., Okun, E., Tang, S.C., *et al.* (2009). Toll-like receptors in ischemia-reperfusion injury. *Shock*, **32**: 4–16.

Baccala, R., Kono, D.H., and Theofilopoulos, A.N. (2005). Interferons as pathogenic effectors in autoimmunity. *Immunol Rev*, **204**: 9–26.

Barton, G.M. and Medzhitov, R. (2003). Toll-like receptor signaling pathways. *Science*, **300**: 1524–5.

Behbehani, A.M. (1983). The smallpox story: life and death of an old disease. *Microbiol Rev*, **47**: 455–509.

Blyszczuk, P., Kania, G., Dieterle, T., *et al.* (2009). Myeloid differentiation factor-88/interleukin-1 signaling controls cardiac fibrosis and heart failure progression in inflammatory dilated cardiomyopathy. *Circ Res* **105**: 912–20.

Brem, W.V. (1914). Case of paroxysmal hemoglobinuria treated with salvarsan. *Cal State J Med*, **12**: 452–4.

Caspi, R. (2008). Autoimmunity in the immune privileged eye: pathogenic and regulatory T cells. *Immunol Res*, **42**: 41–50.

Cihakova, D. and Rose, N.R. (2008). Pathogenesis of myocarditis and dilated cardiomyopathy. *Adv Immunol*, **99**: 95–114.

Curtiss, L.K. and Tobias, P.S. (2009). Emerging role of Toll-like receptors in atherosclerosis. *J Lipid Res*, **50**(Suppl.): S340–5.

Davidson, A. and Diamond, B. (2001). Autoimmune diseases. *N Engl J Med*, **345**: 340–50.

Donermeyer, D.L., Beisel, K.W., Allen, P.M., and Smith, S.C. (1995). Myocarditis-inducing epitope of myosin binds constitutively and stably to I-A^k on antigen-presenting cells in the heart. *J Exp Med*, **182**: 1291–300.

Dunne, A. and O'Neill, L.A. (2005). Adaptor usage and Toll-like receptor signaling specificity. *FEBS Lett*, **579**: 3330–5.

Ercolini, A.M. and Miller, S.D. (2009). The role of infections in autoimmune disease. *Clin Exp Immunol*, **155**: 1–15.

Fairweather, D. and Rose, N.R. (2007). Coxsackievirus-induced myocarditis in mice: a model of autoimmune disease for studying immunotoxicity. *Methods*, **41**: 118–22.

Fairweather, D., Kaya, Z., Shellam, G.R., *et al.* (2001). From infection to autoimmunity. *J Autoimmun*, **16**: 175–86.

Fairweather, D., Yusung, S., Frisancho, S., *et al.* (2003). IL-12 receptor β1 and Toll-like receptor 4 increase IL-1β- and IL-18-associated myocarditis and coxsackievirus replication. *J Immunol*, **170**: 4731–7.

Fairweather, D., Frisancho-Kiss, S., and Rose, N.R. (2005). Viruses as adjuvants for autoimmunity: evidence from Coxsackievirus-induced myocarditis. *Rev Med Virol*, **15**: 17–27.

Fousteri, G., Dave, A., Morin, B., *et al.* (2011). Nasal cardiac myosin peptide treatment and OX40 blockade protect mice from acute and chronic virally-induced myocarditis. *J Autoimmun*, **36**: 210–20.

Freund, J. (1951). The effect of paraffin oil and mycobacteria on antibody formation and sensitization; a review. *Am J Clin Pathol*, **21**: 645–56.

Fujinami, R.S., von Herrath, M.G., Christen, U. and Whitton, J.L. (2006). Molecular mimicry, bystander activation, or viral persistence: infections and autoimmune disease. *Clin Microbiol Rev*, **19**: 80–94.

Fuse, K., Chan, G., Liu, Y., *et al.* (2005). Myeloid differentiation factor-88 plays a crucial role in the pathogenesis of Coxsackievirus B3-induced myocarditis and influences type I interferon production. *Circulation*, **112**: 2276–85.

Garaude, J., Kent, A., van Rooijen, N., and Blander, J.M. (2012). Simultaneous targeting of toll- and nod-like receptors induces effective tumor-specific immune responses. *Sci Transl Med*, **4**: 120ra16.

Gaspari, A, Tyring, S.K., and Rosen, T. (2009). Beyond a decade of 5% imiquimod topical therapy. *J Drugs Dermatol*, **8**: 467–74.

Gay, N.J. and Keith, F.J. (1991). Drosophila Toll and IL-1 receptor. *Nature*, **351**: 355–6.

Germain, R.N. (2004). An innately interesting decade of research in immunology. *Nat Med*, **10**: 1307–20.

Goldstein, D.R. (2006). Toll like receptors and acute allograft rejection. *Transpl Immunol*, **17**: 11–15.

Gorbea, C., Makar, K.A., Pauschinger, M., *et al.* 2010. A role for Toll-like receptor 3 variants in host susceptibility to enteroviral myocarditis and dilated cardiomyopathy. *J Biol Chem*, **285**: 23 208–23.

Grabar, P. and Miescher, P.A. (1959). *Immunopathology – Immunopathologie: 1st International Symposium*. Basel: Benno Schwabe.

Hoffmann, J. and Akira, S. (2013). Innate immunity. *Curr Opin Immunol*, **25**: 1–3.

Holm, C.K., Paludan, S.R., and Fitzgerald, K.A. (2013). DNA recognition in immunity and disease. *Curr Opin Immunol*, **25**: 13–18.

Hoption Cann, S.A., van Netten, J.P., and van Netten, C. (2003). Dr William Coley and tumour regression: a place in history or in the future. *Postgrad Med J*, **79**: 672–80.

Huber, S.A., Polgar, J., Schultheiss, P., and Schwimmbeck, P. (1994). Augmentation of pathogenesis of coxsackievirus B3 infections in mice by exogenous administration of interleukin-1 and interleukin-2. *J Virol*, **68**: 195–206.

Ishii, K.J., Koyama, S., Nakagawa, A., *et al.* (2008). Host innate immune receptors and beyond: making sense of microbial infections. *Cell Host Microbe*, **3**: 352–63.

Iwasaki, A. and Medzhitov, R. (2004). Toll-like receptor control of the adaptive immune responses. *Nat Immunol*, **5**: 987–95.

Janeway, C.A. Jr., (1989). Approaching the asymptote? Evolution and revolution in immunology. *Cold Spring Harb Symp Quant Biol*, **54**(Pt 1): 1–13.

Kaufmann, S.H. (2008). Immunology's foundation: the 100-year anniversary of the Nobel Prize to Paul Ehrlich and Elie Metchnikoff. *Nat Immunol*, **9**: 705–12.

Kawai, T. and Akira, S. (2006). Innate immune recognition of viral infection. *Nat Immunol*, **7**: 131–7.

Kawai, T. and Akira, S. (2009). The roles of TLRs, RLRs and NLRs in pathogen recognition. *Int Immunol*, **21**: 317–37.

Kaya, Z., Dohmen, K.M., Wang, Y., *et al.* (2002). Cutting edge: a critical role for IL-10 in induction of nasal tolerance in experimental autoimmune myocarditis. *J Immunol*, **168**: 1552–6.

Kerfoot, S.M., Long, E.M., Hickey, M.J., *et al.* (2004). TLR4 contributes to disease-inducing mechanisms resulting in central nervous system autoimmune disease. *J Immunol*, **173**: 7070–7.

Kono, H. and Rock, K.L. (2008). How dying cells alert the immune system to danger. *Nat Rev Immunol*, **8**: 279–89.

Kuchroo, V.K., Ohashi, P.S., Sartor, R.B., and Vinuesa, C.G. (2012). Dysregulation of immune homeostasis in autoimmune diseases. *Nat Med*, **18**: 42–7.

Lamm, D.L., Thor, D.E., Harris, S.C., *et al.* 1980. Bacillus Calmette-Guerin immunotherapy of superficial bladder cancer. *J Urol*, **124**: 38–40.

Lane, J.R., Neumann, D.A., Lafond-Walker, A., *et al.* (1991). LPS promotes CB3-induced myocarditis in resistant B10.A mice. *Cell Immunol*, **136**: 219–33.

Lane, J.R., Neumann, D.A., Lafond-Walker, A., *et al.* (1992). Interleukin 1 or tumor necrosis factor can promote Coxsackie B3-induced myocarditis in resistant B10.A mice. *J Exp Med*, **175**: 1123–9.

Lane, J.R., Neumann, D.A., Lafond-Walker, A., *et al.* (1993). Role of IL-1 and tumor necrosis factor in coxsackie virus-induced autoimmune myocarditis. *J Immunol*, **151**: 1682–90.

Lemaitre, B., Nicolas, E., Michaut, L., *et al.* (1996). The dorsoventral regulatory gene cassette *spätzle/Toll/cactus* controls the potent antifungal response in Drosophila adults. *Cell*, **86**: 973–83.

Lim, B.K., Choe, S.C., Shin, J.O., *et al.* (2002). Local expression of interleukin-1 receptor antagonist by plasmid DNA improves mortality and decreases myocardial inflammation in experimental coxsackieviral myocarditis. *Circulation*, **105**: 1278–81.

Marshak-Rothstein, A. (2006). Toll-like receptors in systemic autoimmune disease. *Nat Rev Immunol*, **6**: 823–35.

Marshak-Rothstein, A. and Rifkin, I.R. (2007). Immunologically active autoantigens: the role of Toll-like receptors in the development of chronic inflammatory disease. *Annu Rev Immunol*, **25**: 419–41.

Marty, R.R., Dirnhofer, S., Mauermann, N., *et al.* (2006). MyD88 signaling controls autoimmune myocarditis induction. *Circulation*, **113**: 258–65.

Mata-Haro, V., Cekic, C., Martin, M., *et al.* (2007). The vaccine adjuvant monophosphoryl lipid A as a TRIF-biased agonist of TLR4. *Science*, **316**: 1628–32.

Matzinger, P. (1994). Tolerance, danger, and the extended family. *Annu Rev Immunol*, **12**: 991–1045.

McKee, A.S., Munks, M.W., and Marrack, P. (2007). How do adjuvants work? Important considerations for new generation adjuvants. *Immunity*, **27**: 687–90.

Medzhitov, R. (2009). Approaching the asymptote: 20 years later. *Immunity*, **30**: 766–75.

Medzhitov, R. and Janeway, C.A. Jr., (2002). Decoding the patterns of self and nonself by the innate immune system. *Science*, **296**: 298–300.

Medzhitov, R., Preston-Hurlburt, P., and Janeway, C.A. Jr., (1997). A human homologue of the Drosophila Toll protein signals activation of adaptive immunity. *Nature*, **388**: 394–7.

Morales, A., Eidinger, D., and Bruce, A.W. (1976). Intracavitary Bacillus Calmette-Guerin in the treatment of superficial bladder tumors. *J Urol*, **116**: 180–3.

Morgan, A.J. and Parker, S. (2007). Translational mini-review series on vaccines: the Edward Jenner Museum and the history of vaccination. *Clin Exp Immunol*, **147**: 389–94.

Munz, C., Lunemann, J.D., Getts, M.T., and Miller, S.D. (2009). Antiviral immune responses: triggers of or triggered by autoimmunity? *Nat Rev Immunol*, **9**: 246–58.

Negishi, H., Osawa, T., Ogami, K., *et al.* (2008). A critical link between Toll-like receptor 3 and type II interferon signaling pathways in antiviral innate immunity. *Proc Natl Acad Sci USA*, **105**: 20 446–51.

Netea, M.G., van de Veerdonk, F.L., Kullberg, B.J., *et al.* (2008). The role of NLRs and TLRs in the activation of the inflammasome. *Expert Opin Biol Ther*, **8**: 1867–72.

Neu, N., Rose, N.R., Beisel, K.W., *et al.* (1987). Cardiac myosin induces myocarditis in genetically predisposed mice. *J Immunol*, **139**: 3630–6.

Neumann, D.A., Lane, J.R., Allen, G.S., *et al.* (1993). Viral myocarditis leading to cardiomyopathy: do cytokines contribute to pathogenesis? *Clin Immunol Immunopathol*, **68**: 181–90.

Neumann, D.A., Rose, N.R., Ansari, A.A., and Herskowitz, A. (1994). Induction of multiple heart autoantibodies in mice with coxsackievirus B3- and cardiac myosin-induced autoimmune myocarditis. *J Immunol*, **152**: 343–50.

Nhu, Q.M., Shirey, K., Teijaro, J.R., *et al.* (2010). Novel signaling interactions between proteinase-activated receptor 2 and Toll-like receptors *in vitro* and *in vivo*. *Mucosal Immunol*, **3**: 29–39.

Nishida, M., Suda, R., Nagamatsu, Y., *et al.* (2010). Pertussis toxin up-regulates angiotensin type 1 receptors through Toll-like receptor 4-mediated Rac activation. *J Biol Chem*, **285**: 15 268–77.

O'Hagan, D.T. and Valiante, N.M. (2003). Recent advances in the discovery and delivery of vaccine adjuvants. *Nat Rev Drug Discov*, **2**: 727–35.

O'Neill, L.A. (2008). When signaling pathways collide: positive and negative regulation of Toll-like receptor signal transduction. *Immunity*, **29**: 12–20.

Palm, N.W. and Medzhitov, R. (2009). Pattern recognition receptors and control of adaptive immunity. *Immunol Rev*, **227**: 221–33.

Pasare, C. and Medzhitov, R. (2003). Toll pathway-dependent blockade of $CD4^+CD25^+$ T cell-mediated suppression by dendritic cells. *Science*, **299**: 1033–6.

Pearson, C.M. (1956). Development of arthritis, periarthritis and periostitis in rats given adjuvants. *Proc Soc Exp Biol Med*, **91**:95–101.

Pearson, C.M., Waksman, B.H., and Sharp, J.T. (1961). Studies of arthritis and other lesions induced in rats by injection of mycobacterial adjuvant. *V. Changes affecting the skin and mucous membranes. Comparison of the experimental process with human disease. J Exp Med*, **113**: 485–510.

Pellegrini, M., Nicolay, U., Lindert, K., *et al.* (2009). MF59-adjuvanted versus non-adjuvanted influenza vaccines: integrated analysis from a large safety database. *Vaccine*, **27**: 6959–65.

Poltorak, A., He, X., Smirnova, I., *et al.* (1998). Defective LPS signaling in C3H/HeJ and C57BL/10ScCr mice: mutations in *Tlr4* gene. *Science*, **282**: 2085–8.

Pummerer, C.L., Luze, K., Grassl, G., *et al.* (1996). Identification of cardiac myosin peptides capable of inducing autoimmune myocarditis in BALB/c mice. *J Clin Invest*, **97**: 2057–62.

Qureshi, S.T., Lariviere, L., Leveque, G., *et al.* (1999). Endotoxin-tolerant mice have mutations in Toll-like receptor 4 (*Tlr4*). *J Exp Med*, **189**: 615–25.

Rangachari, M. and Kuchroo, V.K. (2013). Using EAE to better understand principles of immune function and autoimmune pathology. *J Autoimmun*, **45**: 31–9.

Reynolds, J.M., Gustavo, J., Martinez, Y.C., and Chen, D. (2012). Toll-like receptor 4 signaling in T cells promotes autoimmune inflammation. *Proc Natl Acad Sci USA*, **109**: 13 064–9.

Riad, A., Westermann, D., Escher, F., *et al.* (2010). Myeloid differentiation factor-88 contributes to TLR9-mediated modulation of acute coxsackievirus B3-induced myocarditis *in vivo*. *Am J Physiol Heart Circ Physiol*, **298**: H2024–31.

Riad, A., Westermann, D., Zietsch, C., *et al.* (2011). TRIF is a critical survival factor in viral cardiomyopathy. *J Immunol*, **186**: 2561–70.

Richer, M.J., Fang, D., Shanina, I., and Horwitz, M.S. (2006). Toll-like receptor 4-induced cytokine production circumvents protection conferred by TGF-β in coxsackievirus-mediated autoimmune myocarditis. *Clin Immunol*, **121**: 339–49.

Rock, F.L., Hardiman, G., Timans, J.C., *et al.* (1998). A family of human receptors structurally related to *Drosophila* Toll. *Proc Natl Acad Sci USA*, **95**: 588–93.

Ronald, P.C. and Beutler, B. (2010). Plant and animal sensors of conserved microbial signatures. *Science*, **330**: 1061–4.

Rose, N.R. (2008a). The adjuvant effect in infection and autoimmunity. *Clin Rev Allergy Immunol*, **34**: 279–82.

Rose, N.R. (2008b). Autoimmunity in coxsackievirus infection. *Curr Top Microbiol Immunol*, **323**: 293–314.

Rose, N.R. and Afanasyeva, M. (2003). From infection to autoimmunity: the adjuvant effect. *ASM News*, **69**: 132–7.

Rose, N.R. and Griffin, D.E. (1991). Virus-induced autoimmunity. In: Talal, N. (ed.). *Molecular Autoimmunity*. London, San Diego: Academic Press, pp. 247–72.

Rose, N.R., Wolfgram, L.J., Herskowitz, A., and Beisel, K.W. (1986). Postinfectious autoimmunity: two distinct phases of coxsackievirus B3-induced myocarditis. *Ann NY Acad Sci*, **475**: 146–56.

Rose, N.R., Beisel, K.W., Herskowitz, A., *et la.* (1987). Cardiac myosin and autoimmune myocarditis. In: Evered, D. and Whelan, J. (eds). *Ciba Foundation Symposium 129 – Autoimmunity and Autoimmune Disease*. Chichester, UK: John Wiley & Sons, Ltd., pp. 3–24.

Sansonetti, P.J. (2006). The innate signaling of dangers and the dangers of innate signaling. *Nat Immunol*, **7**: 1237–42.

Shpacovitch, V., Feld, M., Bunnett, N.W., and Steinhoff, M. (2007). Protease-activated receptors: novel PARtners in innate immunity. *Trends Immunol*, **28**: 541–50.

Shpacovitch, V., Feld, M., Hollenberg, M.D., *et al.* (2008). Role of protease-activated receptors in inflammatory responses, innate and adaptive immunity. *J Leukoc Biol*, **83**: 1309–22.

Sozzani, S., Bosisio, D., Scarsi, M., and Tincani, A. (2010). Type I interferons in systemic autoimmunity. *Autoimmunity*, **43**: 196–203.

Takeuchi, O., Hoshino, K., Kawai, T., *et al.* (1999). Differential roles of TLR2 and TLR4 in recognition of gram-negative and gram-positive bacterial cell wall components. *Immunity*, **11**: 443–51.

Thotathil, Z. and Jameson, M.B. (2007). Early experience with novel immunomodulators for cancer treatment. *Expert Opin Investig Drugs*, **16**: 1391–403.

Toshchakov, V., Jones, B.W., Perera, P.Y., *et al.* (2002). TLR4, but not TLR2, mediates IFN-β-induced STAT1α/β-dependent gene expression in macrophages. *Nat Immunol*, **3**: 392–8.

Toubi, E. and Shoenfeld, Y. (2004). Toll-like receptors and their role in the development of autoimmune diseases. *Autoimmunity*, **37**: 183–8.

Triantafilou, K., Orthopoulos, G., Vakakis, E., *et al.* (2005). Human cardiac inflammatory responses triggered by Coxsackie B viruses are mainly Toll-like receptor (TLR) 8-dependent. *Cell Microbiol*, **7**: 1117–26.

Trinchieri, G. and Sher, A. (2007). Cooperation of Toll-like receptor signals in innate immune defence. *Nat Rev Immunol*, **7**: 179–90.

van de Veerdonk, F.L., Teirlinck, A.C., Kleinnijenhuis, J., *et al.* (2010). Mycobacterium tuberculosis induces IL-17A responses through TLR4 and dectin-1 and is critically dependent on endogenous IL-1. *J Leukoc Biol*, **88**: 227–32.

Vogel, S.N., Fitzgerald, K.A., and Fenton, M.J. (2003). TLRs: differential adapter utilization by Toll-like receptors mediates TLR-specific patterns of gene expression. *Mol Interv*, **3**: 466–77.

Waksman, B.H. (2002). Immune regulation in adjuvant disease and other arthritis models: relevance to pathogenesis of chronic arthritis. *Scand J Immunol*, **56**: 12–34.

Wang, Z.Y., Yang, D., Chen, Q., *et al.* (2006). Induction of dendritic cell maturation by pertussis toxin and its B subunit differentially initiate Toll-like receptor 4-dependent signal transduction pathways. *Exp Hematol*, **34**: 1115–24.

Wang, J.P., Cerny, A., Asher, D.R., *et al.* (2010). MDA5 and MAVS mediate type I interferon responses to coxsackie B virus. *J Virol*, **84**: 254–60.

Weithauser, A., Bobbert, P., Antoniak, S., *et al.* (2013). Protease-activated receptor 2 regulates the innate immune response to viral infection in a CVB3-induced myocarditis. *J Am Coll Cardiol*, **62**(19): 1737–45.

White, R.G. (1976). The adjuvant effect of microbial products on the immune response. *Annu Rev Microbiol*, **30**: 579–600.

Whitehouse, M.W. (2007). Adjuvant arthritis 50 years on: the impact of the 1956 article by C. M. Pearson, "Development of arthritis, periarthritis and periostitis in rats given adjuvants." *Inflamm Res*, **56**: 133–8.

Whitham, S., Dinesh-Kumar, S.P., Choi, D., *et al.* (1994). The product of the tobacco mosaic virus resistance gene *N*: similarity to toll and the interleukin-1 receptor. *Cell*, **78**: 1101–15.

Wolfgram, L.J., Beisel, K.W., Herskowitz, A., and Rose, N.R. (1986). Variations in the susceptibility to Coxsackievirus B3-induced myocarditis among different strains of mice. *J Immunol*, **136**: 1846–52.

Yamamoto, M., Takeda, K., and Akira, S. (2004). TIR domain-containing adaptors define the specificity of TLR signaling. *Mol Immunol*, **40**: 861–8.

3 Experimental Models of Adjuvants

Nicola Bassi, Mariele Gatto, Anna Ghirardello, and Andrea Doria
Division of Rheumatology, Department of Medicine, University of Padua, Padua, Italy

Adjuvant models

Many studies have been carried out to investigate the immune pathways stimulated by adjuvants, the role of adjuvants in enhancing pathogen-specific immune responses, and the potential noxious effects of adjuvants for the recipient, both in animal models and in humans. Mounting evidence demonstrates that the adjuvants and preservatives included in vaccines may not only enhance antigenic stimulation, but also be capable of inducing autoantibodies, inflammation, and aberrant immune responses triggering overt autoimmune manifestations, such as arthritis, neuronal damage, encephalitis, myocarditis, and vasculitis (Geier and Geier, 2005; Abu-Shakra, 2009). In fact, mineral oils may induce sclerosing lipogranulomas (Di Benedetto *et al.*, 2002); silicone has been implicated in the induction of scleroderma, systemic lupus erythematosus (SLE), and rheumatoid arthritis (RA) (Spiera *et al.*, 1994); and alum, aluminum hydroxide [$Al(OH)_3$], and squalene (2,6,10,15,19,23-hexamethyl-2,6,10,14,18,22-tetracosahexaene) are all reported as cofactors in chronic fatigue syndrome, polymyalgia, macrophagic myofasciitis (MMF), and Gulf War syndrome (GWS) (Asa *et al.*, 2002; Gherardi, 2003). Adjuvants may also induce the production of antibodies against themselves, as reported in US military personnel with GWS, in whom high levels of antisqualene circulating antibodies were found (Asa *et al.*, 2000; Gronseth, 2005; Lippi *et al.*, 2010).

Such autoimmune effects induced by adjuvants have been extensively evaluated using different animal models, ranging from rats to primates, though rodents have been the most extensively studied (Table 3.1). Interestingly, rodents models include both autoimmune-prone strains (i.e. strains spontaneously developing autoimmune diseases, e.g. NZB/NZW mice) and models of experimental autoimmune diseases (in which autoimmunity may be triggered, e.g. BALB/c mice, Dark Agouti mice, and C5BL/6), and even autoimmune-resistant strains (e.g. Sprague–Dawley mice) (Germolec *et al.*, 2012).

Murine models

Rats

Rats are mainly useful for studies on RA, since they develop arthritis following the injection of oil adjuvants, such as complete Freund's adjuvant (CFA), pristane, squalene, or avridine. Notably, these adjuvants are not immunogenic, because they do not contain major histocompatibility complex (MHC)-binding peptides.

Dark Agouti rats

Females belonging to the Dark Agouti (DA) strain display a defective bile acid transport, and might function as an animal model for debrisoquine hydroxylation in humans (Reichen *et al.*, 1986). This strain is widely used by immunologists, because it is susceptible to development of autoimmune thyroiditis (Rose, 1975) or severe

Vaccines and Autoimmunity, First Edition. Edited by Yehuda Shoenfeld, Nancy Agmon-Levin, and Lucija Tomljenovic.

collagen-induced arthritis following immunization with bovine, chick, or rat type II collagens, which is exacerbated by infection caused by rat cytomegalovirus (Griffiths *et al.*, 1994).

Injection of CFA, incomplete Freund's adjuvant (IFA), pristane (2,6,10,14-tetramethylpentadecane), or squalene (2,6,10,15,19,23-hexamethyl-2,6,10, 14,18,22-tetracosahexaene) alone has been demonstrated to induce chronic arthritis in this model (Kleinau *et al.*, 1991, 1994; Holmdahl and Kvick, 1992; Holmdahl *et al.*, 1992, 2001; Cannon *et al.*, 1993; Carlson *et al.*, 2000).

Sprague–Dawley rats

Sprague–Dawley rats are characterized by a naturally well-balanced immunity (Authier *et al.*, 2006). This rat model is especially sensitive to CFA-induced arthritis (Bersani-Amado *et al.*, 1990). The intraplantar injection of CFA also induces behavioral hyperalgesia, along with glucocorticoid inducible kinase 1 phosphorylation in the ipsilateral dorsal horn (Peng *et al.*, 2012), an enhanced mechanical allodynia, thermal hyperalgesia, a static weight-bearing deficit, and notably pronounced spontaneous foot lifting (Allchorne *et al.*, 2012). Furthermore, CFA induces temporomandibular joint inflammation after injection into the unilateral temporomandibular joint (Wang *et al.*, 2012). Finally, it has been demonstrated that this model can develop MMF induced by the injection of an $Al(OH)_3$-adjuvanted vaccine (Authier *et al.*, 2006).

Mice

BALB/c

The BALB/c model is used as a general-purpose strain in many disciplines and is well known for the development of plasmacytomas by the injection of mineral oil and pristane mixtures (Anderson and Potter, 1969). Moreover, subcutaneous injection of mineral oil induces sclerosing lipogranulomas, a chronic local inflammatory reaction (Di Benedetto *et al.*, 2002), and a single injection of pristane, CFA, or squalene has been shown to induce lupus-related autoantibodies toward nRNP/Sm and Su in nonautoimmune BALB/c mice (Satoh and Reeves, 1994; Satoh *et al.*, 1995, 2003; Kuroda *et al.*, 2004; Reeves *et al.*, 2009).

Notably, this model was used to study the effects of MF59, a safe and potent oil/water emulsion adjuvant for human use, eliciting both humoral and cellular responses; no collateral effects were observed (Dupuis *et al.*, 1998, 2000).

Finally, it has been shown that phytol compounds and adjuvants from porcine small-intestinal submucosa (SIS-H) are as effective as alum and have few deleterious effects; similar evidence has been provided from C57BL/6 and NZB/NZW mice (Aachoui and Ghosh, 2011).

C57BL/6

This model has been used to explore the effects of aluminum adjuvants. It has been shown that NALP3 inflammasome is a crucial element in the adjuvant effect of aluminum and that the innate inflammasome pathway can elicit a humoral adaptive immune response (Eisenbarth *et al.*, 2008; McKee *et al.*, 2009). Moreover, the immunization of transgenic factor V Leiden-mutated C57/BL6-back-crossed mice with CFA or IFA induced high levels of pathogenic antiphospholipid antibodies (Katzav *et al.*, 2012), leading to the development of an antiphospholipid-like syndrome, whose ontogenetic mechanisms are similar to those underlying the so-called "autoimmuune (autoinflammatory) syndrome induced by adjuvants" (ASIA) (Shoenfeld and Agmon-Levin, 2011).

NZB/NZWF1

This is a model of human SLE. It develops a lupus-like glomerulonephritis (GLN) within 5–7 months of age. As in humans, anti-double-stranded DNA (dsDNA) antibodies are found at high levels in the circulation and deposited as immune complexes in glomeruli of NZB/NZWF1.

In this model, CFA injection has been demonstrated to accelerate proteinuria and GLN onset and to worsen systemic organ involvement (Bassi *et al.*, 2012a). Notably, it has also been used to demonstrate that alum is as effective as and safer than CFA (Bassi *et al.*, 2012b).

Salmon

In order to prevent several costly infectious diseases, farmed salmon are intraperitoneally injected with vaccines containing adjuvant oil and a number of different antigens (Sommerset *et al.*, 2005). The amount of vaccine used, adjusted for salmon body weight, is considerably higher than in mammals (Koppang *et al.*, 2008), which may account for the higher frequency and severity of vaccine-induced side effects in salmon compared to farmed mammals and humans. Recipient fish may over time develop mild to severe pathological changes as a consequence of vaccination; observed

Table 3.1 Autoimmune and inflammatory manifestations in animal models induced by different adjuvants

Model	Characteristics/aims	Model for	Adjuvants	Symptoms
Dark Agouti (DA) rats	Defective bile acid transport Autoimmune thyroiditis	Debrisoquine hydroxilation Collagen-induced arthritis	CFA, IFA, pristane, squalene	Arthritis
Sprague–Dawley Rats	Well-balanced immunity	Arthritis	CFA	Arthritis, hyperalgesia, mechanical allodynia, weight-bearing deficit, foot lifting, joint inflammation
			$Al(OH)_3$	MMF
BALB/c Mice	General-purpose strain	Plasmacytomas	Mineral-oil adjuvants	Plasmacytomas, sclerosing lipogranulomas
			Pristane, CFA, squalene	nRNP/Sm, Su autoantibodies
			MF59, SIS-H	No side effects
C57BL/6 mice	General-purpose strain	Inflammasome	Aluminum	Inflammasome activation
			CFA, IFA	Antiphospholipid autoantibodies, ASIA syndrome
NZB/NZWF1 mice	GLN lupus-like	SLE	CFA	GLN acceleration
			Alum	No side effects
Salmon	Vaccination		Oil adjuvants	Impaired growth rate, spinal deformities, uveitis, inflammatory reactions, rheumatoid factor, immunocomplex GLN, antinuclear and anticytoplasmatic autoantibodies
Rabbits	Vaccine production		Gerbu, Montanide, $Al(OH)_3$, $AlPO_4$	No side effects
			(AS)01, AS03, AS15	Increased CRP, increased fibrinogen
Swine	Vaccination		Penaut oil, $Al(OH)_3$	Granulomatous inflammation
			CFA, IFA, montanide ISA 206	Local adverse reactions
Primates	Preclinical vaccine testing		Methylmercury, ethylmercury	No persistent side effects
			$Al(OH)_3$	Potential delayed acquisition of neonatal reflexes

CFA, complete Freund's adjuvant; IFA, incomplete Freund's adjuvant; $Al(OH)_3$, aluminum hydroxide; MMF, macrophagic myofasciitis syndrome; GLN, glomerulonephritis; SLE, systemic lupus erythematosus; $AlPO_4$, aluminium phosphate; CRP, C-reactive protein

adverse reactions include impaired growth rate, decreased carcass quality, spinal deformities, uveitis, and inflammatory reactions in the abdominal cavity, due to activation of the MHC, probably caused by adjuvants in the vaccines (Koppang *et al.*, 2003, 2004, 2005; Evensen *et al.*, 2005).

In vaccinated farmed Atlantic salmon, rheumatoid factor, antinuclear and anticytoplasmic autoantibodies, immune-complex GLN, and chronic granulomatous inflammation have all been detected (Koppang *et al.*, 2008; Haugarvoll *et al.*, 2010). These systemic autoimmune reactions have been potentially linked with the oil adjuvants contained in vaccines; however, whether this is due to oil redistribution of injected vaccine components or to a generalized inflammatory response, or both, cannot be directly inferred (Haugarvoll and Koppang, 2005; Koppang *et al.*, 2005, 2008; Haugarvoll *et al.*, 2010).

Rabbits

The rabbit model is often used for vaccines intended for intramuscular injection. Indeed, a full human dose volume can be injected via this route at a single site in this animal, and the full human dose rather than a lower dose provides a more representative view of the reactogenicity of the vaccine (i.e. its capacity to induce certain manifestations of an inflammatory response at the injection site, such as redness/erythema and swelling/edema), as combined effects of volume and quantity of antigen(s) may affect the potential toxic effects at the injection site.

Rabbits have been used to evaluate new adjuvants. Gerbu adjuvant produces a similar or enhanced immune response without the undesirable side effects associated with CFA and IFA, demonstrating that it produces some of the most sensitive antibodies at relatively low titers and with common adverse effects at injection sites. Montanide adjuvant produces no adverse effects and the related antibodies levels are comparable to those produced with both CFA and IFA, suggesting it might be a suitable replacement for CFA and IFA in the production of polyclonal antibodies to low-molecular-weight compounds in rabbits (Fodey *et al.*, 2008).

Recently, in order to find new inflammatory markers by which to evaluate side effects in adjuvant and vaccine toxicity studies, rabbits were injected with aluminum phosphate, $Al(OH)_3$, AS01, AS03, or AS15. This study demonstrated that C reactive protein (CRP) and fibrinogen levels increased after the injection of AS01, AS03, and AS15, peaking at day 1 after injection. All increases in CRP and fibrinogen serum levels at day 1 after injection were accompanied by increases in the number of circulating heterophils. In contrast, after the injection of aluminum phosphate or $Al(OH)_3$, there were no increases in CRP and fibrinogen serum levels, and no increases in circulating heterophils (Destexhe *et al.*, 2013).

Swine

Porcine pleuropneumonia is a worldwide contagious disease of the respiratory tract of pigs that has led to considerable losses in swine herds over recent decades (Bossé *et al.*, 2002; Sebunya and Saunders, 1983). The etiological agent of this infection, *Actinobacillus pleuropneumoniae*, causes severe and fatal fibrinous hemorrhagic necrotizing pneumonia in pigs.

Many studies have been carried out to investigate the efficacy of vaccines and the potential side effects of adjuvants in this disease. In one, the effects of four mineral-oil adjuvant compounds, including one peanut-oil compound and aluminum hydroxide $Al(OH)_3$, were compared (Straw *et al.*, 1985). The adjuvants were inoculated in swine neck, quadriceps, and semitendinosus muscles. The mineral-oil adjuvants were highly irritant and caused extensive areas of granulomatous inflammation that lasted 8 weeks after injection. In another study (Krejci *et al.*, 2013), piglets were intradermally treated with repeated administrations of antigen mixed with different adjuvants (CFA, IFA, $Al(OH)_3$, montanide ISA 206, and emulsigen). Although the benefits of oil adjuvants in terms of improvement of mucosal and cell-mediated immune responses induced by intradermal immunization have been demonstrated, none of the oil adjuvants used in this experiment could be recommended without reservation because adverse local reactions were always found to take place. Given the most favorable ratio between local adverse reactions and immunostimulating effects, montanide ISA 206 appears to be the most promising candidate for further development (Krejci *et al.*, 2013).

Primates

The Rhesus macaque is commonly used in preclinical testing of vaccine neurovirulence and displays complex early neurobehavioral and developmental processes that are well characterized (Ruppenthal and Sackett, 1992).

Macaques have been extensively used in studies on methyl- and ethylmercury toxicokinetics and developmental neurotoxicity, in which persistent side effects were not found (Rice and Gilbert, 1982, 1990; Gunderson *et al.*, 1986, 1988; Burbacher *et al.*, 1990, 2005). Studies on male primates support the mounting evidence for a gender-selective neurotoxicity of organomercurials in both humans and animals (Rossi *et al.*, 1997; Sakamoto *et al.*, 1998; Gao *et al.*, 2007; White *et al.*, 2007; Branch *et al.*, 2009; Malagutti *et al.*, 2009).

Notably, it has been shown that aluminum contained in vaccines is detectable in vaccinated Macaques at the site of injection for at least 6 months, but no evidence of MMF has been found (Verdier *et al.*, 2005); however, a delayed acquisition of neonatal reflexes in newborn primates vaccinated for hepatitis B has been attributed to $Al(OH)_3$ or preservatives contained in vaccines (Hewitson *et al.*, 2010).

Conclusions

Several kinds of animal models have been employed to investigate adjuvant effects and midterm reactions *in vivo*, in an attempt to better unravel adverse events in humans following vaccinations or prosthesis implantations. Higher organisms, such as primates, are rarely used; rather, different mouse models have provided the bulk of the evidence for vaccine and adjuvants effects in living beings. Interestingly, adverse reactions rarely take place in non-overtly autoimmune-prone models, acting as putative triggering agents on an apparent neutral background; further, spontaneous autoimmune-prone strains are rapidly doomed to autoimmunity as they undergo adjuvant administration. Notably, animal models rendered transgenic for a non-autoimmune-related alteration (e.g. mutation of V factor Leiden) show a greater susceptibility in developing an acquired autoimmune condition following adjuvant administration (e.g. antiphospholipid syndrome), suggesting that genetic alterations, though devoid of immune influence, may result in autoimmunity as soon as a triggering agent is encountered. The underlying mechanisms are still unknown, but a role for perturbations in cytokine production via Toll-like receptor (TLR) stimulation is likely, alongside the capability of adjuvants to foster an immune response, whether protective or aberrant. Further studies are warranted, in order to progressively unravel the pathogenic pathways elicited by adjuvants and to definitely state the caution required in treating autoimmunity-bearing patients.

References

Aachoui, Y. and Ghosh, S.K. (2011). Immune enhancement by novel vaccine adjuvants in autoimmune-prone NZB/W F1 mice: relative efficacy and safety. *BMC Immunol*, **12**: 61.

Abu-Shakra, M. (2009). Safety of vaccination of patients with systemic lupus erythematosus. *Lupus*, **18**(13): 1205–8.

Allchorne, A.J., Gooding, H.L., Mitchell, R., and Fleetwood-Walker, S.M. (2012). A novel model of combined neuropathic and inflammatory pain displaying long-lasting allodynia and spontaneous pain-like behaviour. *Neurosci Res*, **74**(3–4): 230–8.

Anderson, P.N. and Potter, M. (1969). Induction of plasma cell tumours in BLAB-c mice with 2,6,10,14-tetramethylpentadecane (pristine). *Nature*, **222**(5197): 994–5.

Asa, P.B., Cao, Y., and Garry, R.F. (2000). Antibodies to squalene in Gulf War syndrome. *Exp Mol Pathol*, **68**(1): 55–64.

Asa, P.B., Wilson, R.B., and Garry, R.F. (2002). Antibodies to squalene in recipients of anthrax vaccine. *Exp Mol Pathol*, **73**(1): 19–27.

Authier, F.J., Sauvat, S., Christov, C., *et al.* (2006). Al(OH)3-adjuvanted vaccine-induced macrophagic myofasciitis in rats is influenced by the genetic background. *Neuromuscul Disord*, **16**(5): 347–52.

Bassi, N., Luisetto, R., Del Prete, D., *et al.* (2012a). Induction of the "ASIA" syndrome in NZB/NZWF1 mice after injection of complete Freund's adjuvant (CFA). *Lupus*, **21**(2): 203–9.

Bassi, N., Luisetto, R., Ghiardello, A., *et al.* (2012b). Vaccination of mice for research purpose: alum is as effective as and safer than complete Freund adjuvant. *Reumatismo*, **64**(6): 380–7.

Bersani-Amado, C.A., Barbuto, J.A., and Jancar, S. (1990). Comparative study of adjuvant induced arthritis in susceptible and resistant strains of rats. *I. Effect of cyclophosphamide. J Rheumatol*, **17**(2): 149–52

Bossé, J.T., Janson, H., Sheehan, B.J., *et al.* (2002). Actinobacillus pleuropneumoniae: pathobiology and pathogenesis of infection. *Microbes Infect*, **4**(2): 225–35.

Branch, D.R. (2009). Gender-selective toxicity of thimerosal. *Exp Toxicol Pathol*, **61**(2): 133–6.

Burbacher, T.M., Sackett, G.P., and Mottet, N.K. (1990). Methylmercury effects on the social behavior of *Macaca fascicularis* infants. *Neurotoxicol Teratol*, **12**(1): 65–71.

Burbacher, T.M., Shen, D.D., Liberato, N., *et al.* (2005). Comparison of blood and brain mercury levels in infant monkeys exposed to methylmercury or vaccines containing thimerosal. *Environ Health Perspect*, **113**(8): 1015–21.

Cannon, G.W., Woods, M.L., Clayton, F., and Griffiths, M.M. (1993). Induction of arthritis in DA rats by incomplete Freund's adjuvant. *J Rheumatol*, **20**(1): 7–11.

Carlson, B.C., Jansson, Å.M., Larsson, A., *et al.* (2000). The endogenous adjuvant squalene can induce a chronic T-cell-mediated arthritis in rats. *Am J Pathol*, **156**(6): 2057–65.

Destexhe, E., Prinsen, M.K., van Schöll, I., *et al.* (2013). Evaluation of C-reactive protein as an inflammatory biomarker in rabbits for vaccine nonclinical safety studies. *J Pharmacol Toxicol Methods*, **68**(3): 367–73.

Di Benedetto, G., Pierangeli, M., Scalise, A., and Bertani, A. (2002). Paraffin oil injection in the body: an obsolete and destructive procedure. *Ann Plast Sur*, **49**(4): 391–6.

Dupuis, M., Murphy, T.J., Higgins, D., *et al.* (1998). Dendritic cells internalize vaccine adjuvant after intramuscular injection. *Cell Immunol*, **186**(1): 18–27.

Dupuis, M., McDonald, D.M., and Ott, G. (2000). Distribution of adjuvant MF59 and antigen gD2 after intramuscular injection in mice. *Vaccine*, **18**(5–6): 434–9.

Eisenbarth, S.C., Colegio, O.R., O'Connor, W., *et al.* (2008). Crucial role for the Nalp3 inflammasome in the immunostimulatory properties of aluminium adjuvants. *Nature*, **453**(7198): 1122–6.

Evensen, Ø., Brudeseth, B., and Mutoloki, S. (2005). The vaccine formulation and its role in inflammatory processes in fish - effects and adverse effects. *Dev Biol (Basel)*, **121**: 117–25.

Fodey, T.L., Delahaut, P., Charlier, C., and Elliott, C.T. (2008). Comparison of three adjuvants used to produce polyclonal antibodies to veterinary drugs. *Vet Immunol Immunopathol*, **122**(1–2): 25–34.

Gao, Y., Yan, C.H., Tian, Y., *et al.* (2007). Prenatal exposure to mercury and neurobehavioral development of neonates in Zhoushan City, China. *Environ Res*, **105**(3): 390–9.

Geier, D.A. and Geier, M.R. (2005). A case–control study of serious autoimmune adverse events following hepatitis B immunization. *Autoimmunity*, **38**(4): 295–301.

Germolec, D., Kono, D.H., Pfau, J.C., and Pollard, K.M. (2012). Animal models used to examine the role of the environment in the development of autoimmune disease: findings from an NIEHS Expert Panel Workshop. *J Autoimmun*, **39**(4): 285–93.

Gherardi, R.K. (2003). Lessons from macrophagic myofasciitis: towards definition of a vaccine adjuvant-related syndrome. *Rev Neurolog*, **159**(2): 162–4.

Griffiths, M.M., Sawitzke, A.D., Harper, D.S., *et al.* (1994). Exacerbation of collagen-induced arthritis in rats by rat cytomegalovirus is antigen-specific. *Autoimmunity*, **18**(3): 177–87.

Gronseth, G.S. (2005). Gulf war syndrome: a toxic exposure? A systematic review. *Neurol Clin*, **23**(2): 523–40.

Gunderson, V.M., Grant, K.S., Burbacher, T.M., *et al.* (1986). The effect of low-level prenatal methylmercury exposure on visual recognition memory in infant crab-eating macaques. *Child Dev*, **57**(4): 1076–83.

Gunderson, V.M., Grant-Webster, K.S., Burbacher, T.M., and Mottet, N.K. (1988). Visual recognition memory deficits in methylmercury-exposed *Macaca fascicularis* infants. *Neurotoxicol Teratol*, **10**(4): 373–9.

Haugarvoll, E. and Koppang, E.O. (2005). Vaksinering av oppdrettsfisk – sjukdomsvern med attåtsmak. *Nor Vet Tidsskr*, **117**: 286–90.

Haugarvoll, E., Bjerkås, I., Szabo, N.J., *et al.* (2010). Manifestations of systemic autoimmunity in vaccinated salmon. *Vaccine*, **28**(31): 4961–9.

Hewitson, L., Houser, L.A., Stott, C., *et al.* (2010). Delayed acquisition of neonatal reflexes in newborn primates receiving a thimerosal-containing hepatitis B vaccine: influence of gestational age and birth weight. *J Toxicol Environ Health A*, **73**(19): 1298–313.

Holmdahl, R. and Kvick, C. (1992). Vaccination and genetic experiments demonstrate that adjuvant-oil-induced arthritis and homologous type II collagen-induced arthritis in the same rat strain are different diseases. *Clin Exp Immunol*, **88**(1): 96–100.

Holmdahl, R., Goldschmidt, T.J., Kleinau, S., *et al.* (1992). Arthritis induced in rats with adjuvant oil is a genetically restricted, alpha beta T-cell dependent autoimmune disease. *Immunology*, **76**(2): 197–202.

Holmdahl, R., Lorentzen, J.C., Lu, S., *et al.* (2001). Arthritis induced in rats with non-immunogenic adjuvants as models for rheumatoid arthritis. *Immunol Rev*, **184**: 184–202.

Katzav, A., Kivity, S., Blank, M., *et al.* (2012). Adjuvant immunization induces high levels of pathogenic antiphospholipid antibodies in genetically prone mice: another facet of the ASIA syndrome. *Lupus*, **21**(2): 210–16.

Kleinau, S., Erlandsson, H., Holmdahl, R., and Klareskog, L. (1991). Adjuvant oils induce arthritis in the DA rat. *I. Characterization of the disease and evidence for an immunological involvement. J Autoimmun*, **4**(6): 871–80.

Kleinau, S., Erlandsson, H., and Klareskog, L. (1994). Percutaneous exposure of adjuvant oil causes arthritis in DA rats. *Clin Exp Immunol*, **96**(2): 281–4.

Koppang, E.O., Bjerkås, E., Bjerkås, I., *et al.* (2003). Vaccination induces major histocompatibility complex class II expression in the Atlantic salmon eye. *Scand J Immunol*, **58**(1): 9–14.

Koppang, E.O., Haugarvoll, E., Hordvik, I., *et al.* (2004). Granulomatous uveitis associated with vaccination in the Atlantic salmon. *Vet Pathol*, **41**(2): 122–30.

Koppang, E.O., Haugarvoll, E., Hordvik, I., *et al.* (2005). Vaccine-associated granulomatous inflammation and melanin accumulation in Atlantic salmon, *Salmo salar* L., white muscle. *J Fish Dis*, **28**(1): 13–22.

Koppang, E.O., Bjerkås, I., Haugarvoll, E., *et al.* (2008). Vaccination-induced systemic autoimmunity in farmed Atlantic salmon. *J Immunol*, **181**(7): 4807–14.

Krejci, J., Nechvatalova, K., Kudlackova, H., *et al.* (2013). Effects of adjuvants on the immune response of pigs after intradermal administration of antigen. *Res Vet Sci*, **94**(1): 73–6.

Kuroda, Y., Nacionales, D.C., Akaogi, J., *et al.* (2004). Autoimmunity induced by adjuvant hydrocarbon oil components of vaccine. *Biomed Pharmacoter*, **58**(5): 325–37.

Lippi, G., Targher, G., and Franchini, M. (2010). Vaccination, squalene and antisqualene antibodies: facts or fiction? *Eur J Intern Med*, **21**(2): 70–3.

Malagutti, K.S., da Silva, A.P., Campos Braga, H., *et al.* (2009). 17ß-estradiol decreases methylmercury-induced neurotoxicity in male mice. *Environ Toxicol Pharmacol*, **27**(2): 293–7.

McKee, A.S, Munks, M.W., MacLeod, M.K., *et al.* (2009). Alum induces innate immune responses through macrophage and mast cell sensors, but these are not required for alum to act as an adjuvant for specific immunity. *J Immunol*, **183**(7): 4403–14.

Peng, H.Y., Chen, G.D., Hsieh, M.C., *et al.* (2012). Spinal SGK1/GRASP-1/Rab4 is involved in complete Freund's adjuvant-induced inflammatory pain via regulating dorsal horn GluR1-containing AMPA receptor trafficking in rats. *Pain*, **153**(12): 2380–92.

Reeves, W.H., Lee, P.Y., Weinstein, J.S., *et al.* (2009). Induction of autoimmunity by pristane and other naturally occurring hydrocarbons. *Trends Immunol*, **30**(9): 455–64.

Reichen, J., Krähenbühl, S., Küpfer, A., *et al.* (1986). Defective bile acid transport in an animal model of defective debrisoquine hydroxylation. *Biochem Pharmacol*, **35**(5): 753–9.

Rice, D.C. and Gilbert, S.G. (1982). Early chronic low-level methylmercury poisoning in monkeys impairs spatial vision. *Science*, **216**(4547): 759–61.

Rice, D.C. and Gilbert, S.G. (1990). Effects of developmental exposure to methyl mercury on spatial and temporal visual function in monkeys. *Toxicol Appl Pharmacol*, **102**(1): 151–63.

Rose, N.R. (1975). Differing responses of inbred rat strains in experimental autoimmune thyroiditis. *Cell Immunol*, **18**(2): 360–4.

Rossi, A.D., Ahlbom, E., Ogren, S.O., *et al.* (1997). Prenatal exposure to methylmercury alters locomotor activity of male but not female rats. *Exp Brain Res*, **117**(3): 428–36.

Ruppenthal, G.C. and Sackett, G.P. (1992). *Research Protocol and Technician's Manual*, 2nd edn. Seattle: Infant Primate Research Laboratory, University of Washington Press.

Sakamoto, M., Wakabayashi, K., Kakita, A., *et al.* (1998). Widespread neuronal degeneration in rats following oral administration of methylmercury during the postnatal developing phase: a model of fetal-type minamata disease. *Brain Res*, **784**(1–2): 351–4.

Satoh, M. and Reeves, W.H. (1994). Induction of lupus-associated autoantibodies in BALB/c mice by intraperitoneal injection of pristane. *J Exp Med*, **180**(6): 2341–6.

Satoh, M., Kumar, A., Kanwar, Y.S., and Reeves, W.H. (1995). Anti-nuclear antibody production and immune-complex glomerulonephritis in BALB/c mice treated with pristine. *Proc Nat Acad Sci USA*, **92**(24): 10 934–8.

Satoh, M., Kuroda, Y., Yoshida, H., *et al.* (2003). Induction of lupus autoantibodies by adjuvants. *J Autoimmun*, **21**(1): 1–9.

Sebunya, T.N.K. and Saunders, J.R. (1983). Haemophilus pleuropneumoniae infection in swine. *J Am Vet Med Assoc*, **182**: 1331–7.

Shoenfeld, Y. and Agmon-Levin, N. (2011). "ASIA" – autoimmune/inflammatory syndrome induced by adjuvants. *J Autoimmun*, **36**(1): 4–8.

Sommerset, I., Krossøy, B., Biering, E., and Frost, P. (2005). Vaccines for fish in aquaculture. *Expert Rev Vaccines*, **4**(1): 89–101.

Spiera, R.F., Gibofsky, A., and Spiera, H. (1994). Silicone gel filled breast implants and connective tissue disease: an overview. *J Rheumatol*, **21**(2): 239–45.

Straw, B.E., MacLachlan, N.J., Corbett, W.T., *et al.* (1985). Comparison of tissue reactions produced by haemophilus pleuropneumoniae vaccines made with six different adjuvants in swine. *Can J Comp Med*, **49**(2): 149–51.

Verdier, F., Burnett, R., Michelet-Habchi, C., *et al.* (2005). Aluminium assay and evaluation of the local reaction at several time points after intramuscular administration of aluminium containing vaccines in the Cynomolgus monkey. *Vaccine*, **23**(11): 1359–67.

Wang, S., Song, L., Tan, Y., *et al.* (2012). A functional relationship between trigeminal astroglial activation and NR1 expression in a rat model of temporomandibular joint inflammation. *Pain Med*, **13**(12): 1590–600.

White, J.F. (2007). Thimerosal: chemistry, toxicology and effects on biological systems. In: Zhao, L.B. (ed.). *Autism Research Challenges*. Hauppauge, NY: Nova Science, pp. 137–67.

4 Answers to Common Misconceptions Regarding the Toxicity of Aluminum Adjuvants in Vaccines

Lucija Tomljenovic[1] and Christopher A. Shaw[2]
[1]Neural Dynamics Research Group, University of British Columbia, Vancouver, BC, Canada
[2]Department of Ophthalmology and Visual Sciences, Program in Experimental Medicine, Program in Neuroscience, University of British Columbia, Vancouver, BC, Canada

Introduction

Numerous studies have documented the appearance of diverse autoimmune disorders following routine vaccinations, the most common of which are those affecting the nervous system. Although the classical explanations for these phenomena have largely centered on vaccine antigens, in recent years attention has shifted to aluminum (Al) compounds, which are the most common commercial adjuvants in current use. Al is both immuno- and neurotoxic, and, in the last decade, studies on animal models and humans have indicated that Al adjuvants have an intrinsic ability to inflict immune and inflammatory responses (Couette *et al.*, 2009; Li *et al.*, 2009; Shaw and Petrik, 2009; Passeri *et al.*, 2011). This research culminated in the delineation of "autoimmune/inflammatory syndrome induced by adjuvants" (ASIA), which encompasses the wide spectrum of adjuvant-triggered medical conditions characterized by a misregulated immune response (Meroni, 2010; Shoenfeld and Agmon-Levin, 2011). Notably, the vast majority of adverse manifestations experimentally triggered by Al in animal models, and of those associated with administration of adjuvanted vaccines in humans, are neurological and neuropsychiatric (Zafrir *et al.*, 2012; Lujan *et al.*, 2013). In this context, recent experiments have revealed that Al adjuvant nanoparticles have a unique capacity to cross the blood–brain barrier (BBB) and blood–cerebrospinal fluid (CSF) barrier and incite deleterious immunoinflammatory responses in neural tissues (Shaw and Petrik, 2009; Khan *et al.*, 2013). These observations may explain in part why vaccines have a predilection to affect the central nervous system (CNS).

In spite of these data, it is currently maintained by both the pharmaceutical industry and drug regulatory agencies that the concentrations at which Al is used in vaccines do not represent a health hazard (Offit and Jew, 2003; Eldred *et al.*, 2006; US FDA, 2014). In this chapter, we aim to answer the most common misconceptions regarding the safety of Al compounds in vaccines. We will do so by providing an overview of what is currently known about Al adjuvants, in particular, and their modes of action and mechanisms of potential toxicity. We first give a brief overview of the crucial role of Al in a variety of neurological disorders and then elaborate on the unresolved controversy about Al adjuvant safety.

Vaccines and Autoimmunity, First Edition. Edited by Yehuda Shoenfeld, Nancy Agmon-Levin, and Lucija Tomljenovic.

Al and neurological disorders

Al is the most abundant neurotoxic element in the earth's crust and is widely bioavailable to humans (ATSDR, 2008). It is present in many sources of drinking water, as a food additive, in many cosmetics, and in many pharmaceuticals, including vaccines. Because of this ubiquity, it is increasingly found in our bodies, too (Gherardi *et al.*, 2001; Walton, 2006; Couette *et al.*, 2009; Passeri *et al.*, 2011). None of this would necessarily be a problem if Al were benign in biological systems. However, in spite of a widely held belief that this is the case (Offit and Jew, 2003; Eldred *et al.*, 2006), it is demonstrably not so. Rather, research shows that Al is toxic on multiple levels. It is a neurotoxin (Joshi, 1990; Shaw and Petrik, 2009; Walton, 2009; Tomljenovic, 2011), a genotoxin (Lukiw, 2001), and an immunotoxin (Gherardi *et al.*, 2001; Israeli *et al.*, 2009; Batista-Duharte *et al.*, 2011), as well as being prooxidant (Verstraeten *et al.*, 1997; Exley, 2004a) and proinflammatory (Lukiw *et al.*, 2005). Additionally, it is an endocrine disruptor (Agarwal *et al.*, 1996; Singla and Dhawan, 2011), depresses glucose metabolism (Joshi, 1990; Singla and Dhawan, 2011), and interferes with many other essential cellular processes, such as calcium homeostasis (Walton, 2012a), various ATP-dependent mechanisms, membrane receptor signaling, and mitochondrial function (Shafer *et al.*, 1994; Tomljenovic, 2011).

The notion that Al is toxic is hardly novel: Dr William Gies, with 7 years of experimental testing in humans and animals on the effects of oral consumption of Al salts used in baking powders and food preservatives, had this to say in 1911 (Gies, 1911):

These studies have convinced me that the use in food of aluminum or any other aluminum compound is a dangerous practice. That the aluminum ion is very toxic is well known. That aluminized food yields soluble aluminum compounds to gastric juice (and stomach contents) has been demonstrated. That such soluble aluminum is in part absorbed and carried to all parts of the body by the blood can no longer be doubted. That the organism can "tolerate" such treatment without suffering harmful consequences has not been shown. It is believed that the facts in this paper will give emphasis to my conviction that aluminum should be excluded from food.

Since Gies' day, epidemiological, clinical, and experimental data have clearly identified the CNS as the most sensitive target of Al's toxic effects, regardless of mode of exposure (oral, injectable as an adjuvant in vaccines, etc.) (Bishop *et al.*, 1997; Rogers and Simon, 1999; Rondeau *et al.*, 2000; ATSDR, 2008; Shaw and Petrik, 2009; Walton, 2012b). The neurotoxicity of Al typically manifests in learning, memory, concentration, and speech deficits, impaired psychomotor control, increased seizure activity, and altered behavior (i.e. confusion, anxiety, repetitive behaviors, and sleep disturbances) (Tomljenovic, 2011).

After 101 years of general ignorance of Gies' prophetic concerns, we need now to reevaluate our increased intake of Al. This need is highlighted in burgeoning evidence that links Al to the spectrum of neurological diseases which plague the 21st century, including Alzheimer's (Tomljenovic, 2011), amyotrophic lateral sclerosis/Parkinsonism dementia (Perl and Moalem, 2006), multiple sclerosis (Authier *et al.*, 2001; Exley *et al.*, 2006), and autism spectrum and neurological impairments in children (Bishop *et al.*, 1997; Tomljenovic and Shaw, 2011; Seneff *et al.*, 2012; Melendez *et al.*, 2013).

In view of this, the question arises: what makes the brain particularly susceptible to Al's toxic impacts? First, it has intrinsically high glucose and oxygen requirements, a high surface area of biological membranes (especially vascular endothelium), a high tubulin content, a high phospholipid content, and a low concentration of antioxidants, compared with other organs (Tomljenovic, 2011). For example, although an adult human brain only weighs ~1.5 kg, it consumes 20% of total body oxygen and 120 g of glucose/day, compared to 190 g for the whole body (Joshi *et al.*, 1994). Given that Al is a known disruptor of glucose metabolism, a prooxidant and proinflammatory agent, and disrupts the assembly of microtubules, it can damage brain function at multiple levels (Tomljenovic, 2011). Moreover, even relatively small amounts of Al (i.e. the equivalent of what is injected via vaccinations) can reach the brain (Redhead *et al.*, 1992; Khan *et al.*, 2013).

In summary, a now abundant literature shows that exposure of humans and animals to Al from various sources can have deleterious consequences on the developing and adult nervous systems. These impacts may depend in large part on various factors, such as the form(s) of Al, the route of administration, and the concentration and duration of exposure. Included in this latter category is the issue of dietary versus injected Al. In addition, the final impact of Al will likely depend on a number of biological variables, including age, gender, and the potential and largely unidentified genetic susceptibility factors enhancing Al toxicity.

Al as adjuvant in vaccines

How is the safety of Al regulated?

Al salts (hydroxide and phosphate) are the most commonly used vaccine adjuvants and were until recently the only adjuvants licensed for use in the United States (Baylor *et al.*, 2002; Eickhoff and Myers, 2002; Brewer, 2006). In the absence of Al, antigenic components of most vaccines (with the exception of live attenuated vaccines) fail to launch an adequate immune response (Brewer, 2006; Israeli *et al.*, 2009). While the US Food and Drug Administration (FDA) does set an upper limit for Al in vaccines at no more than 850 μg/dose (Baylor *et al.*, 2002), it is important to note that this amount was selected empirically from data showing that Al in such amounts enhanced the antigenicity of the vaccine, rather than from existing safety data or on the basis of toxicological considerations (Baylor *et al.*, 2002). In preventative vaccination, where a vaccine is administered to healthy individuals, a compromise in efficacy for the sake of additional margins of safety should not necessarily be viewed as an unreasonable expectation (Batista-Duharte *et al.*, 2011). It is also of note that the FDA Department of Health and Human Services (DHHS) requires limits on Al in parenteral feeding solutions and requires warning labels about potential Al hazards, yet sets no safety limits or required warnings for Al in vaccines (US FDA DHHS, 2005).

The consequence of this view is best reflected in the fact that a large number of vaccine trials use an Al adjuvant-containing placebo or another Al-containing vaccine as the "control group," despite much evidence showing that Al in vaccine-relevant exposures is toxic to humans and animals (Gherardi *et al.*, 2001; Couette *et al.*, 2009; Li *et al.*, 2009; Shaw and Petrik, 2009; Passeri *et al.*, 2011; Shaw *et al.*, 2013) and that, therefore, its use as a placebo in vaccine trials is scientifically untenable (Exley, 2011). That the safety issue of Al in vaccines has indeed been overlooked by the regulators (for more than 90 years while these compounds have been in use) is illustrated by the following statement from the World Health Organization (WHO) Special Committee on the Safety of Vaccines (WHO, 2005): "The Committee considered the safety of adjuvants used in vaccines. This hitherto neglected subject is becoming increasingly important given modern advances in vaccine development and manufacture."

Dietary versus vaccine-derived Al: is there a difference?

Although Al is clearly neurotoxic, a common assertion is that humans obtain much more Al from diet than from vaccines, and that, therefore, the adjuvant form of Al does not represent a toxicological risk (Offit and Jew, 2003). However, this notion contradicts basic toxicological principles. For instance, it should be obvious that the route of exposure which bypasses the protective barriers of the gastrointestinal tract (GIT) and/or the skin will likely require a much lower dose to produce a toxic outcome. In the case of Al, only ~0.25% of dietary Al is absorbed into systemic circulation (Yokel *et al.*, 2008), and it is rapidly filtered by the kidneys in those with healthy kidney function. In contrast, Al hydroxide (the most common adjuvant form) injected intramuscularly may be absorbed at nearly 100% efficiency over time (Yokel and McNamara, 2001) and follows a completely different route in the body (i.e. accumulation in other organs, including the spleen and the brain) (Khan *et al.*, 2013).

What is also not widely known is that current regular human dietary consumption of Al is far from innocuous (Joshi, 1990; Rogers and Simon, 1999; Walton, 2012b). Although average estimates of total daily intakes vary between 2 and 25 mg Al/day (14–175 mg/week), individual intake in urban societies can easily exceed 100 mg/day (700 mg/week), due to a widespread increase in consumption of processed convenience foods, which are typically high in Al-containing additives (Tomljenovic, 2011). In response to increased dietary intake of Al, in 2006 the Food and Agriculture Organization (FAO) WHO Expert Committee amended its provisional tolerable weekly intake (PTWI) for Al from 7 mg/kg/bw (490 mg/week, for an average 70 kg human) to 1 mg/kg/bw (70 mg/week) (FAO/WHO, 2006). The Committee concluded that, "aluminum compounds have the potential to affect the reproductive system and developing nervous system at doses lower than those used in establishing the previous PTWI and therefore [we] revised the PTWI" (FAO/WHO, 2006). The take-home message is that a large proportion of people are unwittingly consuming significantly more Al than is considered safe by the expert food authorities (for more details, refer to Tomljenovic (2011)).

It is further important to note that although the half-life of enterally or parenterally absorbed Al from the body is short (approximately 24 hours), the same cannot be assumed for Al adjuvants in vaccines, as Al is tightly complexed to the vaccine antigen. Although the tightness of bonding between the Al adjuvant and the antigen is considered a desired feature, as it enhances the immunogenicity of vaccines (Egan *et al.*, 2009), this feature represents an additional problem for effective clearance of Al from the body, as the size of most Al-adsorbed antigen complexes is greater than the molecular weight cut-off of the glomerulus (Tomljenovic and Shaw, 2011). Experiments in adult rabbits demonstrate that even in an antigen-free form, Al hydroxide, the most commonly used adjuvant, is poorly excreted. The cumulative amount of Al hydroxide excreted in the urine of adult rabbits as long as 28 days post-intramuscular injection was less than 6%, as measured by accelerator mass spectrometry (Hem, 2002).

Moreover, current research shows that, other than antigens, Al can form complexes with other vaccine excipients. Recently, Lee (2013a) explored the melting profiles of the residual HPV L1 gene DNA contaminant recently detected in the quadrivalent HPV vaccine Gardasil. This quadrivalent vaccine contains genotype-specific L1 capsid proteins of four HPV strains (HPV-16, −18, −6, −11), in the form of virus-like particles (VLPs), as active ingredients, in addition to the Al adjuvant. Because viral DNA fragments, if present in the vaccine, may bind to the insoluble Al adjuvant and cause unintended pathophysiologic effects, Lee undertook experiments to develop a PCR-based test for HPV L1 gene DNA detection in the final products of Gardasil. The results showed that all samples tested (a total of 16 Gardasil vials) contained residues of the synthetic HPV-11 L1 gene DNA and/or HPV-18 L1 gene DNA. At least 7 of the 16 samples also contained HPV-16 L1 gene DNA, which was amplified by a pair of modified nondegenerate primers. While the HPV-11 and HPV-18 L1 gene DNA in Gardasil were readily amplified by the common degenerate consensus primers, the HPV-16 L1 gene DNA needed a specially designed nondegenerate PCR primer and different stringency conditions for amplification and detection (Lee, 2013a).

Notably, the specific melting profile of the HPV-16 L1 gene DNA detected in Gardasil vials was similar to that of the HPV-16 L1 gene DNA recently discovered in the post-mortem blood and spleen of a young woman who suffered unexpected death 6 months following Gardasil vaccination (Lee, 2012a, 2013b). Collectively, these findings suggest that the topological conformational changes in the HPV L1 gene DNA residues bound to the Al adjuvant may be genotype-related. Additionally, the particular conformation of the HPV-16 L1 gene DNA may prevent its degradation by endonucleases and cause the Al-HPV DNA complex to persist in the bodies of vaccine recipients long-term after injection (i.e. up to 6 months), thus increasing the risk for adverse immune responses (Lee, 2012a). Hence, routine testing for the presence of residual viral and microbial DNA bound to Al adjuvant in vaccines should be warranted for studies of vaccination safety, according to the PCR protocol developed by Lee (2013a,b).

Although the WHO Web page and documents provided by the vaccine manufacturers to regulatory agencies for licensing purposes specifically state that Gardasil is a highly purified vaccine and that the VLPs contain no nucleic acids (Merck Research Laboratories, 2010; WHO, 2014), the finding of such DNA residuals in Gardasil vials (Lee, 2012b) shows that the current methods of purification employed by the vaccine manufacturers are not as efficient as claimed. Neither are their current methods of testing of the final product sensitive enough to detect potential contaminant DNA residuals. In this context, the protocol described by Lee (2013a,b) represents the first attempt toward developing a quality assurance test for residual HPV L1 gene DNA fragments in the insoluble fraction of the Gardasil vaccine.

Finally, the latest research shows that peripherally injected Al adjuvant nanoparticles engulfed by macrophages actively spread throughout the body, eventually crossing the BBB and blood–CSF barrier (Khan *et al.*, 2013). Once in the CNS, Al adjuvant nanoparticles incite deleterious inflammatory responses, resulting in a range of neuropathological effects (Petrik *et al.*, 2007; Shaw and Petrik, 2009). It should be noted that Al on its own can alter the properties of the BBB (Banks and Kastin, 1989; Zheng, 2001; Yokel, 2006), making the brain more accessible to inflammatory and immune mediators. Al also increases endothelial adhesion of activated monocytes (Oesterling *et al.*, 2008), which, in the case of Al penetration in the CNS, can likewise facilitate the entry of immune-competent cells into the CNS and lead to adverse manifestations. In accordance with these observations, Zinka *et al.* (2006) reported six cases of sudden infant death that occurred within 48 hours after vaccination with hexavalent vaccines. The post-mortem analysis of the six children, aged 4–17 months (five of whom were vaccinated with Hexavac and one with Infanrix Hexa), revealed abnormal pathologic findings, particularly affecting the nervous system. The overall pathological abnormalities included acute congestion, defective BBB, infiltration of the leptomeninx by macrophages and lymphocytes, perivascular lymphocytic infiltration, diffuse infiltration of the pons, mesencephalon, and cortex by T-lymphocytes, microglia in the hippocampus and pons, and, in one case, necrosis in the cerebellum (Zinka *et al.*, 2006).

Long-term persistence of Al adjuvants in the body and its effects

The prolonged hyperactivation of the immune system and chronic inflammation triggered by repeated exposure and unexpectedly long persistence of Al adjuvants in the human body (up to 11 years post-vaccination: Gherardi *et al.*, 2001; Ryan *et al.*, 2006; Shivane *et al.*, 2012) are thought to be the principal factors underlying the toxicity of these compounds. One of the reasons for this long retention of Al adjuvants in bodily compartments, including systemic circulation, is most likely its tight association with the vaccine antigen or other vaccine excipients (i.e. contaminant DNA), as already explained. Even dietary Al has been shown to accumulate in the CNS over time, producing Alzheimer-type outcomes in experimental animals fed equivalent amounts of Al to what humans consume through a typical Western diet (Walton, 2007; Walton and Wang, 2009).

The long retention of Al adjuvants was first identified, and has since been extensively studied, in macrophagic myofasciitis (MMF) patients. MMF is a condition characterized by highly specific myopathological alterations at deltoid muscle biopsy, first recognized in 1998, and subsequently shown to result from long-term persistence of vaccine-derived Al hydroxide nanoparticles within macrophages at the site of previous vaccine injections (Gherardi *et al.*, 1998, 2001; Couette *et al.*, 2009; Passeri *et al.*, 2011). Patients diagnosed with MMF tend to be female (70%) and middle-aged at time of biopsy (median 45 years), and to have received 1 to 17 intramuscular (i.m.) Al-containing vaccine administrations (mean 5.3) in the 10 years before MMF detection (Gherardi and Authier, 2012). The central histopathological feature in MMF is a granulomatous lesion comprising Al-loaded macrophages at the site of previous intramuscular vaccination. Notably, MMF lesions have been experimentally reproduced by i.m. vaccination in rats and monkeys (Verdier *et al.*, 2005; Authier *et al.*, 2006).

Clinical manifestations in MMF patients include diffuse myalgia, arthralgia, chronic fatigue, muscle weakness, and cognitive dysfunction. In particular, up to 93% of patients suffer from chronic fatigue (over 6 months in duration: Authier *et al.*, 2003), and up to 89% from chronic diffuse myalgias (over 6 months in duration), with or without arthralgias (Gherardi and Authier, 2012). Fatigue is disabling in 87% of cases and affects physical and mental functioning in 53% (Authier *et al.*, 2003). Overt cognitive alterations affecting memory and attention are manifested in 51% of cases (Gherardi and Authier, 2012). In addition to chronic fatigue syndrome, 15–20% of patients with MMF concurrently develop an autoimmune disease, the most frequent being multiple sclerosis-like demyelinating disorders, Hashimoto's thyroiditis, and diffuse autoimmune neuromuscular diseases, such as dermatomyositis, necrotizing autoimmune myopathy, myasthenia gravis, and inclusion body myositis (Gherardi and Authier, 2012). Even in the absence of overt autoimmune disease, low titers of various autoantibodies, increased inflammatory biomarkers, and abnormal iron status are commonly detected (Gherardi and Authier, 2003).

The pathological significance of the MMF lesion has long been ill understood, because of the lack of an obvious link between persistence of Al agglomerates in macrophages at sites of previous vaccination and delayed onset of systemic and neurological manifestations. However, recent experiments in animal models have revealed that a proportion of injected Al adjuvant nanoparticles do not stay localized at a site of injection. In particular, following injection, antigen-presenting cells (APCs) avidly take up Al particles (Morefield *et al.*, 2005) and so become long-lived cells (Hamilton *et al.*, 2000), impeding Al solubilization in the interstitial fluid (Gherardi *et al.*, 2001). Thus, a proportion of Al nanoparticles escape the injected muscle (mainly within immune cells), travel to regional draining lymph nodes, and exit the lymphatic system to reach the bloodstream, eventually gaining access to distant organs, including the spleen and the brain, where Al deposits can still be detected 1 year after injection. Moreover, the Trojan horse mechanism by which Al loaded in macrophages enters the brain results in its slow accumulation due to lack of recirculation, and is likely responsible for the myriad of cognitive deficits associated with administration of

Al-containing vaccines observed in MMF patients (Passeri *et al.*, 2011).

Another point requiring emphasis is that the bioaccumulation of Al in the brain appears to occur at a very low rate in normal conditions, thus potentially explaining the presumably good overall tolerance of this adjuvant despite its strong neurotoxic potential. Nonetheless, according to Khan *et al.* (2013), continuously increasing doses of the poorly biodegradable Al adjuvant may become "insidiously unsafe," especially in cases of repetitive closely-spaced vaccinations (otherwise known as "vaccine rechallenge") and in those with an immature/altered BBB, such as the very young or those suffering past head injuries. In this context, the latest research by Lujan *et al.* (2013), who described a severe neurodegenerative syndrome in commercial sheep, linked to the repetitive inoculation of Al-containing vaccines, is noteworthy. In particular, the "sheep ASIA syndrome" mimics in many aspects human neurological diseases linked to Al adjuvants (Lujan *et al.*, 2013). The adverse chronic phase of this syndrome affects 50–70% of flocks and up to 100% of animals within a flock. It is characterized by severe neurobehavioural outcomes, all of which are consistent with Al toxicity (restlessness, compulsive wool biting, generalized weakness, muscle tremors, loss of response to stimuli, ataxia, tetraplegia, stupor, coma, and death), inflammatory lesions in the brain, and the presence of Al in CNS tissues (Lujan *et al.*, 2013). The main histopathologic change in the chronic phase of sheep ASIA syndrome is located at the spinal cord and consists of multifocal neuronal necrosis and neuron loss in both dorsal and ventral column of the gray matter.

These findings by Lujan *et al.* (2013) are consistent with those of Shaw and Petrik (2009) and Khan *et al.* (2013), who both demonstrated the ability of Al adjuvants to penetrate the blood–CSF barrier and BBB. More significantly, the quoted research also shows that the resulting presence of Al in the brain can trigger severe neurological damage, with devastating consequences. Collectively, these findings also explain in part why the majority of reported adverse reactions following vaccinations with adjuvanted vaccines are neurological and neuropshychiatric with an underlying immunoinflammatory or autoimmune component (Konstantinou *et al.*, 2001; Carvalho and Shoenfeld, 2008; Couette *et al.*, 2009; Passeri *et al.*, 2011; Zafrir *et al.*, 2012). As an example, Zafrir *et al.* (2012) recently reported an analysis of 93 patients who experienced the appearance of a new immune-mediated phenomenon following vaccination with hepatitis B (86% of whom also fulfilled the ASIA criteria). By far the most commonly reported adverse manifestations were neurological (70%), with 25% of patients in this cohort being diagnosed with a specific neurological disease (multiple sclerosis, Guillain–Barré syndrome (GBS), transverse myelitis, etc.) (Zafrir *et al.*, 2012).

In summary, it is clear from the previously quoted research that the toxicity potential of Al will be influenced by its biopersistence and its biodistribution (i.e. whether the bioactive Al adjuvant nanoparticles remain localized at injection sites or scatter and accumulate in distant organs and tissues). All the clinical and experimental evidence collected thus far identifies at least three main risks associated with Al in vaccines:

1. it can persist in the body (up to 11 years following vaccination);
2. it can trigger pathological immunological responses;
3. it can make its way into the CNS, where it can drive further deleterious immunoinflammatory processes, resulting in brain inflammation and long-term neural dysfunction.

Al's activation of the NLRP3 inflammasome pathway and its role in autoimmune-inflammatory diseases

Al adjuvants exert their immunostimulatory effect through many different actions, which impinge on both the innate and adaptive immune systems (Eisenbarth *et al.*, 2008; Exley *et al.*, 2010). These include: (i) activation of the NLRP3 inflammasome pathway, (ii) protection of antigens, resulting in prolonged delivery; (iii) induction of prompt vaccine particle phagocytosis by dendritic cells and macrophages, with upregulation of their antigen-presenting function; (iv) translocation of antigens to lymphoid organs, where the primary activation of naïve T cells takes place; (v) amplification of the inflammatory reaction in the injection site and its draining lymph nodes, through interaction with pattern-recognition receptors (PRRs) and release of inflammatory cytokines; and (vi) priming of B cells in spleen (Marrack *et al.*, 2009; Exley *et al.*, 2010).

There appears to be a fine balance between the efficacy of vaccine adjuvants and their potential toxicity (Batista-Duharte *et al.*, 2011). This is because the same mechanisms that drive the immune-stimulatory effect of adjuvants have the capacity to provoke a variety of autoimmune and/or inflammatory adverse reactions, including those associated with the ASIA syndrome (Tomljenovic and Shaw, 2012; Shaw *et al.*, 2013). A perfect example of this is Al's activation of the NLRP3 inflammasome signaling pathway (and its downstream mediators caspase-1 and IL-18; Li *et al.*, 2007; Eisenbarth *et al.*, 2008; Exley *et al.*, 2010), which is responsible for the immune adjuvant-stimulatory properties of Al. Unfortunately, activation of the NLPR3 inflammasome pathway (which is the principal immunostimulatory pathway through which Al adjuvants operate) is also critically involved in the development of serious autoimmune and inflammatory diseases, including type 2 diabetes, CNS demyelinating diseases (inflammatory bowel disease), colitis, and atherosclerosis (Bauer *et al.*, 2010; Chakraborty *et al.*, 2010; Jha *et al.*, 2010; Rajamaki *et al.*, 2010; Wen *et al.*, 2011).

NLPR3 activation triggers type 2 diabetes through interference with insulin signaling and promotion of insulin resistance. In particular, using NLPR3 knockout mice, Wen *et al.* (2011) demonstrated that the absence of inflammasome components led to a better maintenance of glucose homeostasis and higher insulin sensitivity. Activation of the inflammasome and its downstream components, proinflammatory cytokines IL-1β and IL-18, is also strongly implicated in the promotion of several CNS disorders, including Alzheimer's disease, Parkinson's disease, and multiple sclerosis (Chakraborty *et al.*, 2010), all of which have been previously linked to Al exposure (Authier *et al.*, 2001; Exley *et al.*, 2006; Perl and Moalem, 2006; Tomljenovic, 2011; Walton, 2012b). Experiments in transgenic animal models show that NLRP3 plays a crucial role in multiple sclerosis (a demyelinating autoimmune disease) by exacerbating CNS inflammation, and that this effect is partly mediated by caspase-1 and IL-18 (Jha *et al.*, 2010). Specifically, mice lacking the *Nlrp3* gene ($Nlrp3^{-/-}$) exhibit delayed neuroinflammation, delayed demyelination, and delayed oligodendrocyte loss in the experimental autoimmune encephalomyelitis (model of multiple sclerosis). These mice also show reduced demyelination. This outcome is also observed for $casp1^{-/-}$ and $IL\text{-}18^{-/-}$ mice, whereas $IL\text{-}1\beta^{-/-}$ mice are indistinguishable from wild-type controls, indicating that Nlrp3-mediated function occurs through caspase-1 and IL-18. Additional analyses have revealed that $Nlrp3^{-/-}$ mice do not exhibit delayed remyelination. Interestingly, $IL\text{-}18^{-/-}$ mice show enhanced remyelination, thus providing a possible compensatory mechanism for the lack of a remyelination defect in $\text{Nlrp3}^{-/-}$ mice. Altogether, these results suggest that NLRP3's role in multiple sclerosis is mediated by caspase-1 and IL-18 (Jha *et al.*, 2010).

There is yet another mechanism by which Al disrupts the myelin sheath: oxidation. As shown by Verstraeten *et al.* (1997), Al (due to its lipophilic nature) binds avidly to membrane phospholipids and, by inducing changes in phospholipid rheology, promotes lipid peroxidation. Consequently, myelin (due to its high lipid-to-protein ratio, 70:30, and relatively low ubiquinol content versus synaptic membranes, 30:70) is the preferred target of Al-mediated oxidative damage both *in vitro* and *in vivo* (Verstraeten *et al.*, 1997).

In view of the numerous reports of autoimmune demyelinating pathologies following administration of Al-adjuvanted vaccines (Table 4.1),

Table 4.1 Autoimmune demyelinating diseases associated with Al-adjuvanted vaccines

Disease	Vaccine	References
Multiple sclerosis	Hepatitis A and B, HPV	Authier *et al.* (2001), Hernan *et al.* (2004), Sutton *et al.* (2009)
Acute disseminated encephalomyelitis	Hepatitis B, HPV	Cabrera-Gomez *et al.* (2002), Wildemann *et al.* (2009), Mendoza *et al.* (2010)
GBS	Hepatitis B, HPV	Khamaisi *et al.* (2004), Souayah *et al.* (2011)
Transverse myelitis	Hepatitis B, DPT	Karaali-Savrun *et al.* (2001), Riel-Romero (2006), Agmon-Levin *et al.* (2009)
Neuromyelitis optica/optic neuritis	Hepatitis A and B, DPT, HPV, tetanus–toxoid	Topaloglu *et al.* (1992), Voigt *et al.* (2001), Beyer *et al.* (2007), DiMario *et al.* (2010)
Demyelinating leukoencephalitis	Hepatitis B	Konstantinou *et al.* (2001)

DPT, diphtheria–pertussis–tetanus; GBS, Guillain–Barré syndrome; HPV, human papilloma virus perhaps a move toward reducing the number of Al-containing vaccines that an individual receives throughout their life should be considered. Indeed, the consequences of continuous life-long exposure to this neurotoxic agent can no longer be seen as benign, in view of the current scientific literature.

Al body burden and the health risks associated with life-long exposure to Al adjuvants

As discussed previously, Al adjuvants act as vehicles for the presentation of antigens in nonbenign ways because they are capable of stimulating pathological immune and inflammatory responses even in the absence of an antigen. Moreover, they have also been shown to act as antigens themselves (Levy *et al.*, 1998). Thus, Al is both adjuvant and antigen, and, as noted by Exley *et al.* (2009), this dual activity must raise questions about how the human body reacts to any future exposures to Al. For example, there is evidence that Al in adjuvants also acts as an antigen, as a significant proportion of vaccine recipients retain a memory of their exposure to Al, in that they show delayed hypersensitivity to subsequent exposures to Al (Bergfors *et al.*, 2003; Hindsen, 2005). Thus, vaccination, as well as allergen therapies which incorporate Al-based adjuvants, may sensitize recipients to adverse outcomes from future exposures to Al. Manifestations of such an enhanced sensitivity to Al are probably as diverse as the myriad ways in which humans are exposed to Al in everyday life (Exley, 2004b; Lerner, 2007; Mannello *et al.*, 2011; Shaw and Tomljenovic, 2013; Walton, 2012b). For example, it may manifest as a skin reaction following antiperspirant exposure or as an allergic asthma triggered by Al in tobacco smoke. The response to a systemic Al challenge (i.e. following the injection of Al-adjuvanted vaccinations) might be more severe, and could potentially explain the spectrum of symptoms associated with such conditions as MMF (Gherardi and Authier, 2012) and chronic fatigue syndrome (Couette *et al.*, 2009; Exley *et al.*, 2009).

Exley *et al.* (2009) further note that sensitization to Al may simply be one manifestation of the physiological response to biologically available Al. The biological availability of Al, as defined by its propensity to induce a biochemical response in an affected system, is known to depend upon the establishment over time of a threshold concentration or burden of Al (Exley and Birchall, 1992). The system (i.e. cell or tissue) copes with the burgeoning burden of Al up until a threshold concentration is reached, at which point there is a net biochemical effect. The immunological memory of early exposures to biologically available Al may vary widely within recipients, such that there may be many different biochemical responses to future exposures to Al. In the case of future Al-adjuvant-containing vaccinations, the threshold may be achieved instantaneously in individuals who have retained a memory of their earlier exposure to Al, and could instigate a severe immune response, with wide-ranging health implications (Exley *et al.*, 2009).

The wider cascade of effects might involve the recruitment of Al antigens in other parts of the body or might be mediated through other antigens that have been sensitized through their previous co-administration with Al adjuvant. An example of this is the sensitization to food allergens following their co-administration with Al salts. Notably, the immunostimulatory properties of Al have been routinely exploited for induction of mast cell-dependent allergic sensitization to food proteins, which ultimately results in intestinal inflammation and diarrhea (Brandt *et al.*, 2003; Berin and Mayer, 2009). Mast cells play key roles in a wide range of inflammatory gastrointestinal pathologies, in which they compromise mucosal immunity and increase intestinal permeability (Brandt *et al.*, 2003; Berin and Mayer, 2009; Theoharides *et al.*, 2009). Particularly relevant in the context of this review is the fact that gastrointestinal dysfunction and food allergies appear to be the most common nonneurological comorbidities in both ASIA and disorders of the autism spectrum (Theoharides *et al.*, 2009; Zafrir *et al.*, 2012). These observations provide further compelling evidence in support of the role of Al adjuvant overexposure in both of these syndromes (Meroni 2010; Shoenfeld and Agmon-Levin, 2011; Tomljenovic and Shaw, 2011; Seneff *et al.*, 2012; Melendez *et al.*, 2013; Shaw *et al.*, 2013).

In summary, an individual's susceptibility to an adverse reaction from Al may be dependent upon the combination of a previous sensitization to Al (e.g. via childhood vaccination) and an ongoing Al overload from all sources (Exley *et al.*, 2009). While the body may cope robustly with a mild but persistent exposure to Al, the coping mechanism will be suddenly and dramatically overwhelmed by a new exposure to Al adjuvant. The latter will not only enhance the antigenicity by itself but will raise the level of the immune response against all significant body stores of Al. Under these conditions, an individual's everyday exposure to Al will continue to fuel the response, and many symptoms of associated autoimmunity will take over, leading to adverse responses to Al exposures which previously would not have been sufficient to elicit a biological response (Exley *et al.*, 2009). When it is considered that as many as 1% of recipients of Al-containing adjuvants might be sensitized to future exposures to Al (Bergfors *et al.*, 2003), a cautionary note can be raised in respect of future mass vaccinations that include this form of adjuvant.

Conclusions

Al salts are the most widely used adjuvants today, and have been since the 1920s (Glenny *et al.*, 1926). The fact that they can trigger pathological immunological responses and a cascade of unwanted health effects has been relatively underappreciated to date. The risks associated with vaccine-derived Al are threefold: it can persist in the body, it can trigger pathological immunological responses, and it can make its way into the CNS, where it can drive deleterious immunoinflammatory and excitotoxic processes. Because infants and children may be most at risk for complications following vaccination, a more rigorous evaluation of potential vaccine-related adverse health impacts in pediatric populations is urgently needed. The recognition of ASIA as a vaccine adjuvant-triggered pathology should alert and encourage both physicians and patients to report vaccine adverse conditions, in order to enable a better estimation of the true prevalence of ASIA. It is clear that the role of adjuvants in the pathogenesis of immune-mediated diseases can no longer be ignored, especially in view of the fact that many nonspecific medical conditions that fall under the ASIA spectrum (i.e., chronic fatigue, myalgias, and cognitive impairments) are frequently disabling and negatively impact individuals' private and professional activities. The inclusion of this category of adverse manifestations under ASIA is of special importance because in the past they were frequently ignored or disregarded as irrelevant and non-vaccine-related, not only by physicians and patients, but also by scientists involved in the design of vaccine trials. Finally, the delineation of ASIA further emphasizes the fact that the use of Al adjuvant-containing placebos in vaccine clinical trials can no longer be justified.

Acknowledgments

The authors thank the Dwoskin Family Foundation, the Katlyn Fox Foundation, and the Luther Allyn Shourds Dean Estate for support.

References

Agarwal, S.K., Ayyash, L., Gourley, C.S., *et al.* (1996). Evaluation of the developmental neuroendocrine and reproductive toxicology of aluminium. *Food Chem Toxicol*, **34**: 49–53.

Agmon-Levin, N., Kivity, S., Szyper-Kravitz, M., and Shoenfeld, Y. (2009). Transverse myelitis and vaccines: a multi-analysis. *Lupus*, **18**: 1198–204.

ATSDR (Agency for Toxic substances, Disease Registry). (2008). Toxicological profile for aluminum. Available from: http://www.atsdr.cdc.gov/toxprofiles/tp22.pdf [last accessed 11 December 2014].

Authier, F.J., Cherin, P., Creange, A., *et al.* (2001). Central nervous system disease in patients with macrophagic myofasciitis. *Brain*, **124**: 974–83.

Authier, F.J., Sauvat, S., Champey, J., *et al.* (2003). Chronic fatigue syndrome in patients with macrophagic myofasciitis. *Arthritis Rheum*, **48**: 569–70.

Authier, F.J., Sauvat, S., Christov, C., *et al.* (2006). AlOH3-adjuvanted vaccine-induced macrophagic myofasciitis in rats is influenced by the genetic background. *Neuromuscul Disord*, **16**: 347–52.

Banks, W.A. and Kastin, A.J. (1989). Aluminum-induced neurotoxicity: alterations in membrane function at the blood–brain barrier. *Neurosci Biobehav Rev*, **13**: 47–53.

Batista-Duharte, A., Lindblad, E.B., and Oviedo-Orta, E. (2011). Progress in understanding adjuvant immunotoxicity mechanisms. *Toxicol Lett*, **203**: 97–105.

Bauer, C., Duewell, P., Mayer, C., *et al.* (2010). Colitis induced in mice with dextran sulfate sodium (DSS) is mediated by the NLRP3 inflammasome. *Gut*, **59**: 1192–9.

Baylor, N.W., Egan, W., and Richman, P. (2002). Aluminum salts in vaccines – US perspective. *Vaccine*, **20**(Suppl. 3): S18–23.

Bergfors, E., Trollfors, B., and Inerot, A. (2003). Unexpectedly high incidence of persistent itching nodules and delayed hypersensitivity to aluminium in children after the use of adsorbed vaccines from a single manufacturer. *Vaccine*, **22**: 64–9.

Berin, M.C. and Mayer, L. (2009). Immunophysiology of experimental food allergy. *Mucosal Immunol*, **2**: 24–32.

Beyer, A.M., Wandinger, K.P., Siebert, E., *et al.* (2007). Neuromyelitis optica in a patient with an early onset demyelinating episode: clinical and autoantibody findings. *Clin Neurol Neurosurg*, **109**: 926–30.

Bishop, N.J., Morley, R., Day, J.P., and Lucas, A. (1997). Aluminum neurotoxicity in preterm infants receiving intravenous-feeding solutions. *N Engl J Med*, **336**: 1557–61.

Brandt, E.B., Strait, R.T., Hershko, D., *et al.* (2003). Mast cells are required for experimental oral allergen-induced diarrhea. *J Clin Invest*, **112**: 1666–77.

Brewer, J.M. (2006). (How) do aluminium adjuvants work? *Immunol Lett*, **102**: 10–15.

Cabrera-Gomez, J.A., Echazabal-Santana, N., Garcia Gonzalez, L., *et al.* (2002). A severe episode in a patient with recurrent disseminated acute encephalitis due to vaccination against hepatitis B. For or against vaccination? *Rev Neurol*, **34**: 358–63.

Carvalho, J.F. and Shoenfeld, Y. (2008). Status epilepticus and lymphocytic pneumonitis following hepatitis B vaccination. *European J Int Med*, **19**: 383–5.

Chakraborty, S., Kaushik, D.K., Gupta, M., and Basu, A. (2010). Inflammasome signaling at the heart of central nervous system pathology. *J Neurosci Res*, **88**: 1615–31.

Couette, M., Boisse, M.F., Maison, P., *et al.* (2009). Long-term persistence of vaccine-derived aluminum hydroxide is associated with chronic cognitive dysfunction. *J Inorg Biochem*, **103**: 1571–8.

DiMario, F.J. Jr,, Hajjar, M., and Ciesielski, T. (2010). A 16-year-old girl with bilateral visual loss and left hemiparesis following an immunization against human papilloma virus. *J Child Neurol*, **25**: 321–7.

Egan, P.M., Belfast, M.T., Gimenez, J.A., *et al.* (2009). Relationship between tightness of binding and immunogenicity in an aluminum-containing adjuvant-adsorbed hepatitis B vaccine. *Vaccine*, **27**: 3175–80.

Eickhoff, T.C. and Myers, M. (2002). *Workshop summary. Aluminum in vaccines. Vaccine*, **20**(Suppl. 3): S1–4.

Eisenbarth, S.C., Colegio, O.R., O'Connor, W., *et al.* (2008). Crucial role for the Nalp3 inflammasome in the immunostimulatory properties of aluminium adjuvants. *Nature*, **453**: 1122–6.

Eldred, B.E., Dean, A.J., McGuire, T.M., and Nash, A.L. (2006). Vaccine components and constituents: responding to consumer concerns. *Med J Aust*, **184**: 170–5.

Exley, C. (2004a). The pro-oxidant activity of aluminum. *Free Radic Biol Med*, **36**: 380–7.

Exley, C. (2004b). Aluminum in antiperspirants: more than just skin deep. *Am J Med*, **117**: 969–70.

Exley, C. (2011). Aluminium-based adjuvants should not be used as placebos in clinical trials. *Vaccine*, **29**: 9289.

Exley, C. and Birchall, J.D. (1992). The cellular toxicity of aluminium. *J Theor Biol*, **159**: 83–98.

Exley, C., Mamutse, G., Korchazhkina, O., *et al.* (2006). Elevated urinary excretion of aluminium and iron in multiple sclerosis. *Mult Scler*, **12**: 533–40.

Exley, C., Swarbrick, L., Gherardi, R.K., and Authier, F.J. (2009). A role for the body burden of aluminium in vaccine-associated macrophagic myofasciitis and chronic fatigue syndrome. *Med Hypotheses*, **72**: 135–9.

Exley, C., Siesjo, P., and Eriksson, H. (2010). The immunobiology of aluminium adjuvants: how do they really work? *Trends Immunol*, **31**: 103–9.

FAO/WHO (Food and Agriculture Organization/World Health Organization). (2006). Summary and conclusions of the sixty-seventh meeting of the Joint FAO/WHO Expert Committee on Food Additives (JECFA), sixty-seventh meeting in Rome. Available from: ftp://ftp.fao.org/ag/agn/jecfa/jecfa67_final.pdf [last accessed 11 December 2014].

Gherardi, R.K. and Authier, F.J. (2003). Aluminum inclusion macrophagic myofasciitis: a recently identified condition. *Immunol Allergy Clin North Am*, **23**: 699–712.

Gherardi, R. and Authier, F. (2012). Macrophagic myofasciitis: characterization and pathophysiology. *Lupus*, **21**: 184–9.

Gherardi, R.K., Coquet, M., Cherin, P., *et al.* (1998). Macrophagic myofasciitis: an emerging entity. Groupe d'Etudes et Recherche sur les Maladies Musculaires Acquises et Dysimmunitaires (GERMMAD) de l'Association Francaise contre les Myopathies (AFM). *Lancet*, **352**: 347–52.

Gherardi, R.K., Coquet, M., Cherin, P., *et al.* (2001). Macrophagic myofasciitis lesions assess long-term persistence of vaccine-derived aluminium hydroxide in muscle. *Brain*, **124**: 1821–31.

Gies, W.J. (1911). Some objections to the use of alum baking-powder. *JAMA*, **57**: 816–21.

Glenny, A.T., Pope, C.G., Waddington, H., and Wallace, U. (1926). XXIII – the antigenic value of toxoid precipitated by potassium alum. *J Pathol Bacteriol*, **29**: 38–9.

Hamilton, J.A., Byrne, R., and Whitty, G. (2000). Particulate adjuvants can induce macrophage survival, DNA synthesis, and a synergistic proliferative response to GM-CSF and CSF-1. *J Leukoc Biol*, **67**: 226–32.

Hem, S.L. (2002). Elimination of aluminum adjuvants. *Vaccine*, **20**(Suppl. 3): S40–3.

Hernan, M.A., Jick, S.S., Olek, M.J., and Jick, H. (2004). Recombinant hepatitis B vaccine and the risk of multiple sclerosis: a prospective study. *Neurology*, **63**: 838–42.

Hindsen, M. (2005). Contact allergy to aluminium in patients hyposensitized with aluminium-containing hyposensitizing extracts. *Contact Dermatitis*, **53**: 301–2.

Israeli, E., Agmon-Levin, N., Blank, M., and Shoenfeld, Y. (2009). Adjuvants and autoimmunity. *Lupus*, **18**: 1217–25.

Jha, S., Srivastava, S.Y., Brickey, W.J., *et al.* (2010). The inflammasome sensor, NLRP3, regulates CNS inflammation and demyelination via caspase-1 and interleukin-18. *J Neurosci*, **30**: 15 811–20.

Joshi, J.G. (1990). Aluminum, a neurotoxin which affects diverse metabolic reactions. *Biofactors*, **2**: 163–9.

Joshi, J.G., Dhar, M., Clauberg, M., and Chauthaiwale, V. (1994). Iron and aluminum homeostasis in neural disorders. *Environ Health Perspect*, **102**(Suppl. 3): 207–13.

Karaali-Savrun, F., Altintas, A., Saip, S., and Siva, A. (2001). Hepatitis B vaccine related-myelitis? *Eur J Neurol*, **8**: 711–15.

Khamaisi, M., Shoenfeld, Y., and Orbach, H. (2004). Guillain-Barre syndrome following hepatitis B vaccination. *Clin Exp Rheumatol*, **22**: 767–70.

Khan, Z., Combadiere, C., Authier, F.J., *et al.* (2013). Slow CCL2-dependent translocation of biopersistent particles from muscle to brain. *BMC Med*, **11**: 99.

Konstantinou, D., Paschalis, C., Maraziotis, T., *et al.* (2001). Two episodes of leukoencephalitis associated with recombinant hepatitis B vaccination in a single patient. *Clin Infect Dis*, **33**: 1772–3.

Lee, S.H. (2012a). Detection of human papillomavirus L1 gene DNA fragments in postmortem blood and spleen after Gardasil® vaccination-A case report. *Advances Biosc Biotech*, **3**: 1214–24.

Lee, S.H. (2012b). Detection of human papillomavirus (HPV) L1 gene DNA possibly bound to particulate aluminum adjuvant in the HPV vaccine Gardasil. *J Inorg Biochem*, **112**: 85–92.

Lee, S.H. (2013a). Melting profiles may affect detection of residual HPV L1 gene DNA fragments in Gardasil. *Curr Med Chem*, **21**(7): 932–40.

Lee, S.H. (2013b). Topological conformational changes of human papillomavirus (HPV) DNA bound to an insoluble aluminum salt-A study by low temperature PCR. *Advances Biol Chem*, **3**: 76–85.

Lerner, A. (2007). Aluminum is a potential environmental factor for Crohn's disease induction: extended hypothesis. *Ann NY Acad Sci*, **1107**: 329–45.

Levy, R., Shohat, L., and Solomon, B. (1998). Specificity of an anti-aluminium monoclonal antibody toward free and protein-bound aluminium. *J Inorg Biochem*, **69**: 159–63.

Li, H., Nookala, S., and Re, F. (2007). Aluminum hydroxide adjuvants activate caspase-1 and induce IL-1beta and IL-18 release. *J Immunol*, **178**: 5271–6.

Li, X., Zheng, H., Zhang, Z., *et al.* (2009). Glia activation induced by peripheral administration of aluminum oxide nanoparticles in rat brains. *Nanomedicine*, **5**: 473–9.

Lujan, L., Perez, M., Salazar, E., *et al.* (2013). Autoimmune/autoinflammatory syndrome induced by adjuvants (ASIA syndrome) in commercial sheep. *Immunol Res*, **56**: 317–24.

Lukiw, W.J. (2001). Aluminum and gene transcription in the mammalian central nervous system-implications for Alzheimer's Disease. In: Exley, C. (ed.). *Aluminium and Alzheimer's Disease: The Science that Describes the Link*. Amsterdam: Elsevier Science, pp. 147–69.

Lukiw, W.J., Percy, M.E., and Kruck, T.P. (2005). Nanomolar aluminum induces pro-inflammatory and pro-apoptotic gene expression in human brain cells in primary culture. *J Inorg Biochem*, **99**: 1895–8.

Mannello, F., Tonti, G.A., Medda, V., *et al.* (2011). Analysis of aluminium content and iron homeostasis in nipple aspirate fluids from healthy women and breast cancer-affected patients. *J Appl Toxicol*, **31**: 262–9.

Marrack, P., McKee, A.S., and Munks, M.W. (2009). Towards an understanding of the adjuvant action of aluminium. *Nat Rev Immunol*, **9**: 287–93.

Melendez, L., Santos, D., Luna Polido, L., *et al.* (2013). Aluminium and other metals may pose a risk to children with autism spectrum disorder: biochemical and behavioural impairments. *Clin Exp Pharmacol*, **3**: 120.

Mendoza Plasencia, Z., Gonzalez Lopez, M., Fernandez Sanfiel, M.L., and Muniz Montes, J.R. (2010). Acute disseminated encephalomyelitis with tumefactive lesions after vaccination against human papillomavirus. *Neurologia*, **25**: 58–9.

Merck Research Laboratories. 2010. GARDASIL® (Human Papillomavirus ([Types 6, 11, 16, 18] Recombinant Vaccine). Vaccines and Related Biological Products Advisory Committee (VRBPAC) Briefing Document. Available from: http://www.fda.gov/downloads/AdvisoryCommittees/CommitteesMeetingMaterials/BloodVaccinesandOtherBiologics/VaccinesandRelatedBiologicalProductsAdvisoryCommittee/UCM231522.pdf [last accessed 11 December 2014].

Meroni, P.L. (2010). Autoimmune or auto-inflammatory syndrome induced by adjuvants (ASIA): old truths and a new syndrome? *J Autoimmun*, **36**: 1–3.

Morefield, G.L., Sokolovska, A., Jiang, D., *et al.* (2005). Role of aluminum-containing adjuvants in antigen internalization by dendritic cells in vitro. *Vaccine*, **23**: 1588–95.

Oesterling, E., Chopra, N., Gavalas, V., *et al.* (2008). Alumina nanoparticles induce expression of endothelial cell adhesion molecules. *Toxicol Lett*, **178**: 160–6.

Offit, P.A. and Jew, R.K. (2003). Addressing parents' concerns: do vaccines contain harmful preservatives, adjuvants, additives, or residuals? *Pediatrics*, **112**: 1394–7.

Passeri, E., Villa, C., Couette, M., *et al.* (2011). Long-term follow-up of cognitive dysfunction in patients with aluminum hydroxide-induced macrophagic myofasciitis (MMF). *J Inorg Biochem*, **105**: 1457–63.

Perl, D.P. and Moalem, S. (2006). Aluminum, Alzheimer's disease and the geospatial occurrence of similar disorders. *Rev Mineral Geochem*, **64**: 115–34.

Petrik, M.S., Wong, M.C., Tabata, R.C., *et al.* (2007). Aluminum adjuvant linked to Gulf War illness induces motor neuron death in mice. *Neuromolecular Med*, **9**: 83–100.

Rajamaki, K., Lappalainen, J., Oorni, K., *et al.* (2010). Cholesterol crystals activate the NLRP3 inflammasome in human macrophages: a novel link between cholesterol metabolism and inflammation. *PLoS One*, **5**: e11765.

Redhead, K., Quinlan, G.J., Das, R.G., and Gutteridge, J.M. (1992). Aluminium-adjuvanted vaccines transiently increase aluminium levels in murine brain tissue. *Pharmacol Toxicol*, **70**: 278–80.

Riel-Romero, R.M. (2006). Acute transverse myelitis in a 7-month-old boy after diphtheria-tetanus-pertussis immunization. *Spinal Cord*, **44**: 688–91.

Rogers, M.A. and Simon, D.G. (1999). A preliminary study of dietary aluminium intake and risk of Alzheimer's disease. *Age Ageing*, **28**: 205–9.

Rondeau, V., Commenges, D., Jacqmin-Gadda, H., and Dartigues, J.F. (2000). Relation between aluminum concentrations in drinking water and Alzheimer's disease: an 8-year follow-up study. *Am J Epidemiol*, **152**: 59–66.

Ryan, A.M., Bermingham, N., Harrington, H.J., and Keohane, C. (2006). Atypical presentation of macrophagic myofasciitis 10 years post vaccination. *Neuromuscul Disord*, **16**: 867–9.

Seneff, S., Davidson, R.M., and Liu, J. (2012). Empirical data confirm autism symptoms related to aluminum and acetaminophen exposure. *Entropy*, **14**: 2227–53.

Shafer, T.J., Nostrandt, A.C., Tilson, H.A., and Mundy, W.R. (1994). Mechanisms underlying AlCl3 inhibition of agonist-stimulated inositol phosphate accumulation. Role of calcium, G-proteins, phospholipase C and protein kinase C. *Biochem Pharmacol*, **47**: 1417–25.

Shaw, C.A. and Petrik, M.S. (2009). Aluminum hydroxide injections lead to motor deficits and motor neuron degeneration. *J Inorg Biochem*, **103**: 1555–62.

Shaw, C.A. and Tomljenovic, L. (2013). Aluminum in the central nervous system (CNS): toxicity in humans and animals, vaccine adjuvants, and autoimmunity. *Immunol Res*, **56**: 304–16.

Shaw, A.C., Li, Y., and Tomljenovic, L. (2013). Administration of aluminium in vaccine-relevant amounts in neonatal mice is associated with long-term adverse neurological outcomes. *J Inorg Biochem*, **128**: 237–44.

Shivane, A., Hilton, D.A., Moate, R.M., *et al.* (2012). Macrophagic myofasciitis: a report of second case from UK. *Neuropathol Appl Neurobiol*, **38**: 734–6.

Shoenfeld, Y. and Agmon-Levin, N. (2011). "ASIA" – Autoimmune/inflammatory syndrome induced by adjuvants. *J Autoimmun*, **36**: 4–8.

Singla, N. and Dhawan, D.K. (2011). Regulatory role of zinc during aluminium-induced altered carbohydrate metabolism in rat brain. *J Neurosci Res*, **90**: 698–705.

Souayah, N., Michas-Martin, P.A., Nasar, A., *et al.* (2011). Guillain-Barre syndrome after Gardasil vaccination: data from Vaccine Adverse Event Reporting System 2006–2009. *Vaccine*, **29**: 886–9.

Sutton, I., Lahoria, R., Tan, I.L., *et al.* (2009). CNS demyelination and quadrivalent HPV vaccination. *Mult Sclerosis*, **15**(1): 116–19.

Theoharides, T.C., Kempuraj, D., and Redwood, L. (2009). Autism: an emerging "neuroimmune disorder" in search of therapy. *Expert Opin Pharmacother*, **10**: 2127–43.

Tomljenovic, L. (2011). Aluminum and Alzheimer's disease: after a century of controversy, is there a plausible link? *J Alzheimers Dis*, **23**: 567–98.

Tomljenovic, L. and Shaw, C.A. (2011). Do aluminum vaccine adjuvants contribute to the rising prevalence of autism? *J Inorg Biochem*, **105**: 1489–99.

Tomljenovic, L. and Shaw, C.A. (2012). Mechanisms of aluminum adjuvant toxicity in pediatric populations. *Lupus*, **21**: 223–30.

Topaloglu, H., Berker, M., Kansu, T., *et al.* (1992). Optic neuritis and myelitis after booster tetanus toxoid vaccination. *Lancet*, **339**: 178–9.

US FDA (Food and Drug Administration). (2014). Vaccines, blood & biologics. Common ingredients in U.S. licensed vaccines. Available from: http://www.fda.gov/BiologicsBloodVaccines/SafetyAvailability/VaccineSafety/ucm187810.htm [last accessed 11 December 2014].

US FDA DHHS (Food and Drug Administration Department of Health and Human Services). (2005). Aluminum in large and small volume parenterals used in total parenteral nutrition. Available from: http://www.gpo.gov/fdsys/pkg/CFR-2005-title21-vol4/pdf/CFR-2005-title21-vol4-sec201-323.pdf [last accessed 11 December 2014].

Verdier, F., Burnett, R., Michelet-Habchi, C., *et al.* (2005). Aluminium assay and evaluation of the local reaction at several time points after intramuscular administration of aluminium containing vaccines in the Cynomolgus monkey. *Vaccine*, **23**: 1359–67.

Verstraeten, S.V., Golub, M.S., Keen, C.L., and Oteiza, P.I. (1997). Myelin is a preferential target of aluminum-mediated oxidative damage. *Arch Biochem Biophys*, **344**: 289–94.

Voigt, U., Baum, U., Behrendt, W., *et al.* (2001). Neuritis of the optic nerve after vaccinations against hepatitis A, hepatitis B and yellow fever. *Klin Monbl Augenheilkd*, **218**: 688–90.

Walton, J.R. (2006). Aluminum in hippocampal neurons from humans with Alzheimer's disease. *Neurotoxicology*, **27**: 385–94.

Walton, J.R. (2007). A longitudinal study of rats chronically exposed to aluminum at human dietary levels. *Neurosci Lett*, **412**: 29–33.

Walton, J.R. (2009). Functional impairment in aged rats chronically exposed to human range dietary aluminum equivalents. *Neurotoxicology*, **30**: 182–93.

Walton, J.R. (2012a). Aluminum disruption of calcium homeostasis and signal transduction resembles change that occurs in aging and Alzheimer's disease. *J Alzheimers Dis*, **29**: 255–73.

Walton, J.R. (2012b). Evidence that ingested aluminum additives contained in processed foods and alum-treated drinking water are a major risk factor for Alzheimer's disease. *Curr Inorg Chem*, **2**: 19–39.

Walton, J.R. and Wang, M.X. (2009). APP expression, distribution and accumulation are altered by aluminum in a rodent model for Alzheimer's disease. *J Inorg Biochem*, **103**: 1548–54.

Wen, H., Gris, D., Lei, Y., *et al.* (2011). Fatty acid-induced NLRP3-ASC inflammasome activation interferes with insulin signaling. *Nat Immunol*, **12**: 408–15.

WHO (World Health Organization). (2005). Weekly epidemiological record/Relevé épidémiologique hebdomadaire. Available from: http://www.who.int/wer/2005/wer8001.pdf [last accessed 11 December 2014].

WHO (Word Health Organization). (2014). Human papillomavirus (HPV) and cervical cancer. Fact sheet No 380. Updated November 2014. Available from: http://www.who.int/mediacentre/factsheets/fs380/en/ [last accessed 11 December 2014].

Wildemann, B., Jarius, S., Hartmann, M., *et al.* (2009). Acute disseminated encephalomyelitis following vaccination against human papilloma virus. *Neurology*, **72**: 2132–3.

Yokel, R.A. (2006). Blood–brain barrier flux of aluminum, manganese, iron and other metals suspected to contribute to metal-induced neurodegeneration. *J Alzheimers Dis*, **10**: 223–53.

Yokel, R.A. and McNamara, P.J. (2001). Aluminium toxicokinetics: an updated minireview. *Pharmacol Toxicol*, **88**: 159–67.

Yokel, R.A., Hicks, C.L., and Florence, R.L. (2008). Aluminum bioavailability from basic sodium aluminum phosphate, an approved food additive emulsifying agent, incorporated in cheese. *Food Chem Toxicol*, **46**: 2261–6.

Zafrir, Y., Agmon-Levin, N., Paz, Z., *et al.* (2012). Autoimmunity following Hepatitis B vaccine as part of the spectrum of "Autoimmune (Auto-inflammatory) Syndrome induced by Adjuvants" (ASIA): analysis of 93 cases. *Lupus*, **21**: 146–52.

Zheng, W. (2001). Neurotoxicology of the brain barrier system: new implications. *J Toxicol Clin Toxicol*, **39**: 711–19.

Zinka, B., Rauch, E., Buettner, A., *et al.* (2006). Unexplained cases of sudden infant death shortly after hexavalent vaccination. *Vaccine*, **24**: 5779–80.

5 Allergy and Autoimmunity Caused by Metals: A Unifying Concept

Vera Stejskal
Department of Immunology, University of Stockholm, Stockholm, Sweden

Introduction

Allergy and autoimmunity are caused by an abnormal immune response and have the same clinical outcomes, including local and systemic inflammation resembling autoimmune/inflammatory syndrome induced by adjuvants (ASIA) (Shoenfeld and Agmon-Levin, 2011; Perricone *et al.*, 2013).

This chapter will give an overview of the literature on metal-induced pathologies, such as delayed-type hypersensitivity and autoimmunity. Because of the vast amount of information available on this subject, the focus of this review will be mainly on specific T cell reactivity to mercury, aluminum, nickel, and gold, all of which are known to induce immunotoxic effects in human subjects. Mercury, as a constituent of thimerosal, and aluminum are both used in vaccines.

The immunological effects of metals include immunomodulation, allergy, and autoimmunity. Metals may act as immunosuppressants or as immune adjuvants. One example of immunomodulation is the ability of metals to modify cytokine production *in vitro* and *in vivo*.

In the body, metal ions may firmly bind to cells and proteins. This binding results in the modification of autologous epitopes (i.e. haptenization). In susceptible individuals, T cells falsely recognize the modified proteins as foreign and start an autoimmune attack (Griem and Gleichmann, 1995; Schiraldi and Monestier, 2009; Wang and Dai, 2013). In experimental animals, the recognition of metal haptens is dependent on the genetic makeup: some rodent strains are resistant, while others are susceptible to the induction of autoimmunity by metals (Griem and Gleichmann, 1995; Bigazzi, 1999; Fournié *et al.*, 2001; Schiraldi and Monestier, 2009). Clusters of autoimmunity have been reported in areas of increased exposure to heavy metals (Ingalls, 1986). It has been found that mercury, nickel, cadmium, lead, aluminum, and arsenic can exert immunotoxic effects through epigenetic mechanisms, such as DNA methylation and histone modification (Greer and McCombe, 2012).

In humans, the expression of autoimmune diseases can differ between genetically identical twins. This suggests that, in addition to genetics, environmental factors are involved in the disease process. The genes controlling susceptibility to metals are the subject of intensive studies (Wang *et al.*, 2012; Woods *et al.*, 2013), but no clear conclusion has yet been reached. Genes that might predispose for toxic effects of metals are, for example, those involved in detoxification and synthesis of glutathione. In the case of metal allergy, only a few genetic studies have been performed, such as those on workers occupationally sensitized to beryllium (Wang and Dai, 2013).

Delayed-type hypersensitivity

The type of allergy induced by metals in humans is cellular-type hypersensitivity, also called type IV delayed-type hypersensitivity. "Delayed" refers to

Vaccines and Autoimmunity, First Edition. Edited by Yehuda Shoenfeld, Nancy Agmon-Levin, and Lucija Tomljenovic.

the fact the first symptoms appear 24–48 hours after initial exposure to the allergen, which makes causal connection difficult. Metals such as mercury are low-molecular haptens and only rarely produce antibodies (Wylie *et al.*, 1992). Hence, immunological responses induced by metals are mostly T cell-mediated.

The gold standard for diagnosis of delayed-type hypersensitivity is patch testing. In patch test, the suspected metal allergens are applied under occlusion on the skin of the back. A dermatologist evaluates the reaction after 2–3 days. Another diagnostic approach, one that is becoming more widespread, is the lymphocyte transformation test (LTT), which allows an objective evaluation of memory lymphocytes present in the blood of patients. In this test, blood lymphocytes are cultivated with metals or other allergens for 5 days *in vitro*, after which the number of proliferating lymphocytes is determined by radioisotope incorporation.

A standardized and validated form of LTT is LTT-MELISA (Memory Lymphocyte Stimulation Assay) (Stejskal *et al.*, 1994, 2006; Prochazkova *et al.*, 2004; Valentine-Thon *et al.*, 2007). In addition to objective radioisotope evaluation, morphological confirmation of the presence of activated lymphocytes (lymphoblasts) is also performed (Stejskal *et al.*, 2006).

The allergic and autoimmune effects of metals

Exposure to metals can be external (e.g. through pollution, occupation, cosmetics, and handling of metallic items) or internal (e.g. through foods, dental restorations, orthopaedic implants, and vaccines). Cigarette smoke contains many metals, such as mercury, cadmium, lead, arsenic, and nickel, and increasing evidence is linking it to autoimmune disorders (Arnson *et al.*, 2010).

Mercury

It has been known for decades that exposure to mercury through skin-lightening ointments will, in some individuals, lead to the development of serious side effects, such as kidney disease (Turk and Baker, 1968; Barr *et al.*, 1972; Kibukamusoke *et al.*, 1974), as well as neurological complications such as peripheral polyneuropathy (Kern *et al.*, 1991; Adawe and Oberg, 2013). In a more recent paper, skin-lightening creams induced neuropsychological problems and glomerulonephritis in a patient with juvenile diabetes (Pelcova *et al.*, 2002). After mercury chelation, the symptoms disappeared, confirming a causal relationship. Mercury-containing ointments are still being used in some countries (Weldon *et al.*, 2000).

The main source of inorganic mercury in the general population is mercury released from dental amalgam fillings (Clarkson *et al.*, 1988). Dental amalgam consists of 50% mercury, ~22–32% silver, ~14% tin, ~8% copper, and other trace metals (Ferracane, 2001). Since mercury functions as both adjuvant and allergen, it has no safe dose level (IPCS, 1991). The most common source of methyl mercury is ingested polluted fish. Methyl mercury can also be formed through the conversion of metallic mercury by oral and gastrointestinal bacteria, and vice versa (Liang and Brooks, 1995).

Thimerosal and phenyl mercury are organic mercury compounds used as antiseptics and preservatives in eye drops and vaccines (Rietschel and Fowler, 2001). Like methyl mercury, these organic mercury compounds are decomposed to inorganic mercury in the body (WHO, 1990; Havarinasab and Hultman, 2005).

Inorganic mercury, thimerosal, and nickel are the most common allergens in children, a fact that is not widely recognized. Of 1094 children with skin disease, 10% reacted to thimerosal (ethylmercury thiosalicylate) and 6% to mercury (Seidenari *et al.*, 2005) in patch test. A review of PubMed articles investigating allergens in at least 100 children from the years 1966–2010 showed that among the top five allergens across 49 studies, three were metals: nickel, gold, and thimerosal (Bonitsis *et al.*, 2011).

Sensitization to thimerosal can be demonstrated *in vitro* by LTT-MELISA, as shown previously (Stejskal *et al.*, 1994, 1999; Stejskal, 2014). In a large study of over 3000 patients, tested by LTT-MELISA in three different laboratories, the prevalence of thimerosal-specific lymphocyte responses was around 7% (Stejskal *et al.*, 1999). As shown in Table 5.1, LTT-MELISA can identify thimerosal-specific responses in patients who have experienced side effects after exposure to thimerosal-containing products.

According to one paper (Westphal *et al.*, 2000), thimerosal sensitization depends on homozygous gene deletion of the glutathione S-transferases, indicating the role of genetics in detoxification capacity.

It is important to note that memory lymphocytes induced by various mercury compounds do not crossreact, as shown by Italian dermatologists (Tosti *et al.*, 1989; Santucci *et al.*, 1998) and by

Table 5.1 Lymphocyte responses in LTT-MELISA to thimerosal and other metals in patients with side effects following exposure to thimerosal-containing products

Patient number	Sex	Age	Health status	Thimerosal exposure	Symptoms after exposure	Positive thimerosal responses (SI)	Other positive responses
1	F	45	CFS	Hepatitis-B vaccine, gamma globulin	Flu-like symptoms after hepatitis B vaccine	**20**	Cadmium, palladium, phenyl mercury, tin
2	F	52	Skin/eye irritation, fatigue	Anti-D globulin × 3, eye drops, TB vaccine, patch test	Worsening of symptoms after thimerosal patch testing	**19**	Ethyl mercury, inorganic mercury, methyl mercury
3	F	58	CFS	Vaccines	Flu-like symptoms post-vaccination	5.9	Inorganic mercury, phenyl mercury
4	F	53	CFS, oral lichen planus	Gamma globulin × 8, cosmetics	Eyelid eczema and edema from cosmetics	**41**	None
5	F	48	CFS	Vaccines	Not known	7.3	None
6	F	18	Heart problems	Vaccines	Not known	**16.3**	Cadmium, copper, inorganic mercury, lead, methyl mercury, phenyl mercury
7	F	57	CFS	Gamma globulin, TB vaccine	Not known	**65**	None
8	F	45	CFS	Vaccines	Not known	**12.4**	Ethyl mercury, gold, inorganic mercury, lead, methyl mercury, nickel, phenyl mercury, tin
9	M	47	CFS	Gamma globulin, eye drops	Not known	4.4	Cadmium, ethyl mercury, gold, inorganic mercury, lead, methyl mercury, nickel, palladium, phenyl mercury, tin
10	F	53	CFS	Gamma globulin, eye drops	Not known	4.4	Cadmium, ethyl mercury, methyl mercury, nickel

Lymphocytes were isolated from human blood and cultivated for 5 days with a wide range of metal salts, including thimerosal, inorganic mercury, methyl mercury, phenyl mercury, gold, palladium, tin, lead, nickel, and cadmium (Stejskal *et al.*, 1999). Metal-specific responses were measured by 3H thymidine uptake. Lymphocyte responses are shown as stimulation index (SI) = counts per minute (cpm) in metal-treated cultures divided by counts per minute in control cultures. SI ≥ 3 is a positive response and SI ≥ 10 is a strongly positive response (shown in **bold**)

LTT-MELISA testing (Stejskal *et al.*, 1994). However, sensitization to several mercury compounds, as well as to other metals, is frequently observed.

Clinical observations accumulated over many years indicate that exposure to mercury can induce multiple sclerosis and other autoimmune diseases. As early as 1966, Baasch suggested that multiple sclerosis is caused by a neuroallergic reaction to mercury released from amalgam fillings, comparing it to an adult form of acrodynia (pink disease) (Baasch, 1966). Acrodynia occurred in some children who were treated with a mercury-containing teething powder (Warkany and Hubbard, 1953). The same conclusion – that dental and environmental exposure to mercury could be one of the factors leading to multiple sclerosis – was also reached by Ingalls (1983, 1986).

Recent research supports these early clinical observations. Prochazkova *et al.* (2004), at Charles University in Prague, studied the impact of amalgam replacement on health in patients with various autoimmune diseases who showed increased mercury-specific responses *in vitro*. After the replacement of mercury-containing amalgam with metal-free materials, 71% of the patients showed health improvement by 6 months later. In the group of patients that did not undergo dental treatment, no health improvement occurred.

Other studies seemingly contradict the hypothesis that mercury might be one of the causes of neurodegenerative diseases. Saxe *et al.* (1999) measured the concentration of mercury in the brains of Alzheimer's patients and controls. Since there were no statistically significant differences in brain mercury levels between the two groups, the authors concluded that mercury does not appear to be a neurotoxic factor in the pathogenesis of Alzheimer's disease. Similar findings were published by Clausen (1993), who studied mercury levels in the brains of patients with multiple sclerosis. The conclusions drawn from these studies may be questioned. In mercury-sensitized patients, even mercury concentrations within the normal range might provoke neuroallergic reactions in the brain.

The protocol of identification of metal hypersensitivity and removal of sensitizing metals has been successfully used in patients with fibromyalgia (Stejskal *et al.*, 2013) and autoimmune thyroid diseases (Sterzl *et al.*, 1999, 2006; Hybenova *et al.*, 2010). In the latter group, the removal of mercury-containing amalgam not only downregulated mercury-specific responses *in vitro*, but also resulted in a significant decrease of antithyroid peroxidase and antithyreoglobulin antibodies compared to levels prior to treatment.

Another disease of autoimmune origin is oral lichen planus. In one study, 72% of patients with oral lichen planus showed a positive response to mercury *in vitro* (Stejskal *et al.*, 1996). In addition to oral symptoms, the patients suffered from arthralgia, myalgia, eczema, and chronic ill health. After removal of amalgams, both local and systemic symptoms significantly decreased.

Finally, a study was recently published which showed successful treatment of orofacial granulomatosis on removal of amalgam in patients with a hypersensitivity to mercury (Tomka *et al.*, 2011).

Gold

The autoimmune potential of gold compounds has been known for many years. Serious side effects, such as nephropathy, were observed in some patients after the use of colloidal gold as a treatment for rheumatoid arthritis (Palosuo *et al.*, 1976), and the possible mechanisms behind these side effects have been discussed (Stejskal *et al.*, 1999). According to some studies, gold allergy is more common in patients who have developed autoimmune side effects after treatment with gold, indicating the existence of both allergy and autoimmunity induced by gold in the same patient (Möller *et al.*, 1996). It is important to emphasize that, as with other metals, gold allergy is not only expressed on the mucosa or skin, but also inside the body. For example, the rate of restenosis after implantation of gold-stented plates is high in patients suffering from gold allergy (Ekqvist *et al.*, 2007).

Nickel

Nickel is the most common sensitizer, and also the most studied (Thyssen and Menné, 2010). In Swedish patients with chronic fatigue syndrome (CFS), the frequency of nickel allergy was around 40%, as diagnosed by LTT-MELISA (Stejskal *et al.*, 1999). The coexistence of both allergic and autoimmune symptoms, induced by nickel, has been published, suggesting the autoimmune potential of nickel compounds (Kosboth *et al.*, 2007; Niedziela and Bluvshteyn-Walker, 2012). Direct evidence of nickel-induced autoimmunity was observed in susceptible rats that developed scleroderma-related autoantibodies and cutaneous sclerosis after exposure to nickel (Al-Mogairen *et al.*, 2010). Since nickel can also induce Toll-like receptors (TLRs) (Schmidt *et al.*, 2010), the autoimmune potential of this metal is plausible and should be studied in the future.

Aluminum

Aluminum is a ubiquitous metal, widely occurring in the environment and used in many everyday objects, foods, and pharmaceuticals. Aluminum is a well-known adjuvant in vaccines, despite its neurotoxic properties (Shaw and Tomljenovic, 2013). As described by Shoenfeld *et al.* (Shoenfeld and Agmon-Levin, 2011; Perricone *et al.*, 2013), adjuvants can promote ASIA in susceptible patients. Allergy to aluminum is rare, but has been described. Delayed-type hypersensitivity to aluminum and itching nodules were found in children exposed to aluminum-containing vaccines (Bergfors *et al.*, 2003). Exley *et al.* (2009) described a patient who developed CFS after multiple vaccinations with aluminum-containing vaccines. A muscle biopsy confirmed the presence of aluminum-containing macrophages; the aluminum content in the patient's urine was also increased. Macrophagic myofasciitis (MMF) has been described by Gherardi and Authier (2012) as a systemic disease whose main histopathological feature is a granulomatous lesion comprising aluminum-loaded macrophages at the site of previous intramuscular vaccination. Typical clinical manifestations in MMF patients include myalgias, arthralgias, marked asthenia, weakness, cognitive dysfunction, and CFS. In addition, 15–20% of MMF patients may also have coexistent autoimmune diseases, the most frequent of which are multiple sclerosis, Hashimoto's thyroiditis, and diffuse autoimmune neuromuscular diseases, such as dermatomyositis, necrotizing autoimmune myopathy, myasthenia gravis, and inclusion body myositis (Authier *et al.*, 2001; Guis *et al.*, 2002).

Conclusions

Scientific literature and clinical experience show that metals play a key role in the development of autoimmune diseases. Whether metals induce autoimmunity or whether they aggravate existing disease, the removal of sensitizing metals upon identification of metal triggers has improved patient health.

Larger randomized studies in susceptible individuals, selected on the basis of genotypic or phenotypic biomarkers, should be pursued in the future. As suggested by Weiss and Liff (1983), studies of phenotypic markers may be suitable for the elucidation of causal pathways and identification of specific risk factors. The limited power of epidemiological studies to detect minor susceptible populations, such as those susceptible to mercury, has been discussed by Wallach *et al.* (2003). The benefits of this approach for patients can be monitored not only by the decrease in antibody titers (Sterzl *et al.*, 1999), but also by downregulation of metal-specific lymphocyte responses *in vitro* (Stejskal *et al.*, 1999, 2006, 2013; Yaqob *et al.*, 2006).

Finally, the identification of sensitized T cells in human blood can be made use of in future studies of vaccine-induced side effects. Elucidation of the possible mechanisms will contribute not only to successful treatment of affected individuals but also to the development of safer vaccines. The use of human blood lymphocytes in vaccine research has recently been suggested (Brookes *et al.*, 2014).

References

Adawe, A. and Oberg, C. (2013). Skin-lightening practices and mercury exposure in the Somali community. *Minn Med*, **96**(7): 48–9.

Al-Mogairen, S.M., Meo, S.A., Al-Arfaj, A.S., *et al.* (2010). Nickel-induced allergy and contact dermatitis: does it induce autoimmunity and cutaneous sclerosis? An experimental study in Brown Norway rats. *Rheumatol Int*, **30**(9): 1159–64.

Arnson, Y., Shoenfeld, Y., and Amital, H. (2010). Effects of tobacco smoke on immunity, inflammation and autoimmunity. *J Autoimmun*, **34**(3): J258–65.

Authier, F.J., Cherin, P., Creange, A., *et al.* (2001). Central nervous system disease in patients with macrophagic myofasciitis. *Brain*, **124**(Pt 5): 974–83.

Baasch, E. (1966). Theoretical considerations on the etiology of multiple sclerosis. Is multiple sclerosis a mercury allergy? *Schweiz Arch Neurol Neurochir Psychiatr*, **98**(1): 1–19.

Barr, R.D., Rees, P.H., Cordy, P.E., *et al.* (1972). Nephrotic syndrome in adult Africans in Nairobi. *BMJ*, **2**(806): 131–4.

Bergfors, E., Trollfors, B., and Inerot, A. (2003). Unexpectedly high incidence of persistent itching nodules and delayed type hypersensitivity to aluminum in children after the use of adsorbed vaccines from a single manufacturer. *Vaccine*, **22**: 64–9.

Bigazzi, P.E. (1999). Metals and kidney autoimmunity. *Environ Health Perspect*, **107**(Suppl. 5): 753–65.

Bonitsis, N.G., Tatsioni, A., Bassioukas, K., and Ioannidis, J.P. (2011). Allergens responsible for allergic contact dermatitis among children: a systematic review and meta-analysis. *Contact Dermatitis*, **64**(5): 245–57.

Brookes, R.H., Hakimi, J., Ha, Y., *et al.* (2014). Screening vaccine formulations for biological activity using fresh human whole blood. *Hum Vaccin Immunother*, **8**: 10(4).

Clarkson, T.W., Friberg, L., Nordberg, G.F., and Sager, P. (1988). *Biological Monitoring of Metals*. New York: Plenum Press.

Clausen, J. (1993). Mercury and multiple sclerosis. *Acta Neurol Scand*, **87**(6): 461–4.

Ekqvist, S., Svedman, C., Möller, H., *et al.* (2007). High frequency of contact allergy to gold in patients with endovascular coronary stents. *Br J Dermatol*, **157**(4): 730–8.

Exley, C., Swarbrick, K., Gherardi, R.K., and Authier, F.J. (2009). Medical A role for the body burden of Al in vaccine-associated macrophagic myofasciitis and chronic fatigue syndrome. *Med Hypotheses*, **72**: 135–9.

Ferracane, J.L. (2001). *Materials in Dentistry: Principles and Applications*. Philadelphia: Lippincott Williams & Wilkin.

Fournié, G.J., Mas, M., Cautain, B., *et al.* (2001). Induction of autoimmunity through bystander effects. Lessons from immunological disorders induced by heavy metals. *J Autoimmun*, **16**(3): 319–26.

Gherardi, R.K. and Authier, F.J. (2012). Macrophagic myofasciitis: characterization and pathophysiology. *Lupus*, **21**(2): 184–9.

Greer, J.M. and McCombe, P.A. (2012). The role of epigenetic mechanisms and processes in autoimmune disorders. *Biologics*, **6**: 307–27.

Griem, P. and Gleichmann, E. (1995). Metal ion induced autoimmunity. *Curr Opin Immunol*, **7**(6): 831–8.

Guis, S., Pellissier, J.F., Nicoli, F., *et al.* (2002). HLA-DRB1 01 and macrophagic myofasciitis. *Arthritis Rheum*, **46**: 2535–2537.

Havarinasab, S. and Hultman, P. (2005). Organic mercury compounds and autoimmunity. *Autoimmun Rev*, **4**(5): 270–5.

Hybenova, M., Hrda, P., Procházková, J., *et al.* (2010). The role of environmental factors in autoimmune thyroiditis. *Neuro Endocrinol Lett*, **31**(3): 283–9.

Ingalls, T.H. (1983). Epidemiology, etiology, and prevention of multiple sclerosis. Hypothesis and fact. *Am J Forensic Med Pathol*, **4**(1): 55–61.

Ingalls, T.H. (1986). Endemic clustering of multiple sclerosis in time and place, 1934–1984. Confirmation of a hypothesis. *Am J Forensic Med Pathol*, **7**(1): 3–8.

IPCS (International Programme on Chemical Safety). (1991). *Environmental Health Criteria 118: Inorganic Mercury*. Geneva: World Health Organization.

Kern, F., Roberts, L., Osterle, L., *et al.* (1991). Ammoniated mercury ointment as a cause of peripheral neuropathy. *Dermatologica*, **183**: 280–2.

Kibukamusoke, J.W., Davies, D.R., and Hutt, M.S.R. (1974). Membranous nephropathy due to skin lightening cream. *BMJ*, **2**(920): 646–7.

Kosboth, M., Chin-Loy, A., Lyons, R., *et al.* (2007). Malar rash caused by metal allergy in a patient with systemic lupus erythematosus. *Nat Clin Pract Rheumatol*, **3**(4): 240–5.

Liang, L. and Brooks, R.J. (1995). Mercury reactions in the human mouth with dental amalgams. *Water, Air and Soil Poll*, **80**: 103–7.

Möller, H., Larsson, A., and Björkner, B. (1996). Flare-up at contact allergy sites in a gold-treated rheumatic patient. *Acta Derm Venereol (Stockh)*, **76**: 55–8.

Niedziela, M. and Bluvshteyn-Walker, S. (2012). Autoimmune thyroid disease and allergic contact dermatitis: two immune-related pathologies in the same patient. *J Pediatr Endocrinol Metab*, **25**(1–2): 31–2.

Palosuo, T., Provost, T.T., and Milgrom, F. (1976). Gold nephropathy: serologic data suggesting an immune complex disease. *Clin Exp Immunol*, **25**(2): 311–18.

Pelcova, D., Lukas, E., Urban, P., *et al.* (2002). Mercury intoxication from skin ointment containing mercuric ammonium chloride. *Int Arch Occup Environ Health*, **75**(Suppl.): 54–9.

Perricone, C., Colafrancesco, S., Mazor, R.D., *et al.* (2013). Autoimmune/inflammatory syndrome induced by adjuvants (ASIA): unveiling the pathogenic, clinical and diagnostic aspects. *J Autoimmun*, **47**: 1–16.

Prochazkova, J., Sterzl, I., Kucerova, H., *et al.* (2004). The beneficial effect of amalgam replacement on health in patients with autoimmunity. *Neuro Endocrinol Lett*, **25**: 211–18.

Rietschel, R.L. and Fowler, J.F. (2001). *Fisher's Contact Dermatitis*, 5th edn. Philadelphia: Lippincott, Williams & Wilkins.

Santucci, B., Cannistraci, C., Cristaudo, A., *et al.* (1998). Thimerosal sensitivities: the role of organo mercury alkyl compounds. *Contact Derm*, **38**: 325–8.

Saxe, S.R., Wekstein, M.W., Kryscio, R.J., *et al.* (1999). Alzheimer's disease, dental amalgam and mercury. *J Am Dent Assoc*, **130**(2): 191–9.

Schiraldi, M. and Monestier, M. (2009). How can a chemical element elicit complex immunopathology? Lessons from mercury-induced autoimmunity. *Trends Immunol*, **30**(10): 502–9.

Schmidt, M., Raghavan, B., Müller, V., *et al.* (2010). Crucial role for human Toll-like receptor 4 in the development of contact allergy to nickel. *Nat Immunol*, **11**(9): 814–19.

Seidenari, S., Giusti, F., Pepe, P., and Mantovani, L. (2005). Contact sensitization in 1094 children undergoing patch testing over a 7-year period. *Pediatr Dermatol*, **22**: 1–5.

Shaw, C.A. and Tomljenovic, L. (2013). Aluminum in the central nervous system (CNS): toxicity in humans and animals, vaccine adjuvants, and autoimmunity. *Immunol Res*, **56**(2–3): 304–16.

Shoenfeld, Y. and Agmon-Levin, N. (2011). "ASIA" – autoimmune/inflammatory syndrome induced by adjuvants. *J Autoimmun*, **36**(1): 4–8.

Stejskal, V.D., Cederbrant, K., Lindvall, A., and Forsbeck, M. (1994). MELISA – an *in vitro* tool for the study of metal allergy. *Toxicol In Vitro*, **8**: 991–1000.

Stejskal, V.D., Forsbeck, M., Cederbrant, K.E., and Asteman, O. (1996). Mercury-specific lymphocytes: an indication of mercury allergy in man. *J Clin Immunol*, **16**(1): 31–40.

Stejskal, V.D., Danersund, A., Lindvall, A., *et al.* (1999). Metal-specific lymphocytes: biomarkers of sensitivity in man. *Neuro Endocrinol Lett*, **20**(5): 289–98.

Stejskal, V., Hudecek, R., Stejskal, J., and Sterzl, I. (2006). Diagnosis and treatment of metal-induced side-effects. *Neuro Endocrinol Lett*, **27**(Suppl. 1): 7–16.

Stejskal, V., Ockert, K., and Bjørklund, G. (2013). Metal-induced inflammation triggers fibromyalgia in metal-allergic patients. *Neuro Endocrinol Lett*, **34**(6): 559–65.

Stejskal V. (2014). Metals as a common trigger of inflammation resulting in non-specific symptoms: diagnosis and treatment. *IMAJ*, **16**: 2–7.

Sterzl, I., Procházková, J., Hrdá, P., *et al.* (1999). Mercury and nickel allergy: risk factors in fatigue and autoimmunity. *Neuro Endocrinol Lett*, **20**(3–4): 221–8.

Sterzl, I., Prochazkova, J., Hrda, P., *et al.* (2006). Removal of dental amalgam decreases anti-TPO and anti-Tg autoantibodies in patients with autoimmune thyroiditis. *Neuro Endocrinol Lett*, **27**: 25–30.

Thyssen, J.P. and Menné, T. (2010). Metal allergy – a review on exposures, penetration, genetics, prevalence, and clinical implications. *Chem Res Toxicol*, **23**(2): 309–18.

Tomka, M., Machovcová, A., Pelclová, D., *et al.* (2011). Orofacial granulomatosis associated with hypersensitivity to dental amalgam. *Oral Surg Oral Med Oral Pathol Oral Radiol Endod*, **112**(3): 335–41.

Tosti, A., Guerra, L., and Bardazzi, F. (1989). Hyposensitizing therapy with standard antigenic extracts: an important source of thimerosal sensitization. *Contact Dermat*, **20**: 173–6.

Turk, J.L. and Baker, H. (1968). Nephrotic syndrome due to ammoniated mercury. *Br J Dermatol*, **80**: 623–4.

Valentine-Thon, E., Ilsemann, K., and Sandkamp, M. (2007). A novel lymphocyte transformation test (LTT-MELISA) for Lyme borreliosis. *Diagn Microbiol Infect Dis*, **57**(1): 27–34.

Wallach, H., Nauman, J., Mutter, J., and Daschner, F. (2003). No difference between self reported amalgam sensitivities and non-sensitivities? Listen carefully to the data. *Int J Hyg Environ Health*, **206**: 139–41.

Wang, Y. and Dai, S. (2013). Structural basis of metal hypersensitivity. *Immunol Res*, **55**(1–3): 83–90.

Wang, Y., Goodrich, J.M., Gillespie, B., *et al.* (2012). An investigation of modifying effects of metallothionein single-nucleotide polymorphisms on the association between mercury exposure and biomarker levels. *Environ Health Perspect*, **120**(4): 530–4.

Warkany, J. and Hubbard, D.M. (1953). Acrodynia and mercury. *J Pediatr*, **42**: 365–86.

Weiss, N.S. and Liff, J.M. (1983). Accounting for the multicausal nature in disease in the design and analysis of epidemiological studies. *Am Epidemiol*, **117**: 14–18.

Weldon, M.M., Smolinski, M.S., Maroufi, A., *et al.* (2000). Mercury poisoning associated with a Mexican beauty cream. *West J Med*, **173**(1): 15–18

Westphal, G.A., Schnuch, A., Schulz, T.G., *et al.* (2000). Homozygous gene deletions of the glutathione S-transferases M1 and T1 are associated with thimerosal sensitization. *Int Arch Occup Environ Health*, **73**(6): 384–8.

WHO (World Health Organization). (1990). Methylmercury, Environmental Health Criteria 61, 1–197. Geneva: World Health Organization.

Woods, J.S., Heyer, N.J., Russo, J.E., *et al.* (2013). Modification of neurobehavioral effects of mercury by genetic polymorphisms of metallothionein in children. *Neurotoxicol Teratol*, **39**: 36–44.

Wylie, D.E., Lu, L., Carlson, S.D., *et al.* (1992). Monoclonal antibodies specific for mercuric ions. *Proc Nat Acad Sci USA*, **89**: 4104–8.

Yaqob, A., Danersund, A., Stejskal, V.D., *et al.* (2006). Metal-specific lymphocyte reactivity is downregulated after dental metal replacement. *Neuro Endocrinol Lett*, **27**(1–2): 189–97.

6 Genetics and Vaccinology

John Castiblanco and Juan-Manuel Anaya
Center for Autoimmune Diseases Research (CREA), School of Medicine and Health Sciences, Del Rosario University, Bogotá, Colombia

Introduction

Vaccines are the most effective and sustainable means of preventing infectious diseases (Rappuoli *et al.*, 2002). A vaccine, typically containing one or several antigens from or similar to a disease-causing microorganism, improves immunity to a particular disease upon administration, by inducing specific immune responses. Over the last century, the availability of vaccines has reduced the incidence and mortality of smallpox, polio, pertussis diphtheria, tetanus, polio, measles, mumps, rubella, pneumococcus, hepatitis B, and meningitis (Rappuoli *et al.*, 2002).

Although the field has been successful, it still lacks safe and effective vaccines against pathogens – such as neglected tropical diseases, tuberculosis, the human immunodeficiency virus (HIV), and malaria – affecting humans worldwide (Thomas and Moridani, 2010). Moreover, the historic development of the field is usually divided into generations: the first generation stands for the administration of inactivated pathogens in whole or live attenuated forms (e.g. bacillus Calmette–Guérin (BCG), plague, pertussis, polio, rabies, and smallpox) (Rappuoli *et al.*, 2002; Kaushik and Sehgal, 2008); the second refers to vaccines assembled from purified microbial cell components, also referred to as "subunit vaccines" (e.g. polysaccharides or protein antigens) (Plotkin, 2005). This approach has exploited recombinant DNA technology and polysaccharide chemistry.

Conventional methods of vaccine development deal with such obstacles as noncultivable *in vitro* pathogens (e.g. hepatitis C, papilloma virus types 16 and 18, and *Mycobacterium leprae*), pathogens with antigen hypervariability (e.g. serogroup B meningococcus, gonococcus, malaria, and HIV) (Rappuoli, 2004), opportunistic pathogens (e.g. *Staphylococcus aureus*) (Projan *et al.*, 2006), and rapidly evolving pathogens (e.g. HIV).

The main goal of every vaccination is to initiate protective and efficient immune responses in the entire receiving population; however, this is rarely observed, reflecting host and pathogen immune system variability. On top of this, tools by which to accurately identify vaccine outcomes are currently lacking (Thomas and Moridani, 2010). The main known factors influencing the observed heterogeneity for immune responses induced by vaccines are gender, age, ethnicity, comorbidity, immune system, and genetic background. The effect of genetic status in defining the response generated directly or indirectly by an innate or adaptive immune response has been demonstrated across multiple viral vaccines (e.g. smallpox, influenza, measles, mumps, and rubella) (Poland *et al.*, 2013).

Vaccine research gained new momentum with the blooming of the genomics field over the last several decades. When the first complete genome sequence of a living microorganism became available in 1995, vaccine research was reinvigorated (Fleischmann *et al.*, 1995). Now, more than 300 bacterial genomes have been sequenced and analyzed, including some of major effect in humans (Fraser and Rappuoli, 2005). The advent of high-throughput sequencing technologies has enabled new and more sophisticated approaches to further expand genomic information, becoming key drivers in disentangling vaccine-induced immune response and hopefully ushering in an era of personalized and predictive vaccinology,

Vaccines and Autoimmunity, First Edition. Edited by Yehuda Shoenfeld, Nancy Agmon-Levin, and Lucija Tomljenovic.

instead of a one-size-fits-all approach (Poland *et al.*, 2013).

The study of pathogen and host genomes by both computational and experimental approaches is broadening the field to mechanistic and functional insights, due to their significant potential to aid in the development of novel diagnostics, therapeutics, and vaccines (Serruto and Rappuoli, 2006). Concepts such as "vaccinomics" describe the common ground upon which a systems biology approach takes into account immunogenetic, immunogenomic, metagenomics, immune profiling, and functional studies in order to understand and predict vaccine-induced immune responses, and uses this information to engineer and test new vaccine candidates (Pulendran *et al.*, 2010).

Vaccine development has largely been focused on identifying a specific immune response that might be exploited to develop a vaccine capable of eliciting long-lived protection against a pathogen (Plotkin, 2008). The power of a systems approach in unraveling novel mechanisms of vaccine action, thus enabling the prediction of the immunogenicity and efficiency of vaccines, has been highlighted by recent studies (Querec *et al.*, 2009). High-throughput approaches allow exploration of the interconnected networks that control and drive the immune response, shifting the field from the search for a single correlate (or multiple, independent correlates) to the identification of multifactorial signatures associated with immunological protection.

Novel diagnostics would help customize the use of vaccines in subpopulations in which they would display enhanced safety and efficacy. Moreover, reframing of the field of public health to include "therapeutics," instead of the classic model of merely "prevention," is giving vaccines application in the treatment of chronic noncommunicable diseases, such as cancer and obesity, that are impacting health worldwide (Daar, 2010). A fresh new look at how we design vaccines and apply them judiciously to benefit global health is essential and timely in the present age of data-enabled science and post-genomics integrative biology. This chapter will focus on giving a glimpse of the genetic status effect of vaccine immune response and how this could contribute to the development of novel vaccine candidates that are better directed and predicted relative to the genetic history of an individual and/or population.

Brief history of vaccinology

Vaccines antedate the vaccinology field by a long time. Their origins can be traced to Asia, where smallpox lesions were used to transmit a mild infection to induce protection (Fenner, 1980). Documented smallpox vaccinations date back as early as the 17th century in the United States and England. These were administered through the use of variolation: the purposeful infection of a person with smallpox (Variola). Variolation became more accepted and safer when Edward Jenner demonstrated protection against smallpox infection through the inoculation of cowpox in 1796, leading to the formulation of the vaccine concept (Artenstein and Poland, 2012). By the 19th century – once the germ theory of disease had been proven and several bacteria species related to infection and viruses discovered – Louis Pasteur described the process of microbial attenuation and its implications for immunization. Then, he further developed the rabies vaccine, the first human vaccine created in a laboratory (Artenstein and Poland, 2012). The principles established by Louis Pasteur (i.e. isolation, inactivation, and administration of disease-causing microorganisms) began the rational development of vaccines and established the basic rules of vaccinology (Rappuoli *et al.*, 2002). Ever since, the field has focused on vaccination as the best defense against numerous bacterial and viral pathogens, with a profound effect in human health. By the mid-20th century, toxoid-based vaccines brought diphtheria and tetanus under control; these were followed by partially successful killed bacterial vaccines for cholera and typhoid and the first inactivated viral vaccine against influenza; and later by a successful attenuated yellow fever vaccine (Norrby, 2007). After this, a series of developments in tissue culture techniques culminated in the first *ex vivo* cultivation of poliovirus, leading to an effective polio vaccine (Enders *et al.*, 1949), and eventually to vaccines against other important childhood diseases, such as measles, mumps, rubella, and varicella.

All existing vaccines are based on killed or live-attenuated microorganisms or subunits purified from the microorganisms, such as toxins detoxified by chemical treatment, purified antigens, or polysaccharide conjugated to proteins (Table 6.1). These vaccines were developed using Pasteur's principles and have become landmarks and tools that have led to the elimination of some of the most devastating infectious diseases

Table 6.1 Approaches to vaccine design in the pre-genomic era: application of Pasteur's principles. Serruto, D. and Rappuoli, R. Post-genomic vaccine development. *FEBS Lett* 580(12): 2985–2992. Copyright © 2006, Elsevier

Microorganism status	Pathogen treatment	Advantages	Drawbacks	Vaccine examples
Killed	Agent is inactivated	Efficacious	Difficult to cultivate in a scalable setting	Polio virus, influenza, rabies, oral cholera
Live attenuated	Live agent does not cause disease			Polio virus, intranasal influenza vaccine, measles, mumps, rubella (MMR)
Subunit	Purified portions of agents	No risk of disease No need to culture	Identification of components is complex and time-consuming	Diphtheria toxoid, tetanus toxoid, pertussis toxoid, hepatitis B vaccine (HBV)
Subunit – conjugated	Polysaccharide component agent is linked to a protein carrier	The conjugated polysaccharide that is poorly immunogenic on its own becomes immunogenic	Need to culture *in vitro* to obtain capsule	Haemophlius influenza, meningococcus A, C, Y, W135, pneumococcus

worldwide. Vaccine development takes time, especially for noncultivable pathogens and for those where there is not an obvious antigen or structure to use as a candidate. On top of this, variation between individuals in vaccine responses remains a complex trait that needs further investigation, due to the fact that a high proportion of vaccinated individuals lack complete protection after routine immunizations (e.g. 10–15% of adults fail to respond to three doses of hepatitis B virus (HBV) vaccine) (Roome *et al.*, 1993).

By the end of the 20th century, traditional vaccine development technologies were becoming unused, since all plausible vaccines that could be developed were described already; the field needed new approaches to counteract the remaining problematic pathogens. Remarkable progress was made by the introduction of new such technologies as recombinant DNA and chemical conjugation of proteins to polysaccharides, as well as novel adjuvants. In 1995, Craig Venter reported the first draft genome of a microorganism (Fleischmann *et al.*, 1995), leading the way to a technological revolution, allowing the use of computational approaches to the design of vaccines by extracting the information from the genome without the need to ever grow the pathogen; this is known as "reverse vaccinology" (Sette and Rappuoli, 2010).

Reverse vaccinology has been applied to many bacterial pathogens, in order to develop protein-based vaccines. The first pathogen addressed was group B streptococcus. Eight genomes were analyzed, leading to the expression of 312 candidate antigens and the development of a vaccine composed of four proteins capable of protecting against all serotypes (Sette and Rappuoli, 2010). For group A streptococcus, another vaccine was developed by crossmatching homology to make sure the selected antigens for the vaccine differed from human encoded proteins. Reverse vaccinology uses the entire protein repertoire of each pathogen to select the best candidate vaccine antigens. This allows for the development of vaccines that were previously difficult or impossible to make, and can lead to the discovery of unique antigens that may improve existing vaccines. The main differences between conventional and reverse vaccinology are summarized in Table 6.2.

Vaccine response and genetics

The immune system is responsible for surveying, recognizing, and generating a response to a presented exposure. Recognition of foreign and hazard signals stems from the ability of antigen-presenting cells (APCs) to expose specific pathogen-derived peptides in the context of the HLA peptide binding grooves determined by the genetic constitution of the individual; this provides

Table 6.2 Comparison between traditional and reverse vaccinology. Sette, A. and Rappuoli, R. Reverse vaccinology: developing vaccines in the era of genomics. *Immunity* 33(4): 530–541. Copyright © 2010, Elsevier

	Traditional	Reverse
Available antigens	Only 10–25 identified	Virtually all antigens encoded by the genome
Antigen properties	Most abundant antigens, immunogenic during disease only from cultivable microorganisms	Antigens from noncultivable microorganisms can be identified
Antigen immunology	Highly immunogenic antigens Some may contain domains mimicking self-antigens and may induce autoimmunity	Conserved protective antigens can be identified The novel antigens are screened against the human genome to avoid homology
Polysaccharide antigens	A major target of traditional bacterial vaccines	Cannot be identified by reverse vaccinology; however, operons coding for the biosynthesis of polysaccharides can be identified
T cell epitopes	Known epitopes limited to the known antigens	Virtually every single T cell epitope is available

a useful model for understanding variability in immune responses (Poland, 1998).

Population genetic studies provide the tools for understanding the underlying genetic factors responsible for the variation in susceptibility to pathogen infection, and also provide further clues to the interactions between host and pathogen that define the host response. However, diversity and heterogeneity in immune responses to vaccines remain obstacles to designing and offering vaccines to the general public. This variability stems from the genetic history of each individual and is believed to be related mostly to polymorphisms in the immune-response genes (Poland *et al.*, 2007). There are a growing number of reports documenting clinically relevant infectious differences in clinical outcomes, depending on the status of genes related to the immune response. Just recently, the idea of genetics influencing the response to vaccine exposure began to be further explored. It is not for lack of trying that a response has not been attained. A vaccine response is defined by the articulation of a plethora of genetic and environmental components, such as genes promoting/suppressing a response due to the presence of a polymorphism, environmental modifications (including epigenetic modifications), and interaction of host and non-host genes at the genetic and environmental levels (Poland *et al.*, 2008).

Research is increasingly focusing on understanding the influence of genetic polymorphisms and their effect on humoral, cell-mediated, and innate immune responses to vaccines at the individual and population levels. This area of study is growing, with genomic tools and technological advances – such as high-throughput, lower-cost platforms and methodologies – coming close to deciphering the roles of genetic variants involved from the time of exposure to such receptors as Toll-like receptors (TLRs) and downstream molecules, cytokines and their receptors, and human leukocyte antigen (HLA) molecules. The introduction of genetics, epidemiology, and genomics to vaccine design has been termed "vaccinomics" (Poland, 2007). Perhaps this path could offer important allelic gene variants, which would allow us to define how likely an individual is to respond to a vaccine challenge. Still, vaccine development for multifactorial complex traits (i.e. complex diseases), including HIV, malaria, dengue fever, and tuberculosis, is in its infancy and will require a shift in vaccine strategies (Ovsyannikova and Poland, 2011; Patarroyo *et al.*, 2011). Several reports present data for the effect of genetic factors in vaccine-induced immunity. In the rest of this section, we provide a summary of genetic factors associated with HLA alleles and single nucleotide polymorphisms (SNPs) in multiple classes of genes that provide immune response to vaccines.

Epidemiology and genetics

Due to the response heterogeneity, vaccines can elicit either partial or complete protection, or can fail to protect individuals treated under the same conditions. Vaccines, such as measles, mumps, and rubella (MMR) and HBV, fail to induce life-long protective levels of antibody in approximately 5–10% of healthy recipients. For this reason, outbreaks continue to take place worldwide (Poland and Jacobson, 1994).

Twin studies support the role of genetics in vaccine response. For measles, mumps, and rubella, 89, 39, and 46% of the variation of IgG titers in humoral immunity after vaccination is attributed to genetic factors rather than chance, respectively (Tan *et al.*, 2001). Moreover, early vaccination in twins shows high heritability for antibody responses in hepatitis B (77%), oral polio (60%), tetanus (44%), and diphtheria (49%) vaccines (Newport *et al.*, 2004). Of the total contribution to this variability, about 40% is attributed to HLA genes and 60% to non-HLA genes (Hohler *et al.*, 2002).

Significant heritability is also observed for interferon-gamma (IFN-g) and interleukin-13 (IL-13) responses to tetanus, pertussis, and some BCG vaccine antigens (39–65%). Twin and sibling studies that examine the heritability of vaccine response and reactivity profiles strongly support the rationale for a genomics approach to inter-individual variations in vaccine immune response (Jacobson *et al.*, 2007).

Epidemiological and family vaccine studies have shown familial aggregation. Subsequently, many association studies have identified both HLA and non-HLA candidate gene markers, including genes in close linkage disequilibrium (LD) with a putative causative marker (Ovsyannikova *et al.*, 2011). These HLA and SNP findings emphasize the importance of identifying and replicating initial reports of genetic associations with vaccine-induced immune responses, as well as of understanding the functional consequences of each gene/SNP association. The most common approaches to evaluating the effect between vaccination and variation in immune response-related genes are the candidate-gene approach and genome-wide association studies (GWAS).

An additional component in host variability includes the multiplicity of immune response genes, as well as the diversity of HLA haplotypes, which gives human populations an almost limitless immune-response repertoire (Brusic and August, 2004). Vaccine efficacy can be impacted by a number of host factors as possible confounders (Mooney *et al.*, 2013). It is now clear that pathogen and host variability, as well as the interactions between them, must be considered in vaccine design.

Polymorphism of the HLA region

Immune responses following exposure to vaccination by MMR, influenza, HBV, and vaccinia vaccines are influenced by the HLA region and other immune-regulatory genes (Ovsyannikova *et al.*, 2011). The HLA region, located on the short arm of chromosome 6 (6p21.3), is by far the most polymorphic region of the human genome, with more than 220 genes contributing significantly to genetic susceptibility to infectious diseases and variations in immune responses to vaccines (Poland *et al.*, 2008). Genes in this region are usually taken as candidate genes in association studies of infectious diseases, due to their role in immune function. The HLA region is divided into three subregions: the class I region, where the *HLA-A*, *–B*, and *-C* genes are located, which are involved in antigen presentation to cytotoxic CD8+ T cells, defining the induction and maintenance of cell-mediated immune responses; the class II region, where genes like *HLA-DR*, *–DQ*, and *–DP* are located, which are associated with the presentation of exogenous antigens to helper CD4+ T cells (active players in humoral immune responses); and the class III region, where immune non-HLA-related genes are located (Figure 6.1). The HLA genes play a key role in determining the immune response to T cell antigens, whereas other genes fine-tune this response profile (Sinha *et al.*, 2007). HLA class I and class II genes represent one of the main focal points, due to their biologic role of presenting pathogen-derived peptide epitopes to T cells and their extraordinary polymorphism.

Genes located in the HLA region are inherited as a block (haplotype), making them codominantly expressed for each individual. A heterozygous human inherits one paternal and one maternal haplotype, each containing three class I (A, B, and C) and three class II (DP, DQ, and DR) loci. Each individual inherits a maximum of two alleles for each locus. The maximum number of class I major histocompatibility complex (MHC) gene products expressed in an individual is six. Thus, as each chromosome is found twice (diploid) in each individual, a normal tissue type will involve at least 12 HLA antigens. Haplotypes, normally, are inherited intact, and hence antigens encoded by different loci are inherited together. However, on occasion, there is crossing over between parental chromosomes, resulting in new recombinant haplotypes (Figure 6.1) (Shiina *et al.*, 2009).

Likewise, a large and growing family of immune-response genes has been identified as critical to immune response, including classical and nonclassical HLA genes, cytokine and cytokine-receptor genes, chemokine and chemokine-receptor genes, killer cell immunoglobulin-like receptors, genes of the leukocyte receptor cluster, signaling molecules, vitamin D and receptor genes, the mannose-binding lectin gene,

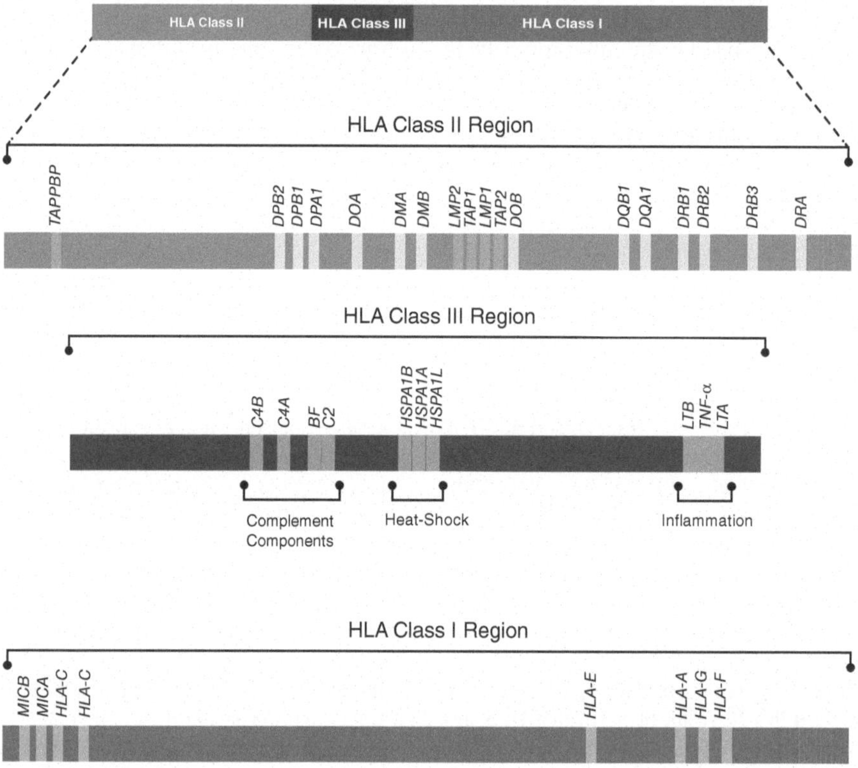

Figure 6.1 Map of the human HLA. The region is conventionally divided into three subregions: the class I, II, and III regions. Each contains numerous genes – only a few of the most relevant are shown here. Abbreviations: *TAPBP*, Tapasin; *LMP1* and *LMP2*, large multifunctional proteases 1 and 2; *TAP1* and *TAP2*, transporter associated with antigen processing 1 and 2; *C2*, *C4A*, and *C4B*, complement components 2, 4A, and 4B; *BF*, complement factor B; *HSPA1A* and *HSPA1B*, heat-shock protein 1A A-type and B-type; *HSPA1L*, heat-shock protein 1A-like; *LTA* and *LTB*, lymphotoxins A and B; *TNFA*, tumor necrosis factor α; and *MICA* and *MICB*, major histocompatibility complex class I chain genes A and B. (For a color version of this figure, please see color plate section.)

genes associated with innate immune responses, TLR genes, and a host of other gene families. There follows a description of the genetic associations reported for HLA class I and II and non-HLA.

Functions of HLA class I and II

Class I and class II molecules are essential for T cell-mediated adaptive immunity. The foreign antigens recognized by the T cell receptor (TCR) are peptides produced by intracellular protein degradation that are bound to class I or class II molecules at the surface of human cells. Degradation of foreign proteins to produce peptides is referred to as "antigen processing," while the binding of peptides by HLA molecules to form ligands for binding to the TCR is called "antigen presentation." When the TCR recognizes HLA-associated peptides on an APC, several T cell surface proteins and intracellular signaling molecules are rapidly mobilized to the site of T cell and APC contact.

Antigen processing in the class I pathway

For the most part, endogenous antigens presented by class I molecules are derived from intracellular infection caused by viruses, proteins synthesized in the cytosol, mature proteins, or defective ribosomal products (Cresswell *et al.*, 2005). Assembly of class I molecules with antigenic peptides requires the coordination of multiple processes, in order to generate, transport, and load the peptides into the peptide-binding groove structure of nascent class I molecules in the endoplasmic reticulum (ER) (Heemels and Ploegh, 1995; Williams *et al.*, 2002). Many of these polypeptides are ubiquitinated and thus are degraded by the proteasome (Figure 6.2) (Cresswell *et al.*, 2005).

Peptides to be presented are transported by transporter associated with antigen processing

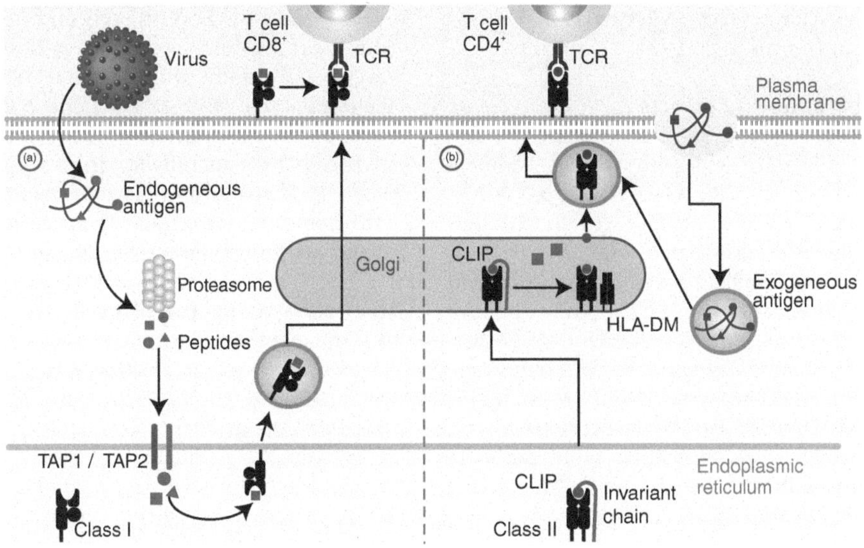

Figure 6.2 Antigen processing by HLA class I and II molecules. (a) Class I antigen processing and presentation occurs when proteins in the cytosol are degraded by the proteosome into small peptides and then are transported by transporter associated with antigen processing (TAP) into the endoplasmic reticulum (ER) lumen. HLA class I molecules are synthesized, translocated, and assembled into the lumen of the ER, where they load the peptide; HLA class I–peptide complexes then leave the ER and move through the Golgi apparatus to the plasma membrane, where they present the joined peptide to the T cell receptor (TCR) of CD8+ T cells. (b) Class II presentation occurs when extracellular proteins are phagocytized and then degraded into small peptides. These peptides are then sorted into vesicles, where they interact with the HLA class II molecules. HLA class II α and β chains, class II-associated invariant peptide (CLIP), and the invariant chain (Ii) molecules are located and assembled in the lumen of the ER, where they cannot bind peptides because the complex occupies the peptide-binding site. Heterotrimers leave the ER and pass through the Golgi apparatus to fuse with vesicles. The Ii is degraded and, with the help of HLA-DM and HLA-DO, a peptide can be joined. Complexes of HLA class II and peptide are relocated to the plasma membrane, where they can be recognized by CD4+ T cells. (For a color version of this figure, please see color plate section.)

(TAP) into the ER, where they associate with heterodimers of HLA class I heavy chain and β2-microglobulin. Tapasin loads the HLA class I molecules to TAP, in association with chaperone molecules (calreticulin and ERp57), to form the peptide loading complex. Once the peptide is loaded, the HLA class I–peptide complex is transported to the cell surface via the ER and Golgi network to be recognized by the specific TCR CD8+ T cell (Figure 6.2) (Williams *et al.*, 2002).

Antigen processing in the class II pathway

Usually, exogenous antigens are presented by class II molecules and derived from pathogens located in the extracellular spaces. APCs have specialized receptors to bind and internalize microorganisms into phagosomes, which fuse with lysosomes, producing phagolysosomes or secondary lysosomes.

Less often, cytoplasmic and membrane proteins may be processed and displayed by HLA class II molecules. In this pathway, cytoplasmic proteins are trapped within membrane-bound vesicles called autophagosomes; these vesicles fuse with lysosomes, and the cytoplasmic proteins are degraded by proteolysis. In both cases, degraded proteins are then able to bind to HLA class II molecules (Figure 6.2) (Cresswell *et al.*, 2005).

HLA class II α and β chains assemble in the ER with a nonpolymorphic protein called invariant chain (Ii). The interaction with the Ii stabilizes the structure of the HLA class II molecule and prevents the binding of peptides within the ER. Ii is anchored in the ER membrane, and the cytosolic portion of the molecule directs intracellular sorting of class II molecules through the Golgi to the MHC class II compartment; Ii is degraded and can be replaced by a peptide derived from degradation in the endosomes or lysosomes of endocytosed material. Proteolytic enzymes, such as cathepsins, that generate peptides from internalized proteins degrade the Ii, leaving only a 24 amino acid remnant called class II-associated invariant peptide (CLIP), which sits in the peptide-binding groove (Ghosh *et al.*, 1995). CLIP is removed by the action of the HLA-DM. Complexes of HLA II and peptide are then taken to the plasma membrane, where they can be recognized by CD4+ T cells (Figure 6.2) (Morris *et al.*, 1994).

HLA class I vaccine effects

In the measles vaccine, *HLA-B*8*, *HLA-B*13*, and *HLA-B*44* alleles are associated with IgG seronegativity after a single dose (Jacobson *et al.*, 2003). In the rubella vaccine response, low-rubella IgG antibody levels are associated with *HLA-B*27:05*, while *HLA-B*45:01* alleles are associated with high antibody levels after two doses (Ovsyannikova *et al.*, 2009a). *HLA-B*35:03* and *HLA-C*15:02* alleles are associated with high levels of lymphocyte proliferation to rubella virus, while *HLA-B*13:02*, *HLA-B*37:01*, and *HLA-B*38:01* alleles are associated with high levels of cellular proliferation to mumps virus following two doses of MMR vaccine (Ovsyannikova *et al.*, 2004).

The association between HLA alleles and rubella-specific IFN-g (Th1) and IL-10 (Th2) cytokine responses among healthy children following two doses of rubella vaccine has been studied. Several class I *HLA-A* (**02:01, *24:02, *68:01*) alleles are significantly associated with rubella vaccine-induced IFN-g secretion (Ovsyannikova *et al.*, 2007a). Both *HLA-A*02:01* and *HLA-A*68:01* alleles are associated with IFN-g and IL-10 secretion.

Normally, vaccines provide immunity by simulating a natural infection; thus, polymorphisms in genes that play a role in the pathogenesis of the infection might also influence or regulate vaccine immune response. Genes involved in antigen processing for HLA class I presentation, such as *TAP1*, *TAP2*, proteasome subunits *LMP2*, *LMP7*, and *Tapasin* are suggested to contribute to susceptibility to human papillomavirus (HPV) type-16-associated cervical cancer (Gostout *et al.*, 2003; Deshpande *et al.*, 2008). Recently, *IL-10* gene polymorphisms have been associated with the clearance of infection and with high-risk HPV types among immunosuppressed adolescent females with varying degrees of HIV-1-induced CD4 immunosuppression (Shrestha *et al.*, 2007). Such information creates a compelling argument for the importance of cytokine gene regions and/or a cluster of genes in the HLA region that regulates host immune responses to HPV infection in a manner that results in inherited susceptibility or resistance to the transforming properties of oncogenic papillomaviruses (Poland *et al.*, 2008).

The variability of immune responses modulated by HLA genes is also a significant factor in the immune response to rubella vaccine. In a recent study, HLA genotyping was performed in a group of 346 healthy schoolchildren and young adults who had previously received two doses of MMR vaccine (Ovsyannikova *et al.*, 2004). Among the alleles analyzed, *HLA-B*35:03* and *HLA-Cw*15:02* were positively associated with lymphoproliferative responses to rubella virus, suggesting that class I HLA alleles may have limited associations with humoral and cellular immune responses to rubella vaccine (Poland *et al.*, 2007).

HLA class II vaccine effects

In the mumps vaccine, *HLA-DQB1*02:01*, *HLA*-DQB1*04:02, *HLA-DQA1*04:01*, *HLA-DRB1*03:01*, *HLA-DRB1*08:01*, *HLA-DRB1*12:01*, and *HLA-DRB1*13:02* alleles are significantly associated with low cellular proliferative responses in healthy children (Ovsyannikova *et al.*, 2008), while those alleles positively associated with rubella-specific lymphocyte proliferation are *HLA-DQB1*05:01*, *HLA-DRB1*01:01*, and *HLA-DRB1*11:04*. Conversely, the *HLA-DQB1*02:02* and *HLA-DRB1*07:01* alleles are negatively associated with rubella-induced cellular proliferation (Ovsyannikova *et al.*, 2005a).

HLA-DQA1 (**01:03, *03:01, *03:03*), *HLA-DQB1* (**02:02, *03:02, *06:03*), and inter-individual variations are associated with rubella virus-induced IL-2 (Ovsyannikova *et al.*, 2009b). *HLA-DPA1*02:01* is associated with low levels of rubella-induced antibodies, whereas *HLA-DPB1*04:01* alleles are associated with high antibody levels (Ovsyannikova *et al.*, 2009a). Specific HLA class II alleles have also been shown to have a significant influence on the immune response to other vaccines (Gelder *et al.*, 2002). In particular, *HLA-DRB1*07* alleles are overrepresented in individuals failing to respond to any component of the trivalent influenza vaccine, compared with responders to the vaccine (Lambkin *et al.*, 2004). Finally, multiple studies have shown relationships between HLA gene polymorphisms and nonresponsiveness to the HBV vaccine. In addition, studies suggest that the immune response to HBV vaccine is independently controlled by both HLA and cytokine polymorphisms (Poland *et al.*, 2007).

HLA haplotypes

Associations between HLA haplotypes have been reported following a second dose of the MMR vaccine (Poland *et al.*, 2007). The class I *HLA-A*29-Cw*16-B*44* haplotype is associated with lower IgG antibody levels to measles and mumps vaccine viruses (Ovsyannikova *et al.*, 2006). The *HLA-A*26-Cw*12-B*38* haplotype is associated with higher cellular immune responses to measles and mumps. The class II *HLA-DRB1*03-DQB1*02-DPB1*04* haplotype presents higher levels of cellular proliferation to measles and to mumps (Ovsyannikova *et al.*, 2006). The *HLA-DRB1*15/16-DQB1*06-DPB1*03* haplotype is associated with low IgG antibody levels to rubella virus, whereas the *HLA-DRB1 *04-DQB1*03-DPB1*03* haplotype is associated with high levels of cellular proliferation to measles and rubella vaccine viruses (Jacobson *et al.*, 2003). The *HLA-A*26-Cw*12-B*38* haplotype is associated with mumps-specific humoral and cell-mediated immune responses. Better characterization of such HLA profiles could inform and advance the design of novel epitope-based vaccines and help to predict both individual- and population-level immune coverage to vaccines (Poland *et al.*, 2007).

Immune responses to vaccines are also affected by extended haplotypes in the class III region. Associations involving haplotypes expanding across the HLA class I region, 10 polymorphisms for *LTA-TNF-LST1*, and the HLA class II region with rubella-specific antibodies are also reported (Ovsyannikova *et al.*, 2010a). Likewise, HLA allele supertypes are grouped according to their shared peptide-binding specificities (Sette and Sidney, 1998). The role of HLA supertypes in immune responses to the MMR vaccine was examined based on the shared sequence property in the peptide-binding pockets of HLA molecules (Ovsyannikova *et al.*, 2007b). HLA class I B44 and B58 supertypes were strongly associated with lower measles vaccine-induced antibody levels, while the most common HLA supertypes, B7 and DR, were associated with higher measles antibody response. Moreover, the DR supertype was significantly associated with lower mumps-specific cellular immune responses. The A3 supertype was associated with higher levels of measles virus-induced IFN-g and IL-4 immune responses. Genetic associations between rubella vaccine immune responses and HLA supertypes were not as strong as those observed for measles and mumps, suggesting that HLA molecules may be less effective in the presentation of rubella cross-reactive peptides than for measles and mumps (Ovsyannikova *et al.*, 2007b).

By identifying naturally processed epitopes, combined with knowledge of HLA supertypes, adjuvanted vaccines can be devised using the combinations of those peptides most likely to be optimally immunogenic (Ovsyannikova *et al.*, 2007b). As an example, using a mass-spectrometry approach, 13 naturally processed and presented measles virus peptides were identified from the class II HLA-DRB1 peptide-binding groove of human cells. These peptides, capable of binding across the common population HLA supertypes, could be used to fine-tune new candidate vaccines (Johnson *et al.*, 2005; Ovsyannikova *et al.*, 2005b).

Non-HLA genetic polymorphisms and vaccine response

Cytokines and their receptors give shape to the immune response (Jin and Wang, 2003). Therefore, knowledge of such a polymorphism might facilitate development of vaccine candidates that incorporate cytokines to "replace" those not made natively (i.e. a cytokine plasmid or cytokine adjuvants) and restore an optimal Th1/Th2 balance that would facilitate a protective immune response (Poland *et al.*, 2008).

Cytokine genes

IL-2 and *IL-10* gene associations are reported with measures of measles vaccine-induced immunity. *IL-2* rs2069762 and rs2069763 SNPs are associated with higher antibody and higher cellular immune responses to measles (Dhiman *et al.*, 2007b). On the other hand, rs1800890, rs1800871, and rs1800872, from the *IL-10* gene, are associated with lower antibody and cellular immune responses to measles vaccine. Polymorphisms proximal to the promoter region of *IL-10* are known to contribute to a lower production of secreted IL-10 (Yilmaz *et al.*, 2005).

Genetic variants rs3790567 and rs372889 at the *IL-12RB2* are associated with both lower antibody and lower cellular immune responses following two doses of measles vaccine (Dhiman *et al.*, 2007b). Conversely, minor allele A of rs372889 within the *IL-12RB1* gene is associated with significantly lower mumps vaccine-induced cellular responses (Ovsyannikova *et al.*, 2008).

SNPs at the *IL12B* promoter have been associated with nonresponsiveness to HBV vaccination in North American adolescents. However, HLA and cytokine gene polymorphisms were found to be independently associated with immune response to HBV vaccination (Wang *et al.*, 2004). A functional polymorphism in the *IL-10* promoter was found to be an important modulator of the immune response to HBsAg and hepatitis A vaccination (Hohler *et al.*, 2005).

In a recent report, when 346 vaccinated individuals received primary smallpox vaccine, fever following live vaccinia virus vaccination was associated with specific haplotypes in the *IL-1* gene complex and with haplotypes within the *IL-18* gene. Conversely, a haplotype in the *IL-4* gene was highly significant for reduced susceptibility to the development of fever following smallpox vaccination, suggesting genetic predisposition to adverse events after vaccination (Poland *et al.*, 2007).

*HLA-DRB1*07* and cytokine SNPs at the *IL-2* and *IL-4* loci, along with insertion/deletion variants at the *IL-12B* locus, have been found to be independently associated with HBV vaccine nonresponse (Wang *et al.*, 2004). Development of a new HBV vaccine consisting of a peptide "cocktail" (novel epitopes identified from chronic carriers) with cytokine adjuvants could circumvent these immunogenetic restrictions. In fact, such vaccine development is underway (Poland *et al.*, 2008).

Rubella vaccine-induced humoral and cytokine responses are significantly modulated by cytokine and cytokine-receptor genetic variants. For example, an increased representation of minor alleles for two promoter SNPs (rs2844482 and rs2857708) of the *TNFA* gene is associated with twofold increases in rubella-specific IgG levels. Furthermore, IL-6 production is associated with intronic SNPs (rs5745993, rs17882988, rs472093, rs5746059, and rs590977) in the *TNFRSF1B* gene, while several promoter and intronic SNPs in the *IL12B* gene are significantly associated with higher IL-6 production after rubella vaccination (Dhiman *et al.*, 2010).

Thus, cytokines play an essential role in the modulation of immune responses, and cytokine production is influenced by the rate of transcription of their cytokine and cytokine-receptor genes. As an example, SNPs in these cytokine genes can affect mRNA splicing and stability, the structure of RNA molecules, and protein folding (Jin and Wang, 2003).

Innate and cell-surface receptor genes

The TLR family of receptors plays an essential role in the primary recognition of pathogens and in the initiation of adaptive immunity. TLRs are pattern-recognition receptors (PRRs) and can contribute to viral detection by sensing RNA and viral proteins, leading to the induction of cytokines and interferon response (Bowie and Haga, 2005). The poxvirus, herpesvirus, retrovirus, and paramyxovirus families activate T cells through TLRs, triggering antiviral innate immune responses (Rassa and Ross, 2003).

For the *TLR3* gene, SNPS rs3775291 and rs5743305 are associated with low antibody and cellular proliferation responses after measles vaccination (Dhiman *et al.*, 2008). Moreover, rs5743305 is associated with rubella-induced granulocyte-macrophage colony stimulating factor (GM-CSF) secretion, lower measles-specific antibody titer, and lower cellular proliferation to measles vaccine (Ovsyannikova *et al.*, 2010a). Both SNPs in the *TLR2* (rs3804100) and *TLR4* (rs5030710) genes are associated with variations in measles vaccine-induced humoral immunity in an allele dose-dependent manner (Ovsyannikova *et al.*, 2010a).

Other associations of innate related genes have been identified between the vitamin A (*RARA, RARB*, and *RARG*), *RIG-I/DDX58*, *TRIM* (*TRIM5* and *TRIM22*), vitamin D receptor, and *RXRA* genes and rubella vaccine-specific immunity (Ovsyannikova *et al.*, 2010b). The *TRIM5* gene variants have been associated with rubella-specific humoral response TNF-a secretion (rs3740996) and IL-2/GM-CSF production (rs10838525) (Ovsyannikova *et al.*, 2011). Genetic variants (rs3741981, rs1051042, rs2660) mapping to the *OAS* gene are associated with rubella-induced IL-2, IL-10, IL-6 secretion and antibody levels. These three variants present a haplotype associated with an increase in rubella-specific IL-2 production (Haralambieva *et al.*, 2010).

Upregulation of TLRs following infection with vaccine strains of measles virus triggers activation of TLR-responsive genes such as IL-1a/b, IL-6, IFN-a/b, and IL-12 and induction of measles' own receptor, signaling lymphocyte activation molecule (SLAM) (Hahm *et al.*, 2007). Measles virus binds to SLAM and CD46. SLAM and CD46 act as cellular receptors for vaccine and laboratory-adapted strains. Wild-type measles virus strains are known to preferentially use SLAM as a receptor. *SLAM* minor allele T for rs3796504 correlates with an allele dose-related decrease of measles antibody levels (Dhiman *et al.*, 2007a). In addition, polymorphisms in *CD46* (rs11118580 and rs2724384) correlate with allele dose-dependent reduction in measles antibody levels (Dhiman *et al.*, 2007a).

Conclusions

These are exciting times in which to be doing research, given the rapid pace of development of high-throughput technologies for clinical and basic use. Methodological approaches are maturing toward a systems view for identifying and characterizing immune responses by inspecting different -omics layers of information (e.g. proteomics, transcriptomics, metabolomics, genomics). The ultimate goal for the application of these new technologies would be to identify biomarker signatures, which would show how innate and adaptive responses are to be integrated into a unified network.

The immune response network theory, in its simplest form, is based on the premise "the response to a vaccine is the cumulative result of interactions driven by a host of genes and their interactions, and is theoretically predictable" (Poland *et al.*, 2013). Scientists are nurturing this definition by recognizing and including the impact of epigenetics, metagenomics, and other factors that might influence or play a role in defining the onset of a vaccine response (Poland *et al.*, 2013). The main obstacles impairing our ability to predict a response and to develop effective vaccines or treatments are an increased genetic variability in the human population and the constant evolution of pathogens. These two effects produce a wide spectrum of possible host–pathogen interactions and compel the use of a systemic approach that can disentangle mechanisms and provide a definition of targeted-population and personalized vaccines, hopefully in the near future.

References

Artenstein, A.W. and Poland, G.A. (2012). Vaccine history: the past as prelude to the future. *Vaccine*, **30**(36): 5299–301.

Bowie, A.G. and Haga, I.R. (2005). The role of Toll-like receptors in the host response to viruses. *Mol Immunol*, **42**(8): 859–67.

Brusic, V. and August, J.T. (2004). The changing field of vaccine development in the genomics era. *Pharmacogenomics*, **5**(6): 597–600.

Cresswell, P., Ackerman, A.L., Giodini, A., *et al.* (2005). Mechanisms of MHC class I-restricted antigen processing and cross-presentation. *Immunol Rev*, **207**: 145–57.

Daar, A.S. (2010). Chronic non-communicable diseases as a threat for all: recent global initiatives. *Sultan Qaboos Univ Med J*, **10**(3): 306–9.

Deshpande, A., Wheeler, C.M., Hunt, W.C., *et al.* (2008). Variation in HLA class I antigen-processing genes and susceptibility to human papillomavirus type 16-associated cervical cancer. *J Infect Dis*, **197**(3): 371–81.

Dhiman, N., Poland, G.A., Cunningham, J.M., *et al.* (2007a). Variations in measles vaccine-specific humoral immunity by polymorphisms in SLAM and CD46 measles virus receptors. *J Allergy Clin Immunol*, **120**(3): 666–72.

Dhiman, N., Ovsyannikova, I.G., Cunningham, J.M., *et al.* (2007b). Associations between measles vaccine immunity and single-nucleotide polymorphisms in cytokine and cytokine receptor genes. *J Infect Dis*, **195**(1): 21–9.

Dhiman, N., Ovsyannikova, I.G., Vierkant, R.A., *et al.* (2008). Associations between SNPs in toll-like receptors and related intracellular signaling molecules and immune responses to measles vaccine: preliminary results. *Vaccine*, **26**(14): 1731–6.

Dhiman, N., Haralambieva, I.H., Kennedy, R.B., *et al.* (2010). SNP/haplotype associations in cytokine and cytokine receptor genes and immunity to rubella vaccine. *Immunogenetics*, **62**(4): 197–210.

Enders, J.F., Weller, T.H., and Robbins, F.C. (1949). Cultivation of the lansing strain of poliomyelitis virus in cultures of various human embryonic tissues. *Science*, **109**(2822): 85–7.

Fenner, F. (1980). Smallpox and its eradication. *Aus NZ J Med*, **10**(4): 459.

Fleischmann, R.D., Adams, M.D., White, O., *et al.* (1995). Whole-genome random sequencing and assembly of Haemophilus influenzae Rd. *Science*, **269**(5223): 496–512.

Fraser, C.M. and Rappuoli, R. (2005). Application of microbial genomic science to advanced therapeutics. *Annu Rev Med*, **56**: 459–74.

Gelder, C.M., Lambkin, R., Hart, K.W., *et al.* (2002). Associations between human leukocyte antigens and non-responsiveness to influenza vaccine. *J Infect Dis*, **185**(1): 114–17.

Ghosh, P., Amaya, M., Mellins, E., and Wiley, D.C. (1995). The structure of an intermediate in class II MHC maturation: CLIP bound to HLA-DR3. *Nature*, **378**(6556): 457–62.

Gostout, B.S., Poland, G.A., Calhoun, E.S., *et al.* (2003). TAP1, TAP2, and HLA-DR2 alleles are predictors of cervical cancer risk. *Gynecol Oncol*, **88**(3): 326–32.

Hahm, B., Cho, J.H., and Oldstone, M.B. (2007). Measles virus-dendritic cell interaction via SLAM inhibits innate immunity: selective signaling through TLR4 but not other TLRs mediates suppression of IL-12 synthesis. *Virology*, **358**(2): 251–7.

Haralambieva, I.H., Dhiman, N., Ovsyannikova, I.G., *et al.* (2010). 2′-5′-oligoadenylate synthetase single-nucleotide polymorphisms and haplotypes are associated with variations in immune responses to rubella vaccine. *Hum Immunol*, **71**(4): 383–91.

Heemels, M.T. and Ploegh, H. (1995). Generation, translocation, and presentation of MHC class I-restricted peptides. *Annu Rev Biochem*, **64**: 463–91.

Hohler, T., Reuss, E., Evers, N., *et al.* (2002). Differential genetic determination of immune responsiveness to hepatitis B surface antigen and to hepatitis A virus: a vaccination study in twins. *Lancet*, **360**(9338): 991–5.

Hohler, T., Reuss, E., Freitag, C.M., and Schneider, P.M. (2005). A functional polymorphism in the IL-10 promoter influences the response after vaccination with HBsAg and hepatitis A. *Hepatology*, **42**(1): 72–6.

Jacobson, R.M., Poland, G.A., Vierkant, R.A., *et al.* (2003). The association of class I HLA alleles and antibody levels after a single dose of measles vaccine. *Hum Immunol*, **64**(1): 103–9.

Jacobson, R.M., Ovsyannikova, I.G., Targonski, P.V., and Poland, G.A. (2007). Studies of twins in vaccinology. *Vaccine*, **25**(16): 3160–4.

Jin, P. and Wang, E. (2003). Polymorphism in clinical immunology – from HLA typing to immunogenetic profiling. *J Transl Med*, **1**(1): 8.

Johnson, K.L., Ovsyannikova, I.G., Poland, G.A., and Muddiman, D.C. (2005). Identification of class II HLA-DRB1*03-bound measles virus peptides by 2D-liquid chromatography tandem mass spectrometry. *J Proteome Res*, **4**(6): 2243–9.

Kaushik, D.K. and Sehgal, D. (2008). Developing antibacterial vaccines in genomics and proteomics era. *Scand J Immunol*, **67**(6): 544–52.

Lambkin, R., Novelli, P., Oxford, J., and Gelder, C. (2004). Human genetics and responses to influenza vaccination: clinical implications. *Am J Pharmacogenomics*, **4**(5): 293–8.

Mooney, M., Mcweeney, S., and Sekaly, R.P. (2013). Systems immunogenetics of vaccines. *Semin Immunol*, **25**(2): 124–9.

Morris, P., Shaman, J., Attaya, M., *et al.* (1994). An essential role for HLA-DM in antigen presentation by class II major histocompatibility molecules. *Nature*, **368**(6471): 551–4.

Newport, M.J., Goetghebuer, T., Weiss, H.A., *et al.* (2004). Genetic regulation of immune responses to vaccines in early life. *Genes Immun*, **5**(2): 122–9.

Norrby, E. (2007). Yellow fever and Max Theiler: the only Nobel Prize for a virus vaccine. *J Exp Med*, **204**(12): 2779–84.

Ovsyannikova, I.G. and Poland, G.A. (2011). Vaccinomics: current findings, challenges and novel approaches for vaccine development. *AAPS J*, **13**(3): 438–44.

Ovsyannikova, I.G., Jacobson, R.M., Vierkant, R.A., *et al.* (2004). The contribution of HLA class I antigens in immune status following two doses of rubella vaccination. *Hum Immunol*, **65**(12): 1506–15.

Ovsyannikova, I.G., Jacobson, R.M., Vierkant, R.A., *et al.* (2005a). Human leukocyte antigen class II alleles and rubella-specific humoral and cell-mediated immunity following measles-mumps-rubella-II vaccination. *J Infect Dis*, **191**(4): 515–19.

Ovsyannikova, I. G., Johnson, K. L., Naylor, S. and Poland, G. A. (2005b). Identification of HLA-DRB1-bound self-peptides following measles virus infection. *J Immunol Methods* **297**(1–2): 153–167.

Ovsyannikova, I.G., Pankratz, V.S., Vierkant, R.A., *et al.* (2006). Human leukocyte antigen haplotypes in the genetic control of immune response to measles-mumps-rubella vaccine. *J Infect Dis*, **193**(5): 655–63.

Ovsyannikova, I.G., Jacobson, R.M., Ryan, J.E., *et al.* (2007a). Relationship between HLA polymorphisms and gamma interferon and interleukin-10 cytokine production in healthy individuals after rubella vaccination. *Clin Vaccine Immunol*, **14**(2): 115–22.

Ovsyannikova, I.G., Jacobson, R.M., Vierkant, R.A., *et al.* (2007b). HLA supertypes and immune responses to measles-mumps-rubella viral vaccine: findings and implications for vaccine design. *Vaccine*, **25**(16): 3090–100.

Ovsyannikova, I.G., Jacobson, R.M., Dhiman, N., *et al.* (2008). Human leukocyte antigen and cytokine receptor gene polymorphisms associated with heterogeneous immune responses to mumps viral vaccine. *Pediatrics*, **121**(5): e1091–9.

Ovsyannikova, I.G., Jacobson, R.M., Vierkant, R.A., *et al.* (2009a). Replication of rubella vaccine population genetic studies: validation of HLA genotype and humoral response associations. *Vaccine*, **27**(49): 6926–31.

Ovsyannikova, I.G., Ryan, J.E., Vierkant, R.A., *et al.* (2009b). Influence of host genetic variation on rubella-specific T cell cytokine responses following rubella vaccination. *Vaccine*, **27**(25–6): 3359–66.

Ovsyannikova, I.G., Dhiman, N., Haralambieva, I.H., *et al.* (2010a). Rubella vaccine-induced cellular immunity: evidence of associations with polymorphisms in the Toll-like, vitamin A and D receptors, and innate immune response genes. *Hum Genet*, **127**(2): 207–21.

Ovsyannikova, I.G., Vierkant, R.A., Pankratz, V.S., *et al.* (2010b). Extended LTA, TNF, LST1 and HLA gene haplotypes and their association with rubella vaccine-induced immunity. *PLoS One*, **5**(7): e11806.

Patarroyo, M.E., Bermudez, A., and Patarroyo, M.A. (2011). Structural and immunological principles leading to chemically synthesized, multiantigenic, multistage, minimal subunit-based vaccine development. *Chem Rev*, **111**(5): 3459–507.

Plotkin, S.A. (2005). Six revolutions in vaccinology. *Pediatr Infect Dis J*, **24**(1): 1–9.

Plotkin, S.A. (2008). [New vaccination strategies]. *Bull Acad Natl Med*, **192**(3): 511–18; disc. 518–19.

Poland, G.A. (1998). Variability in immune response to pathogens: using measles vaccine to probe immunogenetic determinants of response. *Am J Hum Genet*, **62**(2): 215–20.

Poland, G.A. (2007). Pharmacology, vaccinomics, and the second golden age of vaccinology. *Clin Pharmacol Ther*, **82**(6): 623–6.

Poland, G.A. and Jacobson, R.M. (1994). Failure to reach the goal of measles elimination. Apparent paradox of measles infections in immunized persons. *Arch Intern Med*, **154**(16): 1815–20.

Poland, G.A., Ovsyannikova, I.G., Jacobson, R.M., and Smith, D.I. (2007). Heterogeneity in vaccine immune response: the role of immunogenetics and the emerging field of vaccinomics. *Clin Pharmacol Ther*, **82**(6): 653–64.

Poland, G.A., Ovsyannikova, I.G., and Jacobson, R.M. (2008). Personalized vaccines: the emerging field of vaccinomics. *Expert Opin Biol Ther*, **8**(11): 1659–67.

Poland, G.A., Kennedy, R.B., Mckinney, B.A., *et al.* (2013). Vaccinomics, adversomics, and the immune response network theory: individualized vaccinology in the 21st century. *Semin Immunol*, **25**(2): 89–103.

Projan, S.J., Nesin, M., and Dunman, P.M. (2006). Staphylococcal vaccines and immunotherapy: to dream the impossible dream? *Curr Opin Pharmacol*, **6**(5): 473–9.

Pulendran, B., Li, S., and Nakaya, H.I. (2010). Systems vaccinology. *Immunity*, **33**(4): 516–29.

Querec, T.D., Akondy, R.S., Lee, E.K., *et al.* (2009). Systems biology approach predicts immunogenicity of the yellow fever vaccine in humans. *Nat Immunol*, **10**(1): 116–25.

Rappuoli, R. (2004). From Pasteur to genomics: progress and challenges in infectious diseases. *Nat Med*, **10**(11): 1177–85.

Rappuoli, R., Miller, H.I., and Falkow, S. (2002). Medicine. The intangible value of vaccination. *Science*, **297**(5583): 937–9.

Rassa, J.C. and Ross, S.R. (2003). Viruses and Toll-like receptors. *Microbes Infect*, **5**(11): 961–8.

Roome, A.J., Walsh, S.J., Cartter, M.L., and Hadler, J.L. (1993). Hepatitis B vaccine responsiveness in Connecticut public safety personnel. *JAMA*, **270**(24): 2931–4.

Serruto, D. and Rappuoli, R. (2006). Post-genomic vaccine development. *FEBS Lett*, **580**(12): 2985–92.

Sette, A. and Rappuoli, R. (2010). Reverse vaccinology: developing vaccines in the era of genomics. *Immunity*, **33**(4): 530–41.

Sette, A. and Sidney, J. (1998). HLA supertypes and supermotifs: a functional perspective on HLA polymorphism. *Curr Opin Immunol*, **10**(4): 478–82.

Shiina, T., Hosomichi, K., Inoko, H., and Kulski, J.K. (2009). The HLA genomic loci map: expression, interaction, diversity and disease. *J Hum Genet*, **54**(1): 15–39.

Shrestha, S., Wang, C., Aissani, B., *et al.* (2007). Interleukin-10 gene (IL10) polymorphisms and human papillomavirus clearance among immunosuppressed adolescents. *Cancer Epidemiol Biomarkers Prev*, **16**(8): 1626–32.

Sinha, P., Snyder, J.A., Kim, E.Y., and Moudgil, K.D. (2007). The major histocompatibility complex haplotypes dictate and the background genes fine-tune the dominant versus the cryptic response profile of a T-cell determinant within a native antigen: relevance to disease susceptibility and vaccination. *Scand J Immunol*, **65**(2): 158–65.

Tan, P.L., Jacobson, R.M., Poland, G.A., *et al.* (2001). Twin studies of immunogenicity--determining the genetic contribution to vaccine failure. *Vaccine*, **19**(17–19): 2434–9.

Thomas, C. and Moridani, M. (2010). Interindividual variations in the efficacy and toxicity of vaccines. *Toxicology*, **278**(2): 204–10.

Wang, C., Tang, J., Song, W., *et al.* (2004). HLA and cytokine gene polymorphisms are independently associated with responses to hepatitis B vaccination. *Hepatology*, **39**(4): 978–88.

Williams, A., Peh, C.A., and Elliott, T. (2002). The cell biology of MHC class I antigen presentation. *Tissue Antigens*, **59**(1): 3–17.

Yilmaz, V., Yentur, S.P., and Saruhan-Direskeneli, G. (2005). IL-12 and IL-10 polymorphisms and their effects on cytokine production. *Cytokine*, **30**(4): 188–94.

7 Silicone and Autoimmune/Inflammatory Syndrome Induced by Adjuvants (ASIA)

Yair Levy and Rotem Baytner-Zamir
Department of Medicine E, Meir Medical Center, Kfar Saba, Israel, affiliated with the Sackler Faculty of Medicine, Tel Aviv University, Tel Aviv, Israel

Introduction

Silicone

Silicones are a family of synthetic polymers that share a backbone of repeating silicon–oxygen bonds. In addition to their links to oxygen, which form the polymeric chain, silicon atoms are also bonded to organic groups, typically methyl groups. Many other groups, such as phenyl, vinyl, and trifluoropropyl, can be substituted for the methyl groups along the chain (Ratner *et al.*, 1996).

The combination of "organic" groups attached to an "inorganic" chain gives silicones a combination of unique attributes, which allows their use as fluids, emulsions, compounds, resins, and elastomers in numerous applications and diverse fields. Silicone was long considered a biologically inert substance and was therefore incorporated into a variety of medical applications, including joint implants, artificial heart valves, catheters, drains, shunts, intraocular lenses, and manymore.

One of the most publically known uses for silicone is as the main component in aesthetic implants, the most popular of which are breast implants, first introduced in the early 1960s. From 1997 to 2012, there was an almost 227% increase in the total number of breast augmentations performed in the United States. Breast augmentation was the most popular cosmetic surgical procedure in the United States in 2012. More than 330 000 breast augmentation surgeries were performed that year, of which 72% used silicone implants (ASAPS, 2013).

Silicone breast implants are most commonly composed of a silicone elastomer envelope filled with silicone gel. Saline implants, which use a silicone elastomer envelope filled with a saline solution, are an alternative. In the majority of cases, silicone implants are preferred over saline-filled implants, due to better overall aesthetic results and a decreased chance of long-term rupture – a more common occurrence with saline-filled implants (Lidar *et al.*, 2012).

Autoimmune/inflammatory syndrome induced by adjuvants (ASIA)

Autoimmune/inflammatory syndrome induced by adjuvants (ASIA) is a new syndrome that assembles a spectrum of immune-mediated conditions triggered by an adjuvant stimulus. Currently, ASIA incorporates four conditions: siliconosis, Gulf War syndrome (GWS), macrophagic myofasciitis (MMF) syndrome, and post-vaccination phenomena (Shoenfeld and Agmon-Levin, 2011).

Substances that have been found to induce an immune adjuvant activity in various animal models include silicone, aluminum, pristine, and infectious components. Exposure to these factors is presumed to provoke autoimmune or autoinflammatory disease in humans (Zandman-Goddard

Vaccines and Autoimmunity, First Edition. Edited by Yehuda Shoenfeld, Nancy Agmon-Levin, and Lucija Tomljenovic.

et al., 1999; Bar-Meir *et al.*, 2003; Agmon-Levin *et al.*, 2009; Israeli *et al.*, 2009).

Interestingly, even though exposure to adjuvants is quite common in the medical setting, adjuvant-related diseases occur relatively rarely. Current theories suggest that the manifestation of a clinically overt adjuvant disease requires the presence of additional risk factors, such as genetic susceptibility or co-exposure to other environmental factors (Meroni 2011; Shoenfeld and Agmon-Levin, 2011).

Silicone and autoimmunity

The possibility of adverse systemic reactions to silicone implants was suggested as early as 1964, when two patients developed connective tissue diseases (CTDs) several years after having breast augmentation surgery (Miyoshi *et al.*, 1964). As additional cases and publications suggesting a link between silicone and CTDs (and autoimmune diseases) surfaced and public concerns grew, the US Food and Drug Administration (FDA) suspended the use of silicone-filled breast implants (Hajdu *et al.*, 2011). In January 2004, the FDA announced it would no longer approve silicone gel breast implants, due to the lack of long-term safety data. However, in a public meeting held by the FDA in April 2005, it announced it would again consider their approval. The following year, in November 2006, the FDA approved silicone breast implants for women over 22 years of age, on the condition that the approved manufacturers would each study 40 000 women with silicone gel implants for a period of 10 years (Levy *et al.*, 2009).

Local adverse effects of silicone

A tissue response occurs following breast augmentation surgery using silicone implants. It reaction is usually limited to a mild foreign-body reaction, followed by encapsulation of the implant in the surrounding tissue. Capsular tissue is formed around any nondegradable material too large to be engulfed by macrophages and is viewed as a normal tissue reaction in the field of aesthetic surgery (McCarty, 1990).

One of the most common local adverse effects linked to breast implants is capsular contracture, which is estimated to occur in around 50% of patients with silicone gel-filled implants, and in 16% with saline-filled implants (Gylbert *et al.* 1990). Capsular contractures are characterized by breast pain, stiffness, a change in the appearance and position of the implant, and possibly implant rupture. Capsular contractures are reported six times more frequently in cases of ruptured implants (Hölmich *et al.*, 2003). This phenomenon might be explained by a correlation between the degree of capsular contracture and the amount of silicone and silicone-laden macrophages in the capsular tissue (Prantl *et al.*, 2006). A contributing mechanism to the risk of capsular contracture relates to low-grade bacterial infection from skin flora, which further stimulates the local immune response around the implant (Netscher, 2004).

Another possible local side effect of implants is an allergic reaction to either silicone or platinum, a catalyst used in silicone polymerization that is found in minute quantities in implants (Hajdu *et al.*, 2011; Lidar *et al.*, 2012). Several case reports describe granulomatous nodular cellulitis and scarring dystrophy following breast implants and other silicone prostheses (Teuber *et al.*, 1995; Mastruserio *et al.*, 1996; Rapaport *et al.*, 1996; Marcusson and Bjarnason, 1999). It has been speculated that regular exposure to silicone in household products such as makeup and baby bottles, and to platinum in the form of automotive catalytic converters, could potentially sensitize patients, thus predisposing them to hypersensitivity reactions when receiving silicone breast implants (Santucci *et al.*, 2000).

The systemic influence of silicone

Diffusion of silicone, more commonly termed "bleeding," through the silicone elastomer envelope into the surrounding tissues, even in the absence of implant rupture, is another possible complication of breast implants (Barker 1978). Silicone bleeding increases with time and increases the inflammatory response around the capsule (Lidar *et al.*, 2012). This bleeding phenomenon suggests that, in addition to local activation of the immune system, silicone may give rise to systemic effects as it degrades and fragments in tissue. The non-inert fragments can spread throughout the body and potentially lead to the development of cancer or autoimmune phenomena (Brown *et al.*, 2000).

Microscopic evidence of silicone has been discovered in body tissues other than the breast tissue of women with breast implants, and silicone compounds have been identified in the blood and liver of patients with silicone implants (Levy *et al.*, 2009). Pain and chronic fatigue symptoms are reported more frequently in patients with ruptured compared to unruptured implants. Such reports suggest the presence of silicone-mediated systemic autoimmunity (Vermeulen and Scholte, 2003). These findings have led implant manufacturers to switch to harder elastomer shells in order to reduce the chance of implant rupture (Vasey *et al.*, 2003). Nevertheless, systemic silicone exposure is not limited to silicone-filled implants, as it can also occur with saline-filled implants. In addition, the phenomenon of chronic silicone bleeding through the elastomer shell facilitates the extracapsular spread of silicone particles and, thus, systemic silicone exposure (Lidar *et al.*, 2012). When silicone migrates outside the capsule of scar tissue that surrounds the implant, patients are significantly more likely ($p = 0.008$) to be diagnosed with an autoimmune or CTD (Brown *et al.*, 2001).

Antibody response following silicone exposure

Anti-silicone antibodies

Several reports demonstrating the presence of antisilicone antibodies in human sera have been published. Patients with severe immune-mediated reactions to implanted silicone devices were found to have increased IgG in the surrounding tissue and higher levels of antisilicone antibodies compared to asymptomatic patients with implanted silicone devices (Goldblum *et al.*, 1992). Direct visualization by immunofluorescence demonstrated the presence of antisilicone antibodies in the capsular tissue of implants ($p < 0.001$) compared to controls (Bekerecioglu *et al.*, 2008). Additionally, serum antisilicone antibodies were detected more frequently in patients following silicone implantation than in women who did not have breast implants (Wolf *et al.*, 1993; Bar-Meir *et al.*, 1995; Levy *et al.*, 2009). Furthermore, antibody titers were statistically higher ($p < 0.001$) in patients with ruptured as opposed to nonruptured implants (Wolf *et al.*, 1993). There is some debate as to whether the immune reaction to silicone is specific or simply an interaction between hydrophobic molecules (IgGs) and hydrophobic surfaces (silicones) in an aqueous-based system (White and Klykken, 1998). Other authors have speculated about the likelihood of a relationship between silicone implants and activation of the immune system (Levy *et al.*, 2009).

Autoantibodies

A prominent feature of autoimmunity is the presence of autoantibodies. In a study in which mice that spontaneously developed murine lupus were implanted with silicone, significant elevations of anti-dsDNA antibodies ($p < 0.02$), modest elevations of rheumatoid factor, and a strong presence of silicone-bound autoantibodies were demonstrated (Schaefer and Wooley, 1999).

Several studies have demonstrated an increase in several types of autoantibodies in patients with exposure to silicone. The presence of autoantibodies was found to be increased in symptomatic compared to asymptomatic women with silicone breast implants (Zandman-Goddard *et al.*, 1999). Increased titers among 122 asymptomatic women studied ranged from 2 to 13%, mostly directed against dsDNA (8%), ssDNA (9%), SSB/La (13%), silicone (9%), and collagen II (9%), whereas 20% of the 86 symptomatic women demonstrated more than four of these autoantibodies and 8% harbored six autoantibodies. It was noted that the mean duration of implants in the asymptomatic group was significantly less than in the symptomatic group ($p < 0.01$), which led to the assumption that the development of autoantibodies might be related to implant duration. Another study found a statistically significant incidence of antibodies to collagen in women with silicone breast implants. De facto, 35% of women with silicone breast implants had such antibodies. One can deduce from these findings that silicone implants might pose a significant risk for immunopathology in genetically susceptible hosts (Teuber *et al.*, 1993). Additionally, one can also deduce that the association of autoantibodies and implants suggests an adjuvant action of silicone and silicone byproducts (Bar-Meir *et al.*, 1995).

Silicone and defined autoimmune diseases

The association between silicone exposure and specific autoimmune diseases has been suggested before. In 1984, researchers described a spectrum of defined autoimmune diseases, including rheumatoid arthritis (RA), systemic lupus erythematosus (SLE), polymyositis, and systemic sclerosis (SSc), in 24 patients with silicone implants. They concluded that the overrepresentation of SSc in their case series corresponded to a threefold increase in relative risk (Kumagai *et al.*, 1984). Another report described three patients with silicone implants: one with both RA and Sjøgren's syndrome, one with SLE, and one with mixed CTD (Van Nunen *et al.*, 1982). During the following decade, over 60 cases of CTDs were reported in patients with silicone implants (Spiera *et al.*, 1994). A large cohort study published in 1996, based on self-reported symptoms and autoimmune diseases among approximately

11 800 women with breast implants, showed a relative risk of 1.24 (95% CI: 1.08–1.41, p = 0.0015) for any defined CTD investigated (Hennekens *et al.*, 1996). A meta-analysis using data from 20 cohort, case–control, and cross-sectional studies, which excluded the previous report, concluded that the adjusted risk between breast implants and CTD was 0.80 (95% CI: 0.62–1.04) (Janowsky *et al.*, 2000). Hence, the authors deduced that there was no evidence of an association between breast implants and any connective tissue diseases or other autoimmune or rheumatic conditions. Furthermore, in a comprehensive study of 20 epidemiological studies, the National Science Panel found no consistent association between silicone breast implants and defined autoimmune diseases (Diamond *et al.*, 1998).

Silicone and scleroderma

Scleroderma

SSc, also known as scleroderma, is a systemic inflammatory autoimmune disease of unknown etiology. It is characterized by essential vasomotor disturbances and fibrosis, with subsequent atrophy of the skin, subcutaneous tissue, muscles, and internal organs (e.g. alimentary tract, lungs, heart, kidney, central nervous system, CNS). These clinical findings are accompanied by immunological disturbances (Levy *et al.*, 2009).

SSc is an acquired, sporadic disease, with a worldwide distribution that affects all races (Aminoff and Asbury, 2012). A conservative estimate of the incidence rate of scleroderma ranges from 2 to 10 cases per million, although rates of up to 20 cases per million have been reported (Lidar *et al.*, 2012). SSc shows a female predominance, with a disease onset in the range of 30–50 years (Aminoff and Asbury, 2012).

The environmental risk factors for scleroderma are difficult to assess, due to the limited number of validated exposure biomarkers, combined with the disease's rarity (Lidar *et al.*, 2012). Nevertheless, case reports and case series have identified clusters of the disease among certain occupational groups, in addition to a link to a variety of environmental agents, such as crystalline silica, rapeseed oil, polyvinyl chloride, and several drugs and vaccines (Levy *et al.*, 2009; Lidar *et al.*, 2012; Aminoff and Asbury, 2012).

The environmental pathogenesis of scleroderma may include the following mechanisms: loss of immune tolerance, immune system activation, and molecular mimicry (Lidar *et al.*, 2012).

Connection between silicone and scleroderma

In the 1980s, case reports of women with silicone breast implants who developed scleroderma began to appear in English medical literature (Levy *et al.*, 2009). During the 1980s and 1990s, more case reports and case series depicting an association between breast implants and scleroderma (as well as other connective tissue disorders) were published (Van Nunen *et al.*, 1982; Okano *et al.*, 1984; Sahn *et al.*, 1990; Spiera *et al.*, 1994; Sanchez-Guerrero *et al.*, 1995). Based on these publications, the possibility of a causal relationship between breast implants and the development of scleroderma was suggested (Lidar *et al.*, 2012).

Because scleroderma is a rare disease, case–control studies are the most suitable epidemiological approach to evaluating the connection between it and breast implants. A population-based case–control study was conducted among women in Michigan, in the United States. A total of 274 individuals with confirmed cases of SSc diagnosed between 1985 and 1991 and 1184 controls completed a telephone-based questionnaire. Two cases and fourteen controls had undergone breast augmentation. Both cases with silicone breast implants were diagnosed with SSc 1 and 12 years after the procedure. The results showed no significant association between silicone breast implants and SSc, with an odds ratio (OR) of 1.3 (95% CI: 0.27–6.23) (Burns *et al.*, 1996).

A study of 4229 rheumatology patients, followed from 1986 to 1992, identified 150 with breast implants. Of these, 12 had rheumatoid arthritis and 1 had an undifferentiated CTD; none of the patients had been diagnosed with scleroderma. None of the 64 scleroderma patients in the study had breast augmentation, corresponding to an OR of zero (Goldman *et al.*, 1995). The study found no evidence that women with breast implants are at an increased risk for having rheumatoid arthritis or other diffuse CTDs.

Other studies published in the medical literature have arrived at similar conclusions: that there is no statistically significant association between silicone breast implants and scleroderma (Englert and Brooks, 1994; Gabriel *et al.*, 1994; Hochberg *et al.*, 1996; Wong, 1996; Stein, 1999; Kjoller *et al.*, 2001; Breiting *et al.*, 2004). Currently, the general belief is that there is no association between silicone breast implants and defined autoimmune diseases.

However, it should be mentioned that the follow-up period in most studies was too short for the development of an autoimmune disease and that the studies focused on specific, defined autoimmune diseases rather than on autoimmune disease-related symptoms (Lidar *et al.*, 2012). One example of the need for a longer follow-up period when assessing the possible connection between silicone breast implants and scleroderma is provided by a report published in 2009, which described four women who developed scleroderma 5, 14, 15, and 20 years after having undergone silicone breast implantation (Levy *et al.*, 2009). Another study that demonstrated the importance of a longer follow-up period evaluated the impact of silicone-filled breast implants on the immune system in 32 patients attending a specialized autoimmunity clinic. All patients fulfilled the diagnostic criteria for ASIA, with a median time between implantation and onset of complaints of 10 years (2–24 years). Median time for the diagnosis of ASIA was 16 years (2–40 years). In 17 of the 32 patients, a systemic autoimmune disease was diagnosed, and 15 of the 32 had an impaired humoral immune system (hypogammaglobulinemia of an IgG subclass deficiency) (Cohen Tervaert and Kappel, 2013). The appearance of signs and symptoms after long-term follow-up suggests that they are connected to implant aging and/or rupture.

Silicone and nondefined autoimmune phenomena

Often, symptomatic breast implant patients do not match the criteria of a specific category of autoimmune disease. Several studies and a meta-analysis have demonstrated a relationship between silicone implants and a particular constellation of symptoms that do not fulfill the diagnostic criteria for any recognized CTDs (Vasey *et al.*, 2003; Hajdu *et al.*, 2011).

A study comparing 1546 patients with silicone breast implants to a control group of 2496 women who underwent reduction mammoplasty showed that the group who underwent breast augmentation manifested significantly higher prevalence of 16 of 28 investigated symptoms compared to controls (Fryzek *et al.*, 2001). Many of these symptoms met several of the criteria for fibromyalgia and chronic fatigue syndrome. This information is congruent with the findings of another study conducted by FDA scientists, which found a statistically significant link between ruptured silicone gel implants and fibromyalgia ($p = 0.004$) (Brown *et al.*, 2001). A limitation of such studies is their reliance on self-reported symptoms – a major weakness when compared to studies based on physician evaluations (Levy *et al.*, 2009; Hajdu *et al.*, 2011).

As early as 1994, there were reports of a new entity: siliconosis, a silicone implant-related disease, which was later incorporated into ASIA (Borenstein, 1994; Solomon, 1994). The US Institute of Medicine (IOM) published a report in 1999 in which it concluded that siliconosis lacked sufficient evidence and was based on selected case series, which 'lacked consistent and reproducible syndrome' (Bondurant *et al.*, 1999). A meta-analysis of symptoms from six different studies of women with silicone breast implants showed a statistically significant increase in 27 signs and symptoms associated with silicone breast implants, including body ache, abnormal fatigue, impaired cognition, depression, dry eyes, dry mouth, skin abnormalities, parasthesias, swollen and tender axillary glands, unexplained fever, hair loss, headache, and morning stiffness, among others (Vasey *et al.*, 2003). The authors proposed major criteria for siliconosis, which included a history of silicone breast implant with local problems, a 6 month history of chronic fatigue and myalgias (RR of 2.7 (95% CI: 1.4–5.2) and RR 1.4 (95% CI: 1.1–1.7), respectively), and symptom exacerbations during exercise, in contrast to patients with fibromyalgia, who experience attenuation of symptoms during exercise. Other studies disputed these phenomena and argued that these symptoms might only be related to capsular contracture or implant rupture, rather than systemic disease (Hölmich *et al.*, 2007). Nonetheless, it appears that the link between silicone and autoimmunity should not be limited to defined diseases, but rather should include undefined symptoms (Hajdu *et al.*, 2011).

Conclusions

It is currently widely believed that there is no association between silicone, most commonly in the form of silicone breast implants, and specific autoimmune diseases. However, silicone-induced inflammatory fibroproliferative response, namely capsular formation around silicone breast implants, is an irrefutable, well-documented occurrence, and the presence of antisilicone antibodies and nondefined silicone-associated autoimmune phenomena seems plausible.

It has been suggested that many studies conducted on the connection between silicone and autoimmunity had several important limitations (Levy *et al.*, 2009):

• Most studies were based on medical records and self-reported data, rather than clinical exams.
• Many studies focused on classically defined autoimmune diseases and did not study other diseases, atypical types of autoimmune diseases, or nondefined immune phenomena.
• Most study samples were too small to meaningfully detect increases in the rare disease they were studying.
• Connective tissue and autoimmune diseases and phenomena may take years to develop and be diagnosed. Most studies had a follow-up period that was too short to allow for the development of immune phenomena or autoimmune diseases. Studies that include women who underwent breast augmentation just a few months or years prior to enrollment cannot determine whether breast implants increase the long-term risk for such diseases.
• Many studies did not specify whether participants had had their implants removed prior to or during the study. This affects the exposure time to silicone, with women who had their implants removed having a shorter exposure than those who kept them.

Future studies that generate long-term data on a wider scale and with end-points that include specific autoimmune diseases and nondefined autoimmune phenomena should help clarify the presumed association between silicone and autoimmunity.

References

Agmon-Levin, N., Paz, Z., Israeli, E., and Shoenfeld, Y. (2009). Vaccines and autoimmunity. *Nat Rev Rheumatol*, **5**(11): 648–52.

Aminoff, M.J. and Asbury, A.K. (2012). Numbness, tingling, and sensory loss. In: Longo, D.L., Fauci, A.S., Kasper, D.L., *et al.* (eds). *Harrison's Principles of Internal Medicine*, 18th edn, Vol. **2**. New York: McGraw-Hill, pp. 2757–69.

ASAPS (American Society for Aesthetic Plastic Surgery). (2013). Cosmetic Surgery National Data Bank: statistics 2012. New York: American Society for Aesthetic Plastic Surgery.

Bar-Meir, E., Teuber, S.S., Lin, H.C., *et al.* (1995). Multiple autoantibodies in patients with silicone breast implants. *J Autoimmun*, **8**(2): 267–77.

Bar-Meir, E., Eherenfeld, M., and Shoenfeld, Y. (2003). Silicone gel breast implants and connective tissue disease – a comprehensive review. *Autoimmunity*, **36**(4): 193–7.

Barker, D.E., Retsky, M.I., and Schultz, S. (1978). "Bleeding" of silicone from bag-gel breast implants, and its clinical relation to fibrous capsule reaction. *Plast Reconstr Surg*, **61**(6): 836–41.

Bekerecioglu, M., Onat, A.M., Tercan, M., *et al.* (2008). The association between silicone implants and both antibodies and autoimmune diseases. *Clin Rheumatol*, **27**(2): 147–50.

Bondurant, S., Ernster, V., and Herdman, R. (1999). Safety of Silicone Breast Implants: Report of the Committee on the Safety of Silicone Breast Implants (IOM). Washington, DC: National Academies Press.

Borenstein, D. (1994). Siliconosis: a spectrum of illness. *Semin Arthritis Rheum*, **24**(1): 1–7.

Breiting, V.B., Hölmich, L.R., Brandt, B., *et al.* (2004). Long-term health status of Danish women with silicone breast implants. *Plast Reconstr Surg*, **114**(1): 217–26.

Brown, S.L., Middleton, M.S., Berg, W.A., *et al.* (2000). Prevalence of rupture of silicone gel breast implants revealed on MR imaging in a population of women in Birmingham, Alabama. *Am J Roentgenol*, **175**(4): 1057–64.

Brown, S.L., Pennello, G.E.N.E., Berg, W.A., *et al.* (2001). Silicone gel breast implant rupture, extracapsular silicone, and health status in a population of women. *J Rheumatol*, **28**(5): 996–1003.

Burns, C.J., Laing, T.J., Gillespie, B.W., *et al.* (1996). The epidemiology of scleroderma among women: assessment of risk from exposure to silicone and silica. *J Rheumatol*, **23**(11): 1904–11.

Cohen Tervaert, J.W. and Kappel, R.M. (2013). Silicone implant incompatibility syndrome (SIIS): a frequent cause of ASIA (Shoenfeld's syndrome). *Immunol Res*, **56**(2–3): 293–8.

Diamond, B.A.H., Kerkvliet, B.S., and Tugwell, N.I.P. (1998). Silicone breast implants in relation to connective tissue diseases and immunologic dysfunction: a report by a National Science Panel to the Hon. Sam C. Pointer Jr., Coordinating Judge for the Federal Breast Implant Multi-District Litigation. Available from: http://www.fjc.gov/BREIMLIT/SCIENCE/chap1.pdf [last accessed 11 December 2014].

Englert, H.J. and Brooks, P. (1994). Scleroderma and augmentation mammoplasty-a causal relationship? *Aus NZ J Med*, **24**(1): 74–80.

Fryzek, J.P., Signorello, L.B., Hakelius, L., *et al.* (2001). Self-reported symptoms among women after cosmetic breast implant and breast reduction surgery. *Plast Reconstr Surg*, **107**(1): 206–13.

Gabriel, S.E., O'Fallon, W.M., Kurland, L.T., *et al.* (1994). Risk of connective-tissue diseases and other disorders after breast implantation. *N Engl J Med*, **330**(24): 1697–702.

Goldblum, R.M., O'Donell, A.A., Pyron, D., *et al.* (1992). Antibodies to silicone elastomers and reactions to ventriculoperitoneal shunts. *Lancet*, **340**(8818): 510–13.

Goldman, J.A., Greenblatt, J., Joines, R., *et al.* (1995). Breast implants, rheumatoid arthritis, and connective tissue diseases in a clinical practice. *J Clin Epidemiol*, **48**(4): 571–82.

Gylbert, L., Asplund, O., and Jurell, G. (1990). Capsular contracture after breast reconstruction with silicone-gel and saline-filled implants: a 6-year follow-up. *Plast Reconstr Surg*, **85**(3): 373–7.

Hajdu, S.D., Agmon-Levin, N., and Shoenfeld, Y. (2011). Silicone and autoimmunity. *Eur J Clin Invest*, **41**(2): 203–11.

Hennekens, C.H., Lee, I.M., Cook, N.R., *et al.* (1996). Self-reported breast implants and connective-tissue diseases in female health professionals. *JAMA*, **275**(8): 616–21.

Hochberg, M.C., Perlmutter, D.L., Medsger, T.A., *et al.* (1996). Lack of association between augmentation mammoplasty and systemic sclerosis (scleroderma). *Arthritis Rheum*, **39**(7): 1125–31.

Hölmich, L.R., Kjøller, K., Fryzek, J.P., *et al.* (2003). Self-reported diseases and symptoms by rupture status among unselected Danish women with cosmetic silicone breast implants. *Plast Reconstr Surg*, **111**(2): 723–32.

Hölmich, L.R., Lipworth, L., McLaughlin, J.K., and Friis, S. (2007). Breast implant rupture and connective tissue disease: a review of the literature. *Plast Reconstr Surg*, **120**(7): S62–9.

Israeli, E., Agmon-Levin, N., Blank, M., and Shoenfeld, Y. (2009). Adjuvants and autoimmunity. *Lupus*, **18**(13): 1217–25.

Janowsky, E.C., Kupper, L.L., and Hulka, B.S. (2000). Meta-analyses of the relation between silicone breast implants and the risk of connective-tissue diseases. *N Engl J Med*, **342**(11): 781–90.

Kjoller, K., Friis, S., Mellemkjær, L., *et al.* (2001). Connective tissue disease and other rheumatic conditions following cosmetic breast implantation in Denmark. *Arch Internal Med*, **161**(7): 973.

Kumagai, Y., Shiokawa, Y., Medsger, T.A., and Rodnan, G.P. (1984). Clinical spectrum of connective tissue disease after cosmetic surgery. *Arthritis Rheum*, **27**(1): 1–12.

Levy, Y., Rotman-Pikielny, P., Ehrenfeld, M., and Shoenfeld, Y. (2009). Silicone breast implantation-induced scleroderma: description of four patients and a critical review of the literature. *Lupus*, **18**(13): 1226–32.

Lidar, M., Agmon-Levin, N., Langevitz, P., and Shoenfeld, Y. (2012). Silicone and scleroderma revisited. *Lupus*, **21**(2): 121–7.

Marcusson, J.A. and Bjarnason, B. (1999). Unusual skin reaction to silicone content in breast implants. *Acta Dermato-venereologica*, **79**: 136–8.

Mastruserio, D.N., Pesqueira, M.J., and Cobb, M.W. (1996). Severe granulomatous reaction and facial ulceration occurring after subcutaneous silicone injection. *J Am Acad Dermatol*, **34**(5): 849–52.

McCarty J.G. (1990). *Plastic Surgery*. Philadelphia: Saunders.

Meroni, P.L. (2011). Autoimmune or auto-inflammatory syndrome induced by adjuvants (ASIA): old truths and a new syndrome? *J Autoimmunity*, **36**(1): 1–3.

Miyoshi, K., Miyaoka, T., Kobayashi, Y., *et al.* (1964). Hypergammaglobulinemia by prolonged adjuvanticity in man: disorders developed after augmentation mammoplasty. *Jap Med J*, **2122**: 9–14.

Netscher, D.T. (2004). Subclinical infection in breast capsules. *Plast Reconstr Surg*, **114**(3): 818–20.

Okano, Y., Nishikai, M., and Sato, A. (1984). Scleroderma, primary biliary cirrhosis, and Sjögren's syndrome after cosmetic breast augmentation with silicone injection: a case report of possible human adjuvant disease. *Ann Rheum Dis*, **43**(3): 520.

Prantl, L., Angele, P., Schreml, S., *et al.* (2006). Determination of serum fibrosis indexes in patients with capsular contracture after augmentation with smooth silicone gel implants. *Plast Reconstr Surg*, **118**(1): 224–9.

Rapaport, M.J., Vinnik, C., and Zarem, H. (1996). Injectable silicone: cause of facial nodules, cellulitis, ulceration, and migration. *Aesthetic Plast Surg*, **20**(3): 267–76.

Ratner, B.D., Hoffman, A.S., Schoen, F.J., and Lemons, J.E. (1996). *Biomaterials Science: An Introduction to Materials in Medicine*, 2nd edn. New York: Academic Press.

Sahn, E.E., Garen, P.D., Silver, R.M., and Maize, J.C. (1990). Scleroderma following augmentation mammoplasty: report of a case and review of the literature. *Arch Dermatol*, **126**(9): 1198.

Sanchez-Guerrero, J., Colditz, G.A., Karlson, E.W., *et al.* (1995). Silicone breast implants and the risk of connective-tissue diseases and symptoms. *N Engl J Med*, **332**(25): 1666–70.

Santucci, B., Valenzano, C., Rocco, M.D., and Cristaudo, A. (2000). Platinum in the environment: frequency of reactions to platinum-group elements in patients with dermatitis and urticaria. *Contact Dermatitis*, **43**(6): 333–8.

Schaefer, C.J. and Wooley, P.H. (1999). The influence of silicone implantation on murine lupus in MRL lpr/lpr mice. *J Rheumatol*, **26**(10): 2215–21.

Shoenfeld, Y. and Agmon-Levin, N. (2011). "ASIA" – autoimmune/inflammatory syndrome induced by adjuvants. *J Autoimmun*, **36**(1): 4–8.

Solomon, G. (1994). A clinical and laboratory profile of symptomatic women with silicone breast implants. *Semin Arthritis Rheum*, **24**: 29–37.

Spiera, R.F., Gibofsky, A., and Spiera, H. (1994). Silicone gel filled breast implants and connective tissue disease: an overview. *J Rheumatol*, **21**(2): 239.

Stein, Z.A. (1999). Silicone breast implants: epidemiological evidence of sequelae. *Am J Public Health*, **89**(4): 484–7.

Teuber, S.S., Rowley, M.J., Yoshida, S.H., *et al.* (1993). Anti-collagen autoantibodies are found in women with silicone breast implants. *J Autoimmunity*, **6**(3): 367–77.

Teuber, S.S., Ito, L.K., Anderson, M., and Gershwin, M.E. (1995). Silicone breast implant – associated scarring dystrophy of the arm. *Arch Dermatol*, **131**(1): 54.

Van Nunen, S.A., Gatenby, P.A., and Basten, A. (1982). Post-mammoplasty connective tissue disease. *Arthritis Rheum*, **25**(6): 694–7.

Vasey, F.B., Zarabadi, S.A., Seleznick, M., and Ricca, L. (2003). Where there's smoke there's fire: the silicone breast implant controversy continues to flicker: a new disease that needs to be defined. *J Rheumatol*, **30**(10): 2092–4.

Vermeulen, R.C. and Scholte, H.R. (2003). Rupture of silicone gel breast implants and symptoms of pain and fatigue. *J Rheumatol*, **30**(10): 2263–7.

White, K.L. and Klykken, P.C. (1998). The non-specific binding of immunoglobulins to silicone implant materials: the lack of a detectable silicone specific antibody. *Immunol Invest*, **27**(4–5): 221–35.

Wolf, L.E., Lappe, M., Peterson, R.D., and Ezrailson, E.G. (1993). Human immune response to polydimethylsiloxane (silicone): screening studies in a breast implant population. *FASEB J*, **7**(13): 1265–8.

Wong, O. (1996). A critical assessment of the relationship between silicone breast implants and connective tissue diseases. *Reg Toxicol Pharmacol*, **23**(1): 74–85.

Zandman-Goddard, G., Blank, M., Ehrenfeld, M., *et al.* (1999). A comparison of autoantibody production in asymptomatic and symptomatic women with silicone breast implants. *J Rheumatol*, **26**(1): 73–7.

8 Silicone Breast Implants and Autoimmune/Inflammatory Syndrome induced by Adjuvants (ASIA): A Literature Search

Elisabetta Borella,[1,2] Eitan Israeli,[2] and Yehuda Shoenfeld[2,3]

[1]Division of Rheumatology, Department of Medicine, University of Padua, Padua, Italy
[2]Zabludowicz Center for Autoimmune Diseases, Sheba Medical Center, Tel Hashomer, Israel
[3]Incumbent of the Laura Schwarz-Kipp Chair for Research of Autoimmune Diseases, Sackler Faculty of Medicine, Tel Aviv University, Tel Aviv, Israel

Introduction

The hypothesis that silicone implants or injections may play a role in the etiology of autoimmune diseases has been suggested since the 1980s (Baldwin and Kaplan, 1983). An extensive review of the literature published in 1998 confirms the data showing that silicone triggers the immune system (Shanklin and Smalley, 1998). Thus, at the end of the 20th century, a new disease presenting rheumatic symptoms and arising following silicone breast implantation was defined: the siliconosis. Siliconosis is a hyperactive immune response induced by silicone exposure, which leads to chronic inflammation, granulomas, and fibrosis, and eventually to a fully autoimmune disease (Shanklin and Smalley, 1998; Shoenfeld and Agmon-Levin, 2011).

However, since no parameters were listed by which to define siliconosis, criteria from several different systemic autoimmune diseases were used to prove the existence of the pathology in several epidemiological studies, although most of them failed in this respect (Stein, 1999; Jensen *et al.*, 2001; Henriksen *et al.*, 2003; Holmich *et al.*, 2003; Kjoller *et al.*, 2004; Cohen Tervaert and Kappel, 2013).

In 2011, Shoenfeld and Agmon-Levin argued that siliconosis is a well-defined disease, although it shares some characteristics with connective tissue diseases (CTDs). In particular, the authors suggested that siliconosis is part of a group of diseases induced by adjuvant, which they termed "autoimmune/inflammatory syndrome induced by adjuvant" (ASIA) (Shoenfeld and Agmon-Levin, 2011).

ASIA and siliconosis

ASIA is a recently defined syndrome, characterized by an autoimmune or immune-mediated condition following exposure to external stimuli (Shoenfeld and Agmon-Levin, 2011). Exposure to adjuvant stimuli may occur through vaccines (e.g. aluminium) (Agmon-Levin *et al.*, 2009), breast implants (e.g. silicone) (Hajdu *et al.*, 2011), and

Vaccines and Autoimmunity, First Edition. Edited by Yehuda Shoenfeld, Nancy Agmon-Levin, and Lucija Tomljenovic.

Table 8.1 Major and minor criteria proposed for the ASIA syndrome. Shoenfeld, Y. and N. Agmon-Levin. ASIA- autoimmune/inflammatory Syndrome Induced by Adjuvants. *Journal of Autoimmunity* 36 (1): 4–8. Copyright © 2011, Elsevier

Major criteria
- Exposure to an external stimulus (infection, vaccine, silicone, adjuvant) prior to clinical manifestation
- The appearance of "typical" clinical manifestations:
  - o Myalgia, myositis, or muscle weakness
  - o Arthralgia and/or arthritis
  - o Chronic fatigue, un-refreshing sleep, or sleep disturbances
  - o Neurological manifestations (especially associated with demyelization)
  - o Cognitive impairment, memory loss
  - o Pyrexia, dry mouth
- Removal of the inciting agent induces improvement
- Typical biopsy of involved organs

Minor criteria
- The appearance of autoantibodies or antibodies directed at the suspected adjuvant
- Other clinical manifestations (e.g. irritable bowel syndrome, IBS)
- Specific HLA (HLA DRB1, HLA DQB1)
- Evolvement of an autoimmune disease (multiple sclerosis, MS; systemic sclerosis, SSc)

infectious agents (e.g. Epstein–Barr virus, EBV) or environmental substances found in building models (e.g. mycotoxins) (Israeli and Pardo, 2011). Although these adjuvants were formerly believed not to elicit significant adverse immune responses, studies in animals (Israeli *et al.*, 2009; Batista-Duharte *et al.*, 2011; Zivkovic *et al.*, 2012; Lujan *et al.*, 2013) and humans (Couette *et al.*, 2009; Passeri *et al.*, 2011; Gherardi and Authier, 2012) seem to suggest otherwise.

However, despite the common exposure, adjuvant disease is rare. Therefore, it is possible that the exposure to an external stimulus with an immune adjuvant effect, such as silicone or aluminium in vaccinations, is not in itself sufficient to trigger the development of an autoimmune disease. Rather, only people with additional genetic or environmental risk factors are susceptible to developing ASIA (Agmon-Levin *et al.*, 2009). The major and minor criteria proposed for ASIA (Agmon-Levin *et al.*, 2009) are listed in Table 8.1.

Numerous case reports have hinted at a possible correlation between silicone breast implant and CTDs. Thus, studies in this field have been performed for 30 years. The conjecture that silicone might trigger an autoimmune response led the US Food and Drugs Administration (FDA) to ban the use of silicone gel-filled breast implants in 1992 (Podolsky and Newman, 1992). Although in the following years several cases of rheumatic diseases developing after adjuvant exposure were reported, the relationship between silicone implants and CTDs has since generally been refuted by the consistent evidence from published large-scale cohort and case–control studies (Shoaib *et al.*, 1994; Vasey *et al.*, 2003; Asherson *et al.*, 2004; Nancy and Shoenfeld, 2008; Levy *et al.*, 2009; Caldeira and Ferreira, 2012; Jara *et al.*, 2012; Kivity *et al.*, 2012), as well as from critical reviews (Bar-Meir *et al.*, 2003; Lipworth *et al.*, 2004, 2011; Holmich *et al.*, 2007; Levy *et al.*, 2009). However, since the criteria of ASIA were defined only recently, few studies have looked for a link between silicone exposure and adjuvant disease (Englert *et al.*, 2004; Vera-Lastra *et al.*, 2012; Cohen Tervaert and Kappel, 2013).

Methods

We searched the PubMed database using the terms "silicone breast implants" and "connective tissue disease," or "undifferentiated connective tissue disease," as well as various combinations of different keywords, such as "ASIA syndrome," "siliconosis," "autoantibodies," and "animal model." Manual searches using the related link facility extended the number of references identified, and additional references were identified by cross-checking the reference lists of the identified publications. We found that only a few epidemiological studies had tried to demonstrate whether silicone acts as an adjuvant in ASIA, and all of them actually did demonstrate this conjecture (Englert *et al.*, 2004; Vera-Lastra *et al.*, 2012; Cohen Tervaert and Kappel, 2013).

Results

In one relevant study (Cohen Tervaert and Kappel, 2013), all patients referred to the Clinical

Immunology Clinic of the Maastricht University Medical Center (Netherlands) between January 2008 and January 2012 were evaluated for the presence of breast implants, with 32 of the 600 patients evaluated presenting with implants. These were classified into subgroups: those with ASIA, those with non-Hodgkin's lymphoma, those suffering from immune-mediated inflammatory diseases (IMIDs) fulfilling the diagnostic criteria for CTDs, and those with silicone breast implant incompatibility (SIIS) and symptoms or signs of silicone allergy, capsular contracture, and/or systemic manifestations such as chronic fatigue, arthralgia, myalgias, asthenia, and/or fever. All patients met the diagnostic criteria for ASIA. Most of them presented with arthralgias and fatigue. The period between start of complaints and implantation of silicone prosthesis varied significantly: median time was 10 years (2–24 years). Of all of the patients, 20% reported signs and symptoms within 2 years after operation; 33% developed signs and symptoms between 2 and 10 years after operation; 28% between 10 and 20 years; and 19% more than 20 years after the breast implant operation. As a consequence, it might be inferred that ASIA can occur even several years after silicone implantation. This may be due to the fact that silicone acts as a trigger for autoimmune disease: in order to develop these diseases, other genetic, epigenetic, or environmental risk factors must to be involved. However, the authors suggested that the long-term follow-up required to develop an autoimmune reaction may be due to aging and/or rupture of silicone implant. It has to be noted that this study has a limitation: the population consisted of women who were referred to an autoimmunity clinic with complaints, so the number of patients who had undergone mammary prosthesis without any complaints is unknown. A review of case reports and clinical studies about systemic autoimmune disease after silicone exposure led Vasey *et al.* (2003) to observe a higher prevalence of undefined rheumatic symptoms in patients with breast implants. Taken together, these data confirm the hypothesis that mild rheumatic symptoms following silicone exposure are only the tip of the iceberg of a systemic autoimmune disorder.

Englert *et al.* (2004) performed a retrospective study comparing 458 silicone breast-implanted women with a control group of 687 women who had undergone other types of non-silicone-associated plastic surgery. Subjects were evaluated using a questionnaire (specifically designed for this study, in order to understand symptoms reported by patients), standardized clinical examination, nailfold capillaroscopy photography, and serological assessment. The authors observed a higher prevalence of nonspecific rheumatologic symptoms in the experimental group, which they recognized as part of the siliconosis disease. Another study investigating the general symptoms reported by patients following cosmetic surgery and breast implantation reported different clinical and serological data fulfilling the diagnostic criteria for ASIA. It found that mineral oil and silicone fluid act as adjuvants, inducing rheumatic symptoms like myalgia, stiffness, arthralgia, memory loss, and elevated titers of autoantibodies in patients' sera (Vera-Lastra *et al.*, 2012).

There are numerous case reports that point to a relationship between silicone and siliconosis, the adverse immune condition triggered by this type of adjuvant (Shoaib *et al.*, 1994; Vasey *et al.*, 2003; Asherson *et al.*, 2004; Nancy and Shoenfeld, 2008; Levy *et al.*, 2009; Caldeira and Ferreira, 2012; Jara *et al.*, 2012; Kivity *et al.*, 2012). Moreover, numerous reviews performed in order to clarify whether CTDs can be elicited by silicone implants show a significant relationship between silicone exposure and undefined rheumatic symptoms, such as chronic fatigue and myalgia (Peters *et al.*, 1999; Bar-Meir *et al.*, 2003; Lipworth *et al.*, 2004, 2011; Holmich *et al.*, 2007; Levy *et al.*, 2009).

Breiting *et al.* (2004) conducted a study on a Danish cohort of 525 women exposed to silicone breast implant and compared this cohort to two control groups. The first consisted of 186 women who underwent other cosmetic surgeries, while the second consisted of 149 women who were not exposed to any surgery. The authors observed that, in the latter group, there was a lower use of antidepressant and hypnotics. In addition, they observed a significant difference in the prevalence of cognitive syndrome, Reynaud, and fatigue between the silicone-exposed group and the two non-exposed control groups.

In another study, 2761 Danish women with cosmetic breast implants were compared with 8807 women who chose other types of cosmetic surgery. A statistical prevalence of rheumatic disorders not fulfilling the CTD criteria was described, with greater prevalence of myalgia and fibromyalgia in the silicone-exposed breast implant group (Fryzek *et al.*, 2007).

Mechanisms of siliconosis

In parallel to the clinical trials, several studies were performed to identify increased levels of autoantibodies in asymptomatic women who had undergone mammary prosthesis implantation. Unfortunately, different studies present conflicting results (Miller *et al.*, 1998; Park *et al.*, 1998; Karlson *et al.*, 1999; Zandman-Goddard *et al.*, 1999; Contant *et al.*, 2000; Karlson *et al.*, 2001; Pastor *et al.*, 2001; De Jong *et al.*, 2004; Bekerecioglu *et al.*, 2008; Wolfram *et al.*, 2008; Silva *et al.*, 2011). However, a single study evaluating the capsular tissue of silicone implants and the sera of implant patients and controls for antisilicone antibodies and nonspecific immunoglobulins (IgG, IgA, IgM, and IgE) using immunofluorescence methods observed increased levels of antibodies, especially of the IgG class, bound to the capsule in the silicone-exposed group (Bekerecioglu *et al.*, 2008). The high concentration of immunoglobulins around the capsule may explain the low sera levels of IgG observed in the silicone breast-implanted population described by Cohen Tervaert and Kappel (2013). Moreover, high levels of Th1/Th17 cells and cytokines in the capsule were observed in a study of 33 women undergoing implant change or removal due to capsular fibrosis or implant deflation, or for aesthetic reasons (Wolfram *et al.*, 2012). These observations confirm the data showing that silicone breast implants cause a foreign-body reaction characterized by infiltration of immune cells (macrophages and T cells), fibrosis reaction, and granuloma-containing silicone. These cells may be involved in the development of autoimmune reactions. As reported by Bassetto *et al.* (2012), the formation of the periprosthetic capsule leads to capsular contracture, but it may also stimulate an autoimmune response, especially in genetically predisposed individuals. According to Bassetto *et al.* (2012), the stimulation of macrophages induces the production of cytokines, the activation of T and B cells, and consequently the production of autoantibodies against silicone. TGFβ induces the activation of fibroblasts, the formation of the fibrotic capsule, and the induction of T-reg cells. In genetically predisposed individuals, IL-1 and IL-6 prevail and induce the formation of Th17, which seems to play an important role in the pathogenesis of autoimmune diseases.

Experimental models

There are few preclinical studies supporting the hypothesis of the existence of siliconosis (McDonald *et al.*, 1998; Schaefer and Wooley, 1999; Naim *et al.*, 2000). For example, Naim *et al.* (2000) performed an experimental study in which ASW mice, which are not predisposed to autoimmune diseases, showed higher levels of gammaglobulins and macrophages compared to controls following intraperitoneal injection of silicone.

Subsequently, Schaefer and Wooley (1999) demonstrated that, compared to saline, the injection of silicone gel or silicone oil in MRL lpr/lpr mice induced higher titers of anti-DNA antibodies, rheumatic factor, and anti-silicone-binding protein (anti-SBP) and higher levels of serum cytokines, especially IL-1. However, no atypical clinical manifestation was observed in mice treated with the adjuvant, suggesting that in this particular case, silicone did exhibit an immune-stimulating action, but without an effect on the clinical manifestation of the disease.

Genetic evidence

Since only a few subjects develop rheumatic symptoms after silicone exposure, it is believed that silicone elicits an autoimmune response only in genetically predisposed individuals. Consistent with this theory, Meier *et al.* (1997) reported the case of two HLA-identical sisters who developed polyarticular arthritis and neurological symptoms after silicone implant. HLA typing revealed alleles typically associated with rheumatic diseases: HLA-DRB 1*0405 and HLA-DQB1*0302. Thus, these women were genetically predisposed to develop an autoimmune disease, which arose only after silicone exposure. In addition, the improvements of rheumatic and neurological symptoms and radiological erosions after the removal of the implants in both the sisters suggested that silicone exposure and rheumatic symptoms were related (Meier *et al.*, 1997). It was further shown that symptomatic women with mammary prosthesis share a common HLA pattern that is characterized by the presence of HLA-DR5 and HLA-DQ2 (Young *et al.*, 1995). Moreover, the increased proportion of patients with HLA-DR5 is similar to that of patients suffering from fibromyalgia (Young *et al.*, 1995).

Conclusions

In conclusion, clinical, experimental and genetic studies allude to the immunogenic role of silicone in susceptible people. This means that silicone is safe and well tolerated in the general population, but may trigger ASIA in patients with a genetically susceptible background.

References

Agmon-Levin, N., Paz, Z., Israeli, E., and Shoenfeld, Y. (2009). Vaccines and autoimmunity. *Nat Rev Rheumatol*, **5**(11): 648–52.

Asherson, R.A., Shoenfeld, Y., Jacobs, P., and Bosman, C. (2004). An unusually complicated case of primary Sjögren's syndrome: development of transient lupus-type autoantibodies following silicone implant rejection. *J Rheumatol*, **31**(1): 196–7.

Baldwin, C.M. Jr, and Kaplan, E.N. (1983). Silicone-induced human adjuvant disease? *Ann Plastic Surg*, **10**(4): 270–3.

Bar-Meir, E., Eherenfeld, M., and Shoenfeld, Y. (2003). Silicone gel breast implants and connective tissue disease – a comprehensive review. *Autoimmunity*, **36**(4): 193–7.

Bassetto, F., Scarpa, C., Vindigni, V. and Doria, A. (2012). The periprosthetic capsule and connective tissue diseases: a piece in the puzzle of autoimmune/autoinflammatory syndrome induced by adjuvants. *Exp Biol Med (Maywood, NJ)*, **237**(10): 1117–22.

Batista-Duharte, A., Lindblad, E.B., and Oviedo-Orta, E. (2011). Progress in understanding adjuvant immunotoxicity mechanisms. *Toxicol Lett*, **203**: 97–105.

Bekerecioglu, M., Onat, A.M., Tercan, M., *et al.* (2008). The association between silicone implants and both antibodies and autoimmune diseases. *Clin Rheumatol*, **27**(2): 147–50.

Breiting, V.B., Holmich, L.R., Brandt, B., *et al.* (2004). Long-term health status of Danish women with silicone breast implants. *Plastic Reconstruct Surg*, **114**(1): 217–26; disc. 227–8.

Caldeira, M. and Ferreira, A.C. (2012). Siliconosis: autoimmune/inflammatory syndrome induced by adjuvants (ASIA). *IMAJ*, **14**(2): 137–8.

Cohen Tervaert, J.W. and Kappel, R.M. (2013). Silicone implant incompatibility syndrome (SIIS): a frequent cause of ASIA (Shoenfeld's syndrome). *Immunol Res*, **56**(2–3): 293–8.

Contant, C.M., Swaak, A.J., Wiggers, T., *et al.* (2000). First evaluation study of the Dutch working party on silicone breast implants (SBI) and the silicone-related symptom complex (SRSC). *Clin Rheumatol*, **19**(6): 458–63.

Couette, M., Boisse, M.F., Maison, P., *et al.* (2009). Long-term persistence of vaccine-derived aluminum hydroxide is associated with chronic cognitive dysfunction. *J Inorg Biochem*, **103**: 1571–8.

De Jong, W.H., Kallewaard, M., Goldhoorn, C.A., *et al.* (2004). Long-term exposure to silicone breast implants does not induce antipolymer antibodies. *Biomaterials*, **25**(6): 1095–103.

Englert, H., Joyner, E., Thompson, M., *et al.* (2004). Augmentation mammoplasty and silicone-osis. *Int Med J*, **34**(12): 668–76.

Fryzek, J.P., Holmich, L., McLaughlin, J.K., *et al.* (2007). A nationwide study of connective tissue disease and other rheumatic conditions among Danish women with long-term cosmetic breast implantation. *Ann Epidemiol*, **17**(5): 374–9.

Hajdu, S.D., Agmon-Levin, N., and Shoenfeld, Y. (2011). Silicone and autoimmunity. *Eur J Clin Invest*, **41**(2): 203–11.

Henriksen, T.F., Holmich, L.R., Fryzek, J.P., *et al.* 2003. Incidence and severity of short-term complications after breast augmentation: results from a nationwide breast implant registry. *Ann Plastic Surg*, **51**(6): 531–9.

Holmich, L.R., Kjoller, K., Fryzek, J.P., *et al.* 2003. Self-reported diseases and symptoms by rupture status among unselected Danish women with cosmetic silicone breast implants. *Plastic Reconstruct Surg*, **111**(2): 723–32; disc. 733–4.

Holmich, L.R., Lipworth, L., McLaughlin, J.K., and Friis, S. (2007). Breast implant rupture and connective tissue disease: a review of the literature. *Plastic Reconstruct Surg*, **120**(7 Suppl. 1): S62–9.

Gherardi, R. and Authier, F. (2012). Macrophagic myofasciitis: characterization and pathophysiology. *Lupus*, **21**: 184–9.

Israeli, E., Agmon-Levin, N., Blank, M., and Shoenfeld, Y. (2009). Adjuvants and autoimmunity. *Lupus*, **18**: 1217–25.

Israeli, E. and Pardo, A. (2011). The sick building syndrome as a part of the autoimmune (auto-inflammatory) syndrome induced by adjuvants. *Mod Rheumatol*, **21**(3): 235–9.

Jara, L.J., Medina, G., Gomez-Banuelos, E., *et al.* (2012). Still's disease, lupus-like syndrome, and silicone breast implants. A case of "ASIA" (Shoenfeld's syndrome). *Lupus*, **21**(2): 140–5.

Jensen, B., Bliddal, H., Kjoller, K., *et al.* (2001). Rheumatic manifestations in Danish women with silicone breast implants. *Clin Rheumatol*, **20**(5): 345–52.

Karlson, E.W., Hankinson, S.E., Liang, M.H., *et al.* (1999). Association of silicone breast implants with immunologic abnormalities: a prospective study. *Am J Med*, **106**(1): 11–19.

Karlson, E.W., Lee, I.M., Cook, N.R., *et al.* (2001). Serologic evaluations of women exposed to breast implants. *J Rheumatol*, **28**(7): 1523–30.

Kivity, S., Katz, M., Langevitz, P., *et al.* (2012). Autoimmune syndrome induced by adjuvants (ASIA) in the Middle East: morphea following silicone implantation. *Lupus*, **21**(2): 136–9.

Kjoller, K., Holmich, L.R., Fryzek, J.P., *et al.* (2004). Self-reported musculoskeletal symptoms among Danish women with cosmetic breast implants. *Ann Plastic Surg*, **52**(1): 1–7.

Levy, Y., Rotman-Pikielny, P., Ehrenfeld, M., and Shoenfeld, Y. (2009). Silicone breast implantation-induced scleroderma: description of four patients and a critical review of the literature. *Lupus*, **18**(13): 1226–32.

Lipworth, L., Holmich, L.R., and McLaughlin, J.K. (2011). Silicone breast implants and connective tissue disease: no association. *Semin Immunopathol*, **33**(3): 287–94.

Lipworth, L., Tarone, R.E., and McLaughlin, J.K. (2004). Silicone breast implants and connective tissue disease: an updated review of the epidemiologic evidence. *Ann Plastic Surg*, **52**(6): 598–601.

Lujan, L., Perez, M., Salazar, E., *et al.* (2013). Autoimmune/autoinflammatory syndrome induced by adjuvants (ASIA syndrome) in commercial sheep. *Immunol Res*, **56**: 317–24.

McDonald, A.H., Weir, K., Schneider, M., *et al.* (1998). Silicone gel enhances the development of autoimmune disease in New Zealand black mice but fails to induce it in BALB/cAnPt mice. *Clin Immunol Immunopathol*, **87**(3): 248–55.

Meier, L.G., Barthel, H.R., and Seidl, C. (1997). Development of polyarthritis after insertion of silicone breast implants followed by remission after implant removal in 2 HLA – identical sisters bearing rheumatoid arthritis susceptibility genes. *J Rheumatol*, **24**(9): 1838–41.

Miller, A.S. 3rd,, Willard V., Kline K., *et al.* (1998). Absence of longitudinal changes in rheumatologic parameters after silicone breast implantation: a prospective 13-year study. *Plastic Reconstruct Surg*, **102**(7): 2299–303.

Naim, J.O., Satoh, M., Buehner, N.A., *et al.* (2000). Induction of hypergammaglobulinemia and macrophage activation by silicone gels and oils in female ASW mice. *Clin Diagnostic Lab Immunol*, **7**(3): 366–70.

Nancy, A.L. and Shoenfeld, Y. (2008). Chronic fatigue syndrome with autoantibodies – the result of an augmented adjuvant effect of hepatitis-B vaccine and silicone implant. *Autoimmun Rev*, **8**(1): 52–5.

Park, A.J., Black, R.J., Sarhadi, N.S., *et al.* (1998). Silicone gel-filled breast implants and connective tissue diseases. *Plastic Reconstruct Surg*, **101**(2): 261–8.

Passeri, E., Villa, C., Couette, M., *et al.* (2011). Long-term follow-up of cognitive dysfunction in patients with aluminum hydroxide-induced macrophagic myofasciitis (MMF). *J Inorg Biochem*, **105**: 1457–63.

Pastor, J.C., Puente, B., Telleria, J., *et al.* (2001). Anti-silicone antibodies in patients with silicone implants for retinal detachment surgery. *Ophthalm Res*, **33**(2): 87–90.

Peters, W., Smith, D., and Lugowski, S. (1999). Silicon assays in women with and without silicone gel breast implants – a review. *Ann Plastic Surg*, **43**(3): 324–30.

Podolsky, D. and Newman, R.J. (1992). A ban on silicone. *US News World Rep*, **112**(2): 61–2.

Schaefer, C.J. and Wooley, P.H. (1999). The influence of silicone implantation on murine lupus in MRL lpr/lpr mice. *J Rheumatol*, **26**(10): 2215–21.

Shanklin, D.R. and Smalley, D.L. (1998). The immunopathology of siliconosis. History, clinical presentation, and relation to silicosis and the chemistry of silicon and silicone. *Immunol Res*, **18**(3): 125–73.

Shoaib, B.O., Patten, B.M., and Calkins, D.S. (1994). Adjuvant breast disease: an evaluation of 100 symptomatic women with breast implants or silicone fluid injections. *Keio J Med*, **43**(2): 79–87.

Shoenfeld, Y. and Agmon-Levin, N. (2011). "ASIA" - autoimmune/inflammatory syndrome induced by adjuvants. *J Autoimmunity*, **36**(1): 4–8.

Silva, M.M., Modolin, M., Faintuch, J., *et al.* (2011). Systemic inflammatory reaction after silicone breast implant. *Aesthet Plastic Surg*, **35**(5): 789–94.

Stein, Z. A. (1999). Silicone breast implants: epidemiological evidence of sequelae. *Am J Public Health*, **89**(4): 484–7.

van Assen, S., Elkayam, O., Agmon-Levin, N., *et al.* (2011). Vaccination in adult patients with auto-immune inflammatory rheumatic diseases: a systematic literature review for the European League Against Rheumatism evidence-based recommendations for vaccination in adult patients with auto-immune inflammatory rheumatic diseases. *Autoimmun Rev*, **10**(6): 341–52.

Vasey, F.B., Zarabadi, S.A., Seleznick, M., and Ricca, L. (2003). Where there's smoke there's fire: the silicone breast implant controversy continues to flicker: a new disease that needs to be defined. *J Rheumatol*, **30**(10): 2092–4.

Vera-Lastra, O., Medina G., Cruz-Dominguez Mdel, P., *et al.* 2012. Human adjuvant disease induced by foreign substances: a new model of ASIA (Shoenfeld's syndrome). *Lupus*, **21**(2): 128–35.

Wolfram, D., Oberreiter, B., Mayerl, C., *et al.* (2008). Altered systemic serologic parameters in patients with silicone mammary implants. *Immunol Lett*, **118**(1): 96–100.

Wolfram, D., Rabensteiner, E., Grundtman, C., *et al.* (2012). T regulatory cells and TH17 cells in peri-silicone implant capsular fibrosis. *Plastic Reconstruct Surg*, **129**(2): 327e–37e.

Young, V.L., Nemecek, J.R., Schwartz, B.D., *et al.* (1995). HLA typing in women with breast implants. *Plastic Reconstruct Surg*, **96**(7): 1497–519; disc. 1520.

Zandman-Goddard, G., Blank, M., Ehrenfeld, M., *et al.* (1999). A comparison of autoantibody production in asymptomatic and symptomatic women with silicone breast implants. *J Rheumatol*, **26**(1): 73–7.

Zivkovic, I., Petrusic, V., Stojanovic, M., *et al.* (2012). Induction of decreased fecundity by tetanus toxoid hyper-immunization in C57BL/6 mice depends on the applied adjuvant. *Innate Immun*, **18**: 333–42.

9 Autoantibodies Induced by Vaccine

Nataša Toplak and Tadej Avčin

Department of Allergology, Rheumatology and Clinical Immunology, University Children's Hospital, University Medical Centre Ljubljana, Ljubljana, Slovenia

Introduction

Autoantibodies, in combination with a clinical picture, are a very helpful tool in establishing diagnosis of certain autoimmune diseases and can also be used to monitor disease activity. However, autoantibodies can also be found in healthy individuals and are not a pathognomonic sign of disease per se. Increased frequencies of autoantibodies have been reported in the healthy elderly population, as compared with healthy adults, presumably due to the loss of ability to recognize self and foreign antigens (Prelog, 2006); as a consequence, autoantibodies may be produced secondary to thymus involution in the elderly, with a decline of naive T cells and accumulation of clonal T cells. Moreover, it has also been shown that development of the immune system during childhood may produce autoantibodies (Avčin *et al.*, 2001). In a study of 61 apparently healthy children, frequency of IgG anticardiolipin antibodies (aCL) was similar to that found in healthy adults.

It is well known that autoantibodies can be induced by infection, and studies have also shown that the induction of autoantibodies is possible following vaccinations in animals, healthy people, and patients with autoimmune diseases. In this chapter, we review research data on autoantibody production and its clinical consequences following vaccination. Case presentations and case series reporting autoantibody induction in connection with autoimmune adverse events and diseases following vaccinations are mentioned only where relevant.

Autoantibodies induced by vaccines in animals

As in humans, induction of autoantibodies is possible following vaccination in animals. Autoantibodies are more frequently present in older than in younger dogs (Papini *et al.*, 2005). Consequently, when older dogs are vaccinated and tested for autoantibodies following vaccination, it is difficult to define what is a direct consequence of the vaccine and what has accumulated during life. In the first controlled experimental study to investigate the production of autoantibodies after routine vaccination, the researchers enrolled young dogs that had not been previously vaccinated (Hogenesch *et al.*, 1999). They showed that mandatory vaccination against rabies, canine distemper virus, and canine parvovirus 9 triggered the production of various autoantibodies. Five beagles were vaccinated with commercially available multivalent vaccine at 8, 10, 12, 16, and 20 weeks of age. The control group of five dogs received subcutaneous injections of sterile saline at the same time points. Blood samples were collected before each vaccination and 2, 5, 7, and 14 days after vaccination. Blood samples collected 7 days after vaccination were used for detection of autoantibodies. Two weeks after the last vaccination, there was an increase in the titer of IgG antibodies reactive with 10 of 17 antigens in the study group. A significant increase

Vaccines and Autoimmunity, First Edition. Edited by Yehuda Shoenfeld, Nancy Agmon-Levin, and Lucija Tomljenovic.

was observed for antibodies against laminin and fibronectin in all dogs in the study group after the last vaccination. The concentration of antifibronectin antibodies began to rise after the second vaccination in three of the dogs and after the third vaccination in the remaining two. Maximum level was reached after the fourth vaccination in all dogs. No increase was observed in the control group. The study was terminated 2 weeks after the last vaccination. At the time of study termination, there were no signs of autoimmune diseases in vaccinated dogs. However, the 2-week follow-up time interval was too short to conclude that there were no autoimmune adverse events following vaccination. Another study evaluated antithyroglobulin antibodies after routine vaccination in pet and research dogs (Scott-Moncrieff *et al.*, 2002) and found a significant increase in two groups of dogs that received the rabies vaccine.

Pathogenic anti-β_2-GPI antibodies were induced in mice vaccinated with tetanus toxoid (Blank *et al.*, 2002). It was shown that tetanus toxoid conformationally mimics hexapeptide TLRVYK epitope, against which pathogenic anti-β_2-GPI were directed. In this study, mice were also immunized with a panel of microbial preparations and studied for the development of anti-β_2-GPI antibodies. Pathogenic anti-β_2-GPI antibodies directed against the hexapeptide TLRVYK epitope were synthesized in mice that were immunized with *Haemophilus influenzae* or *Neisseria Gonorrhoeae*, which both exhibit the TLRVYK sequence. This study provided the first experimental evidence for an infectious origin of antiphospholipid syndrome (APS) induced by molecular mimicry. Another study focused on tetanus toxoid as a model for microbial antigen (Stojanović *et al.*, 2009). Detection of antibodies specific for tetanus toxoid, β_2-GPI, and laminin was performed at different time points after vaccination. In a group of BALB mice pretreated with complete Freund's adjuvant (CFA) and vaccinated with high doses of tetanus toxoid mixed with aluminum hydroxide as adjuvant, the concentrations of antilaminin IgG found 1 week after the last vaccine dose were significantly higher than those in normal sera or in other study groups. Preliminary screening for IgM and IgG anti-β_2-GPI antibodies showed the presence of these antibodies, but neither pretreatment nor adjuvants per se induced an increase in their levels. Affinity of anti-β_2-GPI IgG was tested. A statistically significant difference in dissociation profiles between identically pretreated groups was registered only after CFA pretreatment. It was also shown that hydrocarbon pristane and other hydrocarbons, namely mineral oil Bayol F, incomplete Freund's adjuvant (IFA), and squalene, which have been used in human and veterinary vaccines as adjuvants, can trigger induction of lupus-specific autoantibodies following an intraperitoneal injection in mice (Satoh *et al.*, 2003). Anti-nRNP/Sm and -Su autoantibodies were induced in 20 and 25% of mice, respectively.

Polyclonal hypergammaglobulinemia, autoantibody production, and vaccination-induced systemic autoimmunity have been found in farmed Atlantic salmon (Koppang *et al.*, 2008). In a controlled study, farmed Atlantic salmon were vaccinated with vaccines containing either mineral oil or animal/vegetable oil adjuvant, and in some cases also emulsifiers. Salmon in vaccinated groups (10 in the experimental group; all together, 154 salmon in 6 different groups of randomly vaccinated fish from salmon farms) received a single intraperitoneal injection of vaccine. A significant induction of antinuclear antigen (ANA) antibodies was found in vaccinated groups of salmon. In the experimental group, 60% of the vaccinated salmon were positive for ANA, whereas none of the unvaccinated controls had an ANA titer >1/160 (1/160 was the limit for a positive result). In the farmed groups, 5% of unvaccinated and 36–85% of vaccinated salmon from the six farms had ANA. No wild salmon were positive for ANA. The sera were also tested for rheumatoid factor (RF), antichromatin, anti-single-stranded (ss) and anti-double-stranded DNA (ds-DNA), antithyroglobulin, anti-β_2-GPI, and antiferritin. Higher levels of these autoantibodies were found in vaccinated salmon. Lupus-related specific autoantibodies, like those found in adjuvant oil-injected normal mice, were not detected in salmon. However, granulomatous inflammation similar to the peritoneal granuloma of mice with adjuvant oil-induced lupus, immune complex glomerulonephritis, and liver thrombosis were found in salmon. Later, the same group of researchers tested whether vaccination-induced autoantibody production in farmed Atlantic salmon was simply the result of polyclonal B cell activation (Satoh *et al.*, 2011). Total IgM levels and autoantibodies against salmon blood cell (SBC) extract were measured and the relationship between hypergammaglobulinemia and autoantibody production was analyzed. Both total IgM and anti-SBC antibodies were increased in vaccinated salmon, but associations were not always strong. In 155 vaccinated farmed fish, many samples had high IgM levels but relatively low anti-SBC levels, or vice versa. Antigen-specific mechanism could be speculated to explain this finding.

Autoantibodies induced by vaccine in apparently healthy people

Induction of autoantibodies following vaccinations in apparently healthy people was studied following hepatitis B and annual influenza vaccination in adults and following hepatitis B and hepatitis A vaccination in children (Belloni *et al.*, 2002; Martinuč Porobič *et al.*, 2005; Toplak *et al.*, 2008; Karali *et al.*, 2011). Autoantibodies were also tested in 1-year-old children after vaccinations recommended in the first year of life (tetanus, diphtheria, polio, mumps, rubella, measles, tuberculosis (BCG), *Haemophilus influenzae B* (Hib), and pertussis) (Wahlberg *et al.*, 2003). It was shown that induction of autoantibodies following hepatitis B, influenza, hepatitis A, BCG, and Hib vaccinations was possible, but was without clinical significance during the 6–12 months observation period following the vaccination. In a few cases, antibodies, mainly antiphospholipid antibodies, were elevated even 6 months after vaccination.

Induction of autoantibodies after hepatitis B vaccination was studied in a group of 85 medical students who were vaccinated with three doses of recombinant DNA hepatitis B vaccine (Martinuč Porobič *et al.*, 2005). ANA, anti-extractable nuclear antigens (anti-ENAs), IgG, and IgM aCL/anti-β_2-GPI antibodies were tested before and 1 and 6 months after the first dose. No significant changes were found in the levels of autoantibodies. However, three participants were transiently positive for autoantibodies 1 month after the first dose: in two participants, aCL levels reached medium positivity, with a drop 5 months later, and in one participant, a similar transient increase was observed for anti-β_2-GPI. In one participant, progressively increased levels of anti-β_2-GPI were found. None of the participants developed any signs of autoimmune disease 6 months after vaccination.

Induction of autoantibodies after influenza vaccination was studied in a group of 92 apparently healthy medical workers who were vaccinated with annual influenza vaccine without adjuvant (Toplak *et al.*, 2008). ANA, anti-ENA, IgG, IgM, IgA aCL/anti-β_2-GPI antibodies, and lupus anticoagulants (LAs) were tested before and 1 and 6 months after vaccination. No significant changes were found in the levels of autantibodies and none of the participants developed any signs of autoimmune disease 6 months after vaccination. However, 1 and 6 months after vaccination, 15 and 13% of participants, respectively, demonstrated increased levels of autoantibodies or the appearance of new antibodies. Persistently elevated levels of autoantibodies were observed in 8% of participants, and two participants showed progressively increased levels of IgM aCL or IgA anti-β_2-GPI, while 11 participants had a transient increase in autoantibodies.

Although this chapter deals with autoantibody induction after vaccination, it is worth mentioning that, in healthy adults, ASO3 squalene-based adjuvanted influenza vaccine, in contrast with nonadjuvanted influenza vaccine, can trigger high nonspecific T lymphocyte responses (Korošec *et al.*, 2012). Although no autoimmune adverse events were reported in the study that monitored the absolute count of T cells and immunogenicity after vaccination with the 2009 adjuvanteded H1N1 vaccine and the 2010 nonadjuvanted H1N1/H3N2/B-Brisbane vaccine, there was a significant increase in local and short-term adverse events following the administration of adjuvanted influenza vaccine compared to the nonadjuvanted vaccine.

Autoantibody induction was studied in healthy children following hepatitis B and hepatitis A vaccination. Induction of autoantibodies 6 years after hepatitis B vaccination was studied in 210 children immunized at birth with recombinant hepatitis B vaccine (Belloni *et al.*, 2002). ANA, anti-DNA, antimitochondrial, anti-liver/kidney microsomal, antireticulin, anti-smooth muscle, and antiribosomal antibodies were tested. The presence of antithyroid antibodies (ant-thyroglobulin and antiperoxidase) and antibodies found in type 1 diabetes (T1D) (thyrosine phosphtase, IA-2A; glutamic acid decarboxylase, GADA) were also tested. The presence of autoantibodies was compared against a group of 109 unvaccinated children: no significant differences in the levels of autoantibodies between the two groups and no signs of autoimmunity were found. Induction of autoantibodies after hepatitis A vaccination was studied in 40 healthy children (mean age 6.3 years, range 2–18 years) (Karali *et al.*, 2011), who were vaccinated with two doses of the hepatitis A vaccine within a 6-month interval. ANA, anti-smooth muscle antibodies (ASMAs), anti-nDNA, antithyroid microsomal antibodies, IgG and IgM aCL, anti-dsDNA, and antineutrophil cytoplasmic antibody (ANCA) profiles were tested before vaccination, 1 month after the first dose, and 1 month after the second dose. During the study period, 10 children were positive for one or two autoantibodies. Three children were transiently positive for ANA and two were persistently positive for ANA. One of the two persistently ANA-positive children was also transiently positive for ANCA. Four children were transiently positive for IgM or IgG aCL antibodies and one child was transiently positive for antithyroid microsomal antibodies. At the end of the study, after 12 months, none of the children had signs of autoimmune disease. Two children continued to be ANA-positive at the same titer as after the first vaccine dose.

Induction of autoantibodies following vaccinations in the first year of life (tetanus, diphtheria, polio, mumps, rubella, measles, BCG, Hib, pertussis) was studied in the ABIS (All Babies in Southeast Sweden) study – a prospective cohort study that followed an unselected birth cohort of the general population (Wahlberg *et al.*, 2003). Not all included children were vaccinated with all recommended vaccines; in particular, BCG, Hib, and pertussis vaccinations were frequently missed. The blood samples of 4400 randomly selected 1-year-old children were tested for T1D-related autoantibodies, namely antibodies against GADA and against IA-2A. These two autoantibodies were shown to be good markers for the disease process sometimes leading to T1D. A 99th percentile cut-off for positivity of IA-2A levels was used in connection with BCG vaccination and a 90th percentile cut-off for positivity of IA-2A and GADA levels in connection with Hib vaccination. The study showed that BCG and Hib vaccinations may induce T1D-related autoantibodies in 1-year-old children. Other vaccinations were not related to the appearance of GADA and IA-2A. However, it is uncertain whether this early stimulation of the immune system has any link to the risk of T1D. Epidemiological studies to date have not proved the connection between Hib vaccination and T1D (Kravonen *et al.*, 1999; DeStefano *et al.*, 2001).

A summary of antibodies that were found to be induced after various vaccinations is presented in Table 9.1.

Autoantibodies induced by vaccine in patients with autoimmune diseases

Induction of autoantibodies in patients with autoimmune diseases was studied following annual influenza and pneumococcal vaccinations and vaccination against human papilloma virus (HPV) in adults (Abu-Shakra *et al.*, 2002, 2007; Tarjan *et al.*, 2002, 2006; Elkayam *et al.*, 2005; Del Porto *et al.*, 2006; Urowitz *et al.*, 2011; Perdan-Pirkmajer *et al.*, 2012; Vista *et al.*, 2012; Mok *et al.*, 2013; Pasoto *et al.*, 2013). Autoantibody induction after HPV vaccination was also studied in adolescent girls with systemic lupus erythematosus (SLE) (Soybilgic *et al.*, 2013). In children, autoantibody induction was studied after annual influenza and pneumococcal vaccinations (Kostinov *et al.*, 2009; Aikawa *et al.*, 2011; Toplak *et al.*, 2012).

Induction of autoantibodies following influenza vaccination was mainly studied in patients with rheumatoid arthritis (RA), SLE, Sjögren syndrome, and mixed connective tissue disease (MCTD) (Abu-Shakra *et al.*, 2002, 2007; Tarjan *et al.*, 2002, 2006; Del Porto *et al.*, 2006; Urowitz *et al.*, 2011; Perdan-Pirkmajer *et al.*, 2012; Vista *et al.*, 2012; Miossi *et al.*, 2013; Pasoto *et al.*, 2013).

In RA patients, two studies investigating induction of autoantibodies after influenza vaccination were published (Del Porto *et al.*, 2006; Perdan-Pirmajer *et al.*, 2012). In the first, only 10 RA patients were included, together with 14 SLE patients (Del Porto *et al.*, 2006). ANA, anti-dsDNA, anti-ENA, RF, aCL, anti-β_2-GPI, and LA were tested before and 90 days after influenza vaccination. No significant changes were detected in either RA or SLE patients. In the second, a prospective cohort study with 6-month follow-up, 218 patients with autoimmune inflammatory rheumatic disease (AIRD) were included (176 RA, 14 psoriatic arthritis (PsA), 16 ankylosing spondylitis (AS), and 12 with other autoimmune diseases) (Perdan Pirkmajer *et al.*, 2012). Among them, 50 were vaccinated against seasonal influenza, 6 against H1N1 pandemic influenza, 104 against both, and 58 were nonvaccinated controls. ANA, anti-ENA, aCL, and anti-β_2-GPI were tested before each vaccination, 1 month after the last vaccination, and 6 months after inclusion. After administration of seasonal influenza vaccine, transient changes in the levels of autoantibodies were detected in patients with AIRD and in healthy controls. A tendency toward anti-ENA development was detected in patients who were ANA-positive before vaccination. Induction of IgM and IgG aCL was significantly higher after vaccination with pandemic influenza vaccine containing the squalene adjuvant than after vaccination against seasonal influenza without an adjuvant. The changes were mainly transient. It is interesting to note that in this study, the long-term effect of vaccination was significantly lower IgG aCL in

Table 9.1 Reported induction of autoantibodies following various vaccinations in selected apparently healthy persons

Vaccine/Ab	ANA	ANCA	aCL	Anti- β_2GPI	LA	GADA	IA-2A
Influenza	+	+	+	+	+		
Hepatitis B	+		+	+			
Hepatitis A	+	+	+				
Hib						+	+
BCG							+

Ab, antibody; ANA, antinuclear antigen antibody; ANCA, antineutrophil cytoplasmic antibody; aCL, anticardiolipin antibody; anti-β_2GPI, anti-β_2 glycoprotein I antibody; LA, lupus anticoagulant; GADA, glutamic acid decarboxylase; IA-2A, thyrosine phosphtase; Hib, *Haemophilus influenzae* B; BCG, bacille Calmette–Guérin

most patients after 6 months. However, almost all vaccinated patients in this cohort were RA patients, and there were only three SLE patients.

In SLE patients, several studies have been published. The most inducible autoantibodies following influenza vaccination are aCL and anti-β_2-GPI (Abu-Shakra *et al.*, 2002, 2007; Tarjan *et al.*, 2002, 2006; Elkayam *et al.*, 2005; Del Porto *et al.*, 2006; Urowitz *et al.*, 2011; Vista *et al.*, 2012). In a study of 24 women with SLE, anti-dsDNA, aCL, anti-Sm, anti-Sm/RNP, anti-Ro, and anti-La were tested before and 6 and 12 weeks after influenza vaccination (Abu-Shakra *et al.*, 2002, 2007). There was mainly transient induction of anti-Sm, anti-RNP, anti-Ro, and anti-La antibodies. However, two patients who were negative for anti-La before vaccination were positive after 12 weeks, and aCL induction for a prolonged period was observed. A higher percentage of positive sera was also present 12 weeks after vaccination. No thrombotic events were observed in this study. The induced autoantibodies did not have any clinical significance. In another study, which included 18 SLE patients with mild disease activity, anti-dsDNA, aCL, and anti-β_2-GPI antibodies were tested before and 4 weeks after a third annually repeated influenza vaccination (Tarjan *et al.*, 2006). Eight patients were positive for aCL and/or anti-β_2-GPI before vaccination (aPL-positive group) and 10 were negative (aPL-negative group). Four weeks after vaccination, IgG aCL levels decreased significantly in both groups. However, levels of anti-β_2-GPI significantly increased during the same period in the aPL-positive group. The measurement of anti-dsDNA revealed significant elevation, especially in the aPL-positive group, without clinical side effects. In contrast, another recently published study which included 101 SLE patients and the same number of healthy controls showed that aCL, but not anti-β_2-GPI, can be induced after influenza vaccination (Vista *et al.*, 2012). Patients were tested for aCL before and 2, 6, and 12 weeks after vaccination. However, only patients positive for aCL were further tested for anti-β_2-GPI antibodies. Patients with SLE (12/101) and healthy controls (7/101) developed new aCL after vaccination. No new anti-β_2-GPI was detected among patients who developed aCL after vaccination. In another study, which included 103 SLE patients, the possible difference in inducing autoantibodies between the adjuvanted and nonadjuvanted influenza vaccine was tested; 51 patients received the adjuvanted and 52 the nonadjuvanted influenza vaccine (Urowitz *et al.*, 2011). RF, ANA, anti-dsDNA, anti-RNP, anti-Sm, anti-Ro, anti-La, anti-Scl-70, and anti-Jo-1 were tested before and 1 and 3 months after vaccination. The percentage of patients with changes in antibodies after vaccination was mostly not significant. Only RF and anti-Ro showed significant changes in frequency between visits. However, the actual mean values did not change significantly. There was no difference between the adjuvanted and the nonadjuvanted vaccine.

Induction of autoantibodies after annual influenza vaccination was studied in 36 patients with Sjögren syndrome and 36 healthy controls (Pasoto *et al.*, 2013). The aim was to evaluate the influence of the nonadjuvanted A/California/7/2009/H1N1-like virus vaccine on autoantibody profile and on clinical manifestations. ANA, anti-dsDNA, anti-RNP/anti-Sm, anti-Ro(SSA)/La(SSB), RF, anti-alpha-fodrin, aCL, and anti-β_2-GPI were tested before and 21 days and 1 year after vaccination. No short-term changes in the levels of autoantibodies were detected, but during 1-year follow-up, four patients developed positivity to anti-Ro/anti-La, anti-alpha-fodrin, or IgM aCL. None developed SLE-specific autoantibodies. A significant increase in the mean levels of anti-Ro/anti-La was detected

1 year after influenza vaccination, with no change in other autoantibodies. The induced autoantibodies were not clinically relevant.

Induction of autoantibodies after annual influenza vaccination was studied in 69 MCTD patients who were vaccinated with nonadjuvant influenza vaccine (Miossi *et al.*, 2013). Anti-RNP were tested prior to and 21 days after vaccination. No significant changes were detected in the levels of anti-RNP after vaccination.

Induction of autoantibodies after pneumococcal vaccination was investigated in two studies of SLE patients (Tarjan *et al.*, 2002; Elkayam *et al.*, 2005). The first included 18 SLE patients (Tarjan *et al.*, 2002) and tested ANA and anti-dsDNA before and 4 weeks after vaccination. No changes were detected. The second included 24 SLE patients (Elkayam *et al.*, 2005) and tested anti-dsDNA, aCL, anti-Sm, anti-RNP, anti-Ro/SSA, and anti-La/SSB before and 2 months after vaccination. No significant changes were detected, but one patient developed IgG aCL antibodies and another turned anti-RNP-negative.

Induction of autoantibodies after HPV vaccination was studied in 50 females with SLE (Mok *et al.*, 2013). Anti-dsDNA and anti-C1q were tested. No significant changes in the levels of autoantibodies were found during an observation period of 12 months.

Induction of autoantibodies after HPV vaccination was also studied in 27 female SLE patients, including children (mean age 20.5 years, range 12–26 years) (Soybilgic *et al.*, 2013). Anti-ds-DNA, aCL, anti-β_2-GPI, LA, anti-RNP, and anti-Smith antibodies were evaluated before and 7 months after vaccination. Two patients who were negative for LA before vaccination developed LA 7 months after vaccination. One patient, who was initially positive for LA, became negative 7 months after vaccination. None of the patients developed *de novo* anti-RNP, anti-Sm, aCL, or anti-β_2-GPI antibodies during the observation period of 7 months. There was no significant change in anti-ds-DNA antibody titers post-vaccination.

ANCA-associated vasculitis following influenza vaccination has been reported in eight cases so far. In three, ANCA-associated vasculitis appeared *de novo* (Uji *et al.*, 2005; Birck *et al.*, 2009; Duggal *et al.*, 2013). Data from studies of ANCA induction after influenza vaccination are lacking: to date, ANCA induction following influenza vaccination has been studied only in 31 children with juvenile idiopathic arthritis (JIA), showing no changes during an observation period of 6 months (Toplak *et al.*, 2012). ANCA was studied following hepatitis A vaccination in 40 healthy children. In one case, a transient increase in cytoplasmic ANCA (c-ANCA) was noticed, without clinical significance (Karali *et al.*, 2011).

In children, autoantibody induction following annual influenza vaccination has been studied mainly in patients with JIA, although a few children with other autoimmune diseases were included in one study (Kanakoudi-Tsakalidou *et al.*, 2001; Aikawa *et al.*, 2011; Toplak *et al.*, 2012). Another study was performed in children with T1D (Kostinov *et al.*, 2009); autoantibodies after vaccination against pneumococcal and influenza infections were determined. In a study of 70 children with different rheumatic diseases (49 JIA, 11 SLE, 10 others), no significant changes were found in the levels of autoantibodies (Kanakoudi-Tsakalidou *et al.*, 2001). ANA, anti-DNA, anti-ENA, RF, SS-A, and SS-B were tested. Blood samples were collected before and 1 month after vaccination. In a prospective controlled longitudinal cohort study of 31 children with JIA, receiving various therapies, ANA, anti-ENA, ANCA, IgG/IgM aCL, IgG/IgM/IgA anti-β_2-GPI, and LA were tested before and 1 and 6 months following annual influenza vaccination (Toplak *et al.*, 2012); this study took place in 2008/09. Children in the JIA and control group were vaccinated with nonadjuvanted influenza vaccine. Viral infections were followed during an observation period of 6 months following vaccination by nasal and oral swabs taken during febrile illness. No significant changes were found in the number of children with positive autoantibodies before and 1 and 6 months after vaccination in the JIA or control group. However, a tendency for progressively increased mean values of IgG aCL after influenza vaccination was found in the JIA group, but the difference before and 6 months after vaccination did not reach statistical significance ($p = 0.05$). Before vaccination, four children in the JIA group and none in the control group were positive for ANA; 1 month after vaccination, seven children with JIA were positive for ANA. In three children, ANA developed *de novo*; two children developed low-titer ANA 1 : 80; and in one child, ANA was positive in a titer 1 : 320 (author's unpublished data). ANA titers in children who were positive before vaccination remained the same. One child in the control group, who was negative for ANA before vaccination, developed ANA titer 1 : 80 1 month after vaccination, but he was infected with influenza A virus 10 days following vaccination. Infection was proven by virus isolation from a nasal and oral swab. ANA was positive in four children 6 months after vaccination. One child who was negative for ANA before and 1 month after vaccination developed ANA titer 1 : 80 *de novo*. One child who had ANA 1 : 80 before and 1 month after vaccination remained positive at the same titer. Two children who were negative for ANA before vaccination and positive 1 month after vaccination were persistently positive for ANA 6 months after vaccination. However, in one child the ANA titer was lower after 6 months (1 : 160) compared to 1 month after vaccination (1 : 320). A child in the control group who was infected with influenza A virus and developed ANA 1 month after vaccination was negative for ANA 6 months after vaccination. All children in the study were negative for anti-ENA and ANCA before and 1 and 6 months after vaccination (Toplak *et al.*, 2012). Before vaccination, three children in the JIA group and three in the control group were positive for IgG aCL. All but one child in the control group, whose levels of aCL were intermediately positive, were low-positive for aCL. Six children in the JIA group were positive for aCL 1 month after vaccination, while four developed aCL *de novo*. Two patients who were positive before vaccination remained positive 1 month after vaccination. In one of them, the levels of aCL were higher than before vaccination. In the control group, the number of children who were positive for aCL before vaccination remained the same 1 month after vaccination. In the JIA group, 10 children were low-positive for aCL, four developed aCL *de novo*, and six were persistently positive 6 months after vaccination (author's unpublished data). However, even in the control group, two children developed *de novo* aCL 6 months after vaccination, but two who were only transiently positive for aCL after 1 month were negative after 6 months, and the number of children with positive aCL after 6 months remained the same as before vaccination. The mean value of IgG aCL after influenza vaccination progressively increased in the JIA group during the observation period of 6 months (Figure 9.1). No such effect was observed in the control group. No connection between infections and aCL presence was found. Anti-β_2-GPI was not induced by vaccination, but three children with JIA were transiently positive for LA 1 month after vaccination. Induced antibodies (ANA, aCL) and LA were not clinically significant.

In a study of children with new-onset JIA who were not vaccinated, the opposite trend was noted for aCL during 12 months of study (Avčin *et al.*, 2002). At study entry, 46% of the children were positive for aCL, mainly IgG aCL; after 12 months, only 28% of the children were still positive. However, a direct comparison between these two studies is not possible. In the study of influenza vaccination in children with JIA, mean disease duration at study entry was 4.3 years (Toplak *et al.*, 2012), while the study of children with JIA who were not vaccinated included participants at disease onset (Avčin *et al.*, 2002). In neither of these two studies were changes in the anti-β_2-GPI levels noticed. This is one possible explanation for the limited prothrombotic potential of aCL in children with JIA. ACL induced by influenza vaccination in children with JIA was not clinically important (Toplak *et al.*, 2012).

Another study that tested autoantibodies in children with JIA found no significant changes in the levels of autoantibodies (Aikawa *et al.*, 2011). ANA, aCL, and RF were tested in 58 JIA patients before and 3 weeks after pandemic influenza vaccination. One patient became low-positive for IgG aCL antibody. Autoantibodies induction was studied in 100 children (2–18 years) with T1D

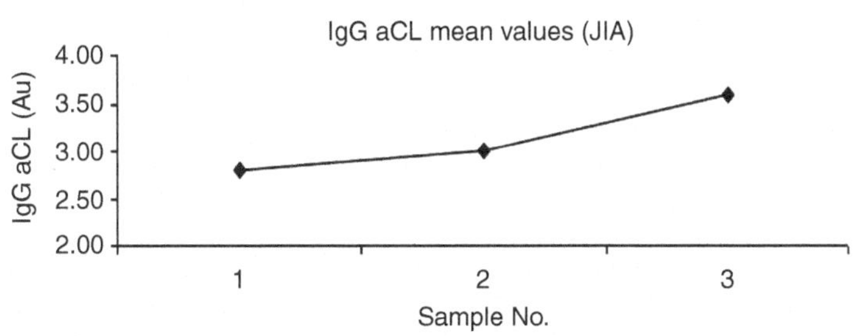

Figure 9.1 Mean values of IgG aCL before and 1 and 6 months after influenza vaccination in patients with juvenile idiopathic arthritis (JIA). The difference between the mean values of IgG aCL before and 6 months after the vaccination was not statistically significant (p = 0.05) (Author's unpublished data). aCL, anticardiolipin antibodies; AU, arbitrary units; 1, sample before vaccination; 2, sample 1 months after vaccination; 3, sample 6 months after vaccination. (For a color version of this figure, please see color plate section.)

Table 9.2 Reported induction of autoantibodies following various vaccinations in selected patients with autoimmune diseases

Vaccine/Ab	ANA	Anti-dsDNA	Anti-ENA	ANCA	aCL	Anti- β_2GPI	LA	Anti-Sm/RNP	Anti-Ro	Ani-La	RF
Influenza[a]	+	+	+	+	+	+	+	+	+	+	+
Pneumococcus					+						
HPV							+				

[a]Influenza vaccine was the most frequently studied for induction of autoantibodies; others were tested in only a few studies
Ab, antibody; ANA, antinuclear antigen antibody; anti-dsDNA, anti-double-stranded DNA; anti-ENA, antibody against extractable nuclear antigens; ANCA, antineutrophil cytoplasmic antibody; aCL, anticardiolipin antibody; anti-β_2GPI, anti-β_2 glycoprotein I antibody; LA, lupus anticoagulant; RF, rheumatoid factor; HPV, human papilloma virus

after pneumococcal vaccination with Pneumo 23 vaccine (Kostinov *et al.*, 2009). Of these, 28 were also vaccinated with annual influenza vaccine. Anti-dsDNA and antibodies against pancreatic and adrenal tissue were tested before and 1 year after vaccination. No significant increase in the levels of autoantibodies was noticed; on the contrary, a decrease of levels to normal was observed in at least half of patients who tested positive before vaccination.

A summary of the antibodies that were found to be induced following various vaccinations is presented in Table 9.2.

Conclusions

Studies that have investigated autoantibody production following various vaccinations show that induction of various antibodies is possible, but so far none has been able to prove that the induced antibodies have any clinical consequence. In contrast, several case reports and case series report patients developing autoantibodies and autoimmune adverse events following various vaccinations, emphasizing that clinical consequences could develop in isolated cases and should not be overlooked. Putting all the published evidence together, it seems prudent to conclude that there are certain individuals, most likely with a genetic predisposition, who could develop autoimmune adverse events or diseases following vaccination under certain unfavorable circumstances, including infection, trauma, psychical stress, and any other event that might disturb the complex immune system balance.

We must bear in mind that autoimmunity, including production of autoantibodies following infections, vaccinations, and other environmental triggers, is a feature of a healthy immune system and that, fortunately, the immune system has developed safety mechanisms to prevent the development of an overt autoimmune disease in the majority of cases. The long-term clinical consequences of autoantibodies induced by vaccinations are not yet known. Due to several environmental factors with influence over the immune system, it seems at present almost impossible to define the importance of vaccine-induced autoantibodies several years after vaccination.

References

Abu-Shakra, M., Press, J., Sukenik, S., and Buskila, D. (2002). Influenza virus vaccination of patients with SLE: effects on generation of autoantibodies. *Clin Rheumatol*, **21**: 369–72.

Abu-Shakra, M., Press, J., Buskila, D., and Sukenik, S. (2007). Influenza vaccination of patients with systemic lupus erythematosus: safety and immunogenecity issues. *Autoimmun Rev*, **6**: 543–6.

Aikawa, N.E., Goldenstein-Schainberg, C., Vendramini, M., *et al.* (2011). Autoimmune response following influenza H1N1 vaccination in patients with juvenile idiopathic arthritis. *Pediatric Rheumatol*, **9**(Suppl. 1): P133.

Avčin, T., Ambrozic, A., Kuhar, M., *et al.* (2001). Anticardiolipin and anti-beta2 glycoprotein I antibodies in sera of 61 apparently healthy children at regular preventive visits. *Rheumatology*, **40**: 565–73.

Avčin, T., Ambrožič, A., Božič, B., *et al.* (2002). Estimation of anticardiolipin antibodies, anti-beta2 glycoprotein I antibodies and lupus anticoagulant in a prospective longitudinal study of children with juvenile idiopathic arthritis. *Clin Exp Rheumatol*, **20**: 101–8.

Belloni, C., Avanzino, M.A., De Silvestri, A., *et al.* (2002). No evidence of autoimmunity in 6-year old children immunized at birth with recombinant hepatitis B vaccine. *Pediatrics*, **110**: e4.

Blank, M., Krause, I., Fridkin, M., *et al.* (2002). Bacterial induction of autoantibodies to beta2-glycoprotein-I accounts for the infectious etiology of antiphospholipid syndrome. *J Clin Invest*, **109**: 797–804.

Birck, R., Kaelsch, I., Schnuelle, P., *et al.* (2009). ANCA-associated vasculitis following influenza vaccination: causal association or mere coincidence? *J Clin Rheumaol*, **15**: 289–91.

Del Porto, F., Lagana, B., Biselli, R., *et al.* (2006). Influenza vaccine administration in patients with systemic lupus erythematosus and rheumatoid arthritis. Safety and immunogenicity. *Vaccine*, **24**: 3217–23.

DeStefano, F., Mullooly, J.P., Okoro, C.A., *et al.* (2001). Childhood vaccinations, vaccination timing and risk of type I diabetes mellitus. *Pediatrics*, **108**: E112.

Duggal, T., Segal, P., Shah, M., *et al.* (2013). Antineutrophil cytoplasmic antibody vasculitis associated with influenza vaccination. *Am J Nephrol*, **38**: 174–8.

Elkayam, O., Paran, D., Burke, M., *et al.* (2005). Pneumococcal vaccination of patients with systemic lupus erythematosus: effects on generation of autoantibodies. *Autoimmunity*, **38**: 493–6.

Hogenesch, H., Azcona-Olivera, J., Scott-Montcrieff, C., *et al.* (1999). Vaccine induced autoimmunity in the dog. *Adv Vet Med*, **41**: 733–47.

Kanakoudi-Tsakalidou, F., Trachana, M., Pratsidou-Gertsi, P., *et al.* (2001). Influenza vaccination in children with chronic rheumatic disease and long-term immunosuppressive therapy. *Clin Exp Rheumatol*, **19**: 589–94.

Karali, Z., Basaranoglu, S.T., Karali, Y., *et al.* (2011). Autoimmunity and hepatitis A vaccine in children. *J Investig Allergol Clin Immunol*, **21**: 389–93.

Koppang, E.O., Bjerkås, I., Haugarvoll, E., *et al.* (2008). Vaccination-induced systemic autoimmunity in farmed Atlantic salmon. *J Immunol*, **181**: 4807–14.

Korošec, P., Petrovec, M., Lainšč scak, M., *et al.* (2012). High non-specific T lymphocyte response to the adjuvant H1N1 vaccine in comparison with the H1N1/H3N2/B-Brisbane vaccine without adjuvant. *Scand J Immunol*, **76**: 497–504.

Kostinov, M.P., Skochilova, T.V., Vorob'eva, V.A., *et al.* (2009). [Autoantibodies after vaccination against pneumococcal and influenza infections in children and adolescents with type I diabetes mellitus.] *Zh Mikrobiol Epidemiol Immunobiol*, **2**: 53–7.

Kravonen, M., Cepaitis, Z., and Tuomilehto, J. (1999). Association between type 1 diabetes and *Haemophilus influezae* type B vaccination: birth cohort study. *BMJ*, **318**: 1169–72.

Martinuč, P., Jana, A.T., Božič, B., *et al.* (2005). Anti-phospholipid antibodies following vaccination with recombinant hepatitis B vaccine. *Clin Exp Immunol*, **142**: 377–80.

Miossi, R., Fuller, R., Moares, J.C., *et al.* (2013). Immunogenicity of influenza vaccination in mixed connective tissue disease: effect of disease and therapy. *Clinics*, **68**: 129–34.

Mok, C.C., Ho, L.Y., Fong, L.S., and To, C.H. (2013). Immunogenicity and safety of a quadrivalent human papillomavirus vaccine in patients with systemic lupus erythematosus: a case control study. *Ann Rheum Dis*, **72**: 659–64.

Papini, R., Pellegrini, O., and Guidi, G. (2005). Anticardiolipin antibodies in healthy and diseased dogs. *Vet Rec*, **4**(156): 732–4.

Pasoto, S.G., Ribeiro, A.C., Viana, V.S., *et al.* (2013). Short and long-term effects of pandemic unadjuvant influenza A(H1N1)pdm09 vaccine on clinical manifestations and autoantibody profile in primary Sjögren's syndrome. *Vaccine*, **31**: 1793–8.

Perdan-Pirkmajer, K., Thallinger, G.G., Snoj, N., *et al.* (2012). Autoimmune response following influenza vaccination in patients with autoimmune inflammatory rheumatic disease. *Lupus*, **21**: 175–83.

Prelog, M. (2006). Aging of the immune system: a risk factor for autoimmunity? *Autoimmun Rev*, **5**: 136–9.

Satoh, M., Kuroda, Y., Yoshida, H., *et al.* (2003). Induction of lupus autoantibodies by adjuvants. *J Autoimmun*, **21**: 1–9.

Satoh, M., Bjerkås, I., Haugarvoll, E., *et al.* (2011). Polyclonal hypergammaglobulinemia and autoantibody production induced by vaccination in farmed Atlantic salmon. *Fish Shellfish Immunol*, **30**: 1080–6.

Scott-Moncrieff, J.C., Azcona-Olivera, J., Glickman, N.W., *et al.* (2002). Evaluation of antithyroglobulin antibodies after routine vaccination in pet and research dogs. *J Am Vet Med Assoc*, **221**: 515–21.

Soybilgic, A., Onel, K.B., Utset, T., *et al.* (2013). Safety and immunogenicity of the quadrivalent HPV vaccine in female Systemic Lupus Erythematosus patients aged 12 to 26 years. *Pediatr Rheumatol Online J*, **11**: 29.

Stojanović, M., Živković, I., Inić-Kanda, A., *et al.* (2009). The context of tetanus toxoid application influences the outcome of antigen-specific and self-directed humoral immune response. *Microbiol Immunol*, **53**: 89–100.

Tarjan, P., Sipka, S., Marodi, L., *et al.* (2002). No short-term immunological effects of Pneumococcus vaccination in patients with systemic lupus erythemtosus. *Scand J Rheumatol*, **31**: 211–15.

Tarjan, P., Sipka, S., Lakos, G., *et al.* (2006). Influenza vaccination and the production of anti-phospholipid antibodies in patients with systemic lupus erythematosus. *Scand J Rheumatol*, **35**: 241–9.

Toplak, N., Kveder, T., Trampuš-Bakija, A., *et al.* (2008). Autoimmune response following annual influenza vaccination in 92 apparently healthy adults. *Autoimmun Rev*, **8**: 134–8.

Toplak, N., Šubelj, V., Kveder, T., *et al.* (2012). Safety and efficacy of influenza vaccination in a prospective longitudinal study of 31 children with juvenile idiopathic arthritis. *Clin Exp Rheumatol*, **30**: 436–44.

Uji, M., Matsushita, H., and Iwata, S. (2005). Microscopic polyangiitis after influenza vaccination. *Intern Med*, **44**: 892–6.

Urowitz, M.B., Anton, A., Ibanez, D., and Gladman, D.D. (2011). Autoantibody response to adjuvant and non-adjuvant H1N1 vaccination in systemic lupus erithematosus. *Arthritis Care Res (Hoboken)*, **63**: 1517–20.

Vista, E.S., Crowe, S.R., Thompson, L.F., *et al.* (2012). Influenza vaccination can induce new-onset anticardiolipins but not β2-glycoprotein-I antibodies among patients with systemic lupus erythematosus. *Lupus*, **21**: 168–74.

Wahlberg, J., Fredriksson, J., Vaarala, O., *et al.* (2003). Vaccinations may induce diabetes-related autoantibodies in one-year-old children. *Ann NY Acad Sci*, **1005**: 404–8.

10 The ASIA Syndrome Registry

Ignasi Rodriguez-Pintó[1] and Yehuda Shoenfeld[2,3]
[1]Department of Autoimmune Disease, Hospital Clínic de Barcelona, Barcelona, Spain
[2]Zabludowicz Center for Autoimmune Diseases, Sheba Medical Center, Tel Hashomer, Israel
[3]Incumbent of the Laura Schwarz-Kipp Chair for Research of Autoimmune Diseases, Sackler Faculty of Medicine, Tel Aviv University, Tel Aviv, Israel

Introduction

The autoimmune/inflammatory syndrome induced by adjuvants (ASIA) was defined by Shoenfeld and Agmon-Levin (2011) as a syndrome in which exposure to an adjuvant leads to the development of a disease characterized by a hyperactive immune response. An adjuvant is defined as a substance that enhances an antigen-specific immune response, preferably without triggering one of its own. (Although several studies in animal models and humans have demonstrated the ability of what we now recognize as adjuvants to trigger an autoimmunity phenomenon and autoimmune disease themselves: Israeli *et al.*, 2009; Cruz-Tapias *et al.*, 2013; Perricone *et al.*, 2013).

A registry per se is not a study. It is an organized collection of data related to patients with a specific diagnosis, or other health conditions assembled and stored in insured repositories. Although the most ancient form of registration of clinical cases is a handwriting tabulation, modern registries are files stored in computers that collect case information.

Registries are often the first approach to a new disease or area of inquiry (Grimes and Schulz, 2002a). For instance, the first report of AIDS was the case-series report of pneumocystis pneumonia in Los Angeles (CDC, 1981). Even when the editors of journals that picked up the torch for evidence-based medicine fall out of love with descriptive studies, they continue to play several important roles in medical research. Indeed, the case series has been used intuitively by physicians for over 2 centuries and continues to have a role in defining new disease. Patient registries are the only means in rare disease by which to pool data in order to achieve a sufficient sample size for epidemiological and/or clinical research. For instance, analysis of catastrophic antiphospholipid syndrome (CAPS) registry data enabled the characterization of the clinical and laboratory features of patients with CAPS, as well as the establishment of preliminary criteria for its classification and guidelines for its management (Cervera *et al.*, 2007). Further, registries have been used to monitor health conditions in the general population in several countries (Carstensen and Borch-Johnsen, 2011).

Sometimes, registries are seen as primitive forms of case–control studies, in which the controls are only implied (Cummings and Weiss, 1998; Grimes and Schulz, 2002a). In a registry, the cases and their exposures are described explicitly, while the frequency of exposure in non-cases is implied but not collected (Cummings and Weiss, 1998). The implied control group is everyone else in the target population (i.e. the population from which the cases were acquired). The information gained is used to generate hypotheses that can be further analyzed in observational studies.

ASIA is probably an underreported disease, due to general unawareness of its existence and failure to attribute symptoms to it. The ASIA Registry gives us a channel through which to highlight

Vaccines and Autoimmunity, First Edition. Edited by Yehuda Shoenfeld, Nancy Agmon-Levin, and Lucija Tomljenovic.

the importance of the syndrome to the scientific community.

Purpose

The ASIA Registry was built to provide clues about the clinical picture of ASIA across the world and to better define the disease and its causes.

The first goal is to describe the disease itself. This will help build up preliminary criteria, which can then be further validated. The registry will then allow us to define the demographic features of patients who develop the syndrome, including sex, age at presentation, and smoking history. Next, we can define the syndrome's epidemiology. This will provide an idea of the prevalence of ASIA in the general population, which can be contrasted with the information supplied through public databases. Lab results will show any possible association with the clinical manifestations. Finally, and hopefully, the ASIA Registry will point us toward possible treatments, in order to improve patient quality of life.

Structure

Using definitions of ASIA established in previous studies (Agmon-Levin *et al.*, 2012; Perricone *et al.* 2013), and in agreement with the ASIA Syndrome Registry Group, the set of variables to be collected was listed. A software package was licensed and used to build a new database Web site, containing appropriate privacy and security features. On 1 December 2014, the ASIA Registry became fully operational. It can be found at https://ontocrf.costaisa.com/web/asia/home1.

Only cases reported by physicians are accepted, but any non-physician with a case to report is encouraged to contact a physician and ask them to fill up the standardized form, which can be sent on request to any physician. The Web site contains information on how to get in touch with the ASIA Syndrome Registry Project Group; submissions should be sent by email to ASIASyndromeRegistry@gmail.com.

Since one of the objectives of this registry is to validate proposed criteria, no strict criteria have yet been established. Every submitted case is reviewed and, if the ASIA Registry Project Group members agree it qualifies, included in the database.

Anyone can consult the registry and see the results. However, only ASIA Registry Project Group members can log in and gain access to the raw data. The main table includes demographic data, including previous personal medical and family history; allergic history, including all reported allergies and known allergens; the variables defining the frame of the disease; information on exposure to adjuvants, including date of exposure; major histocompatibility complex (MHC) data and laboratory tests and biopsies performed, including their time-relation with the disease and their main results; and information on treatment and follow-up.

Analyses and uses of ASIA Registry data

The data are being reviewed and are published periodically to the scientific community in the form of articles in indexed journals. Forthcoming results are searchable on the Web site.

Limitations

There are several intrinsic limitations to data gathered through spontaneous reporting systems (Grimes and Schulz, 2002b).

All registry systems, even those owned by public health services for which reporting is mandated, suffer from underreporting (Krumholz, 2009). Cases reports to the registry are not a random sample of cases found in the general population and therefore may have some bias.

As in any study in which some information is collected retrospectively, reports are rarely thorough. At best, the quality relies on the information recorded in clinical records, which is not always full; alternatively, it relies on the memory of the physician filling out the form.

Since the clinician or researcher selects which cases to register, registered cases may be subject to selection bias. This affects what conclusions can be drawn, as missing cases may have had very different outcomes (Krumholz, 2009).

Unlike other registries, for which a distinct case definition exists, the cases included in the ASIA Registry are clinical pictures, with a weak and poorly defined definition.

The information recorded in the registry should be analyzed with caution. Patients may be exposed to different triggers at any one time, so attributing the disease to a single trigger is difficult.

Caution is thus needed in extrapolating conclusions from ASIA Registry data to the general population.

Conclusions

A new registry has been created in order to increase our knowledge of a recent proposed syndrome: the ASIA syndrome. This registry will be a scientific tool through which to cluster information on patients seen by many physicians, which until now has not been grouped in a distinct entity. It will enable us to define the disease criteria and discover any association of clinical manifestations, laboratory features, and pathology findings.

References

Agmon-Levin, N., Hughes, G.R.V, and Shoenfeld, Y. (2012). The spectrum of ASIA: "Autoimmune (Auto-inflammatory) Syndrome induced by Adjuvants". *Lupus*, **21**: 118–20.

Carstensen, B. and Borch-Johnsen, K. (2011). Register-based studies of diabetes. *Scand J Public Health*, **39**: 175–9.

CDC (Centers for Disease Control and Prevention). (1981). Pneumocystis pneumonia – Los Angeles. *MMWR*, **30**: 250–2.

Cervera, R., Bucciarelli, S., Espinosa, G., et al. (2007). Catastrophic antiphospholipid syndrome: lessons from the "CAPS registry" – a tribute to the late Josep Font. *Ann NY Acad Sci*, **1108**: 448–56.

Cruz-Tapias, P., Agmon-Levin, N., Israeli, E., *et al.* (2013). Autoimmune (auto-inflammatory) syndrome induced by adjuvants (ASIA) – animal models as a proof of concept. *Curr Med Chem*, **20**: 4030–6.

Cummings, P. and Weiss, N.S. (1998). Case series and exposure series: the role of studies without controls in providing information about the etiology of injury or disease. *Inj Prev*, **4**: 54–7.

Grimes, D.A. and Schulz, K.F. (2002a). Descriptive studies: what they can and cannot do. *Lancet*, **359**: 145–9.

Grimes, D.A. and Schulz, K.F. (2002b). An overview of clinical research: the lay of the land. *Lancet*, **359**: 57–61.

Israeli, E., Agmon-Levin, N., Blank, M., and Shoenfeld, Y. (2009). Adjuvants and autoimmunity. *Lupus*, **18**: 1217–25.

Krumholz, H.M. (2009). Registries and selection bias: the need for accountability. *Circ Cardiovasc Qual Outcomes*, **2**: 517–18.

Perricone, C., Colafrancesco, S., Mazor, R.D., *et al.* (2013). Autoimmune/inflammatory syndrome induced by adjuvants (ASIA) 2013: unveiling the pathogenic, clinical and diagnostic aspects. *J Autoimmun*, **47C**: 1–16.

Shoenfeld, Y. and Agmon-Levin, N. (2011). "ASIA" – autoimmune/inflammatory syndrome induced by adjuvants. *J Autoimmun*, **36**: 4–8.

11 Vaccination in Autoimmune Diseases

Carla Gonçalves,[1] Schahin Saad,[1] Clóvis A. Silva,[2] and Eloisa Bonfá[1]

[1]Division of Rheumatology, Children's Institute, Faculty of Medicine, University of São Paulo, São Paulo, Brazil

[2]Pediatric Rheumatology Unit, Children's Institute, Faculty of Medicine, University of São Paulo, São Paulo, Brazil

Introduction

Patients with autoimmune rheumatic diseases (ARDs) are at increased risk of infection attributed to the underlying disease immunosuppression and treatment immunomodulatory effect (Wolfe *et al.*, 1994; Doran *et al.*, 2002; Bosch *et al.*, 2006; Falagas *et al.*, 2007). Vaccination is an attractive method by which to prevent such infections, including influenza, invasive pneumococcal diseases, herpes zoster, and human papillomavirus (HPV) (van Assen *et al.*, 2011a). However, efficacy in patients with ARD may be reduced, and there is a potential risk of flares following vaccination. In addition, adjuvanted vaccines have been reported to trigger autoantibodies and autoimmune/inflammatory syndrome induced by adjuvants (ASIA) (Shoenfeld and Agmon-Levin, 2011).

This chapter will update our knowledge of the efficacy and safety of vaccination in patients with ARD and provide vaccine recommendations for such patients (Table 11.1).

Vaccination of patients with ARDs

Live vaccines

Live vaccines, including bacillus Calmette–Guérin (BCG) vaccine, vaccine against herpes zoster, vaccine against yellow fever (YF), and measles, mumps, and rubella (MMR) triple vaccine, are generally contraindicated in immunosuppressed ARD patients, due to the risk of an uncontrolled vaccine viral replication (van Assen *et al.*, 2011a, 2011b; Bijl *et al.*, 2012).

BCG vaccine

Patients with rheumatoid arthritis (RA) on synthetic disease-modifying antirheumatic drugs (DMARDs) have an increased incidence of tuberculosis (TB), while those under anti-tumor necrosis factor alpha (TNF-α) therapy have an even higher frequency of this complication (Gómez-Reino *et al.*, 2003; Brassard *et al.*, 2006). Screening and treatment for latent TB infection (LTBI) is recommended before the use of anti-TNF-α therapy. In adults, most TB cases are due to disease reactivation or new infection, while the efficacy of the BCG vaccine has solely been demonstrated in infants, so BCG is not recommended for ARD patients (Bijl *et al.*, 2012).

Vaccine against herpes zoster

Patients with ARD are at higher risk of developing herpes zoster infection than the general population, particularly in those under corticosteroids and biologics therapies (Kahl, 1994; Smitten *et al.*, 2007). The vaccine against varicella contains live attenuated viruses derived from the Oka strain. The vaccine against herpes zoster has had its efficacy demonstrated, reducing the number of

Vaccines and Autoimmunity, First Edition. Edited by Yehuda Shoenfeld, Nancy Agmon-Levin, and Lucija Tomljenovic.

Table 11.1 Vaccine card recommendation by age for adult patients with autoimmune rheumatic diseases (ARDs)

Immunization	Protocol
Non-live vaccines (highly recommended)	
Influenza	1 annual dose
Tetanus, diphtheria, pertussis (Tdap) #	Complete basic vaccination schedule: booster shot with Tdap and then one dose of dT every 10 years
HPV	3 doses of quadrivalent vaccine 0–2–6 months (up to 26 years)
Pneumococcal 23	1 or 2 doses (1 booster >65 years)
Conjugated pneumococcal 13	1 dose or more
Conjugated meningococcal	1 dose, even for individuals vaccinated during childhood or more than 5 years previously
Hepatitis A	2 doses, minimum interval 6 months
Hepatitis B	3 doses (0, 1, and 6 months)
Live vaccines (generally contraindicated)	
Yellow fever	1 dose every 10 years for those living in endemic areas or travelling to such areas
Herpes zoster	1 dose >50 years

Dt, adult combined vaccine against diphtheria and tetanus; Tdap, combination vaccine with acellular pertussis of the adult type; HPV, human papillomavirus

infections and complications in adults over the age of 60 years and in patients with chronic inflammatory diseases over the age of 50 years (Zhang *et al.*, 2011).

The vaccine against herpes zoster can be indicated for ARD patients with a positive serology for varicella and/or previous varicella vaccination, preferably before starting immunosuppression. It is contraindicated when patients are highly immunosuppressed: high doses of corticosteroids (>20 mg prednisone/day or equivalent) for 2 weeks or longer, pulse therapy, cytotoxic or alkylating agents, synthetic DMARDs at doses above those recommended, or immunobiological therapy. A rigorous follow-up after vaccination is recommended for early diagnosis and appropriate treatment with acyclovir (van Assen *et al.*, 2011a).

Vaccine against YF

YF is a noncontagious viral hemorrhagic febrile disease that is transmitted by the bite of insects, such as *Aedes* and *Haemagogus genera*. The lethality ranges from 5 to 10% and there is no specific treatment for the disease (Vasconcelos 2003). The 17D vaccine against YF provides protection for at least 10 years (Barret and Teuwen, 2009). The lack of effective treatment for YF vaccine-induced viral replication precludes a specific recommendation for ARD patients, although occasional use of the YF vaccine in the population of endemic areas may be considered (van Assen *et al.*, 2011a). In this regard, a case series of 70 ARDs who had been inadvertently immunized with the YF vaccine reported only minor adverse events (Mota *et al.*, 2009).

MMR triple vaccine

The triple viral vaccine is a combined vaccine containing live, attenuated viruses that protects against measles, mumps, and rubella. Usually, the MMR vaccine causes few adverse events, being well tolerated. Studies on the safety of the MMR vaccine in adults with ARD are still lacking (Heijstek *et al.*, 2007).

Non-live vaccines

The major advantage of inactivated vaccines is the total lack of infectious potential of the pathogenic agents. Such vaccines do not induce disease, but maintain the immunologic characteristics of the agent. However, the inactivated or recombinant vaccines have the disadvantage of inducing a suboptimal immune response, sometimes requiring the association of adjuvants or transporting proteins and the administration of booster shots. According to EULAR guidelines, the following vaccines are recommended: influenza vaccine, pneumococcal vaccine (13V-conjugated and 23-polysaccharide), HPV vaccine, meningococcal vaccine, *Haemophilus influenzae* type B

(Hib) vaccine, hepatitis A (HVA) and B (HVB) vaccine, and tetanus vaccine. Such vaccines can be safely administered, preferentially before starting DMARDs, in an attempt to reach the expected immunogenicity. When the vaccine card cannot be updated prior to the beginning of treatment, all of these vaccines can be administered to patients with ARDs, even those on corticosteroids and/or synthetic or biological DMARDs, based on their safety in several studies; however, the vaccine immunogenicity might be impaired (van Assen *et al.*, 2011a).

Influenza vaccine

Respiratory infections are common among patients with ARD and have a high mortality rate. Vaccination against influenza has been shown to reduce the number of hospital admissions and mortality due to respiratory infections in elderly patients, being effective even in patients on DMARDs (van Assen *et al.*, 2011a; Saad *et al.*, 2011). Response to the influenza vaccine seems be impaired in patients with systemic lupus erythematosus (SLE) and RA, which has been associated with corticosteroid and methotrexate use, respectively (Ribeiro *et al.*, 2011; Borba *et al.*, 2012a). Disease activity is also associated with poor vaccine response in SLE (Borba *et al.*, 2012b).

There is also evidence of an impaired vaccine-induced antibody production in response to pneumococcal and influenza vaccines when administered to patients on rituximab (Bingham *et al.*, 2010; van Assen *et al.*, 2011a). The response to the influenza vaccine (including vaccine against influenza A and H1N1) is particularly impaired when administered early, 4–8 weeks after the administration of rituximab. Thus, influenza vaccines should be administered before starting rituximab or 6 months after its first infusion and 4 weeks before its next dose (Buch *et al.*, 2011). With regard to abatacept in association with traditional DMARDs in RA patients, immune response has been reported to be severely hampered (Ribeiro *et al.*, 2013).

The influenza vaccine is considered safe, and has been used in annual campaigns for the population aged 60 years and over and for adults and children over the age of 6 months with ARD (Fiore *et al.*, 2010; van Assen *et al.*, 2011a, 2011b). It is contraindicated only in patients with a history of allergy to egg or to the vaccine itself, as well as in those who had Guillain–Barré syndrome up to 6 weeks after receiving it.

Pneumococcal vaccine

Bacterial infections of the respiratory tract are more common in patients diagnosed with ARD than in the general population and contribute to increase morbidity and mortality (Doran *et al.*, 2002), particularly invasive pneumococcal infection (Naveau and Houssiau, 2005). Thus, vaccination against *Streptococcus pneumoniae* (*pneumococcus*) is highly relevant for patients with ARD (Van Assen *et al.*, 2011a).

The pneumococcal vaccine available for adults is the 23-valent polysaccharide vaccine (Pn23), a polyvalent vaccine prepared from purified polysaccharides of the bacterial capsule, containing 23 serotypes of *Streptococcus pneumoniae*. A single dose of the vaccine is administered, with only one booster shot 5 years after the initial dose. However, it is associated with low immune response when compared with conjugated formulations (pneumo 7, 10, and 13). In RA, both similar and lower responses to pneumococcal vaccination have been reported. Two studies reported an impaired response to pneumococcal vaccination in RA patients on the combination of methotrexate and anti-TNF-α (Kapetanovic *et al.*, 2006; Kaine *et al.*, 2007; Visvanathan *et al.*, 2007). Finally, a lower response after rituximab was demonstrated (Bingham *et al.*, 2010).

The pneumococcal vaccine should be indicated for all patients with ARD, and can be more effective when administered before beginning synthetic or biological DMARDs.

HPV vaccine

HPV is a sexually transmitted virus and is highly prevalent around the world. HPV infection is the major risk factor for uterine cervix cancer, being associated with tumors of the penis, anus, mouth, and throat. HPV also causes genital warts or condyloma acuminatum. The quadrivalent vaccine is highly effective in preventing infections by subtypes 16 and 18 (the most oncogenic subtypes) and 6 and 11 (responsible for genital warts). Several countries recommend vaccination against HPV in young women, ideally before they initiate sexual activity. The vaccine is administered intramuscularly, in three doses in months 0, 1–2, and 6. Few adverse events have been described, although some patients can have mild local reactions. In SLE, the incidence of HPV infection is known to be increased. Data on the efficacy and safety of the HPV vaccine in patients with ARD are scarce (Tam *et al.*, 2010; Gatto *et al.*, 2013; Mok *et al.*, 2013). Recently, the quadrivalent HPV vaccine was studied in women with SLE aged 18–35 years. It was well tolerated and reasonably effective in patients with stable SLE and did not induce an increase in lupus activity or flare (Mok *et al.*, 2013). The HPV vaccine should be considered for adolescents and young women with ARD, preferably before they initiate their sexual life.

Meningococcal and Hib vaccines

Meningococcal vaccine is indicated to prevent invasive disease caused by *Neisseria meningitidis*, especially in patients with asplenia and complement deficiency. When hyposplenic/asplenic patients with ARDs plan to travel to or live in endemic areas vaccination is indicated (van Assen *et al.*, 2011a). The quadrivalent meningococcal conjugate vaccine (types A.C, W135, and Y) should be considered an option when immunizing adolescents and adults (Zonneveld-Huijssoon *et al.*, 2007). Studies on the efficacy and safety of the meningococcal vaccine in patients with ARD are still lacking.

Patients with rheumatic diseases are at greater risk of developing infections related to Hib and have indication for immunization. ARD patients should be immunized as soon as their diagnosis is made, preferentially before immunosuppressive therapy begins, because of the possible interference with vaccine response. In an uncontrolled study, Hib vaccination resulted in protection in 88% of SLE patients, and a trend toward a lower response was observed in patients on immunosuppressive drugs (Battafarano *et al.*, 1998). Asplenic adults with ARD should also be immunized.

HVA and HVB vaccine

Data on the incidence of hepatitis A and B infection in patients with ARD are still lacking. Vaccination is recommended in accordance with national vaccination guidelines.

The safety and efficacy of the HVB vaccine in RA have been assessed in a prospective study, which showed HVB vaccination is safe and produces antibodies in 68% of such patients (Elkayam *et al.*, 2002). In lupus, one study showed that HVB vaccination was safe in inactive patients between 18 and 50 years of age with an adequate vaccine response rate (Kuruma *et al.*, 2007).The HVB vaccine should be indicated for patients with ARD when their serology against HBsAg is negative, preferably before beginning treatment with biological DMARDs.

Combined vaccine against diphtheria, tetanus, and acellular pertussis (DTaP/Tdap) and combined vaccine against diphtheria and tetanus (dT)

The DTaP is a combined vaccine against diphtheria, tetanus, and pertussis, in which the pertussis component is acellular. Adult and elderly individuals with a complete basic vaccination schedule should receive a booster shot with Tdap (combination vaccine with acellular pertussis of the adult type) every 10 years (van Assen *et al.*, 2011a). The combined vaccine against diphtheria and tetanus (dT) is indicated for adolescents and adults. Individuals with an incomplete basic vaccination schedule (who have received fewer than their doses of the tetanus component during their lifetime) should complete their three-dose schedule, receiving one dose of Tdap and one or two doses of dT according to the 0–2–6-month schedule. In patients with RA and SLE, efficacy for tetanus toxoid vaccination has been demonstrated to be comparable with that in healthy controls (Abe and Homma, 1971; Devey *et al.*, 1987). Tetanus toxoid vaccination led to adequate immune response 24 weeks after rituximab administration. Vaccines should ideally be administered before B cell-depleting biological therapy is started or, when patients are on such a treatment already, at least 6 months after the start but 4 weeks before the next course, and patients should undergo passive immunization with tetanus immunoglobulin in case of exposure (Bingham *et al.*, 2010; van Assen *et al.*, 2011a).

References

Abe, T. and Homma, M. (1971). Immunological reactivity in patients with systemic lupus erythematosus. Humoral antibody and cellular immune responses. *Acta Rheumatol Scand*, **17**(1): 35–46.

Barrett, A.D. and Teuwen, D.E. (2009). Yellow fever vaccine – how does it work and why do rare cases of serious adverse events take place? *Curr Opin Immunol*, **21**(3): 308–13.

Battafarano, D.F., Battafarano, N.J., Larsen, L., *et al.* (1998). Antigen-specific antibody responses in lupus patients following immunization. *Arthritis Rheum*, **41**(10): 1828–34.

Bijl, M., Agmon-Levin, N., Dayer, J.M., *et al.* (2012). Vaccination of patients with auto-immune inflammatory rheumatic diseases requires careful benefit-risk assessment. *Autoimmun Rev*, **11**(8): 572–6.

Bingham, C.O. 3rd,, Looney, R.J., Deodhar, A., *et al.* (2010). Immunization responses in rheumatoid arthritis patients treated with rituximab: results from a controlled clinical trial. *Arthritis Rheum*, **62**(1): 64–74.

Borba, E.F., Saad, C.G., Pasoto, S.G., *et al.* (2012a). Influenza A/H1N1 vaccination of patients with SLE: can antimalarial drugs restore diminished response under immunosuppressive therapy? *Rheumatology (Oxford)*, **51**(6): 1061–9.

Borba, E.F., Pasoto, S.G., Calich, A.L., *et al.* (2012b). Lupus disease activity severely impairs pandemic influenza A/H1N1 vaccine immune response in patients without therapy. *Arthritis Rheum*, **64**(10): 2618.

Bosch, X., Guilabert, A., Pallarés, L., *et al.* (2006). Infections in systemic lupus erythematosus: a prospective and controlled study of 110 patients. *Lupus*, **15**(9): 584–9.

Brassard, P., Kezouh, A., and Suissa, S. (2006). Antirheumatic drugs and the risk of tuberculosis. *Clin Infect Dis*, **43**(6): 717–22.

Buch, M.H., Smolen, J.S., Betteridge, N., *et al.* (2011). Updated consensus statement on the use of rituximab in patients with rheumatoid arthritis. *Ann Rheum Dis*, **70**(6): 909–20.

Devey, M.E., Bleasdale, K., and Isenberg, D.A. (1987). Antibody affinity and IgG subclass of responses to tetanus toxoid in patients with rheumatoid arthritis and systemic lupus erythematosus. *Clin Exp Immunol*, **68**(3): 562–9.

Doran, M.F., Crowson, C.S., Pond, G.R., *et al.* (2002). Frequency of infection in patients with rheumatoid arthritis compared with controls: a population-based study. *Arthritis Rheum*, **46**(9): 2287–93.

Elkayam, O., Yaron, M., and Caspi, D. (2002). Safety and efficacy of vaccination against hepatitis B in patients with rheumatoid arthritis. *Ann Rheum Dis*, **61**(7): 623–5.

Falagas, M.E., Manta, K.G., Betsi, G.I., and Pappas, G. (2007). Infection-related morbidity and mortality in patients with connective tissue diseases: a systematic review. *Clin Rheumatol*, **26**(5): 663–70.

Fiore, A.E., Uyeki, T.M., Broder, K., *et al.* (2010). Prevention and control of influenza with vaccines: recommendations of the Advisory Committee on Immunization Practices (ACIP), 2010. *MMWR Recomm Rep*, **59**(RR-8): 1–62.

Gatto, M., Agmon-Levin, N., Soriano, A., *et al.* (2013). Human papillomavirus vaccine and systemic lupus erythematosus. *Clin Rheumatol*, **32**(9): 1301–7.

Gómez-Reino, J.J., Carmona, L., Valverde, V.R., *et al.* (2003). Treatment of rheumatoid arthritis with tumor necrosis factor inhibitors may predispose to significant increase in tuberculosis risk: a multicenter active-surveillance report. *Arthritis Rheum*, **48**(8): 2122–7.

Heijstek, M.W., Pileggi, G.C., Zonneveld-Huijssoon, E., *et al.* (2007). Safety of measles, mumps and rubella vaccination in juvenile idiopathic arthritis. *Ann Rheum Dis*, **66**(10): 1384–7.

Kahl, L.E. (1994). Herpes zoster infections in systemic lupus erythematosus: risk factors and outcome. *J Rheumatol*, **21**(1): 84–6.

Kaine, J.L., Kivitz, A.J., Birbara, C., and Luo, A.Y. (2007). Immune responses following administration of influenza and pneumococcal vaccines to patients with rheumatoid arthritis receiving adalimumab. *J Rheumatol*, **34**(2): 272–9.

Kapetanovic, M.C., Saxne, T., Sjöholm, A., *et al.* (2006). Influence of methotrexate, TNF blockers and prednisolone on antibody responses to pneumococcal polysaccharide vaccine in patients with rheumatoid arthritis. *Rheumatology (Oxford)*, **45**(1): 106–11.

Kuruma, K.A., Borba, E.F., Lopes, M.H., *et al.* (2007). Safety and efficacy of hepatitis B vaccine in systemic lupus erythematosus. *Lupus*, **16**(5): 350–4.

Mok, C.C., Ho, L.Y., Fong, L.S., and To, C.H. (2013). Immunogenicity and safety of a quadrivalent human papillomavirus vaccine in patients with systemic lupus erythematosus: a case–control study. *Ann Rheum Dis*, **72**(5): 659–64.

Mota, L.M., Oliveira, A.C., Lima, R.A., *et al.* (2009). Vaccination against yellow fever among patients on immunosuppressors with diagnoses of rheumatic diseases. *Rev Soc Bras Med Trop*, **42**(1): 23–7.

Naveau, C. and Houssiau, F.A. (2005). Pneumococcal sepsis in patients with systemic lupus erythematosus. *Lupus*, **14**(11): 903–6.

Ribeiro, A.C., Guedes, L.K., Moraes, J.C., *et al.* (2011). Reduced seroprotection after pandemic H1N1 influenza adjuvant-free vaccination in patients with rheumatoid arthritis: implications for clinical practice. *Ann Rheum Dis*, **70**(12): 2144–7.

Ribeiro, A.C., Laurindo, I.M., Guedes, L.K., *et al.* (2013). Abatacept and reduced immune response to pandemic 2009 influenza A/H1N1 vaccination in patients with rheumatoid arthritis. *Arthritis Care Res (Hoboken)*, **65**(3): 476–80.

Saad, C.G., Borba, E.F., Aikawa, N.E., *et al.* (2011). Immunogenicity and safety of the 2009 non-adjuvanted influenza A/H1N1 vaccine in a large cohort of autoimmune rheumatic diseases. *Ann Rheum Dis*, **70**(6): 1068–73.

Shoenfeld, Y. and Agmon-Levin, N. (2011). "ASIA" – autoimmune/inflammatory syndrome induced by adjuvants. *J Autoimmun*, **36**(1): 4–8.

Smitten, A.L., Choi, H.K., Hochberg, M.C., *et al.* (2007). The risk of herpes zoster in patients with rheumatoid arthritis in the United States and the United Kingdom. *Arthritis Rheum*, **57**(8): 1431–8.

Tam, L.S., Chan, P.K., Ho, S.C., *et al.* (2010). Natural history of cervical papilloma virus infection in systemic lupus erythematosus – a prospective cohort study. *J Rheumatol*, **37**(2): 330–40.

van Assen, S., Agmon-Levin, N., Elkayam, O., *et al.* (2011a). EULAR recommendations for vaccination in adult patients with autoimmune inflammatory rheumatic diseases. *Ann Rheum Dis*, **70**(3): 414–22.

van Assen, S., Elkayam, O., Agmon-Levin, N., *et al.* (2011b). Vaccination in adult patients with auto-immune inflammatory rheumatic diseases: a systematic literature review for the European League Against Rheumatism evidence-based recommendations for vaccination in adult patients with auto-immune inflammatory rheumatic diseases. *Autoimmun Rev,* **10**(6): 341–52.

Vasconcelos, P.F. (2003). Yellow fever. *Rev Soc Bras Med Trop,* **36**(2): 275–93.

Visvanathan, S., Keenan, G.F., Baker, D.G., *et al.* (2007). Response to pneumococcal vaccine in patients with early rheumatoid arthritis receiving infliximab plus methotrexate or methotrexate alone. *J Rheumatol,* **34**(5): 952–7.

Wolfe, F., Mitchell, D.M., Sibley, J.T., *et al.* (1994). The mortality of rheumatoid arthritis. *Arthritis Rheum,* **37**(4): 481–94.

Zhang, J., Delzell, E., Xie, F., *et al.* (2011). The use, safety, and effectiveness of herpes zoster vaccination in individuals with inflammatory and autoimmune diseases: a longitudinal observational study. *Arthritis Res Ther,* **13**(5): R174.

Zonneveld-Huijssoon, E., Ronaghy, A., Van Rossum, M.A., *et al.* (2007). Safety and efficacy of meningococcal vaccination in juvenile idiopathic arthritis. *Arthritis Rheum,* **56**(2): 639–46.

12 Vaccination in Patients with Autoimmune Inflammatory Rheumatic Diseases

Abdulla Watad,[1,2] Alessandra Soriano,[1,3] and Yehuda Shoenfeld[1,4]

[1]Zabludowicz Center for Autoimmune Diseases, Sheba Medical Center, Tel Hashomer, Israel
[2]Department of Internal Medicine B, Sheba Medical Center, Tel Hashomer, Israel
[3]Department of Clinical Medicine and Rheumatology, Campus Bio-Medico University, Rome, Italy
[4]Incumbent of the Laura Schwarz-Kipp Chair for Research of Autoimmune Diseases, Sackler Faculty of Medicine, Tel Aviv University, Tel Aviv, Israel

Introduction

Vaccines represent the most effective way of preventing morbidity and mortality associated with infections, in healthy as well as immunocompromised subjects. Patients with autoimmune inflammatory rheumatic diseases (AIRDs) are at increased risk of contracting infections, especially in the course of immunosuppressive treatment. Thus, in the last few decades, several studies have summarized the risk of infection in AIRD and the related benefits of immunization.

On the other hand, several observations underline that the real efficacy of vaccines in subjects with AIRD may be reduced. Moreover, vaccine safety, which is a pivotal issue, may also be reduced, because of the risk of AIRD flare following vaccinations.

In this chapter, we reviewed the main evidence for the risk of infections in patients affected by AIRD, the safety and efficacy of vaccines in this category of patients, and the current recommendations about who should be vaccinated.

Risk of infection in patients with AIRD

Infections are still one of the leading causes of morbidity and mortality for AIRD patients, due to both the disease processes and medications. Risk of infection in patients with AIRD is increased, especially in those with active disease and those using immunosuppressive treatments (van Assen *et al.*, 2011a).

With regard to influenza infection, two retrospective studies have shown an increased risk of influenza in elderly patients (≥65 years) with rheumatic diseases, including vasculitides (Nichol *et al.*, 1998; Hak *et al.*, 2002). In the first (Nichol *et al.*, 1998), the odds ratio (OR) was 1.56 (95% confidence interval (CI): 1.23–2.02) for hospital admissions for either pneumonia or influenza, and 2.67 (95% CI: 2.26–3.16) for death. In the second study (Hak *et al.*, 2002), 4.5–7.0% of AIRD patients who were unvaccinated for influenza were admitted for pneumonia/influenza or death, compared to 0.8% of unvaccinated healthy controls.

Vaccines and Autoimmunity, First Edition. Edited by Yehuda Shoenfeld, Nancy Agmon-Levin, and Lucija Tomljenovic.

Streptococcus pneumoniae is considered one of the main causative pathogens of pulmonary infections in patients with AIRD (van Assen *et al.*, 2011a). A recent study by Wotton and Goldacre (2012) analyzed the risk of invasive pneumococcal disease in people hospitalized with selected immune-mediated disorders. Significantly high risk of pneumococcal disease following hospital admission was found for many of the immune-mediated diseases, including rheumatoid arthritis (RA), systemic sclerosis (SSc), Sjögren syndrome (SjS), and systemic lupus erythematosus (SLE).

In three studies focused on SLE patients (Klippel *et al.*, 1979; Yee *et al.*, 1997; Naveau and Houssiau, 2005), invasive pneumococcal infection was documented in 1.34% of patients over an unknown period of time, in 1.9% over 25 years, and in 2.4% over 9 years, respectively.

With regard to tuberculosis (TB), its incidence strongly differs with geographic area. A multicenter active surveillance report performed by Gomez-Reino *et al.* (2003) showed that the highest incidence of TB was detected in Spanish RA patients treated with infliximab. The patients had been screened for latent TB before therapy began.

RA patients have increased risk for contracting TB. The relative risk is estimated at from 2.0 to 10.9 when patients are treated with nonbiological disease-modifying antirheumatic drugs (DMARDs) (Wolfe *et al.*, 2004; Askling *et al.*, 2005; Sichletidis *et al.*, 2006; Yamada *et al.*, 2006; Seong *et al.*, 2007; Baronnet *et al.*, 2008; Brassard *et al.*, 2009) and at from 4.0 to 90.1 when they are treated with tumor necrosis factor alpha (TNF-α) biological agents (Gomez-Reino *et al.*, 2003; Yamada *et al.*, 2006; Seong *et al.*, 2007; Baronnet *et al.*, 2008). Moreover, it has been observed that TB occurs more often in SLE patients than in the general population (7.9 vs 2.3/1000 patients/year) (Yun *et al.*, 2002).

Steroid therapy is also a risk factor (hazard ratio (HR) estimates 1.4–2.4), increasing the risk for TB 1.5–8.7-fold (Wolfe *et al.*, 2004). In light of this, in recent studies on biological therapies, all patients have been screened for latent TB before therapy begins.

Herpes zoster (HZ) infection is another important issue in patients with AIRD. In a study by Smitten *et al.* (2007) aimed at evaluating the risk of HZ infection in patients with RA in the United States and the United Kingdom, it was shown that RA per se is a risk factor for HZ infection (adjusted HR 1.65 and 1.91 compared to healthy controls, respectively).

Finally, the risk of HZ infection in patients with SLE has been investigated in recent decades: compared to the general population, this risk appears to be increased, by a factor ranging from 5- to 16-fold (Kahl, 1994; Manzi *et al.*, 1995; Kang *et al.*, 2005). It correlates with the use of cyclophosphamide, azathioprine, and steroids in combination with other immunosuppressive drugs.

Safety and efficacy of vaccines in patients with AIRD

Influenza vaccine

In general, trivalent inactivated influenza vaccine (TIV) can be used for any patient, including those with high-risk conditions. Live-attenuated influenza vaccine (LAIV) can be used for healthy nonpregnant persons aged from 2 to 49 years. No preference is indicated for TIV or LAIV in such patients (CDC, 2012; Murdaca *et al.*, 2014). According to the European League Against Rheumatism (EULAR), influenza vaccine is strongly recommended in patients with AIRD (Guissa *et al.*, 2012).

Saad *et al.* (2011) studied 1668 patients with AIRD vaccinated with nonadjuvanted influenza A/California/7/2009(H1N1) virus-like flu strain. After immunization, all the evaluated parameters (seroprotection and seroconversion rates, as well as the increase in the geometric mean titer (GMT) of antibodies) resulted in significantly lower in AIRD patients, particularly in SLE patients, when compared to healthy controls. However, about 50% of the patients were under glucocorticoid and/or immunosuppressive therapy. In this study, the nonadjuvanted preparation was chosen, in order to prevent triggering an "adjuvant disease" in genetically susceptible individuals.

Indeed, as adjuvants may act as ligands for Toll-like receptors (TLRs) and stimulate innate immune responses, their use is a potential risk in autoimmune-prone subjects (Agmon-Levin *et al.*, 2009). In addition, aluminum adjuvants commonly used in human vaccines were found to be associated with macrophagic myofasciitis (MMF), an immune-mediated disease (Agmon-Levin *et al.*, 2009).

Autoimmune/inflammatory syndrome induced by adjuvants (ASIA) has recently been described by Shoenfeld and Agmon-Levin (2011), both in animal models and in humans, following injection of adjuvanted vaccines or exposure to other substances with immune adjuvant properties

(Agmon-Levin *et al.*, 2012; Bassi *et al.*, 2012; Jara *et al.*, 2012; Zafrir *et al.*, 2012).

Vaccine efficacy in patients with AIRD undergoing immunosuppressive treatment has been reported in a few studies; except for rituximab, treatment with glucocorticoids, DMARDs, and TNF-α inhibitors did not seem to have a negative effect upon the immunogenicity of seasonal influenza vaccine (Conti *et al.*, 2008; Gelinck *et al.*, 2008a; Bingham *et al.*, 2009; Elkayam *et al.*, 2010; Salemi *et al.*, 2010; van Assen *et al.*, 2010).

Patients who receive TNF-α inhibitors, who are at risk of developing an inadequate serologic response, could receive booster immunizations to favor a better immune response. Finally, it needs to be underlined that, beyond the common adverse reactions (i.e. fever, irritability, headache, pain and swelling in the injection site, itching), severe allergic and anaphylactic reactions may occur in response to a number of influenza vaccine components, but such reactions remain rare (CDC, 2012).

Influenza vaccine in SLE

Influenza vaccination in patients with SLE has been a matter of debate for many years, because of the potential risk of favoring the onset or exacerbation of the disease (Aron-Maor and Shoenfeld, 2001; Holvast *et al.*, 2007; CDC, 2012). Indeed, several cases of SLE precipitation have been reported following immunization (Perdan-Pirkmajer *et al.*, 2012). Influenza vaccine may also favor a transient increase of the levels of different autoantibodies in about 10–15% of patients, even in the absence of disease exacerbation (Older *et al.*, 1999; Abu-Shakra *et al.*, 2000; Perdan-Pirkmajer *et al.*, 2012; Vista *et al.*, 2012). Nevertheless, in most SLE patients with quiescent disease, any increase in both clinical and laboratory parameters has been detected following influenza vaccination (Holvast *et al.*, 2007).

In one of the first studies on this issue, Hess and Hahn (1978) reported 2 SLE patients out of 109 immunized with A/NewJersey/76 vaccine who developed renal failure; one presented an active urinary sediment, while the other had a diffuse proliferative nephritis. In another study, performed by Del Porto *et al.* (2006), two cases of modest flare among 14 SLE patients vaccinated with nonadjuvanted TIV (A/New Caledonia/20/99 [H1N1], A/Moscow/10/99 [H3N2], B/Hong Kong/330/2001) were reported.

Other studies have observed that SLE patients vaccinated against influenza did not develop SLE-flares (Abu-Shakra *et al.*, 2000, Holvast *et al.*, 2006). Abu-Shakra *et al.* (2000) studied the potential effect of influenza vaccination on SLE by evaluating the SLE Disease Activity Index (SLEDAI) score in 24 SLE patients who received influenza vaccine and in 24 SLE patients who did not. The SLEDAI score was similar in both groups, confirming the safety of the influenza vaccine in SLE patients.

Some studies have reported that the immune response to the influenza vaccine in SLE patients is significantly lower than that in the general population. As a matter of fact, an effective immune response to this vaccine depends on both humoral and cell-mediated responses, which may be compromised in SLE patients (Holvast *et al.*, 2007).

A significantly lower immune response following immunization with A/NewJersey/76(Hsw1N1) and A/Victoria3/75(H3N2) strains has been reported in two different studies. Williams *et al.* (1978) reported that, among 19 evaluated SLE patients, 47% presented seroconversion, as compared with 94% in an age-matched control group. Ristow *et al.* (1978) found only 48% of SLE patients and 62% of controls presented a fourfold increase in antibody titers following vaccination.

Holvast *et al.* (2009) reported that 56 SLE patients presented fewer seroconversions compared to healthy controls: 43% of patients versus 94% of controls for A (H1N1), 39% of patients versus 88% of controls for A(H3N2), and 41% of patients versus 71% of controls for B/Hong Kong.

In other studies, a similar (or modestly reduced) response to the influenza vaccination has been reported in SLE patients compared to healthy controls (Gross *et al.*, 1995; De Jong *et al.*, 2003; Mercado *et al.*, 2004; Holvast *et al.*, 2006).

The immunogenicity of the influenza vaccine has long been evaluated through assessment of the humoral response to the vaccine, where a hemagglutination inhibition assay titer (HA) $\geq$ 40 is considered protective in healthy adults (Wiesik-Szewczyk *et al.* 2010). However, it has been reported that the median titer of 28 seems to protect 50% of vaccinated healthy adults (Del Porto *et al.*, 2006). The cell-mediated response seems to play a key role in protecting against symptomatic influenza in elderly patients, in whom hemagglutination inhibition assay titers $\geq$ 40 are not always protective. SLE patients seem to have similar humoral responses to influenza vaccine to healthy controls (Holvast *et al.*, 2009).

A booster influenza vaccination 4 weeks after first vaccination did not seem to increase the seroprotection rates and the GMT of antibodies in SLE. However, Heijstek *et al.* (2011) confirmed the efficacy of influenza vaccination even in the presence of lower levels of protective serum antibodies.

Influenza vaccine in RA

Influenza vaccination in RA patients has been a subject of debate for many years. Chalmers *et al.* (1994) stratified 126 RA patients into three groups: patients who were receiving the usual therapy for RA and had a history of influenza vaccine within 24 months; patients who were receiving the usual therapy without prior vaccine; and patients who were receiving prednisone >7.5 mg/day or other immunosuppressive drugs. In each group, patients were randomized to receive vaccine or placebo, and during the first month of follow-up, adverse reactions occurred equally among RA patients and healthy controls. Similar and significant increases in antibody titers to the vaccine were obtained in all groups of patients with RA and in healthy controls.

Fomin *et al.* (2006) studied the effect of influenza vaccination in 82 RA patients and 30 healthy controls using a split-virion inactivated vaccine containing 15 mcg hemagglutinin (HA)/dose of each of B/Hongkong/330 and A/New caledonian/20/90.

Six weeks after vaccination, a significant increase in GMT of antibodies for each antigen was observed in both groups. The activity of the disease was assessed by different parameters, including the number of tender and swollen joints, morning stiffness, pain level, erythrocyte sedimentation rate (ESR), and C reactive protein (CRP); all of these parameters were found to be unchanged. The study showed clearly that influenza vaccine induced a good humoral response in RA patients, although lower than that in healthy controls (Fomin *et al.*, 2006).

Other studies have shown that glucocorticoids do not significantly affect the development of protective antibodies to influenza vaccine (Malleson *et al.*, 1993; Chalmers *et al.*, 1994; Kanakoudi-Tsakalidou *et al.*, 2001). Similar results were observed for gold, azathioprine, and methotrexate (Chalmers *et al.*, 1994; Fomin *et al.*, 2006).

Influenza vaccine in dermatopolymyositis

The effect of vaccinations in patients with inflammatory myopathies has also been investigated. A single institution-controlled, prospective study (Shinjo *et al.*, 2012) showed that vaccination with nonadjuvanted, non-live H1N1/2009 was not associated with short-term (21 days) harmful effects in adults affected (37 dermatomyositis and 21 polymyositis; mean age 43.1 years).

In addition, an adequate immunogenicity in patients with inflammatory myopathies concurrently treated with various maintenance immunoregulatory drugs was observed. Clinical and laboratory parameters (serum level of creatine kinase and aldolase) remained stable through the study. The long-term effects of the vaccination were not evaluated.

Only limited data exist on the safety and efficacy of vaccines in patients with juvenile dermatomyositis. A single-institution, prospective, controlled study was performed in patients with juvenile dermatomyositis (30 patients with a mean age of 15.5 years) to evaluate the safety and the efficacy of H1N1 vaccine and revealed no short-term harmful effects. The seroconversion rates were particularly hampered by chronic disease and concomitant immunosuppressive therapy (Guissa *et al.*, 2012).

Influenza vaccine in fibromyalgia

A possible role of vaccination in the pathogenesis of fibromyalgia has been posited, following the description of considerable overlap between Gulf War syndrome (GWS), ASIA, and fibromyalgia.

Ablin *et al.* (2013) evaluated the efficacy and safety of influenza vaccine in patients with fibromylagia. Nineteen patients who fulfilled the diagnostic criteria for fibromylagia according to the 1990 American College of Rheumatology criteria were recruited and asked to fill out the Fibromyalgia Impact Questionnaire and the Widespread Pain Index and Symptoms Severity Scale, which are also components of the 2010 fibromylagia diagnostic criteria. All of the patients had received the inactivated split-virion influenza vaccine. Six weeks after vaccination, patients were reevaluated and sera were tested by hemagglutination inhibition assay for antibodies against the three antigens included in the vaccine: A/California (H1N1), A/Perth (H3N2), and B/Brisbane. The patients displayed significant increases in GMT of hemagglutination inhibition assay of antibodies against H1N1 and B/Bri viruses: from 29.9 to 387.9 ($p = 0.0011$), from 82.9 to 460.9 ($p = 0.0007$), and from 28.8 to 96.0 ($p = 0.08$), respectively. The rates of seroprotection (defined as antibody levels above 1/40) increased from 22.9% for H1N1 to 89.5% post-vaccination. Significant increases in hemagglutination inhibition assay of antibody titers were also demonstrated among healthy controls: H1N1 ($p = 0.000435$), B/Bri ($p = 0.000331$), and Perth ($p = 0.004953$). Influenza vaccination proved both safe and effective in patients affected by fibromyalgia. Neither severe adverse reactions nor significant worsening of symptoms were recorded following vaccination and serological evidence of seroconversion was observed, as in healthy controls (Ablin *et al.*, 2013).

Pneumococcal vaccine in AIRD patients

As recently underlined by van Assen *et al.* (2011b), the efficacy of pneumococcal vaccination is difficult to determine because of the lack of accepted response criteria and because of the different numbers of pneumococcal serotypes contained in different vaccine formulations (polysaccharide and conjugate vaccines).

The first study aimed at evaluating the immunogenicity and safety of vaccination against *Streptococcus pneumoniae* was published by Elkayam *et al.* (2002a), who analyzed the antibody response 1 month after the administration of a single dose of 23-valent pneumococcal vaccine in 42 patients with RA, 24 with SLE, and 20 healthy control subjects matched for age and sex. Pneumococcal vaccination was not associated with a significant worsening of any clinical or laboratory parameter of disease activity in either RA or SLE patients.

Minor acute adverse effects were observed in two patients, one affected with RA, who experienced transient diffuse musculoskeletal pain the day after the vaccine, and one with SLE, who developed clinical pleuritic pain lasting 5 days following the vaccine.

Concerning the immunogenicity, neither demographic data nor the clinical and laboratory measures of disease activity were able to predict poor vaccine responders. In a study by Elkayam *et al.* (2002a), antibody levels to seven pneumococcal serotypes were tested before and 1 month after vaccination, and there was evidence of significant increase after 1 month. Nevertheless, the 23-valent vaccine did not appear to be uniformly immunogenic in either population, and a substantial proportion of patients (30% of RA and 20% of SLE patients) had an extremely poor response, as demonstrated by the absence of response in some individuals and the production of an effective autoantibody response to only one of the seven tested polysaccharides in others.

Elkayam *et al.* (2005) also showed that immunization of SLE patients with 23-valent pneumococcal vaccine did not induce generation of autoantibodies such as anti-dsDNA, anti-cardiolipin, anti-Sm, anti-RNP, anti-Ro/SSA, and anti-La/SSB.

Other controlled and uncontrolled studies following the first study performed by Elkayam *et al.* (2002a) showed a similar or reduced autoantibody response independent of the concomitant therapy with steroids, azathioprine, or cyclophosphamide in healthy controls (Elkayam *et al.*, 2005; Tarjan *et al.*, 2002).

Regarding the influence of the immunosuppressive therapy (TNF-α blocking agents and methotrexate) on the autoantibody response, there are still controversial results in RA and SLE population studies. TNF-α blocking agents did not reduce the efficacy of pneumococcal vaccination, except in two studies, in which a reduced response was shown. In particular, the combination of methotrexate and anti-TNF-α agents – as well as monotherapy with methotrexate - has been shown to impair the response to pneumococcal vaccine in RA patients (Kapetanovic *et al.*, 2006; Gelinck *et al.*, 2008b).

Finally, patients treated with anti-CD20 monoclonal antibody therapy (rituximab) showed a reduced response to pneumococcal polysaccharide vaccine, as recently demonstrated in a study on 69 RA patients vaccinated 28 weeks after rituximab administration (Bingham *et al.*, 2009).

Hepatitis B vaccine

Vaccination against hepatitis B virus (HBV) is a universal recommendation of the World Health Organization (WHO, 2004). The HBV vaccine is a recombinant vaccine that contains viral surface antigen emulsified within aluminum hydroxide, serving as an adjuvant. It was introduced into the market in the early 1980s and, like other vaccines, it has been associated with immune and nonimmune adverse events in post-marketing and surveillance studies, in rare cases (CDC, 2006; Mikaeloff *et al.*, 2009; Stubgen, 2010).

According to EULAR recommendations, the HBV vaccine should be considered in selected patients with AIRD (grade of evidence II–III; strength of recommendation B–D) (van Assen *et al.*, 2011b). The main studies of the safety and efficacy of the HBV vaccine, focusing on SLE and RA populations, are reviewed in this section.

HBV vaccine in SLE

The only study aimed at investigating the safety and efficacy of the HBV vaccine in SLE patients was performed by Kuruma *et al.* (2007). This study looked 28 female SLE patients, all fulfilling the ACR classification criteria for SLE (Hochberg, 1997). The main inclusion criteria were: age between 18 and 50 years, SLEDAI < 4, negative anti-dsDNA and anti-cardiolipin antibodies, and a recent negative HBV serology (HBsAg, anti-HBc, anti-HBs) at entry. All patients received three doses of the vaccine. Clinical and laboratory parameters (complete blood count, urine analysis, ANA, anti-dsDNA, anti-HBsAg antibody) were also evaluated at entry. Treatment with prednisone >20 mg daily dose and/or immunosuppressive treatment and previous history of vaccination allergy were considered as exclusion criteria.

The development of protective anti-HBV antibodies after immunization was observed in 26 out of the 28 SLE patients (93%) at the end of the study. A lower than expected frequency of seroconversion was observed after the first (4%) and second (54%) doses. The two nonresponsive patients received an additional dose (fourth dose) of vaccine and positive anti-HBV antibody titers were detected after 1 month, without evidence of clinical or laboratory flares (Kuruma *et al.*, 2007). Only three patients flared up during the vaccination period (11%). One patient flared up 2 months after the second dose, with cutaneous manifestations (rash and alopecia) associated with a decrease in complement levels and an increase of the anti-dsDNA antibodies (1/160), requiring an increase in prednisone dose to 30 mg/day, associated with azathioprine (100 mg/day). The two additional patients flared up 5 days and 2 weeks after the third dose. One experienced a mild photosensitivity and discoid rash, associated with a decrease in complement levels, and was treated with low-dose prednisone (10 mg/day). The other experienced polyarthritis, associated with a decrease in complement levels, which required the introduction of prednisone (20 mg/day) and methotrexate (10 mg/week).

In summary, with regard to safety, in this study, three disease flares (11%) occurred during the follow-up, which were limited to cutaneous and articular manifestations associated with laboratory findings. Thus, the safety of the HBV vaccine in SLE patients was comparable to that described for other vaccines (influenza and pneumococcal) (Williams *et al.*, 1978; Battafarano *et al.*, 1998), and, according to the authors, a direct causal association between HBV vaccine and flares is not supported, since the rate of post-vaccination flares was low and was similar to that observed in the previous year (before the vaccination).

Two-thirds of the patients with disease flares had a history of renal disease. Furthermore, although more than one-third of the patients were formerly anti-dsDNA antibody-positive during their disease, only one of them had a positive test after vaccination, suggesting that the vaccine does not induce this specific antibody.

In the same manner, none of the SLE patients was positive for IgG and IgM anticardiolipin antibodies at baseline, and they all remained negative during the study. These results contrast with those of a previous study by Abu Shakra *et al.* (2002), in which antiphospholipid antibodies were detected in half of all SLE patients after influenza vaccination.

In conclusion, according to the current evidence, the response to the HBV immunizing antigen does not seem to induce flares and autoantibodies in SLE. Although the role of individual genetic predisposition cannot be excluded, inactive lupus patients may benefit from HBV vaccination with successful protective antibody production, and HBV vaccine may be considered as part of prevention programs in SLE patients.

HBV vaccine in RA

In RA, the safety and efficacy of vaccination against HBV were investigated in a prospective study by Elkayam *et al.* (2002b). The aim of this study was to evaluate the humoral immune response of RA patients to recombinant HBV vaccine, the short-term adverse effects of the disease, and the disease's rate of exacerbation.

The study group consisted of 22 patients who were vaccinated with three doses of recombinant HBV vaccine (ENGERIX B, GlaxoSmithKline) intramuscularly in the deltoid region; the second and the third doses were given 1 and 6 months after the first. The control group included 22 other RA patients who did not receive the vaccine. Physical examination, clinical assessment (morning stiffness, daytime pain with visual analog scale 0–10, number of tender and swollen joints), routine laboratory tests (complete blood count, serum chemistry panel, erythrocyte sedimentation rate, CRP, urine analysis) were performed before and 2 and 7 months after the immunization.

Enzyme-linked immunosorbent assay (ELISA) was performed to determine the antibodies to HBsAg and 68% of patients who received the vaccine responded with antibody levels above 10 IU/l after 6 months. None of the patients reported any adverse effect after the immunization, and there was no association with a significant worsening in any clinical or serological parameter.

Nevertheless, a subgroup of patients remained exposed to infection with HBV despite the vaccination. Different factors were reported to be associated with the lack of humoral response, including older age (mean (SD) age 59 (10.5) in nonresponders versus 46 (17.5) in responders) and increased daytime pain at vaccination (4.8 in nonresponders versus 2.1 in responders).

Quadrivalent HPV vaccine

Very few studies have been evaluated the safety and efficacy of the quadrivalent human papillomavirus (HPV) vaccine in AIRD patients.

It has been observed that SLE patients are at higher risk of persistent HPV infection compared to healthy females, and that they also have higher risk of developing abnormal cervical smears and squamous intraepithelial lesions (SILs) of the cervix (Nyberg *et al.*, 1981; Blumenfeld *et al.*, 1994; Klumb *et al.*, 2010).

In this regard, two studies have been performed to evaluate the HPV vaccine in SLE women (Soybilgic *et al.* 2013; Mok *et al.*, 2013). The first was an open-label, prospective, pre-post-intervention study, which recruited 27 SLE patients (age 12–26 years), of whom 22 had juvenile SLE (diagnosed before the age of 16 years). Laboratory parameters, SLEDAI scores, and medications in the year prior to study entry were also evaluated, and used as control data for the same patients. Of the 27 total subjects, 2, 4, and 20 received one, two, and three doses of HPV vaccine, respectively. Only 20 of the 22 juvenile SLE patients completed the study. There was a significant reduction in the mean SLEDAI scores in this population, from 6.14 pre-vaccination to 4.49 post-vaccination at month 7 ($p = 0.010$; 95% CI: −2.85 to −0.44). Two patients withdrew from the study because of an increase in arthralgia, without a complementary increase in their SLEDAI scores. There was no change in the anti-dsDNA titers, and the levels of C3, C4, erythrocyte sedimentation rate and CRP remained stable. Of the 27 patients, 9 (33.3%) had mild to moderate flare-ups during the study period, typically with symptoms similar to those they experienced in flares prior to vaccination; five had arthralgia, four had rash, two had pleuritis, and one had peripheral neuropathy. One of the patients with rash had a severe cutaneous flare following sun exposure, 3 months after receiving the second dose of the vaccine.

Nonetheless, the authors concluded that quadrivalent HPV vaccine is immunogenic, generally safe, and well tolerated in adolescent and young women with SLE. The seropositivity to HPV immunization was greater than 94% in all four HPV types, which was considered an excellent response, because the majority of the patients were on prednisone therapy (Soybilgic *et al.*, 2013).

In the other study, by Mok *et al.* (2013), 50 women (mean age 18–35 years) with stable SLE were recruited to receive quadrivalent HPV vaccine, together with an equal number of age-matched healthy women. The mean age and disease duration were 25.8 ± 3.9 years and 6.6 ± 4.5 years, respectively. At month 12, the seroconversion rates of anti-HPV serotypes 6, 11, 16, and 18 in SLE patients and healthy women were 82, 89, 95, and 76% and 98, 98, 98, and 80%, respectively. In SLE patients, there were no significant changes in the titers of anti-dsDNA, levels of complements, or SLEDAI score through a follow-up period of 12 months. One mild to moderate flare at months 0–2, two mild to moderate flares at months 3–5, and six mild to moderate and two severe flares at months 7–12 were observed.

All flares were treated with the usual regime, and the causal relationship between the vaccination and the flares was considered unclear; no withdrawals were observed because of the flares.

Even in this study, the seroconversion of the anti-HPV exceeded 93% in healthy controls and 76% in patients with SLE, and the vaccine did not lead to an increase in SLEDAI or disease flares 12 months after immunization (Mok *et al.*, 2013).

Nevertheless, cases of AIRD onset and exacerbations (including SLE) have been reported following HPV vaccination (Soldevilla *et al.*, 2012; Gatto *et al.*, 2013; Melo Gomes *et al.*, 2013), further emphasizing the fact that in a small number of individuals, such as those with a familial or personal history of autoimmunity or those with previous adverse manifestations following vaccines, a higher risk of full-blown autoimmunity may be present if they undergo further immunizations.

As a matter of fact, a systematic case–control study on the incidence of autoimmune diseases associated with HPV vaccination in young women in France recently analyzed a total of 269 HPV-vaccinated cases (14–26 years of age) in comparison to 1096 age- and sex-matched controls who did not receive the vaccine (Grimaldi *et al.*, 2014). The incidences of six types of autoimmune diseases were evaluated in this study, including SLE. The results showed that the vaccinated cases had approximately 9 times higher incidence of previous personal history of autoimmune disease than the controls (4.5 versus 0.6 %) and 2.5 times higher incidence of familial history of autoimmunity (14.1 versus 5.6%). Both of these differences were highly statistically significant ($p = 0.001$ and $p < 0.001$), indicating that a personal or familial history of autoimmunity confers a significantly elevated risk for autoimmune manifestations following HPV vaccination.

Tetanus and diphtheria vaccines

Only a few studies have evaluated the efficacy of diphtheria and tetanus vaccination in patients with AIRD. One recent study performed by Csuka *et al.* (2013) recruited 279 SLE patients (205 females, aged 45 ± 13.8), 158 myasthenia gravis (MG) patients (101 females, 55 ± 18.7), and 208 healthy controls (122 females, 48 ± 14.6) in order to assess the efficacy of the vaccine by determining the serum concentration of IgG antidiphtheria-antitoxin and tetanus-antitoxoid-IgG through ELISA.

The study showed that equal proportions of healthy controls, SLE patients, and MG patients had proper responses and immunogenicity against diphtheria and tetanus. The serum concentration of antidiphtheria antibodies in all three groups decreased significantly with age throughout the study population, while the antitetanus antibodies dropped only in elderly subjects (>60 years old). The only particular finding was that antidiphtheria antibody serum concentration was significantly lower in SLE patients (<40 years old) than in healthy controls.

In summary, this study suggested that the immunogenicity induced by diphtheria and tetanus vaccine in patients with SLE and MG is overall comparable to that in the healthy population.

Historically, one of the first studies on this issue was performed by Fawcett *et al.* (1984), who recruited 18 female (mean age 52) patients with Hashimoto's disease (all of them on replacement therapy and euthyroid) and 10 healthy female controls. All individuals received an aluminum adjuvanted diphtheria-tetanus vaccine. Serum samples were obtained at the time of immunization and 4 and 10 weeks later. No significant differences between Hashimoto's patients and the control group in terms of kinetics or the magnitude of their humoral immune response to tetanus toxoid were observed, in agreement with previous observations (Barr *et al.*, 1964). Nevertheless, a marked rise of tetanus toxoid antibodies (IgG) was observed in 7 of 10 normal individuals and 9 of 18 patients with Hashimoto's disease, suggesting a difference in the ability of normal subjects and Hashimoto's patients to regulate their response to tetanus toxoid.

Normal individuals showed a significant negative correlation between the pre-immunization levels of antibody and the increase in antibody levels following the immunization, in agreement with a study by Looney *et al.* (1956). In Hashimoto's patients, the magnitude of increase in tetanus toxoid antibody levels following immunization appeared to be unrelated to tetanus toxoid antibody levels at the time of immunization. It was suggested that this inability to regulate antibody production in response to antigens that are continuously present is likely to have serious consequences, and such a defect could be involved in the initiation and perpetuation of the autoimmune response in Hashimoto's disease.

EULAR recommendations for immunizations in AIRD patients

In 2011, a EULAR task force comprising 11 experts representing 11 European countries (eight rheumatologists, four clinical immunologists, one rheumatologist/clinical immunologist, one infectious disease physician, one nephrologist, one pediatrician/rheumatologist, and one clinical epidemiologist) developed recommendations for the vaccination of patients with AIRD, combining evidence from clinical studies with expert opinions where sufficient evidence was lacking (van Assen *et al.*, 2011b). The task force formulated 13 recommendations for adult patients with AIRD, as summarized in Table 12.1.

There are some issues which remain a matter of debate regarding these recommendations:

1. No studies have been performed comparing efficacy and harms between patients with AIRD and with stable and unstable disease; thus, no grade of evidence has been produced in regard to the recommendation that vaccines be administered during stable disease – the theoretical risk of disease flare following vaccination in unstable patients has always to be considered.

2. As the authors underlined, no randomized controlled studies were available that addressed the efficacy of vaccinations in patients with AIRD on clinical end points. This is why the highest strength of the recommendations is B, reached only in 3 out of 13. Also, a category of evidence of II is reached only in the case of the fourth

Table 12.1 Recommendations for vaccination in adult patients with AIRD, showing level of evidence, strength of recommendation, and results of Delphi voting for each. Strength of recommendations is graded in categories A–D, evidence in categories I–IV. van Assen S et al. Humoral responses after influenza vaccination are severely reduced in patients with rheumatoid arthritis treated with rituximab. Arthritis Rheum 62:75–81. Copyright © 2010, John Wiley & Sons, Inc

Recommendation	Efficacy of vaccination	Harm of vaccination	Strength of recommendation
The vaccination status should be assessed in the initial investigation of patients with AIRD			D
Vaccination in patients with AIRD should ideally be administered during stable disease			D
Live attenuated vaccines should be avoided whenever possible in immunosuppressed patients with AIRD			D
Vaccination in patients with AIRD can be administered during the use of DMARDs and TNF-α blocking agents, but should ideally be administered before B cell-depleting biological therapy is started			B
Influenza vaccination should be strongly considered for patients with AIRD	Ib	Ib	B–C
23-valent polysaccharide pneumococcal vaccination should be strongly considered for patients with AIRD	Ib	Ib	B–C
Patients with AIRD should receive tetanus toxoid vaccination, in accordance with recommendations for the general population. In case of major and/or contaminated wounds in patients who received rituximab within the last 24 weeks, passive immunization with tetanus immunoglobulin should be administered	II	II	B–D
HZ vaccination may be considered in patients with AIRD	—	IV	C–D
HPV vaccination should be considered in selected patients with AIRD	—	—	C–D
In hyposplenic/asplenic patients with AIRD, influenza, pneumococcal, *Haemophilus influenzae* b, and meningococcal C vaccinations are recommended			D
Hepatitis A and/or B vaccination is only recommended in patients with AIRD at risk	II[a]	III[a]	D
Patients with AIRD who plan to travel are recommended to receive their vaccinations according to general rules, except for live attenuated vaccines, which should be avoided whenever possible in immunosuppressed patients			D
BCG vaccination is not recommended in patients with AIRD	—	—	D

[a]For hepatitis B only.

AIRD, autoimmune inflammatory disease; BCG, bacillus Calmette–Guérin; DMARD, disease-modifying antirheumatic drug; HZ, herpes zoster; HPV, human papillomavirus; TNF, tumor necrosis factor

Levels of evidence: Ia, evidence from meta-analysis of randomized controlled trials; Ib, evidence from at least one randomized controlled trial; IIa, evidence from at least one controlled study without randomization; IIb, evidence from at least one other type of quasi-experimental study; III, evidence from nonexperimental descriptive studies, such as comparative studies, correlation studies, and case–control studies; IV, evidence from expert committee reports or opinions, or clinical experience of respected authorities, or both

Grades of recommendation: A, directly based on Level I evidence; B, directly based on Level II evidence or extrapolated recommendations from Level I evidence; C, directly based on Level III evidence or extrapolated recommendations from Level I or II evidence; D, directly based on Level IV evidence or extrapolated recommendations from Level I, II, or III evidence

recommendation, referring to the indication for vaccines in the course of therapy with DMARDs and TNF-α blocking agents.

Thus, further studies are needed to clearly establish the safety and the efficacy of several vaccines, especially the new ones (HZ, H1N1 influenza A, HPV vaccine, and new conjugated pneumococcal vaccines). The efficacy and safety of adjuvants also require further investigations, in order to clarify their effects on subjects with AIRD.

Table 12.2 Summary of the main studies evaluating the efficacy and safety of influenza, pneumococcal, HVB, tetanus, and HPV vaccines

	Author	Study design	No. cases	Efficacy	Influence of ID on efficacy	Safety
Influenza vaccination	Abu-Shakra *et al.* (2002)	Controlled	24 SLE 24 SLE-DC	Reduced in SLE	Reduced on AZA	No flares
	Holvast *et al.* (2009)	RCT	49 WG 23 WG-DC 49 HC	No difference	No	No difference
	Elkayam *et al.* (2010)	Controlled	20 RA-anti-TNF 23 RA-DC 18 SpA-anti-TNF	No difference	No	No difference
	van Assen *et al.* (2010)	Controlled	29 RA 23 RA-RTX 21 HC	Reduced in RA-RTX	Reduced on RTX	No flares
Pneumococcal vaccination	Elkayam *et al.* (2002a)	Controlled	24 SLE 42 RA 29 HC	Reduced in SLE and RA	No	No difference
	Elkayam (2004)	Controlled	11 RA-anti-TNF 5 AS-anti-TNF 17 RA-DC	No difference	Reduced anti-TNF	Not assessed
	Kapetanovic *et al.* (2006)	Controlled	50 RA-MTX/ anti-TNF 62 RA-anti-TNF 37 RA-MTX 47 HC	No difference in RA-anti-TNF Reduced in RA-MTX	Reduced on MTX and MTX/anti-TNF	Not assessed
	Bingham *et al.* (2009)	Controlled	69 RA-RTX 34 RA-DC	Reduced in RA-RTX	Reduced on RTX	Not assessed
Hepatitis B vaccination	Elkayam *et al.* (2002b)	Controlled	22 RA 22 RA-DC	68% protection	No	No flares
	Kuruma *et al.* (2007)	Uncontrolled	28 SLE 20 SpA-anti-TNF	93% protection	Not addressed	11% flares
	Franco Salinas *et al.* (2009)	Controlled	20 SpA-anti-TNF 10 SpA-DC	Reduced in SpA-anti-TNF	Reduced on anti-TNF	Not assessed
Tetanus vaccination	Battafarano *et al.* (1998)	Uncontrolled	73 SLE	90% protection	Trend lower response on PRED and AZA	Not assessed
	Kashef *et al.* (2008)	Controlled	40 SLE 60 HC	No difference	No	Not assessed
	Bingham *et al.* (2010)	Controlled	69 RA-RTX 34 RA-DC	No difference	No	Not assessed
Quadrivalent HPV vaccination	Mok *et al.* (2013)	Controlled	50 SLE 50 HC	Reduced in SLE	Not addressed	10% flares
	Soybilgic *et al.* (2013)	Uncontrolled	20 SLE	No difference	Not addressed	No difference

ID, immunosuppressive drug; RCT, randomized control trial; HC, healthy control; RA, rheumatoid arthritis; SLE, systemic lupus erythematosus; SpA, spondyloarthritis; WG, Wegener granulomatosis; DC, disease control; RTX, rituximab; NA, not assessed; MTX, methotrexate; AZA, azathioprine; PRED, prednisone

Finally, the impact of multiple therapeutic schedules on the prevalence of infections and on the efficacy of vaccines must be further investigated, in order to better define the correct therapeutic strategies in patients affected with AIRD (van Assen *et al.*, 2011b).

Vaccination in AIRD and ASIA patients

As already mentioned, when vaccines are administered in patients with AIRD, they may exacerbate the disease by acting as an adjuvant. Nevertheless, the rate of flares following vaccination in all studies reported is very low (Table 12.2), which can be explained by three main reasons:

1. Most patients evaluated in these studies were already on several treatment schedules, which included immunosuppressive drugs in many cases. Thus, the low seroprotection rates observed in several studies are widely explained by the concomitant or previous immunosuppressive therapy. Indeed, the latest preventive therapy can explain both the low rates of ASIA and the flares in patients with AIRD.

2. The long-term period of observations following the vaccines does not exceed 1 year in several studies. As extensively demonstrated in the literature, the time interval between the administration of an immunogenic stimulus and clinical manifestations of AIRD may be longer (Shoenfeld and Agmon-Levin, 2011)

3. It is possible that in AIRD, the immune machinery is already fully activated via the presence of increased cytokine occupation of receptors, and so on. Thus, the addition of vaccine and adjuvants cannot activate more of the immune system components. Such competition can be noted when different stimuli (i.e. two adjuvants) are given in parallel.

References

Ablin, J.N., Aloush, V., Brill, A., *et al.* (2013). Safety and efficacy of influenza vaccination in patients suffering from fibromyalgia. *Arthritis Rheum*, **65**: 1098.

Abu-Shakra, M., Zalmanson, S., Neumann, L., *et al.* (2000). Influenza virus vaccination of patients with systemic lupus erythematosus: effects on disease activity. *J Rheumatol*, **27**: 1681–5.

Abu-Shakra, M., Press, J., Sukenik, S., and Buskila, D. (2002). Influenza virus vaccination of patients with SLE: effects on generation of autoantibodies. *Clin Rheumatol*, **21**: 369–72.

Agmon-Levin, N., Paz, Z., Israeli, E., and Shoenfeld, Y. (2009). Vaccines and autoimmunity. *Nat Rev Rheumatol*, **5**: 648–52.

Agmon-Levin, N., Hughes, G.R.V., and Shoenfeld, Y. (2012). The spectrum of ASIA: "Autoimmune (Auto-inflammatory) Syndrome induced by Adjuvants." *Lupus*, **21**: 118–20.

Askling, J., Fored, C.M., Brandt, L., *et al.* (2005). Risk and case characteristics of tuberculosis in rheumatoid arthritis associated with tumor necrosis factor antagonists in Sweden. *Arthritis Rheum*, **52**: 1986–92.

Aron-Maor, A. and Shoenfeld, Y. (2001). Vaccination and systemic lupus erythematosus: the bidirectional dilemmas. *Lupus*, **10**: 237–40.

Baronnet, L., Kahn, V., Lacoin, C., *et al.* (2008). Tuberculosis incidence in rheumatoid arthritis affected patients. *Ann Rheum Dis*, **67**: 544.

Barr, M., Buchanan, W.W., Doniach, D., and Roitt, I.M. (1964). The antibody response to tetanus toxoid in patients with autoimmune thyroiditis. *Scott Med J*, **9**: 295.

Bassi, N., Luisetto, R., Del Prete, D., *et al.* (2012). Induction of the "ASIA" syndrome in NZB/NZWF1 mice after injection of complete Freund's adjuvant (CFA). *Lupus*, **21**: 203–9.

Battafarano, D.F., Battafarano, N.J., Larsen, L., *et al.* (1998). Antigen-specific antibody responses in lupus patients following immunization. *Arthritis Rheum*, **14**: 1828–34.

Bingham, C.O. III, Looney, R.J., Deodhar, A., *et al.* (2009). Immunization responses in rheumatoid arthritis patients treated with rituximab: results from a controlled clinical trial. *Arthritis Rheum*, **62**: 64–74.

Blumenfeld, Z., Lorber, M., Yoffe, N., and Sharf, Y. (1994). Systemic lupus erythematosus: predisposition for uterine cervical dysplasia. *Lupus*, **3**: 59–61.

Brassard, P., Lowe, A.M., Bernatsky, S., *et al.* (2009). Rheumatoid arthritis, its treatments, and the risk of tuberculosis in Quebec, Canada. *Arthritis Rheum*, **61**: 300–4.

CDC (Centers for Disease Control and Prevention). (2006). Surveillance for acute viral hepatitis, United States, 2006. Surveillance Summary, 2008. Vol. 57, No. SS-2.

CDC (Centers for Disease Control and Prevention). (2012). Prevention and control of influenza with vaccines: recommendations of the Advisory Committee on Immunization Practices (ACIP) – United States, 2012–13 influenza season. *MMWR*, **61**: 613–18.

Chalmers, A., Scheifele, D., Patterson, C., *et al.* (1994). Immunization of patients with rheumatoid arthritis against influenza: a study of vaccine safety and immunogenicity. *J Rheumatol*, **21**: 1203–6.

Conti, F., Rezai, S., and Valesini, G. (2008). Vaccination and autoimmune rheumatic diseases. *Autoimmun Rev*, **8**: 124–8.

Csuka, D., Czirjàk, L., Hobor, R., *et al.* (2013). Effective humoral immunity against diphtheria and tetanus in patients with systemic lupus erythematosus or myasthenia gravis. *Mol Immunol*, **54**: 453–6.

Del Porto, F., Laganà, B., Biselli, R., *et al.* (2006). Influenza vaccine administration in patients with systemic lupus erythematosus and rheumatoid arthritis Safety and immunogenicity. *Vaccine*, **24**: 3217–23.

De Jong, J.C., Palache, A.M., Beyer, W.E., *et al.* (2003). Haemagglutination-inhibiting antibody to influenza virus. *Dev Biol (Basel)*, **115**: 63–73.

Elkayam, O., Paran, D., Caspi, D., *et al.* (2002a). Immunogenicity and safety of pneumococcal vaccination in patients with rheumatoid arthritis or systemic lupus erythematosus. *Clin Infec Dis*, **34**: 147–53.

Elkayam, O., Yaron, M., and Caspi, D. (2002b). Safety and efficacy of vaccination against hepatitis B in patients with rheumatoid arthritis. *Ann Rheum Dis*, **61**: 623–5.

Elkayam, O., Caspi, D., Reitblatt, T., *et al.* (2004). The effect of tumor necrosis factor blockade on the response to pneumococcal vaccination in patients with rheumatoid arthritis and ankylosing spondylitis. *Semin Arthritis Rheum*, **33**: 282–8.

Elkayam, O., Paran, D., Burke, M., *et al.* (2005). Pneumococcal vaccination of patients with systemic lupus erythematosus: effects on generation of autoantibodies. *Autoimmunity*, **38**: 493–6.

Elkayam, O., Bashkin, A., Mandelboim, M., *et al.* (2010). The effect of infliximab and timing of vaccination on the humoral response to influenza vaccination in patients with rheumatoid arthritis and ankylosing spondylitis. *Semin Arthritis Rheum*, **39**: 442–7.

Fawcett, J., Hutton, C., Mclachlan, S.M., *et al.* (1984). Defective regulation of the immune response to tetanus toxoid in Hashimoto's disease. *Immunology*, **52**: 525.

Fomin, I., Caspi, D., Levy, V., *et al.* (2006). Vaccination against influenza in rheumatoid arthritis: the effect of the disease modifying drugs, including TNF-alpha blockers. *Ann Rheum Dis*, **65**: 191–4.

Franco Salinas, G., de Rycke, L., Cantaert, T., *et al.* (2009). TNF blockade impairs T cell dependent antibody responses. *Ann Rheum Dis*, **68**: 238.

Gatto, M., Agmon-Levin, N., Soriano, A., *et al.* (2013). Human papilloma virus vaccine and systemic lupus erythematosus. *Clin Rheumatol*, **32**: 1302–7.

Gelinck, L.B., van der Bijl, A.E., Beyer, W.E., *et al.* (2008a). The effect of anti-tumour necrosis factor alpha treatment on the antibody response to influenza vaccination. *Ann Rheum Dis*, **67**: 713–16.

Gelinck, L.B., van der Bijl, A.E., Visser, L.G., *et al.* (2008b). Synergistic immunosuppressive effect of anti-TNF combined with methotrexate on antibody responses to the 23-valent pneumococcal polysaccharide vaccine. *Vaccine*, **26**: 3528–33.

Gomez-Reino, J.J., Carmona, L., Valverde, V.R., *et al.* (2003). Treatment of rheumatoid arthritis with tumor necrosis factor inhibitors may predispose to significant increase in tuberculosis risk: a multicenter active-surveillance report. *Arthritis Rheum*, **48**: 2122–7.

Grimaldi-Bensouda, L., Guillemot, D., Godeau, B., *et al.* (2014). Autoimmune disorders and quadrivalent human papillomavirus vaccination of young female subjects. *J Int Med*, **275**: 398–408.

Gross, P.A., Hermogenes, A.W., Sacks, H.S., *et al.* (1995). The efficacy of influenza vaccine in elderly persons. A meta-analysis and review of the literature. *Ann Intern Med*, **123**: 518–27.

Guissa, V.R., Pereira, R.M., Sallum, A.M., *et al.* (2012). Influenza A H1N1/2009 vaccine in juvenile dermatomyositis: reduced immunogenicity in patients under immunosuppressive therapy. *Clin Exp Rheumatol*, **30**: 583–8.

Hak, E., Nordin, J., Wei, F., *et al.* (2002). Influence of high-risk medical conditions on the effectiveness of influenza vaccination among elderly members of 3 large managed-care organizations. *Clin Infect Dis*, **35**: 370–7.

Hess, E.V. and Hahn, B. (1978). Influenza immunization in lupus erythematosus: safe, effective? *Ann Intern Med*, **88**: 833–4.

Heijstek, M.W., Ott de Bruin, L.M., Borrow, R., *et al.* (2011). Vaccination in paediatric patients with auto-immune rheumatic diseases: a systemic literature review for the European League against Rheumatism evidence based recommendations. *Autoimmun Rev*, **11**: 112–22.

Hochberg, M.C. (1997). Updating the American College of Rheumatology revised criteria for the classification of systemic lupus erythematosus. *Arthritis Rheum*, **40**: 1725.

Holvast, A., Huckriede, A., Wilschut, J., *et al.* (2006). Safety and efficacy of influenza vaccination in systemic lupus erythematosus patients with quiescent disease. *Ann Rheum Dis*, **65**: 913–18.

Holvast, B., Huckriede, A., Kallenberg, C.G., and Bijl, M. (2007). Influenza vaccination in systemic lupus erythematosus: safe and protective? *Autoimmun Rev*, **6**: 300–5.

Holvast, A., van Assen, S., de Haan, A., *et al.* (2009). Effect of a second, booster, influenza vaccination on antibody responses in quiescent systemic lupus erythematosus: an open, prospective, controlled study. *Rheumatology (Oxford)*, **48**: 1294–9.

Jara, L.J., Medina, G., Gómez-Bañuelos, E., *et al.* (2012). Still's disease, lupus-like syndrome, and silicone breast implants. A case of "ASIA" (Shoenfeld's syndrome). *Lupus*, **21**: 140–5.

Kahl, L.E. (1994). Herpes zoster infections in systemic lupus erythematosus: risk factors and outcome. *J Rheumatol*, **21**: 84–6.

Kanakoudi-Tsakalidou, F., Trachana, M., Pratsidou-Gertsi, P., *et al.* (2001). Influenza vaccination in children with chronic rheumatic diseases and long-term immunosuppressive therapy. *Clin Exp Rheumatol*, **19**: 589–94.

Kang, T.Y., Lee, H.S., Kim, T.H., *et al.* (2005). Clinical and genetic risk factors of herpes zoster in patients with systemic lupus erythematosus. *Rheumatol Int*, **25**: 97–102.

Kapetanovic, M.C., Saxne, T., Sjoholm, A., *et al.* (2006). Influence of methotrexate, TNF blockers and prednisolone on antibody responses to pneumococcal polysaccharide vaccine in patients with rheumatoid arthritis. *Rheumatology (Oxford)*, **45**: 106–11.

Kashef, S., Ghazizadeh, F., Derakhshan, A., *et al.* (2008). Antigen-specific antibody response in juvenile-onset SLE patients following routine immunization with tetanus toxoid. *Iran J Immunol*, **5**: 181–4.

Klippel, J.H., Karsh, J., Stahl, N.I., *et al.* (1979). A controlled study of pneumococcal polysaccharide vaccine in systemic lupus erythematosus. *Arthritis Rheum*, **22**: 1321–5.

Klumb, E.M., Pinto, A.C., Jesus, G.R., *et al.* (2010). Are women with lupus at higher risk of HPV infection? *Lupus*, **19**: 1485–91.

Kuruma, K., Borba, E.F., Lopes, M.H., *et al.* (2007). Safety and efficacy of hepatitis B vaccine in systemic lupus erythematosus. *Lupus*, **16**: 350–4.

Looney, J.M., Edsall, G., Ipsen, J., and Chansen, W.H. (1956). Persistence of antitoxin levels after tetanus toxoid inoculation in adults and effect of a booster dose after various intervals. *N Engl J Med*, **254**: 6.

Malleson, P.N., Tekano, J.L., Scheifele, D.W., and Weber, J.M. (1993). Influenza immunization in children with chronic arthritis: a prospective study. *J Rheumatol*, **20**: 1769–73.

Manzi, S., Kuller, L.H., Kutzer, J., *et al.* (1995). Herpes zoster in systemic lupus erythematosus. *J Rheumatol*, **22**: 1254–8.

Melo Gomes, S., Glover, M., Malone, M., and Brogan, P. (2013). Vasculitis following HPV immunization. *Rheumatology (Oxford)*, **52**: 581–2.

Mercado, U., Acosta, H., and Avendano, L. (2004). Influenza vaccination of patients with systemic lupus erythematosus. *Rev Invest Clin*, **56**: 16–20.

Mikaeloff, Y., Caridade, G., Suissa, S., and Tardieu, M. (2009). Hepatitis B vaccine and the risk of CNS inflammatory demyelination in childhood. *Neurology*, **72**: 873–80.

Mok, C.C., Ho, L.Y., Fong, L.S., and To, C.H. (2013). Immunogenicity and safety of a quadrivalent human papillomavirus vaccine in patients with systemic lupus erythematosus: a case–control study. *Ann Rheum Dis*, **72**: 659–64.

Murdaca, G., Orsi, A., Spanò, F., *et al.* (2014). Influenza and pneumococcal vaccinations of patients with systemic lupus erythematosus: current views upon safety and immunogenicity. *Autoimmun Rev*, **13**: 75–84.

Naveau, C. and Houssiau, F.A. (2005). Pneumococcal sepsis in patients with systemic lupus erythematosus. *Lupus*, **14**: 903–6.

Nichol, K.L., Wuorenma, J., and von Sternberg, T. (1998). Benefits of influenza vaccination for low, intermediate, and high risk senior citizens. *Arch Intern Med*, **158**: 1769–76.

Nyberg, G., Eriksson, O., and Westberg, N.G. (1981). Increased incidence of cervical atypia in women with systemic lupus erythematosus treated with chemotherapy. *Arthritis Rheum*, **24**: 648–50.

Older, S.A., Battafarano, D.F., Enzenauer, R.J., and Krieg, A.M. (1999). Can immunization precipitate connective tissue disease? Report of five cases of systemic lupus erythematosus and review of the literature. *Semin Arthritis Rheum*, **29**: 131–9.

Perdan-Pirkmajer, K., Thallinger, G.G., Snoj, N., *et al.* (2012). Autoimmune response following influenza vaccination in patients with autoimmune inflammatory rheumatic disease. *Lupus*, **21**: 175–83.

Ristow, S.C., Gouglas, G., and Condemi, J.J. (1978). Influenza vaccination of patients with systemic lupus erythematosus. *Ann Intern Med*, **88**: 786–9.

Saad, C.G., Borba, E.F., Aikawa, N.E., *et al.* (2011). Immunogenicity and safety of the 2009 non-adjuvanted influenza A/H1N1 vaccine in a large cohort of autoimmune rheumatic diseases. *Ann Rheum Dis*, **70**: 1068–73.

Salemi, S., Picchianti-Diamanti, A., Germano, V., *et al.* (2010). Influenza vaccine administration in rheumatoid arthritis patients under treatment with TNFalpha blockers: safety and immunogenicity. *Clin Immunol*, **134**: 113–20.

Seong, S.S., Choi, C.B., Woo, J.H., *et al.* (2007). Incidence of tuberculosis in Korean patients with rheumatoid arthritis (RA): effects of RA itself and of tumor necrosis factor blockers. *J Rheumatol*, **34**: 706–11.

Shoenfeld, Y. and Agmon-Levin, N. (2011). "ASIA" – autoimmune/inflammatory syndrome induced by adjuvants. *J Autoimmun*, **36**: 4–8.

Shinjo, S.K., de Moraes, J.C., Levy-Neto, M., *et al.* (2012). Pandemic unadjuvanted influenza A (H1N1) vaccine in dermatomyositis and polymyositis: immunogenicity independent of therapy and no harmful effect in disease. *Vaccine*, **31**: 202–6.

Sichletidis, L., Settas, L., Spyratos, D., *et al.* (2006). Tuberculosis in patients receiving anti-TNF agents despite chemoprophylaxis. *Int J Tuberc Lung Dis*, **10**: 1127–32.

Smitten, A.L., Choi, H.K., Hochberg, M.C., *et al.* (2007). The risk of herpes zoster in patients with rheumatoid arthritis in the United States and the United Kingdom. *Arthritis Rheum*, **57**: 1431–8.

Soldevilla, H.F., Briones, S.F., and Navarra, S.V. (2012). Systemic lupus erythematosus following HPV immunization or infection? *Lupus*, **21**: 158–61.

Soybilgic, A., Onel, K.B., Utset, T., *et al.* (2013). Safety and immunogenicity of the quadrivalent HPV vaccine in female systemic lupus erythematosus patients aged 12 to 26 years. *Pediatr Rheumatol Online J*, **11**: 29.

Stubgen, J.P. (2010). Neuromuscular disorders associated with hepatitis B vaccination. *J Neurol Sci*, **292**: 1–4.

Tarjan, P., Sipka, S., Marodi, L., *et al.* (2002). No short-term immunological effects of Pneumococcus vaccination in patients with systemic lupus erythematosus. *Scand J Rheumatol*, **31**: 211–15.

van Assen, S., Holvast, A., Benne, C.A., *et al.* (2010). Humoral responses after influenza vaccination are severely reduced in patients with rheumatoid arthritis treated with rituximab. *Arthritis Rheum*, **62**: 75–81.

van Assen, S., Elkayam, O., Agmon-Levin, N., *et al.* (2011a). Vaccination in adult patients with auto-immune inflammatory rheumatic diseases: a systematic literature review for the European League Against Rheumatism evidence-based recommendations for vaccination in adult patients with auto-immune inflammatory rheumatic diseases. *Autoimmun Rev*, **10**: 341–52.

van Assen, S., Agmon-Levin, N., Elkayam, O., *et al.* (2011b). EULAR reccomendations for vaccination in adult patients with autoimmune inflammatory rheumatic diseases. *Ann Rheum Dis*, **70**: 414–22.

Vista, E.S., Crowe, S.R., Thompson, L.F., *et al.* (2012). Influenza vaccination can induce new-onset anticardiolipins but not β2-glycoprotein-I antibodies among patients with systemic lupus erythematosus. *Lupus*, **21**: 168–74.

Wiesik-Szewczyk, E., Romanowska, M., Mielnik, P., *et al.* (2010). Anti-influenza vaccination in systemic lupus erythematosus patients: an analysis of specific humoral response and vaccination safety. *Clin Rheumatol*, **29**: 605–13.

Williams, G.W., Steinberg, A.D., Reinertsen, J.L., *et al.* (1978). Influenza immunization in systemic lupus erythematosus. A double-blind trial. *Ann Intern Med,* **88**: 729–34.

Wolfe, F., Michaud, K., Anderson, J., and Urbansky, K. (2004). Tuberculosis infection in patients with rheumatoid arthritis and the effect of infliximab therapy. *Arthritis Rheum,* **50**: 372–9.

Wotton, C.J. and Goldacre, M.J. (2012). Risk of invasive pneumococcal disease in people admitted to hospital with selected immune-mediated diseases: record linkage cohort analyses. *J Epidemiol Community Health,* **66**: 1177–81.

WHO (World Health Organization). (2004). Hepatitis B vaccine. *Weekly Epidemiol Rec,* **79**: 255–63.

Yamada, T., Nakajima, A., Inoue, E., *et al.* (2006). Increased risk of tuberculosis in patients with rheumatoid arthritis in Japan. *Ann Rheum Dis,* **65**: 1661–3.

Yee, A.M., Ng, S.C., Sobel, R.E., and Salmon, J.E. (1997). Fc gamma RIIA polymorphism as a risk factor for invasive pneumococcal infections in systemic lupus erythematosus. *Arthritis Rheum,* **40**: 1180–2.

Yun, J.E., Lee, S.W., Kim, T.H., *et al.* (2002). The incidence and clinical characteristics of Mycobacterium tuberculosis infection among systemic lupus erythematosus and rheumatoid arthritis patients in Korea. *Clin Exp Rheumatol,* **20**: 127–32.

Zafrir, Y., Agmon-Levin, N., Paz, Z., *et al.* (2012). Autoimmunity following Hepatitis B vaccine as part of the spectrum of "Autoimmune (Auto-inflammatory) Syndrome induced by Adjuvants" (ASIA): analysis of 93 cases. *Lupus,* **21**: 146–52.

II

Studies on Autoimmune Conditions Induced by Vaccination

13 Measles, Mumps, and Rubella Vaccine: A Triad to Autoimmunity

Carlo Perricone,[1] Guido Valesini,[1] and Yehuda Shoenfeld[2,3]

[1]Rheumatology, Department of Internal and Specialized Medicine, Sapienza University of Rome, Rome, Italy

[2]Zabludowicz Center for Autoimmune Diseases, Sheba Medical Center, Tel Hashomer, Israel

[3]Incumbent of the Laura Schwarz-Kipp Chair for Research of Autoimmune Diseases, Sackler Faculty of Medicine, Tel Aviv University, Tel Aviv, Israel

Introduction

Vaccines against measles, mumps, and rubella (MMR) are indicated for simultaneous vaccination in individuals ≥12 months of age. These viruses can infect most exposed subjects and for a long time they led to high morbidity and mortality. Today, they have been eradicated in large geographic areas; this was achieved through the release of the first trivalent MMR vaccine in 1971 (CDC, 2011a,b).

Seven years later, in 1978, MMR II, with the HPV-77 DE strain of live attenuated rubella replaced by the Wistar RA 27/3 live attenuated rubella strain, was licensed; since then, this has been the only version used (Plotkin *et al.*, 1973). More recently, recombinant human albumin replaced human-derived serum albumin, allowing for the elimination of any human-derived substances. As of 2010, over 575 million doses of MMR II had been administered. The efficacy of the vaccine was recently reviewed in a Cochrane meta-analysis: data from 14 700 000 children assessed in different studies, including five randomized controlled trials (RCTs), suggest that one MMR vaccine dose is at least 95% effective in preventing clinical measles and 92% effective in preventing secondary cases among household contacts. The efficacy in preventing clinical mumps ranges from 69 to 81%, while data on rubella are less consistent (Demicheli *et al.*, 2012).

The MMR vaccine has been generally considered as safe. In a long-term post-marketing surveillance, 17 536 adverse events were voluntarily reported, with an overall rate of 30.5 adverse events per million doses (Lievano *et al.*, 2012). Looking at the controlled clinical trials, the most frequently reported clinical adverse effects were injection-site reactions, consisting of erythema, pain, and swelling. Transient fever and upper-respiratory infection have also been documented. Serious adverse events (SAEs) are quite rare, with a risk ratio (RR) of 8.4 serious events per 106 doses distributed. However, 136 temporally associated deaths have been reported following MMR vaccination, mostly in children <5 years of age, and mostly caused by bacterial and viral infections. Approximately 1 in 10 of the fatal cases involved immunocompromised patients; six of these developed measles infections. Nonetheless, it is well known that post-marketing surveillance has limitations, because it is self-reported and the reports are often incomplete. The former aspect is of particular relevance, since the subject may not

Vaccines and Autoimmunity, First Edition. Edited by Yehuda Shoenfeld, Nancy Agmon-Levin, and Lucija Tomljenovic.

recognize an adverse event that is not included in the vaccine data sheet and that may be temporally distant from time of vaccination. Indeed, there are concerns that such vaccinations may be a trigger for adverse events of an autoimmune nature, which manifest themselves as a clinically recognizable disease only many months or even years after administration (Perricone *et al.*, 2012).

Central nervous system

The involvement of the central nervous system (CNS) following infections with measles and rubella viruses is rather uncommon. The onset of post-infectious encephalitis has been described with a rate of 1:1000 to 1:5000 acute cases. The time of latency in these acute forms is 1–3 weeks with monophasic clinical course. Acute disseminated encephalomyelitis (ADEM), a clinical entity that is immune-mediated and similar to experimental autoimmune encephalomyelitis, has been associated with several viral infections, including measles, mumps, and rubella (Altman *et al.*, 2012). These viruses may rarely provoke the development of another inflammatory condition, often associated with autoimmune diseases, such as systemic lupus erythematosus (SLE), which is transverse myelitis (Agmon-Levin *et al.*, 2009a) Sensorineural hearing loss, optic neuritis, and polyneuritis have also been reported in association with the MMR (Asatryan *et al.*, 2008). No causal association seems to prevail between MMR vaccination and Guillain–Barré syndrome (GBS) (Arshi *et al.*, 2004), and only very recently, a fatal case of encephalitis associated with MMR vaccine was reported in Brazil (Patja *et al.*, 2001). In this patient, symptoms developed within 3 days of vaccination. Histopathology confirmed encephalitis, immunohistochemistry, and qPCR were positive for rubella virus on brain tissue. Virus was also isolated from cerebrospinal fluid (CSF) and other clinical samples. The sequence obtained from the isolated virus was identical to that of the RA 27/3 vaccine strain, suggesting a causal link (Gualberto *et al.*, 2013).

Thrombocytopenia

There are several lines of evidence that thrombocytopenia is related to the MMR vaccination (Molina and Shoenfeld, 2005). In the US Vaccine Adverse Event Reporting System (VAERS), there were 250 reports (259 events) of thrombocytopenia within 77 days post-vaccination (Mantadakis *et al.*, 2010). Usually, the onset appears to be sudden, with a median of 13 days. Patient age was largely variable, ranging from 16 weeks to 53 years.

In a Finnish report, acute idiopathic thrombocytopenia (ITP) developed shortly after MMR vaccination in 23 out of 700 000 children (Nieminen *et al.*, 1993). Usually the event is not serious, but it should be remembered that the risk for thrombocytopenia during infection with these viruses is much greater than the vaccine-associated risk (Bayer *et al.*, 1965). MMR vaccine was first associated with ITP in small studies (Miller *et al.*, 2001), and this association was later confirmed in a review of more data (Black *et al.*, 2003). An increased risk of ITP within 6 weeks of MMR immunization in children aged 12–23 months was assessed in one case–control study (RR 6.3; 95% CI: 1.3–30.1) and in one small self-controlled case series (incidence rate ratio (IRR) 5.38; 95% CI: 2.72–10.62). Increased risk of thrombocytopenic purpura within 6 weeks of MMR exposure was also assessed in one other case–control study involving 2311 children and adolescents between 1 month and 18 years of age (odds ratio (OR) 2.4; 95% CI: 1.2–4.7) (Demicheli *et al.*, 2012).

In another study looking at a total of 1 036 689 children who cumulatively received 1 107 814 MMR vaccinations, 259 developed ITP. Exposed patients aged 12–23 months had lower median platelet counts than those who were unexposed and had similar median duration of illness (11 versus 10 days) (France *et al.*, 2008). The IRR was highest for children aged 12–15 months, at 7.10; the IRR for boys aged 12–15 months was 14.6, and that for girls in the same age group was 3.2. In children aged 12–23 months, 76% of ITP cases were attributable to the MMR vaccination. In this study, the vaccine caused 1 case of ITP for every 40 000 doses, which is a relatively low incidence rate (France *et al.*, 2008).

A recent study evaluated the consistency of the association between drug and vaccine use and ITP onset in children. This was part of an ongoing Italian multicenter study on adverse drug reactions in children, coordinated by the Italian National Institute of Health, which began in November 1999 (Bertuola *et al.*, 2010). Up to December 2007, the study population included 387 cases of thrombocytopenia and 1924 controls. After statistical analyses, it was found that, despite the low platelet count, ITP was generally a mild disease, without serious bleeding in the majority of cases, and associated with a short length of hospital stay. After adjusting for concurrent potential confounding factors, MMR vaccination was associated with an increased risk of developing ITP (OR 2.4). The results of this study provide evidence for the association between ITP and exposure to MMR vaccination. However, in this case, MMR-associated ITP was self-limited and non-life-threatening. Thus, it seems that MMR vaccination is justified in children, because its benefits clearly outweigh the potential adverse secondary autoimmune phenomena (Mantadakis *et al.*, 2010).

Arthritis

Joint reactions following rubella vaccination have been well recognized since the late 1960s. The vaccine may cause viremia and the development of transient arthralgia (about 25%), acute arthritis (<10%), and even chronic arthritis (although rarely) (Tingle *et al.*, 1986; Howson and Fineberg, 1992; Geier and Geier, 2002). Female gender, older age, prior seronegativity, and certain HLA types appear to be risk factors. More recent studies, however, have found no evidence of increased risk for chronic arthropathy among women vaccinated against rubella (Slater *et al.*, 1995; Ray *et al.*, 1997).

The incidence of joint manifestations was assessed 6 weeks after immunization with the MMR vaccine (Benjamin *et al.*, 1992). The study included 2658 immunized and 2359 nonimmunized (but immunization-eligible) children. There was an increased risk of joint symptoms (arthralgia or arthritis) in the immunized children 6 weeks after immunization. It was concluded that the rubella component of the vaccine was the most likely to have caused the joint symptoms reported (Abedi *et al.*, 2012). The risk of arthritis was less than after wild rubella infection.

Arthritis-related complications occur less frequently in children. In most cases, arthritis resolves completely without any permanent sequelae. Only rare cases may develop a system-chronic polyarthritis resembling the features of rheumatoid arthritis. More recently, a comprehensive search for adverse events following MMR vaccination was conducted using physician records in the US VAERS. Just 1.8% reported "joint or muscle aches," but no SAEs (Tishler and Shoenfeld, 2004).

Type 1 diabetes

Several studies show a growing rate of type 1 diabetes mellitus (T1DM) in childhood coincident with increasing diagnosis of viral infections. In an Italian cohort study, the authors found a significant association between the incidence of T1DM and that of mumps ($p = 0.034$) and rubella ($p = 0.014$). Notably, data from the Sardinian registry were excluded from the statistical analysis in this study, because of the different genetic background of this population (Ramondetti *et al.*, 2012). In another study, it was shown that previous exposure to MMR, but not vaccination during adolescence, correlated to the prevalence of pancreatic and thyroid autoantibodies (Lindberg *et al.*, 1999). Indeed, the vaccination did not change the prevalence or the level of autoantibodies: those children with rubella antibodies before vaccination had higher levels of immune complexes than those who were seronegative.

Nonetheless, there seems to be no association between the MMR vaccination and T1DM. Indeed, in an epidemiological study looking at 4 720 517 persons/year of follow-up, T1DM was diagnosed in 681 children (Hviid *et al.*, 2004). The rate ratio for T1DM among children who received at least one dose of vaccine, as compared to unvaccinated children, was 0.91 (95% CI: 0.74–1.12), and 1.14 (95% CI: 0.90–1.45) for MMR vaccine. Furthermore, the development of type 1 diabetes in genetically predisposed children (defined as those who had siblings with type 1 diabetes) was not significantly associated with vaccination, and there was no evidence of any clustering of cases 2–4 years after vaccination with any vaccine (Hviid *et al.*, 2004). Interestingly, the OR for the association between T1DM and MMR vaccine was the same (1.14; 95% CI: 4 (0.51–2.57)) in another case–control study conducted within four health-maintenance organizations that participate in the US Centers for Disease Control and Prevention (CDC)'s Vaccine Safety Datalink project (DeStefano *et al.*, 2001). Such results are further confirmed by data prospectively collected from questionnaires obtained at birth, at 9 months of age, and at 2 years of age for 823 children from parents with type 1 diabetes in Germany (Hummel *et al.*, 2000). Even the effect of vaccinations on antibody development was similar in children with T1DM high-risk human leukocyte antigen (HLA) genotypes. However, the presence of low levels of IgG class mumps virus vaccination-induced antibodies was lower in diabetic patients than in their nondiabetic siblings ($p < 0.0005$), suggesting a selective decrease in mumps antibody levels in type 1 diabetic children (Hiltunen *et al.*, 1994). Moreover, in Finland, the risk for T1DM seemed to be decreased following the introduction of the MMR vaccination (Hyöty *et al.*, 1993). A retrospective cohort study of US military personnel, consisting of 2 385 102 individuals followed for approximately 7 644 098 person years of service, included 1074 T1DM cases. MMR vaccine did not increase the risk of T1DM (RR 0.71; 95% CI: 0.61, 0.83). The main limitations of the study were the retrospective analysis and the method of data collection (Duderstadt *et al.*, 2012). Indeed, a clustering of diabetes 2–4 years after measles and mumps infections was observed (Lipman *et al.*, 2002). The presence of distinct rises in the incidence of T1DM occurring 2–4 years following the introduction of the MMR pertussis vaccines was suggested. Such a temporal period may allow for the establishment of the disease, with progression to T1DM in vaccinated patients developing with antipancreatic autoantibodies (Classen and Classen, 2003).

Other autoimmune conditions

Infections with the MMR viruses are seldom associated with the later development of autoimmune damage (Shoenfeld, 2009). Measles virus was at one time linked to some cases of idiopathic autoimmune hepatitis (Robertson *et al.*, 1987), but the evidence for this remains scarce. Postnatal rubella infection has been suggested (along with varicella, influenza, and other viruses) to potentially be linked to myocarditis, and to some cases of chronic arthritis, which may in fact be related to viral persistence (Chantler *et al.*, 1985; Schattner, 2005). Meanwhile, other important studies have found no support for an increased risk of conditions such as asthma, leukemia, hay fever, and Crohn's disease (Miller, 2002) following exposure to the MMR vaccine.

Conclusions

It is clear that, excluding exceptional reports (two cases each) of anterior uveitis, retinopathy, vasculitis, and myositis, most adverse events following MMR vaccination involve the nervous system, joints, and blood-related disorders (i.e. ITP) (Table 13.1). It is noteworthy that such kinds of disease reflect the spectrum of autoimmune manifestations that may occur following MMR virus infections (Agmon-Levin *et al.*, 2009b). There are a number of plausible mechanisms that might account for vaccine-induced autoimmunity, including the fact that the MMR vaccines are composed of infectious antigens, immune adjuvants, preservatives, and other ingredients that may trigger the development or exacerbation of autoimmune phenomena (Colafrancesco *et al.*, 2013). Vaccines are an essential part of preventive modern medicine, but the potential for them to trigger the ASIA syndrome in susceptible individuals should not be ignored. Further efforts should also be made to better investigate the long-term safety of routinely used vaccines (Shoenfeld and Agmon-Levin, 2011; Perricone *et al.*, 2013).

Table 13.1 Autoimmune manifestations following measles, mumps, and rubella (MMR) vaccination

Neurologic
• Aseptic meningitis
• Encephalitis (ADEM type)
• Optic neuritis
• Guillain–Barré syndrome
Hematologic
• Thrombocytopenia/acute thrombocytopenic purpura
• Hemolytic–uremic syndrome
• Hemolytic anemia
Other
• Acute (transient) and chronic arthritis
• Sensorineural hearing loss

References

Abedi, G.R., Mutuc, J.D., Lawler, J., *et al.* (2012). Adverse events following a third dose of measles, mumps, and rubella vaccine in a mumps outbreak. *Vaccine*, **30**(49): 7052–8.

Agmon-Levin, N., Kivity, S., Szyper-Kravitz, M., and Shoenfeld, Y. (2009a). Transverse myelitis and vaccines: a multi-analysis. *Lupus*, **18**(13): 1198–204.

Agmon-Levin, N., Paz, Z., Israeli, E., and Shoenfeld, Y. (2009b). Vaccines and autoimmunity. *Nat Rev Rheumatol*, **5**(11): 648–52.

Altman, A., Szyper-Kravitz, M., Agmon-Levin, N., *et al.* (2012). Prevalence of rubella serum antibody in autoimmune diseases. *Rev Bras Reumatol*, **52**(3): 307–18.

Arshi, S., Sadeghi-Bazargani, H., Ojaghi, H., *et al.* (2004). The first rapid onset optic neuritis after measles-rubella vaccination: case report. *Vaccine*, **22**(25–6): 3240–2.

Asatryan, A., Pool, V., Chen, R.T., *et al.* (2008). Live attenuated measles and mumps viral strain-containing vaccines and hearing loss: Vaccine Adverse Event Reporting System (VAERS), United States, 1990–2003. *Vaccine*, **26**(9): 1166–72.

Bayer, W.L., Sherman, F.E., Michaels, R.H., *et al.* (1965). Purpura in congenital and acquired rubella. *N Engl J Med*, **273**: 1362–6.

Benjamin, C.M., Chew, G.C., and Silman, A.J. (1992) Joint and limb symptoms in children after immunisation with measles, mumps, and rubella vaccine. *BMJ*, **304**(6834): 1075–8.

Bertuola, F., Morando, C., Menniti-Ippolito, F., *et al.* (2010). Association between drug and vaccine use and acute immune thrombocytopenia in childhood: a case-control study in Italy. *Drug Saf*, **33**: 65–72.

Black, C., Kaye, J.A., and Jick, H. (2003). MMR vaccine and idiopathic thrombocytopaenic purpura. *Br J Clin Pharmacol*, **55**(1):107-11.

CDC (Centers for Disease Control and Prevention). (2011a). Notifiable diseases and mortality tables. *MMWR*, **60**: 1362–75.

CDC (Centers for Disease Control and Prevention). (2011b). Summary of notifiable diseases – United States 2009. *MMWR*, **58**: 1–100.

Chantler, J.K., Tingle, A.J., and Petty RE. (1985). Persistent rubella virus infection associated with chronic arthritis in children. *N Engl J Med*, **313:** 1117–23.

Classen, J.B. and Classen, D.C. (2003) Clustering of cases of type 1 diabetes mellitus occurring 2-4 years after vaccination is consistent with clustering after infections and progression to type 1 diabetes mellitus in autoantibody positive individuals. *J Pediatr Endocrinol Metab*, **16**(4): 495–508.

Colafrancesco, S., Agmon-Levin, N., Perricone, C., and Shoenfeld, Y. (2013). Unraveling the soul of autoimmune diseases: pathogenesis, diagnosis and treatment adding dowels to the puzzle. *Immunol Res*, **56**(2–3): 200–5.

Demicheli, V., Rivetti, A., Debalini, M.G., and Di Pietrantonj, C. (2012). Vaccines for measles, mumps and rubella in children. *Cochrane Database Syst Rev*, **2**: CD004407.

DeStefano, F., Mullooly, J.P., Okoro, C.A., *et al.* (2001). Childhood vaccinations, vaccination timing, and risk of type 1 diabetes mellitus. *Pediatrics*, **108**(6): E112.

Duderstadt, S.K., Rose, C.E. Jr, Real, T.M., *et al.* (2012). Vaccination and risk of type 1 diabetes mellitus in active component U.S. Military, 2002–2008. *Vaccine*, **30**(4): 813–19.

France, E.K., Glanz, J., Xu, S., *et al.* (2008). Risk of immune thrombocytopenic purpura after measles-mumps-rubella immunization in children. *Pediatrics*, **121**: 687–92.

Geier, D.A. and Geier, M.R. (2002). A one year follow-up of chronic arthritis following rubella and hepatitis B vaccination based upon analysis of the Vaccine Adverse Events Reporting System (VAERS) database. *Clin Exp Rheumatol*, **20**: 767–71.

Gualberto, F.A., de Oliveira, M.I., Alves, V.A., *et al.* (2013). Fulminant encephalitis associated with a vaccine strain of rubella virus. *J Clin Virol*, **58**(4): 737–40.

Hiltunen, M., Hyöty, H., Leinikki, P., *et al.* (1994). Low mumps antibody levels induced by mumps-measles-rubella vaccinations in type 1 diabetic children. *Diabet Med*, **11**(10): 942–6.

Howson, C.P. and Fineberg, H.V. (1992). Adverse events following pertussis and rubella vaccines. *Summary of a report of the Institute of Medicine. JAMA*, **267**: 392–6.

Hummel, M., Füchtenbusch, M., Schenker, M., and Ziegler, A.G. (2000). No major association of breast-feeding, vaccinations, and childhood viral diseases with early islet autoimmunity in the German BABYDIAB Study. *Diabetes Care*, **23**(7): 969–74.

Hviid, A., Stellfeld, M., Wohlfahrt, J., and Melbye, M. (2004). Childhood vaccination and type 1 diabetes. *N Engl J Med*, **350**(14): 1398–404.

Hyöty, H., Hiltunen, M., Reunanen, A., *et al.* (1993). Decline of mumps antibodies in type 1 (insulin-dependent) diabetic children and a plateau in the rising incidence of type 1 diabetes after introduction of the mumps-measles-rubella vaccine in Finland. Childhood Diabetes in Finland Study Group. *Diabetologia*, **36**(12): 1303–8.

Lievano, F., Galea, S.A., Thornton, M., *et al.* (2012). Measles, mumps, and rubella virus vaccine (M-M-R™II): a review of 32 years of clinical and postmarketing experience. *Vaccine*, **30**(48): 6918–26.

Lindberg, B., Ahlfors, K., Carlsson, A., *et al.* (1999). Previous exposure to measles, mumps, and rubella – but not vaccination during adolescence – correlates to the prevalence of pancreatic and thyroid autoantibodies. *Pediatrics*, **104**(1):e12.

Lipman, T.H., Chang, Y., and Murphy, K.M. (2002). The epidemiology of type 1 diabetes in children in Philadelphia 1990-1994: evidence of an epidemic. *Diabetes Care*, **25**(11), 1969–75.

Mantadakis, E., Farmaki, E., and Buchanan, G.R. (2010). Thrombocytopenic purpura after measles-mumps-rubella vaccination: a systematic review of the literature and guidance for management. *J Pediatr*, **156**: 623–8.

Miller, E. (2002). MMR vaccine: review of benefits and risks. *J Infect*, **44**(1): 1–6.

Miller, E., Waight, P., Farrington, C.P., *et al.* (2001). Idiopathic thrombocytopenic purpura and MMR vaccine. *Arch Dis Child*, **84**(3): 227–9.

Molina, V. and Shoenfeld, Y. (2005). Infection, vaccines and other environmental triggers of autoimmunity. *Autoimmunity*, **38**(3): 235–45.

Nieminen, U., Peltola, H., Syrjala, M.T., *et al.* (1993). Acute thrombo-cytopenic purpura following measles, mumps, and rubella vaccination. A report on 23 patients. *Acta Paediatr*, **82**: 267–70.

Patja, A., Paunio, M., Kinnunen, E., *et al.* (2001). Risk of Guillain-Barré syndrome after measles-mumps-rubella vaccination. *J Pediatr*, **138**(2): 250–4.

Perricone, C., Agmon-Levin, N., Valesini, G., and Shoenfeld, Y. (2012). Vaccination in patients with chronic or autoimmune rheumatic diseases: the ego, the id and the superego. *Joint Bone Spine*, **79**(1): 1–3.

Perricone, C., Colafrancesco, S., Mazor, R.D., *et al.* (2013). Autoimmune/inflammatory syndrome induced by adjuvants (ASIA) 2013: unveiling the pathogenic, clinical and diagnostic aspects. *J Autoimmun*, **47**: 1–16.

Plotkin, S.A., Farquhar, J.D., and Ogra, P.L. (1973). Immunologic properties of RA27/3 rubella virus vaccine: a comparison with strains presently licensed in the United States. *JAMA*, **225**: 585–90.

Ramondetti, F., Sacco, S., Comelli, M., *et al.* (2012). Type 1 diabetes and measles, mumps and rubella childhood infections within the Italian Insulin-dependent Diabetes Registry. *Diabet Med*, **29**(6): 761–6.

Ray, P., Black, S., Shinefield, H., *et al.* (1997). Risk of chronic arthropathy among women after rubella vaccination. *JAMA*, **278**: 551–6.

Robertson, D.A., Zhang, S.L., Guy, E.C., and Wright, R. (1987). Persistent measles virus genome in autoimmune chronic active hepatitis. *Lancet*, **2**: 9–11.

Schattner, A. (2005). Consequence or coincidence? The occurrence, pathogenesis and significance of autoimmune manifestations after viral vaccines. *Vaccine*, **23**(30): 3876–86.

Shoenfeld, Y. (2009). Infections, vaccines and autoimmunity. *Lupus*, **18**(13): 1127–8.

Shoenfeld, Y. and Agmon-Levin, N. (2011). "ASIA" – autoimmune/inflammatory syndrome induced by adjuvants. *J Autoimmun*, **36**(1): 4–8.

Slater, P.E., Ben-Zvi, T., Fogel, A., *et al.* (1995). Absence of an association between rubella vaccination and arthritis in underimmune postpartum women. *Vaccine*, **13**: 1529–32.

Tingle, A.J., Allen, M., Petty, R.E., *et al.* (1986). Rubella-associated arthritis. Comparative study of joint manifestations associated with natural rubella infection and RA27/3 rubella immunization. *Ann Rheum Dis*, **45**: 110–14.

Tishler, M. and Shoenfeld, Y. (2004). Vaccination may be associated with autoimmune diseases. *Isr Med Assoc J*, **6**(7): 430–2.

14 Yellow Fever Vaccine and Autoimmunity

Roger A. Levy[1] and Rodrigo Poubel V. Rezende[1,2]
[1]Faculty of Medical Sciences, Rio de Janeiro State University, Rio de Janeiro, Brazil
[2]Brazilian Society of Rheumatology, Rio de Janeiro, Brazil

Introduction

Yellow fever (YF) is an acute febrile illness caused by a mosquito-borne flavivirus that afflicts humans and other primates and is currently endemic in 44 countries in the tropical regions of Africa and South America. Three types of transmission cycle have been observed: sylvatic (or jungle), intermediate (African savanna), and urban. The sylvatic is the most common form of transmission in Central and South America, as infected *Haemagogus* and *Aedes* mosquitoes incidentally bite young men (70–90%) working in or near the rainforest. Large outbreaks in urban settings occur when infected people are bitten by mosquitoes and then transmit the virus (*Aedes aegypti*) to noninfected people living in communities with little or no immunity to YF (WHO, 2013). Scientists have estimated that unvaccinated travelers bound for Africa are 10 times more likely to acquire and die from YF than those heading to South America. Higher YF vaccination coverage in the latter region and the natural virus transmission in the forest canopy (i.e. away from human contact) are the chief reasons for this finding (Monath and Cetron, 2002; Monath *et al.*, 2002).

Infection with the YF virus can go unnoticed or can result in a wide spectrum of clinical manifestations. Severe forms usually present with hepatorenal failure, hemorrhagic diathesis, and cardiovascular instability. The brain is frequently spared, as the virus exhibits a viscerotropic rather than neurotropic affinity. Laboratory diagnosis of YF is generally accomplished by serological detection of YF virus-specific immunoglobulin M (IgM)- and immunoglobulin G (IgG)-neutralizing antibodies (WHO, 2013). The infected host mounts an immune response normally characterized by IgM peak levels at week 2, followed by a decline in titers over the next 1–2 months. Interestingly, these antibodies may eventually persist for more than a year (Gibney *et al.*, 2012). However, individuals previously infected with other flaviviruses may not induce a detectable IgM response subsequent to YF virus infection (Niedrig *et al.*, 2008). Specific IgG-neutralizing antibodies – those which confer lifelong protection against YF – may become detectable in the serum at the end of the first week and last for at least 35 years, or the entire lifespan. Determination and titration through plaque-reduction neutralization test (PRNT) remains the most sensitive and specific diagnostic test by which to ascertain immunity to YF. At present, treatment is based only on supportive clinical measures (WHO, 2013).

Yellow fever vaccine

Overview

Current YF vaccines are live attenuated viral substrains derived from the original wild-type YF virus (lineage 17D) isolated in Ghana in 1927. The two viral variants (17D-204 and 17DD) used for vaccine preparation share 99.9% sequence homology, and both are cultured in embryonic

Vaccines and Autoimmunity, First Edition. Edited by Yehuda Shoenfeld, Nancy Agmon-Levin, and Lucija Tomljenovic.

chicken eggs. The YF vaccine is lyophilized, contains sorbitol and/or gelatin as stabilizers, and lacks preservatives (WHO, 2013).

The YF vaccination has three main goals: to protect inhabitants living in endemic areas, to protect travelers to these at-risk areas, and to prevent international virus spread by viremic travelers. According to the World Health Organization (WHO)'s most recently issued position paper on vaccination against YF (WHO, 2013), a single dose of the YF vaccine is sufficient to provide lifelong protective immunity against the YF disease, removing the need for a booster dose every 10 years. The paper also states that, unless contraindicated, YF vaccination should be offered to every unvaccinated person aged ≥9 months living in or travelling to any area with reported YF cases (WHO, 2013).

The YF vaccine is currently contraindicated for persons with a history of hypersensitivity to any of the vaccine components, including eggs, egg products, chicken proteins, and gelatin; infants aged <6 months; persons with a thymic disorder that is associated with abnormal immune cell function (e.g. thymoma or myasthenia gravis); persons with primary immunodeficiencies, malignant neoplasms, or transplantation; persons with symptomatic HIV infection or $CD4^+$ T cell values < 200/mm^3; and persons on immunosuppressive or immunomodulatory therapies. Additionally, caution is recommended in vaccinating asymptomatic HIV-infected individuals with a $CD4^+$ T lymphocyte count of 200–499/mm^3, persons aged ≥60 years, infants aged 6–8 months, and pregnant or breastfeeding women (Staples *et al.*, 2010).

Immunogenicity and effectiveness

Clinical trials have shown that almost 100% of healthy YF vaccine recipients develop protective levels of neutralizing antibodies within 30 days of vaccination (Monath and Cetron, 2002; Monath *et al.*, 2002). A recent systematic review of the need for a YF booster dose every 10 years has verified that a significant proportion (>90%) of YF vaccinees still have detectable serum levels of neutralizing antibodies up to 20 years after YF vaccination (Gotuzzo, 2013). In view of the proven efficacy and safety record of the YF vaccine, the WHO's latest statement on the YF vaccine recommends that a single dose is enough to confer sustained lifelong protective immunity against YF disease, with no need for additional doses every 10 years (WHO, 2013).

Safety

Since the introduction of YF vaccination in the 1930s (and the subsequent administration of more than 540 million doses), there have been only 12 suspected cases of YF disease post-vaccination (WHO, 2013). In two of these cases, nucleotide sequencing identified wild-type YF virus rather than vaccine virus (Filippis *et al.*, 2004).

Studies of YF vaccination have shown that roughly 25% of vaccinees report some type of mild adverse event, mainly headache, myalgia, low-grade fever, pruritus, urticaria, rash, and discomfort at the injection site (Lindsey *et al.*, 2008). With regard to the occurrence of serious adverse events (SAEs) following YF immunization, three categories have been established:

1. Anaphylactic reactions: estimated to occur in 0.8 of every 100 000 vaccinations, most commonly in individuals with allergies to eggs or gelatin.

2. YF vaccine-associated neurologic disease (YEL-AND): a group of neurologic conditions resulting from either direct viral invasion of the central nervous system (CNS) by the vaccine virus, leading meningitis or encephalitis, or an autoimmune reaction, leading to such conditions as Guillain–Barré syndrome (GBS) or acute disseminated encephalomyelitis (ADEM).

3. YF vaccine-associated viscerotropic disease (YEL-AVD): caused by replication and dissemination of the vaccine virus, mimicking the natural viral infection. Approximately 60% of these cases evolve with multiple-organ dysfunction syndrome.

To date, all reported cases in the literature of YEL-AND and YEL-AVD have been described in primary vaccinees, with recorded incidence rates of 0.25–0.80/100 000 vaccine doses and 0.25–0.40/100 000 vaccine doses, respectively (Khromava *et al.*, 2005; Lindsey *et al.*, 2008). The viremia experienced by many primary YF vaccinees may be one of the reasons for this association, although it is not described in recipients of a booster dose (WHO, 2013).

The risk of YEL-AND is inversely proportional to age (Gotuzzo, 2013). For this reason, YF vaccine is contraindicated in children aged <6 months and, except during epidemics, is not recommended for children aged 6–8 months (WHO, 2013). A WHO systematic review on the risk of YEL-AVD among the elderly (≥60 years) found high numbers of reported cases (n = 19) in this group compared to all other age groups combined (n = 22) (Rafferty *et al.*, 2013). A recent prospective controlled cohort study showed that elderly subjects, in contrast to younger primary YF vaccine recipients, exhibited a delayed antibody response and higher viremia following YF vaccination (Roukens *et al.*, 2011). The authors have hypothesized that these findings entail an increased risk of development of SAEs in the elderly. Despite this higher risk in persons aged ≥60 years, the overall risk remains low (WHO, 2013).

Autoimmune manifestations post-vaccination

It has been suggested that neurologic syndromes such as GBS and ADEM represent autoimmune conditions in which antibodies and/or T cells produced in response to the YF vaccine crossreact with neuronal epitopes and lead to central or peripheral nerve damage (Staples *et al.*, 2010). Robust data were gathered by an analysis of the development of neurologic adverse events within 30 days of YF vaccine administration in the United States from January 1990 to April 2005. The investigators searched for reports received by the Vaccine Adverse Event Reporting System (VAERS), a passive surveillance system jointly managed by the US Centers for Disease Control and Prevention (CDC) and Food and Drug Administration (FDA) that monitors adverse events subsequent to vaccination (McMahon *et al.*, 2007). Of 97 reports assigned with neurologic coding terms, only three fit case definitions for ADEM, and six for GBS. The ADEM patients' median age was 19 years, and the onset of neurologic symptoms ranged from 7 to 20 days after vaccination. Based on defined criteria for levels of causal relationship to YF vaccine, one case of ADEM was deemed probable, due to the presence of YF vaccine virus-specific IgM in the cerebrospinal fluid (CSF). The remaining ADEM cases temporally associated with YF vaccination were considered suspect. Regarding those vaccinees who developed GBS, median age was 52 years and the neurologic syndrome manifested between 7 and 27 days after YF vaccination. According to the same causality criteria, all six cases of GBS were classified as suspect. Overall, the estimated reporting rate for ADEM and GBS following YF vaccination was 0.4 and 1.9 reported cases, respectively, per 10^6 YF vaccine doses distributed in the study period among US civilians. It is worth emphasizing that no death was recorded and that all nine individuals with autoimmune neurologic syndromes received other vaccines concurrently with YF vaccine administration.

Due to the increased awareness of the link between YF vaccine and post-vaccination autoimmune adverse reactions, other reports have begun to appear in the literature. Neurologists recently cared for an otherwise healthy 23-year-old man who developed severe ADEM 3 weeks after receiving a single dose of the YF 17D-204 vaccine as he was about to visit a YF endemic region in South America (Miravalle *et al.*, 2009). The authors found higher titers of YF vaccine-specific IgM in CSF than in serum, suggesting intrathecal antibody production. The assumption of a YF vaccine-associated autoimmune reaction was further corroborated by a dramatic clinical and neuroimaging improvement following a 5-day course of intravenous methylprednisolone pulse therapy.

In Argentina, physicians have lately dealt with an atypical case of a previously healthy 56-year-old man who evolved longitudinal myelitis without encephalitis 45 days after receipt of the YF 17D-204 substrain (Chaves *et al.*, 2009). Besides ruling out infectious causes, including other South American flaviviruses, the authors identified a high concentration of YF vaccine-specific IgM in the CSF through enzyme-linked immunosorbent assay. Despite the absence of controlled studies on the presence of YF vaccine-specific IgM in the CSF of YF vaccinees without subsequent adverse reactions, experts consider that the detection of YF vaccine-specific IgM in the CSF in cases with clinically compatible illness and temporally associated with YF vaccination does constitute sufficient evidence for a causal relationship between the vaccine and the subsequent disease, particularly when there is neither clinical, epidemiological, nor serological evidence of concurrent infection with a different flavivirus and there is no apparent alternative etiology (McMahon *et al.*, 2007; Lindsey *et al.*, 2008).

Post-vaccinal peripheral nervous system involvement following YF administration has been rarely reported in the literature. There is a single case of a 66-year-old man who presented with progressive motor and sensory neuropathy involving the four limbs and bilateral facial paralysis 15 days after being given the YF vaccine (Vital *et al.*, 2002). Diagnostic work-up, including electrophysiological studies and immunohistochemistry analyses of the superficial peroneal nerve and the peroneus brevis muscle, yielded a diagnosis of chronic inflammatory demyelinating polyneuropathy (CIDP). The patient needed tracheostomy and exhibited a very slow recovery process.

There is one published study in the literature of the association between YF immunization and multiple sclerosis (MS) relapse risk, which employed a self-controlled case-series method (Farez and Correale, 2011). Seven patients diagnosed as having clinical relapsing–remitting MS each received a single dose of the 17D-204 YF vaccine and were prospectively followed-up every 3 months for 2 years. The control subjects were selected contemporaneously with the YF-vaccinated MS patients and formed the following groups: seven age- and sex-matched healthy individuals; seven age- and sex-matched unvaccinated MS patients; and seven influenza-vaccinated MS patients. The follow-up period was divided into an initial at-risk period (ARP), which began 1 week after YF immunization and ended 5 weeks later, and a non-risk period following this. Five MS patients had neurological exacerbations during the ARP. The calculated annual exacerbation rate during the ARP was 8.57, whereas the relapse rate outside this period was 0.67, giving a favorable relapse risk ratio of 12.77 (95% confidence interval (CI): 4.28–38.13; $p < 0.001$). However, other physicians have noted methodological issues with this paper, arguing that the selected study design did not support the inference of a causal relationship between YF vaccination and MS relapses (Pool *et al.*, 2012).

Aside from central and peripheral nervous system involvement, investigators have also reported the development of autoimmune hepatitis (AIH) in a previously healthy 31-year-old woman who had received concomitantly hepatitis A virus vaccine and YF vaccine 11 days before visiting Nigeria (Perumalswami *et al.*, 2009). Her symptoms started on the first day of travel and she sought medical care 2 days following arrival. After extensive investigation, including viral serologies, YF DNA polymerase chain reaction (PCR), and needle liver biopsy, the authors made a presumptive diagnosis of AIH, even though the autoimmune serum markers were negative. The patient was started on prednisone 40 mg daily and showed a dramatic improvement in her serum liver function and transaminase tests. Although vaccination and the onset of AIH may have been unrelated events, the evidence points to an idiosyncratic response to either hepatitis A or YF vaccination.

Yellow fever vaccine and autoimmune inflammatory rheumatic diseases

Since most patients with autoimmune inflammatory rheumatic diseases (AIRDs) require permanent use of disease-modifying antirheumatic drugs (DMARDs) in order to control the symptoms and slow or halt disease progression, they are not eligible candidates to receive any live attenuated vaccine, due to the fear of post-vaccinal life-threatening infection (van Assen *et al.*, 2011). In spite of their immunosuppressed status, many are inadvertently immunized against YF during vaccination campaigns in endemic areas or when travelling to at-risk regions. Taking benefit of this situation, Brazilian researchers performed a retrospective study of 70 patients with AIRDs who were being regularly followed-up in an endemic region of the country and had inadvertently received YF vaccine between November 2007 and January 2008 (Mota *et al.*, 2009). Through application of a standardized questionnaire and chart review, the authors collected data on the demographics and disease characteristics of the YF patients vaccinated up till 60 days before the medical appointment with a rheumatologist (Table 14.1).

Most of the participants were women (90%) and their mean age was 46 years. The rheumatic disorders sampled were rheumatoid arthritis (RA), systemic lupus erythematosus (SLE), spondyloarthropathies (SpA), systemic sclerosis (SSc), and overlap SLE and RA. Overall, only 16 subjects reported side effects post-vaccination, and all were deemed mild. Out of eight patients on biologics, only one treated with infliximab

Table 14.1 Patients demographics and disease characteristics. Adapted from Mota *et al.* (2009)

	Patients			Patients with mild adverse events	
Rheumatic disorder	No.	%	Mean diagnosis (years)	No.	%
RA	52	74.3	9	9	12.8
SLE	9	12.8	14.7	3	4.3
SpA	5	7.1	12.2	2	2.9
SSc	2	2.9	17	1	1.4
RA + SLE	2	2.9	11.5	1	1.4
Total	**70**	**100**		**16**	**22.8**

RA, rheumatoid arthritis; SLE, systemic lupus erythematosus; SPA, spondyloarthropathies; SSc, systemic sclerosis

Table 14.2 Characteristics of patients who suffered adverse events following YF vaccination. Adapted from Mota *et al.* (2009)

Patient	Gender	Age (years)	Diagnosis	Diagnosis (years)	Treatment (mg)	Treatment years	Adverse event reported
1	F	62	RA	3	MTX 5/week + IFX 200 8/8w	5.25	exanthema
2	F	43	RA	11	CyP (NI)	NI	headache, myalgia, arthralgia, fever
3	M	35	ReA	2.16	LFN 20/day + SSZ 2000/day	0.67/2	local pain, myalgia, arthralgia
4	F	46	RA	NI	MTX 15/week + PDN 5/day	0.42/0.42	myalgia, fever
5	F	17	SLE	3	PDN 10/day + AZA 50/day	NI	fever
6	F	59	SLE	14	PDN 10/day	14	headache, arthralgia, myalgia
7	F	66	RA	1.16	MTX 20/week + PDN 10/day	1.16/1.16	headache
8	F	39	RA	3	MTX 20/week + SSZ 1500/day	3 + 1	arthralgia
9	F	55	RA	9	PDN 10/day	5	myalgia
10	F	67	RA	10	LFN 20/day	4	erythema
11	F	41	RA + SLE	8/12	MTX 15/week	10	headache, diarrhea
12	F	42	SS	27	CyP 2,5 mg/kg/day	1	fever
13	F	54	RA	22	MTX 15/week	10	Local pain, erythema
14	F	59	RA	7	MTX 7,5/week	0.25	myalgia
15	F	50	SLE	17	PDN 5/day	10	myalgia
16	F	26	ReA	1	MTX 10/week + SSZ 1000/day	0.83/0.25	elevated liver enzymes

ReA, reactive arthritis; MTX, methotrexate; IFX, infliximab; CyP, cyclophosphamide, SSZ, sulfasalazine; LFN, leflunomide; PDN, prednisone; NI, not informed

experienced a minor adverse event (Table 14.2). There was no positive correlation between the onset of side effects and any specific disease or treatment schedule.

Table 14.3 Serologic respone before and after YF revaccination (values indicated by number). Scheinberg, M. Yellow fever revaccination during infliximab therapy. Arthritis Care Res (Hoboken). Jun;62(6):896–8. Copyright © 2010, John Wiley & Sons, Inc

	Titer				
	1:800	1:400	1:200	1:100	Negative
Before vaccination					
Controls	0	0	3	12	0
Patients	0	0	3	10	2
After vaccination					
Controls	6	6	2	0	1
Patients	0	6	6	4	1

Two patients had only post-vaccination results available

Lately, Brazilian scientists have also examined the YF antibody profile pre- and post-revaccination in 15 RA patients on infliximab maintenance therapy (range 1–3 years) and 15 paired healthy controls (Scheinberg *et al.*, 2010). None of the patients were receiving steroids and all used methotrexate (range 15–20 mg/week) at the time of revaccination. The authors utilized an indirect immunofluorescence assay for the detection of IgM and IgG antibodies against YF vaccine and the results were reported from negative to positive with multiple dilutions (Table 14.3). A comparison of the magnitude of the antibody response post-revaccination between the groups showed a trend toward lower response in RA patients, but, due to the small number of participants, a formal statistical analysis could not be performed.

Conclusions

Large-scale YF immunization has been very effective, due to its excellent immunogenicity and safety profile. Every nonvaccinated person living in or travelling to an endemic YF area should be offered the vaccine. Vaccination risk assessment should weigh the odds of acquiring YF (e.g. location, season, duration of exposure, occupational and recreational activities, and local rate of virus transmission) against the odds of experiencing an SAE following YF vaccination. One rational approach would be to measure the serum levels of specific neutralizing IgG antibodies when the lag time after the primary YF vaccination has exceeded 10 years and then revaccinate those who did not sustain protective values. However, this does not seem mandatory, as studies have demonstrated that the majority of YF-vaccinated people exhibit immunological response for decades.

References

Chaves, M., Riccio, P., Patrucco, L., *et al.* (2009). Longitudinal myelitis associated with yellow fever vaccination. *J Neurovirol,* **15**(4): 348–50.

Farez, M.F. and Correale, J. (2011). Yellow fever vaccination and increased relapse rate in travellers with multiple sclerosis. *Arch Neurol,* **68**(10): 1267–71.

Filippis, A.M., Noqueira, R.M., Jabor, A.V., *et al.* (2004). Isolation and characterization of wild type yellow fever in cases temporally associated with 17DD vaccination during an outbreak of yellow fever in Brazil. *Vaccination,* **22**: 1073–78.

Gibney, K.B., Edupuganti, S., Panella, A.J., *et al.* (2012). Detection of anti-yellow fever virus immunoglobulin M antibodies at 3–4 years following yellow fever vaccination. *Am J Trop Med Hyg,* **87**(6): 1112–15.

Gotuzzo, E. (2013). Efficacy and duration of immunity following yellow fever vaccine: a systematic review on the need of yellow fever booster every 10 years. Background paper prepared for the SAGE meeting of April 2013. Available from: http://www.who.int/immunization/sage/meetings/2013/april/1_Background_Paper_Yellow_Fever_Vaccines.pdf [last accessed 11 December 2014].

Khromava, A.Y., Eidex, R.B., Weld, L.H., *et al.* (2005) Yellow fever vaccine: an updated assessment of advanced age as a risk factor for serious adverse events. *Vaccine,* **23**(25): 3256–63.

Lindsey, N.P., Schroeder, B.A., Miller, E.R., *et al.* (2008). Adverse event reports following yellow fever vaccination. *Vaccine,* **26**(48): 6077–82.

McMahon, A.W., Eidex, R.B., Marfin, A.A., *et al.* (2007). Neurologic disease associated with 17D-204 yellow fever vaccination: a report of 15 cases. *Vaccine,* **25**(10): 1727–34.

Miravalle, A., Biller, J., Silva, E., *et al.* (2009). Acute disseminated encephalomyelitis following yellow fever vaccination. *Arq Neuropsiquiatr,* **67**(3A): 710–11.

Monath, T.P. and Cetron, M.S. (2002) Prevention of yellow fever in persons traveling to the tropics. *Clin Infect Dis,* **34**(10): 1369–78.

Monath, T.P., Nichols, R., Archambault, W.T., *et al.* (2002). Comparative safety and immunogenicity of two yellow fever 17D vaccines(ARILVAX and YF-VAX) in a phase III, multicentre, double blind clinical trial. *Am J Trop Med Hyg,* **66**(5): 533–41.

Mota, L.M., Oliveira, A.C., Lima, R.A., *et al.* (2009). [Vaccination against yellow fever among patients on immunosuppressors with diagnoses of rheumatic diseases]. *Rev Soc Bras Med Trop,* **42**(1): 23–7.

Niedrig, M., Kürsteiner, O., Herzog, C., and Sonnenberg, K. (2008). Evaluation of an indirect immunofluorescence assay for detection of immunoglobulin M (IgM) and IgG antibodies against yellow fever virus. *Clin Vacc Immunol,* **15**: 177–81.

Perumalswami, P., Peng, L., and Odin, J.A. (2009). Vaccination as a triggering event for autoimmune hepatitis. *Semin Liver Dis,* **29**(3): 331–4.

Pool, V., Gordon, D.M., and Decker, M. (2012). Methodological issues with the risk of relapse study in patients with multiple sclerosis after yellow fever vaccination. *Arch Neurol,* **69**(1): 144; author reply 144–5.

Rafferty, E., Duclos, P., Yactayo, S., and Schuster, M. (2013). Risk of yellow-fever vaccine associated viscerotropic disease among the elderly: systematic review. *Vaccine,* **31**(49): 5798–805.

Roukens, A.H., Soonawala, D., Joosten, S.A., *et al.* (2011). Elderly subjects have a delayed antibody response and prolonged viraemia following yellow fever vaccination: a prospective controlled cohort study. *PLoS One,* **6**(12): e27753.

Scheinberg, M., Guedes-Barbosa, L.S., Manqueira, C., *et al.* (2010). Yellow fever revaccination during infliximab therapy. *Arthritis Care Res (Hoboken),* **62**(6): 896–8.

Staples, J.E., Gershman, M., Fischer, M., and Centers for Disease Control and Prevention. (2010). Yellow fever vaccine: recommendations of the Advisory Committee on Immunization Practices (ACIP). *MMWR Recomm Rep,* **59**(RR-7): 1–27.

van Assen, S., Agmon-Levin, N., Elkayam, O., *et al.* (2011). EULAR recommendations for vaccination in adult patients with autoimmune inflammatory rheumatic diseases. *Ann Rheum Dis,* **70**: 414–22.

Vital, C., Vital, A., Gbikpi-Benissan, G., *et al.* (2002). Post-vaccinal inflammatory neuropathy: peripheral nerve biopsy in 3 cases. *J Peripher Nerv Syst,* **7**(3): 163–7.

WHO (World Health Organization). (2013). Weekly Epidemiological Record (WER), *No. 27,* Vol. **88**, pp. 269–84.

15 Antiphospholipid Syndrome and Vaccines

Miri Blank and Paola Cruz-Tapias

[1]Zabludowicz Center for Autoimmune Diseases, Sheba Medical Center, Tel Hashomer, Israel

[2]Doctoral Program in Biomedical Sciences, Del Rosario University, Bogotá, Colombia

Introduction

Antiphospholipid syndrome (APS) is an autoimmune multisystemic disease associated with recurrent fetal loss, thromboembolic phenomena, thrombocytopenia, and neurological, cardiac, and dermatological involvement (Cervera *et al.*, 2002; Shoenfeld, 2003). APS is characterized by the presence of antiphospholipid antibodies, which bind negatively charged phospholipids, mainly through β2-glycoprotein I (β2GPI) (Galli *et al.*, 1990). The factors causing production of anti-β2GPI antibodies remain undefined, but there is evidence that APS is one of those autoimmune conditions that can be induced by infections and can be related to vaccination (Cruz-Tapias *et al.*, 2012).

β2GPI and anti-β2GPI antibodies in APS

β2GPI has been identified as the most important antigen in APS (Schwarzenbacher *et al.*, 1999). It is an anionic phospholipid-binding glycoprotein that belongs to the complement control protein (CCP) superfamily. β2GPI consists of 326 amino acids organized in five CCP domains. The first four domains have the regular, conserved sequences, but the fifth contains 20 amino acids with C-terminal loop endings, which constitute a large positively charged patch that determines affinity for anionic phospholipids (Bouma *et al.*, 1999; Schwarzenbacher *et al.*, 1999). It has been shown that the interaction between β2GPI and anionic phospholipids undergoes conformational changes that result in the exposure of a cryptic epitope recognized by anti-β2GPI antibodies; this crypticity can be an important characteristic of the epitopes recognized by autoantibodies in APS (Pengo *et al.*, 1995; Agar *et al.*, 2011). β2GPI has several properties *in vitro* which define it as an anticoagulant (e.g. inhibition of prothrombinase activity, ADP-induced platelet aggregation, platelet factor IX production) (Blank *et al.*, 1991; Pengo *et al.*, 1995; Bouma *et al.*, 1999; Schwarzenbacher *et al.*, 1999; Agar *et al.*, 2011). Passive transfer of anti-β2GPI antibodies induces experimental APS in naive mice and thrombus formation in *ex vivo* models (Blank *et al.*, 1991; Pierangeli *et al.*, 1996). Immunization with human β2GPI molecule induces experimental APS in naive mice, which is manifested by high titers of mouse anti-β2GPI, an increased percentage of fetal resorptions (the equivalent of fetal loss in human APS), thrombocytopenia, and prolonged activated partial thromboplastin time (Blank *et al.*, 1994). Oral administration of β2GPI to APS mice results in tolerance induction (Blank *et al.*, 1998). The disease can be treated by β2GPI-related synthetic peptides (Blank *et al.*, 1999, 2011).

APS and tetanus toxoid vaccine

Tetanus toxoid (TTd) is a potent exotoxin produced by the bacterium *Clostridium tetani*. It is

Vaccines and Autoimmunity, First Edition. Edited by Yehuda Shoenfeld, Nancy Agmon-Levin, and Lucija Tomljenovic.

produced as an inactive single polypeptide chain of 150 kDa, composed of three 50 kDa domains connected by protease-sensitive loops. The toxin is activated upon selective proteolytic cleavage, which generates two disulfide-linked chains: a heavy chain (100 kDa) and a light chain (50 kDa). The carboxyl terminus of the heavy chain binds to neural cell membrane and the amino terminus facilitates cell entry. The toxin has a predominant effect on inhibitory neurons, inhibiting release of gamma aminobutiric acid (GABA). When spinal inhibitory interneurones are affected, symptoms appear (Cook *et al.*, 2001). The infection may be prevented through use of a vaccine, which is composed of TTd and an adjuvant (usually aluminum hydroxide). However, the vaccine can induce the production of antibodies against the toxoid, which can react with other autoantigens.

A human hybridoma-derived IgM antibody (H3 mAb) was produced from a healthy individual boosted with diphtheria and TTds and was found to react with both toxoids and human cardiolipins (in a β2GPI-dependent manner). Further support for this cross-reactivity was given by analysis of serum antibodies from patients with systemic lupus erythematosus (SLE) and normal individuals (Sutjita *et al.*, 1988). Individual sera were passed through a diphtheria toxoid-sepharose affinity column and bound antibodies were examined by ELISA. Crossreactivity of H3 mAb to TTd was found in preparations from both normal and SLE sera (Sutjita *et al.*, 1988). Anti-H3 antiidiotype antibody was able to block the binding of H3 to cardiolipin, diphtheria, and TTds. Further, high levels of H3 idiotype were found in 83% of SLE patients' sera and there was a good correlation between the amount of H3 idiotype and the total (IgG, IgA, and IgM) cardiolipin antibodies (rs = 0.70, $p < 0.001$)(Sutjita *et al.*, 1989; Hohmann *et al.*, 1991).

Naive mice immunized with H3 mAb developed experimental APS exemplified by the presence of high titers of circulating anticardiolipin antibodies, elevated fetal loss, thrombocytopenia, and prolonged coagulation time (Bakimer *et al.*, 1992). H3 mAb binds anticardiolipin via β2GPI. In order to identify the β2GPI epitope target of H3 mAb, the H3 mAb was introduced to a hexapeptide phage display library. TLRVYK peptide mimics the H3/β2GPI binding epitope. TLRVYK peptide is located in the third domain of the β2GPI molecule (amino acids 133–138 on domain III). Not only was the binding of H3 mAb to β2GPI inhibited by TLRVYK peptide, but so was the activation of endothelial cells. Therefore, mice infused with pathogenic H3 mAb and then treated with the TLRVYK peptide were protected from induction of experimental APS (Blank *et al.*, 1999).

Using the Swiss Protein Database, a high homology between the TLRVYK hexapeptide and peptidic domain of various bacteria and viruses and TTd was found (Blank *et al.*, 2002). The TLRVYK peptide is located on β2GPI and TTd molecules (it appears three times in the TTd, not as linear peptides, but as conformational mimotopes) (Figure 15.1) (Blank *et al.*, 2002). Then, naive BALB/c mice, immunized with TTd, developed antibodies directed to TTd, dsDNA, and β2GPI and were extremely sick. Therefore, in order to assess the pathogenic potential of antibodies directed to the common peptide related to TTd and β2GPI antibodies from the TTd-immunized mice, the crossreactive antibodies were affinity-purified on a column composed of TLRVYK synthetic peptide. Anti-TTd/ β2GPI antibodies bound β2GPI and TTd with high affinity, in a dose-dependent manner. Passive transfer of the affinity-purified anti-TLRVYK Abs to naive mice induced an experimental model of APS, manifested by a high

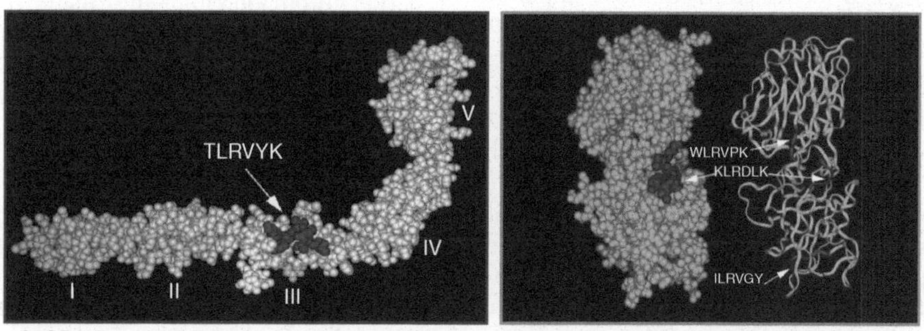

▸ β2GP1: Peptide TLRVYK is located on domain III ▸ Tetanus toxoid: One mimotope is external and 2 mimotopes are cryptic

Figure 15.1 Homology between β2GP1-related peptide and TTd. (For a color version of this figure, please see color plate section.)

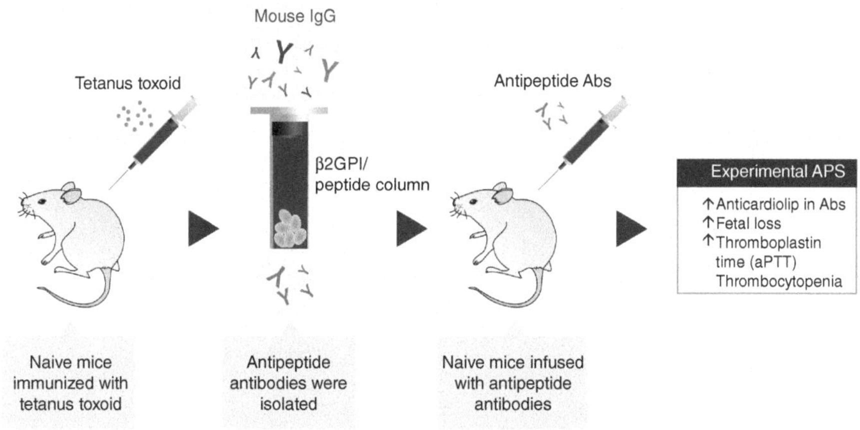

Figure 15.2 Active immunization of mice with TTd induces anti-β2GPI antibodies. The anti-β2GPI antibodies, when passively infused into another set of naive mice, induce experimental APS. (For a color version of this figure, please see color plate section.)

significant percentage of fetal loss, prolonged coagulation time, and thrombocytopenia similar to that of a control group of mice immunized with a pathogenic anti-β2GPI monoclonal antibody (Figure 15.2) (Blank *et al.*, 2002).

All together, the evidence presented here supports the molecular mimicry mechanism between β2GPI and TTd as one of the possible causes of APS. In particular, it appears that specific TTd epitopes can induce an immune response that breaks tolerance to the host epitopes. The crossreactive T or B cell is then able to induce a pathogenic autoimmune response that leads to disease (Figure 15.3). Autoimmunity resulting from epitope mimicry may be an unfortunate side effect of the immune response against TTd epitopes that can lead to disease in predisposed individuals.

Further support for epitope mimicry as a possible mechanism for APS development comes from other authors (Inic-Kanada *et al.*, 2009; Zivkovic *et al.*, 2011, 2012). Induction of APS in two different non-autoimmune-prone mouse strains, BALB/c and C57BL/6, was achieved by TTd hyperimmunization using different adjuvants (Inic-Kanada *et al.*, 2009; Zivkovic *et al.*, 2011). APS had different manifestations of reproductive pathology in BALB/c and C57BL/6 mice: fetal resorption (as a consequence of extreme T cell activation obtained in the course of pretreatment) and lowering of fecundity (as a consequence of polyclonal B cell stimulation), respectively (Zivkovic *et al.*, 2011). Hyperimmunization of BALB/c mice with TTd in aluminum hydroxide, glycerol, or complete Freund's adjuvant (CFA) resulted in elevated circulating antibodies to TTd, β2GPI, gangliosides, and laminin and induced fetal loss (Zivkovic *et al.*, 2011). A decrease in fecundity was recorded only in C57BL/6 mice immunized with aluminum hydroxide adjuvant, associated with an increase in low-affinity anti-β2GPI IgG antibodies and Th1 prevalence. These observations suggest that the immune response to TTd in mice depends on the genetic background and the specific adjuvant used for immunization.

There are some case reports showing the possibility of triggering APS following tetanus vaccination (Meyer *et al.*, 2010). Active or passive immunization with vaccines or sera can cause lesions in the central and peripheral nervous systems. Tezzon *et al.* (1994) reported a case of transverse myelitis with radicular component that occurred acutely following administration of TTd, with the patient having a partially favorable outcome. Komolafe *et al.* (2008) reported a patient who presented neck pain, paroxysmal tonic spasms, a positive Lhermitte's sign, and spastic quadriplegia and later developed bilateral optic neuritis and had clinical and biochemical features of APS. As far as central nervous system (CNS) complications are concerned, there is one case of a woman who, 5 days after receiving tetanus vaccine, manifested a serious clinical outcome produced by CNS and peripheral nervous system involvement, which had a favorable prognosis (Schlenska, 1977). In 1992, two new cases were reported in *The Lancet* (Topaloglu *et al.*, 1992). The first was characterized by optic neuritis and acute myelitis, appearing 3 days after the vaccination, with a favorable resolution over a period of 1 year. The second presented severe acute encephalomyelitis, with liquor pleocytosis and an increased protein titer,

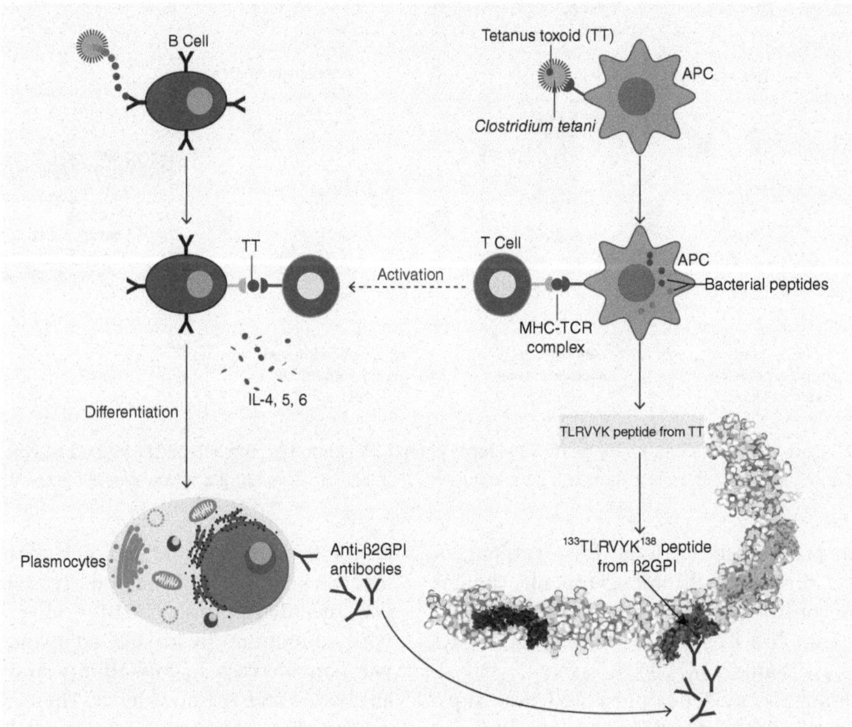

Figure 15.3 Molecular mimicry between β2GPI and TTd. (For a color version of this figure, please see color plate section.)

10 days after the prophylactic administration of the tetanus vaccine.

APS post-hepatitis B virus vaccination

Enhanced circulating autoantibodies directed to phospholipid (cardiolipin, β2GPI) and lupus anticoagulant were identified in the sera of hepatitis B virus (HBV)-infected APS patients (Huh *et al.*, 2011). The most frequently identified antiphospholipid antibodies in these patients were of the IgM isotype, which were persistently detected over a 12-week period (Huh *et al.*, 2011). However, patients with chronic hepatitis develop β2GPI-independent antiphospholipid antibodies (Zachou *et al.*, 2003). In all cases of antiphospholipid antibodies related to the HBV vaccine, they were not associated with thrombosis after hematological manifestations of the APS. DNA HBV vaccination was given to 85 healthy students (Martinuc Porobic *et al.*, 2005) and, 1 month post-vaccination, a minority of individuals showed changes in IgG or IgM anticardiolipin and anti-β2GPI antibodies or lupus anticoagulant ($p < 0.001$). Among subjects in whom changes of IgG anti-β2GPI were observed, a significantly higher number of increased (8/85) than decreased (2/85) values were found ($p < 0.01$). One of the explanations for the HBV vaccination-induced anti-β2GPI antibodies may be β2GPI binding to recombinant HBV surface antigen (rHBsAg) (Huh *et al.*, 2011). The phospholipid binding site on the β2GPI fifth domain is targeted by rHBsAg (Mehdi *et al.*, 2008).

APS post-influenza virus vaccination

Reports on the association between influenza vaccine and APS are scarce. Rare cases of APS clinical manifestations have been reported, such as a patient presenting ischemic stroke 4 days after influenza vaccination, associated with secondary APS to lupus (Vainer-Mossel *et al.*, 2009). Toplak *et al.* (2008) reported the presence of anti-β2GPI antibodies in 15% of 92 healthy medical workers up to 6 months post-influenza vaccination; persistently elevated levels of autoantibodies were observed in seven (8%) participants, and two showed progressively increased levels of IgM anticardiolipin and IgA anti-β2GPI antibodies, respectively (Toplak *et al.*, 2008). So far, no APS clinical picture has been documented, although no one can predict the future clinical presentation of the second hit trigger in individuals genetically predisposed for APS. Tarján *et al.* (2013) found that repeated influenza vaccination in clinically stable SLE patients with low disease activity may result in increased production of anti-β2GPI antibodies, and therefore may increase the risk of thrombotic manifestations.

Conclusions

All of the evidence points to the association between diverse vaccines and APS. Molecular mimicry has been proposed as one of the mechanisms by which experimental APS can occur in association with pathogens. Sequence similarities between foreign proteins (i.e., TTd) and self-proteins are sufficient to trigger a loss of immune tolerance, either by molecular mimicry or by bystander-effect mechanisms, resulting in the formation of pathogenic autoantibodies related to APS.

References

Agar, C., de Groot, P.G., Mörgelin, M., *et al.* (2011). β_2-glycoprotein I: a novel component of innate immunity. *Blood*, **117**(25): 6939–47.

Bakimer, R., Fishman, P., Blank, M., *et al.* (1992). Induction of primary antiphospholipid syndrome in mice by immunization with a human monoclonal anticardiolipin antibody (H-3). *J Clin Invest*, **89**(5): 1558–63.

Blank, M., Cohen, J., Toder, V., and Shoenfeld, Y. (1991). Induction of anti-phospholipid syndrome in naive mice with mouse lupus monoclonal and human polyclonal anti-cardiolipin antibodies. *Proc Nat Acad Sci USA*, **88**(8): 3069–73.

Blank, M., Faden, D., Tincani, A., *et al.* (1994). Immunization with anticardiolipin cofactor (beta-2-glycoprotein I) induces experimental antiphospholipid syndrome in naive mice. *J Autoimmun*, **7**(4): 441–55.

Blank, M., George, J., Barak, V., *et al.* (1998). Oral tolerance to low dose beta 2-glycoprotein I: immunomodulation of experimental antiphospholipid syndrome. *J Immunol*, **161**(10): 5303–12.

Blank, M., Shoenfeld, Y., Cabilly, S., *et al.* (1999). Prevention of experimental antiphospholipid syndrome and endothelial cell activation by synthetic peptides. *Proc Nat Acad Sci USA*, **96**(9): 5164–8.

Blank, M., Baraam, L., Eisenstein, M., *et al.* (2011). β2-glycoprotein-I based peptide regulate endothelial-cells tissue-factor expression via negative regulation of pGSK3β expression and reduces experimental-antiphospholipid-syndrome. *J Autoimmun*, **37**(1): 8–17.

Blank, M., Krause, I., Fridkin, M., *et al.* (2002). Bacterial induction of autoantibodies to beta2-glycoprotein-I accounts for the infectious etiology of antiphospholipid syndrome. *J Clin Invest*, **109**(6): 797–804.

Bouma, B., de Groot, P.G., van den Elsen, J.M., *et al.* (1999). Adhesion mechanism of human beta(2)-glycoprotein I to phospholipids based on its crystal structure. *EMBO J*, **18**(19): 5166–74.

Cervera, R., Piette, J.C., Font, J., *et al.* (2002). Antiphospholipid syndrome: clinical and immunologic manifestations and patterns of disease expression in a cohort of 1,000 patients. *Arthritis Rheum*, **46**(4): 1019–27.

Cook, T.M., Protheroe, R.T., and Handel, J.M. (2001). Tetanus: a review of the literature. *Br J Anaesth*, **87**(3): 477–87.

Cruz-Tapias, P., Blank, M., Anaya, J.M., and Shoenfeld, Y. (2012). Infections and vaccines in the etiology of antiphospholipid syndrome. *Curr Opin Rheumatol*, **24**(4): 389–93.

Galli, M., Comfurius, P., Maassen, C., *et al.* (1990). Anticardiolipin antibodies (ACA) directed not to cardiolipin but to a plasma protein cofactor. *Lancet*, **335**(8705): 1544–7.

Hohmann, A., Comacchio, R., Boswarva, V., *et al.* (1991). The H3 anti-phospholipid idiotype is found in patients with systemic lupus erythematosus (SLE) but not in patients with syphilis. *Clin Exp Immunol*, **86**(2): 207–11.

Huh, J.Y., Yi, D.Y., Hwang, S.G, *et al.* (2011). Characterization of antiphospholipid antibodies in chronic hepatitis B infection. *Korean J Hematol*, **46**(1): 36–40.

Inic-Kanada, A., Stojanovic, M., Zivkovic, I., *et al.* (2009). Murine monoclonal antibody 26 raised against tetanus toxoid cross-reacts with beta2-glycoprotein I: its characteristics and role in molecular mimicry. *Am J Reprod Immunol*, **61**(1): 39–51.

Komolafe, M.A., Komolafe, E.O., Sunmonu, T.A., *et al.* (2008). New onset neuromyelitis optica in a young nigerian woman with possible antiphospholipid syndrome: a case report. *J Med Case Rep*, **2**: 348.

Martinuc Porobic, J., Avcin, T., Bozic, B., *et al.* (2005). Anti-phospholipid antibodies following vaccination with recombinant hepatitis B vaccine. *Clin Exp Immunol*, **142**(2): 377–80.

Mehdi, H., Naqvi, A., and Ilyas Kamboh, M. (2008). Recombinant hepatitis B surface antigen and anionic phospholipids share a binding region in the fifth domain of beta2-glycoprotein I (apolipoprotein H). *Bioch Biophys Acta*, **1782**(3): 163–8.

Meyer, A., Rotman-Pikielny, P., Natour, A., and Levy, Y. (2010). Antiphospholipid syndrome following a diphtheria-tetanus vaccination: coincidence vs. causality. *IMAJ*, **12**(10): 638–9.

Pengo, V., Biasiolo, A., and Fior, M.G. (1995). Autoimmune antiphospholipid antibodies are directed against a cryptic epitope expressed when beta 2-glycoprotein I is bound to a suitable surface. *Thromb Haemost*, **73**(1): 29–34.

Pierangeli, S.S. and Harris, E.N. (1996). In vivo models of thrombosis for the antiphospholipid syndrome. *Lupus*, **5**(5): 451–5.

Schlenska, G.K. (1977). Unusual neurological complications following tetanus toxoid administration. *J Neurol*, **215**(4): 299–302.

Schwarzenbacher, R., Zeth, K., Diederichs, K., *et al.* (1999). Crystal structure of human beta2-glycoprotein I: implications for phospholipid binding and the antiphospholipid syndrome. *EMBO J*, **18**(22): 6228–39.

Shoenfeld, Y. (2003). Systemic antiphospholipid syndrome. *Lupus*, **12**(7): 497–8.

Sutjita, M., Hohmann, A., Comacchio, R., *et al.* (1989). A common anti-cardiolipin antibody idiotype in autoimmune disease: identification using a mouse monoclonal antibody directed against a naturally-occurring anti-phospholipid antibody. *Clin Exp Immunol*, **75**(2): 211–16.

Sutjita, M., Hohmann, A., Comacchio, R., and Bradley, J. (1988). Polyspecific human and murine antibodies to diphtheria and tetanus toxoids and phospholipids. *Clin Exp Immunol*, **73**(2): 191–7.

Tarján, P., Sipka, S., Lakos, G., *et al.* (2013). Influenza vaccination and the production of anti-phospholipid antibodies in patients with systemic lupus erythematosus. *Scand J Rheumatol*, **35**(3): 241–3.

Tezzon, F., Tomelleri, P., Ferrari, G., and Sergi, A. (1994). Acute radiculomyelitis after antitetanus vaccination. *Ital J Neurol Sci*, **15**(4): 191–3.

Topaloglu, H., Berker, M., Kansu, T., *et al.* (1992). Optic neuritis and myelitis after booster tetanus toxoid vaccination. *Lancet*, **339**(8786): 178–9.

Toplak, N., Kveder, T., Trampus-Bakija, A., *et al.* (2008). Autoimmune response following annual influenza vaccination in 92 apparently healthy adults. *Autoimmun Rev*, **8**(2): 134–8.

Vainer-Mossel, E.D., Mekori, Y.A., and Mor, A. (2009). Ischemic stroke in a patient with lupus following influenza vaccination: a questionable association. *IMAJ*, **11**(3): 186–7.

Zachou, K., Liaskos, C., Christodoulou, D.K., *et al.* (2003). Anti-cardiolipin antibodies in patients with chronic viral hepatitis are independent of beta2-glycoprotein I cofactor or features of antiphospholipid syndrome. *Eur J Clin Invest*, **33**(2): 161–8.

Zivkovic, I., Petrusic, V., Stojanovic, M., *et al.* (2012). Induction of decreased fecundity by tetanus toxoid hyper-immunization in C57BL/6 mice depends on the applied adjuvant. *Innate Immun*, **18**(2): 333–42.

Zivkovic, I., Stojanovic, M., Petrusic, V., *et al.* (2011). Induction of APS after TTd hyper-immunization has a different outcome in BALB/c and C57BL/6 mice. *Am J Repr Immunol*, **65**(5): 492–502.

16 Hepatitis B Vaccination and Autoimmunity

Daniel S. Smyk,[1] Lazaros I. Sakkas,[2] Yehuda Shoenfeld,[3,4] and Dimitrios P. Bogdanos,[1,2]

[1]Institute of Liver Studies, King's College London School of Medicine at King's College Hospital, London, UK
[2]Department of Medicine, School of Health Sciences, University of Thessaly, Larissa, Greece
[3]Zabludowicz Center for Autoimmune Diseases, Sheba Medical Center, Tel Hashomer, Israel
[4]Incumbent of the Laura Schwarz-Kipp Chair for Research of Autoimmune Diseases, Sackler Faculty of Medicine, Tel Aviv University, Tel Aviv, Israel

Introduction

Autoimmune diseases represent a complex group of conditions that are believed to arise from the interaction between genetic susceptibility and environmental exposures (Shoenfeld *et al.*, 2008a,b,c; Smyk *et al.*, 2011). These environmental exposures include inorganic and organic compounds, and the latter include infectious agents (Smyk *et al.*, 2011; Bogdanos *et al.*, 2013). Infectious agents include bacteria, viruses, and parasites, and may also involve alterations in the normal flora within the human body (Bogdanos *et al.*, 2013). A variety of mechanisms have been proposed to explain how infectious agents may contribute to the development of autoimmune disease, and several of these are well described in other chapters of this book. Briefly, these mechanisms include molecular mimicry (Vial and Descotes, 2004; Olson *et al.*, 2005; Fujinami *et al.*, 2006), epitope spreading (Miller *et al.*, 1997; Katz-Levy *et al.*, 1999), bystander effect (McCoy *et al.*, 2006; Roner *et al.*, 2008), microbial superantigens (Wucherpfennig, 2001), immune complex formation (Ram and Shoenfeld, 2008), major histocompatibility complex (MHC) class II expression on nonimmune cells (Ravel *et al.*, 2004), direct inflammatory damage (Ram and Shoenfeld, 2008), high levels of proinflammatory cytokines such as interferon gamma (IFN-γ) (Vial and Descotes, 2004), and immune dysregulation in the form of T-regulatory impairment. The same immune system mechanisms that are involved in the induction of autoimmunity during infections may also lead, to some extent, to the loss of tolerance to self-antigens during vaccinations (Shoenfeld *et al.*, 2000; Geier and Geier, 2002a; Vial and Descotes, 2004). Vaccine components include both the microorganismal and the adjuvant parts of vaccines (Shoenfeld *et al.*, 2000). To date, it is not clear whether vaccine-contained adjuvants, microbial components, or both, are responsible for the induction of autoimmune-related adverse reactions following vaccination. As has been described in other chapters of this book, most viral or bacterial components are moderately immunogenic, and for that reason adjuvants are added to vaccine regimens in order to boost antigen-specific immune responses. Adjuvant components include inorganic (such as alum or other mineral salts) and organic (such as yeast cells, viral/bacterial components, and antigens, including CpG DNA and lipopolysaccharide) materials, which may mimic conserved molecules that activate an innate immune response (Gavin *et al.*, 2006; Kool *et al.*,

Vaccines and Autoimmunity, First Edition. Edited by Yehuda Shoenfeld, Nancy Agmon-Levin, and Lucija Tomljenovic.

2008; Agmon-Levin *et al.*, 2009a,b; Israeli *et al.*, 2009; Marrack *et al.*, 2009; Cooper *et al.*, 2010). Very few of the adjuvants so far identified are considered safe for vaccine use in humans, and some of them can induce immune-mediated destruction against the host. Such adjuvant-induced autoimmunity is not directly relevant to the unwanted effects triggered by the main antigenic component of the vaccine; it has recently been termed "autoimmune/inflammatory syndrome induced by adjuvants," or ASIA (see Chapter 7) (Rosenblum *et al.*, 2011; Shoenfeld and Agmon-Levin, 2011; Agmon-Levin *et al.*, 2012; Blank *et al.*, 2012; Zafrir *et al.*, 2012; Perricone and Shoenfeld, 2013). Vaccine-related autoimmune phenomena have already been documented, including Guillain–Barré syndrome (GBS) (after the 1976 swine influenza vaccine), immune thrombocytopenia purpura (after the measles, mumps, and rubella (MMR) vaccine), and myocarditis (after smallpox vaccination) (Salemi and D'Amelio, 2010).

Among vaccinations implicated in the pathogenesis of autoimmune disease, the hepatitis B virus (HBV) vaccine has received particular attention in regards to several autoimmune diseases (see Table 16.1) (Perricone and Shoenfeld, 2013), of which multiple sclerosis (MS) is most notable (Geier and Geier, 2004; Ram and Shoenfeld, 2008; Stubgen 2012). Case reports of post-HBV vaccine twins affected by the same autoimmune disease (in particular, arthritis) have been reported, shedding light on the potential implication of genetic makeup.

Hepatitis B virus

HBV is a DNA virus of the Hepadnaviridae family, responsible for acute and chronic liver disease (CDC, 1990). Current estimations indicate that more than 2 billion people worldwide have been exposed to HBV, including 360 million who are chronically infected. Among these, a significant proportion (including those who are treated with antiviral therapies) will develop serious complications, cirrhosis, and hepatocellular cancer. In fact, 600 000 individuals lose their lives each year as a result of HBV infection and its complications (Goldstein *et al.*, 2005). In the United States, there are about 300 000 new HBV infection cases annually, which represent patients who have not received vaccinations (CDC, 1991). HBV transmission is by mucosal and/or percutaneous exposure to blood via sexual contact, needle stick or medical device injury, and exposure to contaminated blood products (Francis *et al.*, 1981). The incubation period of HBV is approximately 2–3 months, and two-thirds of patients are either subclinical/asymptomatic or have a mild illness (Stubgen, 2012). Chronic HBV infection occurs in 5–10% of adults (Liang, 2009).

The first virological marker of HBV infection is hepatitis B surface antigen (HBsAg), which can be detected in the serum 2–12 weeks following infection with HBV. Its persistence varies widely among individuals and its presence indicates that a person is potentially infectious. The endemicity of hepatitis B is defined by the extent of the prevalence of HBsAg in the general population, and it varies largely between geographic regions. The most widely used classification is that based on the following three groups: high endemic areas (>8% of the population is HBsAg-positive), intermediate prevalence (2~8%), and low prevalence (<2%). At the molecular level, HBsAg is a glycoprotein constituent of the outer envelope of HBV, but it is also found as 22 nm spheres and tubular forms in the sera of acutely and chronically infected patients.

Hepatitis B vaccine

Fortunately, HBV vaccines have reduced the risk of developing chronic infection and provided prophylaxis (Stubgen, 2012). Seminal studies, mainly from Taiwan and other highly endemic countries, show that HBV vaccine markedly reduces the incidence of liver cancer in children (Chang *et al.*, 1997). These data largely explain why HBV vaccine is considered the first efficient vaccine against a major human cancer. As of 2001, 142 countries were using HBV vaccine in routine immunization, while in 2011 this number increased to 179 countries. By the end of 2002, more than 2 billion doses of HBV vaccine had been administered.

The first vaccine against HBV became available in 1982. It was prepared from particles of HBsAg from the plasma of chronically infected individuals. More recent advances in HBV vaccine development include recombinant HBsAg, developed by inserting the HBsAg gene into yeast cells. The vaccination scheme includes a three-dose series (Tron *et al.*, 1989). The amount of HBsAg per dose is estimated to vary from 2.5 to 40 μg.

Aluminum phosphate or aluminum hydroxide is added to these vaccines as an adjuvant, sometimes preserved with thiomersal. Unwanted reactions

Table 16.1 Autoimmune reactions associated with hepatitis B vaccine (HBVacc). Many of the entries are based solely on case reports, and the evidence is not strong, but larger studies have been conducted in relation to some other conditions. Please see the text for more in-depth discussion

	Autoimmune condition	References
Neuromuscular disorders	Multiple sclerosis	Positive findings: Herroelen *et al.* (1991); Matsui *et al.* (1996); Gout and Lyon-Caen (1998); Marshall (1998); Shoenfeld and Aron-Maor (2000); Touze *et al.* (2000); Karaali-Savrun *et al.* (2001); Geier and Geier (2004); Hernan *et al.* (2004) Negative findings: Gout and Lyon-Caen (1998); Ascherio *et al.* (2001a, 2001b), Confavreux *et al.* (2001); DeStefano *et al.* (2003)
	Myelitis	Mahassin *et al.* (1993); Trevisani *et al.* (1993); Tartaglino *et al.* (1995); Senejoux *et al.* (1996); Song *et al.* (1997); Renard *et al.* (1999); Iniguez *et al.* (2000); Karaali-Savrun *et al.* (2001); Fonseca *et al.* (2003); Agmon-Levin *et al.* (2009a); Stubgen (2012)
	Optic neuritis	Shaw *et al.* (1988); Geier and Geier (2004)
	Guillain–Barré syndrome	Shaw *et al.* (1988); Tuohy (1989); Geier and Geier (2004); Haber *et al.* (2004); Khamaisi *et al.* (2004); Souayah *et al.* (2009); van Doorn (2009); Salemi and D'Amelio (2010)
	Neuropathy	Shaw *et al.* (1988); McMahon *et al.* (1992); Maillefert *et al.* (1997); Creange *et al.* (1999); Dano and Korczyn (2001); DeJonckere and de Surgeres (2001); Sindern *et al.* (2001); Stubgen (2010)
	Myopathy	Positive findings: Fernandez-Funez and Polo Romero (1998); Ramirez-Rivera *et al.* (2003); Capasso *et al.* (2006); Agmon-Levin and Shoenfeld (2008); Altman *et al.* (2008); Buskila *et al.* (2008); Orbach and Tanay (2009); Stubgen (2010) Negative findings: Maillefert *et al.* (1999); Geier and Geier (2005)
	Myasthenia gravis	Biron *et al.* (1988); Domigo *et al.* (1999)
	Chronic fatigue syndrome	Delage *et al.* (1993); Agmon-Levin and Shoenfeld (2008); Ortega-Hernandez and Shoenfeld (2009); Rosenblum *et al.* (2011)
	Gulf War syndrome	
	Juvenile dermatomyositis	Fernandez-Funez and Polo Romero (1998); Ramirez-Rivera *et al.* (2003); Altman *et al.* (2008)
Autoimmune rheumatic disorders	Arthritides	Positive findings: Gross *et al.* (1995); Pope *et al.* (1998); Maillefert *et al.* (1999); Geier and Geier (2002a, 2004) Negative findings: Ray *et al.* (2011)
	General vasculitides	Masse and Descoffres (1998); Le Hello *et al.* (1999); Zaas *et al.* (2001)
	Pulmonary and cutaneous vasculitis	Allen *et al.* (1993); Bui-Quang *et al.* (1998)
	Churg–Strauss vasculitis	Vanoli *et al.* (1998); Beretta *et al.* (2001)
	Henoch–Schonlein purpura	Chave *et al.* (2003)
	Still's disease	Grasland *et al.* (1998)
	Kawasaki's disease	Miron *et al.* (2003)
	Polyarteritis nodosa	Le Goff *et al.* (1988); Kerleau *et al.* (1997); De Keyser *et al.* (2000); Saadoun *et al.* (2001); Bourgeais *et al.* (2003); Begier *et al.* (2004); de Carvalho *et al.* (2008); Ventura *et al.* (2009)

(*continued*)

Table 16.1 *(Continued)*

	Autoimmune condition	References
	Systemic lupus erythematosus	Positive findings: Tudela *et al.* (1992); Mamoux and Dumont (1994); Grezard *et al.* (1996); Guiserix (1996); Finielz *et al.* (1998); Maillefert *et al.* (1999); Shapiro and Kopicky (2000); Shoenfeld and Aron-Maor (2000); Shoenfeld *et al.* (2000); Aron-Maor and Shoenfeld (2001); Chen *et al.* (2001); Fineschi (2001); Geier and Geier (2004); Ravel *et al.* (2004); Choffray *et al.* (2007); Santoro *et al.* (2007); Agmon-Levin *et al.* (2009b); Cooper *et al.* (2010) Negative findings: Cooper *et al.* (2002)
	Antiphospholipid syndrome	Martinuc Porobic *et al.* (2005); Blank *et al.* (2012)
	Bullous pemphigoid	Erbagci (2002); Merida *et al.* (2005)
Autoimmune skin conditions	Lichen planus	Ciaccio and Rebora (1990); Trevisan and Stinco (1993); Aubin *et al.* (1994); Lefort *et al.* (1995); Gisserot *et al.* (1997); Ferrando *et al.* (1998); Rybojad *et al.* (1998); Rebora *et al.* (1999); Schupp and Vente (1999); Agrawal *et al.* (2000); Al-Khenaizan (2001); Daramola *et al.* (2002); Limas and Limas (2002); Schuh *et al.* (2002); Agrawal and Shenoi (2004); Calista and Morri (2004); Criado *et al.* (2004); Kanwar and De (2010)
	Erythema multiforme	Loche, Schwarze *et al.* (2000); Bergstrom and Lindh (2008); Kaur and Handa (2008); Cho *et al.* (2011)
	Gianotti–Crosti syndrome	Andiran *et al.* (2002); Karakas *et al.* (2007)
	Alopecia	Wise *et al.* (1997)
Haemotological	Thrombocytopenia and pancytopenia	Meyboom *et al.* (1995); Neau *et al.* (1998); Viallard *et al.* (2000); Geier and Geier (2004, 2005); Nuevo *et al.* (2004); Schattner (2005)
Miscallaneous autoimmune conditions	Graves' disease	Chen *et al.* (2001); Yu *et al.* (2007)
	Glomerulonephritis	Geier and Geier (2004)

due to these adjuvants have been considered for a series of vaccines, and are not limited to HBV vaccine. Thus, a clear differentiation between autoimmune reactions caused by the HBV vaccine-contained HBAg immunogenicity and those caused by the other constituents of the vaccine is difficult to establish.

Several studies have indicated a potential link between HBV vaccine and autoimmunity, by crossreactivity with HBsAg epitopes, yeast antigens, or other adjuvants included in the vaccine (Geier and Geier, 2002a,b, 2004, 2005; Geier *et al.*, 2003; Ravel *et al.*, 2004; Schattner, 2005; Cooper *et al.*, 2010; Blank *et al.*, 2012; Perricone and Shoenfeld, 2013). HBV vaccine has also been implicated in ASIA, as well as in undifferentiated connective tissue diseases (Perricone and Shoenfeld, 2013). This section will discuss the role of the HBV vaccine in the development of autoimmune disease, including MS and other neuromuscular disorders, autoimmune skin and vascular disease, and other autoimmune diseases. It will become apparent that a large proportion of the data herein is based on case reports, as few large studies examining the role of HBV vaccine in autoimmunity exist. In Table 16.2, we summarize the potential pathogenetic mechanisms by which HBV vaccine might trigger autoimmune manifestations.

Neuromuscular disorders

Neuromuscular disorders, including MS, make up the largest group of autoimmune conditions to which HBV vaccine has been linked (Geier and Geier, 2004, 2005; Ram and Shoenfeld, 2008; Salemi and D'Amelio, 2010; Stubgen, 2010, 2012). A 2004 study examining adverse reports in the

Table 16.2 Potential pathogenetic mechanisms inducing HBVacc-related autoimmunity. Mechanisms are given as either working hypotheses (WH) or experimental data (ED)

	Pathogenetic mechanisms	Working hypotheses (WH)/experimental data (ED)	References
A	Vaccine-induced T cell dysregulation	HBVvacc responders have lower frequency of $CD4^+CD25^+$ (T regs) compared to nonresponders (ED) and this may make them prone to development of autoimmune phenomena (WH)	Li *et al.* (2010); Perricone and Shoenfeld (2013)
B	ASIA Aluminum-related	HBVacc-contained Al-related immunological phenomena Al salt activates dendritic cells and complement and increases levels of chemokine secretion (ED) Al regulates T-helper 2 response (ED) Al modulates expression of proinflammatory cytokines such as IL-12 (ED) Al immunoregulates IL-17 production (ED)	Agmon-Levin *et al.* (2012); Blank *et al.* (2012); Hogenesch (2013); Perricone and Shoenfeld (2013)
	(Plasma HBVacc-containing) thiomersal related (before 1999)[a] Yeast-induced dysregulation of regulatory T cells HBVaccs can be based on novel autoimmunity-sparing adjuvants	Immunoneutotoxicity related to thimerosal-containing mercury (ED) Numerical and functional impairment of T regs is induced by maturation of dendritic cells with yeast (ED) HBVaccs not containing conservative adjuvants (such as polysaccharides based on inulin) do not induce immune-mediated manifestations in vaccinated individuals (ED)	Wilcock *et al.* (2004); Cereda *et al.* (2011); Mori *et al.* (2012); Perricone and Shoenfeld (2013); Saade *et al.* (2013)
C	Molecular mimicry between hepatitis B surface antigen and self-antigens (e.g. myelin antigens in the case of MS)	Post-HBVacc patients develop antibodies crossrecognizing HBSag and MOG mimicking pairs (ED)	Bogdanos *et al.* (2005)

[a]Thiomersal removed from the HBVacc preparation in 1999

Al, aluminum; HBsAg, hepatitis B surface antigen; IL, interleukin; MOG, myelin oligodendrocyte glycoprotein; T regs, T regulatory cells; MS, multiple sclerosis

Vaccine Adverse Event Reporting System (VAERS) and PubMed found several cases of autoimmune neuromuscular disorders following HBV vaccine, including 130 cases of myelitis, 100 of optic neuritis, 101 of GBS, and 183 of MS (this study also noted several other cases of nonneuromuscular autoimmune diseases) (Geier and Geier, 2004). Although the volume of literature examining this link is larger for these disorders than for others, it still largely comprises individual case reports. This section will discuss the role of HBV vaccine in MS, myelitis, neuropathy (including GBS), myasthenia gravis (MG), and myopathy. It should be noted that, despite these reports, it is generally agreed that the benefits of vaccination outweigh the potential risks, and that reported adverse autoimmune events are infrequent, occurring in individuals with an underlying genetically determined predisposition.

Multiple sclerosis

Most studies linking MS to HBV vaccine describe neurological signs and symptoms, with MRI evidence of demyelination occurring days to weeks post-vaccination (Shoenfeld and Aron-Maor, 2000). Recombinant HBV vaccine was first associated with MS onset in 1991, in a case report of two individuals who received HBV vaccine and had human leukocyte antigen (HLA) haplotypes demonstrating susceptibility to MS (Herroelen *et al.*, 1991). This was followed by several other such case reports (Matsui *et al.*, 1996; Gout and Lyon-Caen, 1998; Marshall, 1998). A larger, hospital-based, case–control study of 121 MS patients who were administered a questionnaire regarding vaccination history could not exclude a link between HBV vaccine and a first demyelinating episode (Touze *et al.*, 2000). A prospective study also found an increased risk of developing MS 3 years after receiving HBV vaccine (odds ratio (OR) 3.1, 95% confidence interval (CI): 1.5, 6.3), with no increased risk being found with tetanus or influenza vaccinations (Hernan *et al.*, 2004). However, other studies have not confirmed such associations, and suggest that there is no link between HBV vaccine and the development of MS (Gout and Lyon-Caen, 1998; Ascherio *et al.*, 2001a,b; Confavreux *et al.*, 2001; DeStefano *et al.*, 2003). The current evidence does suggest that there may be a small increase in the risk of developing MS following HBV vaccine in predisposed individuals.

The mechanisms responsible for HBV vaccine-induced MS remain elusive. The role of molecular mimicry involving myelin antigens and their HBsAg mimics has been studied in healthy subjects receiving the HBV vaccine (Bogdanos *et al.*, 2005). There is amino acid similarity between HBsAg and myelin oligodendrocyte glycoprotein (MOG) or other myelin antigens, and peptidyl sequences corresponding to the viral/myelin pairs become targets of antibody responses soon after vaccination. Approximately 60% of vaccinees in this study had HBsAg/MOG double reactivity on at least one occasion, compared to none before vaccination, and inhibition studies confirmed the crossreactive nature of this immune response (Bogdanos *et al.*, 2005). Notably, at 6 months post-vaccination, reactivity was lost in 29% of the cases that were anti-MOG reactive at 3 months, and none of the vaccinees reported symptoms of demyelinating disorders. The extent of crossreactive immune responses in HBV vaccine vaccinees with demyelinating disorders has not yet been studied.

Myelitis

Several reports of HBV vaccine-associated myelitis appear in the literature (Mahassin *et al.*, 1993; Trevisani *et al.*, 1993; Tartaglino *et al.*, 1995; Senejoux *et al.*, 1996; Song *et al.*, 1997; Renard *et al.*, 1999; Iniguez *et al.*, 2000; Karaali-Savrun *et al.*, 2001; Fonseca *et al.*, 2003). One of these reports consists of four patients, one of whom developed relapsing–remitting disease that was eventually diagnosed as MS (Karaali-Savrun *et al.*, 2001). The majority of reports involve females, although age ranges vary from 3 to 64 years (Stubgen, 2012). One systematic review found that HBV vaccine is the most common form of immunization associated with acute myelitis (Agmon-Levin *et al.*, 2009a). That analysis located 37 cases of transverse myelitis associated with different vaccines, including, but not limited to, HBV, MMR, and diphtheria, tetanus, and pertussis (dTP), which were administered to infants, children, and adults (Agmon-Levin *et al.*, 2009a). The temporal association in most cases was between several days and 3 months, although several years was also noted in some cases (Agmon-Levin *et al.*, 2009a).

A recent systematic review assessed the potential association between HBV vaccine and MS (or other central nervous system (CNS) disorders) (Martinez-Sernandez and Figueiras, 2013). The authors identified a total of 216 published reports in a period between 1981 and 2011, including 12 meeting the criteria for proper analysis. Among these, six were case–control studies (sample size ranged from 121 to 14 362 individuals in the case group and from 121 to 7671 in the control group; three were nested into one or two cohorts), two were cohort studies (sample size varied from 11 to 27 229 subjects in the study group and from 71 to 107 469 in the comparison group), and one was a case–crossover study (containing 643 patients). The analysis concluded that there was no association between HBV vaccine and MS, but stressed the relatively low quality of the observational studies, which prevents vigorous conclusions from being drawn (Martinez-Sernandez and Figueiras, 2013).

Neuropathy

To date, there is no conclusive evidence linking HBV vaccine to neuropathy (Stubgen, 2010). However, several case reports of neuropathy (including several of GBS) developing after recombinant and plasma-derived HBV vaccine (Tuohy, 1989; McMahon *et al.*, 1992; Kakar and Sethi, 1997; Maillefert *et al.*, 1997; Creange *et al.*, 1999; Sinsawaiwong and Thampanitchawong, 2000; Dano and Korczyn, 2001; DeJonckere and de Surgeres, 2001; Sindern *et al.*, 2001; Seti *et al.*, 2002; Khamaisi *et al.*, 2004). The number of injections prior to development of symptoms ranged from one to four, with time between injection and symptoms ranging from 1 day to 8 weeks (Tuohy, 1989; McMahon *et al.*, 1992; Kakar and Sethi, 1997; Maillefert *et al.*, 1997; Creange *et al.*, 1999; Sinsawaiwong and Thampanitchawong, 2000; Dano and Korczyn, 2001; DeJonckere and de Surgeres, 2001; Sindern *et al.*, 2001; Seti *et al.*, 2002; Khamaisi *et al.*, 2004). In addition to case reports, post-marketing surveillance following HBV vaccine in 850 000 patients revealed multiple neurological events, including five cases of convulsions, ten of Bell's palsy, nine of GBS, five of lumbar radiculopathy, three of brachial plexus neuropathy, five of optic neuritis, and four of transverse myelitis (Shaw *et al.*, 1988). Further studies utilizing VAERS have also noted that HBV vaccine is linked to autoimmune events, including GBS (Geier and Geier, 2004; Haber *et al.*, 2004; Souayah *et al.*, 2009; van Doorn, 2009).

Myasthenia gravis

The current evidence linking HBV vaccine with MG is not strong, with only a few case reports (Biron *et al.*, 1988; Domigo *et al.*, 1999). Although all case reports noted in this chapter had positive antibodies to acetylcholine receptors, there is no clear link between HBV vaccine and the development of symptoms. One case involved the development of MG in a 48-year-old male following HBV vaccine, although this patient had also received general anesthesia at the time (Biron *et al.*, 1988). The lack of larger studies, in addition to the low volume of case reports, indicates that MG following HBV vaccine in unlikely or exceedingly rare.

Myopathy and macrophagic myofascitiis

Currently, there is insufficient evidence to strongly link HBV vaccine with myopathy (Buskila *et al.*, 2008; Agmon-Levin and Shoenfeld, 2008; Orbach and Tanay, 2009; Ortega-Hernandez and Shoenfeld, 2009; Stubgen, 2010), despite the presumed ability of HBV to infect muscle fibers, with muscle damage resulting from an immune-mediated response to viral antigens (Capasso *et al.*, 2006). Like other aluminum-containing vaccines, HBV vaccine has been considered responsible for macrophagic myofascilitis (MMF), a pathological disorder first described in France in 1993, and seen in muscle biopsies from a small number of patients suffering with various complaints (Cherin *et al.*, 1999; Fischer *et al.*, 2003). It is characterized by inflammatory macrophagic lesions (periodic acid–Schiff stain positive for glycogen) and aluminum salt deposition at the site of injection, and possibly at other sites (Fischer *et al.*, 2003). Over the years, several hundred MMF cases have been recognized, and their apparent association with aluminum-containing vaccines has been documented.

Several cases have associated juvenile dermatomyositis (DM) with HBV vaccine (Fernandez-Funez and Polo Romero, 1998; Ramirez-Rivera *et al.*, 2003; Altman *et al.*, 2008), with one study noting HLA-DR3 as a possible predisposing factor (Fernandez-Funez and Polo Romero, 1998). Other studies have found no association between HBV vaccine and myopathy (Maillefert *et al.*, 1999; Geier and Geier, 2005), although one reported a possible link between HBV vaccine and other rheumatic conditions (rheumatoid arthritis (RA), SLE, reactive arthritis, polyarthralgia-myalgia, and vasculitis) (Maillefert *et al.*, 1999).

Chronic fatigue syndrome

The term "chronic fatigue syndrome" (CFS) was first used in the early 1980s to describe a severe condition of primary persistent or relapsing fatigue of unknown etiology. The pathogenesis of CFS remains elusive, although neuroendocrine, hormonal, and immunoglial factors are likely mediators accounting for the induction of this disorder (Agmon-Levin and Shoenfeld, 2008; Ortega-Hernandez and Shoenfeld, 2009; Rosenblum *et al.*, 2011). In early 1990, some media coverage in Canada raised concern as to whether HBV vaccine is responsible for CFS (Delage *et al.*,

1993), but such an association has not been confirmed by the epidemiological data thus far (Public Health Agency of Canada, 1993).

Autoimmune rheumatic disorders

A variety of autoimmune rheumatic disorders have been linked with HBV vaccine administration. A considerable number of reports link HBV vaccine with arthritides, vasculitides, systemic lupus erythematosus (SLE), and antiphospholipid syndrome (APS).

Arthritis

Several case reports have found an association between the development of arthritis and/or RA and HBV vaccine (Pope *et al.*, 1998; Maillefert *et al.*, 1999), including which also notes HLA-DR4 as a potential predisposing factor (Gross *et al.*, 1995). A study from the United States conducted in 2000 reported an association between HBV vaccine and arthritis (as well as pharyngitis, nasopharyngitis, and ear infection) in a pediatric cohort (Fisher *et al.*, 2001). However, another larger study did not find any association between RA and HBV vaccine (Ray *et al.*, 2011). A 2004 study that noted adverse events following HBV vaccine, using VAERS and PubMed, found 166 cases of RA and 415 cases of arthritis (non-RA) (Geier and Geier, 2004). A VAERS analysis in 2002 by the same authors found adult rubella vaccines and HBV vaccine were statistically associated with chronic arthritis, which persisted for a minimum of 1 year (Geier and Geier, 2002a).

Vasculitides

There is a large body of evidence regarding an association between HBV vaccine and vasculitis, although largely in the form of case reports (Le Goff *et al.*, 1988; Allen *et al.*, 1993; Bui-Quang *et al.*, 1998; Grasland *et al.*, 1998; Masse and Descoffres, 1998; Vanoli *et al.*, 1998; Le Hello *et al.*, 1999; De Keyser *et al.*, 2000; Beretta *et al.*, 2001; Zaas *et al.*, 2001; Bourgeais *et al.*, 2003; Chave *et al.*, 2003; Miron *et al.*, 2003). There are multiple case reports of polyarteritis nodosa (PAN) (some involving pediatric patients), with symptoms developing after the first dose and booster doses, and onset of symptoms ranging from 1 week to 2 months (Le Goff *et al.*, 1988; Kerleau *et al.*, 1997; De Keyser *et al.*, 2000; Saadoun *et al.*, 2001; Bourgeais *et al.*, 2003; de Carvalho *et al.*, 2008; Ventura *et al.*, 2009). Of 10 definite or probable PAN cases reported to the VAERS, nine (90%) were post-HBV vaccine, with an average onset of symptoms 2 weeks after vaccination (Begier *et al.*, 2004). However, it should be noted that other infectious triggers were not ruled out, and that identification of vaccine antigens was not attempted in clinical samples (Begier *et al.*, 2004).

Systemic lupus erythematosus

SLE has also been linked to HBV vaccine in a number of case series and reports (Tudela *et al.*, 1992; Mamoux and Dumont, 1994; Grezard *et al.*, 1996; Guiserix, 1996; Finielz *et al.*, 1998; Maillefert *et al.*, 1999; Shapiro and Kopicky, 2000; Aron-Maor and Shoenfeld, 2001; Fineschi, 2001; Ravel *et al.*, 2004; Agmon-Levin *et al.*, 2009b; Cooper *et al.*, 2010). One case series reported SLE in 10 patients after they received HBV vaccine (Agmon-Levin *et al.*, 2009b), while an analysis of VAERS and PubMed found four cases of SLE to be associated with HBV vaccine (Geier and Geier, 2004). Another study reported exacerbations of previously undiagnosed SLE in two individuals following HBV vaccine (Maillefert *et al.*, 1999), and SLE flares following HBV vaccine have been noted in other studies, too (Shoenfeld *et al.*, 2000; Shoenfeld and Aron-Maor, 2000; Chen *et al.*, 2001; Choffray *et al.*, 2007), as has lupus nephritis (Santoro *et al.*, 2007). However, a case–control study of 265 SLE patients did not find an association between SLE and HBV vaccine (Cooper *et al.*, 2002). It currently appears that HBV vaccine may induce SLE or SLE flares in a small number of cases, but there is no evidence indicating a significant risk.

Antiphospholipid syndrome

The classical APS is clinically associated with recurrent fetal loss, thrombocytopenia thromboembolic phenomena, and multiorgan involvement, including heart, kidney, CNS, and skin (Cervera *et al.*, 2002). The serological marker of APS is the presence of antiphospholipid antibodies (aPLs), which bind negatively charged phospholipids, platelets, and endothelial cells, mainly through the plasma protein beta-2-glycoprotein-I (b2GPI) (Cervera *et al.*, 2002). The presence of IgG and IgM anticardiolipin antibodies (aCLs) and lupus anticoagulant (LA) is associated with thrombosis in patients with APS.

Chronic HBV infection (Zachou *et al.*, 2003; Blank *et al.*, 2004), and indeed HBV vaccine, has been considered a trigger of APS-related autoantibodies. Mean absorbance values of IgG aCL and IgG anti-β_2GPI levels were increased 1 month following vaccination with three doses of recombinant HBV vaccine in 8 of 85 students (9%) and decreased in 2. Two vaccinees showed transient increase for aCL and one showed transient increase for anti-β_2GPI. None of the vaccinees developed APS during the study period. Another student was initially low-positive for IgG anti-β2GPI, and the levels increased progressively during 6-month follow-up after vaccination. Blank *et al.* (2012) suggested that the induction of anti-b2GPI antibodies in HBV vaccine vaccinees may be caused by the binding of the fifth domain of b2GPI to recombinant HBsAgI (Mehdi *et al.*, 2008), which is probably involved in the initiation of autoimmune responses.

Thrombocytopenia and pancytopenia

Thrombocytopenia/pancytopenia following HBV vaccine has been reported in case series and larger studies (Meyboom *et al.*, 1995; Neau *et al.*, 1998; Viallard *et al.*, 2000; Geier and Geier, 2004, 2005; Nuevo *et al.*, 2004; Schattner, 2005). An analysis of VAERS and PubMed reports between 1966 and 2003 revealed 283 cases of thrombocytopenia/pancytopenia (Geier and Geier, 2004) and a case–control study found a significantly increased OR for thrombocytopenia in adults receiving HBV vaccine (Geier and Geier, 2005). Several cases are also present in the literature (Meyboom *et al.*, 1995; Neau *et al.*, 1998; Viallard *et al.*, 2000; Nuevo *et al.*, 2004). Although rare, there does appear to be an increased incidence of thrombocytopenia/pancytopenia following HBV vaccine.

Autoimmune skin conditions

Although several autoimmune skin conditions have been associated with HBV vaccine, most are only represented by a small number of case reports, and there is insufficient evidence to definitively link most conditions to HBV vaccine. Lichen planus was first reported post-HBV vaccine in 1990 (Ciaccio and Rebora, 1990), and it has received a relatively large number of case reports and series since then (Trevisan and Stinco, 1993; Aubin *et al.*, 1994; Lefort *et al.*, 1995; Gisserot *et al.*, 1997; Ferrando *et al.*, 1998; Rybojad *et al.*, 1998; Rebora *et al.*, 1999; Schupp and Vente, 1999; Agrawal *et al.*, 2000; Al-Khenaizan, 2001; Daramola *et al.*, 2002; Limas and Limas, 2002; Schuh *et al.*, 2002; Agrawal and Shenoi, 2004; Calista and Morri, 2004; Criado *et al.*, 2004; Kanwar and De, 2010). One case series reports lichen planus in 100 children, with 16 of these developing the condition post-HBV vaccine (Kanwar and De, 2010). To our knowledge, only two case reports of bullous pemphigoid following HBV vaccine exist (Erbagci, 2002; Merida *et al.*, 2005), although one of these involved multiple vaccines, including polio and *Haemophilus influenzae B* (Merida *et al.*, 2005). A small number of cases of erythema multiforme following HBV vaccine have also been reported, and a majority of these also consisted of multiple vaccinations (Loche *et al.*, 2000; Bergstrom and Lindh, 2008; Kaur and Handa, 2008; Cho *et al.*, 2011). Only a single case report of Henoch–Schönlein purpura appears in the literature (Chave *et al.*, 2003), and only a small number link Gianotti–Crosti syndrome to HBV vaccine (Andiran *et al.*, 2002; Karakas *et al.*, 2007), although the patient in one of them also received a measles vaccination with the HBV vaccine (Andiran *et al.*, 2002).

Alopecia has been reported following routine HBV vaccine (Wise *et al.*, 1997), but analyses of WHO and CDC HBV vaccine safety databases do not support a causal link between hair loss and HBV vaccine.

Miscellaneous autoimmune conditions

Two other autoimmune conditions have also been linked to HBV vaccine. As they have only few reports, we include them here for reference purposes. These conditions are Graves' disease (mixed results) (Chen *et al.*, 2001; Yu *et al.*, 2007) and glomerulonephritis (Geier and Geier, 2004).

Conclusions

Although rare, autoimmune conditions related to HBV vaccine do appear to occur in some predisposed individuals. Data regarding HBV vaccine and most autoimmune conditions are scarce, which does not allow definitive conclusions to be drawn. However, several conditions do have a large number of case reports, in addition to larger studies. Links appear to be strongest in regards to MS and myelitis, vasculitis, chronic arthritis, SLE, and thrombocytopenia/pancytopenia. However, it must be noted that even these incidents are exceedingly rare, and vaccination is still recommended, especially in at-risk individuals. Development of vaccines lacking autoimmunity-prone adjuvants (Saade *et al.*, 2013) (such as aluminum) could further minimize the risk for development of HBV vaccine-induced autoimmune phenomena.

References

Agmon-Levin, N. and Shoenfeld, Y. (2008). Chronic fatigue syndrome with autoantibodies – the result of an augmented adjuvant effect of hepatitis-B vaccine and silicone implant. *Autoimmun Rev*, **8**(1): 52–5.

Agmon-Levin, N., Kivity, S., Szyper-Kravitz, M., and Shoenfeld, Y. (2009a). Transverse myelitis and vaccines: a multi-analysis. *Lupus*, **18**(13): 1198–204.

Agmon-Levin, N., Zafrir, Y., Paz, Z., *et al.* (2009b). Ten cases of systemic lupus erythematosus related to hepatitis B vaccine. *Lupus*, **18**(13): 1192–7.

Agmon-Levin, N., Hughes, G.R.V., and Shoenfeld, Y. (2012). The spectrum of ASIA: "Autoimmune (Auto-inflammatory) Syndrome induced by Adjuvants." *Lupus*, **21**(2): 118–20.

Agrawal, A. and Shenoi, S.D. (2004). Lichen planus secondary to hepatitis B vaccination. *Indian J Dermatol Venereol Leprol*, **70**(4): 234–5.

Agrawal, S., Garg, V.K., Joshi, A., *et al.* (2000). Lichen planus after HBV vaccination in a child: a case report from Nepal. *J Dermatol*, **27**(9): 618–20.

Al-Khenaizan, S. (2001). Lichen planus occurring after hepatitis B vaccination: a new case. *J Am Acad Dermatol*, **45**(4): 614–15.

Allen, M.B., Cockwell, P., and Page, R.L. (1993). Pulmonary and cutaneous vasculitis following hepatitis B vaccination. *Thorax*, **48**(5): 580–1.

Altman, A., Szyper-Kravitz, M., and Shoenfeld, Y. (2008). HBV vaccine and dermatomyositis: is there an association? *Rheumatol Int*, **28**(6): 609–12.

Andiran, N., Senturk, G.B., and Bukulmez, G. (2002). Combined vaccination by measles and hepatitis B vaccines: a new cause of Gianotti-Crosti syndrome. *Dermatology*, **204**(1): 75–6.

Aron-Maor, A. and Shoenfeld, Y. (2001). Vaccination and systemic lupus erythematosus: the bidirectional dilemmas. *Lupus*, **10**(3): 237–40.

Ascherio, A., Munger, K.L., Lennette, E.T., *et al.* (2001a). Epstein-Barr virus antibodies and risk of multiple sclerosis: a prospective study. *JAMA*, **286**(24): 3083–8.

Ascherio, A., Zhang, S.M., Hernan, M.A., *et al.* (2001b). Hepatitis B vaccination and the risk of multiple sclerosis. *N Engl J Med*, **344**(5): 327–32.

Aubin, F., Angonin, R., Humbert, P., and Agache, P. (1994). Lichen planus following hepatitis B vaccination. *Arch Dermatol*, **130**(10): 1329–30.

Begier, E.M., Langford, C.A., Sneller, M.C., *et al.* (2004). Polyarteritis nodosa reports to the vaccine adverse event reporting system (VAERS): implications for assessment of suspected vaccine-provoked vasculitis. *J Rheumatol*, **31**(11): 2181–8.

Beretta, L., Caronni, M., Vanoli, M., and Scorza, R. (2001). Churg-Strauss vasculitis with brain involvement following hepatitis B vaccination. *Clin Exp Rheumatol*, **19**(6): 757.

Bergstrom, H. and Lindh, J. (2008). [A connection between erythema multiforme and Twinrix vaccination?]. *Lakartidningen*, **105**(30–1): 2114–15.

Biron, P., Montpetit, P., Infante-Rivard, C., and Lery, L. (1988). Myasthenia gravis after general anesthesia and hepatitis B vaccine. *Arch Intern Med*, **148**(12): 2685.

Blank, M., Asherson, R.A., Cervera, R., and Shoenfeld, Y. (2004). Antiphospholipid syndrome infectious origin. *J Clin Immunol*, **24**(1): 12–23.

Blank, M., Israeli, E., and Shoenfeld, Y. (2012). When APS (Hughes syndrome) met the autoimmune/inflammatory syndrome induced by adjuvants (ASIA). *Lupus*, **21**(7): 711–14.

Bogdanos, D., Smith, H., Ma, Y., *et al.* (2005). A study of molecular mimicry and immunological cross-reactivity between hepatitis B surface antigen and myelin mimics. *Clin Dev Immunol*, **12**(3): 217.

Bogdanos, D.P., Smyk, D.S., Invernizzi, P., *et al.* (2013). Infectome: a platform to trace infectious triggers of autoimmunity. *Autoimmun Rev*, **12**(7): 726–40.

Bourgeais, A.M., Dore, M.X., Croue, A., *et al.* (2003). [Cutaneous polyarteritis nodosa following hepatitis B vaccination]. *Ann Dermatol Venereol*, **130**(2 Pt 1): 205–7.

Bui-Quang, D., Thomas, E., Riols, M., *et al.* (1998). [Cutaneous vasculitis after hepatitis B vaccination with recombinant vaccine in a renal transplant patient]. *Presse Med*, **27**(26): 1321–3.

Buskila, D., Atzeni, F., and Sarzi-Puttini, P. (2008). Etiology of fibromyalgia: the possible role of infection and vaccination. *Autoimmun Rev*, **8**(1): 41–3.

Calista, D. and Morri, M. (2004). Lichen planus induced by hepatitis B vaccination: a new case and review of the literature. *Int J Dermatol*, **43**(8): 562–4.

Capasso, M., Di Muzio, A. Comar, M., *et al.* (2006). The association of chronic hepatitis B and myopathy. *Neurology*, **67**(8): 1467–9.

CDC (Centers for Disease Control and Prevention). (1990).Protection against viral hepatitis. Recommendations of the Immunization Practices Advisory Committee (ACIP). *MMWR*, **39**(RR-2): 1–26.

CDC (Centers for Disease Control and Prevention). (1991). Hepatitis B virus: a comprehensive strategy for eliminating transmission in the United States through universal childhood vaccination. Recommendations of the Immunization Practices Advisory Committee (ACIP). *MMWR*, **40**(RR-13): 1–25.

Cereda, V., Vergati, M., Huen, N.Y., *et al.* (2011). Maturation of human dendritic cells with *Saccharomyces cerevisiae* (yeast) reduces the number and function of regulatory T cells and enhances the ratio of antigen-specific effectors to regulatory T cells. *Vaccine*, **29**(31): 4992–9.

Cervera, R., Piette, J.C., Font, J., *et al.* (2002). Antiphospholipid syndrome: clinical and immunologic manifestations and patterns of disease expression in a cohort of 1,000 patients. *Arthritis Rheum*, **46**(4): 1019–27.

Chang, M.H., Chen, C.J., Lai, M.S., *et al.* (1997). Universal hepatitis B vaccination in Taiwan and the incidence of hepatocellular carcinoma in children. Taiwan Childhood Hepatoma Study Group. *N Engl J Med*, **336**(26): 1855–9.

Chave, T., Neal, C., and Camp, R. (2003). Henoch-Schonlein purpura following hepatitis B vaccination. *J Dermatolog Treat*, **14**(3): 179–81.

Chen, R.T., Pless, R., and Destefano, F. (2001). Epidemiology of autoimmune reactions induced by vaccination. *J Autoimmun*, **16**(3): 309–18.

Cherin, P., Laforet, P., Gheradi, R.K., *et al.* (1999). [Macrophagic myofasciitis: description and etiopathogenic hypotheses. Study and Research Group on Acquired and Dysimmunity-related Muscular Diseases (GERMMAD) of the French Association against Myopathies (AFM)]. *Rev Med Interne*, **20**(6): 483–9.

Cho, Y.J., Huh, S.Y., Hong, J.S., *et al.* (2011). Neonatal erythema multiforme: a case report. *Ann Dermatol*, **23**(3): 382–5.

Choffray, A., Pinquier, L., and Bachelez, H. (2007). Exacerbation of lupus panniculitis following anti-hepatitis-B vaccination. *Dermatology*, **215**(2): 152–4.

Ciaccio, M. and Rebora, A. (1990). Lichen planus following HBV vaccination: a coincidence? *Br J Dermatol*, **122**(3): 424.

Confavreux, C., Suissa, S., Saddier, P., *et al.* (2001). Vaccinations and the risk of relapse in multiple sclerosis. Vaccines in Multiple Sclerosis Study Group. *N Engl J Med*, **344**(5): 319–26.

Cooper, G.S., Dooley, M.A., Treadwell, E.L., *et al.* (2002). Risk factors for development of systemic lupus erythematosus: allergies, infections, and family history. *J Clin Epidemiol*, **55**(10): 982–9.

Cooper, G.S., Wither, J., Bernatsky, S., *et al.* (2010). Occupational and environmental exposures and risk of systemic lupus erythematosus: silica, sunlight, solvents. *Rheumatology*, **49**(11): 2172–80.

Creange, A., Temam, G., and Lefaucheur, J.P. (1999). Lumbosacral acute demyelinating polyneuropathy following hepatitis B vaccination. *Autoimmunity*, **30**(3): 143–6.

Criado, P.R., de Oliveira Ramos, R., Vasconcellos, C., *et al.* (2004). Two case reports of cutaneous adverse reactions following hepatitis B vaccine: lichen planus and granuloma annulare. *J Eur Acad Dermatol Venereol*, **18**(5): 603–6.

Dano, M. and Korczyn, A.D. (2001). Precipitation of hereditary motor-sensory neuropathy following hepatitis B vaccination. *J Clin Neuromuscul Dis*, **3**(2): 75–6.

Daramola, O.O., Ogunbiyi, A.O., and George, A.O. (2002). Lichen planus following hepatitis B vaccination in an African girl. *Trop Doct*, **32**(2): 117–18.

de Carvalho, J.F., Pereira, R.M., and Shoenfeld, Y. (2008). Systemic polyarteritis nodosa following hepatitis B vaccination. *Eur J Intern Med*, **19**(8): 575–8.

De Keyser, F., Naeyaert, J.M., Hindryckx, P., *et al.* (2000). Immune-mediated pathology following hepatitis B vaccination. Two cases of polyarteritis nodosa and one case of pityriasis rosea-like drug eruption. *Clin Exp Rheumatol*, **18**(1): 81–5.

DeJonckere, P.H. and de Surgeres, G.G. (2001). Acute tinnitus and permanent audiovestibular damage after hepatitis B vaccination. *Int Tinnitus J*, **7**(1): 59–61.

Delage, G., Salit, I., Pennie, R., *et al.* (1993). [The possible relation between hepatitis B vaccination and chronic fatigue syndrome]. *Union Med Can*, **122**(4): 278–9.

DeStefano, F., Verstraeten, T., Jackson, L.A., *et al.* (2003). Vaccinations and risk of central nervous system demyelinating diseases in adults. *Arch Neurol*, **60**(4): 504–9.

Domigo, V., Authier, F.J., Bachoud-Levi, A.C., *et al.* (1999). Should hepatitis B vaccine be contra-indicated in myasthenia gravis? *Autoimmunity*, **29**(2): 139–40.

Erbagci, Z. (2002). Childhood bullous pemphigoid following hepatitis B immunization. *J Dermatol*, **29**(12): 781–5.

Fernandez-Funez, A. and Polo Romero, F.J. (1998). [Juvenile dermatomyositis concomitant with hepatitis B vaccination]. *Med Clin (Barc)*, **111**(17): 675.

Ferrando, M.F., Doutre, M.S., Beylot-Barry, M., *et al.* (1998). Lichen planus following hepatitis B vaccination. *Br J Dermatol*, **139**(2): 350.

Fineschi, S. (2001). Can recombinant anti-hepatitis B vaccine be a cause of systemic lupus erythematosus? *Lupus*, **10**(11): 830.

Finielz, P., Lam-Kam-Sang, L.F., and Guiserix, J. (1998). Systemic lupus erythematosus and thrombocytopenic purpura in two members of the same family following hepatitis B vaccine. *Nephrol Dial Transplant*, **13**(9): 2420–1.

Fischer, D., Reimann, J., and Schroder, R. (2003). [Macrophagic myofasciitis: inflammatory, vaccination-associated muscular disease]. *Dtsch Med Wochenschr*, **128**(44): 2305–8.

Fisher, M.A., Eklund, S.A., James, S.A., and Lin, X. (2001). Adverse events associated with hepatitis B vaccine in U.S. children less than six years of age, 1993 and 1994. *Ann Epidemiol*, **11**(1): 13–21.

Fonseca, L.F., Noce, T.R., Teixeira, M.L., *et al.* (2003). Early-onset acute transverse myelitis following hepatitis B vaccination and respiratory infection: case report. *Arq Neuropsiquiatr*, **61**(2A): 265–8.

Francis, D.P., Favero, M.S. *et al.* (1981). Transmission of hepatitis B virus. *Semin Liver Dis*, **1**(1): 27–32.

Fujinami, R.S., von Herrath, M.G., Christen, U., and Whitton, J.L. (2006). Molecular mimicry, bystander activation, or viral persistence: infections and autoimmune disease. *Clin Microbiol Rev*, **19**(1): 80–94.

Gavin, A.L., Hoebe, K., Duong, B., *et al.* (2006). Adjuvant-enhanced antibody responses in the absence of toll-like receptor signaling. *Science*, **314**(5807): 1936–8.

Geier, D.A. and Geier, M.R. (2002a). A one year followup of chronic arthritis following rubella and hepatitis B vaccination based upon analysis of the Vaccine Adverse Events Reporting System (VAERS) database. *Clin Exp Rheumatol*, **20**(6): 767–71.

Geier, M.R. and Geier, D.A. (2002b). Hepatitis B vaccination safety. *Ann Pharmacother*, **36**(3): 370–4.

Geier, M.R. and Geier, D.A. (2004). A case-series of adverse events, positive re-challenge of symptoms, and events in identical twins following hepatitis B vaccination: analysis of the Vaccine Adverse Event Reporting System (VAERS) database and literature review. *Clin Exp Rheumatol*, **22**(6): 749–55.

Geier, D.A. and Geier, M.R. (2005). A case-control study of serious autoimmune adverse events following hepatitis B immunization. *Autoimmunity*, **38**(4): 295–301.

Geier, M.R., Geier, D.A., and Zahalsky, A.C. (2003). A review of hepatitis B vaccination. *Expert Opin Drug Saf*, **2**(2): 113–22.

Gisserot, O., Carsuzaa, F., Marlier, S., *et al.* (1997). [Lichen planus after hepatitis B vaccination. 3 new cases]. *Presse Med*, **26**(16): 760.

Goldstein, S.T., Zhou, F., Hadler, S.C., *et al.* (2005). A mathematical model to estimate global hepatitis B disease burden and vaccination impact. *Int J Epidemiol*, **34**(6): 1329–39.

Gout, O. and Lyon-Caen, O. (1998). [Sclerotic plaques and vaccination against hepatitis B]. *Rev Neurol (Paris)*, **154**(3): 205–7.

Grasland, A., Le Maitre, F., Pouchot, J., *et al.* (1998). [Adult-onset Still's disease after hepatitis A and B vaccination?]. *Rev Med Interne*, **19**(2): 134–6.

Grezard, P., Chefai, M., Philippot, V., *et al.* (1996). [Cutaneous lupus erythematosus and buccal aphthosis after hepatitis B vaccination in a 6-year-old child]. *Ann Dermatol Venereol*, **123**(10): 657–9.

Gross, K., Combe, C., Kruger, K., and Schattenkirchner, M. (1995). Arthritis after hepatitis B vaccination. Report of three cases. *Scand J Rheumatol*, **24**(1): 50–2.

Guiserix, J. (1996). Systemic lupus erythematosus following hepatitis B vaccine. *Nephron*, **74**(2): 441.

Haber, P., DeStefano, F., Angulo, F.J., *et al.* (2004). Guillain-Barre syndrome following influenza vaccination. *JAMA*, **292**(20): 2478–81.

Hernan, M.A., Jick, S.S., Olek, M.J., and Jick, H. (2004). Recombinant hepatitis B vaccine and the risk of multiple sclerosis: a prospective study. *Neurology*, **63**(5): 838–42.

Herroelen, L., de Keyser, J., and Ebinger, G. (1991). Central-nervous-system demyelination after immunisation with recombinant hepatitis B vaccine. *Lancet*, **338**(8776): 1174–5.

Hogenesch, H. (2013). Mechanism of immunopotentiation and safety of aluminum adjuvants. *Front Immunol*, **3**: 406.

Iniguez, C., Mauri, J.A., Larrode, P., *et al.* (2000). [Acute transverse myelitis secondary to hepatitis B vaccination]. *Rev Neurol*, **31**(5): 430–2.

Israeli, E., Agmon-Levin, N., Blank, M., and Shoenfeld, Y. (2009). Adjuvants and autoimmunity. *Lupus*, **18**(13): 1217–25.

Kakar, A. and Sethi, P.K. (1997). Guillain Barre syndrome associated with hepatitis B vaccination. *Indian J Pediatr*, **64**(5): 710–12.

Kanwar, A.J. and De, D. (2010). Lichen planus in childhood: report of 100 cases. *Clin Exp Dermatol*, **35**(3): 257–62.

Karaali-Savrun, F., Altintas, A., Saip, S., and Siva, A. (2001). Hepatitis B vaccine related-myelitis? *Eur J Neurol*, **8**(6): 711–15.

Karakas, M., Durdu, M., Tuncer, I., and Cevlik, F. (2007). Gianotti-Crosti syndrome in a child following hepatitis B virus vaccination. *J Dermatol*, **34**(2): 117–20.

Katz-Levy, Y., Neville, K.L., Girvin, A.M., *et al.* (1999). Endogenous presentation of self myelin epitopes by CNS-resident APCs in Theiler's virus-infected mice. *J Clin Invest*, **104**(5): 599–610.

Kaur, S. and Handa, S. (2008). Erythema multiforme following vaccination in an infant. *Indian J Dermatol Venereol Leprol*, **74**(3): 251–3.

Kerleau, J.M., Levesque, H., Lair, G., *et al.* (1997). [Is hepatitis B vaccination a new cause of necrotizing vasculitis?]. *Rev Med Interne*, **18**(6): 491–2.

Khamaisi, M., Shoenfeld, Y., and Orbach, H. (2004). Guillain-Barre syndrome following hepatitis B vaccination. *Clin Exp Rheumatol*, **22**(6): 767–70.

Kool, M., Soullie, T., van Nimwegen, M., *et al.* (2008). Alum adjuvant boosts adaptive immunity by inducing uric acid and activating inflammatory dendritic cells. *J Exp Med*, **205**(4): 869–82.

Le Goff, P., Fauquert, P., Youinou, P., and Hoang, S. (1988). [Periarteritis nodosa following vaccination against hepatitis B]. *Presse Med*, **17**(34): 1763.

Le Hello, C., Cohen, P. Bousser, M.G., *et al.* (1999). Suspected hepatitis B vaccination related vasculitis. *J Rheumatol*, **26**(1): 191–4.

Lefort, A., Dachary, D., Vergier, B., and Boiron, G. (1995). [Lichen planus and vaccination against hepatitis B]. *Ann Dermatol Venereol*, **122**(10): 701–3.

Li, J.C., Tan, D.M., Liu, H.B., and Li, K.C. (2010). [The roles of CD4+CD25+ T regulatory cells and FoxP3 mRNA expression in the subjects with and without responses to hepatitis B virus vaccination]. *Zhonghua Gan Zang Bing Za Zhi*, **18**(10): 726–30.

Liang, T.J. (2009). Hepatitis B: the virus and disease. *Hepatology*, **49**(5 Suppl.): S13–21.

Limas, C. and Limas, C.J. (2002). Lichen planus in children: a possible complication of hepatitis B vaccines. *Pediatr Dermatol*, **19**(3): 204–9.

Loche, F., Schwarze, H.P., Thedenat, B., *et al.* (2000). Erythema multiforme associated with hepatitis B immunization. *Clin Exp Dermatol*, **25**(2): 167–8.

Mahassin, F., Algayres, J.P., Valmary, J., *et al.* (1993). [Acute myelitis after vaccination against hepatitis B]. *Presse Med*, **22**(40): 1997–8.

Maillefert, J.F., Farge, P., Gazet-Maillefert, M.P., and Tavernier, C. (1997). Mental nerve neuropathy as a result of hepatitis B vaccination. *Oral Surg Oral Med Oral Pathol Oral Radiol Endod*, **83**(6): 663–4.

Maillefert, J.F., Sibilia, J., Vignon, E., *et al.* (1999). Rheumatic disorders developed after hepatitis B vaccination. *Rheumatology*, **38**(10): 978–83.

Mamoux, V. and Dumont, C. (1994). [Lupus erythematosus disseminatus and vaccination against hepatitis B virus]. *Arch Pediatr*, **1**(3): 307–8.

Marrack, P., McKee, A.S., and Munks, M.W. (2009). Towards an understanding of the adjuvant action of aluminium. *Nat Rev Immunol*, **9**(4): 287–93.

Marshall, E. (1998). A shadow falls on hepatitis B vaccination effort. *Science*, **281**(5377): 630–1.

Martinez-Sernandez, V. and Figueiras, A. (2013). Central nervous system demyelinating diseases and recombinant hepatitis B vaccination: a critical systematic review of scientific production. *J Neurol*, **260**(8): 1951–9.

Martinuc Porobic, J., Avcin, T., Bozic, B., *et al.* (2005). Anti-phospholipid antibodies following vaccination with recombinant hepatitis B vaccine. *Clin Exp Immunol*, **142**(2): 377–80.

Masse, I. and Descoffres, M.C. (1998). [Hypersensitivity vasculitis after hepatitis B vaccination]. *Presse Med*, **27**(20): 965–6.

Matsui, M., Kakigi, R., Watanabe, S., and Kuroda, Y. (1996). Recurrent demyelinating transverse myelitis in a high titer HBs-antigen carrier. *J Neurol Sci*, **139**(2): 235–7.

McCoy, L., Tsunoda, I., and Fujinami, R.S. (2006). Multiple sclerosis and virus induced immune responses: autoimmunity can be primed by molecular mimicry and augmented by bystander activation. *Autoimmunity*, **39**(1): 9–19.

McMahon, B.J., Helminiak, C., Wainwright, R.B., *et al.* (1992). Frequency of adverse reactions to hepatitis B vaccine in 43,618 persons. *Am J Med*, **92**(3): 254–6.

Mehdi, H., Naqvi, A., and Kamboh, M.I. (2008). Recombinant hepatitis B surface antigen and anionic phospholipids share a binding region in the fifth domain of beta2-glycoprotein I (apolipoprotein H). *Biochim Biophys Acta*, **1782**(3): 163–8.

Merida, C., Martinez-Escribano, J.A., Frias, J.F., *et al.* (2005). [Bullous pemphigoid in an infant after vaccination]. *Actas Dermosifiliogr*, **96**(4): 255–7.

Meyboom, R.H., Fucik, H., and Edwards, I.R. (1995). Thrombocytopenia reported in association with hepatitis B and A vaccines. *Lancet*, **345**(8965): 1638.

Miller, S.D., Vanderlugt, C.L., Begolka, W.S., *et al.* (1997). Persistent infection with Theiler's virus leads to CNS autoimmunity via epitope spreading. *Nat Med*, **3**(10): 1133–6.

Miron, D., Fink, D., and Hashkes, P.J. (2003). Kawasaki disease in an infant following immunisation with hepatitis B vaccine. *Clin Rheumatol*, **22**(6): 461–3.

Mori, A., Oleszycka, E., Sharp, F.A., *et al.* (2012). The vaccine adjuvant alum inhibits IL-12 by promoting PI3 kinase signalling while chitosan does not inhibit IL-12 and enhances Th1 and Th17 responses. *Eur J Immunol*, **42**(10): 2709–19.

Neau, D., Bonnet, F., Michaud, M., *et al.* (1998). Immune thrombocytopenic purpura after recombinant hepatitis B vaccine: retrospective study of seven cases. *Scand J Infect Dis*, **30**(2): 115–18.

Nuevo, H., Nascimento-Carvalho, C.M., Athayde-Oliveira, C.P., *et al.* (2004). Thrombocytopenic purpura after hepatitis B vaccine: case report and review of the literature. *Pediatr Infect Dis J*, **23**(2): 183–4.

Olson, J.K., Ercolini, A.M., and Miller, S.D. (2005). A virus-induced molecular mimicry model of multiple sclerosis. *Curr Top Microbiol Immunol*, **296**: 39–53.

Orbach, H. and Tanay, A. (2009). Vaccines as a trigger for myopathies. *Lupus*, **18**(13): 1213–16.

Ortega-Hernandez, O.D. and Shoenfeld, Y. (2009). Infection, vaccination, and autoantibodies in chronic fatigue syndrome, cause or coincidence? *Ann NY Acad Sci*, **1173**: 600–9.

Perricone, C. and Shoenfeld, Y. (2013). Hepatitis B vaccination and undifferentiated connective tissue disease: another brick in the wall of the autoimmune/inflammatory syndrome induced by adjuvants (Asia). *J Clin Rheumatol*, **19**(5): 231–3.

Pope, J.E., Stevens, A., Howson, W., and Bell, D.A. (1998). The development of rheumatoid arthritis after recombinant hepatitis B vaccination. *J Rheumatol*, **25**(9): 1687–93.

Public Health Agency of Canada. (1993). Report of the working group on the possible relationship between hepatitis B vaccination and the chronic fatigue syndrome. *CMAJ*, **149**(3): 314–19.

Ram, M. and Shoenfeld, Y. (2008). Hepatitis B: infection, vaccination and autoimmunity. *IMAJ*, **10**(1): 61–4.

Ramirez-Rivera, J., Vega-Cruz, A.M., and Jaume-Anselmi, F. (2003). Polymyositis: rare complication of hepatitis B vaccination. An unusual cause of toxic shock syndrome. *Bol Asoc Med P R*, **95**(6): 13–16.

Ravel, G., Christ, M., Horand, F., and Descotes, J. (2004). Autoimmunity, environmental exposure and vaccination: is there a link? *Toxicology*, **196**(3): 211–16.

Ray, P., Black, S., Shinefield, H., *et al.* (2011). Risk of rheumatoid arthritis following vaccination with tetanus, influenza and hepatitis B vaccines among persons 15–59 years of age. *Vaccine*, **29**(38): 6592–7.

Rebora, A., Rongioletti, F., Drago, F., and Parodi. (1999). Lichen planus as a side effect of HBV vaccination. *Dermatology*, **198**(1): 1–2.

Renard, J.L., Guillamo, J.S., Ramirez, J.M., *et al.* (1999). [Acute transverse cervical myelitis following hepatitis B vaccination. Evolution of anti-HBs antibodies]. *Presse Med*, **28**(24): 1290–2.

Roner, S., Zinser, E., Menges, M., *et al.* (2008). Minor role of bystander tolerance to fetal calf serum in a peptide-specific dendritic cell vaccine model against autoimmunity: comparison with serum-free cultures. *J Immunother*, **31**(7): 656–64.

Rosenblum, H., Shoenfeld, Y., and Amital, H. (2011). The common immunogenic etiology of chronic fatigue syndrome: from infections to vaccines via adjuvants to the ASIA syndrome. *Infect Dis Clin North Am*, **25**(4): 851–63.

Rybojad, M., Moraillon, I., Laglenne, S., *et al.* (1998). [Lichen planus in children: 12 cases]. *Ann Dermatol Venereol*, **125**(10): 679–81.

Saade, F., Honda-Okubo, Y., Trec, S., and Petrovsky, N. (2013). A novel hepatitis B vaccine containing Advax, a polysaccharide adjuvant derived from delta inulin, induces robust humoral and cellular immunity with minimal reactogenicity in preclinical testing. *Vaccine*, **31**(15): 1999–2007.

Saadoun, D., Cacoub, P., Mahoux, D., *et al.* (2001). Postvaccine vasculitis: a report of three cases. *Rev Med Interne*, **22**(2): 172–6.

Salemi, S. and D'Amelio, R. (2010). Could autoimmunity be induced by vaccination? *Int Rev Immunol*, **29**(3): 247–69.

Santoro, D., Stella, M., Montalto, G., and Castellino, S. (2007). Lupus nephritis after hepatitis B vaccination: an uncommon complication. *Clin Nephrol*, **67**(1): 61–3.

Schattner, A. (2005). Consequence or coincidence? The occurrence, pathogenesis and significance of autoimmune manifestations after viral vaccines. *Vaccine*, **23**(30): 3876–86.

Schuh, T., Rocken, M., Schmoeckel, C., and Degitz, K. (2002). [Lichen ruber planus after hepatitis B vaccination]. *Hautarzt*, **53**(10): 650–1.

Schupp, P. and Vente, C. (1999). Lichen planus following hepatitis B vaccination. *Int J Dermatol*, **38**(10): 799–800.

Senejoux, A., Roulot, D., Belin, C., *et al.* (1996). [Acute myelitis after immunization against hepatitis B with recombinant vaccine]. *Gastroenterol Clin Biol*, **20**(4): 401–2.

Seti, N.K., Reddi, R., Anand, I., and Sethi, P.K. (2002). Gulliane Barre syndrome following vaccination with hepatitis B vaccine. *J Assoc Physicians India*, **50**: 989.

Shapiro, E. and Kopicky, J. (2000). Comment on the article can immunization precipitate connective tissue disease? Report of 5 cases of systemic lupus erythematosus and review of the literature. *Semin Arthritis Rheum*, **30**(3): 215–16.

Shaw, F.E. Jr, Graham, D.J., Guess, H.A., *et al.* (1988). Postmarketing surveillance for neurologic adverse events reported after hepatitis B vaccination. Experience of the first three years. *Am J Epidemiol*, **127**(2): 337–52.

Shoenfeld, Y. and Aron-Maor, A. (2000). Vaccination and autoimmunity – vaccinosis: a dangerous liaison? *J Autoimmunity*, **14**(1): 1–10.

Shoenfeld, Y. and Agmon-Levin, N. (2011). ASIA – autoimmune/inflammatory syndrome induced by adjuvants. *J Autoimmun*, **36**(1): 4–8.

Shoenfeld, Y., Aharon-Maor, A., and Sherer, Y. (2000). Vaccination as an additional player in the mosaic of autoimmunity. *Clin Exp Rheumatol*, **18**(2): 181–4.

Shoenfeld, Y., Gilburd, B., Abu-Shakra, M., *et al.* (2008a). The mosaic of autoimmunity: genetic factors involved in autoimmune diseases – 2008. *IMAJ*, **10**(1): 3–7.

Shoenfeld, Y., Zandman-Goddard, G., Stojanovich, L., *et al.* (2008b). The mosaic of autoimmunity: hormonal and environmental factors involved in autoimmune diseases – 2008. *IMAJ*, **10**(1): 8–12.

Shoenfeld, Y., Blank, M., Abu-Shakra, M., *et al.* (2008c). The mosaic of autoimmunity: prediction, autoantibodies, and therapy in autoimmune diseases –2008. *IMAJ*, **10**(1): 13–19.

Sindern, E., Schroder, J.M., Krismann, M., and Malin, J.P. (2001). Inflammatory polyradiculoneuropathy with spinal cord involvement and lethal [correction of letal] outcome after hepatitis B vaccination. *J Neurol Sci*, **186**(1–2): 81–5.

Sinsawaiwong, S. and Thampanitchawong, P. (2000). Guillain–Barré syndrome following recombinant hepatitis B vaccine and literature review. *J Med Assoc Thai*, **83**(9): 1124–6.

Smyk, D., Rigopoulou, E.I., Baum, H., *et al.* (2011). Autoimmunity and environment: am I at risk? *Clin Rev Allerg Immunol*, **42**(2): 199–212.

Song, H.K., Kim, H.C., and Yun, Y.H. (1997). Acute myelitis after hepatitis B vaccination. *J Korean Med Sci*, **12**(3): 249–51.

Souayah, N., Nasar, A., Suri, M.F., and Qureshi, A.I. (2009). Guillain-Barre syndrome after vaccination in United States: data from the Centers for Disease Control and Prevention/Food and Drug Administration Vaccine Adverse Event Reporting System (1990–2005). *J Clin Neuromuscul Dis*, **11**(1): 1–6.

Stubgen, J.P. (2010). Neuromuscular disorders associated with Hepatitis B vaccination. *J Neurol Sci*, **292**(1–2): 1–4.

Stubgen, J.P. (2012). Immune-mediated myelitis following hepatitis B vaccination. *Autoimmun Rev*, **12**(2): 144–9.

Tartaglino, L.M., Heiman-Patterson, T., Friedman, D.P., and Flanders, A.E. (1995). MR imaging in a case of postvaccination myelitis. *AJNR Am J Neuroradiol*, **16**(3): 581–2.

Touze, E., Gout, O., Verdier-Taillefer, M.H., *et al.* (2000). [The first episode of central nervous system demyelinization and hepatitis B virus vaccination]. *Rev Neurol (Paris)*, **156**(3): 242–6.

Trevisan, G. and Stinco, G. (1993). Lichen ruber planus following HBV vaccination. *Acta Derm Venereol*, **73**(1): 73.

Trevisani, F., Gattinara, G.C., Caraceni, P., *et al.* (1993). Transverse myelitis following hepatitis B vaccination. *J Hepatol*, **19**(2): 317–18.

Tron, F., Degos, F., Brechot, C., *et al.* (1989). Randomized dose range study of a recombinant hepatitis B vaccine produced in mammalian cells and containing the S and PreS2 sequences. *J Infect Dis*, **160**(2): 199–204.

Tudela, P., Marti, S., and Bonal, J. (1992). Systemic lupus erythematosus and vaccination against hepatitis B. *Nephron*, **62**(2): 236.

Tuohy, P.G. (1989). Guillain–Barre syndrome following immunisation with synthetic hepatitis B vaccine. *N Z Med J*, **102**(863): 114–15.

van Doorn, P.A. (2009). What's new in Guillain–Barre syndrome in 2007–2008? *J Peripher Nerv Syst*, **14**(2): 72–4.

Vanoli, M., Gambini, D., and Scorza, R. (1998). A case of Churg-Strauss vasculitis after hepatitis B vaccination. *Ann Rheum Dis*, **57**(4): 256–7.

Ventura, F., Antunes, H., Brito, C., *et al.* (2009). Cutaneous polyarteritis nodosa in a child following hepatitis B vaccination. *Eur J Dermatol*, **19**(4): 400–1.

Vial, T. and Descotes, J. (2004). Autoimmune diseases and vaccinations. *Eur J Dermatol*, **14**(2): 86–90.

Viallard, J.F., Boiron, J.M. *et al.* (2000). Severe pancytopenia triggered by recombinant hepatitis B vaccine. *Br J Haematol*, **110**(1): 230–3.

Wilcock, L.K., Francis, J.N., and Durham, S.R. (2004). Aluminium hydroxide down-regulates T helper 2 responses by allergen-stimulated human peripheral blood mononuclear cells. *Clin Exp Allergy*, **34**(9): 1373–8.

Wise, R.P., Kiminyo, K.P., and Salive, M.E. (1997). Hair loss after routine immunizations. *JAMA*, **278**(14): 1176–8.

Wucherpfennig, K.W. (2001). Mechanisms for the induction of autoimmunity by infectious agents. *J Clin Invest*, **108**(8): 1097–104.

Yu, O., Bohlke, K., Hanson, C.A., *et al.* (2007). Hepatitis B vaccine and risk of autoimmune thyroid disease: a Vaccine Safety Datalink study. *Pharmacoepidemiol Drug Saf*, **16**(7): 736–45.

Zaas, A., Scheel, P., Venbrux, A., and Hellmann, D.B. (2001). Large artery vasculitis following recombinant hepatitis B vaccination: 2 cases. *J Rheumatol*, **28**(5): 1116–20.

Zachou, K., Liaskos, C., Christodoulou, D.K., *et al.* (2003). Anti-cardiolipin antibodies in patients with chronic viral hepatitis are independent of beta2-glycoprotein I cofactor or features of antiphospholipid syndrome. *Eur J Clin Invest*, **33**(2): 161–8.

Zafrir, Y., Agmon-Levin, N., Paz, Z., *et al.* (2012). Autoimmunity following hepatitis B vaccine as part of the spectrum of "Autoimmune (Auto-inflammatory) Syndrome induced by Adjuvants" (ASIA): analysis of 93 cases. *Lupus*, **21**(2): 146–52.

17 Adverse Reactions to Human Papillomavirus Vaccines

Lucija Tomljenovic[1] and Christopher A. Shaw[2]

[1]Neural Dynamics Research Group, University of British Columbia, Vancouver, BC, Canada

[2]Department of Ophthalmology and Visual Sciences, Program in Experimental Medicine, Program in Neuroscience, University of British Columbia, Vancouver, BC, Canada

Introduction

Cervical cancer is a serious disease that affects millions of women around the world. Almost 90% of cervical cancer deaths occur in developing countries, which have an insufficient medical infrastructure to fully implement regular Papanicolaou (Pap) screening programs. In contrast, in developed countries, due to successful screening efforts, deaths from cervical cancer have been reduced by more than 75% and rarely occur in females who undergo regular screening. The annual death rate from cervical cancer in developed countries is 1.4–2.3 per 100 000 (Tomljenovic and Shaw, 2013).

There are at least 100 types of human papillomavirus (HPV) strains, 15 of which have been associated with various types of cancer (Bosch and de Sanjose, 2003). In the last decade, two vaccines (Gardasil and Cervarix) were developed by different pharmaceutical companies for the prevention of HPV infection and its associated morbidity. Both HPV vaccines contain as antigens the L1 capsid proteins of several HPV types. Cervarix is a bivalent vaccine (directed at HPV serotypes 16 and 18; De Carvalho *et al.*, 2010), while Gardasil is quadrivalent (directed at HPV serotypes 16, 18, 6, 11; Villa *et al.*, 2005). HPV-16 and −18 are associated with approximately 70% (Munoz *et al.*, 2003) of all cervical cancers, while HPV-6 and −11 cause approximately 90% of external genital warts in both men and women (Garland *et al.*, 2009). Beside antigenic constituents, the two HPV vaccines also differ in the type of adjuvant complex used. Namely, Merck's Gardasil uses aluminum (Al) hydroxyphosphate sulfate, while Glaxo's Cervarix contains a combination of the oil-based adjuvant monophosphoryl lipid A (MPL) and Al hydroxide (a proprietary brand of the vaccine manufacturer's, otherwise known as AS04; De Carvalho *et al.*, 2010).

The pre- and post-licensure epidemiological studies on both HPV vaccines showed no autoimmune safety concerns (Chao *et al.*, 2011; Lu *et al.*, 2011; Klein *et al.*, 2012) and they have both been considered to have a remarkably good safety profile (Gostin, 2011). In the current post-licensure literature, however, there are numerous reports substantiating the link between adverse immune reactions and HPV vaccines, including fatal reactions. We shall here review these data and discuss the results from safety clinical trials on HPV vaccines.

Pre- and post-licensure overall safety data

Vaccine safety surveillance programs and case reports

Since the introduction of both vaccines (Gardasil in 2006 and Cervarix in 2009), 21 301 adverse drug reactions (ADRs) have been reported to

Vaccines and Autoimmunity, First Edition. Edited by Yehuda Shoenfeld, Nancy Agmon-Levin, and Lucija Tomljenovic.

Table 17.1 Age-adjusted rate of ADRs related to HPV compared with all other vaccines in the United States reported to VAERS as of 12 September 2013. The VAERS Internet database (USVAERS) was searched using the following criteria: (i) Vaccine Products: HPV4 (human papillomavirus types 6,11,16,18), HPV2 (human papillomavirus bivalent), HPVX (human papillomavirus vaccine unspecified), and All Vaccine Products; (ii) Gender: Female; (iii) Age: 6–29 Years, the target age group for HPV vaccines; (iv) Territory: All Locations; and (v) Date Vaccinated: January 2007–September 2013, the HPV vaccine post-licensure period

Events	HPV	All vaccines	% ADRs from HPV
All	21 301	43 274	42.2
Serious	3310	5020	65.9
Death	69	111	62.1
Life-threatening	472	714	66.1
Permanent disability	760	946	80.3
Prolonged hospitalization	198	273	72.5
Emergency room visit	7834	14 984	52.3

the US Vaccine Adverse Event Reporting System (VAERS), 3310 of which were serious (15.5%), including 69 deaths, 760 permanent disabilities, and 7834 emergency room visits in women aged 6–29 years (Table 17.1). Notably, compared with all other vaccines given to the same age group, Gardasil alone was associated with 65.9% of all serious ADRs, including 62% of all deaths, 66% of all life-threatening reactions, and 80% of all permanent disabilities (Table 17.1).

It needs to be emphasized that, although VAERS shares the inherent limitations of all passive surveillance systems, it is national in scope and can thus provide important signals that may require further attention (Slade *et al.*, 2009). Moreover, similar trends to those observed in the VAERS database have also been seen in other countries, including an unusually high proportion of ADRs being reported in association with HPV vaccines compared to other vaccines in the various vaccination schedules (Tomljenovic and Shaw, 2013).

In 2008, Australia reported an annual ADR rate of 7.3/100 000 – the highest since 2003 – representing an 85% increase compared with 2006 (Lawrence *et al.*, 2008). This increase was almost entirely due to ADRs reported following the commencement of the national HPV vaccination program for females aged 12–26 years in April 2007 (705 out of a total of 1538 ADR records). Thus, nearly 50% of all ADRs reported during 2007 were related to the HPV vaccine. Moreover, HPV vaccine was the only suspected vaccine in 674 (96%) records, while 203 (29%) had causality ratings of "certain" or "probable," and 43 (6%) were defined as "serious." During 2008, the HPV vaccine was still the number one vaccine on the list of ADRs in Australia, with 497 records (32% of all ADRs), and was accountable for nearly 30% of convulsions (13 out of 43) (Menzies *et al.*, 2009). During 2009, the Australian reported ADR rate for adolescents decreased by almost 50% (from 10.4 to 5.6/100 000) (Mahajan *et al.*, 2010). This decline was directly attributed to a reduction in the number of HPV vaccine-related reports, following cessation of the catch-up component of the HPV program (Mahajan *et al.*, 2010). The percentage of ADRs related to HPV vaccines was only 6.4 in 2009 (Mahajan *et al.*, 2010), compared to 50 in 2007 (Lawrence *et al.*, 2008). In spite of the overall significant decrease in ADR rate, the percentage of convulsions attributable to the HPV vaccine remained comparable between 2007 and 2009 (51% (Lawrence *et al.*, 2008) and 40% (Mahajan *et al.*, 2010), respectively).

It has to be noted that reports to passive vaccine ADR surveillance systems do not prove that the vaccine caused the ADR. However, given that there is an unusually high frequency of ADRs related to HPV vaccines reported worldwide, and that they fit a consistent pattern (i.e. nervous system-related disorders rank the highest in frequency; Tomljenovic *et al.*, 2013), it is somewhat difficult to dismiss these events as coincidental occurrences. Substantiating this notion is the fact that the pattern of ADRs reported worldwide for HPV vaccines through various vaccine ADR surveillance systems matches the data from a large number of case reports documenting similar serious ADRs associated with Gardasil administration, with nervous system disorders of autoimmune origin being the most frequently reported (Table 17.2). In particular, out of 28 total cases identified via PubMed, 12 were related to neuro-ophthalmologic disorders (43%). Cumulatively, these data suggest that the risks of HPV

Table 17.2 Summary of cases of autoimmune and inflammatory-like manifestations following HPV vaccination

Age	Medical history	Dose preceding first symptom manifestation	Onset of symptoms post-vaccination	Symptoms/main clinical features	Final diagnosis	References
17	No relevant history	2nd	<15 days	Visual impairments	ADEM	Mendoza *et al.* (2010)
15	No relevant history	2nd	<23 days	Headache, nausea, fever, vertigo, diplopia	ADEM	Schaffer *et al.* (2008)
20	No relevant history	2nd	<28 days	Headache, nausea, vomiting, diplopia	ADEM	Wildemann *et al.* (2009)
16	No relevant history	3rd	<3 weeks	Upper limb pseudoathetosis	CIS	Sutton *et al.* (2009)
25	No relevant history	2nd	<3 weeks	Acute hemiparesis	CIS	Sutton *et al.* (2009)
21	Previous history of MS	2nd	<3 weeks	Incomplete TM, left optic neuritis	MS	Sutton *et al.* (2009)
26	Previous history of MS	3rd	<3 weeks	Headache, incomplete TM	MS	Sutton *et al.* (2009)
16	Previous history of MS	2nd	<3 weeks	Incomplete TM, brainstem syndrome	MS	Sutton *et al.* (2009)
19	No relevant history	2nd	<30 days	Leg numbness, midthoracic back pain.	Demyelinating disease unspecified	Chang *et al.* (2011)
18	Motor vehicle accident without head trauma	1st	<6 weeks	Blurriness, paraesthesias, optic neuritis	Demyelinating disease unspecified	Chang *et al.* (2011)
11	Seasonal allergies and mild asthma, family history of lupus, epilepsy, and fibromyalgia	1st	15 days	Mood swings, abnormal eye movements, dizziness, leg weakness, myoclonic jerks	OMS	McCarthy and Filiano (2009)
16	No relevant history	2nd	<10 days	Visual loss, headaches, left hemiparesis	Optic neuritis	DiMario *et al.* (2010)
14	No relevant history	1st	2 days	Skin rash, fever, nausea, stomach aches, headache, insomnia, night sweats, arthralgia, anxiety, depression, amenorrhea	POF	Colafrancesco *et al.* (2013)
13	No relevant history	1st	10 days	Depression, sleep disturbance, lightheadedness, tremulousness, anxiety, panic, attacks, cognitive dysfunction, amenorrhea	POF	Colafrancesco *et al.* (2013)
21	No relevant history	3rd	<3 months	Amenorrhea preceded by oligomenorrhea	POF	Colafrancesco *et al.* (2013)
16	No relevant history	Not reported	Not reported	5 months amenorrhea preceded by 12 months oligomenorrhea	POF	Little and Ward (2012)
15	No relevant history	2nd	3 days	Vasculitic rash, soft tissue swellings of ankles and forearms, arthralgia, lethargy, epistaxis	Vasculitis	Melo Gomes *et al.* (2013)
15	Previous history of vasculitis	1st	3 days	Severe flare of cutaneous vasculitis	Vasculitis	Melo Gomes *et al.* (2013)
16	No relevant history	2nd	Not reported	Fatigue associated with prolonged menorrhagia, antiplatelet autoantibodies	Thrombocytopenic purpura	Pugnet *et al.* (2009)

(continued)

Table 17.2 *(Continued)*

Age	Medical history	Dose preceding first symptom manifestation	Onset of symptoms post-vaccination	Symptoms/main clinical features	Final diagnosis	References
11	No relevant history	Not reported	<36 days	Jaundice, hepatosplenomegaly elevated serum aminotransferases	Autoimmune hepatitis	Della Corte *et al.* (2011)
26	No relevant history	1st	4 days	Severe constant epigastric pain, vomiting, fever	Pancreatitis	Das *et al.* (2008)
32	Family history of autoimmune thyroid diseases	1st	<4 weeks	Fatigue, severe myalgia, polyarthralgia, anorexia, severe skin rash, malar rash, aphtous stomatitis, pharyngodynia, cervical lymphadenopathy, alopecia, severe weight loss, anemia	SLE	Gatto *et al.* (2013)
29	Immune thrombocytopenia	2nd	3 weeks	Weakness, diarrhea, malar rash, photosensitivity, arthritis, alopecia, severe weight loss	SLE	Gatto *et al.* (2013)
16	Personal and family history of Raynaud's syndrome and family syndrome of systemic sclerosis	1st	8 days	High-grade fever, generalized asthenia, diffuse polyarthralgia, multiple erythematous annular cutaneous lesions on the face, trunk, and lower limbs	SLE	Gatto *et al.* (2013)
16	Personal and family history of Raynaud's syndrome	2nd	3 weeks	Fever, pharyngodynia, erythematous skin lesions of elbows and knees, generalized asthenia, anorexia, polyarthralgia	Fever-APLA	Gatto *et al.* (2013)
19	Personal history of lupus	1st	10 days	Mild arthralgia, dyspnea, cervical lymphadenopathy, skin rash	SLE flare	Gatto *et al.* (2013)
13	Family history of autoimmunity, including SLE	2nd	3 weeks	Erythematous facial rash, fever, periorbital edema, weight loss, malaise, fatigue, alopecia, cervical, axillary, and inguinal lymphadenopathy, anemia, thrombocytopenia	SLE	Gatto *et al.* (2013)
20	No relevant history	1st	2 weeks	Persistent dizziness, exercise intolerance, fatigue, nausea, loss of appetite, severe weight loss	POTS	Blitshteyn (2010)

ADEM, acute disseminated encephalomyelitis; APLA, antiphospholipid antibodies; CIS, clinically isolated syndrome; MS, multiple sclerosis; POF, primary ovarian failure; OMS, opsoclonus–myoclonus syndrome; POTS, postural orthostatic tachycardia syndrome; SLE, systemic lupus erythematosus; TM, transverse myelitis vaccination may not have been properly evaluated in pre-licensure clinical trials (Tomljenovic and Shaw, 2012a,b,c; Tomljenovic *et al.*, 2013).

In addition, the observations from the Australian vaccine ADR surveillance database point to a possible causal relationship between these events, given that the rate of ADRs increased dramatically following introduction of the HPV vaccine (by 85%) and then dropped by 50% upon cessation of one part of the HPV vaccination program. It is a well known principle in toxicology that one type of evidence supporting causality is when, upon the removal of the suspected agent, the adverse event is mitigated or ameliorated.

With regard to autoimmune diseases, since licensure, HPV vaccination has been linked to Guillain–Barré syndrome (GBS) (Souayah *et al.*, 2011), other demyelinating neuropathies (multiple sclerosis, MS; acute disseminated encephalomyelitis, ADEM; transverse myelitis, TM), postural orthostatic tachycardia syndrome (POTS), systemic lupus erythematosus (SLE), primary ovarian failure (POF), pancreatitis, vasculitis, thrombocytopenic purpura, and autoimmune hepatitis (Table 17.2).

Of special note, in their analysis of VAERS data, Souayah *et al.* (2011) identified 69 reports of GBS following Gardasil vaccination in the United States between 2006 and 2009. The onset of symptoms was within 6 weeks after vaccination in 70% of the patients in whom the date of vaccination was known. The estimated weekly reporting rate of post-Gardasil GBS within the first 6 weeks (6.6/10 000 000) was higher than that of the general population, and higher than post-Menactra (meningococcal C vaccine) and post-influenza vaccinations. In particular, there was a nearly 10 times greater risk of acquiring GBS within 6 weeks after Gardasil vaccination when compared with the general population. Additionally, Gardasil vaccination was associated with approximately 8.5 times more emergency department visits, 12.5 times more hospitalizations, 10 times more life-threatening events, and 26.5 times more disability than the Menactra vaccination (Souayah *et al.*, 2011).

Of the 34 patients who developed GBS within 6 weeks post-vaccination, 25 (74%) developed symptoms within the first 2 weeks. The probability of observing an asymmetrical distribution over the 6 weeks by chance alone was low (p = 0.0 002). Hospitalization after Gardasil™ vaccination occurred in 42 (61%) subjects. Disability, defined by a substantial disruption of the ability to conduct normal life functions, occurred in 12 (17%) subjects (Souayah *et al.*, 2011).

Pre- and post-licensure vaccine manufacturer's safety data

Given these data, the question arises as to why the pre-licensure manufacturer's epidemiological studies did not seem to identify any significant safety concerns. One possible explanation is that all published Gardasil and Cervarix safety trials used either an Al-adjuvant containing placebo or hepatitis B vaccine as the "control" group (Harper *et al.*, 2004, 2006; Villa *et al.*, 2005; Mao *et al.*, 2006; Garland *et al.*, 2007a,b; Verstraeten *et al.*, 2008; Munoz *et al.*, 2009). This practice persists in vaccine trials despite considerable data showing that Al in vaccine-relevant exposures is neurotoxic and can have unintended adverse immunological effects (Petrik *et al.*, 2007; Couette *et al.*, 2009; Li *et al.*, 2009; Passeri *et al.*, 2011; Khan *et al.*, 2013), and that therefore it cannot constitute a valid placebo.

The results of the pre-licensure safety evaluation of Gardasil that are presented in the vaccine manufacturer's package insert and the US Food and Drug Administration (FDA) product approval information (Merck & Co, 2006) show that, compared to one small study population that received the saline placebo, those women receiving the Al-containing placebo reported approximately two to five times more injection-site ADRs. On the other hand, the proportion of injection-site ADRs reported in the Gardasil treatment group was comparable to that of the Al "control" group. Thus, the vaccine manufacturer's own data seem to indicate that a large proportion of ADRs from the HPV vaccine were caused by the effect of the Al adjuvant. In spite of these observations, in assessing serious ADRs of autoimmune etiology, the manufacturer pooled the results from the study participants who received the saline placebo with those who received the Al-containing placebo and presented them as one "control" group. The outcome of this procedure was that Gardasil and the Al "control" group had exactly the same rate of serious ADRs (2.3%) (Tomljenovic and Shaw, 2013). Thus, at best, Gardasil was shown to be as safe as its potentially neuroimmunotoxic constituent Al.

The unusual frequency of ADRs following HPV vaccination cannot be attributed solely to the Al adjuvant, as many other vaccines also contain Al (e.g. tetanus, diphtheria, etc.) but are not apparently associated with as many ADRs. However, it is the adjuvant that evokes the enhanced immune reaction necessary to induce the production of the elevated titers of antibodies (Israeli *et al.*, 2009). The antigen on its own is not capable of evoking this strong immune response. Because of this, any adverse effect arising from the antigen (or other constituents in the vaccine) is ultimately linked to the action of the adjuvant. For example, Zivkovic *et al.* (2012) showed that induction of antiphospholipid syndrome (APS) and its associated decreased fecundity by tetanus toxoid (TTd) hyperimmunization in C57BL/6 mice critically depends on the Al adjuvant. In particular, in this study, Zivkovic *et al.* (2012) investigated reproductive pathology induced in C57BL/6 mice by TTd hyperimmunization using a combination of different pretreatments (complete Freund's adjuvant (CFA) or glycerol) and adjuvants (Al-hydrogel or glycerol). A decrease in fecundity was recorded in only C57BL/6 mice immunized with Al-hydrogel adjuvant, irrespective of the kind of pretreatment applied.

In another example, Lee (2012a) reported the case of a teenage girl with no relevant previous medical history who suddenly died approximately 6 months after her third Gardasil HPV vaccine booster. According to the documents presented at the coroner's inquest, the patient experienced a range of nonspecific symptoms shortly after her first dose of Gardasil injection, including dizziness spells, paresthesia in her hands, and memory lapses. After the second injection, her condition worsened, and she further developed an intermittent weak arm, frequent tiredness requiring daytime naps, increased pins-and-needles feelings in her hands (causing her to drop things), appetite increase with no weight gain, night sweats, loss of ability to use common objects, intermittent chest pain, and sudden unexpected "racing heart." A full autopsy analysis revealed no anatomical, histological, toxicological, genetic, or microbiological findings that might be linked to a potential cause of death. In contrast, Lee's post-mortem blood and spleen tissue analysis revealed HPV-16 L1 gene DNA fragments corresponding to those previously found in 16 separate Gardasil vials from different vaccine lots (presumably representing contaminants from the vaccine manufacturing process) (Lee, 2012b, 2013). Specifically, the HPV DNA fragments detected in Gardasil vials were bound to the Al adjuvant used in the vaccine formulation and thus likely protected against enzymatic degradation by the endogenous nucleases (Lee, 2013). Lee's findings indicate that contaminant HPV DNA fragments, if present in the vaccine, may bind to the insoluble Al adjuvant and cause unintended pathophysiologic effects (Lee, 2012a).

Further scrutiny of post-licensure HPV vaccine manufacturer's safety trials reveals additional problems in study design that may account for the negative findings with regards to potential immune and autoimmune risks. We shall briefly refer to some of these studies in the following sections.

Cervarix

In their 2008 pre-licensure analysis of ADRs of potential autoimmune etiology in a large integrated safety database of AS04 adjuvanted vaccines (including Cervarix), Verstraeten *et al.* (2008) pointed out that "It is important to note that none of these studies were set up primarily to study autoimmune disorders." If the purpose of the study was indeed to assess ADRs of "potential autoimmune aetiology," as the title itself clearly states (Verstraeten *et al.*, 2008), then the study should have been designed to detect them. All of the eight authors of the AS04 safety study were employees of Cervarix's vaccine manufacturer, Glaxo. The authors also noted that "our search of the literature found no studies conducted by independent sources on this subject," and, "All studies included in this analysis were funded by GSK Biologicals, as was the analysis itself. GSK Biologicals was involved in the study design, data collection, interpretation and analysis, preparation of the manuscript and decision to publish" (Verstraeten *et al.*, 2008). Given that vaccines can trigger autoimmune disorders (Shoenfeld and Aron-Maor, 2000; Agmon-Levin *et al.*, 2009; Israeli *et al.*, 2012), a more rigorous safety assessment than that provided by the GSK-sponsored study would appear to have been warranted.

Gardasil

Large post-licensure epidemiological studies assessing the safety of Gardasil likewise failed to identify any significant autoimmune safety concerns (Chao *et al.*, 2011; Klein *et al.*, 2012). For example, the study by Chao *et al.* (2011), which was sponsored by the vaccine manufacturer, was conducted as a post-licensure commitment to the FDA, the European Medicines Agency (EMA), and other regulatory authorities in order to help evaluate the autoimmune safety of the vaccine. There were two specific biases that influenced the outcome of the safety analysis by Chao *et al.* (2011). First, the study included all women who received at least one dose of Gardasil, thus making the population sample less sensitive for detection of serious ADRs, as such events are likely to occur less frequently if fewer doses of the vaccine are administered. Since the authors did not report how many women actually completed the recommended three-dose HPV vaccination regimen, it is impossible to know what proportion of the study population was actually at high risk from vaccine-related serious ADRs. Second, the Safety Review Committee (SRC) which reviewed all safety data included a general pediatrician/clinical epidemiologist, a perinatologist/teratologist, a vaccinologist, a pediatric rheumatologist, and a pharmacoepidemiologist. Given that the autoimmune conditions of interest examined by this expert Committee included (i) rheumatologic/autoimmune disorders; (ii) autoimmune endocrine conditions; and (iii) autoimmune neurological/ophthalmic disorders, the question arises as to why the vaccine manufacturer's research team failed to recruit a panel with expertise that more closely matched their study's task. It is surprising to note the absence of an immunologist/autoimmunologist, neurologist, and ophthalmologist, especially since such experts were in fact present at a later stage, in the analysis of case reports selected by the SRC (Chao *et al.*, 2011).

In addition, Chao *et al.* (2011) only looked for the presence of clinically apparent systemic autoimmune diseases (e.g. MS, ADEM, GBS, etc.), failing to take into account that post-vaccine autoimmune manifestation may not always fit into the category of a well-defined autoimmune disorder (e.g. arthralgia, myalgia, paraesthesia, weakness, and cognitive disturbances – typical ASIA symptoms: Shoenfeld and Agmon-Levin 2011; Zafrir *et al.*, 2012). These "nonspecific" manifestations are all too easily ignored or disregarded as irrelevant and non-vaccine-related (Poser and Behan, 1982; Zafrir *et al.*, 2012). However, failure to register such cases represents an obvious impediment to an objective estimate of the frequency and severity of vaccine-associated risks. In this regard, it is worth emphasizing the fact that many such ill-defined medical conditions that fall under the ASIA spectrum are frequently disabling and thus of significant clinical relevance (Authier *et al.*, 2003; Couette *et al.*, 2009; Passeri *et al.*, 2011). Further, these nonspecific, but nonetheless clinically relevant, manifestations can progress to a full-blown autoimmune disease upon subsequent vaccine rechallenge (Poser and Behan, 1982; Gatto *et al.*, 2013).

A more recent analysis by Klein *et al.* (2012) concluded that the only new safety concerns following Gardasil were same-day syncope and skin infections within 2 weeks of vaccination. However, as in Chao *et al.*'s analysis, the pool of study participants received on average fewer than two doses of the vaccine. Only 23% of the study participants received all three HPV vaccine injections, and they were furthermore preselected based on undisclosed criteria (Klein *et al.*, 2012). This is clearly not how the vaccine is intended to be administered. Not completing the series of three doses not only distorts the safety outcomes but may considerably lower the efficacy of the vaccine. Indeed, a recent report by Dobson *et al.* (2013) shows that the administration of two doses of the vaccine results in inferior protection against HPV-18 by 24 months and HPV-6 by 36 months after vaccination, indicating diminished protection against both cancer and genital warts. In addition, the study by Klein *et al.* (2012) used an interval distant from vaccination as control, instead of an interval prior to the administration of the vaccine. The authors acknowledged the limitations of their study by stating the following: "This study has limitations. First, we could only detect new-onset conditions requiring ED [emergency department] visits or hospitalizations within 60 days after vaccination; it was not designed to investigate long-term safety outcomes or risk of HPV4-associated recurrence/progression of disease."

Long onset of autoimmunity post-vaccination

An interval of 6 weeks between vaccine exposure and adverse outcome is often used as evidence in favor of a plausible causal association (Burwen *et al.*, 2010; Andrews *et al.*, 2011). However, immune and autoimmune diseases are chronic diseases which evolve slowly and therefore, more often than not, have a long incubation time (Tomljenovic and Shoenfeld, 2013). For example, Arbuckle *et al.* (2003) reported that SLE evolves slowly and progressively over many years, and only when enough auto-antibodies are present. In particular, they found autoantibodies were found in 88% of SLE patients up to 9.4 years before the clinical diagnosis of the syndrome (mean 3.3 years) (Arbuckle *et al.*, 2003). Thus, long-term persistence of elevated titers of autoantibodies was necessary for the emergence of clinically overt signs and symptoms for the diagnosis of SLE. Notably, the accumulation of autoantibodies occurred while patients were still asymptomatic.

Similarly, post-vaccination adverse immune phenomena can have long latency periods (i.e. months to years following vaccination) (Poser and Behan, 1982; Authier *et al.*, 2001; Hernan *et al.*, 2004; Mikaeloff *et al.*, 2009; Gherardi and Authier, 2003, 2012).

There are two studies that are particularly worth mentioning in this regard. First is the nested case–control study within the UK General Practice Research Database (GPRD) by Hernan *et al.* (2004). Cases were patients with a diagnosis of MS confirmed through examination of medical records, and with at least 3 years of continuous recording in the GPRD before their date of first symptoms (index date). Up to 10 controls per case were randomly selected, matched on age, sex, practice, and date of joining the practice. Overall, the analyses included 163 cases of MS and 1604 controls. The odds ratio (OR) of MS for vaccination within 3 years before the index date compared to no vaccination was 3.1 (95% confidence interval (CI): 1.5, 6.3). No increased risk of MS was associated with tetanus and influenza vaccinations.

The second study, conducted by Mikaeloff *et al.* (2009), is a population-based case–control study of children with a first episode of acute central nervous system (CNS) inflammatory demyelination in France (1994–2003). Each case was matched on age, sex, and geographic location to up to 12 controls, randomly selected from the general population. In total, in comparing 349 children with a first episode of acute CNS inflammatory demyelination matched with 2941 healthy controls, Mikaeloff *et al.* (2009) showed that hepatitis B vaccine Engerix B administered 3 years earlier was associated with an increased risk for CNS demyelinating events and a higher risk of having a confirmed diagnosis of MS.

Interestingly, as early as 1982, compelling evidence from epidemiological, clinical, and animal research emerged to show that autoimmune neuropathies (i.e. GBS, MS, and ADEM, all of which have been associated with administration of HPV vaccines; Table 17.2 and Souayah *et al.*, 2011) can occur 4–10 months following vaccination (Poser and Behan, 1982). In such cases, the disease first manifests with vague symptoms (i.e. arthralgia, myalgia, paraesthesia, weakness – typical ASIA symptoms), which are frequently deemed minor and thus not further investigated. These symptoms, otherwise known as "bridging symptoms," and consistent with a mild subclinical disease, progress slowly and insidiously up to the point of exposure to a secondary immune stimulus. This then triggers the rapid and acute clinical manifestation of the disease (Poser and Behan, 1982). In other words, it is the secondary anamnestic response that brings about the acute overt manifestation of an already present subclinical long-term persisting disease.

Consistent with these observations, Gatto *et al.* (2013) recently described several cases of autoimmunity (SLE) following Gardasil (Table 17.2) in which the nonspecific ASIA-related manifestations eventually progressed to a full-blown immune disease following subsequent vaccine re-exposure. In all cases, several common features were observed, namely, a personal or familial susceptibility to autoimmunity and an adverse response to a prior dose of the vaccine, both of which were associated with a higher risk of post-vaccination full-blown autoimmunity (Gatto *et al.*, 2013).

One of the cases was a 32-year-old woman who was admitted to the hospital 5 days following her third vaccination with Gardasil (Gatto *et al.*, 2013). On admission, she suffered from general weakness, severe myalgia, polyarthralgia, anorexia, severe skin rash (urticarialike), malar rash, aphtous stomatitis, pharyngodynia, cervical lymphadenopathy (>3.5 cm), and hair loss. In addition, in the 4 weeks prior to her hospitalization, she lost 10 kg of body weight. Laboratory tests demonstrated an elevated erythrocyte sedimentation rate (ESR) and C-reactive protein (CRP), anemia (without evidence of hemolysis), leucopenia, and lymphopenia. Autoantibody screening showed positive antinuclear antibodies with very high titers of anti-Ro (SSA) and anti-La (SSB) antibodies and high positive anti-dsDNA antibodies. Antiphospholipid antibodies were undetectable, complement (C3) levels were very low, urine analysis showed no active sediment, and no evidence of infections was documented (i.e. blood and urine culture were normal and there was negative serology for hepatitis, Epstein–Barr virus (EBV), cytomegalovirus, parvovirus). CPK and TSH levels were normal, as were chest radiography and endoscopy. The patient was diagnosed as having SLE and treatment with high-dose prednisone (PDN) and hydroxychloroquine (HCQ) was commenced, producing gradual clinical improvement. PDN therapy was tapered slowly up to 5 mg/day, and HCQ was continued at 400 mg/day along with supplementation of calcium and vitamin D. After 8 months the patient was in remission, with normalization of inflammatory laboratory parameters (CRP, ESR) and of blood counts and complement levels. Notably, her medical history was unremarkable prior to vaccination. However, mild weakness, facial malar rash, and hair loss were observed following the first vaccination (6 months prior to hospitalization). Local reaction to vaccination, fever, fatigue, mild rash, and arthralgia were documented following the second dose but were misinterpreted as a "common cold." Her family history was remarkable for autoimmune thyroid diseases.

In many of the other cases of autoimmunity following HPV vaccination presented in Table 17.2, although the first manifestations of symptoms normally occurred within the first 3 weeks (68% of all cases) following the first or second HPV vaccine injection, the definite clinical manifestation and final diagnosis of the disease occurred months later (McCarthy and Filiano, 2009; Little and Ward, 2012; Colafrancesco *et al.*, 2013).

Possible mechanisms of HPV vaccine-induced autoimmunity

The ramifications of the anamnestic response caused by the secondary immune challenge (i.e. subsequent vaccine boosters) may depend on the genetic immune makeup of the host (Agmon-Levin *et al.*, 2009) and the immunological similarity between the vaccine antigens and those previously encountered (Poser and Behan, 1982). In particular, in individuals who have already been presensitized by previous exposure to HPV (via either previous vaccination or natural infection) or similar viruses, subsequent vaccinations or infections can trigger pathological anamnestic responses, leading to acute clinically diagnosable full-blown autoimmune diseases. Of further note, compared with infections, vaccines induce potentiated immunological responses, due to the presence of adjuvants or multiple antigens (four, in the case of Gardasil). Moreover, vaccines are often administered over fairly short periods. Thus, the full ramifications of such closely spaced vaccinations might include a greater risk for autoimmune manifestations to the recipient than is caused by infections.

Apart from the hyperstimulation of the immune system, it is also possible that molecular mimicry may play a role in mediating autoimmunity following HPV vaccination. Gardasil, for example, may be more likely to trigger autoimmune adverse manifestation than other vaccines, due to the high antigenicity of its recombinant proteins. For instance, Gardasil vaccination induces a 40-fold increase in anti-HPV antibodies compared with the physiological antibody level triggered by a natural HPV infection (Harro *et al.*, 2001). The antibody titer against HPV-16 and −18 may remain 11 times higher than those induced by natural infections 5.5 years after vaccination (Bayas *et al.*, 2008). Similarly, Cervarix has induced sustained antibody titers for HPV-18 more than fourfold higher than natural infection titers at 8.4 years after initial vaccination, maintaining 100% seropositivity as measured by the pseudovirion-based assay, and more than tenfold higher than natural infection titer measured by ELISA, with similar 100% seropositivity maintained (Harper and Williams, 2010). The fact that vaccines are designed to hyperstimulate antibody production (thus producing much higher antibody levels than occur following natural infection), which is accomplished via the immunostimulatory properties of adjuvants, suggests that vaccination may indeed carry a much higher risk of autoimmunity than do natural infections.

It is clear that more research is needed to identify those individuals who may develop autoimmune diseases following vaccinations. Although genetic predisposition and personal and familial history of autoimmunity represent clear risk factors, we have noted with interest that 57% of the case reports of autoimmunity in the currently available literature had no such susceptibilities (Table 17.2). Thus, further investigations should be aimed at identifying other risk factors, which may include previous exposure to medications (including oral contraceptives, other vaccinations, antipsychotic drugs, smoking, etc.).

Conclusions

Vaccines are given to healthy people for the prevention of diseases they may never encounter in their lifetimes, and, as such, they need to be held to the highest safety standards possible. The human clinical trial data for the two HPV vaccines currently on the market reveal a troubling safety profile that requires an accurate reevaluation of the risks and benefits. The HPV vaccines do not replace currently available methods for screening or treating cervical cancer, and their effectiveness in preventing any cancer death will not be known for several decades. Given that the death rate from cervical cancer in 9–20-year-old girls is zero, the short-term risks from the vaccine to otherwise healthy individuals seem to significantly outweigh the as yet unproven long-term benefits (Gerhardus and Razum, 2010; Tomljenovic and

Shaw, 2012b; Tomljenovic *et al.*, 2013). Recommendations for the continued use of HPV vaccines should be urgently and accurately reassessed and new guidelines should be requested on the use of appropriate placebos in vaccine safety trials.

Acknowledgments

The authors thank the Dwoskin Family Foundation, the Katlyn Fox Foundation, and the Luther Allyn Shourds Dean Estate Foundation for support

References

Agmon-Levin, N., Paz, Z., Israeli, E., and Shoenfeld, Y. (2009). Vaccines and autoimmunity. *Nat Rev Rheumatol*, **5**: 648–52.

Andrews, N., Stowe, J., Al-Shahi Salman, R., and Miller, E. (2011). Guillain-Barre syndrome and H1N1 (2009) pandemic influenza vaccination using an AS03 adjuvanted vaccine in the United Kingdom: self-controlled case series. *Vaccine*, **29**: 7878–82.

Arbuckle, M.R., McClain, M.T., Rubertone, M.V., *et al.* (2003). Development of autoantibodies before the clinical onset of systemic lupus erythematosus. *N Engl J Med*, **349**: 1526–33.

Authier, F.J., Cherin, P., Creange, A., *et al.* (2001). Central nervous system disease in patients with macrophagic myofasciitis. *Brain*, **124**: 974–83.

Authier, F.J., Sauvat, S., Champey, J., *et al.* (2003). Chronic fatigue syndrome in patients with macrophagic myofasciitis. *Arthritis Rheum*, **48**: 569–70.

Bayas, J.M., Costas, L., and Munoz, A. (2008). Cervical cancer vaccination indications, efficacy, and side effects. *Gynecol Oncol*, **110**: S11–14.

Blitshteyn, S. (2010). Postural tachycardia syndrome after vaccination with Gardasil (letter to the editor). *Eur J Neurol*, **17**: e52.

Bosch, F.X. and de Sanjose, S. (2003). Chapter 1: Human papillomavirus and cervical cancer – burden and assessment of causality. *J Natl Cancer Inst Monogr*, **31**: 3–13.

Burwen, D.R., Ball, R., Bryan, W.W., *et al.* (2010). Evaluation of Guillain-Barre syndrome among recipients of influenza vaccine in 2000 and 2001. *Am J Prev Med*, **39**: 296–304.

Chang, J., Campagnolo, D., Vollmer, T.L., and Bomprezzi, R. (2011). Demyelinating disease and polyvalent human papilloma virus vaccination. *J Neurol Neurosurg Psychiatry*, **82**(11): 1296–8.

Chao, C., Klein, N.P., Velicer, C.M., *et al.* (2011). Surveillance of autoimmune conditions following routine use of quadrivalent human papillomavirus vaccine. *J Intern Med*, **271**: 193–203.

Colafrancesco, S., Perricone, C., Tomljenovic, L., and Shoenfeld, Y. (2013). HPV vaccines and primary ovarian failure: another facet of the autoimmune/inflammatory syndrome induced by adjuvants (ASIA). *Am J Reprod Immunol*, **70**: 309–16.

Couette, M., Boisse, M.F., Maison, P., *et al.* (2009). Long-term persistence of vaccine-derived aluminum hydroxide is associated with chronic cognitive dysfunction. *J Inorg Biochem*, **103**: 1571–8.

Das, A., Chang, D., Biankin, A.V., and Merrett, N.D. (2008). Pancreatitis following human papillomavirus vaccination. *Med J Aust*, **189**: 178.

De Carvalho, N., Teixeira, J., Roteli-Martins, C.M., *et al.* (2010). Sustained efficacy and immunogenicity of the HPV-16/18 AS04-adjuvanted vaccine up to 7.3 years in young adult women. *Vaccine*, **28**: 6247–55.

Della Corte, C., Carlucci, A., Francalanci, P., *et al.* (2011). Autoimmune hepatitis type 2 following anti-papillomavirus vaccination in a 11-year-old girl. *Vaccine*, **29**: 4654–6.

DiMario, F.J. Jr, Hajjar, M., and Ciesielski, T. (2010). A 16-year-old girl with bilateral visual loss and left hemiparesis following an immunization against human papilloma virus. *J Child Neurol*, **25**: 321–7.

Dobson, S.R., McNeil, S., Dionne, M., *et al.* (2013). Immunogenicity of 2 doses of HPV vaccine in younger adolescents vs 3 doses in young women: a randomized clinical trial. *JAMA*, **309**: 1793–802.

Garland, S.M., Hernandez-Avila, M., Wheeler, C.M., *et al.* (2007a). Quadrivalent vaccine against human papillomavirus to prevent anogenital diseases. *N Engl J Med*, **356**: 1928–43.

Garland, S.M., Hernandez-Avila, M., Wheeler, C.M., *et al.* (2007b). Supplementary appendix. Supplement to: Garland, S.M., Hernandez-Avila, M., Wheeler, C.M., *et al.* (FUTURE I Investigators). Quadrivalent vaccine against human papillomavirus to prevent anogenital diseases. *N Engl J Med*, **356**: 1928–43.

Garland, S.M., Steben, M., Sings, H.L., *et al.* (2009). Natural history of genital warts: analysis of the placebo arm of 2 randomized phase III trials of a quadrivalent human papillomavirus (types 6, 11, 16, and 18) vaccine. *J Infect Dis*, **199**: 805–14.

Gatto, M., Agmon-Levin, N., Soriano, A., *et al.* (2013). Human papillomavirus vaccine and systemic lupus erythematosus. *Clin Rheumatol*, **32**: 1301–7.

Gerhardus, A. and Razum, O. (2010). A long story made too short: surrogate variables and the communication of HPV vaccine trial results. *J Epidemiol Community Health*, **64**: 377–8.

Gherardi, R. and Authier, F. (2012). Macrophagic myofasciitis: characterization and pathophysiology. *Lupus*, **21**: 184–9.

Gherardi, R.K. and Authier, F.J. (2003). Aluminum inclusion macrophagic myofasciitis: a recently identified condition. *Immunol Allergy Clin North Am*, **23**: 699–712.

Gostin, L.O. (2011). Mandatory HPV vaccination and political debate. *JAMA*, **306**: 1699–700.

Harper, D.M. and Williams, K.B. (2010). Prophylactic HPV vaccines: current knowledge of impact on gynecologic premalignancies. *Discov Med*, **10**: 7–17.

Harper, D.M., Franco, E.L., Wheeler, C., *et al.* (2004). Efficacy of a bivalent L1 virus-like particle vaccine in prevention of infection with human papillomavirus types 16 and 18 in young women: a randomised controlled trial. *Lancet*, **364**: 1757–65.

Harper, D.M., Franco, E.L., Wheeler, C.M., *et al.* (2006). Sustained efficacy up to 4.5 years of a bivalent L1 virus-like particle vaccine against human papillomavirus types 16 and 18: follow-up from a randomised control trial. *Lancet*, **367**: 1247–55.

Harro, C.D., Pang, Y.Y., Roden, R.B., *et al.* (2001). Safety and immunogenicity trial in adult volunteers of a human papillomavirus 16 L1 virus-like particle vaccine. *J Natl Cancer Inst*, **93**: 284–92.

Hernan, M.A., Jick, S.S., Olek, M.J., and Jick, H. (2004). Recombinant hepatitis B vaccine and the risk of multiple sclerosis: a prospective study. *Neurology*, **63**: 838–42.

Israeli, E., Agmon-Levin, N., Blank, M., *et al.* (2012). Guillain-Barre syndrome –a classical autoimmune disease triggered by infection or vaccination. *Clin Rev Allergy Immunol*, **42**: 121–30.

Israeli, E., Agmon-Levin, N., Blank, M., and Shoenfeld, Y. (2009). Adjuvants and autoimmunity. *Lupus*, **18**: 1217–25.

Khan, Z., Combadiere, C., Authier, F.J., *et al.* (2013). Slow CCL2-dependent translocation of biopersistent particles from muscle to brain. *BMC Med*, **11**: 99.

Klein, N.P., Hansen, J., Chao, C., *et al.* (2012). Safety of quadrivalent human papillomavirus vaccine administered routinely to females. *Arch Pediatr Adolesc Med*, **166**: 1140–8.

Lawrence, G., Gold, M.S., Hill, R., *et al.* (2008). Annual report: surveillance of adverse events following immunisation in Australia, 2007. *Commun Dis Intell Q Rep*, **32**(4): 371–87.

Lee, S.H. (2012a). Detection of human papillomavirus L1 gene DNA fragments in postmortem blood and spleen after Gardasil® vaccination – a case report. *Advances Biosc Biotech*, **3**: 1214–24.

Lee, S.H. (2012b). Detection of human papillomavirus (HPV) L1 gene DNA possibly bound to particulate aluminum adjuvant in the HPV vaccine Gardasil. *J Inorg Biochem*, **112**: 85–92.

Lee, S.H. (2013). Topological conformational changes of human papillomavirus (HPV) DNA bound to an insoluble aluminum salt – a study by low temperature PCR. *Advances Biol Chem*, **3**: 76–85.

Li, X., Zheng, H., Zhang, Z., *et al.* (2009). Glia activation induced by peripheral administration of aluminum oxide nanoparticles in rat brains. *Nanomedicine*, **5**: 473–9

Little, D.T. and Ward, H.R. (2012). Premature ovarian failure 3 years after menarche in a 16-year-old girl following human papillomavirus vaccination. *BMJ Case Rep*, doi: 10.1136/bcr-2012-006879.

Lu, B., Kumar, A., Castellsague, X., and Giuliano, A.R. (2011). Efficacy and safety of prophylactic vaccines against cervical HPV infection and diseases among women: a systematic review & meta-analysis. *BMC Infect Dis*, **11**: 13.

Mahajan, D., Roomiani, I., Gold, M.S., *et al.* (2010). Annual report: surveillance of adverse events following immunisation in Australia, 2009. *Comm Dis Intell Q Rep*, **34**(3): 259–76.

Mao, C., Koutsky, L.A., Ault, K.A., *et al.* (2006). Efficacy of human papillomavirus-16 vaccine to prevent cervical intraepithelial neoplasia: a randomized controlled trial. *Obstet Gynecol*, **107**: 18–27.

McCarthy, J.E. and Filiano, J. (2009). Opsoclonus myoclonus after human papilloma virus vaccine in a pediatric patient. *Parkinsonism Relat Disord*, **15**: 792–4.

Melo Gomes, S., Glover, M., Malone, M., and Brogan, P. (2013). Vasculitis following HPV immunization. *Rheumatology (Oxford)*, **52**: 581–2.

Mendoza Plasencia, Z., Gonzalez Lopez, M., Fernandez Sanfiel, M.L., and Muniz Montes, J.R. (2010). Acute disseminated encephalomyelitis with tumefactive lesions after vaccination against human papillomavirus. *Neurologia*, **25**: 58–9.

Menzies, R., Mahajan, D., Gold, M.S., *et al.* (2009). Annual report: surveillance of adverse events following immunisation in Australia, 2008. *Commun Dis Intell Q Rep*, **33**(4): 365–81.

Merck & Co. (2006). Gardasil product sheet, *Date of Approval* **2006**. Available from: http://www.fda.gov/downloads/BiologicsBloodVaccines/Vaccines/ApprovedProducts/UCM111263.pdf [last accessed 11 December 2014].

Mikaeloff, Y., Caridade, G., Suissa, S., and Tardieu, M. (2009). Hepatitis B vaccine and the risk of CNS inflammatory demyelination in childhood. *Neurology*, **72**: 873–80.

Munoz, N., Bosch, F.X., de Sanjose, S., *et al.* (2003). Epidemiologic classification of human papillomavirus types associated with cervical cancer. *N Engl J Med*, **348**: 518–27.

Munoz, N., Manalastas, R. Jr, Pitisuttithum, P., *et al.* (2009). Safety, immunogenicity, and efficacy of quadrivalent human papillomavirus (types 6, 11, 16, 18) recombinant vaccine in women aged 24–45 years: a randomised, double-blind trial. *Lancet*, **373**: 1949–57.

Passeri, E., Villa, C., Couette, M., *et al.* (2011). Long-term follow-up of cognitive dysfunction in patients with aluminum hydroxide-induced macrophagic myofasciitis (MMF). *J Inorg Biochem*, **105**: 1457–63.

Petrik, M.S., Wong, M.C., Tabata, R.C., *et al.* (2007). Aluminum adjuvant linked to Gulf War illness induces motor neuron death in mice. *Neuromolecular Med*, **9**: 83–100.

Poser, C.M. and Behan, P.O. (1982). Late onset of Guillain-Barre syndrome. *J Neuroimmunol*, **3**: 27–41.

Pugnet, G., Ysebaert, L., Bagheri, H., *et al.* (2009). Immune thrombocytopenic purpura following human papillomavirus vaccination. *Vaccine*, **27**: 3690.

Schaffer, V., Wimmer, S., Rotaru, I., *et al.* (2008). HPV vaccine: a cornerstone of female health a possible cause of ADEM? *J Neurol*, **255**: 1818–20.

Shoenfeld, Y. and Agmon-Levin, N. (2011). "ASIA" – autoimmune/inflammatory syndrome induced by adjuvants. *J Autoimmun*, **36**: 4–8.

Shoenfeld, Y. and Aron-Maor, A. (2000). Vaccination and autoimmunity – "vaccinosis": a dangerous liaison? *J Autoimmun*, **14**: 1–10.

Slade, B.A., Leidel, L., Vellozzi, C., *et al.* (2009). Postlicensure safety surveillance for quadrivalent human papillomavirus recombinant vaccine. *JAMA*, **302**: 750–7.

Souayah, N., Michas-Martin, P.A., Nasar, A., *et al.* (2011). Guillain-Barre syndrome after Gardasil vaccination: data from Vaccine Adverse Event Reporting System 2006–2009. *Vaccine*, **29**: 886–9.

Sutton, I., Lahoria, R., Tan, I.L., *et al.* (2009). CNS demyelination and quadrivalent HPV vaccination. *Multiple Scler*, **15**: 116–19.

Tomljenovic, L. and Shaw, C.A. (2012a). No autoimmune safety signal after vaccination with quadrivalent HPV vaccine Gardasil? *J Intern Med*, **272**: 514–15.

Tomljenovic, L. and Shaw, C.A. (2012b). Too fast or not too fast: the FDA's approval of Merck's HPV vaccine Gardasil. *J Law Med Ethics*, **40**: 673–81.

Tomljenovic, L. and Shaw, C.A. (2012c). Who profits from uncritical acceptance of biased estimates of vaccine efficacy and safety? *Am J Public Health*, **102**: e13–14.

Tomljenovic, L. and Shaw, C.A. (2013). Human papillomavirus (HPV) vaccine policy and evidence-based medicine: Are they at odds? *Ann Med*, **45**: 182–93.

Tomljenovic, L. and Shoenfeld, Y. (2013). Association between vaccination and Guillain-Barre syndrome. *Lancet Infect Dis*, **13**: 730–1.

Tomljenovic, L., Spinosa, J.P., and Shaw, C.A. (2013). Human papillomavirus (HPV) vaccines as an option for preventing cervical malignancies: (how) effective and safe? *Curr Pharm Des*, **19**: 1466–87.

Verstraeten, T., Descamps, D., David, M.P., *et al.* (2008). Analysis of adverse events of potential autoimmune aetiology in a large integrated safety database of AS04 adjuvanted vaccines. *Vaccine*, **26**: 6630–8.

Villa, L.L., Costa, R.L., Petta, C.A., *et al.* (2005). Prophylactic quadrivalent human papillomavirus (types 6, 11, 16, and 18) L1 virus-like particle vaccine in young women: a randomised double-blind placebo-controlled multicentre phase II efficacy trial. *Lancet Oncol*, **6**: 271–8.

Wildemann, B., Jarius, S., Hartmann, M., *et al.* (2009). Acute disseminated encephalomyelitis following vaccination against human papilloma virus. *Neurology*, **72**: 2132–3.

Zafrir, Y., Agmon-Levin, N., Paz, Z., *et al.* (2012). Autoimmunity following hepatitis B vaccine as part of the spectrum of "Autoimmune (Auto-inflammatory) Syndrome induced by Adjuvants" (ASIA): analysis of 93 cases. *Lupus*, **21**: 146–52.

Zivkovic, I., Petrusic, V., Stojanovic, M., *et al.* (2012). Induction of decreased fecundity by tetanus toxoid hyper-immunization in C57BL/6 mice depends on the applied adjuvant. *Innate Immun*, **18**: 333–42.

18 Influenza Vaccine and Autoimmune Diseases

Luis J. Jara,[1] Gabriela Medina,[2] Pilar Cruz Dominguez,[3] Olga Vera-Lastra,[4] Miguel A. Saavedra,[5] Mónica Vázquez del Mercado,[6] and Minoru Satoh[7]

[1]Direction of Education and Research
[2]Clinical Epidemiological Research Unit
[3]Research Division
[4]Department of Internal Medicine
[5]Department of Rheumatology, Hospital de Especialidades "Dr Antonio Fraga Mouret," Mexican Social Security Institute, National Autonomous University of Mexico, Mexico City, Mexico
[6]Institute of Research in Rheumatology and Musculoloeskeletal System, Hospital Civil JIM. University of Guadalajara, Jalisco, Mexico
[7]School of Health Sciences, University of Occupational and Environmental Health, Kitakyushu, Japan

Introduction

Influenza is an acute viral infection caused by influenza type A, B, and C viruses of the Orthomyxoviridae family that affects the respiratory tract (Steinhauer and Skehel, 2002). In 24 March–24 April 2009, 18 cases of pneumonia caused by a novel swine-origin influenza (A/H1N1) virus were identified in Mexico City. Notably, the majority of patients were between 13 and 47 years of age, had no preexisting disease, and 16 of the 18 patients were hospitalized for the first time. These cases showed that A/H1N1 infection could cause acute respiratory distress syndrome (ARDS) and death in young, healthy individuals (Perez-Padilla *et al.*, 2009). By 31 July 2009, 63 479 cases of influenza-like illness were reported in patients belonging to the Mexican Institute for Social Security, 6945 (11%) caused by A/H1N1 virus, leading to 63 (<1 %) deaths. The most affected patients had an age between 10 and 39 years (56%). Mortality rates showed high risk in those aged 70 years and older, delayed admission, and presence of chronic diseases. Risk of infection was lower in those who had been vaccinated for seasonal influenza with 2008–9 trivalent inactivated vaccine (Echevarría-Zuno *et al.*, 2009). As in chronic diseases, underlying medical conditions such as diabetes and immune suppression increased the risk of death in patients hospitalized with A/H1N1 infection (Chowell *et al.*, 2012).

It has long been known that influenza vaccine can stimulate the formation of autoantibodies (Endoh *et al.*, 1984). Nevertheless, in view of the morbidity and mortality caused by the 2009 H1N1 influenza and the effectiveness of the vaccine, clinicians and patients should be assured that the benefits of inactivated pandemic vaccines greatly outweigh the risks (Salmon *et al.*, 2013).

The purpose of this chapter is to analyze the relationship between influenza, vaccination, autoimmunity, and autoimmune diseases, with special emphasis on the A/H1N1 influenza pandemic that occurred in Mexico.

Vaccines and Autoimmunity, First Edition. Edited by Yehuda Shoenfeld, Nancy Agmon-Levin, and Lucija Tomljenovic.

Influenza and the immune system

The 2009 influenza A pandemic virus (H1N1) was the result of a mutation that allowed the virus to be transmitted between species, from pigs to humans. The H1N1 influenza A virus pandemic began in Mexico in early April 2009; by October, 191 countries had reported more than 375 000 confirmed cases of H1N1/09, with more than 4500 deaths (Patel *et al.*, 2010). The majority of cases were mild and self-limiting disease, but a subset was characterized by serious illness, often requiring hospitalization and mechanical ventilatory support (Perez-Padilla *et al.*, 2009).

Between 10 April and 28 May 2009, 100 people died in Mexico from the A/H1N1 infection. The clinical severity and mortality in younger age groups was higher than that in the 1918 influenza pandemic. The main clinical symptoms were fever (84%), cough (85%), dyspnea (75%), and myalgia (30%) (Fajardo-Dolci *et al.*, 2010). Some factors, such as obesity, have been associated with immunodysregulation and adversely affected pulmonary function. Another susceptibility factor was pregnancy, which has been associated with cytokine dysregulation, without placenta or fetus involvement, mainly in IgG2 subclass deficiency (Chan *et al.*, 2011b).

Following this emergency, investigations revealed that the severe disease was a consequence of reactive hemophagocytosis, thrombotic phenomena, lymphoid atrophy, diffuse alveolar damage, and multiorgan dysfunction. Patients who died frequently had bacterial coinfections (30.4%), myocarditis (21.7%), or viremia (13.0%). This pathological change was associated with a delayed viral clearance and high plasma levels of proinflammatory cytokines and chemokines (To *et al.*, 2010). A recent study found interleukin (IL)-6 to be an important feature of the host response in both humans and mice infected with A/Mexico/4108/2009 (H1N1pdm strain). Elevated serum levels of IL-6 were associated with severe disease in patients hospitalized with H1N1pdm infection. These findings suggest that IL-6 may be a potential disease severity biomarker (Paquette *et al.*, 2012). The host immune response during the severe 2009 A/H1N1 infection is poorly understood. In an investigation of the viral load, immune response, and apoptosis in lung tissues from 50 fatal cases of 2009 H1N1 virus infection, 7 of the 27 cytokines/chemokines had notably high expression: IL-1 receptor antagonist protein, IL-6, tumor necrosis factor alpha (TNF-α), IL-8, monocyte chemoattractant protein-1, macrophage inflammatory protein 1-β, and interferon-inducible protein-10 in lung tissues. Viral load was positively correlated with mRNA levels of cytokines/chemokines. Apoptosis was found in lung tissues stained by the TUNEL assay (Gao *et al.*, 2013).

Influenza virus infection and vaccination activate the human adaptive immune system, which reacts via either humoral response with antibody production or cell-mediated response with T and B lymphocyte activation. The differential antibody response has recently been analyzed in the sera of patients with natural infection by pandemic H1N1 2009 influenza virus and the sera of recipients of the vaccine. Surprisingly, the overall seropositivity was low (40%). Lower antibody levels were found in both naturally infected patients and immunized recipients, but naturally infected patients exhibited higher titers. This finding may be related to differences between antigen presentation by the intramuscular route of vaccination and viral replication in mucosal cells of the respiratory tract (Chan *et al.*, 2011a).

Alterations in T cell immunity occur with aging, affecting the function and proportions of T cell subsets. The frequency of naive $CD4^+$ and $CD8^+$ T cells decreases with age, whereas the frequency of memory $CD4^+$ and $CD8^+$ T cells increases. Also, changes in T cell proliferation, cytokine production, memory response, and cytotoxicity, as well as in regulatory T cell number and function, have been reported with aging. On the other hand, T cells play a major role in defending the host against influenza virus infection, which has a high morbidity and mortality in the elderly. Thus, altered T cell immunity may account in part for the development of such respiratory problems with aging (Lee *et al.*, 2012).

In this regard, severe A/H1N1-associated pneumonia may result from the emergence of particular T cell subsets and cytokines/chemokines, as well as from distinct responses to infection. A recent study determined the T cell subset distribution and cytokine/chemokine levels in peripheral blood and bronchoalveolar lavage (BAL) in patients with severe A/H1N1 infection, asymptomatic household contacts, and healthy controls. Several inflammatory mediators were upregulated in peripheral blood and lung samples from A/H1N1-infected patients who developed severe pneumonia. The A/H1N1 strain induced higher levels of proinflammatory cytokines (such as IL-6) and chemokines than the seasonal H1N1 strain. These findings suggest the presence of biomarkers in severe pneumonia and point to the therapeutic use of immunomodulatory drugs in severe pneumonia associated with A/H1N1 infection (Zúñiga *et al.*, 2011).

Statin treatment reduced the 30 day mortality (due to sepsis and pneumonia) in 41% of patients hospitalized with laboratory-confirmed seasonal influenza. Other immunomodulatory agents also reduced mortality in patients with pneumonia, with no increase of virus replication. Randomized controlled trials (RCTs) will be needed to provide convincing evidence of the beneficial effects of these immunomodulatory agents (Fedson, 2013).

Influenza, vaccination, and pregnancy

Pregnant patients experience more frequent and severe complications than nonpregnant women of the same age or the general population (Van Kerkhove *et al.*, 2011). Physiological modulation of the immune system is required for fetal tolerance during pregnancy. Nonetheless, this regulation might lead to impaired self-defense against pathogens, as seen by the prevalence of severe complications in pregnant women infected with the pandemic influenza virus in 2009. In addition to increased death rates during pregnancy, severe influenza can affect pregnancy outcome, including preterm delivery, low birth weight, and fetal loss. A study revealed that the modulation of CD54 and CD86 expression in peripheral blood plasmacytoid dendritic cells during pregnancy may diminish the onset of adaptive antiviral immune response (Cordeau *et al.*, 2012). On the other hand, the severity of diseases that are ameliorated by inflammatory responses, such as influenza infection, is increased during pregnancy. The bidirectional interactions between hormones and the immune system contribute to both the outcome of pregnancy and female susceptibility to disease (Robinson and Klein, 2012).

Gordon *et al.* (2010) found that pregnancy-related reductions in IgG2 level may explain the increased severity of H1N1 in some patients, probably due to IgG2 deficiency being associated with an inability to mount an early effective immune response.

Responses to MF59-adjuvanted vaccination in pregnant women were slightly altered compared with those in nonpregnant women, as demonstrated by a lower rate of seroconversion in the former group. As in other studies, the adjuvanted vaccine had more local reactions but did not increase systemic adverse reactions (Bischoff *et al.*, 2013). Vaccinating pregnant women against influenza can protect both mother and child and reduces the complications associated with H1N1 infection in this priority group.

Influenza, vaccination, and autoimmune rheumatic diseases

Patients with autoimmune rheumatic disease (ARD) are at increased risk of influenza infection, due to the underlying immunological abnormalities and the effects of immunosuppressive therapy. There is some evidence to link influenza virus with autoimmunity. Although the pathogenetic mechanisms of autoimmune response are not yet fully elucidated, Agmon-Levin *et al.* (2009a) have suggested a role of the molecular mimicry mechanism.

Systemic lupus erythematosus (SLE) is an autoimmune disease characterized by a defect in B lymphocytes, with abnormalities in apoptosis and persistent production of autoantibodies. Dysregulation of B cells and immunosuppressive therapy in SLE patients might be responsible for impaired humoral immune responses to the A/H1N1 influenza virus (Yurasov *et al.*, 2005).

Other mechanisms, such as bystander activation and exposure to hidden epitopes due to tissue destruction, may be involved in the induction of antiphospholipid antibodies by viral infections and vaccinations (Tarjan *et al.*, 2006).

Studies have demonstrated that influenza vaccine is safe and immunogenic in patients with SLE or rheumatoid arthritis (RA), diminishing the risk of respiratory infections (Rahier *et al.*, 2010). The incidence of viral infection, especially influenza and bacterial complications, was lower in vaccinated patients, without exacerbation of the underlying disease. Thus, the benefits from vaccination outweigh the possible adverse effects. Influenza vaccines are considered safe and effective in the healthy population, as well as in ARD patients. The recommendations of the European League Against Rheumatism (EULAR) favor annual influenza vaccination in patients with ARDs. Major recommendations include the following: (i) vaccination should ideally be administered during stable disease; (ii) influenza vaccination and pneumococcal vaccination should be strongly considered; (iii) vaccination can be administered during the use of disease-modifying antirheumatic drugs (DMARDs) and TNF inhibitors, but before starting rituximab; (iv) live-attenuated vaccines should be avoided whenever possible in immunosuppressed patients; and (v) bacillus Calmette–Guérin (BCG) vaccination should be avoided (van Assen *et al.*, 2011).

Influenza, vaccination, and underlaying conditions

People with underlying medical conditions are known to be at increased risk of severe influenza infection, and sometimes fatal illness. During the A/H1N1 influenza pandemic, defective T cell responses were noted in severe cases of this infection. Immune alterations associated with comorbid conditions such as obesity, diabetes mellitus, chronic obstructive pulmonary disease, cancer, asthma, and chronic heart failure, among others, may interfere with the normal development of the specific response to the virus (Angeles-Garay *et al.*, 2011; Chowell *et al.*, 2012).

Annual vaccination of older adults and other risk groups is the most effective measure to reduce morbidity and mortality associated with A/H1N1 infection. So far, it is unknown whether vaccination increases the incidence of autoimmune diseases, especially in high-risk groups; meanwhile, the recommendations suggest that the benefits of vaccination outweigh the possible harmful events. Prospective studies are needed on this subject.

Influenza vaccine and autoimmunity

Vaccines are a prototypic source for natural immune stimulation, but may be involved in pathogenic disease in the setting of aberrant immune system function. However, a causal relationship between influenza vaccines and induction of autoimmune diseases remains unproved (Shoenfeld *et al.*, 2008). Comparing infection- versus vaccine-induced autoimmune reactions, the latter generally show a lower incidence, with the majority displaying a milder and more self-limiting clinical course (Schattner, 2005). In fact, the prevalence of autoimmune phenomena due to vaccination is notably lower than the prevalence of autoimmune diseases (3%) (Wraith *et al.*, 2003). Clinical syndromes described following influenza vaccination mainly include neurological pictures such as Guillain–Barré syndrome (GBS). The strongest association was with the vaccine used in an epidemic of influenza virus of swine origin in the United States in the mid-1970s, probably due to crossreacting antibodies against peripheral nerve gangliosides, which may develop after vaccination with this influenza virus (Nachamkin *et al.*, 2008). However, in general, the risk is low (Vera-Lastra *et al.*, 2013). In fact, recent studies found the incidence of GBS following contemporary H1N1 influenza vaccine is low (Shaikh *et al.*, 2012).

It has been reported that influenza vaccine in general does not alter the prevalence of autoantibodies in healthy adults (Toplak *et al.*, 2008). However, a more recent study showed that influenza vaccination has the potential to induce not only transient but also long-term autoantibody changes, including the development of new antibodies (especially anticardiolipin antibodies, aCLs) in some susceptible autoimmune inflammatory rheumatic disease (AIRD) patients and healthy subjects (Tarjan *et al.*, 2006). Further, Cerpa-Cruz *et al.* (2013) confirmed the link between vaccinations and increased risk of autoimmune responses suggestive of autoimmune/inflammatory syndrome induced by adjuvants (ASIA). In this study, 43 out of 120 patients experienced the new onset of a nonspecific inflammatory response (fever, myalgia, arthralgia) or signs or symptoms of an immune-mediated condition following vaccination, including by influenza vaccines. Adjuvants used in the vaccines included aluminum salts, thimerosal, and squalene.

The mechanisms by which vaccines may induce autoimmune diseases are mainly extrapolated from the known capacity of the infectious agents they target, both in animal models and in *in vitro* studies (Salemi and D'Amelio, 2010). In consequence, several mechanisms may be involved in the development of autoimmune phenomena caused by the use of vaccines. For example, the molecular mimicry of microbial antigens may play a role in the development of post-vaccination autoimmunity, especially through the presence of an adjuvant in the vaccine (Rose, 2010). Adjuvants optimize the immune response against the coadministered antigen, but also contribute to the toxicity of vaccines, which may increase the recognition and activation of autoreactive lymphocytes in genetically predisposed individuals (Batista-Duharte *et al.*, 2011). In fact, induction of lupus-related autoantibodies such as anti-U1RNP and Su/Argonaute2 by a single intraperitoneal injection of adjuvant hydrocarbon oils, such as squalene, a main component of an adjuvant MF59, and incomplete Freund's adjuvant (IFA) (mineral oil), has been reported in mice (Kuroda *et al.*, 2004).

The formation of immune complexes, causing vasculitis or exacerbation of latent autoimmune disorders, has also been proposed as a potential mechanism (Orbach, *et al.*, 2010). Another possibility is bystander activation: the release of sequestered self-antigens from the infected host tissue, leading to activation of antigen-presenting cells (APCs), which are capable of stimulating preprimed dormant autoreactive T cell clones. In addition, bystander activation is a situation in which enhanced cytokine production appers to promote the expansion of autoreactive T cells. In the case of polyclonal activation of B cells, the increased B cell proliferation, antibody production, and generation of circulating immune complexes may eventually damage self-tissues (Agmon-Levin *et al.*, 2009b). This mechanism has also been suggested in swine influenza vaccine (Nachamkin *et al.*, 2008).

Post-influenza vaccination syndrome: clinical spectrum

The post-vaccination phenomena are rare conditions that have been associated with clinical manifestations of specific AIRD and nonspecific autoimmune manifestations. They can induce transient or long-term autoantibodies in some susceptible AIRD patients and healthy subjects (Table 18.1). All these clinical manifestations are included in ASIA (Vera-Lastra *et al.*, 2013).

Vasculitis

Several types of vasculitis have been associated with post-influenza vaccination syndrome (P-IVS). Antineutrophil cytoplasmic antibody-associated vasculitis has been reported: its main clinical manifestations are purpura, mononeuritis, glomerulonephritis, and necrotizing vasculitis involving small blood vessels (Hull *et al.*, 2004; Urso *et al.*, 2011; Duggal *et al.*, 2013).

Microscopic polyangiitis

Microscopic polyangiitis (MPA) affects different organs, such as lung, kidney, and skin, and produces a high level of myeloperoxidase antineutrophil cytoplasmic antibody (Uji *et al.*, 2005; Konishi *et al.*, 2011).

Leucocytoclastic vasculitis

Patients present cutaneous vasculitis and abnormal urinalysis suggestive of renal involvement. As influenza vaccination is increasingly used, physicians should be aware of the development of P-IVS leucocytoclastic vasculitis (LV) (Ulm *et al.*, 2006).

Henoch–Schonlein purpura

Preexisting Henoch–Schönlein purpura (HSP) may be exacerbated by influenza vaccination (Mormile *et al.*, 2004).

Giant cell arteritis

Giant cell arteritis (GCA) and polymyalgia rheumatica (PMR) are associated with the application of the influenza vaccine in elderly people. Mild transitory side effects after vaccination are common, while severe and systemic complications, such as vasculitis and rheumatic disorders, remain rare (Marti and Anton, 2004). Recently, 10 cases of GCA and PMR were reported within 3 months of influenza vaccination (Inf-V), in addition to 11 previous cases of GCA/PMR associated with Inf-V. Therefore, a systematic review of previous vaccination should be conducted in patients with recent onset of GCA/PMR. Subjects at higher risk of developing GCA/PMR should be followed up for 2–6 months after Inf-V (Soriano *et al.*, 2012).

Systemic lupus erythematous and antiphospholipid syndrome

A transient increase of aCLs, but not anti-β2 glycoprotein-1 (β2 GP1), was reported in SLE patients who received influenza vaccination. The induction of antibodies against β2-GP1 has been shown to act as a cofactor among aCL-positive SLE patients, increasing thrombosis risk (Vista *et al.*, 2012). In rare cases, post-influenza vaccination induces clinical manifestation of antiphospholipid syndrome (APS) (Blank *et al.*, 2012; Cruz-Tapias *et al.*, 2012).

Inflammatory myopathy

Immune mechanisms play an important role in dermatomyositis. Sporadic associations between inflammatory myopathies and vaccinations have been described in the literatura. Recently, it was reported that three patients developed polymyositis complicated by interstitial lung disease (two cases) and dermatomyositis (one case) following influenza A (H1N1) vaccination (Ferri *et al.*, 2012).

Still's disease

Some cases of adult-onset Still's disease (AOSD) in which a causal relationship is suggested have been published. Bystander activation may play an

Table 18.1 Clinical spectrum of post-influenza vaccination syndrome

1.1 Vasculitis
a. Microscopic polyangiitis (MPA)
b. Leucocytoclastic vasculitis (LV)
c. Henoch–Schönlein purpura (HSP)
d. Giant cell arteritis (GCA)
1.2 Systemic lupus erythematosus (SLE) and antiphospholipid syndrome (APS)
1.3 Inflammatory myopathy
1.4 Still's disease
1.5 Juvenile idiopathic arthritis (JIA)
1.6 Neurological syndromes
a. Guillain–Barré syndrome (GBS)
b. Lambert–Eaton myasthenic syndrome (LEMS)
c. Narcolepsy
d. Multiple sclerosis (MS)
e. Acute disseminated encephalomyelitis (ADEM)
1.7 Idiopathic thrombocytopenic purpura (ITP)

important role in inducing the immune reaction in this condition (Yoshioka *et al.*, 2011).

Juvenile idiopathic arthritis

The relapse of systemic juvenile idiopathic arthritis (JIA) has recently been reported following influenza vaccination in a patient receiving tocilizumab (Shinoki *et al.*, 2012).

Neurological syndromes

Neurological manifestations associated with vaccines are diverse. They include GBS, Lambert–Eaton myasthenic syndrome (LEMS), narcolepsy, multiple sclerosis (MS), and acute disseminated encephalomyelitis (ADEM), among others.

Guillain–Barré syndrome

The onset of GBS was associated with the swine influenza vaccine in 1976–77 (Stowe *et al.*, 2009). However, subsequent studies of influenza vaccination detected no significant increase in the overall risk for GBS (Roscell *et al.*, 1991; Lasky *et al.*, 1998), but recent studies of contemporary H1N1 influenza vaccine have again found an association with GBS.

Five patients presented "atypical" GBS variants after 4 weeks of 2010/11 H1N1 influenza vaccine. These patients showed sensory ataxia, areflexia, extremity and oropharyngeal paresthesias, numbness, pain, weakness, sphincteric disturbances, dysautonomia, and Miller Fisher syndrome. However, epidemiological studies with large cohorts are necessary to confirm excess cases of atypical GBS following H1N1 influenza vaccination (Shaikh *et al.*, 2012). In this regard, Pandemrix vaccine against influenza seems to be safe, causing no change in the risk for GBS, MS, type 1 diabetes, or RA. However, relative risks were significantly increased for Bell's palsy, paresthesia, and inflammatory bowel disease (IBS) after vaccination, especially in the early phase of the vaccination campaign (Bardage *et al.*, 2011). In Quebec, the 2009 influenza A (H1N1) vaccine was associated with a slightly increased risk of GBS. However, it is likely that the benefits of immunization outweigh the risks (Poland and Jacobsen, 2012).

Lambert–Eaton myasthenic syndrome

Nonparaneoplastic LEMS has been reported following Pandemrix vaccination (Ansakorpi *et al.*, 2012).

Narcolepsy

A sudden increase in childhood narcolepsy was observed in Finland soon after an influenza epidemic and vaccination with ASO3-adjuvanted Pandemrix (Melén *et al.*, 2013). A ninefold increased risk of narcolepsy in children and adolescents following H1N1 vaccination with Pandemrix was found (Zarocostas, 2011). In a few individuals, the onset of narcolepsy occurred approximately 8 weeks following vaccination (Dauvilliers *et al.*, 2010). In Sweden, a relative risk of narcolepsy of 6.6 was found in vaccinated children and adolescents (Eurosurveillance Editorial Team, 2011).In China, narcolepsy onset correlated with seasonal and annual patterns of upper airway infections, including H1N1 influenza. The correlation was independent of H1N1 vaccination (Han *et al.*, 2011).

Multiple sclerosis

Influenza vaccines should be recommended as part of treatment practice in MS, because influenza infections are associated with increased risk of exacerbations of MS. Vaccines containing viable pathogens should not be used during immunosuppressive therapy (Farez and Correale, 2011).

Acute disseminated encephalomyelitis

ADEM includes several categories of primary inflammatory demyelinating disorders of the central nervous system (CNS), including MS, optic neuropathy, acute transverse myelitis, neuromyelitis optica, and Devic's disease. Post-infectious and post-immunization encephalomyelitis make up about 75% of cases; post-vaccination ADEM has been associated with several vaccines, including influenza vaccin. Recently, one patient was found to have bilateral optic neuropathy within 3 weeks of "inactivated" influenza vaccination, followed by a delayed onset of ADEM 3 months post-vaccination (Huynh *et al.*, 2008).

Idiopathic thrombocytopenic purpura

As the risk of thrombocytopenia after natural influenza seems to be much higher than that after immunization, annual influenza vaccination is advised for patients with a personal history of idiopathic thrombocytopenic purpura (ITP) (Mantadaki *et al.*, 2010). Refractory ITP has been reported following influenza vaccination (Tsuji *et al.*, 2009).

Recent findings

Recent studies suggested a reduced immunogenicity of the influenza AH1N1/2009 vaccine in juvenile rheumatic diseases. In this regard, 118 juvenile SLE patients and 102 healthy controls of a similar age were vaccinated; 3 weeks after immunization, seroprotection rate, seroconversion rate, geometric mean titer (GMT), and factor increase in GMT were substantially lower in juvenile SLE patients than in controls. The only significant factor for nonseroconversion was a SLEDAI-2K score greater than or equal to 8. This study demonstrated adequate disease safety. However, high disease activity impairs influenza A H1N1/2009 vaccine antibody production in juvenile SLE, in spite of an overall immune response within recommended levels (Campos *et al.*, 2013). Another study assesed the efficacy and safety of pandemic 2009 influenza A/H1N1 vaccination in SLE. Patients with and without therapy and healthy controls were followed after vaccination. The SLE group with treatment had lower seroconversion than healthy controls, and the concomitant use of chloroquine was associated with seroconversion responses comparable to those of patients not receiving treatment. These finding suggest that vaccine response is diminished even in patients under immunosuppressive treatment; antimalarials seems to help restore this immunogenicity (Borba *et al.*, 2012).

Finally, new studies have shown the efficiency, immunogenicity, and safety of trivalent inactivated split influenza vaccine (vaccine without adjuvant) in patients with AIRDs such as SLE, RA, and Sjögren syndrome (SjS). Therefore, the use of nonadjuvanted vaccine is recommended in order to reduce the risk of respiratory infecctions and of exacerbations of autoimmune diseases, including the onset of ASIA (Milanovic *et al.*, 2013; Murdaca *et al.*, 2014).

Conclusions

1. Influenza infection can be mild or severe and can even cause death, especially in vulnerable population groups. The complications observed in patients with influenza are caused not only by the severity of the infection, but also by the development of opportunist infections, as well as by an inadequate and untimely immune response, especially in AIRD.
2. The best preventive measure that we currently have is a vaccine. In order to avoid complications or onset of ASIA, the use of nonadjuvanted vaccine is recommended.
3. The medical community should be alert and report any side effects, including the onset or activation of AIRDs associated with or related to the influenza vaccine.
4. Even though the autoimmune diseases and ASIA have been described in patients with the influenza A (H1N1) vaccine, the benefits of immunization still outweigh the risks.

References

Agmon-Levin, N., Paz, Z., Israeli, E., and Shoenfeld, Y. (2009a). Vaccines and autoimmunity. *Nat Rev Rheumatol*, **5**: 648–52.

Agmon-Levin, N., Kivity, S., Shoenfeld, Y. (2009b). Influenza vaccine and autoimmunity. *IMAJ*, **11**: 183–5.

Angeles-Garay, U., Gayosso Rivera, J.A., Zacate-Palacios, Y., *et al.* (2011). Clinical features and contagiousness of influenza A(H1N1) in health care workers and hospitalized patients in a Mexico City hospital. *Enferm Infecc Microbiol Clin*, **29**: 679–82.

Ansakorpi, H., Rusanen, H., Rytky, S., and Färkkilä, M. (2012) Lambert-Eaton myasthenic syndrome following H1N1-influenza vaccination: a case report. *Acta Neurol Scand*, **126**(5): e25–8.

Blank, M., Israeli, E., Shoenfeld, Y. (2012). When APS (Hughes syndrome) met the autoimmune/inflammatory syndrome induced by adjuvants (ASIA). *Lupus*, **21**: 711.

Bardage, C., Persson, I., Ortqvist, A., *et al.* (2011). Neurological and autoimmune disorders after vaccination against pandemic influenza A (H1N1) with a monovalent adjuvanted vaccine: population based cohort study in Stockholm, Sweden. *BMJ*, **343**: d5956.

Batista-Duharte, A., Lindblad, E.B., and Oviedo-Orta, E. (2011). Progress in understanding adjuvant immunotoxicity mechanisms. *Toxicol Lett*, **203**: 97–105.

Borba, E.F., Saad, C.G., Pasoto, S.G., *et al.* (2012). Influenza A/H1N1 vaccination of patients with SLE: can antimalarial drugs restore diminished response under immunosuppressive therapy? *Rheumatology (Oxford)*, **51**: 1061–9.

Bischoff, A.L., Følsgaard, N.V., Carson, C.G., *et al.* (2013). Altered response to A(H1N1)pnd09 vaccination in pregnant women: a single blinded randomized controlled trial. *PloS One*, **8**: e56700.

Campos, L.M., Silva, C.A., Aikawa, N.E., *et al.* (2013). High disease activity: an independent factor for reduced immunogenenicity of the pandemic influenza a vaccine in patients with juvenile systemic lupus erythematosus. *Arthritis Care Res (Hoboken)*, **65**: 1121–7.

Cerpa-Cruz, S., Paredes-Casillas, P., Landeros Navarro, E., *et al.* (2013). Adverse events following immunization with vaccines containing adjuvants. *Immunol Res*, **56**: 299–303.

Chan, K.H., To, K.K., Hung, I.F., *et al.* (2011a). Differences in antibody responses of individuals with natural infection and those vaccinated against pandemic H1N1 2009 influenza. *Clin Vaccine Immunol*, **18**: 867–73.

Chan, J.F., To, K.K., Tse, H., *et al.* (2011b). The lower serum immunoglobulin G2 level in severe cases than in mild cases of pandemic H1N1 2009 influenza is associated with cytokine dysregulation. *Clin Vaccine Immunol*, **18**: 305–10.

Chowell, G., Echevarría-Zuno, S., Viboud, C., *et al.* (2012). Epidemiological characteristics and underlying risk factors for mortality during the autumn 2009 pandemic wave in Mexico. *PLoS One*, **7**: e41069.

Cordeau, M., Herblot, S., Charrier, E., *et al.* (2012). Defects in CD54 and CD86 up-regulation by plasmacytoid dendritic cells during pregnancy. *Immunol Invest*, **41**: 497–506.

Cruz-Tapias, P., Blank, M., Anaya, J.M., and Shoenfeld, Y. (2012). Infections and vaccines in the etiology of antiphospholipid syndrome. *Curr Opin Rheumatol*, **24**: 389–93.

Dauvilliers, Y., Montplaisir, J., Cochen, V., *et al.* (2010). Post-H1N1 narcolepsy-cataplexy. *Sleep*, **33**: 1428–30.

Duggal, T., Segal, P., Shah, M., *et al.* (2013). Antineutrophil cytoplasmic antibody vasculitis associated with influenza vaccination. *Am J Nephrol*, **38**: 174–8.

Echevarría-Zuno, S., Mejía-Aranguré, J.M., Mar-Obeso, A.J., *et al.* (2009). Infection and death from influenza A H1N1 virus in Mexico: a retrospective analysis. *Lancet*, **374**: 2072–9.

Endoh, M., Suga, T., Miura, M., *et al.* (1984). In vivo alteration of antibody production in patients with IgA nephropathy. *Clin Exp Immunol*, **57**: 564–70.

Eurosurveillance Editorial Team. (2011). Swedish Medical Products Agency publishes report from a case inventory study on Pandemrix vaccination and development of narcolepsy with cataplexy. *Euro Surveill*, **16**(26). pii: 19904.

Fajardo-Dolci, G., Gutierrez-Vega, R., Arboleya-Casanova, H., *et al.* (2010). Clinical characteristics of fatalities due to influenza A (H1N1) virus in Mexico. *Thorax*, **65**: 505–9.

Farez, M.F. and Correale, J. (2011). Immunizations and risk of multiple sclerosis: systematic review and meta-analysis. *J Neurol*, **258**: 1197–206.

Fedson, D.S. (2013). Treating influenza with statins and other immunomodulatory agents. *Antiviral Res*, **99**: 417–35.

Ferri, C., Colaci, M., Manzini, C.U., *et al.* (2012). Polymyositis following Pandemic Influenza A (H1N1) and 2009–10 Seasonal Trivalent Vaccines. *Case Rep Rheumatol*, **2012**: 836930.

Gao, R., Bhatnagar, J., Blau, D., *et al.* (2013). Cytokine and chemokine profiles in lung tissues from fatal cases of 2009 pandemic influenza A (H1N1): role of the host immune response in pathogenesis. *Am J Pathol*, **183**(4): 1258–68.

Gordon, C.L., Johnson, P.D., Permezel, M., *et al.* (2010). Association between severe pandemic 2009 influenza A (H1N1) virus infection and immunoglobulin G-(2) subclass deficiency. *Clin Infect Dis*, **50**: 672–8.

Han, F., Lin, L., Warby, S.C., *et al.* (2011). Narcolepsy onset is seasonal and increased following the 2009 H1N1 pandemic in China. *Ann Neurol*, **70**: 410–17.

Hull, J.H., Mead, S.H., Foster, O.J., and Modarres-Sadeghi, H. (2004). Severe vasculitic neuropathy following influenza vaccination. *J Neurol Neurosurg Psychiatry*, **75**: 1507–8.

Huynh, W., Cordato, D.J., Kehdi, E., *et al.* (2008). Post-vaccination encephalomyelitis: literature review and illustrative case. *J Clin Neurosci*, **15**: 1315–22.

Konishi, M., Koarada, S., Yamaguchi, K., *et al.* (2011). Case of microscopic polyangiitis and giant cell arteritis after influenza vaccination. *Nihon Rinsho Meneki Gakkai Kaishi*, **34**: 154–61.

Kuroda, Y., Nacionales, D.C., Akaogi, J., *et al.* (2004). Autoimmunity induced by adjuvant hydrocarbon oil components of vaccine. *Biomed Pharmacother*, **58**: 325–37.

Lasky, T., Terracciano, G.J., Magder, L., *et al.* (1998). The Guillain–Barre syndrome and the 1992–1993 and 1993–1994 influenza vaccines. *N Engl J Med*, **339**: 1797–802.

Lee, N., Shin, M.S., and Kang, I. (2012). T-cell biology in aging, with a focus on lung disease. *J Gerontol A Biol Sci Med Sci*, **67**: 254–63.

Mantadakis, E., Farmaki, E., Thomaidis, S., *et al.* (2010). A case of immune thrombocytopenic purpura after influenza vaccination: consequence or coincidence? *J Pediatr Hematol Oncol*, **32**: e227–9.

Marti, J. and Anton, E. (2004). Polymyalgia rheumatica complicating influenza vaccination. *J Am Geriatr Soc*, **52**: 1412.

Melén, K., Partinen, M., Tynell, J., *et al.* (2013). No serological evidence of influenza A H1N1pdm09 virus infection as a contributing factor in childhood narcolepsy after pandemrix vaccination campaign in Finland. *PLoS One*, **8**(8): e68402.

Milanovic, M., Stojanovich, L., Djokovic, A., *et al.* (2013). Influenza vaccination in autoimmune rheumatic disease patients. *Tohoku J Exp Med*, **229**: 29–34.

Mormile, R., D'Alterio, V., Treccagnoli, G., and Sorrentino, P. (2004). Henoch-Schonlein purpura with antiphospholipid antibodies after influenza vaccination: how fearful is it in children? *Vaccine*, **23**: 567–8.

Murdaca, G., Orsi, A., Spano, F., *et al.* (2014). Influenza and pneumococcal vaccinations of patients with systemic lupus erythematosus: current views upon safety and immunogenicity. *Autoimmun Rev*, **13**: 75–84.

Nachamkin, I., Shadomy, S.V., Moran, A.P., *et al.* (2008). Anti-ganglioside antibody induction by swine (A/NJ/1976/H1N1) and other influenza vaccines: insights into vaccine associated Guillain-Barré syndrome. *J Infect Dis*, **198**: 226–33.

Orbach, H., Agmon-Levin, N., and Zandman-Goddard, G. (2010). Vaccines and autoimmune diseases in the adult. *Discov Med*, **9**: 90–7.

Patel, M., Dennis, A., Flutter, C., and Khan, Z. (2010). Pandemic (H1N1) 2009 influenza. *Br J Anaesth*, **104**: 128–42.

Paquette, S.G., Banner, D., Zhao, Z., *et al.* (2012). Interleukin-6 is a potential biomarker for severe pandemic H1N1 influenza A infection. *PLoS One*, **7**(6): e38214.

Perez-Padilla, R., Rosa-Zamboni, D., Ponce, D.L., *et al.* (2009). Pneumonia and respiratory failure from swine-origin influenza A (H1N1) in Mexico. *N Engl J Med*, **361**: 680–9.

Poland, G.A. and Jacobsen, S.J. (2012). Influenza vaccine, Guillain-Barré syndrome, and chasing zero. *Vaccine*, **30**: 5801–3.

Rahier, J.-F., Moutschen, M., Van Gompel, A., *et al.* (2010). Vaccinations in patients with immune-mediated inflammatory diseases. *Rheumatology (Oxford)*, **49**: 1815–27.

Robinson, D.P. and Klein, S.L. (2012). Pregnancy and pregnancy-associated hormones alter immune responses and disease pathogenesis. *Horm Behav*, **62**: 263–71.

Rose, N.R. (2010). Autoimmunity, infections and adjuvants. *Lupus*, **19**: 354–8.

Roscelli, J.D., Bass, J.W., and Pang, L. (1991). Guillain-Barré syndrome and influenza vaccination in the US Army, 1980–1988. *Am J Epidemiol*, **133**: 952–5.

Salemi, S. and D'Amelio, R. (2010). Could autoimmunity be induced by vaccination? *Int Rev Immunol*, **29**: 247–69.

Salmon, D.A., Proschan, M., Forshee, R., *et al.* (2013). Association between Guillain-Barré syndrome and influenza A (H1N1) 2009 monovalent inactivated vaccines in the USA: a meta-analysis. *Lancet*, **38**: 1461–8.

Schattner, A. (2005). Consequence or coincidence? The occurrence, pathogenesis and significance of autoimmune manifestations after viral vaccines. *Vaccine*, **23**: 3876–86.

Shaikh, A.G., Termsarasab, P., Nwankwo, C., *et al.* (2012). Atypical forms of Guillain–Barré syndrome and H1N1-influenza vaccination. *Vaccine*, **30**: 3251–4.

Shinoki, T., Hara, R., Kaneko, U., *et al.* (2012). Safety and response to influenza vaccine in patients with systemic-onset juvenile idiopathic arthritis receiving tocilizumab. *Mod Rheumatol*, **22**: 871–6.

Shoenfeld, Y., Zandman-Goddard, G., Stojanovich, L., *et al.* (2008). The mosaic of autoimmunity: hormonal and environmental factors involved in autoimmune diseases. *IMAJ*, **10**: 8–12.

Soriano, A., Verrecchia, E., Marinaro, A., *et al.* (2012). Giant cell arteritis and polymyalgia rheumatica after influenza vaccination: report of 10 cases and review of the literature. *Lupus*, **21**: 153–7.

Steinhauer, D.A. and Skehel, J.J. (2002). Genetics of influenza viruses. *Annu Rev Genet*, **36**: 305–32.

Stowe, J., Andrews, N., Wise, L., and Miller, E. (2009). Investigation of the temporal association of Guillain-Barre syndrome with influenza vaccine and influenza like illness using the United Kingdom General Practice Research Database. *Am J Epidemiol*, **169**: 382–8.

Tarjan, P., Sipka, S., Lakos, G., *et al.* (2006). Influenza vaccination and the production of anti-phospholipid antibodies in patients with systemic lupus erythematosus. *Scand J Rheumatol*, **35**: 241–9.

To, K.K., Hung, I.F., Li, I.W., *et al.* (2010). Delayed clearance of viral load and marked cytokine activation in severe cases of pandemic H1N1 influenza virus infection. *Clin Infect Dis*, **50**: 850–9.

Toplak, N., Kveder, T., Trampuš-Bakija, A., *et al.* (2008). Autoimmune response following annual influenza vaccination in 92 apparently healthy adults. *Autoimmun Rev*, **8**: 134–8.

Tsuji, T., Yamasaki, H., and Tsuda, H. (2009). Refractory idiopathic thrombocytopenic purpura following influenza vaccination. *Rinsho Ketsueki*, **50**: 57.

Uji, M., Matsushita, H., and Iwata, S. (2005). Microscopic polyangiitis after influenza vaccination. *Intern Med*, **44**: 892–6.

Ulm, S., Hummel, M., Emig, M., *et al.* (2006). Leukocytoclastic vasculitis and acute renal failure after influenza vaccination in an elderly patient with myelodysplastic syndrome. *Onkologie*, **29**: 470–2.

Urso, R., Bevilacqua, N., Gentile, M., *et al.* (2011). Pandemic 2009 H1N1 virus infection associated with purpuric skin lesions: a case report. *J Med Case Rep*, **5**: 132.

Van Kerkhove, M.D., Vandemaele, K.A., Shinde, V., *et al.* (2011). Risk factors for severe outcomes following 2009 influenza A (H1N1) infection: a global pooled analysis. *PLoS Med*, **8**(7): e1001053.

van Assen, S., Agmon-Levin, N., Elkayam, O., *et al.* (2011). EULAR recommendations for vaccination in adult patients with autoimmune inflammatory rheumatic diseases. *Ann Rheum Dis*, **70**(3): 414–22.

Vera-Lastra, O., Medina, G., Cruz-Dominguez, M.P., *et al.* (2013). Autoimmune/inflammatory syndrome induced by adjuvants (Shoenfeld's syndrome): clinical and immunological spectrum. *Exp Rev Clin Immunol*, **9**: 361–73.

Vista, E.S., Crowe, S.R., Thompson, L.F., *et al.* (2012). Influenza vaccination can induce new-onset anticardiolipins but not β2-glycoprotein-I antibodies among patients with systemic lupus erythematosus. *Lupus*, **21**: 168–74.

Wraith, D.C., Goldman, M., and Lambert, P.H. (2003). Vaccination and autoimmune diseases: what is the evidence? *Lancet*, **362**: 1659–66.

Yoshioka, K., Fujimoto, S., Oba, H., *et al.* (2011). Onset of adult-onset Still's disease following influenza vaccination. *Mod Rheumatol*, **21**: 432–5.

Yurasov, S., Wardemann, H., Hammersen, J., *et al.* (2005). Defective B cell tolerance checkpoints in systemic lupus erythematosus. *J Exp Med*, **201**: 703–11.

Zarocostas, J. (2011). WHO backs further probes into possible link between H1N1 vaccine and narcolepsy in children. *BMJ*, **342**: d909.

Zúñiga, J., Torres, M., Romo, J., *et al.* (2011). Inflammatory profiles in severe pneumonia associated with the pandemic influenza A/H1N1 virus isolated in Mexico City. *Autoimmunity*, **44**: 562–70.

19 Vaccines and Autoimmunity: Meningococcal Vaccines

Giovanna Passaro,[1] Alessandra Soriano,[2,3] and Raffaele Manna[1]

[1]Periodic Fevers Research Centre, Department of Internal Medicine, Catholic University of the Sacred Heart, Rome, Italy

[2]Zabludowicz Center for Autoimmune Diseases, Sheba Medical Center, Tel Hashomer, Israel

[3]Department of Clinical Medicine and Rheumatology, Campus Bio-Medico University, Rome, Italy

Introduction

Meningococcal disease can refer to any illness that is caused by *Neisseria meningitidis*, also known as meningococcus. Meningococcal disease remains a serious global health threat, associated with high mortality and morbidity, despite advances in antibiotic therapy, organ support techniques, and modern vaccination strategies (Nadel, 2012). The recent incidence of meningococcal disease is 0.35–1.01 cases per 100 000 inhabitants in most Western countries. It is a leading cause of bacterial meningitis in children aged 2–18 years in the United States. *N. meningitidis* can cause meningitis, sepsis, and focal infections, both as an endemic and an epidemic infection. Children are particularly vulnerable to meningococcal disease because of the relative immaturity of their immune system, particularly their impaired immunity to the meningococcus polysaccharide capsule. Worldwide, the epidemiology of bacterial meningitis has changed dramatically in the last 2 decades, following the introduction of new, highly effective conjugate protein/polysaccharide vaccines (Nadel, 2012).

N. meningitidis

N. meningitidis is an obligate human commensal living in the upper respiratory tract, an encapsulated, aerobic Gram-negative diplococcus. Thirteen antigenically and chemically distinct polysaccharide capsules have been described. Almost all invasive disease is caused by one of five serogroups: A, B, C, Y, and W-135.

Vaccines

The modern era of meningococcal vaccines began in the late 1960s, triggered by the evolution of resistance to available chemotherapies (Maiden, 2013). There are two main types of vaccines used for protection against meningococcal infection: pure polysaccharide vaccines and protein/polysaccharide conjugate vaccines. These are based on the capsular polysaccharide of the bacteria, which is a major virulence factor and is responsible for the prevention of host-mediated bacterial killing (Nadel, 2012). The first monovalent (group C) polysaccharide vaccine was licensed

Vaccines and Autoimmunity, First Edition. Edited by Yehuda Shoenfeld, Nancy Agmon-Levin, and Lucija Tomljenovic.

in the United States in 1974, while a quadrivalent polysaccharide vaccine was licensed in 1978. The invention of conjugate vaccines in the 1980s was a major breakthrough in polysaccharide vaccine development: these vaccines contain a polysaccharide molecule chemically conjugated to a T cell-stimulating antigen, such as the diphtheria toxoids or tetanus toxoids (TTds). This has the effect of recruiting T cell help and, therefore, results in the generation of affinity-matured immunological responses and immunological memory (Maiden, 2013). Meningococcal conjugate vaccine (MCV) has been licensed in the United Kingdom since 1999 and has had a major impact on the incidence of type C meningococcal disease. A quadrivalent conjugate vaccine was first licensed in the United States in 2005. In January 2013, the European Commission approved a meningococcal serogroup B vaccine (Bexsero) for use in individuals over 2 months of age in the European Union (Novartis, 2013).

Several different meningococcal vaccines are available in the United States:

- The current quadrivalent A, C, Y, W-135 polysaccharide vaccine (Menomune, Sanofi Pasteur) was licensed in 1978. Each dose consists of 50 mcg of each of the four purified bacterial capsular polysaccharides. The vaccine contains lactose as a stabilizer. It is the only meningococcal vaccine licensed for people older than 55 (CDC, 2012). It does not generate memory T cells, and attempts to boost protection with repeated vaccination may result in a diminished antibody response. Also, like other polysaccharide vaccines, this meningococcal vaccine does not prevent mucosal colonization and therefore does not provide herd immunity through the interrupted transmission of *N. Meningitidis* (Gardner, 2006).
- A quadrivalent meningococcal polysaccharide vaccine (MPV) conjugated to diphtheria toxoid (Menactra – Sanofi Pasteur) was licensed in 2005. The vaccine contains *N. meningiditis* serogroups A, C, Y, and W-135 capsular polysaccharide antigens conjugated to diphtheria toxoid protein (CDC, 2012).
- A combination conjugate vaccine against meningococcus serogoups C and Y and *Haemophilus influenza* type b (MenHibrix, Hib/MenCY) was approved in 2012 for infants and children aged 6 weeks to 18 months.

N. meningitidis serogroup B (MenB), a leading cause of bacterial meningitis in industrialized countries, remains a serious unresolved public health challenge. Global incidence of the MenB infection is estimated to be between 20 000 and 80 000 cases per year, with a 10% fatality rate even with appropriate treatment. MenB glycoconjugate vaccines are not immunogenic and, hence, vaccine design has focused on subcapsular antigens (Findlow, 2013). Bexsero (multicomponent meningococcal group B vaccine (rDNA, adsorbed)) is designed to provide protection against MenB disease for children aged 2 months and older (Novartis, 2013).

Immunopathology

The MCVs elicit a T cell-dependent memory response that increases their effectiveness, resulting in an improved primary response to vaccination and a strong anamnestic response at reexposure (Keyserling *et al.*, 2005). Some researchers suggested that the antigenic similarities between brain components and group B meningococci capsular polysaccharide could induce immunopathology (Stein *et al.*, 2006). MenB capsular polysaccharide is composed of a linear homopolymer of $\alpha(2\rightarrow8)$ N-acetyl-neuroaminic acid (polysialic acid, PSA). The MenB PSA and PSA found on neural cell-adhesion molecules are structurally identical; it has been proposed that infection with MenB or vaccination with PSA may be associated with subsequent autoimmune or neurological disease (Howitz *et al.*, 2007; Gottfredsson *et al.*, 2011). In a recent study, no evidence of increased autoimmunity was found to be associated with MenB compared with MenC (Gottfredsson *et al.*, 2011). MenC infections were associated with arthritis and migraine headaches more frequently than were MenB infections. This study does not support the hypothesis that the MenB infection may be predisposed to autoimmunity (Howitz *et al.*, 2007; Gottfredsson *et al.*, 2011). A natural exposure to MenB antigens is not associated with autoimmune diseases in humans. There is no increased risk of autoimmune diseases among persons with previous MenB disease (Howitz *et al.*, 2007). A major challenge in the development of a meningococcal serogroup B vaccine has been that the serogroup B polysaccharide is a very poor immunogen in humans, probably due to the similarity of its immunochemical structure to human intracellular adhesion molecules (Findlow *et al.*, 2010). There is no epidemiological or clinical evidence to associate the pathology with PSA antibodies (Stein *et al.*, 2006).

Menactra and possible association with Guillain–Barré syndrome

The US Vaccine Adverse Event Reporting System (VAERS), operated by the Centers for Disease

Control and Prevention (CDC) and the FDA, is a national passive surveillance system that monitors the safety of vaccines (CDC, 2005). Spontaneous reports to VAERS shortly after the introduction of quadrivalent conjugated meningococcal vaccine (MCV4) raised concerns of a possible association with Guillain–Barré syndrome (GBS) (Haber *et al.*, 2009). In September 2005, the FDA and the CDC issued an advisory regarding the occurrence of GBS in five recipients of MCV4 (Menactra: Sanofi Pasteur, Inc., Swiftwater, PA) (Table 19.1) (CDC, 2005). In September 2006, a total of 15 cases of GBS were reported in persons aged 11–19 year with onset within 6 weeks after vaccination by Menactra; all patients recovered or were recovering at that time. In 2007, the US Advisory Committee on Immunization Practices (ACIP) indicated a personal history of GBS as a limiting condition for meningococcal vaccination. However, the comparison with expected rates of GBS was inconclusive for an increased risk, and the lack of controlled epidemiological studies makes it difficult to draw conclusions about a causal association (Haber *et al.*, 2009). In 2010, the ACIP removed precautionary language from its recommendations, since more recent studies showed no increased risk of GBS in individuals receiving Menactra (CDC, 2011).

Vaccines in patients with complement dysfunction

A significant proportion of individuals with sporadic meningococcal infection have deficiencies of the complement system. A recent study found that 15–23% of adults with meningococcal meningitis had an underlying complement deficiency (Keiser and Broderick, 2012). Current CDC guidelines recommend that individuals with complement deficiencies receive a two-dose primary series of meningococcal conjugate vaccine, followed by boosting every 5 years (CDC, 2011). Vaccination against *N. meningitides* serogroup C (MenC) is essential in patients with dysfunction of the complement system, as induced by eculizumab (Soliris: Alexion Pharmaceuticals, Inc., Cheshire, CT). Patients should undergo meningococcal vaccination at least 2 weeks prior to receiving the first eculizumab treatment and undergo revaccination according to current medical guidelines. Patients must be monitored and evaluated immediately for early signs of meningococcal infections and treated with antibiotics as indicated (Dmytrijuk *et al.*, 2008).

Henoch–Schönlein purpura and meningococcal vaccines

Henoch–Schönlein purpura (HSP) is an autoimmune disease characterized by acute small-vessel vasculitis involving immunoglobulin A (IgA). Several risk factors play important roles in the pathogenesis of HSP (Nikibakhsh *et al.*, 2012). The most convincing finding is the association of HLA-DRB1*01, 07, and 11 with HSP susceptibility (He *et al.*, 2013). Some case reports have linked vaccines to HSP (Goodman *et al.*, 2010). No studies have evaluated this relationship. The association between vaccination and the onset of vasculitis including HSP has been reported before. HSP has been observed following influenza vaccination and measles vaccination, hepatitis B and bacillus Calmette–Guérin (BCG) vaccinations, and viral and streptococcal infections:

- HSP following meningitis C vaccination (Goodman *et al.*, 2010). This case involved a vasculitis occurring 7 days after vaccination in a 17-year-old woman. The patient presented palpable purpuric rash, pyrexia, severe abdominal pain, arthritis, and an elevation of inflammatory markers. The onset of this illness 1 week following meningitis C vaccination (Meningitec) may implicate the vaccine as a possible trigger. The case report contains no reference to other familial autoimmune diseases.
- HSP following a meningococcal vaccine (Lambert *et al.*, 2003). This case involved a leukocytoclastic vasculitis occurring 10 days after Menomune MPV4. The patient presented multiple purpuric papules, polyarthritis, and bilateral lower-extremity edema and pain. The patient developed proteinuria and hematuria 3 months after the immunization. A renal biopsy, performed 8 months after vaccination, showed a proliferative glomerulonephritis with IgA deposits.
- HSP and polysaccharide meningococcal vaccine (Courtney *et al.*, 2001). This case occurred 10 days after administration of MPV4 in a healthy 18-year-old woman. The patient had rash, migrating polyarthralgias, and gastrointestinal symptoms.

A study of the Vaccine Safety Datalink cohort, aimed at estimating the 42-day post-vaccination incidence rate of HSP, revealed that HSP was not statistically associated with MPV4 in the 16–20-year-old age group.

Bullous pemphigoid and meningococcal vaccine

Bullous pemphigoid (BP) is an acquired autoimmune blistering disease that is extremely rare in childhood, characterized by circulating IgG

Table 19.1 Guillain–Barré syndrome (GBS) in five recipients of MCV4 in 2005

Patient	Sex	Age (years)	Onset interval (days)	Signs and symptoms	Familiarity	Concomitant vaccines	Comorbidity	Nerve conduction study consistent with GBS
1	Male	18	15	Tingling in feet and hands	Mother GBS 5 year earlier	None	None	Yes
2	Male	17	25	Difficulty walking, difficulty moving from a standing to a seated position	None	None	Attention deficit hyperactivity disorder Asperger's syndrome	Yes
3	Female	17	14	Numbness of toes and tongue, numbness of thighs and fingertips, arm weakness, inability to run, difficulty walking with falling	None	None	GBS at ages 2 and 5 years 14 days after vaccination with childhood vaccines	Not performed
4	Female	18	31	Numbness of legs and trouble standing on toes	None	None	Mild ulcerative colitis	Yes
5	Female	18	14	Heaviness in legs walking upstairs, bilateral leg pain, headache, back and neck pain, vomiting and tingling in both hands	None	None	None	Yes

antibodies to antigens of the epidermal basement membrane zone. In general, the clinical course of this condition is good, and relapses are rare. Early diagnosis and treatment are fundamental. Valdivielso-Ramos *et al.* (2011) reported a case of 3-month-old girl with a blistering eruption on her palms and soles and urticarial plaques on her trunk and face, which occurred 3 weeks after her vaccinations at 2 months (hepatitis B, diphtheria, tetanus, pertussis, polio, *Haemophilus influenzae B*, MenC, pneumococcus). The clinical course worsened with vaccinations at 4 and 6 months. Control of the lesions was achieved with oral deflazacort 1 mg/kg/day, with a gradual decrease throughout 3 months of therapy. The patient was still in remission after 8 months of follow-up (Valdivielso-Ramos *et al.*, 2011).

In this case, family history was negative for autoimmune disease. There were no antibodies against the epidermal basal membrane. Furthermore, there was a generalization of the booster effect, with worsening of the clinical symptoms 4 days after immunization at 4 months in the vaccination calendar. The corticosteroid therapy obtained a stable remission of the disease. The vaccinations at 15 months (for measles, mumps, and rubella (MMR) and chickenpox), during oral deflazacort 1 mg/kg/day, were totally asymptomatic. Childhood BP is an acquired disorder that occurs as a result of the taking of drugs and vaccines, especially with first vaccination. Lesions appear 1–4 weeks later. Because of the high number of vaccinations that are needed in the first year of life, it is difficult to establish the causal relationship with a specific type. However, the onset and recurrence of symptoms after 3 weeks of vaccination establishes a clear temporal relation to the same. We can conclude vaccination may unmask a subclinical BP or induce an immune activation in individuals without familiarity for autoimmune diseases. However, the rarity of cases can be explained only by a rare genetic predisposition.

Safety

In a study aimed at evaluating immune response, antibody persistence, and the safety of a single dose of quadrivalent meningococcal serogroups A, C, W-135, and Y TTd conjugate (MenACWY-TT) vaccine in 500 adolescents and adults, no new onset of chronic illnesses or serious adverse events reported up to 3 years after vaccination were considered to be related to vaccination, and both vaccines were well tolerated (Borja-Tabora *et al.*, 2013). Injection-site redness and swelling with any intensity were more frequently reported in participants receiving the MenACWY-TT vaccine than those who received the MenACWY polysaccharide vaccine. These observations are likely due to the TTd content of the conjugate vaccine and are consistent with previous studies showing that local reactions are more frequent in individuals vaccinated with quadrivalent meningococcal conjugate vaccines than with plain polysaccharide vaccines. However, we cannot rule out the possibility that the intramuscular administration of the conjugate vaccine (versus subcutaneous for that of polysaccharide) may, in part, explain the higher reactogenicity, as a previous study suggested that intramuscular administration of MPV4 induced higher rates of erythema as compared with subcutaneous administration (Borja-Tabora *et al.*, 2013).

In a study based on the analysis of approximately 63 000 doses of group B meningococcal vaccinations (MeNZB) in New Zealand, no statistically significant increase in incidences of simple febrile seizures was found. That vaccine is unlikely to induce a heightened risk of simple febrile seizures (Stehr-Green *et al.*, 2008).

Vaccines in autoimmune disease

British Society for Rheumatology (BSR) guidelines state that the immune response to the MenC conjugate vaccine in immunosuppressed patients with rheumatic disease may be suboptimal and that therefore these patients may require boosters. The MenC conjugate vaccine does not aggravate juvenile idiopathic arthritis (JIA) activity or increase relapse frequency, and results in adequate antibody levels, even in patients receiving highly immunosuppressive medication. Therefore, patients with JIA can be vaccinated safely and effectively with the MenC conjugate (Zonneveld-Huijssoon *et al.*, 2007), regardless of the immunosuppressive treatment given.

Conclusions

We can conclude that meningococcal vaccines do not cause autoimmune diseases but may unmask autoimmune phenomena in rare individuals, who probably carry an unknown genetic predisposition.

According to the CDC, persons at increased risk for meningococcal disease, for whom immunization is therefore recommended (Cohn 2013), are:

- College freshmen living in dormitories.
- Microbiologists routinely exposed to *N. meningitides*.
- Populations in which an outbreak of meningococcal disease occurs.
- Military recruits.
- Persons with increased susceptibility (those with anatomical or functional asplenia or terminal complement deficiency).
- Travelers to regions where *N. meningitides* is hyperendemic (e.g. sub-Saharan Africa and Saudi Arabia) or epidemic.

People with a history of GBS who are not in a high-risk group for invasive meningococcal disease should not receive MCV4, although, in 2010, the ACIP removed precautionary language from its recommendations, since recent studies had not shown an increased risk of GBS in individuals receiving Menactra (CDC, 2011)

References

Borja-Tabora, C., Montalban, C., Memish, Z.A., *et al.* (2013). Immune response, antibody persistence, and safety of a single dose of the quadrivalent meningococcal serogroups A, C, W-135, and Y tetanus toxoid conjugate vaccine in adolescents and adults: results of an open, randomised, controlled study. *BMC Infect Dis*, **13**: 1–13.

CDC (Centers for Disease Control and Prevention). (2005). Guillain-Barré syndrome among recipients of Menactra® meningococcal conjugate vaccine-United States, June-July 2005. *MMWR*, **54**: 1023–5.

CDC (Centers for Disease Control and Prevention). (2011). Updated recommendations for use of meningococcal conjugate vaccines- Advisory committee on immunization practices (ACIP), 2010. *MMWR*, **60**: 72.

CDC (Centers for Disease Control and Prevention). (2012). Epidemiology and prevention of vaccine-preventable diseases meningococcal disease. *The Pink Book: Course Textbook*, 12th edn. Atlanta, GA: Centers for Disease Control and Prevention.

Cohn, A.C., MacNeil, J.R., Clark, T.A., *et al.* (2013). Prevention and control of meningococcal disease: recommendations of the Advisory Committee on Immunization Practices (ACIP). *MMWR*, **62**: 1–28.

Courtney, P.A., Patterson, R.N., and Lee, R.J.E. (2001). Henoch–Schönlein purpura following meningitis C vaccination. *Rheumatology*, **40**(3): 345–6.

Dmytrijuk, A., Robie-Suh, K., Cohen, M.H., *et al.* (2008). FDA report: eculizumab (Soliris) for the treatment of patients with paroxysmal nocturnal hemoglobinuria. *Oncologist*, **13**(9): 993–1000.

Findlow, J. (2013). Meningococcal group B vaccines. *Hum Vacc Immunother*, **9**(6): 1387–8.

Findlow, J., Borrow, R., Snape, M.D., *et al.* (2010). Multicenter, open-label, randomized phase II controlled trial of an investigational recombinant meningococcal serogroup B vaccine with and without outer membrane vesicles, administered in infancy. *Clin Infect Dis*, **51**(10): 1127–37.

Gardner, P. (2006). Prevention of meningococcal disease. *N Engl J Med*, **355**(14): 1466–73.

Goodman, M.J., Nordin, J.D., Belongia, E.A., *et al.* (2010). Henoch–Schönlein purpura and polysaccharide meningococcal vaccine. *Pediatrics*, **126**(2): e325–9.

Gottfredsson, M., Reynisson, I.K., Ingvarsson, R.F. *et al.* (2011). Comparative long-term adverse effects elicited by invasive group B and C meningococcal infections. *Clin Infect Dis*, **53**(9): e117–24.

Haber, P., Sejvar, J., Mikaeloff, Y., and DeStefano, F. (2009). Vaccines and Guillain–Barré syndrome. *Drug Saf*, **32**(4): 309–23.

He, X., Yu, C., Zhao, P., *et al.* (2013). The genetics of Henoch–Schönlein purpura: a systematic review and meta-analysis. *Rheumatol Int*, **33**(6): 1387–95.

Howitz, M., Krause, T.G., Simonsen, J.B., *et al.* (2007). Lack of association between group B meningococcal disease and autoimmune disease. *Clin Infect Dis*, **45**(10): 1327–34.

Keiser, P.B. and Broderick, M. (2012). Meningococcal polysaccharide vaccine failure in a patient with C7 deficiency and a decreased anti-capsular antibody response. *Hum Vacc Immunother*, **8**(5): 582–6.

Keyserling, H., Papa, T., Koranyi, K., *et al.* (2005). Safety, immunogenicity, and immune memory of a novel meningococcal (groups A, C, Y, and W-135) polysaccharide diphtheria toxoid conjugate vaccine (MCV-4) in healthy adolescents. *Arch Pediatr Adolesc Med*, **159**(10): 907–13.

Lambert, E.M., Liebling, A., Glusac, E., and Antaya, R.J. (2003). Henoch-Schonlein purpura following a meningococcal vaccine. *Pediatrics*, **112**(6 Pt. 1): e491.

Maiden, M.C. (2013). The impact of protein-conjugate polysaccharide vaccines: an endgame for meningitis? *Philos Trans R Soc Lond B Biol Sci*, **368**(1623): 20120147.

Nadel, S. (2012). Prospects for eradication of meningococcal disease. *Arch Dis Child*, **97**(11): 993–8.

Nikibakhsh, A.A., Houshmand, M., Bagheri, M., *et al.* (2012). MEFV gene mutations (M694V, V726A, M680I, and A744S) in Iranian children with Henoch-Schönlein purpura. *Pneumologia*, **61**(2): 84–7.

Novartis. (2013). Novartis receives EU approval for Bexsero®, first vaccine to prevent the leading cause of life-threatening meningitis across Europe. Available from: http://www.novartis.com/newsroom/media-releases/en/2013/1672036.shtml [last accessed 11 December 2014].

Stehr-Green, P., Radke, S., Kieft, C., *et al.* (2008). The risk of simple febrile seizures after immunisation with a new group B meningococcal vaccine, New Zealand. *Vaccine*, **26**(6): 739–42.

Stein, D.M., Robbins, J., Miller, M.A., *et al.* (2006). Are antibodies to the capsular polysaccharide of *Neisseria meningitides* group B and *Escherichia coli* K1 associated with immunopathology? *Vaccine*, **24**(3): 221–8.

Valdivielso-Ramos, M., Velázquez, D., Tortoldeo, A., and Hernanz, J.M. (2011). Penfigoide ampolloso infantil en relación con la vacunación hexavalente, meningococo y neumococo. *An Pediatr*, **75**(3): 199–202.

Zonneveld-Huijssoon, E., Ronaghy, A., Van Rossum, M.A., *et al.* (2007). Safety and efficacy of meningococcal C vaccination in juvenile idiopathic arthritis. *Arthritis Rheum*, **56**(2): 639–46.

20 Pneumococcal Vaccines and Autoimmune Phenomena

Elisabetta Borella,[1,2] Nancy Agmon-Levin,[2,3] Andrea Doria,[1] and Yehuda Shoenfeld[2,4]

[1]Division of Rheumatology, Department of Medicine, University of Padua, Padua, Italy
[2]Zabludowicz Center for Autoimmune Diseases, Sheba Medical Center, Tel Hashomer, Israel
[3]Sackler Faculty of Medicine, Tel Aviv University, Tel Aviv, Israel
[4]Incumbent of the Laura Schwarz-Kipp Chair for Research of Autoimmune Diseases, Sackler Faculty of Medicine, Tel Aviv University, Tel Aviv, Israel

Introduction

Streptococcus pneumoniae (Pneumococcus) is the main cause of bacterial community-acquired pneumonia and meningitis in Western countries (Jit, 2010), as well as of more than 800 000 childhood death in developing countries (O'Brien *et al.*, 2009). The impact of Pneumococcal infections on morbidity and mortality of the general population since the 1900s has forced the medical community to find an effective weapon against this bacterium (Hitoshi *et al.*, 1991). Today, three antipneumococcal vaccines are available: two conjugated to a protein carrier (PCV7 and PCV13) and one not conjugated (PPV23). The latter was licensed in 1983 and consists of the capsular polysaccharides of 23 different *Streptococcus pneumoniae* sero types (1, 2, 3, 4, 5, 6B, 7F, 8, 9N, 9V, 10A, 11A, 12F, 14, 15B, 17F, 18C, 19F, 19A, 20, 22F, 23F, and 33F) that elicit the production of antibodies. However, the humoral response induced by PPV acts in a T cell-independent manner and does not elicit immunological memory, so there is no anamnestic or booster response to revaccination. Moreover, it seems that reimmunization is associated with a decreased response (De Roux *et al.*, 2008). For these reasons, PPV23 is believed to be less effective in younger children and is usually administered to adults (>65 years). Lately, two new conjugated vaccines against Pneumococcus have been introduced: in 2000, a conjugate vaccine against seven of the most frequent serotypes associated with invasive disease (4, 6B, 9V, 14,18C, 19F, and 23F: PCV7) was licensed for use in children under 5 years of age in the United States, while in 2010 a PCV13-valent was licensed directed at serotypes 1, 3, 4, 5, 6A, 6B, 7F, 9V, 14, 18C, 19A, 19F, and 23F.

The two conjugated vaccines differ from the nonconjugated one in that they are composed of polysaccharides that are chemically activated to make saccharides. These saccharides are bound to CRM197, a nontoxic variant of diphtheria toxin, to form a glycol conjugate. The CRM197 derives from cultures of *Corynebacterium diphtheriae* strain C7 (β197) grown in a casamino acid and yeast extract-based medium. The individual glycoconjugates are then purified by ultrafiltration and column chromatography. Another difference is the presence of an adjuvant aluminium (0.125 mg) in PCVs.

Generally, invasive pneumococcal disease (IPD) occurs in children under 2 years of age and it is mainly associated with infection with 14, 23F,19F, 6B, 6A, 9V, 18C, 4, and 19A (Zangwill *et al.*, 1996) serotypes. The introduction of 7-valent vaccine in 2001 had a substantial impact on IPD in all age groups, with a decrease of 69% among children less than 2 years old, 32% among adults 20–39

Vaccines and Autoimmunity, First Edition. Edited by Yehuda Shoenfeld, Nancy Agmon-Levin, and Lucija Tomljenovic.

years old, 8% in adults 40–64 years old, and 18% in older adults. Moreover, since 2004, a drop in hospital admission due to IPD has been observed (Whitney *et al.*, 2003). The 13-valent pneumococcal conjugate vaccine was licensed recently and it has replaced PCV7 for the prevention of otitis media and IPD (American Academy of Pediatrics Committee on Infectious Diseases, 2010). This vaccine contains the same serotypes of PCV7 plus serotypes 1, 3, 5, 6A, 7F, and 19A, thus encompassing 92% of the serotypes that cause IPD in children under 5 years of age in the United States. Studies of its efficacy have not yet been performed, but results from phase III clinical trials in infants show a similar immune response and side effects profile to those seen with PCV7 (Dinleyici and Yargic, 2009; Vanderkooi *et al.*, 2012), with the additional property of eliciting an antibody response against the six additional pneumococcal serotypes. On the other hand, the 23-valent vaccine seems to have mild or no effect on IPDs in adult patients, as reported in a recent matched case–control study (Melegaro and Edmunds, 2004; Vila-Corcoles *et al.*, 2012). In fact, it appears that PPV is protective against IPD in the general elderly population (65%, 95% CI: −49 to +92%) but offers a mild protection in the high-risk elderly (20%; 95% CI: −188 to −78%) (Melegaro and Edmunds, 2004). Pneumococcal vaccination does not significantly alter the risk of overall pneumococcal pneumonia (16% in the general elderly), especially among subjects older than 75 years (Vila-Corcoles *et al.*, 2012).

According to US Center for Disease Control and Prevention (CDC) guidelines (ACIP, 2013a,b), different protocols are recommended for each pneumococcal vaccine. Pneumococcal polysaccharide (PPV23) vaccination is recommended for adults over 65 years old or for adults less than 65 years with chronic diseases (involving lung, heart, kidney, and liver), diabetes mellitus, or addiction to alcohol. It is also recommended to caregivers in nursing homes or long-term care facilities and to smokers (ACIP, 2013a,b). Subjects vaccinated with PPV23 for these diseases require one-time revaccination 5 years after the first dose (ACIP, 2013a,b). In addition, subjects that receive one or two doses of PPV23 before age 65 years should receive another dose of the vaccine at age 65 years or later, if at least 5 years have passed since their previous dose (ACIP, 2013a,b). Pneumococcal conjugate vaccines are considered routine vaccines during childhood (ACIP, 2013a,b). A series of PCV13 vaccine are administered at ages 2, 4, and 6 months, with a booster at age 12–15 months. PCV13 or PPV23 is also administered to adults with immunocompromising conditions, chronic renal failure, nephrotic syndrome and functional or anatomic asplenia, cerebrospinal fluid (CSF) leaks, or cochlear implants. When PCV13 is also indicated, it should be given first (one dose), and at least 8 weeks later a dose of PPV23 should follow (ACIP, 2013a,b).

Pneumococcal vaccine safety

The safety profile of pneumococcal vaccines is reassuring (Paradiso, 2011; Black *et al.*, 2000). The most common complaint after vaccination is mild local reactions, which usually last no more than 1 or 2 days and are present in up to 50% of people immunized with the adjuvanted vaccines (PCVs) (Bryant *et al.*, 2010), and less so in those immunized with the nonadjuvanted PPV. In the latter case, this local reaction is present in 9% of people vaccinated intramuscularly and in 24% of those immunized subcutaneously (Cook *et al.*, 2007). Notably, the incidence of local reactions tends to increase with revaccination (Cook *et al.*, 2007). Among subjects receiving either PCV or PCV, 80–85% complain of systemic reactions, such as fever, irritability, decreased appetite, and increased or decreased sleep (Jackson *et al.*, 2013). In contrast to local reactions, symptoms like arthralgia, arthritis, myalgia, paresthesia, and fatigue are more frequently reported in subjects who have undergone PPV, which may be related to the different age groups immunized with the different vaccines. Serious side effects like anaphylactic and anaphylactoid reactions, serum sickness, and thrombocytopenia are rare. The concurrent administration of other routine childhood vaccinations does not increase the rate of side effects, nor does it reduce the elicited immune response to pneumococcal vaccines (Bryant *et al.*, 2010). However, in recent years, concerns regarding the long-term safety of vaccines in general and those targeting autoimmune phenomena in particular have been expressed. Thus, in the current study, we aimed to search the evidence regarding autoimmune phenomena following immunization with pneumococcal vaccines.

Methods

We searched PubMed using the term "Pneumococcal Vaccine" as well as various combinations of different keywords, such as "Pneumovax," "7-Valent" or "13-Valent," "23-Valent," and "Autoimmune or Rheumatic Disease." We included epidemiological studies performed mainly in patients with systemic lupus erythematosus (SLE) and rheumatoid arthritis (RA).

Manual searches using the related link facility extended the number of references identified, and additional references were identified by cross-checking the reference lists of the identified publications. We excluded from our research reports of reactions following coadministration of pneumococcal vaccine with other vaccines. Notably, only five cases of adverse events following coadministration of vaccines including Pneumovax were reported (Houston, 1983; Thaler, 2008; Hafiji *et al.*, 2010; Valdivielso-Ramos *et al.*, 2011).

Results

We have not found any epidemiological study investigating specifically autoimmune adverse events post-pneumococcal vaccination. However, during a period of 33 years (1980–2013), only 14 case reports addressed the plausibility of the appearance of rheumatic disorders following immunization with a pneumococcal vaccine (Table 20.1). Among these, six reports described reactivation of an autoimmune disorder following immunization with pneumococcal vaccine. Three subjects had some kind of autoimmune reaction after previous vaccination. The mean age of the patients was 50 (the youngest was 17 years old, the oldest was 87), and nine subjects (64.3%) were men. The autoimmune disorders reported in the 14 cases can be divided into dermatological, neurological, and hematological phenomena. Intriguingly, almost all the reported cases are related to the nonadjuvanted 23-valent vaccine, which has been on the market for the last 3 decades.

In addition, several studies of larger cohorts have been performed in order to define the safety of pneumococcal vaccine among patients with autoimmune diseases (Klippel *et al.*, 1979; Battafarano *et al.*, 1998; Elkayam *et al.*, 2002, 2007; Tarján *et al.*, 2002; Pisoni *et al.*, 2003; Heijstek *et al.*, 2011; Shunsuke *et al.*, 2013). These studies followed a total of 313 adults with SLE and 98 with RA. The results showed pneumococcal immunization was safe, with a smaller profile of adverse reactions than the one observed among healthy subjects. Moreover, the last EULAR recommendation about vaccines in pediatric populations concluded that pneumococcal vaccines were safe for children with autoimmune disorders (Heijstek *et al.*, 2011).

Discussion

We have reviewed a period of more than 30 years, during which many millions of pneumococcal immunizations were given. In this period, only 14 cases of autoimmune adverse event post-pneumococcal immunization were reported, as summarized in Table 20.1 – an observation which shows that the risk of autoimmunity following this vaccine is extremely low, if not inconsequential. In these 14 cases, rheumatic diseases were reported a few days following immunization, and in at least six of them, an autoimmune disorder had already been diagnosed in the past, and an exacerbation was suggested (Schulman *et al.*, 1979; Kelton, 1981; Citron and Moss, 1982; Torri *et al.*, 1992; Neil, 1994). However, it should be noted that Pneumovax is regularly recommended to patients with idiopathic thrombocytopenic purpura (ITP) and to other patients prior to performing a splenectomy, while reoccurrence of ITP following splenectomy may occur regardless of the vaccine used. Therefore, these cases seem to describe a rebound of the disease already diagnosed in the patients, rather than the consequence of vaccination.

Recently, great interest has been given to adjuvant effects in humans, since a new disease has been described by Shoenfeld and Agmon-Levin, the autoimmune/inflammatory syndrome induced by adjuvants (ASIA) (Shoenfeld and Agmon-Levin, 2011). According to the authors, exposure to an external stimulus in genetically predisposed individuals may induce the onset of an autoimmune disorder. In fact, initially, patients might develop nonspecific symptoms such as sleeping disturbances, myalgia, arthralgia, extreme fatigue, memory loss, and irritable bowel syndrome (IBS), which may subsequently evolve into a well-defined connective tissue disease (Shoenfeld and Agmon-Levin, 2011). Common exposure to adjuvants follows silicone implantation and vaccination, especially utilizing adjuvanted vaccines. The PPV is a nonadjuvanted vaccine, so it is not surprising that so few cases of autoimmunity following immunization with this vaccine have been reported. Accordingly, local reactions after administration of the pneumococcal vaccines have proven to be more frequent after PCV than PPV, showing that the adjuvanted

Table 20.1 Report cases of autoimmunity following pneumococcal vaccination

Reference	Age (years)	Gender	Concomitant diseases	PPV or PCV	Adverse reactions to vaccines in the past	Disease	Time of onset	Therapy
Citron and Moss (1982)	17	Male	ITP	PPV		Rebound ITP	2 days	CS
de la Monte *et al.* (1986)	59	Male	Hairy-cell leukemia	PPV	Swine flu	Peripheral and central demyelinating disease	72 hours	CS
Fox and Peterson (1998)	57	Male		PPV		Rash, migratory arthralgia, leucocytoclastic vasculitis	2 weeks	NSAID
Friedland and Wittels (1983)	50	Male	Hairy-cell leukemia	PPV	Swine flu	GBS	72 hours	CS
Kelton (1981)	38	Male	COPD and bronchiectasis	PPV	Influenza vaccination	ITP	2 weeks	CS
Kelton (1981)	48	Female	ITP	PPV		Rebound ITP	10 days	CS, CT
Kikuchi *et al.* (2002)	67	Female	RA	PPV		Minimal-change nephritic syndrome	1 week	CS
Kitazawa *et al.* (2012)	87	Male	Prostate adenocarcinoma	PPV		*Neuromyelitis* optica	Few days	CS
Maddox and Motley (1990)	36	Female		PPV		Sweet's syndrome	4 hours	CS
Neil (1994)	29	Male	ITP and autoimmune hemolytic anemia	PPV		Rebound ITP and autoimmune hemolytic anemia	2 days	
Ries and Shemonsky (1981)						SLE		
Schulman *et al.* (1979)	75	Female	AILD	PPV		Reactivation AILD		CT
Schulman *et al.* (1979)	62	Male	AILD	PPV		Reactivation AILD		CT
Torri *et al.* (1992)	27	Male	Autoimmune hemolytic anemia	PPV		Rebound autoimmune hemolytic anemia	8 days	CS

AILD, angioimmunoblastic lymphadenopathy; COPD, *chronic obstructive pulmonary disease;* CS, corticosteroid; NSAID, nonsteroidal antiinflammatory drug; CT, chemotherapy; ITP, idiopathic thrombocytopenic purpura; PCV, pneumococcal conjugate vaccine; PPV, pneumococcal polysaccharide vaccine; GBS, Guillain–Barré syndrome; SLE, systemic lupus erythematosus vaccines have a higher tendency to trigger the immune system (Cook *et al.*, 2007; Bryant *et al.*, 2010). Nevertheless, we did not find data on autoimmune reactions after the administration of PCVs. The lack of data regarding adverse reactions after the 7- and 13-valent vaccines, despite the presence of aluminum in their composition, can be explained by a variety of reasons. In fact, while PPV has been administered for more than 30 years, PCV has been on the market for just 10, and has been mainly used in children. In addition, in all the epidemiological studies, the side effects were evaluated for only a few months following vaccination; therefore, it may be possible that some adverse reactions requiring more time to develop were missed. Moreover, symptoms like sleeping disturbances, fatigue, memory loss, and arthralgia, cannot be easily recognized in the pediatric population, which is commonly immunized with PCVs. Finally, none of the clinical studies reported in the literature focused on the onset of undefined rheumatic phenomena after PCV vaccination, so the absence of a correlation between PCV administration and the development of ASIA remains to be proven.

In conclusion, according to the current data, the pneumococcal vaccines, and particularly PPV, seem to be efficacious and safe for both healthy and diseased subjects (Heijstek *et al.*, 2011).

References

ACIP (Advisory Committee on Immunization Practice). (2013). Recommended immunization schedule for person aged 0 through 18 years – United States, 2013. *MMWR*, **62**(1): 2–8; erratum **62**(13): 256.

ACIP (Advisory Committee on Immunization Practice). (2013). Recommended immunization schedule for adults aged 19 years and older – United States, 2013. *MMWR*, **62**(1): 9–19.

American Academy of Pediatrics Committee on Infectious Diseases. (2010). Recommendations for the prevention of *Streptococcus pneumoniae* infections in infants and children: use of 13 valent pneumococcal conjugate vaccine (PCV13) and pneumococcal polysaccharide vaccine (PPSV23). *Pediatrics*, **126**: 186–90.

Battafarano, D.F., Battafarano, N.J., Larsen, L, *et al.* (1998). Antigen-specific antibody responses in lupus patients following immunization. *Arthritis Rheum*, **41**: 1828–34.

Black, S., Shinefield, H., Fireman, B. *et al.* (2000). Efficacy, safety and immunogenicity of heptavalent pneumococcal conjugate vaccine in children. Northern California Kaiser Permanente Vaccine Study Center Group. *Pediatr Infect Dis J*, **19**(3): 187–95.

Bryant, K.A., Block, S.L., Baker, S.A., *et al.* (2010). Safety and immunogenicity of a 13-valent pneumococcal conjugate vaccine. *Pediatrics*, **125**: 866–75.

Citron, M.L. and Moss, B.M. (1982). Pneumococcal-vaccine-induced thrombocytopenia. *JAMA*, **248**: 1178.

Cook, I.F., Pond, D., and Hartel, G. (2007). Comparative reactogenicity and immunogenicity of 23 valent pneumococcal vaccine administered by intramuscular or subcutaneous injection in elderly adults. *Vaccine*, **25**: 4767–74.

de la Monte, S.M., Ropper, A.H., Dickersin, G.R., *et al.* (1986). Relapsing central and peripheral dmyelinating diseases. Unusual pathologic features. *Arch Neurol*, **43**: 626–9.

de Roux, A., Schmole-Thoma, B., Siber, G.R., *et al.* (2008). Comparison of pneumococcal conjugate polysaccharide and free polysaccharide vaccines in elderly adults: conjugate vaccine elicits improved antibacterial immune responses and immunological memory. *Clin Infect Dis*, **46**: 1015–23.

Dinleyici, E.C. and Yargic, Z.A. (2009). Current knowledge regarding the investigational 13-valent pneumococcal conjugate vaccine. *Expert Rev Vaccines*, **8**: 977–86.

Elkayam, O., Paran, D., Caspi, D., *et al.* (2002). Immunogenicity and safety of pneumococcal vaccination in patients with rheumatoid arthritis and systemic lupus erythematosus. *Clin Infect Dis*, **34**: 147–53.

Elkayam, O., Ablin, J., and Caspi, D. (2007). Safety and efficacy of vaccination against streptococcus pneumonia in patients with rheumatic diseases. *Autoimmun Rev*, **6**: 312–14.

Fox, B.C. and Peterson, A. (1998). Leukocytoclastic vasculitis after pneumococcal vaccination. *Am J Infect Control*, **26**: 365–6.

Friedland, M.L. and Wittels, E.G. (1983). An unusual neurologic reaction following polyvalent pneumococcal vaccine in a patient with hairy cell leukemia. *Am J Hematol*, **14**: 189–91.

Hafiji, J., Bhogal, B., Rytina, E., and Burrows, N.P. (2010). Bullous pemphigoid in infancy developing after the first vaccination. *Clin Exp Dermatol*, **35**: 940–1.

Heijstek, M.W., Ott de Bruin, L.M., Bijl, M., *et al.* (2011). EULAR recommendation for vaccination in paediatric patients with rheumatic diseases. *Ann Rheum Dis*, **70**: 1704–12.

Hitoshi, S., Ito, Y., Takehara, K., *et al.* (1991). A case of malignant hypertension and scleroderma after cosmetic surgery. *Jpn J Med*, **30**: 97–100.

Houston, T.P. (1983). Small-vessel vasculitis following simultaneous influenza and pneumococcal vaccination. *NY State J Med*, **83**: 1182–3.

Jackson, L.A., Gurtman, A., van Cleeff, M., *et al.* (2013). Immunogenicity and safety of a 13 valent pneumococcal conjugate vaccine compared to a 23-valent pneumococcal polysaccharidevaccine in pneumococcal vaccine-naive adults. *Vaccine*, **31**: 3577–84.

Jit, M. (2010). The risk of sequelae due to pneumococcal meningitis in high-income countries: a systematic review and meta-analysis. *J Infect*, **61**: 114–24.

Kelton, J.G. (1981). Vaccination-associated relapse of immune thrombocytopenia. *JAMA*, **245**: 369–70.

Kikuchi, Y., Imakiire, T., Hyodo, T., *et al.* (2002). Minimal change nephrotic syndrome, lymphadenopathy and hyperimmunoglobulinemia after immunization with a pneumococcal vaccine. *Clin Nephrol*, **58**: 68–72.

Kitazawa, Y., Warabi, Y., Bandoh, M., *et al.* (2012). Elderly-onset neuromyelitis optica which developed after the diagnosis of prostate adenocarcinoma and relapsed after a 23-valent pneumococcal polysaccharide vaccination. *Intern Med*, **51**: 103–7.

Klippel, J.H., Karsh, J., Stahl, N., *et al.* (1979). A controlled study of pneumococcal polysaccharide vaccine in systemic lupus erythematosus. *Arthritis Rheum*, **22**: 1321–5.

Maddox, P.R. and Motley, R.J. (1990). Sweet's syndrome: a severe complication of pneumococcal vaccination following emergency splenectomy. *Br J Surg*, **77**: 809–10.

Melegaro, A. and Edmunds, W.J. (2004). The 23-valent pneumococcal polysaccharide vaccine. Part I. Efficacy of PPV in the elderly: a comparison of meta-analyses. *Eur J Epidemiol*, **19**: 353–63.

Neil, V.S. (1994). Long term management after splenectomy: revaccination may cause relapse. *BMJ*, **308**(6924): 339.

Paradiso, P.R. (2011). Advances in pneumococcal disease prevention: 13-valent pneumococcal conjugate vaccine for infants and children. *Clin Infect Dis*, **52**: 1241–7.

O'Brien, K.L., Wolfson, L.J., Watt, J.P., *et al.* (2009). Burden of disease caused by *Streptococcus pneumoniae* in children younger than 5 years: global estimates. *Lancet*, **374**(9693): 893–902.

Pisoni, C., Sarano, J., Benchetrit, G., *et al.* (2003). [Antipneumococcal vaccination in patient with systemic lupus erythematosus]. *Medicina (B Aires)*, **63**: 388–92.

Ries, K. and Shemonsky, N.K. (1981). Acute lupus erythematosus (SLE) following polyvalent pneumococcal vaccine. *S D J Med*, **34**: 27–8.

Schulman, P., Budman, D.R., Vinciguerra, V.P., and Degnan, T.J. (1979). B-cell activation in angioimmunoblastic lymphadenopathy after immunisation with multivalent pneumococcal polysaccharide vaccine. *Lancet*, **2**(8152): 1141.

Shoenfeld, Y. and Agmon-Levin, N. (2011). "ASIA" – autoimmune/inflammatory syndrome induced by adjuvants. *J Autoimmun*, **36**: 4–8.

Shunsuke, M., Yukitaka, U., Yukihiro, A., *et al.* (2013). Pneumococcal polysaccharide vaccination in rheumatoid arthritis patients receiving tocilizumab therapy. *Ann Rheum Dis*, **72**: 1362–6.

Tarján, P., Sipka, S., Maródi, L., *et al.* (2002). No short-term immunological effects of Pneumococcus vaccination in patients with systemic lupus erythematosus. *Scand J Rheumatol*, **31**(4): 211–15.

Thaler, A. (2008). Miller fisher syndrome in a 66-year-old female after flu and pneumovax vaccinations. *J Am Med Dir Assoc*, **9**: 283–4.

Torri, O., Veyssier, P., Kaloustian, E., and Reis-Borges, R. (1992). Acute hemolysis after early pneumococcal revaccination in a patient splenectomized for autoimmune hemolytic anemia. *Presse Med*, **21**: 1873–4.

Valdivielso-Ramos, M., Velazquez, D., Tortoledo, A., and Hernanz, J.M. (2011). Infantile bullous pemphigoid developing after hexavalent, meningococcal and pneumococcal vaccinations. *An Pediatr (Barc)*, **75**: 199–202.

Vanderkooi, O.G., Scheifele, D.W., Girgenti, D., *et al.* (2012). Safety and immunogenicity of a 13-valent pneumococcal conjugate vaccine in healthy infants and toddlers given with routine pediatric vaccinations in Canada. *Pediatr Infect Dis J*, **31**: 72–7.

Vila-Corcoles, A., Ochoa-Gondar, O., Rodriguez-Blanco, T., *et al.* (2012). Clinical effectiveness of 23-valent pneumococcal polysaccharide vaccine against pneumonia in patients with chronic pulmonary diseases: a matched case-control study. *Hum Vaccin Immunother*, **8**: 639–44.

Whitney, C.G., Farley, M.M., Hadler, J., *et al.* (2003). Decline in invasive pneumococcal disease after the introduction of protein-polysaccharide conjugate vaccine. *N Engl J Med*, **348**: 1737–46.

Zangwill, K.M., Vadheim, C.M., Vannier, A.M., *et al.* (1996). Epidemiology of invasive pneumococcal disease in southern California: implications for the design and conduct of a pneumococcal conjugate vaccine efficacy trial. *J Infect Dis*, **174**: 752–9.

21 BCG and Autoimmunity

Luigi Bernini, Carlo Umberto Manzini, and Clodoveo Ferri

Rheumatology Unit, Department of Internal Medicine, University of Modena and Reggio Emilia, Medical School, Modena, Italy

Introduction

The bacillus Calmette–Guérin (BCG) is a live, attenuated vaccine from *Mycobacterium bovis* obtained by Albert Calmette and Camille Guerin through 230 *in vitro* passages between 1908 and 1921, when it was first used in Paris. Since then, BCG has been distributed all over the word as a preventive tool against tuberculosis (TB), although its use has changed over time as different epidemiological conditions have arisen, and despite debate about its preventive effect reaching only 51% risk reduction. Indeed, the BCG vaccination shows a positive protection on extrapulmonary TB and mortality in children but has a variable and only partial prevention against pulmonary disease in adults (Colditz *et al.*, 1994; Trunz *et al.*, 2006). Alongside this current main use, other nonspecific immunological effects of BCG have become evident over the years, leading to its application in the clinical management of several diseases (Table 21.1). Since 1976, intravesical instillation of BCG has represented the chief immunotherapy against bladder cancer (Morales *et al.*, 1976), and at present it is an integral and approved part of the management of non-muscle invasive bladder cancer. Other neoplasms, such as colorectal cancer, lung cancer, and melanoma, have been the targets of BCG immunotherapy, especially in the past. Because of their immunologic background, a variety of diseases (asthma, type 1 diabetes, multiple sclerosis (MS), leprosy) have been evaluated for treatment with BCG (Ritz *et al.*, 2013).

Immune mechanisms of BCG

Although not fully clarified, the protective effect of BCG is produced by a robust cellular immunity response that occurs through the involvement of $CD4^+$ T cells and the Th-1 cytokines IFNγ, TNFβ, and IL17 (Figure 21.1). Subsequently, $CD8^+$ T cells are also involved in the immunization process, while IL-2 has a role in central memory T cell responses (McShane, 2011).

Similar mechanisms, although with different biologic meanings, are evoked when BCG invokes antitumor immunity (Mosolits *et al.*, 2005). In this context, the local treatment of bladder cancer with BCG instillations represents an interesting model, although one that is not yet fully understood. When the BCG is instilled in the bladder, it binds to fibronectin present at the endoluminal layer and forms fibronectin–BCG complexes embedded in both normal and tumor cells. The BCG antigens, expressed at the surface of urothelial, neoplastic, and antigen-presenting cells (APCs), trigger an early, large influx of neutrophils (75%) and macrophages. This in turn induces a Th1 response via $CD4^+$ T cells, producing IFN-γ, TNF-β, IL-2, and IL-12. Finally, this Th1 response activates the NK cells cytotoxic T lymphocytes (CTLs), which destroy the neoplastic cells. A Th2 cytokine profile (IL-4, -5, -6, and -10) is also produced, but Th1 response seems to exert a more favorable antitumor action. Specific effects of some interleukins are as follows: IL-2 has a stimulatory effect on CTLs, IL-18 activates CTLs and NK cells, and IL-8, produced by activated macrophages, has a recruiting effect on PMN granulocytes, which promote

Vaccines and Autoimmunity, First Edition. Edited by Yehuda Shoenfeld, Nancy Agmon-Levin, and Lucija Tomljenovic.

Table 21.1 Clinical applications of BCG. Aside from its current use in vaccination against tuberculosis and leprosy, BCG has been proposed as a possible immunotherapy treatment in both neoplastic and autoimmune diseases

BCG vaccination	BCG immunotherapy	
	Cancer	Autoimmune diseases
Tuberculosis Leprosy	Bladder cancer	Asthma Type 1 diabetes Multiple sclerosis

secretion of TNF-related apoptosis-inducing ligand (TRAIL), a member of TNF family that induces apoptosis in neoplastic but not in normal cells. A functional host immune system is a necessary prerequisite to achieving an effective antitumor response (Kresowik and Griffith, 2009; Zuiverloon *et al.*, 2012; Brincks *et al.*, 2013); however, a limited number of cancer cells, a juxtaposition of BCG and tumor cells, and an appropriate dose of BCG are also essential factors – all fulfilled in bladder cancer – for an effective immunotherapy action (Gandhi *et al.*, 2013).

Reactive arthritis (ReA) is an aseptic synovitis triggered by several pathogens of either the genitourinary or the gastrointestinal tract, including intravesical BCG (Bernini *et al.*, 2013). Indeed, ReA may share many of the immunological mechanisms involved in intravesical BCG therapy. As a first step, a CD4 T cell response occurs at the site of inflammation, stimulated and maintained by bacterial components, with the subsequent participation of CD8 T cells. However, the Th1 cytokine response (IFN-γ, IL-2, -12) is unable to remove the triggering bacterial agent, so a prevalence of Th2 response (IL-4 and -10) is produced, keeping the bacteria in the joints alive. It is conceivable that HLA-B27, often present in patients with ReA, may act as a restriction molecule for antigenic bacterial peptides, presented to and crossrecognized by cytotoxic $CD8^+$ T lymphocytes, which allows the persistence of the inflammatory process even after the elimination of the bacteria (Colmegna *et al.*, 2004). So-called "molecular mimicry" is usually invoked to explain the autoimmune effects of infectious microorganisms; in the case of intravesical BCG, the shared homology between mycobacterial HSP65 and cartilage proteoglycan link protein might be determining. Thus, the

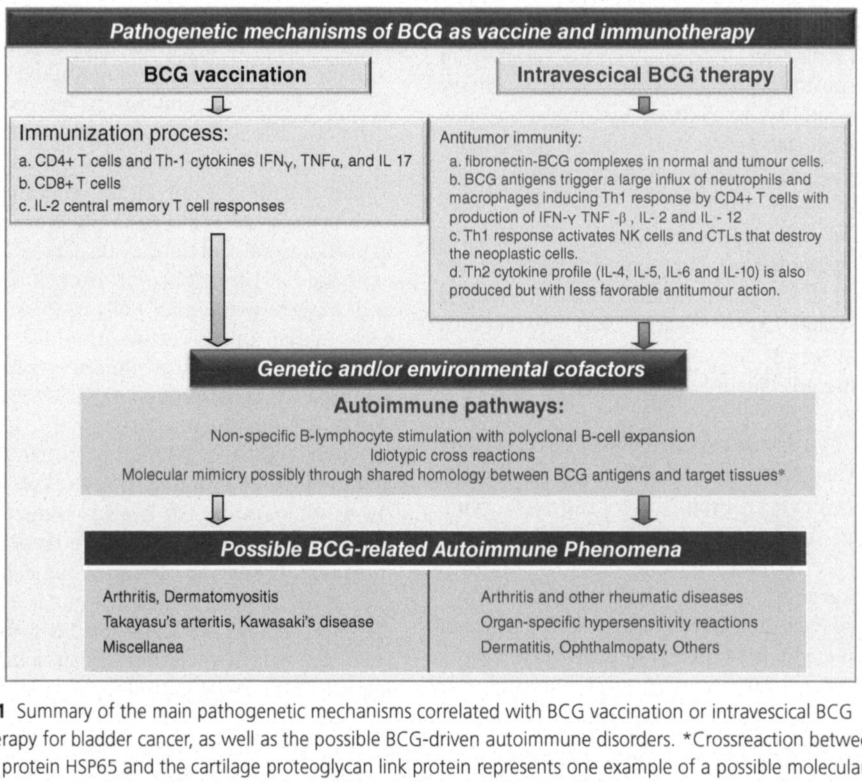

Figure 21.1 Summary of the main pathogenetic mechanisms correlated with BCG vaccination or intravescical BCG immunotherapy for bladder cancer, as well as the possible BCG-driven autoimmune disorders. *Crossreaction between the BCG heat-shock protein HSP65 and the cartilage proteoglycan link protein represents one example of a possible molecular mimicry pathogenetic mechanism. (For a color version of this figure, please see color plate section.)

bacteria or bacterial antigens spreading from the bladder to the circulation may produce a systemic immune-mediated response with the joints as target, especially in the presence of genetic predisposition (positive HLA-B27 antigen) (Tishler and Shoenfeld, 2006).

Clinical applications of BCG

Tuberculosis

The specific, preventive anti-TB use of BCG has revealed some interesting nonspecific effects of its immunogenic action. One of these is a significant reduction in infant and child mortality attributable to causes other than TB. Since 1940, and continuing into the last 10 years, several observations have been made of this effect in high-mortality countries. Similarly, BCG immunization at birth reduces the death from respiratory infections and sepsis. It is possible that BCG immunization at birth favorably influences the antibody response to routine immunizations administered later in infancy (vaccinations against pneumococcal, *Haemophilus influenzae* type B, tetanus toxoid, and hepatitis B surface antigen) (Ritz *et al.*, 2013).

Leprosy

Leprosy is caused by infection with *Mycobacterium leprae*. The most common preventive strategy relies on BCG vaccination, which was initially used for both TB and leprosy (Duthie *et al.*, 2011). BCG's antileprosy effect shares the same protective immune mechanism as its effect against TB, with increased production of Th-1 cytokines: IFN-γ, TNF-α, and IL-17. The protective impact of BCG against leprosy is variable, with an overall protection of 41–76% for multibacillary and 62% for paucibacillary forms, and with 41% for experimental trial, 60% for observational studies, 53% for general population, and 68% for contact subjects. Despite this variability, there is clear-cut evidence that BCG exerts a preventive effect against leprosy, mostly in the populations at highest risk for household contact. Although this effect declines with time, it persists over 30 years or more. Early diagnosis and multidrug therapy are the key points of leprosy control, but BCG vaccine retains its important preventive role (Merle *et al.*, 2010).

Asthma and other hypersensitivity diseases

BCG has a stimulatory effect on Th1 cytokines and inhibits the Th2 immunologic response, and consequently it counteracts the risk of Th2-dependent atopic diseases (Rousseau *et al.*, 2008). Based on this, an epidemiological association between BCG vaccination in early life and asthma and other atopic conditions has been evaluated, with conflicting results. Some recent meta-analyses have failed to prove a significant protective effect of BCG against allergy in general, atopic dermatitis or eczema, allergic rhinoconjunctivitis, urticaria, or food allergy, although they found it may provide some benefits against asthma (Afshar *et al.*, 2011; Arnoldussen *et al.*, 2011). However, other studies did not confirm the suggested protective role of BCG vaccination against asthma and/or other allergic conditions (Flohr *et al.*, 2012).

Type 1 diabetes

Type 1 diabetes results in part from impaired activation of transcription factor NF-κB in a subpopulation of activated T cells, which become more sensitive to TNF-α-induced apoptosis. The cytokine TNF-α selectively kills autoreactive T cells – namely, the only disease-causing cells – allowing pancreas regeneration. The ability of BCG to stimulate the production of TNF-α, producing similar results to those obtained with TNF-α administration, has supported its use as a possible treatment of type 1 diabetes (Kodama *et al.*, 2005).

In the past, immunotherapy with a single, low dose of BCG in patients with late-stage pre-diabetes was shown to induce a clinical remission in some patients (Shehadeh *et al.*, 1994), but these results were not confirmed in recent studies of newly diagnosed diabetic children evaluated for 18 (Elliott *et al.*, 1998) and 24 (Allen *et al.*, 1999) months post-vaccination. However, in a recent controlled trial, two low-dose BCG vaccinations in patients with long-term type 1 diabetes resulted in a rapid release of dead insulin autoreactive T cells, as a consequence of TNF-α activity, and in a significant increase in C-peptide secretion, as an expression of a restored β-cell function. Interestingly, similar results were documented in a placebo patient with unexpected Epstein–Barr virus (EBV) infection, a well-known trigger of innate immunity, by inducing a host TNF-α response (Faustman *et al.*, 2012).

Multiple sclerosis

MS is a neurological autoimmune disorder whose specific mechanisms remain to be fully elucidated (Noseworthy *et al.*, 2000). The rationale for the use of BCG in MS rests on the documented suppression of autoimmune responses in experimental autoimmune encephalomyelitis, an animal model of human MS (Lee *et al.*, 2008). Among the complex immune-mediated processes responsible for MS, there seems to be a conflict between the clinical results of anti-TNF-α treatments and the biology of TNF-α. This cytokine is abnormally increased in the cerebrospinal fluid (CSF) and active lesions of patients with MS, but TNF-α agonists may worsen MS and induce demyelinating syndromes in other autoimmune diseases. This apparent paradoxical effect can be explained by the presence of a single-nucleotide polymorphism (rs1800693) in the TNFRSF1A gene, which encodes tumor necrosis factor receptor 1 (TNFR1), which is associated with MS but not with other autoimmune diseases. The MS risk allele directs the expression of a new, soluble form of TNFR1, an MS-associated TNFR1 variant, that can block TNF, mimicking the effect of TNF-α antagonists in rheumatoid arthritis (RA), psoriasis, and Crohn's disease (Gregory *et al.*, 2012).

A trial of BCG vaccine in MS showed evidence of reduced disease activity over 6 months in 14 patients with relapsing–remitting MS (Ristori *et al.*, 1999). There was a consistent and significant reduction (57%) in the mean number of active MRI lesions from the run-in period to the post-BCG period, without serious side effects.

Cancer

The observation that patients with TB were less frequently affected by cancer led to a study of the effect of live, attenuated BCG as an adjuvant immunotherapy (mostly with autologous tumor cells) for a variety of neoplasms, including colorectal cancer, lung cancer, melanoma, and renal cell carcinoma. The underlying hypothesis is that increasing the immunogenicity of tumor cells with adjuvants will enhance immune responses to the endogenous tumor antigens. The prerequisite for a positive outcome was a limited tumor expansion. However, because of the lack of documented efficacy, in several studies such a use was gradually abandoned for the more concrete prospective of novel and specific vaccines. Indeed, the molecular definition of cancer-associated antigens introduced the possibility of specific vaccines and a new era in genetically engineered whole-cell vaccination has involved the modification of tumor cells through transfer of genes encoding cell-membrane immunostimulatory molecules or cytokines.

To date, BCG is the most common intravesical therapy for the treatment of non-muscle-invasive urothelial carcinoma, and it is able to reduce the recurrence rate and the risk of progression to muscle-invasive disease in patients with carcinoma *in situ* (gold-standard therapy), as well as to treat superficial bladder tumors (Kresowik and Griffith, 2009). The immunologic mechanisms of BCG therapy have already been discussed; it is interesting to note that the local immunogenic action of intravesical BCG is much more powerful if compared with the systemic effect of the intradermal route in other malignancies. Intravesical BCG is commonly administered once weekly for 6 weeks as induction following transurethral resection of the bladder tumor. The optimal maintenance schedules are to be determined. Most include 3-weekly instillations at 3-month intervals for 3 years, but reduced doses and durations have recently been proposed (Oddens *et al.*, 2013).

Autoimmune phenomena produced by BCG

Like any vaccine, BCG can be described as a "double-acting tool" because its own immunogenicity produces a preventive effect in a variety of diseases on one hand and may trigger a number of autoimmune phenomena on the other.

Although several immune-mediated adverse events triggered by the BCG vaccination have been reported thus far, a clear causal link has not been demonstrated. Furthermore, some cases might be correlated to other environmental factors or represent normal responses to foreign proteins. Conceptually, an immune-mediated response against self could result from host response to any component of the vaccine (Koenig *et al.*, 2011). In the presence of genetic and environmental factors, the host immune response may be triggered by various mechanisms, mainly nonspecific B-lymphocyte stimulation with polyclonal B cell expansion, idiotypic crossreaction, and molecular mimicry (Figure 21.1).

BCG vaccination for TB is safe and well tolerated, with a very low rate of side effects (ranging from local reactions to disseminated BCG disease), which can have various degrees of severity and are mostly of an infectious nature. To a large extent, the more serious complications occur in immunocompromised subjects, as in HIV-infected infants. Equally rare, although speculatively more intriguing, are the post-vaccine autoimmune phenomena, which, however, have different clinical significances according to the route of administration (systemic-intradermal or local-intravesical).

Intradermal BCG

Arthritis

Arthritis following cancer immunotherapy has been described in 10 of 159 patients treated for various types of malignancy. The main clinical pattern was a bilateral and symmetric polyarthritis, which was predominantly confined to small joints of the hands, preceded by morning stiffness, and associated with elevated inflammatory markers; it occurred within 1–5 months from first BCG injection and was exacerbated by subsequent BCG injection. The arthritis partially recovered in 1–3 months with nonsteroid antiinflammatory drugs treatment. Synovial fluid analysis and biopsy, when conducted, were negative for acid-fast bacilli and indicative of nonspecific chronic inflammation (Torisu *et al.*, 1978). An acute inflammatory polyarthritis with skin maculopapular rash was reported in a healthy woman following intradermal BCG. The culture of synovial fluid was repeatedly negative; after failure of antibiotic multitherapy, the clinical symptoms promptly and fully subsided with corticosteroid treatment (Kodali and Clague, 1998).

Dermatomyositis

This autoimmune condition is rarely reported as a consequence of BCG vaccination. Two cases (one after revaccination) have been diagnosed by skin and muscle biopsy (Kåss *et al.*, 1978). Generally, a substantial increase in the incidence of dermatomyositis or polymyositis after any kind of vaccine has not been reported.

Takayasu's arteritis

Based on historical, epidemiological, and immunologic background, a link has been proposed between Takayasu's arteritis (TA) and TB/BCG (Kothari, 1995). Both diseases present a granulomatous histological feature: patients with TA produce both humoral and cell-mediated responses against antigens of *M. tuberculosis* and have a higher reactivity of peripheral lymphocytes to heat-shock protein (HSP) 65 kDa, which is remarkably induced in the media and the vasa vasorum of TA biopsies. However, this intriguing linkage has been poorly studied and, at present, is not confirmed (Arnaud *et al.*, 2011).

Kawasaki's disease

A reaction at the BCG inoculation site (erythema, induration, or crust formation) is common, especially in Japan, and occurs in about 30–50% of Kawasaki's disease (KD) patients. Although this sign is more frequent than cervical lymphadenopathy in children with complete KD aged 3–20 months, it is not included among diagnostic criteria. The skin reaction appears early (24–48 hours) after febrile onset and is replaced by a crust immediately after the fever subsides. It is considered an important tool for early diagnosis in febrile children, especially in those younger than 2 years. It has been hypothesized that crossreactivity may occur between specific epitopes of mycobacterial HSP65 and the human homolog, HSP65 (Lai *et al.*, 2013).

Miscellaneous adverse events

Several pathological conditions are associated with BCG vaccination, usually as anecdotal reports. Immediate hypersensitivity reactions are associated with sensitization to dextran; delayed reactions, such as maculopapular exanthems, erythema nodosum, urticarial vasculitis, and neutrophilic dermatoses (Sweet's syndrome and pyoderma gangrenosum), rarely occur (Bellet and Prose, 2005). Cutaneous sarcoidosis (Osborne *et al.*, 2003), psoriasis skin lesions (Takayama *et al.*, 2008), guttate psoriasis-like lesions (Koca *et al.*, 2004), and an extensive ulcerating vasculitis, defined as granulomatous type one (type IV hypersensitivity reaction) (Ghattaura *et al.*, 2009), have all been reported after BCG vaccination. A special mention must be given to the relationship of BCG with MS, a disease in which the specific immunization might be either protective or inductive. To our knowledge, a causal link between BCG vaccination and MS is referred to in only one case (Miller *et al.*, 1967), while reports of MS following vaccine against hepatitis B virus (HBV) are much more common (Aharon-Maor and Shoenfeld, 2000). Moreover, no significantly increased risk of developing MS after vaccination for BCG was found in a recent review (Farez and Correale, 2011). In the past, cases of polyneuropathy following BCG vaccination or revaccination were occasionally reported (Katznelson *et al.*, 1982; Wilmshurst *et al.*, 1999).

Intravesical BCG

Side effects from intravesical BCG are rare, being documented in less than 5% of cases and generally classified as local and systemic, infective, or secondary to hypersensitivity response. Because they mainly occur after systemic absorption of the vaccine, correct administration procedures must be followed, and the vaccine must not be given if contraindicated (Babjuk *et al.*, 2013). An analysis of the septic adverse events, either local or systemic, is not the purpose of this survey, which is focused on the autoimmune reactions induced by BCG. There are some barriers to a correct review of the literature, especially where the data are from a urologic source. For instance, articular symptoms are often referred to simply as "arthralgia" or "arthritis," without any further detail, and are associated to an unspecified "skin rash" and defined as "possible allergic reactions." Similarly, side effects such as granulomatous pneumonia and hepatitis are reported without sufficient data to support the possible link with intravesical BCG and the autoimmune mechanism. In such cases, it may be difficult to clearly establish whether they represent an infectious complication or a hypersensitivity reaction to BCG. Indeed, histological and cultural findings from affected organs are often unable to demonstrate the presence of acid-fast bacilli. In this context, however, some interesting data still arise.

Arthritis and other rheumatic diseases

Inflammation of the joints is a well-known side effect of intravesical BCG, reported with a variable frequency, ranging from 0.5 to 28.1% in various series (Bernini *et al.*, 2013). Its clinical pattern is a polyarthritis comparable to arthritis triggered by other agents and generally characterized by typical clinical features of classical ReA. Intravesical BCG-related ReA is characterized by the evidence of a preceding infection with a known trigger bacterium of manifest source site. The articular inflammation is aseptic, and it invariably shows latency from the antigen exposure. The large joints of the lower limbs are by far the most involved, whether alone or in association, and asymmetry is the prevalent pattern. Inflammatory low back pain is also reported, albeit infrequently. Enthesitis, tendinitis, bursitis, and dactylitis are typically present, although never as a unique reactive feature. Fever is the most common associated symptom, followed by typical hypersensitivity symptoms such as conjunctivitis, urethritis, uveitis, balanitis, and ketatoderma. A definite genetic link is documented by HLA-B27 carriers in about half of the documented cases, supporting the idea that, when major histocompatibility complex (MHC) class I associations occur, failure of immunological tolerance can lead in combination with environmental factors (such as a microbial agent in the bladder) to immune-mediated inflammatory reaction in the joints. Nonsteroidal antiinflammatory drugs (NSAIDs) and/or corticosteroids are the mainstay of pharmacologic treatment, which is largely effective, and arthritis usually has a benign course, with poor tendency to chronicity (Bernini *et al.*, 2013).

It is worth noting the comparison with polyarthritis secondary to immunotherapy with intradermal BCG (Torisu *et al.*, 1978; Kodali and Clague, 1998) reported in a few cases; their different articular phenotypes might be correlated to the distinct routes of administration and quite different immunogenic actions.

Other anecdotal autoimmune rheumatic diseases have been reported following intravesical BCG, particularly Sjögren-like syndrome, confirmed by salivary gland scintigraphy and biopsy of the submandibular gland showing a lymphoplasmacytoid sialoadenitis without granuloma (Narváez *et al.*, 2003); polymyalgia rheumatica with temporal artheritis (Genereau *et al.*, 1996); remitting seronegative symmetrical synovitis pitting oedema (El Mahou *et al.*, 2006); and cryoglobulinemic vasculitis (Granel *et al.*, 2004).

Organ-specific hypersensitivity reactions

Provided that the lack of any microbiological evidence of BCG dissemination does not exclude an infection with *M. tuberculosis* or mycobacteria other than *M. bovis*, several reports of noninfective organ involvement may indicate an underlying hypersensitivity mechanism. Similar cases have been found in intradermal BCG immunotherapy for neoplasms other than bladder cancer, since viable mycobacteria were not detected in the lesions. Pneumonitis/hepatitis was reported with a prevalence of 0.7% in a series of 2602 patients treated with intravesical BCG for superficial bladder cancer, but was ascribed to systemic BCG infection (Lamm *et al.*, 1992). Although it is debated whether these events are really a result of sterile hypersensitivity reaction, the lack of evidence for infection by acid-fast bacilli and the prompt response to steroid treatment (alone or added to antitubercular drugs) supports this hypothesis. Pneumonitis is generally described as interstitial and bilateral, with a radiological appearance of ground glass opacites and multiple pulmonary nodules and with expression of noncaseating sterile granulomas (Israel-Biet *et al.*, 1987; Horinaga *et al.*, 1999; Orikasa *et al.*, 2003; Um *et al.*, 2009; Davies *et al.*, 2012). Moreover, granulomatous (noncaseating epithelioid granulomas with Langhans giant cells) hepatitis may be concurrently associated with lung involvement. Cultures, PCR analysis, and Ziehl-Neelsen staining performed in blood, bone marrow, urine, feces, biopsy samples, and broncho-alveolar lavage failed to isolate *Mycobacterium* species (Molina *et al.*, 1992; Uetsuki *et al.*, 2011; Valentini *et al.*, 2012). A granulomatous, sterile hepatitis is also described as a single complication of intravesical BCG, although much more rarely than an infectious one (Van Outryve *et al.*, 2004). Even the kidney may be the target of hypersensitivity reaction to intravesical BCG, which can induce an interstitial nephritis with non-necrotizing, sterile granulomas responsive to steroid therapy (Kennedy *et al.*, 2006; Manzanera Escribano *et al.*, 2007). Anecdotally, renal involvement may be observed in association with pneumonitis and hepatitis in the same patient (Kiely *et al.*, 2011). In these cases, corticosteroid therapy is always effective, either as monotherapy or as second-line treatment after the failure of antitubercular drugs.

An interesting report suggested a possible involvement of both pathogenetic mechanisms, namely infection and hypersensitivity, in the same patient, with multiorgan involvement and a significant time course. First, granulomatous hepatitis with positive PCR for mycobacterial DNA and aortic valve vegetations occurred following intravesical BCG; these manifestations were successfully treated with three-drug antitubercular therapy. Several days later, cutaneous leucocytoclastic vasculitis, leucopenia, thrombocytopenia, antinuclear antibodies, and rheumatoid factor appeared together with a persistent impaired cholestasis. For a possible hepatotoxic effect, ethambutol and rifampin were stopped while ofloxacine was added to isoniazid. Three months later, the granulomatous lesions were spread to the lungs (interstitial micronodular pattern on chest CT), spleen (multiple foci on abdominal ultrasound), bone marrow, and liver (extensive non-caseating granulomata at liver biopsy, with negative stains and cultures for acid-fast bacilli). In addition, marked hypercalcemia was documented for the first time. Therefore, corticosteroid therapy was given with complete recovery of the systemic features after 15-month follow-up. The authors suggested both an early infective phase (PCR positive and response to anti-TB treatment), probably triggered by too short a time interval between surgery and intravesical BCG, and a late hypersensitivity reaction (vasculitis, cytopenias, antinuclear antibodies, and rheumatoid factor seropositivity) with disseminated, sterile granulomatosis, responsible for hypercalcemia due to activation of the reticuloendothelial system, as in sarcoidosis. Concordant with their pathogenetic hypothesis, this second phase of disease showed a marked response to low-dose corticosteroid therapy (Schattner *et al.*, 2002).

Skin

Not-otherwise-specified "rashes" are reported in 0.3% of patients (Lamm *et al.*, 1992). However, cutaneous complications of intravesical BCG have seldom been described in anecdotal observations (Bernini *et al.*, 2013). Id reaction or auto-eczematization (a dermatitis appearing at a distance from the initial site of infection or sensitization) was noted in a patient as an intensely pruritic, scaly, erythematous eruption involving all extremities 2 weeks after the start of weekly intravesical use of BCG therapy. The skin biopsy showed spongiotic dermatitis with overlying scaling and an eosinophilic infiltrate. The eruption resolved with topical corticosteroids (Lowther *et al.*, 2013). Finally, vitiligo has also been reported as an autoimmune side effect of intravesical BCG (Beisland and Holsen, 2004).

Ophthalmopathy

Uveitis can be caused by intraocular infections or autoimmune stimulation. Several reports, also after BCG vaccination, have focused on ocular complications of intravesical BCG, but, not infrequently, a clear-cut exclusion of infective origin is lacking, probably due to the low sensitivity of the diagnostic tools. However, some cases show a complete recovery with topical or oral steroid therapy. Notably, when searched, HLA-B27 is negative (Wertheim and Astbury, 2002; Garip *et al.*, 2009; Uppal *et al.*, 2010), unlike in patients with uveitis and ReA (7.7%), who are invariably HLA-B27-positive (Bernini *et al.*, 2013). An autoimmune retinopathy associated with intravesical BCG therapy has also been reported (Sharan *et al.*, 2005).

Other

Among other reports is one case of anaphylactoid purpura in the lower legs and one of Henoch–Schönlein purpura occurring after intravesical BCG therapy in the absence of any concomitant possible cause (Nan *et al.*, 2005; Hirayama *et al.*, 2008).

BCG vaccination in autoimmune diseases

In spite of the TB control achieved in Central Europe, the disease still represents a major health problem around the world, particularly in developing countries, high-income countries, Latin America, and the Western Pacific region (WHO, 2009). Although it has variable efficacy and is largely ineffective in the most common lung infection, the BCG vaccination remains the only currently licensed TB vaccine. The risk of contracting TB and the protective effect of BCG must be weighed against the possible harms of vaccination. In patients with RA, corticosteroids, disease-modifying drugs, and, especially, TNF-α-blocking agents are all associated with increased risk of TB, although this is reduced by the screening procedures for latent TB. In ankylosing spondylitis patients treated with infliximab, etanercept, and adalimumab, no cases of active TB have been reported (van Assen *et al.*, 2011a,b). A EULAR evidence-based review recently concluded that BCG vaccination is not recommended in patients with autoimmune inflammatory rheumatic diseases, according to the following considerations: the incidence of active TB is increased in patients with autoimmune rheumatic diseases, especially when treated with immunosuppressive drugs (mainly TNF-α-blocking agents); these cases are mostly reactivations of latent TB that could not be prevented by the vaccine; the protective effect of BCG vaccination in adults is not clearly demonstrated; and BCG vaccine carries attenuated, live mycobacteria, which, in immunocompromised patients, might increase the risk of disseminated TB (van Assen *et al.*, 2011a,b). However, some exceptions should be considered, such as health care workers exposed to resistant TB (Bijl *et al.*, 2012).

References

Afshar, M. and Morris-Jones, R. (2011). A novel approach to the management of inflammatory bacille Calmette-Guérin reactions using topical steroids. *Br J Dermatol*, **164**: 1402–3.

Aharon-Maor, A. and Shoenfeld, Y. (2000). The good, the bad and the ugly of vaccination. *IMAJ*, **2**(3): 225–7.

Allen, H.F., Klingensmith, G.J., Jensen, P., *et al.* (1999). Effect of bacillus Calmette-Guerin vaccination on new-onset type 1 diabetes. A randomized clinical study. *Diabetes Care*, **22**: 1703–7.

Arnaud, L., Haroche, J., Mathian, A., *et al.* (2011). Pathogenesis of Takayasu's arteritis: a 2011 update. *Autoimmun Rev*, **11**(1): 61–7.

Arnoldussen, D.L., Linehan, M., and Sheikh, A. (2011). BCG vaccination and allergy: a systematic review and meta-analysis. *J Allergy Clin Immunol*, **127**(1): 246–53.

Babjuk, M., Burger, M., Zigeuner, R., *et al.* (2013). EAU guidelines on non-muscle-invasive urothelial carcinoma of the bladder: update 2013. *Eur Urol*, **64**(4): 639–53.

Beisland, C. and Holsen, D.S. (2004). Vitiligo – an autoimmune side-effect of intravesical bacillus Calmette-Guérin instillation? *Scand J Urol Nephrol*, **38**(2): 182–3.

Bellet, J.S. and Prose, N.S. (2005). Skin complications of bacillus Calmette-Guérin immunization. *Curr Opin Infect Dis*, **18**(2): 97–100.

Bernini, L., Manzini, C.U., Giuggioli, D., *et al.* (2013). Reactive arthritis induced by intravesical BCG therapy for bladder cancer: our clinical experience and systematic review of the literature. *Autoimmun Rev*, **12**(12): 1150–9.

Bijl, M., Agmon-Levin, N., Dayer, J.M., *et al.* (2012). Vaccination of patients with auto-immune inflammatory rheumatic diseases requires careful benefit-risk assessment. *Autoimmun Rev*, **11**(8): 572–6.

Brincks, E.L., Risk, M.C., and Griffith, T.S. (2013). PMN and anti-tumor immunity – the case of bladder cancer immunotherapy. *Semin Cancer Biol*, **23**(3): 183–9.

Colditz, G.A., Brewer, T.F., Berkey, C.S., *et al.* (1994). Efficacy of BCG vaccine in the prevention of tuberculosis. *Metaanalysis of the published literature. JAMA*, **271**: 698–702.

Colmegna, I., Cuchacovich, R., and Espinoza, L.R. (2004). HLA-B27-associated reactive arthritis: pathogenetic and clinical considerations. *Clin Microbiol Rev*, **17**(2): 348–69.

Davies, B., Ranu, H., and Jackson, M. (2012). Pulmonary complications of intravesicular BCG immunotherapy. *Thorax*, **67**(10): 933–4.

Duthie, M.S., Gillis, T.P., and Reed, S.G. (2011). Advances and hurdles on the way toward a leprosy vaccine. *Hum Vaccin*, **7**(11): 1172–83.

El Mahou, S., Popa, L., Constantin, A., *et al.* (2006). Remitting seronegative symmetrical synovitis pitting oedema after BCG instillation. *Clin Rheumatol*, **25**(4): 566–7.

Elliott, J.F., Marlin, K.L., and Couch, R.M. (1998). Effect of bacille Calmette-Guerin vaccination on C-peptide secretion in children newly diagnosed with IDDM. *Diabetes Care*, **21**: 1691–3.

Farez, M.F. and Correale, J. (2011). Immunizations and risk of multiple sclerosis: systematic review and meta-analysis. *J Neurol*, **258**(7): 1197–206.

Faustman, D.L., Wang, L., Okubo, Y., *et al.* (2012). Proof-of-concept, randomized, controlled clinical trial of Bacillus-Calmette-Guerin for treatment of long-term type 1 diabetes. *PLoS One*, **7**(8): e41756.

Flohr, C., Nagel, G., Weinmayr, G., *et al.* (2012). Tuberculosis, bacillus Calmette-Guérin vaccination, and allergic disease: findings from the International Study of Asthma and Allergies in Childhood Phase Two. *Pediatr Allergy Immunol*, **23**(4): 324–31.

Gandhi, N.M., Morales, A., and Lamm, D.L. (2013). Bacillus Calmette-Guérin immunotherapy for genitourinary cancer. *BJU Int*, **112**(3): 288–97.

Garip, A., Diedrichs-Möhring, M., Thurau, S.R., *et al.* (2009). Uveitis in a patient treated with Bacille-Calmette-Guérin: possible antigenic mimicry of mycobacterial and retinal antigens. *Ophthalmology*, **116**(12): 2457–62.

Genereau, T., Koeger, A.C., Chaïbi, P., and Bourgeois, P. (1996). Polymyalgia rheumatica with temporal arteritis following intravesical Calmette-Guérin bacillus immunotherapy for bladder cancer. *Clin Exp Rheumatol*, **14**(1): 110.

Ghattaura, A., Eley, K.A., Molenaar, E., and Smith, G. (2009). A case of extensive ulcerating vasculitis following a BCG vaccination. *Plast Reconstr Aesthet Surg*, **62**(8): e286–9.

Granel, B., Serratrice, J., Morange, P.E., *et al.* (2004). Cryoglobulinemia vasculitis following intravescical instillations of bacillus Calmette-Guerin. *Clin Exp Rheumatol*, **22**: 481–2.

Gregory, A.P., Dendrou, C.A., Attfield, K.E., *et al.* (2012). TNF receptor 1 genetic risk mirrors outcome of anti-TNF therapy in multiple sclerosis. *Nature*, **488**(7412): 508–11.

Hirayama, T., Matsumoto, K., Tsuboi, T., *et al.* (2008). Anaphylactoid purpura after intravesical therapy using bacillus Calmette-Guerin for superficial bladder cancer. *Hinyokika Kiyo*, **54**(2): 127–9.

Horinaga, M., Nakamura, K., Nishiyama, T., and Murai, M. (1999). A case of interstitial pneumonitis caused by intravesical bacillus Calmette-Guérin instillation. *Hinyokika Kiyo*, **45**(7): 493–5.

Israel-Biet, D., Venet, A., Sandron, D., *et al.* (1987). Pulmonary complications of intravesical Bacille Calmette-Guérin immunotherapy. *Am Rev Respir Dis*, **135**(3): 763–5.

Kåss, E., Straume, S., and Munthe, E. (1978). Dermatomyositis after B.C.G. vaccination. *Lancet*, **1**(8067): 772.

Katznelson, D., Gross, S., and Sack, J. (1982). Polyneuritis following BCG re-vaccination. *Postgrad Med J*, **58**(682): 496–7.

Kennedy, S.E., Shrikanth, S., Charlesworth, J.A. (2006). Acute granulomatous tubulointerstitial nephritis caused by intravesical BCG. *Nephrol Dial Transplant*, **21**(5): 1427–9.

Kiely, B., McLaughlin, A.M., Lynch, T.H., and Keane, J. (2011). Intravesical bacille Calmette-Guérin-induced multiorgan failure after treatment for transitional cell carcinoma. *Scand J Urol Nephrol*, **45**(4): 278–80.

Koca, R., Altinyazar, H.C., Numanoğlu, G., and Unalacak, M. (2004). Guttate psoriasis-like lesions following BCG vaccination. *J Trop Pediatr*, **50**(3): 178–9.

Kodali, V.R. and Clague, R.B. (1998). Arthritis after BCG vaccine in a healthy woman. *J Intern Med*, **244**(2): 183–4.

Kodama, S., Davis, M., and Faustman, D.L. (2005). The therapeutic potential of tumor necrosis factor for autoimmune disease: a mechanistically based hypothesis. *Cell Mol Life Sci*, **562**: 1850–62.

Koenig, H., Sutherland, A., Izurieta, H.S., and McGonagle, D. (2011). Application of the immunological disease continuum to study autoimmune and other inflammatory events after vaccination. *Vaccine*, **29**: 913–19.

Kothari, S.S. (1995). Aetiopathogenesis of Takayasu's arteritis and BCG vaccination: the missing link? *Med Hypotheses*, **45**(3): 227–30.

Kresowik, T.P. and Griffith, T.S. (2009). Bacillus Calmette–Guerin immunotherapy for urothelial carcinoma of the bladder. *Immunotherapy*, **1**(2): 281–8.

Lai, C.C., Lee, P.C., Wang, C.C., *et al.* (2013). Reaction at the bacillus Calmette–Guérin inoculation site in patients with Kawasaki disease. *Pediatr Neonatol*, **54**(1): 43–8.

Lamm, D.L., van der Meijden, P.M., Morales, A., *et al.* (1992). Incidence and treatment of complications of bacillus Calmette-Guerin intravesical therapy in superficial bladder cancer. *J Urol*, **147**(3): 596–600.

Lee, J., Reinke, E.K., Zozulya, A.L., *et al.* (2008). Mycobacterium bovis bacille Calmette-Guérin infection in the CNS suppresses experimental autoimmune encephalomyelitis and Th17 responses in an IFN-gamma-independent manner. *J Immunol*, **181**(9): 6201–12.

Lowther, C., Miedler, J.D., and Cockerell, C.J. (2013). Id-like reaction to BCG therapy for bladder cancer. *Cutis*, **91**(3): 145–6, 151.

Manzanera Escribano, M.J., Morales Ruiz, E., Odriozola Grijalba, M., *et al.* (2007). Acute renal failure due to interstitial nephritis after intravesical instillation of BCG. *Clin Exp Nephrol*, **11**(3): 238–40.

McShane, H. (2011). Tubercolosis vaccines: beyond bacille Calmette-Guerin. *Phil Trans R Soc B*, **366**: 2782–9.

Merle, C.S., Cunha, S.S., and Rodrigues, L.C. (2010). BCG vaccination and leprosy protection: review of current evidence and status of BCG in leprosy control. *Expert Rev Vaccines*, **9**: 209–22.

Miller, H., Cendrowski, W., and Shapira, K. (1967). Multiple sclerosis and vaccination. *Br Med J*, **2**(5546): 210–13.

Molina, J.M., Rabian, C., D'Agay, M.F., and Modai, J. (1992). Hypersensitivity systemic reaction following intravesical bacillus Calmette-Guerin: successful treatment with steroids. *J Urol*, **147**(3): 695–7.

Morales, A., Eidinger, D., and Bruce, A.W. (1976). Intracavitary bacillus Calmette-Guerin in the treatment of superficial bladder tumors. *J Urol*, **116**(2): 180–3.

Mosolits, S., Ullenhag, G., and Mellstedt, H. (2005). Therapeutic vaccination in patients with gastrointestinal malignancies. A review of immunological and clinical results. *Ann Oncol*, **16**: 847–62.

Nan, D.N., Fernández-Ayala, M., García-Ibarbia, C., *et al.* (2005). Henoch-Schönlein purpura after intravesical administration of bacillus Calmette-Guérin. *Scand J Infect Dis*, **37**(8): 613–15.

Narváez, J., Castro-Bohorquez, F.J., and Vilaseca-Momplet, J. (2003). Sjögren's-like syndrome following intravesical bacillus Calmette-Guérin immunotherapy. *Am J Med*, **115**(5): 418–20.

Noseworthy, J.H., Lucchinetti, C., Rodriguez, M., and Weinshenker, B.G. (2000). Multiple sclerosis. *N Engl J Med*, **343**: 938–52.

Oddens, J., Brausi, M., Sylvester, R., *et al.* (2013). Final results of an EORTC-GU cancers group randomized study of maintenance bacillus Calmette-Guérin in intermediate- and high-risk Ta, T1 papillary carcinoma of the urinary bladder: one-third dose versus full dose and 1 year versus 3 years of maintenance. *Eur Urol*, **63**(3): 462–72.

Orikasa, K., Namima, T., Ota, S., *et al.* (2003). Acute eosinophilic pneumonia associated with intravesical bacillus Calmette-Guérin therapy of carcinoma in situ of the bladder. *Int J Urol*, **10**(11): 622–4.

Osborne, G.E., Mallon, E., and Mayou, S.C. (2003). Juvenile sarcoidosis after BCG vaccination. *J Am Acad Dermatol*, **48**(5 Suppl.): S99–102.

Ristori, G., Buzzi, M.G., Sabatini, U., *et al.* (1999). Use of bacilli Calmette-Guèrin (BCG) in multiple sclerosis. *Neurology*, **53**(7): 1588–9.

Ritz, N., Mui, M., Balloch, A., and Curtis, N. (2013). Non-specific effect of bacille Calmette-Guérin vaccine on the immune response to routine immunisations. *Vaccine*, **31**(30): 3098–103.

Rousseau, M.C., Parent, M.E., and St-Pierre, Y. (2008). Potential health effects from non-specific stimulation of the immune function in early age: the example of BCG vaccination. *Pediatr Allergy Immunol*, **19**: 438–48.

Schattner, A., Gilad, A., and Cohen, J. (2002). Systemic granulomatosis and hypercalcaemia following intravesical bacillus Calmette-Guérin immunotherapy. *J Intern Med*, **251**(3): 272–7.

Sharan, S., Thirkill, C.E., and Grigg, J.R. (2005). Autoimmune retinopathy associated with intravesical BCG therapy. *Br J Ophthalmol*, **89**(7): 927–8.

Shehadeh, N., Calcinaro, F., Bradley, B.J., *et al.* (1994). Effect of adjuvant therapy on development of diabetes in mouse and man. *Lancet*, **343**: 706–7.

Takayama, K., Satoh, T., Hayashi, M., and Yokozeki, H. (2008). Psoriatic skin lesions induced by BCG vaccination. *Acta Derm Venereol*, **88**(6): 621–2.

Tishler, M. and Shoenfeld, Y. (2006). BCG immunotherapy – from pathophysiology to clinical practice. *Expert Opin Drug Saf*, **5**(2): 225–9.

Torisu, M., Miyahara, T., Shinohara, K., *et al.* (1978). A new side effetct of BCG immunitherapy – BCG-induced arthritis in man. *Cancer Immunol Immunother*, **5**: 77–83.

Trunz, B.B., Fine, P., and Dye, C. (2006). Effect of BCG vaccination on childhood tuberculous meningitis and miliary tuberculosis worldwide: a meta-analysis and assessment of cost-effectiveness. *Lancet*, **367**: 1173e80.

Uetsuki, H., Hirama, H., Matsuoka, Y., and Modai, J. (2011). A case with hepatitis and interstitial pneumonitis caused by intravesical bacillus Calmette-Guérin (BCG) instillation. *Nihon Hinyokika Gakkai Zasshi*, **102**(5): 691–5.

Um, S.J., Lee, S.K., and Yang, D.K. (2009). Hypersensitivity pneumonitis following intravesical bacille Calmette-Guérin immunotherapy for superficial bladder cancer. *J Investig Allergol Clin Immunol*, **19**(3): 230–2.

Uppal, G.S., Shah, A.N., Tossounis, C.M., and Tappin, M.J. (2010). Bilateral panuveitis following intravesical BCG immunotherapy for bladder carcinoma. *Ocul Immunol Inflamm*, **18**: 292–6.

Valentini, C.G., Bozzoli, V., Larici, A.R., *et al.* (2012). Systemic granulomatous reaction secondary to treatment of bladder cancer with bacillus calmette-guerin. *Mediterr J Hematol Infect Dis*, **4**(1).

van Assen, S., Agmon-Levin, N., Elkayam, O., *et al.* (2011a). EULAR recommendations for vaccination in adult patients with autoimmune inflammatory rheumatic diseases. *Ann Rheum Dis*, **70**(3): 414–22.

van Assen, S., Elkayam, O., Agmon-Levin, N., *et al.* (2011b). Vaccination in adult patients with auto-immune inflammatory rheumatic diseases: a systematic literature review for the European League Against Rheumatism evidence-based recommendations for vaccination in adult patients with auto-immune inflammatory rheumatic diseases. *Autoimmun Rev*, **10**(6): 341–52.

Van Outryve, S.M., Francque, S.M., Gentens, P.A., *et al.* (2004). Bacillus Calmette-Guérin-induced granulomatous hepatitis in a patient with a superficial bladder carcinoma. *Eur J Gastroenterol Hepatol*, **16**(10): 1027–32.

Wertheim, M. and Astbury, N. (2002). Bilateral uveitis after intravesical BCG immunotherapy for bladder carcinoma. *Br J Ophthalmol*, **86**(6): 706.

WHO (World Health Organization). (2009). Global Tuberculosis Control. Geneva: World Health Organization.

Wilmshurst, J.M., Macleod, M.J., Hughes, E., and Hughes, R.A. (1999). Acute sensory neuropathy in an adolescent girl following BCG vaccination. *Eur J Paediatr Neurol*, **3**(6): 277–9.

Zuiverloon, T.C.M., Nieuweboer, A.J.M., Vèkony, H., *et al.* (2012). Markers predicting response to bacillus Calmette-Guerin immunotherapy in high-risk bladder cancer patients: a systematic review. *Eur Urol*, **61**: 128–45.

III Autoimmune Diseases Solicited by Vaccination

22 Systemic Lupus Erythematosus Induced by Vaccines

Nurit Katz-Agranov[1] and Gisele Zandman-Goddard[1,2]

[1]Department of Medicine, Wolfson Medical Center, Tel Aviv, Israel
[2]Sackler Faculty of Medicine, Tel Aviv University, Tel Aviv, Israel

Introduction

Immunization of healthy individuals is the most effective way of protecting the public from infections and epidemics and has proven to decrease morbidity and mortality worldwide. Although licensed vaccinations have been proven to be widely safe and effective, there are cases in which they may be associated with adverse effects. Autoimmune manifestations following immunization, though considered a rare side effect, have been documented in otherwise healthy individuals, as have flares in individuals with known autoimmune diseases. Reduced antibody response to vaccinations has also been documented in individuals with autoimmune conditions, and, while it has been suggested by some that the underlying disease is the cause of this phenomenon, others have argueed that it may result from the concurrent use of medications that affect the immune response in these individuals (Chatham *et al.*, 2012). These autoimmune side effects may be associated with humoral response to self-antigens, due to molecular mimicry, epitope spread, bystander activation, or polyclonal triggering, and suggest a strong link between infectious agents and autoimmunity whose pathogenesis is characterized by a complex interaction between genetic and immune defects, and environmental and hormonal factors (Colafrancesco *et al.*, 2013). Recently, adjuvants found in various vaccinations have also been suggested to be inducers of immune-mediated conditions, aluminum and silicone being the most common (Blank *et al.*, 2012; Colafrancesco *et al.*, 2013). It is important to state that vaccination has been proven to reduce the burden of infectious disease in patients with autoimmune diseases, and, although live-attenuated vaccines are not recommended for profoundly immune-suppressed patients, other vaccines have adequate safety and efficacy profiles in most studies published to date. Moreover, the immune response to live vaccines is variable in these patients but generally adequate, despite concomitant use of immunosuppressive and biological agents. It is thus important to detect individuals who may be at risk of developing adverse effects and to weigh the benefits against the risks, especially in individuals prone to autoimmune conditions. Physicians should be alerted to this potential association, which may have a long latency period and unique presentations, and should be encouraged to report and analyze such cases.

Vaccination-induced systemic lupus erythematosus or lupus-like syndromes

Systemic lupus erythematosus (SLE) is a severe multisystem autoimmune disease serologically characterized by the production of a variety of autoantibodies (von Muhlen and Tan, 1995; Qiao *et al.*, 2005). Among these, antibodies to double-stranded DNA (dsDNA), especially those of the immunoglobulin G (IgG) isotype, are thought to be diagnostic markers in SLE, and their presence often correlates with disease pathogenesis

Vaccines and Autoimmunity, First Edition. Edited by Yehuda Shoenfeld, Nancy Agmon-Levin, and Lucija Tomljenovic.

(Madaio *et al.*, 1987; Vlahakos *et al.*, 1992; Suzuki *et al.*, 1993; Lefkowith and Gilkeson, 1996). Several factors are involved in the induction of autoantibodies and development of this disease (Elkon, 2000), which is characterized by a complex interaction between genetic and immune defects, and environmental and hormonal factors (Agmon-Levin *et al.*, 2009a,b; Colafrancesco *et al.*, 2013). Infectious agents, as well as various vaccine components, have also been suggested as factors that may contribute to the precipitation or exacerbation of lupus.

Post-vaccination SLE and lupus-like syndromes have been reported throughout the years, ranging from the induction of autoantibodies without concurrent clinical manifestations to full-blown clinical disease. Although considered a rare adverse event, like other autoimmune manifestations, the link between vaccines and SLE has been growing, with reports linking lupus to many vaccines and possibly adjuvants, including hepatitis B virus (HBV), measles, mumps, and rubella (MMR), diphtheria, pertussis, and tetanus (dTP), human papillomavirus (HPV), influenza, bacillus Calmette–Guérin (BCG), pneumococcal vaccinations, and even smallpox (Orbach *et al.*, 2010). In addition, although evidence suggests that most vaccines are safe in patients with SLE, with the exception of live viruses, disease exacerbation post-vaccination has been reported, as has decreased antibody response (Chatham *et al.*, 2012), due to the underlying disease and the frequent use of immunosuppressive drugs (Zandman-Goddard and Shoenfeld, 2005). One analysis demonstrated an overall relative risk (RR) for developing an autoimmune disease of 0.98 (post-HPV vaccination), so no direct statistically significant difference between the groups was encountered. However, when each disease was looked at individually, SLE had the second highest relative risk for an individual event (RR 3.00). Further analysis (of the entire vaccination database) demonstrated that the highest relative risk for an individual event was for SLE (RR 2.39) (Verstraeten *et al.*, 2008; Orbach *et al.*, 2010).

It is important to note for all cases of vaccine-induced SLE that the primary component for adverse effects is variable, and while some studies argue that it is the microbial components, not the adjuvant components, which are considered to be of primary importance, others have demonstrated otherwise (Hamilton *et al.*, 1998; Geier and Geier, 2005).

SLE induced by HBV vaccines

Several reports and studies have tried to link HBV vaccines with clinical SLE manifestations, but they have only shown a temporal relationship (Older *et al.*, 1999). SLE related to HBV vaccines may differ from idiopathic SLE in its clinical presentation and may better resemble drug-induced lupus (Agmon-Levin *et al.*, 2009a,b). Reports published to date have demonstrated that some post-HBV SLE manifestations, which are seen at high prevalence, are typical; these include involvement of the joints, skin, muscles, and photosensitivity and thrombocytopenia. Other manifestations differ in this unique group of SLE patients, such as low rates of kidney and hematologic involvement, and a relatively high rate of hepatitis. Neurological and pulmonary symptoms are also common. In most studies, a female predominance is found, mostly young adults (Hamilton *et al.*, 1998). Overall, SLE patients presented post-HBV vaccination with mild to moderate disease and without life-threatening organ involvement. SLE has been reported to the Vaccine Adverse Event Reporting System (VAERS) with an odds ratio (OR) of 9 : 1. According to an analysis of the UK General Practice Research Database (GPRD) from the year 2000, vaccination against HBV is associated with a 1.6-fold increase in the risk of developing lupus erythematosus; however, it was noted that those vaccinated in Britain were not representative of the general population. Another study, by French and Dutch researchers, found that among people aged over 40, lupus risk was elevated 2.6-fold by the vaccine. The analysis was prompted by 274 reports to the French pharmacovigilance system of autoimmune disorders associated with HBV vaccination – including 32 reports of lupus. The study identified 255 patients on the GPRD who had had at least two diagnoses of lupus or discoid lupus, at least one by a specialist, between 1989 and 1998. Each case was matched to up to 10 controls from the same practice. The overall incidence of lupus was 10.7 per 100 000 patients per year. Incidence was highest in those aged 40–59, and was sixfold higher in women than in men. Results were presented to the International Society for Pharmacoepidemiology in Barcelona in 2005 (ChildHealthSafety, 2011). On the other hand, other studies have demonstrated the safety of this vaccine, such as a case–control study of 265 newly diagnosed lupus patients, which did not show that HBV vaccine was a risk factor for developing SLE (OR 1.4) (Schattner, 2005).

The reported latency period from the first HBV immunization and onset of autoimmune symptoms varies from less than 1 week and to up to 2 years. The classical period between vaccination and autoimmunity was considered to be several weeks, similar to the timeframe suggested in the past for post-infectious autoimmunity phenomena (Older *et al.*, 1999; Agmon-Levin *et al.*, 2009a,b).

Interestingly, many studies have shown that the majority of affected subjects continue to be vaccinated, and aggravation of their condition by additional doses has been documented (Agmon-Levin *et al.*, 2009a,b). One study found that, while 60% received all three vaccinations, only 20% received one inoculation and 20% received two doses. Another case series showed that 70% of patients continued their immunization protocol even though adverse events were documented, while yet another found that 47% of patients continued with the immunization program despite adverse events (Zafrir *et al.*, 2012).

Large-scale epidemiological studies are needed to further elucidate the link between HBV vaccines and lupus. In this regard, it is important to note that, in addition to the potential epitopes in the HBV surface antigen vaccine, adjuvants containing aluminum and mercury may provide potential antigenic stimulation (Schattner, 2005).

SLE induced by HPV vaccines

In 2006, the US Food and Drug Administration (FDA) approved a vaccine for HPV types 6, 11, 16, and 18. Its efficacy in preventing infection exceeds 90%. The two HPV vaccines currently available are Gardasil and Cervarix. Both are composed of HPV-like proteins, each utilizing different adjuvants (Balofsky *et al.*, 2010). The HPV vaccine targets young woman, who are not only at risk for HPV but are in the same age group as those at risk for development of SLE. This has led to an association between HPV and SLE, which may be more than circumstantial, because studies have demonstrated an increased prevalence of HPV in individuals with lupus compared to the general population, after adjusting for other potential risk factors – an association which has been attributed to the immune suppression found in the SLE and which has increased awareness for the need to vaccinate this high-risk population (Lee *et al.*, 2010; Rojo-Contreras *et al.*, 2012; Lyrio *et al.*, 2013). Despite the need to immunize this high-risk population, the decision to do so is not a simple one, as studies have found an association between immunization with HPV vaccines and the appearance of a spectrum of SLE-like conditions, as well as flares in known lupus patients. In a recent study describing five cases of post-HPV-vaccine lupus and one post-vaccine flare-up, many common features were found, such as a personal or family history of autoimmune rheumatic conditions, suggesting genetic or epigenetic contributing components. All patients were young females. Five experienced severe adverse events following a boost immunization (second or third vaccination) but did report mild adverse events with a previous dose that were disregarded. The latency period following immunization ranged from 5 days to 3 weeks. All patients responded favorably to immunosuppressive therapy (Gatto *et al.*, 2013). Another report described three cases: two new onsets of lupus, with a favorable response to immunosuppressive therapy, and one exacerbation of a known disease, leading to death. Typical manifestations of lupus, such as arthralgias, malar rash, and renal involvement, were reported, in addition to elevated titers of antinuclear antibody, other typical autoantibodies, and hyocomplementemia. As in the previous study, all patients were female. Each received two doses of the vaccine prior to initial presentation of the adverse effects, with a latency period of 2–4 months (Soldevilla *et al.*, 2012). In contrast, other studies have shown that the vaccine is in fact well tolerated and reasonably effective in patients with SLE and that it does not induce an increase in lupus activity or flares. An example is a recent case–control study which evaluated 50 patients with SLE and 50 healthy controls, who received Gardasil, a quadrivalent HPV vaccine. This study showed that there were no significant changes in the titers of anti-dsDNA, complements, or anti-C1q, or in disease activity in a 12-month follow-up, and that disease flare-ups were comparable to those of 50 matched SLE controls. In addition, the incidence of adverse events was comparable between patients with SLE and controls (Mok *et al.*, 2013). Due to the conflicting data, it seems that, at present, a carful individualized risk assessment of both patient medical history of autoimmune and infectious diseases and of adverse reactions to past vaccination is required (Bijl *et al.*, 2012), in addition to a close assessment following each boost of vaccination (Agmon-Levin *et al.*, 2009a,b). The high incidence of HPV infection found in patients with SLE compared to the general population, in whom the infection may contribute to disease activity, makes it especially important to assess the risk–benefit balance of the vaccine among these individuals, for its benefits may still outweigh the risk.

SLE induced by influenza vaccines

Routine influenza vaccination of SLE patients seems indicated and safe. As with other vaccinations, there is concern regarding the activation of an autoimmune response, which has been inconsistently described in various reports over the years, ranging from increased autoantibody production to full-blown clinical flare-ups. Such was the case in an older study which found a high incidence of clinical SLE exacerbations in 11 of 46 patients; these were all mild and tended to remit spontaneously (Brodman *et al.*, 1978). Another, older paper reports the case of a female patient who, following vaccination, developed a severe disease exacerbation and *de novo* lupus nephritis (Louie *et al.*, 1978). In a more recent review of 10 studies of 265 SLE patients who received influenza vaccines (with a follow-up period of 4–24 weeks), only six were reported to develop a flare, of whom two had renal involvement (Del Porto *et al.*, 2006; Abu-Shakra *et al.*, 2007; Holvast *et al.*, 2007; Conti *et al.*, 2008). Similar results were found in various other studies (Ristow *et al.*, 1978; Abu-Shakra *et al.*, 2000). Evaluation of autoantibody titers in 103 SLE patients receiving adjuvant- and nonadjuvant-containing H1N1 vaccines found that there was no increase in the levels of SLE-specific autoantibodies in patients with SLE (Urowitz *et al.*, 2011). It has also been shown that, in patients with SLE, the immune response to influenza vaccination leads to a blunted humoral response (Holvast *et al.*, 2007; Saad *et al.*, 2011). A study of influenza-specific antibody responses to vaccination in 72 SLE patients, as well as of disease activity post-vaccination, demonstrated that, compared to high responders to the vaccine, low responders were significantly more likely to have hematologic criteria, to have more American College of Rheumatology classification criteria for SLE, and to be receiving concurrent prednisone treatment. Following vaccination, low responders were also more likely to experience disease flares ($p = 0.01$) and to have increased titers of antinuclear antibodies. Another study, which evaluated the efficacy and safety of influenza vaccine in SLE 62 SLE patients and 47 healthy subjects followed up for 12 weeks post-immunization, similarly concluded that immunization with an inactivated anti-influenza vaccine was safe in subjects with SLE and did not significantly change the underlying disease activity, although it was associated with a transient increase of dsDNA and antinuclear antigen (ANA). This study showed that seroconversion rates and seroprotection rates were significantly lower in the SLE group compared to controls (Wiesik-Szewczyk *et al.*, 2010). Another interesting case–control study, which investigated whether influenza vaccination triggers the development of antiphospholipid antibodies (aPLs) in patients with SLE, found that both SLE patients and healthy controls could develop new-onset anticardiolipin antibodies (aCLs) post-vaccination, at rates which did not differ significantly, without anti-β2GPI response. Vaccine response was not different between patients with and without new-onset aCL reactivity (Vista *et al.*, 2012). The elevation of aCL was found to be mostly transient. This study, and others with similar results, implies that the H1N1 vaccine might have a slight tendency toward aCL induction (Toplak *et al.*, 2008; Perdan-Pirkmajer *et al.*, 2012). Sufficient follow-up of such individuals has yet to be carried out. Therefore, to date, the data regarding such hematological findings, and even more scarce data regarding new onset of clinical SLE manifestations after seasonal/influenza A (H1N1) vaccine in previously healthy adults, do not allow confirmation of anything other than coincidental association between the two. As for patients with known lupus: if the disease is in remission, flares, although documented, are infrequent, and influenza vaccine can be administered without harm (Ristow *et al.*, 1978; Abu-Shakra *et al.*, 2000, 2007).

SLE induced by dTP vaccines

Throughout the years, animal studies have demonstrated autoimmune manifestations following dTP vaccines, and more specifically post-toxoid vaccines. Sporadic cases of autoimmune manifestations have also been documented in humans throughout the years, but usually by unidentified sources and with only limited details (Engleman *et al.*, 1981).

The most studied autoimmune syndrome related to the toxoid vaccine is antiphospholipid syndrome (APS), which is on the spectrum of SLE. Animal studies have shown successful induction of this syndrome in non-autoimmune-prone strains, by tetanus toxoid (TTd) hyperimmunization using different adjuvants (glycerol or aluminium hydroxide) and different adjuvant pretreatments (glycerol or complete Freund's adjuvant (CFA)). Both molecular mimicry and polyclonal B cell activation occur in APS induction (Zivković *et al.*, 2011; Dimitrijević *et al.*, 2012).

APS is characterized by the presence of pathogenic autoantibodies against β2-GPI. Throughout the years, the pathogenic potential of anti-TTd antibody crossreactivity with β2-GPI has been demonstrated in animal models, inducing experimental APS (Cruz-Tapias *et al.*, 2012). A molecular mimicry mechanism, resulting from structural homology between TTd and β2GPI, induces the adjuvant-dependent appearance of these crossreactive antibodies, which leads to a β2GPI-specific immune response (Cruz-Tapias *et al.*, 2012; Stojanović *et al.*, 2013; Zivković *et al.*, 2013).

The pertussis vaccine has also been shown to induce a number of autoimmune diseases, but no specific SLE manifestations have been studied. Reports of dTP-induced SLE cannot indicate which of the vaccine components is the actual cause of the adverse events. For example, the TDAP (Boostrix) vaccine had SLE reported as an adverse event in 2010, but, other than the fact that the patient was between 10–18 years, developed lupus within 6 months of immunization, later recovered fully, and was 1 of 11 similar cases, there are no further data (MedAlerts, 2010). Following these reports, an open, prospective, observational study was carried out in a health maintenance organization and monitored safety outcomes among 13 427 10–18-year-old adolescents up to 2 months after receiving TDAP vaccination as part of their normal health care. The study showed no increased risk for medically attended neurological, hematological, or allergic reactions following TDAP vaccination. It must, however, be noted that there was no information available regarding who conducted the study or why (VAERS request made 18 August 2013).

The efficacy of dTP vaccinations in SLE patients was found to be comparable to that in controls in a recent study. No significant age-related differences were found between the groups, except that in subjects younger than 40 years, the antidiphtheria level was significantly ($p = 0.029$) lower in SLE patients than in controls (Csuka *et al.*, 2013). In contrast, an older study did demonstrate lower protective antibody titers in nine SLE patients compared to nine controls. Patients with SLE had a lower mean pre-boost serum titer of antitetanus, and one-third showed a blunted serum antitetanus response. This difference was found to be due to a lack of SLE B cell response, not to abnormalities of SLE-helper or SLE-suppressor T cell function (Nies *et al.*, 1980).

SLE induced by BCG vaccines

There are only scarce data addressing the association between BCG vaccination and autoimmunity. Nevertheless, a link between this vaccine and autoimmunity has long been demonstrated in animal models, primarily nonobese diabetic (NOD) mice, which developed features of non-organ-specific autoimmune rheumatic disease, such as antinuclear antibodies and late-onset hemolytic anemia, after administration of heat-killed BCG. The vaccine precipitated a syndrome similar to SLE, which included hemolytic anemia, anti-DNA, and anti-Sm antinuclear autoantibodies and an increased severity of sialadenitis. Perivascular lymphocytic infiltration in the kidneys and glomerular immune complex deposition were also found (Baxter *et al.*, 1994). A similar study demonstrated the development of a spontaneous autoimmune disease characterized by the appearance of antinuclear antibodies and premature death due to immune-complex glomerulonephritis in mice treated with BCG (Engleman *et al.*, 1981). The pathogenesis is still not clear, but proposed mechanisms for the development of SLE-like manifestations range from adjuvant-like activity of the BCG (as treated mice showed a substantial increase in reticuloendothelial cell function and enhanced antigen-presentation capacity: Baxter *et al.*, 1994) to increased levels of serum type II interferon (found in both BCG-treated mice and patients with active SLE, suggesting its involvement in the pathogenesis: Engleman *et al.*, 1981).

To date, human use of BCG vaccines has been relatively safe. However, autoimmune rheumatic complications must be kept in mind (Toiusu *et al.*, 1978) and it may be prudent for patients with a history of organ-specific autoimmune disease or autoimmune rheumatic diseases who are candidates for BCG immunotherapy to be monitored carefully after treatment.

SLE induced by MMR vaccines

Data linking the measles virus and SLE have long been reported. In various studies, tubular structures resembling paramyxovirus nucleocapsids were detected in the cytoplasm of endothelial cells from SLE patients, measles virus antigens were detected by immunofluorescence in SLE cells, elevated measles virus antibody titers were detected in SLE patients, and (in one study) deoxyribonucleic acid from tissues of SLE patients was found to contain sequences which hybridized with measles RNA (Phillips and Christian, 1973;

Alekberova *et al.*, 1975; Morgan and Rapp, 1977). In addition, a direct relationship between measles and gamma globulin levels was found, suggesting that the elevated antibody levels in SLE result from hyperimmunoglobulinemia (Phillips and Christian, 1972). Virus antibody titers were also found to be elevated in SLE patients, and although nonspecific, this elevation was most marked for measles. In addition, a significant direct correlation was found between measles antibody and gamma globulin levels in SLE (Phillips and Christian, 1973). Although these data emphasize the possible connection between the measles virus and SLE, they are neither specific nor sensitive enough to form a casual association between the two, and they definitely do not provide grounds to form an association between the administration of MMR vaccination and the development of SLE. There are known adverse effects of this vaccination that may be included in the spectrum of rheumatologic diseases, such as arthralgias, rash, and thrombocytopenia, and there are reported cases in the VAERS system describing elevation of lupus-specific markers following MMR vaccinations, but these are not detailed with regards to the latency period or specific symptoms and findings.

The immunogenicity of MMR vaccinations in SLE patients hasn't been evaluated specifically. A single foreign study evaluating anti-MMR titers in rheumatologic pediatric patients found that 30% of patients did not acquire immunity (Tarasova *et al.*, 2008).

Therefore, more data are needed if we are to further form an association between lupus-like manifestations and the MMR vaccination or to elucidate the efficacy and safety of the vaccine in SLE patients, who may be severely immunocompromised. For now, caution should be used when administering the vaccine to these patients.

SLE induced by pneumococcal vaccines

In several studies, pneumococcal vaccines were not associated with an appreciable deterioration in any clinical or laboratory measure of SLE disease activity (Klippel *et al.*, 1979; Tarján *et al.*, 2002). In one, which observed 24 lupus patients who received the 23 serotype pneumococcal vaccine, only a single patient developed anticardiolipin IgG antibodies post-vaccination (Elkayam *et al.*, 2002, 2005). The efficacy of pneumococcal vaccines in SLE patients is comparable to that in the general population, and an older study found it to be unaffected by immunosuppressive agents (e.g. prednisone, cyclophosphamide, azathioprine) (Lipnick *et al.*, 1985; Tarján *et al.*, 2002). One study did, however, find that a subset of patients (20.8%) who had been immunized with pneumococcal vaccinations did not show immunity to more than one of seven polysaccharides tested (Elkayam *et al.*, 2002).

SLE induced by vaccine adjuvants, ASIA, and others

An adjuvant is a substance that enhances the activation of the immune system, both innate and adoptive (Kool *et al.*, 2008; Israeli *et al.*, 2009; Marrack *et al.*, 2009). The weak immunogenicity of many foreign antigens can be overcome through the use of such adjuvants (Audibert and Lise, 1993; Schwartz, 1993). The adjuvant effect encompass physical protection of the antigen from degradation, stimulation of innate immunity by Toll-like receptors (TLRs) and non-TLR sensors, antigen translocation to regional lymph nodes, and activation of the complement system (Israeli *et al.*, 2009). Consequently, adjuvants enable a longer exposure of the immune system to the antigen and prime the system for the production and activation of B and T cells, resulting in a more robust response.

Despite their ability to boost immune responses, both animal studies and reports of human diseases have clearly demonstrated the ability of adjuvants to inflict autoimmunity (e.g. autoantibodies) and even well-defined autoimmune diseases (Agmon-Levin *et al.*, 2009a,b; Israeli *et al.*, 2009; Cervera, 2011; Shoenfeld and Agmon-Levin, 2011a,b). Experimental evidence also shows that simultaneous administration of as few as two to three immune adjuvants can overcome genetic resistance to autoimmunity (Tomljenovic and Shaw, 2012). This "vaccine burden" has therefore been associated with the autoimmune phenomena, as have various vaccine combinations administered simultaneously (Agmon-Levin *et al.*, 2009a,b). Furthermore, the chronicity of the adjuvant effect explains why the timeframe for the development of autoimmunity can vary from 3 weeks to years, again depending on the vaccine burden (Cervera, 2011).

Autoimmune/inflammatory syndrome induced by adjuvants (ASIA)

The acceleration of an autoimmune or immune-mediated condition following exposure to external stimuli has been recently defined as autoimmune/inflammatory syndrome induced by adjuvants (ASIA) (Shoenfeld and Agmon-Levin, 2011a,b; Agmon-Levin *et al.*, 2012). This syndrome assembles a spectrum of immune-mediated diseases triggered by an adjuvant stimulus (Agmon-Levin *et al.*, 2009a,b; Balofsky *et al.*, 2010; Hajdu *et al.*, 2011). The mechanisms by which various adjuvants trigger autoimmunity are diverse, but, as a group, they may adversely incorporate an adjuvant effect. ASIA has been characterized by common and often disabling complaints coincident in many individuals diagnosed with post-vaccination events such as siliconosis, macrophagic myofasciitis (MMF), and Gulf War syndrome (GWS) (Agmon-Levin *et al.*, 2012), which are enigmatic and nondefined medical conditions, as well as in those diagnosed with distinct immune-mediated diseases, such as SLE, which shares some diagnostic criteria with ASIA, including arthritis, neurologic manifestations, serologic markers, and typical biopsy. However, diagnosis of ASIA also encompasses those with a mild or atypical clinical picture and the absence of manifestations specific for any major rheumatologic disease, and is in a way more similar to undefined connective-tissue disease (UCTD), in which arthralgias, arthritis, mucocutaneous involvement, and sicca symptoms and serologic markers are also considered common manifestations (LeRoy *et al.*, 1980; Greer and Panush, 1989; Alarcón *et al.*, 1991; Vilá *et al.*, 2000; Mosca *et al.*, 1999; Swaak *et al.*, 2001).

The noteworthy common denominator in all patients diagnosed with ASIA is that exposure to a component comprising an adjuvant effect can be documented in each of these medical conditions. Aluminum, squalene, silicone, and other "hidden adjuvants" (Israeli and Pardo, 2011; Shoenfeld and Agmon-Levin, 2011a,b) have all been suggested as triggers to these phenomena, which can occur weeks and even years following exposure to a culprit agent. Moreover, genetic links observed in animal models bring about the notion that the adjuvant effect promotes the appearance of an adjuvant disease in subjects who are genetically susceptible (e.g. HLA-DRB1) or in those who encounter an additional trigger, such as the effect of another deleterious environmental factor (e.g. infectious agent) or co-exposure to more than one adjuvant (Cervera, 2011; Shoenfeld and Agmon-Levin, 2011a,b).

Many studies, mostly animal, have been conducted in an attempt to better understand this association and define characteristic features for the population prone to adjuvant-associated autoimmunity.

Oil adjuvants

One of the most studied adjuvants in this context is pristane, which was found to be capable of inducing an autoimmune disease like SLE in a murine model (Reeves *et al.*, 2009). Replicating features of human disease, pristane-induced lupus is characterized by the production of autoantibodies, as well as by organ damage (e.g. renal disease) that depends on the interferon (IFN)-I receptor signaling pathway. Another adjuvant, squalene, can also induce arthritis in rats and the production of SLE-associated autoantibodies in mice (Santoro *et al.*, 2007; Reeves *et al.*, 2009). Another study similarly found that exposure to several oil adjuvants induces lupus-specific autoantibodies in nonautoimmune mice, including pristane, squalene (used in the adjuvant MF59), and incomplete Freund's adjuvant (IFA), and that various mineral oils induced the production of high levels of IL-6, IL-12, and tumor necrosis factor alpha (TNF-α), which appeared to be associated with the ability to induce lupus autoantibodies (Satoh *et al.*, 2003). The ability to induce such antibodies was found, in a later study, to be associated with the low molecular weight of hydrocarbon, since high-molecular-weight medicinal mineral oils did not induce similar antibodies. Further studies found that, in addition to the induction of autoantibodies, clinical manifestations characteristic of SLE, such as immune-complex glomerulonephritis and arthritis, can also be induced in rodents injected with even a single injection of adjuvant hydrocarbon oil. For example, n-Hexadecane (C(16)H(34)), another low-molecular-weight hydrocarbon oil with adjuvant activity, was found to induce arthritis and a limited set of specific autoantibodies, similar to those induced by other lupus-inducing oils (Kuroda *et al.*, 2004, 2006; Koppang *et al.*, 2008). Other studies evaluating pristane demonstrated its ability to induce a lupus-like syndrome characterized by SLE-specific autoantibodies and immune-complex glomerulonephritis; its ability to predominantly induce antiribosomal P antibodies, which exhibited similar fine specificities to anti-P antibodies in human SLE; its ability to induce severe glomerulonephritis characterized by proteinuria, mesangial proliferation, and glomerular immune-complex deposits in Swiss/Jackson laboratory (SJL) mice (Satoh *et al.*, 1996); and its ability to induce lupus associated with marked hypergammaglobulinemia, in the development of which microbial stimulation plays an important role (Hamilton *et al.*, 1998). Additional studies, some more recent than others, found that Freund's adjuvant could induce aPLs in heterozygous factor V Leiden (FVL) mice (Katzav *et al.*, 2012); that immunization of young dogs resulted in the production of autoantibodies, including lupus-associated ones; and that immunization of salmon fish with oil-adjuvanted vaccines resulted in the production of autoantibodies, thromboembolic disease, and immune-mediated glomerlulonephritis (Agmon-Levin *et al.*, 2009a,b).

Metal adjuvants

Other commonly used adjuvants are metals such as aluminum and mercury. Aluminum adjuvants are used at low quantities in numerous non-live vaccines, which are regulated by the Center for Biologics Evaluation and Research (CBER).

Although there are various adjuvants on the market, aluminum salts are the only materials that can be used in the United States, hydrated potassium aluminum sulfate being the most common.

The success of aluminum as a vaccine adjuvant is due to its potent and multifactorial stimulatory effects on the immune system. In fact, with the exception of attenuated viruses, in the absence of aluminum, most antigenic compounds fail to launch an adequate immune response (Dillon *et al.*, 1992; Seubert *et al.*, 2008; Israeli *et al.*, 2009), suggesting that a significant part of the immunostimulatory effect of a vaccine may be driven by the aluminum adjuvant itself.

While the potency and toxicity of aluminum adjuvants should be adequately balanced so that the necessary immune stimulation is achieved with minimal side effects, such a balance is difficult to achieve in practice, because the same mechanisms that drive the immunostimulatory effects of adjuvants have the capacity to provoke a variety of adverse reactions, including those associated with the ASIA syndrome (Agmon-Levin *et al.*, 2009a,b; Shaw and Petrik, 2009). There are many shared aspects between autoimmune/inflammatory conditions and the immunostimulatory properties of aluminum vaccine adjuvants. In SLE, specifically, an excessive Th2 response is seen, leading to an increase of IL-10, -18, and -6, IFN-g, and TNF-α (Elenkov and Chrousos, 1999; Elenkov *et al.*, 2000; Aringer *et al.*, 2004). A similar response has been demonstrated with aluminum, which generally stimulates Th2 responses, as well as the complement cascade and Th1 response in the presence of other Th1 stimulators (Dillon *et al.*, 1992; Davis *et al.*, 1998; Brazolot *et al.*, 1998; Aron-Maor and Shoenfeld, 2001; Lindblad, 2004; Smith *et al.*, 2006; Shoenfeld and Agmon-Levin, 2011a,b). Another possible mechanism through which aluminum adjuvants may trigger autoimmunity is a bystander effect, activating dormant autoreactive T cells in certain individuals (Fournie *et al.*, 2001; Agmon-Levin *et al.*, 2009a,b). In theory, overactivation of the Th2 response demonstrated after exposure to aluminum solutions may contribute to vaccine-induced SLE, whose pathogenesis stems from similar Th2 overactivation, but more studies are needed to elucidate a more specific relation between the two, because to date the literature is lacking large-series animal and human studies that demonstrate aluminum-induced SLE specifically and studies evaluating lupus exacerbation post-aluminum exposure. This may be due to difficulty isolating the effects of aluminum from those of other components found in vaccines, or to the fact that there is no correlation found between the aluminum quantities in vaccines and their ability to induce SLE. For example, HBV and HPV vaccinations, although they share the ability to induce SLE, do not differ in their aluminum quantities from several other vaccines that do not have a similar effect (Table 22.1); on the other hand, vaccines which have been shown to induce

Table 22.1 Quantities of aluminum in vaccines (Grabenstein, 2013)

Vaccine	Aluminum quantity
Pneumococcal vaccine	0.125 mg/dose
Diphtheria-tetanus-acellular pertussis (DTaP) vaccine	<0.17 to <0.625 mg/dose
Haemophilus influenzae type b (Hib) vaccine	0.225 mg/dose
Hib/Hep B vaccine	0.225 mg/dose
Hepatitis A vaccine (Hep A)	0.225–0.250 mg/dose (pediatrics), 0.45–0.50 mg/dose (adults)
Hepatitis B vaccine (Hep B)	0.225–0.500 mg/dose
Hep A/Hep B vaccine	0.45 mg/dose
DTaP/inactivated polio/Hep B vaccine	<0.85 mg/dose
DTaP/inactivated polio/Hib vaccine	0.33 mg/dose
Human papillomavirus (HPV) vaccine	0.225 mg/dose
Japanese encephalitis (JE) vaccine	0.25 mg/dose

Table 22.2 Association between vaccines and SLE

Vaccine type	Vaccine immunogenicity in SLE patients	Induction of naive SLE	Exacerbation of known SLE	Number of cases reported on VAER system (Lupton, 1987)[a]	Special considerations	Number of patients in cases studied in literature	References
HBV	Comparable to controls	Has been documented in several case series	Has been documented in several case series	203		53	Geier and Geier (2005), Zandman-Goddard and Shoenfeld (2005), Agmon-Levin *et al.* (2009), Orbach *et al.* (2010), ChildHealthSafety (2011), Zafrir *et al.* (2012)
HPV	Comparable to controls	Has been documented in several case series	Has been documented in several case series	113	HPV has been shown to have a high prevalence in SLE patients	14	Lee *et al.* (2010), Rojo-Contreras *et al.* (2012), Soldevilla *et al.* (2012), Gatto *et al.* (2013), Lyrio *et al.* (2013), Mok *et al.* (2013)
Influenza	Variable data May be decreased	Has not been reported	Variable data Case reports have been documented	63	An association between poor vaccine immunogenicity and SLE exacerbation was found There may be a tendency of the H1N1 vaccine toward aCL induction in SLE patients, as well as healthy individuals	36	Brodman *et al.* (1978), Louie *et al.* (1978), Del Porto *et al.* (2006), Abu-Shakra *et al.* (2007), Holvast *et al.* (2007), Conti *et al.* (2008), Toplak *et al.* (2008), Wiesik-Szewczyk *et al.* (2010), Saad *et al.* (2011), Urowitz *et al.* (2011), Perdan-Pirkmajer *et al.* (2012), Vista *et al.* (2012)
dTP	Variable data May be decreased	Human case reports are documented	No reports found in literature	32	dTP vaccine is known to induce antiphospholipid syndrome in animal models	1	VAERS

(continued)

Table 22.2 *(Continued)*

Vaccine type	Vaccine immunogenicity in SLE patients	Induction of naive SLE	Exacerbation of known SLE	Number of cases reported on VAER system (Lupton, 1987)[a]	Special considerations	Number of patients in cases studied in literature	References
BCG	No data found	Widely demonstrated in animal studies Nonspecific human case reports	No reports found in literature	0			
MMR	More data required	None specific Scarce reports have been made to VAERS	No reports found in literature	16	MMR may be contraindicated in severely immunosuppressed individuals	0	
Pneumococcal	Comparable to controls	Has not been reported	No reports found in literature	18		0	Lipnick *et al.* (1985), Elkayam *et al.* (2002, 2005)
Adjuvant		Widely demonstrated in animal studies		-			
Others		Single case report of discoid lupus erythematosus on smallpox vaccination scar Nonspecific reports regarding SLE induction by foreign substances: mineral oils, silicone, collagen, etc.		Smallpox: 7 Meningococcal: 15 Antrhrax: 10 H. influenza: 2		1	Verstraeten *et al.* (2008), Vera-Lastra *et al.* (2012)

[a]These reports vary in specificity, essential details, and length of follow-up. VAERS data are current up to 18 August 2013

SLE may not contain aluminum at all (e.g. BCG, influenza, etc.) (Table 22.2).

Other adjuvants

In 1987, a single case report described the development of discoid lupus erythematosus in a smallpox vaccination scar; however, no other cases have been reported in the literature since, probably due to the uncommon use of the smallpox vaccination in the general population (Lupton, 1987).

Another substance injected subcutaneously which was able to induce an SLE-like syndrome in a murine model was highly purified lymphocyte-derived DNA (referred to as "ALD DNA"). The SLE-like syndrome was expressed by high levels of anti-dsDNA antibodies and other autoantibodies, as well as the development of glomerulonephritis, glomerular deposition of IgG and C3, and proteinuria (Qiao *et al.*, 2005).

In summary, although it may be supported by similar pathogenic pathways that aluminum and other adjuvants are the main trigger for post-vaccination onset of lupus, a causal relation will be difficult to prove, due to the complexity of the various vaccine formulations and of the pathogenesis of SLE.

References

Abu-Shakra, M., Zalmanson, S., Neumann, L., *et al.* (2000). Influenza virus vaccination of patients with systemic lupus erythematosus: effects on disease activity. *J Rheumatol*, **27**: 1681–5.

Abu-Shakra, M., Press, J., Buskila, D., and Sukenik, S. (2007). Influenza vaccination of patients with systemic lupus erythematosus: safety and immunogenicity issues. *Autoimmune Rev*, **6**: 543–6.

Agmon-Levin, N., Paz, Z., Israeli, E., and Shoenfeld, Y. (2009a). Vaccines and autoimmunity. *Nat Rev Rheumatol*, **5**: 648–52.

Agmon-Levin, N., Zafrir, Y., Paz, Z., *et al.* (2009b). Ten cases of systemic lupus erythematosus related to hepatitis B vaccine. *Lupus*, **18**: 1192–7.

Agmon-Levin, N., Hughes, G.R.V., and Shoenfeld, Y. (2012). The spectrum of ASIA: "Autoimmune (Auto-inflammatory) Syndrome induced by Adjuvants." *Lupus*, **21**: 118–20.

Alarcón, G.S., Williams, G.V., Singer, J.Z., *et al.* (1991). Early undifferentiated connective tissue disease. I. Early clinical manifestations in a large cohort of patients with undifferentiated connective tissue diseases compared with cohorts of well-established connective tissue disease. *J Rheumatol*, **18**: 1332–9.

Alekberova, Z.S., Parfanovich, M.I., Nassonova, V.A., and Zhdanov, V.M. (1975). Molecular pathogenesis of SLE. *Arch Virol*, **47**: 109–21.

Aringer, M. and Smolen, J.S. (2004). Tumour necrosis factor and other proinflammatory cytokines in systemic lupus erythematosus: a rationale for therapeutic intervention. *Lupus*, **13**: 344–7.

Aron-Maor, A. and Shoenfeld, Y. (2001). Vaccination and systemic lupus erythematosus: the bidirectional dilemmas. *Lupus*, **10**: 237–40.

Audibert, F.M. and Lise, L.D. (1993). Adjuvants: current status, clinical perspectives and future prospects. *Immunol Today*, **14**: 281–4.

Balofsky, A., Agmon-Levin, N., and Shoenfeld, Y. (2010). The new H1N1 and HPV vaccines and old fears. *Curr Opin Rheumatol*, **22**: 431–6.

Baxter, A.G., Horsfall, A.C., Healey, D., *et al.* (1994). Mycobacteria precipitate an SLE-like syndrome in diabetes-prone NOD mice. *Immunology*, **83**: 227–31.

Bijl, M., Agmon-Levin, N., Dayer, J.M., *et al.* (2012). Vaccination of patients with auto-immune inflammatory rheumatic diseases requires careful benefit-risk assessment. *Autoimmun Rev*, **11**: 572–6.

Blank, M., Israeli, E., and Shoenfeld, Y. (2012). When APS (Hughes syndrome) met the autoimmune/inflammatory syndrome induced by adjuvants (ASIA). *Lupus*, **21**: 711–14.

Brazolot Millan, C.L., Weeratna, R., Krieg, A.M., *et al.* (1998). CpG DNA can induce strong Th1 humoral and cell-mediated immune responses against hepatitis B surface antigen in young mice. *Proc Natl Acad Sci USA*, **95**: 15 553–8.

Brodman, R., Gilfillan, R., Glass, D., and Schur, P.H. (1978). Influenzal vaccine response in systemic lupus erythematosus. *Ann Intern Med*, **88**: 735–40.

Cervera, R. (2011). "ASIA": a new systeminc autoimmune syndrome? *Lupus*, **20**: 665–6.

Chatham, W.W., Wallace, D.J., Stohl, W., *et al.* (2012). Effect of belimumab on vaccine antigen antibodies to influenza, pneumococcal, and tetanus vaccines in patients with systemic lupus erythematosus in the BLISS-76 trial. *J Rheumatol*, **39**: 1632–40.

ChildHealthSafety. (2011). US government concedes hep B vaccine causes systemic lupus erythematosus. Available from: http://childhealthsafety.wordpress.com/2011/04/25/us-concedes-hep-b-causes-lupus/ [last accessed 11 December 2014].

Colafrancesco, S., Agmon-Levin, N., Perricone, C., and Shoenfeld, Y. (2013). Unraveling the soul of autoimmune diseases: pathogenesis, diagnosis and treatment adding dowels to the puzzle. *Immunol Res*, **56**: 200–5.

Conti, F., Rezai, S., and Valesini, G. (2008). Vaccination and autoimmune rheumatic diseases. *Autoimmun Rev*, **8**: 124–8.

Cruz-Tapias, P., Blank, M., Anaya, J.M., and Shoenfeld, Y. (2012). Infections and vaccines in the etiology of antiphospholipid syndrome. *Curr Opin Rheumatol*, **24**: 389–93.

Csuka, D., Czirják, L., Hóbor, R., *et al.* (2013). Effective humoral immunity against diphtheria and tetanus in patients with systemic lupus erythematosus or myasthenia gravis. *Mol Immunol*, **54**: 453–6.

Davis, H.L., Weeratna, R., Waldschmidt, T.J., *et al.* (1998). CpG DNA is a potent enhancer of specific immunity in mice immunized with recombinant hepatitis B surface antigen. *J Immunol*, **160**: 870–6.

Del Porto, F., Lagana, B., Biselli, R., *et al.* (2006). Influenza vaccine administration in patients with systemic lupus erythematosus and rheumatoid arthritis: safety and immunogenicity. *Vaccine*, **24**: 3217–23.

Dillon, S.B., Demuth, S.G., Schneider, M.A., *et al.* (1992). Induction of protective class I MHC-restricted CTL in mice by a recombinant influenza vaccine in aluminium hydroxide adjuvant. *Vaccine*, **10**: 309–18.

Dimitrijević, L., Živković, I., Stojanović, M., *et al.* (2012). Vaccine model of antiphospholipid syndrome induced by tetanus vaccine. *Lupus*, **21**: 195–202.

Elenkov, I.J. and Chrousos, G.P. (1999). Stress hormones, Th1/Th2 patterns, pro/anti-inflammatory cytokines and susceptibility to disease. *Trends Endocrinol Metab*, **10**: 359–68.

Elenkov, I.J., Wilder, R.L., Chrousos, G.P., and Vizi, E.S. (2000). The sympathetic nerve–an integrative interface between two supersystems: the brain and the immune system. *Pharmacol Rev*, **52**: 595–638.

Elkayam, O., Paran, D., Caspi, D., *et al.* (2002). Immunogenicity and safety of pneumococcal vaccination in patients with rheumatoid arthritis or systemic lupus erythematosus. *Clin Infect Dis*, **34**: 147–53.

Elkayam, O., Paran, D., Burke, M., *et al.* (2005). Pneumococcal vaccination of patients with systemic lupus erythematosus: effects on generation of autoantibodies. *Autoimmunity*, **38**: 493–6.

Elkon, K.B. (2000). Antibodies in SLE. In: Klippel, J.H. and Dieppe, P.A. (eds). *Rheumatology*. Saint Louis, MO: Mosby-Year Book, pp. 1–10.

Engleman, E.G., Sonnenfeld, G., Dauphinee, M., *et al.* (1981). Treatment of NZB/NZW F1 hybrid mice with *Mycobacterium bovis* strain BCG or type II interferon preparations accelerates autoimmune disease. *Arthritis Rheum*, **24**: 1396–402.

Fournie, G.J., Mas, M., Cautain, B., *et al.* (2001). Induction of autoimmunity through bystander effects. Lessons from immunological disorders induced by heavy metals. *J Autoimmun*, **16**: 319–26.

Gatto, M., Agmon-Levin, N., Soriano, A., *et al.* (2013). Human papillomavirus vaccine and systemic lupus erythematosus. *Clin Rheumatol*, **32**: 1301–7.

Geier, D.A. and Geier, M.R. (2005). A case–control study of serious autoimmune adverse events following hepatitis B immunization. *Autoimmunity*, **38**: 295–301.

Grabenstein, J.D. (2013). *ImmunoFacts: Vaccines and Immunologic Drugs – 2013*, 38th revision. St Louis, MO: Wolters Kluwer Health.

Greer, J.M. and Panush, R.S. (1989). Incomplete lupus erythematosus. *Arch Intern Med*, **149**: 2473–6.

Hajdu, S.D., Agmon-Levin, N., and Shoenfeld, Y. (2011). Silicone and autoimmunity. *Eur J Clin Invest*, **41**: 203–11.

Hamilton, K.J., Satoh, M., Swartz, J., *et al.* (1998). Influence of microbial stimulation on hypergammaglobulinemia and autoantibody production in pristane-induced lupus. *Clin Immunol Immunopathol*, **86**: 271–9.

Holvast, B., Huckriede, A., Kallenberg, C.G., and Bijl, M. (2007). Influenza vaccination in systemic lupus erythematosus: safe and protective? *Autoimmun Rev*, **6**: 300–5.

Israeli, E. and Pardo, A. (2011). The sick building syndrome as a part of the autoimmune (autoinflammatory) syndrome induced by adjuvants. *Mod Rheumatol*, **21**: 235–9.

Israeli, E., Agmon-Levin, N., Blank, M., and Shoenfeld, Y. (2009). Adjuvants and autoimmunity. *Lupus*, **18**: 1217–25.

Katzav, A., Kivity, S., Blank, M., *et al.* (2012). Adjuvant immunization induces high levels of pathogenic antiphospholipid antibodies in genetically prone mice: another facet of the ASIA syndrome. *Lupus*, **21**: 210–16.

Klippel, J.H., Karsh, J., Stahl, N.I., *et al.* (1979). A controlled study of pneumococcal polysaccharide vaccine in systemic lupus erythematosus. *Arthritis Rheum*, **22**: 1321–5.

Kool, M., Soullié, T., van Nimwegen, M., *et al.* (2008). Alum adjuvant boosts adaptive immunity by inducing uric acid and activating inflammatory dendritic cells. *J Exp Med*, **14**(205): 869–82.

Koppang, E.O., Bjerkås, I., Haugarvoll, E., *et al.* (2008). Vaccination-induced systemic autoimmunity in farmed Atlantic salmon. *J Immunol*, **181**: 4807–14.

Kuroda, Y., Nacionales, D.C., Akaogi, J., *et al.* (2004). Autoimmunity induced by adjuvant hydrocarbon oil components of vaccine. *Biomed Pharmacother*, **58**: 325–37.

Kuroda, Y., Ono, N., Akaogi, J., *et al.* (2006). Induction of lupus-related specific autoantibodies by non-specific inflammation caused by an intraperitoneal injection of n-hexadecane in BALB/c mice. *Toxicology*, **218**: 186–96.

Lee, Y.H., Choe, J.Y., Park, S.H., *et al.* (2010). Prevalence of human papilloma virus infections and cervical cytological abnormalities among Korean women with systemic lupus erythematosus. *J Korean Med Sci*, **25**: 1431–7.

Lefkowith, J.B. and Gilkeson, G.S. (1996). Nephritogenic autoantibodies in lupus: current concepts and continuing controversies. *Arthritis Rheum*, **39**: 894–903.

LeRoy, E.C., Maricq, H.R., and Kahaleh, M.B. (1980). Undifferentiated connective tissue syndromes. *Arthritis Rheum*, **23**: 341–3.

Lindblad, E.B. (2004). Aluminium compounds for use in vaccine. *Immunol Cell Biol*, **82**: 497–505.

Lipnick, R.N., Karsh, J., Stahl, N.I., *et al.* (1985). Pneumococcal immunization in patients with systemic lupus erythematosus treated with immunosuppressives. *J Rheumatol*, **12**: 1118–21.

Louie, J.S., Nies, K.M., Shoji, K.T., *et al.* (1978). Clinical and antibody responses after influenza immunization in systemic lupus erythematosus. *Ann Intern Med*, **88**: 790–2.

Lupton, G.P. (1987). Discoid lupus erythematosus occurring in a smallpox vaccination scar. *J Am Acad Dermatol*, **17**: 688–90.

Lyrio, L.D., Grassi, M.F., Santana, I.U., *et al.* (2013). Prevalence of cervical human papillomavirus infection in women with systemic lupus erythematosus. *Rheumatol Int*, **33**: 335–40.

Madaio, M.P., Carlson, J., Cataldo, J., *et al.* (1987). Murine monoclonal anti-DNA antibodies bind directly to glomerular antigens and form immune deposits. *J Immunol*, **138**: 2883–9.

Marrack, P., McKee, A.S., and Munks, M.W. (2009). Towards an understanding of the adjuvant action of aluminium. *Nat Rev Immunol*, **9**: 287–93.

MedAlerts. (2010). *VAERS ID* 394489. Available from: http://medalerts.org/vaersdb/findfield.php?IDNUMBER=394489 [last accessed 11 December 2014].

Mok, C.C., Ho, L.Y., Fong, L.S., and To, C.H. (2013). Immunogenicity and safety of a quadrivalent human papillomavirus vaccine in patients with systemic lupus erythematosus: a case–control study. *Ann Rheum Dis*, **72**: 659–64.

Morgan, E.M. and Rapp, F. (1977). Measles virus and its associated diseases. *Bacteriol Rev*, **41**: 636–66.

Mosca, M., Neri, R., and Bombardieri, S. (1999). Undifferentiated connective tissue diseases (UCTD): a review of the literature and a proposal for preliminary classification criteria. *Clin Exp Rheumatol*, **17**: 615–20.

Nies, K., Boyer, R., Stevens, R., and Louie, J. (1980). Anti-tetanus toxoid antibody synthesis after booster immunization in systemic lupus erythematosus. Comparison of the in vitro and in vivo responses. *Arthritis Rheum*, **23**: 1343–50.

Older, S.A., Battafarano, D.F., Enzenauer, R.J., and Krieg, A.M. (1999). Can immunization precipitate connective tissue disease? Report of five cases of systemic lupus erythematosus and review of the literature. *Semin Arthritis Rheum*, **29**: 131–9.

Orbach, H., Agmon-Levin, N., and Zandman-Goddard, G. (2010). Vaccines and autoimmune diseases of the adult. *Discov Med*, **9**: 90–7.

Phillips, P.E. and Christian, C.L. (1972). The influence of serum immunoglobulin concentration on measles antibody level. *Proc Soc Exp Biol Med*, **140**: 1340–3.

Phillips, P.E. and Christian, C.L. (1973). Virus antibodies in systemic lupus erythematosus and other connective tissue diseases. *Ann Rheum Dis*, **32**: 450–6.

Perdan-Pirkmajer, K., Thallinger, G.G., Snoj, N., *et al.* (2012). Autoimmune response following influenza vaccination in patients with autoimmune inflammatory rheumatic disease. *Lupus*, **21**: 175–83.

Qiao, B., Wu, J., Chu, Y.W., *et al.* (2005). Induction of systemic lupus erythematosus-like syndrome in syngeneic mice by immunization with activated lymphocyte-derived DNA. *Rheumatology (Oxford)*, **44**: 1108–14.

Reeves, W.H., Lee, P.Y., Weinstein, J.S., *et al.* (2009). Induction of autoimmunity by pristane and other naturally occurring hydrocarbons. *Trends Immunol*, **30**: 455–64.

Ristow, S.C., Douglas, R.G. Jr., and Condemi, J.J. (1978). Influenza vaccination of patients with systemic lupus erythematosus. *Ann Intern Med*, **88**: 786–9.

Rojo-Contreras, W., Olivas-Flores, E.M., Gamez-Nava, J.I., *et al.* (2012). Cervical human papillomavirus infection in Mexican women with systemic lupus erythematosus or rheumatoid arthritis. *Lupus*, **21**: 365–72.

Saad, C.G., Borba, E.F., Aikawa, N.E., *et al.* (2011). Immunogenicity and safety of the 2009 non-adjuvanted influenza A/H1N1 vaccine in a large cohort of autoimmune rheumatic diseases. *Ann Rheum Dis*, **70**: 1068–73.

Santoro, D., Stella, M., Montalto, G., and Castelli, S. (2007). Lupus nephritis after hepatitis B vaccination: an uncommon complication. *Clin Nephrol*, **67**: 61–3.

Satoh, M., Hamilton, K.J., Ajmani, A.K., *et al.* (1996). Autoantibodies to ribosomal P antigens with immune complex glomerulonephritis in SJL mice treated with pristane. *J Immunol*, **157**: 3200–6.

Satoh, M., Kuroda, Y., Yoshida, H., *et al.* (2003). Induction of lupus autoantibodies by adjuvants. *J Autoimmun*, **21**: 1–9.

Schattner, A. (2005). Consequence or coincidence? The occurrence, pathogenesis and significance of autoimmune manifestations after viral vaccines. *Vaccine*, **23**: 3876–86.

Schwartz, R.S. (1993). Autoimmunity and autoimmune diseases. In: Paul, W.E. (ed.). *Fundamental Immunology*, 3rd edn. New York: Raven Press, pp. 1033–97.

Seubert, A., Monaci, E., Pizza, M., *et al.* (2008). The adjuvants aluminum hydroxide and MF59 induce monocyte and granulocyte chemoattractants and enhance monocyte differentiation toward dendritic cells. *J Immunol*, **180**: 5402–12.

Shaw, C.A. and Petrik, M.S. (2009). Aluminum hydroxide injections lead to motor deficits and motor neuron degeneration. *J Inorg Biochem*, **103**: 1555–62.

Shoenfeld, Y. and Agmon-Levin, N. (2011a). "ASIA" – autoimmune/inflammatory syndrome induced by adjuvants. *J Autoimmun*, **36**: 4–8.

Shoenfeld, Y. and Agmon-Levin, N. (2011b). ASIA: a new way to put the puzzle together. Autoimmune (autoinflammatory) syndrome induced by adjuvants provides a diagnostic framework for enigmatic conditions. *The Rheumatologist*, June 2011. Available from: http://www.the-rheumatologist.org/details/article/

1081203/ASIA_A_New_Way_to_Put_the_Puzzle_Together.html [last accessed 11 December 2014].

Smith, K.M., Garside, P., McNeil, R.C., and Brewer, J.M. (2006). Analysis of costimulatory molecule expression on antigen-specific T and B cells during the induction of adjuvant-induced Th1 and Th2 type responses. *Vaccine*, **24**: 3035–43.

Soldevilla, H.F., Briones, S.F., and Navarra, S.V. (2012). Systemic lupus erythematosus following HPV immunization or infection? *Lupus*, **21**: 158–61.

Stojanović, M., Petrušić, V., Zivković, I., *et al.* (2013). Role of molecular mimicry and polyclonal cell activation in the induction of pathogenic β2-glycoprotein I-directed immune response in Balb/c mice upon hyperimmunization with tetanus toxoid. *Immunol Res*, **56**: 20–31.

Suzuki, N., Harada, T., Mizushima, Y., and Sakane, T. (1993). Possible pathogenic role of cationic anti-DNA autoantibodies in the development of nephritis in patients with systemic lupus erythematosus. *J Immunol*, **151**: 1128–36.

Swaak, A.J., van de Brink, H., Smeenk, R.J., *et al.* (2001). Incomplete lupus erythematosus: results of a multicenter study under the supervision of the EULAR Standing Committee on International Clinical Studies Including Therapeutic Trials (ESCISIT). *Rheumatology*, **40**: 89–94.

Tarasova, A.A., Kostinov, M.P., and Korovkina, T.I. (2008). Immunity to measles after vaccination of children with rheumatic diseases. *Zh Mikrobiol Epidemiol Immunobiol*, **5**: 95–8.

Tarján, P., Sipka, S., Maródi, L., *et al.* (2002). No short-term immunological effects of Pneumococcus vaccination in patients with systemic lupus erythematosus. *Scand J Rheumatol*, **31**: 211–15.

Toiusu, M., Miyahara, T., Shinohara, H., *et al.* (1978). A new side effect of BCG immunotherapy-BCG induced arthritis in man. *Cancer Immunother*, **5**: 77.

Tomljenovic, L. and Shaw, C.A. (2012). Mechanisms of aluminum adjuvant toxicity and autoimmunity in pediatric populations. *Lupus*, **21**: 223–30.

Toplak, N., Kveder, T., Trampus-Bakija, A., *et al.* (2008). Autoimmune response following annual influenza vaccination in 92 apparently healthy adults. *Autoimmun Rev*, **8**: 134–8.

Urowitz, M.B., Anton, A., Ibanez, D., and Gladman, D.D. (2011). Autoantibody response to adjuvant and non-adjuvant H1N1 vaccination in systemic lupus erythematosus. *Arthritis Care Res (Hoboken)*, **63**: 1517–20.

Vera-Lastra, O., Medina, G., Cruz-Dominguez Mdel, P., *et al.* (2012). Human adjuvant disease induced by foreign substances: a new model of ASIA (Shoenfeld's syndrome). *Lupus*, **21**: 128–35.

Verstraeten, T., Descamps, D., David, M.P., *et al.* (2008). Analysis of adverse events of potential autoimmune aetiology in a large integrated safety database of AS04 adjuvanted vaccines. *Vaccine*, **26**: 6630–8.

Vilá, L.M., Mayor, A.M., Valentín, A.H., *et al.* (2000). Clinical outcome and predictors of disease evolution in patients with incomplete lupus erythematosus. *Lupus*, **9**: 110–15.

Vista, E.S., Crowe, S.R., Thompson, L.F., *et al.* (2012). Influenza vaccination can induce new-onset anticardiolipins but not β2-glycoprotein-I antibodies among patients with systemic lupus erythematosus. *Lupus*, **21**: 168–74.

Vlahakos, D., Foster, M.H., Ucci, A.A., *et al.* (1992). Murine monoclonal anti-DNA antibodies penetrate cells, bind to nuclei, and induce glomerular proliferation and proteinuria in vivo. *J Am Soc Nephrol*, **2**: 1345–54.

von Muhlen, C.A. and Tan, E.M. (1995). Autoantibodies in the diagnosis of systemic rheumatic diseases. *Semin Arthritis Rheum*, **24**: 323–58.

Wiesik-Szewczyk, E., Romanowska, M., Mielnik, P., *et al.* (2010). Anti-influenza vaccination in systemic lupus erythematosus patients: an analysis of specific humoral response and vaccination safety. *Clin Rheumatol*, **29**: 605–13.

Zafrir, Y., Agmon-Levin, N., Paz, Z., *et al.* (2012). Autoimmunity following hepatitis B vaccine as part of the spectrum of "autoimmune (auto-inflammatory) syndrome induced by adjuvants" (ASIA): analysis of 93 cases. *Lupus*, **21**: 146–52.

Zandman-Goddard, G. and Shoenfeld, Y. (2005). Infections and SLE. *Autoimmunity*, **38**: 473–85.

Zivkovic, I., Stojanovic, M., Petrusic, V., *et al.* (2011). Induction of APS after TTd hyper-immunization has a different outcome in BALB/c and C57BL/6 mice. *Am J Reprod Immunol*, **65**: 492–502.

Zivković, I., Petrušić, V., Dimitrijević, R., *et al.* (2013). Adjuvant dependence of APS pathology-related low-affinity antibodies during secondary immune response to tetanus toxoid in BALB/c mice. *Immunol Res*, **56**: 143–9.

23 Vasculitides

Alessandra Soriano,[1,2] Rotem Inbar,[1] Giovanna Passaro,[3] and Raffaele Manna[3]

[1]Zabludowicz Center for Autoimmune Diseases, Sheba Medical Center, Tel Hashomer, Israel
[2]Department of Clinical Medicine and Rheumatology, Campus Bio-Medico University, Rome, Italy
[3]Periodic Fevers Research Center, Department of Internal Medicine, Catholic University of the Sacred Heart, Rome, Italy

Introduction

Vasculitides are a heterogeneous group of autoimmune-mediated inflammatory diseases involving blood vessels of different types and sizes. They show different etiologies, pathogeneses, genetic predispositions, types of vessels affected, organ distributions, types of inflammation, clinical manifestations, and distinctive demographic characteristics; these and other features can be used for disease categorization (Jennette *et al.*, 2013). The 2012 Revised International Chapel Hill Consensus Conference Nomenclature of Vasculitides (CHCC2012) was an international effort to update the previous CHCC1994 nomenclature system according to our new knowledge of these diseases. In particular, the new system separates vasculitides with known causes – such as infections – from those without known causes. Several reports of vasculitis onset following exposure to vaccines have been described since the 1970s. Nevertheless, these phenomena remain rare, so it is difficult to establish whether some specific vaccines should be included among the environmental factors sometimes associated with the development of these diseases. As a matter of fact, to date a clear etiological definition of "post-vaccination" vasculitides is still missing. A causal relationship between autoimmune inflammatory diseases and vaccinations has been accepted by the medical community in the case of Guillain–Barré syndrome (GBS) and the 1976–77 A/New/Jersey/8/76 swine flu vaccine (Marks and Halpin, 1980), idiopathic thrombocytopenia and measles, mumps, and rubella (MMR) vaccination in children (Jonville-Béra *et al.*, 1996), and transverse myelitis and oral polio vaccine (Agmon-Levin *et al.*, 2009a).

In this chapter, we discuss the main etiopathogenic hypotheses for how vasculitic processes may be triggered by vaccines, and we analyze all the relevant Medline records from 1970 through July 2013 relating to the onset of different types of vasculitides following exposure to different vaccines.

Vasculitides following vaccinations: plausible mechanisms

Vaccines contain viral or bacterial antigens, in many cases in combination with adjuvants and with several types of preservatives (e.g. thimerosal, gelatin, formaldehyde, etc.). As each component of a vaccine might induce blood vessels inflammation via several mechanisms, many plausible theories have been proposed for the etiopathogenic link between exposure to such stimuli and the onset of these diseases.

The role of infectious agents

In recent decades, it has become clear that infectious agents and vaccines have many similarities

Vaccines and Autoimmunity, First Edition. Edited by Yehuda Shoenfeld, Nancy Agmon-Levin, and Lucija Tomljenovic.

in their ability to facilitate antibody production, immune reactions, and a wide spectrum of autoimmune phenomena, including vasculitides (Kivity *et al.*, 2009; Shoenfeld, 2009; Orbach *et al.*, 2010). This overlapping of vaccine/infection-induced immune responses is not a surprising feature, given that the essence of a vaccine is, in many cases, a live-attenuated or recombinant pathogenic antigen (Chen *et al.*, 2001; Molina and Shoenfeld, 2005; Piaggio *et al.*, 2005; Shoenfeld *et al.*, 2008; Agmon-Levin *et al.*, 2009b), which is able to trigger blood-vessel inflammation in some individuals with a genetic predisposition (Tishler and Shoenfeld, 2004; Toplak and Avcin, 2009).

Infectious agents are considered the most common triggers (Blank *et al.*, 2002) of autoimmunity, and they have also been implicated in the pathogenesis of many vasculitic syndromes. A causal relationship between a specific infectious agent and the onset of the vasculitic process has been formally established in the cases of polyarteritis nodosa (PAN) and hepatitis B virus (HBV) and of cryoglobulinemia and hepatitis C virus (HCV) (Gocke *et al.*, 1970; Misiani *et al.*, 1992; Trepo and Guillevin, 2001). For others, animal models provide additional validity to the hypothesis of such a link (Mathieson *et al.*, 1993; Weck *et al.*, 1997; Abe *et al.*, 1998; Buonocore *et al.*, 2004; Huugen *et al.*, 2005).

Several mechanisms are thought to be involved in the pathogenesis of infection-related vasculitides, among which are direct microbial invasion of vessel wall endothelial cells, vessel wall damage induced by immune complex formation, and B and T cell stimulation through molecular mimicry and superantigens. In some cases, more than one mechanism is involved (Millikan and Flynn, 1999; Guillevin, 2004; Rodriguez-Pla and Stone, 2006; Lidar *et al.*, 2009).

Vaccines may provide a transient inflammatory setting for bystander activation and autoreactive T cells through antigen nonspecific mechanisms, which can initiate the autoimmunity process.

The vasculitic process associated with cytomegalovirus (CMV), herpes simplex virus (HSV), *Rickettsia*, and *Staphylococcus aureus* is thought to be achieved through direct microbial invasion (Beekhuizen *et al.*, 1997; Witort-Serraglini *et al.*, 1999; Matussek *et al.*, 2005; Bechah *et al.*, 2008). Immune-complex deposition-mediated endothelial damage is the mechanism behind the virulence of HCV-associated cryoglobulinemic vasculitis and of HBV-associated PAN (Agnello *et al.*, 1992; Millikan and Flynn, 1999; Ferri and Zignego, 2000; Strassburg *et al.*, 2003).

Additional viral infections have been found to have an association with vasculitic syndromes. For example, giant cell arteritis (GCA), PAN, Henoch–Schönlein purpura (HSP), Kawasaki's disease (KD), and granulomatosis with polyangiitis (GPA) were all described following Parvovirus B-19 infection (Corman and Dolson 1992; Cioc *et al.*, 2002; Lehmann *et al.*, 2003; Baskan *et al.*, 2007), with the apparent mechanism being direct vascular injury (Finkel *et al.*, 1994).

The role of adjuvants: ASIA syndrome

In addition to the infectious antigen, vaccines contain a variety of other substances, any of which may harbor the ability to trigger an immune response; these include preservatives, adjuvants, and various manufacturing residues. The main role of the adjuvant is to augment the immune response to the antigen. However, in recent years, many adjuvants have been found to trigger autoimmunity themselves (Gherardi and Authier, 2003; Satoh *et al.*, 2003; Agmon-Levin and Shoenfeld, 2008; Shaw and Petrik, 2009; Shoenfeld, 2009). Such adjuvants merge the diverse and multifaced autoimmune and inflammatory reactions caused by different pharmaceutical, industrial, and environmental compounds with the immune-mediating capabilities of an "adjuvant effect" (Shoenfeld and Agmon-Levin, 2011). This category includes not just the adjuvants traditionally used in the vaccines but any substance with this immune-triggering capacity: notable examples are aluminum, silicone, pristane, and infectious agents; but also the yeast *Saccharomyces cerevisiae* (SC), ovalbumin (egg protein), gelatin, neomycin (antibiotic), thimerosal, and formaldehyde. All of these are vectors used in vaccine preparation, manufacturing, and preservation. Allergic reactions, angioedema, hypersensitivity vasculitis (HV), and other cutaneous reactions have been described following exposure to these ingredients (Brightman *et al.*, 1989; Kelso *et al.*, 1993; Nakayama *et al.*, 1999; Leventhal *et al.*, 2012; Barbaud *et al.*, 2013). In particular, immediate hypersensitivity reactions are generally considered to be transient post-vaccination phenomena, and they have been related to hypersensitivity to injected foreign antigens. Delayed reactions include urticarial vasculitides, neutrophilic dermatoses, and maculopapular exanthems.

All of the post-vaccination syndromes, including systemic post-vaccine vasculitides, appear to have surprising similarities in their signs and symptoms to the usual "pattern" of disease presentation.

When there has been a previous exposure to a vaccine, these are believed to be the "epiphenomena" of the "adjuvant effect," and they can be traced back to the common denominator: exposure to an adjuvant substance. Significant evidence is accumulating to support this notion, as highlighted by the recently defined autoimmune/inflammatory syndrome induced by adjuvants (ASIA) (Shoenfeld and Agmon-Levin, 2011; Bassi *et al.*, 2012; Katzav *et al.*, 2012).

Clinical evidence from the literature

We performed a Medline search of all relevant publications, without any date limitation, and with special emphasis on finding each kind of vasculitide linked to exposure to different types of vaccines. All relevant publications were retrieved and critically analyzed.

Large-vessel vasculitides

A total of 18 cases of large-vessel vasculitides have been detected in the literature in a period of over 30 years: 15 cases of GCA, 2 cases of Takayasu disease (TD), and 1 case of large-cell arteritis involving subclavian and renal arteries, which was similar to TD but had atypical onset in a 61-year-old woman and spared other vessels (Table 23.1). Interestingly, all of the cases of GCA were described following seasonal influenza vaccination; many cases were biopsy-proven, and one showed clinical findings of microscopic polyangiitis (MPA) at the same time. The time interval between vaccine administration and onset of clinical manifestations varied from 1 day to 3 months. The two TD cases and the one with atypical onset followed HBV vaccines. A previous diagnosis of ankylosing spondylitis was detected in one case, while in another the patient had received a diagnosis of polymyalgia rheumatica (PMR)-like illness seven years earlier, suggesting the possibility that patients with previous immunologic activation/dysfunction may be more susceptible to developing vasculitis following vaccination (Zaas *et al.*, 2001). Similarly, in their case series of post-vaccination vasculitides, Soriano *et al.* (2012) described a patient with a diagnosis of PMR following influenza vaccine who experienced a relapse of her disease 2 years later, when she underwent seasonal influenza vaccine for the second time. In the case of TD described by Zaas *et al.* (2001), the third dose of HBV vaccine was administered despite the patient having already experienced some symptoms of the disease following the second dose, which finally determined a clear manifestation of TD.

As regards possible mechanisms for the induction of the vasculitic process by influenza vaccine, it must be underlined that all seasonal influenza vaccines in the United States are produced in hen's eggs and do not contain any adjuvants; they are considered safe and well tolerated, and they represent a widely accepted recommendation in high-risk individuals, such as those over 65 years of age. Nevertheless, they may contain thimerosal: in the formulation of multidose vials, it is used as a preservative to prevent bacterial overgrowth, and it may trigger the inflammatory response in susceptible subjects.

Medium-vessel vasculitides

The first case of medium-vessel vasculitis related to vaccine exposure was described in 1983 by Guillevin *et al.* (1983), who reported an exacerbation of pulmonary manifestations in the course of PAN following the administration of tetanus and BCG vaccines in a 19-year-old man.

Since then, with the exception of sporadic cases following influenza vaccine, all cases of PAN detected in the literature have followed the administration of HBV vaccine in adults (Guillevin *et al.*, 1983; Le Goff *et al.*, 1988; Allen *et al.*, 1993; Mader *et al.*, 1993; Kerleau *et al.*, 1997; Bouroz-Joly *et al.*, 1998; De Keyser *et al.*, 2000; Bourgeais *et al.*, 2003; Begier *et al.*, 2004; de Carvalho *et al.*, 2008; Ventura *et al.*, 2009; Gil *et al.*, 2011).

Case reports of medium-vessels vasculitide – both PAN and KD – have also been published in pediatric patients. KD has been described 1 day after the second dose of HBV vaccine and following yellow fever vaccine (Miron *et al.*, 2003; Schmöeller *et al.*, 2009). Two additional cases of PAN have been reported: an 11-year-old child diagnosed with cutaneous PAN 1 week after HBV vaccination (Ventura *et al.*, 2009) and a 14-year-old boy who developed full-blown PAN 2 months after HBV vaccination (de Carvalho *et al.*, 2008). In both cases, the medical histories were unremarkable.

HBV vaccine is administered universally to all age groups, from neonates to adults. It is the second most frequently administered vaccine in the United States (Zhou *et al.*, 2003), and both plasma-derived and recombinant vaccines are considered safe for use. Despite the fact it has been declared "generally well tolerated," few large series on serious autoimmune adverse events have been reported (Grotto *et al.*, 1998): these conditions can be divided into transient conditions, such as HV,

Table 23.1 Vasculitides of large vessels following different types of vaccinations: summary of cases detected in the literature (source: PubMed/Medline)

Diagnosis	Sex, age (years)	Type of vaccine	Time interval	Other relevant data	References
GCA/PMR	F, 80	Inf-V	3 months		Soriano *et al.* (2012)
GCA[a]	F, 78	Inf-V	3 weeks		Soriano *et al.* (2012)
GCA[a]	F, 67	Inf-V	1 month		Soriano *et al.* (2012)
GCA[a]	F, 78	Inf-V	2 months		Soriano *et al.* (2012)
GCA[a]	F, 80	Inf-V	2 months		Soriano *et al.* (2012)
GCA[a]	F, 73	Inf-V	3 months		Soriano *et al.* (2012)
GCA[a]	F, 70	Inf-V	3 months		Soriano *et al.* (2012)
GCA[a]	F, 74	Inf-V	1 month		Soriano *et al.* (2012)
GCA	F, 70	Inf-V	1 day		Wada *et al.* (2011)
GCA[a]/MPA	F, 67	Inf-V	N/A		Konishi *et al.* (2011)
GCA/PMR	M,74	Inf-V	1 week		Pou *et al.* (2008)
GCA	F, 64	Inf-V	3 days		Saadoun *et al.* (2001)
GCA	M,70	Inf-V	Few days later		Finsterer *et al.* (2001)
GCA	M,76	Inf-V	1 week		Perez and Maravi (2000)
GCA	F, 76	Inf-V	1 week		Ghose *et al.* (1976)
TD	F, 19	rHBV-V (Engerix B)	2 months after the second dose[b]	Previous diagnosis of ankylosing spondilytis (AS) and AS history family	Zaas *et al.* (2001)
Large cell arteritis involving subclavian and renal arteries	F, 61	rHBV-V (Engerix B)	1 month after the second dose	PMR-like illness 7 ys before	Zaas *et al.* (2001)
TD	F, 26	HBV-V (plasma-derived)	3 months	Concomitant erythema nodosum and liver granuloma	Castresana-Isla *et al.* (1993)

[a]Biopsy-proven.

[b]Patient received the third dose of the vaccine despite the appearance of symptoms.

GCA, giant cell arteritis; PMR, polymyalgia rheumatica; TD, Takayasu disease; MPA, microscopic polyangiitis; Inf-V, seasonal influenza vaccination; HBV-V, hepatitis B vaccine; rHBV-V, recombinant hepatitis B vaccine; N/A, not available post-vaccinal arthritis, erythema nodosum; and onset or relapse of rheumatic (rheumatoid arthritis, lupus erythematosus, spondyloarthropathies, systemic vasculitides) or nonrheumatic (multiple sclerosis, etc.) chronic diseases. Several pathogenic models can be put forward to explain rheumatic disorders: the transient conditions could be explained by the deposition of circulating immune complexes containing viral antigen and anti-HBs antibodies, such as those observed in some cases of HBV infections, or by hypersensitivity to some components of the vaccines, such as aluminum or yeast proteins (e.g. *Saccharomyces cerevisiae*). On the other hand, the onset of chronic inflammatory or autoimmune diseases might be a result of molecular mimicry, and HBV immunization might trigger the onset, the relapse, or the exacerbation of these diseases in individuals with underlying genetic and immunological susceptibility (Maillefert *et al.*, 1999).

Small-vessel vasculitides: ANCA-associated vasculitides

There are 12 reported cases of antineutrophil cytoplasmic antibody (ANCA)-associated vasculitides following vaccination: three cases of eosinophilic granulomatosis with polyangiitis (EGPA), three cases of MPA, and six cases of GPA, in a period of over 30 years (Table 23.2). Guillevin *et al.* (1983) described the first case of EGPA following tetanus vaccination in 1983; two other cases have been reported following HBV vaccine, by Beretta *et al.* (2001) and Vanoli *et al.* (1998), 2 weeks and 1 month after the third dose of the vaccine, respectively. ANCAs were absent in one case (Vanoli *et al.*, 1998), and they were not available in the first case described by Guillevin *et al.* (1983). All the reported cases of MPA and GPA were related to influenza vaccines, but neither the brand manufacturer nor the ingredients of the vaccines have been specified. Interestingly enough, Birck *et al.* (2009) described a case of a patient with GPA in clinical remission for 2 years who developed an exacerbation of her disease 12 days after the influenza vaccine. Spaetgens *et al.* (2009) reported another case of exacerbation of GPA with fatal relapse in a 20-year-old man, to whom influenza vaccine was administered in the course of active disease (glomerulonephritis). In this case, the authors suggested that further activation of the vasculitic process after influenza vaccination was caused by so-called bystander activation, in which the vaccine caused the activation of antigen-presenting cells (APCs) expressing the autoantigen proteinase 3.

Immune complex small-vessel vasculitides

HSP is the most common vasculitis of childhood. It is generally benign and self-limited. HSP is mediated by IgA immune complex deposition in various tissues, as well as in small-sized blood vessels. Genetic risk factors play an important role in the pathogenesis of the disease: the association with HLA-DRB*01, 07, and 11 is one of the most convincing findings of susceptibility. It is known that immune complex deposition in HSP may be triggered by infections with distinct bacteria and viruses.

As seen with infectious agents, an immune complex may form between vaccine antigens and native antibodies, thereby initiating the vasculitic process. As a matter of fact, HSP has been observed following the administration of several vaccines, including seasonal influenza, influenza A (H1N1), pneumococcal and meningococcal disease, hepatitis A virus (HAV), HBV, anti-human papillomavirus (HPV), and multiple combinations of vaccines, including typhoid, cholera, and yellow fever, which were simultaneously administered in one case (Mastroiacovo, 1976; Damjanov and Amato, 1979; Yoshimoto *et al.*, 1987; Ledermann and Hoffbrand, 1993; Goodman *et al.*, 2010; Jariwala *et al.*, 2011; Pimentel *et al.*, 2011; Watanabe, 2011; Melo Gomes *et al.*, 2013). Mastroiacovo (1976) described the first case following smallpox vaccine in 1976, hypothesizing a causal relationship between vaccines and rheumatic disorders in the pediatric population for the first time. Interestingly, in a recent literature review of HSP cases following influenza vaccine, Watanabe (2011) observed in many cases a past history of food allergy, drug eruptions, or immunologically mediated disease in patients with post-vaccination HSP.

Reported cases of post-vaccine HSP have occurred in subjects aged between 1 and over 50 years; however, the dose sequence and type of vaccine are often not mentioned. The most commonly reported symptoms include arthralgias, rashes, joint swelling, abdominal pain, and proteinuria. The reported time course of symptom onset after vaccine administration ranges from a few hours to 14 days. However, where there is a short delay between vaccine administration and onset of clinical manifestations, the hypothesis of a hypersensitivity reaction to the vaccine or to any of its adjuvants and preservatives should be considered.

HV commonly denotes a form of small-vessel vasculitis that may clinically manifest either as

Table 23.2 ANCA-associated vasculitides following different types of vaccinations: summary of cases detected in literature (source: PubMed/Medline)

Diagnosis	Sex, age (years)	Type of vaccine	Time interval	Other relevant data	References
EGPA	M, 47	HBV-V	2 weeks after third dose		Beretta *et al.* (2001)
EGPA	F, 20	rHBV-V (Engerix B®)	1 month after third dose	ANCA absent	Vanoli *et al.* (1998)
EGPA	N/A, 54	Tetanus	Following days	ANCA N/A	Guillevin *et al.* (1983)
MPA	F, 55	Inf-V	3 weeks		Birck *et al.* (2009)
MPA	F, 83	Inf-V	7 days		Uji *et al.* (2005)
MPA	M, 34	Trivalent Inf-V	2 weeks		Kelsall *et al.* (1997)
GPA	F, 28	Inf-V	2 weeks		Birck *et al.* (2009)
GPA	M, 70	Inf-V	3 weeks		Birck *et al.* (2009)
GPA	F, 67	Inf-V	12 days	Exacerbation GPA in remission for 2 years	Birck *et al.* (2009)
GPA	M, 20	Inf-V, performed in course of active glomerulo nephritis	Not specified ("shortly after")	Exacerbation Fatal relapse	Spaetgens *et al.* (2009)
GPA	F, 75	Inf-V	4 weeks		Duggal *et al.* (2013)
GPA	M, 50	Inf-V	2 weeks		Duggal *et al.* (2013)

EGPA, eosinophilic granulomatosis with polyangiitis; MPA, microscopic polyangiitis; GPA, granulomatosis with polyangiitis; Inf-V, seasonal influenza vaccination; HBV-V, hepatitis B vaccine; rHBV-V, recombinant hepatitis B vaccine; N/A, not available

cutaneous disease only or as systemic vasculitis, when skin disease is associated with organ involvement. Hypersensitivity vasculitides are usually represented histopathologically as leukocytoclastic vasculitis (LV); the etiological cause of LV remains unknown in as many as 50% of cases.

In the literature, LV has been associated with several vaccines, including influenza vaccine (Yanai-Berar *et al.*, 2002; Tavadia *et al.*, 2003; Kasper *et al.*, 2004; Walker *et al.*, 2004; Hyla-Klekot *et al.*, 2005; Famularo *et al.*, 2006; Ulm *et al.*, 2006), HAV vaccine (Bani-Sadr *et al.*, 1996), HBV vaccine (Masse and Descoffres, 1998), pneumococcal vaccine (Fox and Peterson, 1998), varicella (Gerecitano *et al.*, 1997), rubella, smallpox (Hanissian *et al.*, 1973; Somer and Finegold, 1995) and the anthrax vaccine (Muniz, 2003). Following MMR vaccine, dermal vasculitis with panuveitis has also been described (Sedaghat *et al.*, 2007).

An abnormal activation of the immune system and a direct effect of the vaccine itself, involving triggering of autoreactive T cells or a deregulated cytokine network, have been described as possible mechanisms. The final result is the deposition of immune complexes at the vessels, which is the pathophysiologic hallmark of leukocytoclastic vasculitides. In other cases, they can result from hypersensitivity to some components of the vaccine. In some reported cases, there is evidence of systemic involvement. Tavadia *et al.* (2003) described four patients with cutaneous LV following influenza vaccine, with abnormal urinalysis suggesting an associated renal vasculitis. Interestingly, this abnormality resolved as the cutaneous signs improved. Three out of four patients had previously received influenza vaccine without adverse sequelae, pointing to the likelihood that, as the virus strains included in influenza vaccine change each year according to World Health Organization (WHO) recommendations, each vaccine has a unique potential to provoke immunologically mediated reactions in genetically predisposed subjects.

Conclusions

Because the incidence of post-vaccination vasculitides remains very low, vaccinations should not be limited for this reason. However, caution may be required with its use in children with immunologically mediated diseases such as HSP, as well as in adults with a known history of autoimmune disease, for whom a careful risk–benefit assessment is required. Some authors suggest that children with a history of vaccine-induced vasculitides should not be revaccinated, but there are currently no clear guidelines for what is the most appropriate management. It has been suggested that flu vaccines must be contraindicated in subjects with a history of rheumatoid purpura, with previous history of vasculitis after flu vaccination, or with active autoimmune rheumatic disease. Further investigations are needed to clarify the biologic plausibility of post-vaccination phenomena. Thus, surveillance systems and registries can be important tools for retrospective as well as prospective evaluations of cases, and also for establishing research studies aimed at elucidating genetic susceptibility factors.

Nevertheless, in clinical practice, when the suspicion of vasculitis onset is raised, a meticulous history-taking with special emphasis on vaccine history is imperative, so that an appropriate diagnosis can be established early and management can be initiated. Moreover, follow-up is mandatory to verify whether a different prognosis is associated with these diseases. The lack of evidence for other causes of the symptoms and the coincidence regarding the vaccination in most of the cases analyzed here strongly supports a causal relationship between the vaccination and vasculitis onset, especially where a plausible temporal association exists. For this reason, the modal peak in time of the onset has to be carefully examined, in order to detect cases with peak onset within a few days of vaccination (which are more consistent with hypersensitivity reactions) or within weeks or months (which are more consistent with delayed immune reactivity).

References

Abe, Y., Nakano, S., Aita, K., and Sagishima, M. (1998). Streptococcal and staphylococcal superantigen -induced lymphocytic arteritis in a local type experimental model: comparison with acute vasculitis in the Arthus reaction. *J Lab Clin Med*, **131**: 93–102.

Agmon-Levin, N. and Shoenfeld, Y. (2008). Chronic fatigue syndrome with autoantibodies-the result of an augmented adjuvant effect of hepatitis-B vaccine and silicone implant. *Autoimmun Rev*, **8**: 52–5.

Agmon-Levin, N., Kivity, S., Szyper-Kravitz, M., and Shoenfeld, Y. (2009a). Transverse myelitis and vaccines: a multi-analysis. *Lupus*, **18**: 1198–204.

Agmon-Levin, N., Paz, Z., Israeli, E., and Shoenfeld, Y. (2009b). Vaccines and autoimmunity. *Nat Rev Rheumatol*, **5**: 648–52.

Agnello, V., Chung, R.T., and Kaplan, L.M. (1992). A role for hepatitis C virus infection in type II cryoglobulinemia. *N Engl J Med*, **327**: 1490–5.

Allen, M.B., Cockwell, P., and Page, R.L. (1993). Pulmonary and cutaneous vasculitis following hepatitis B vaccination. *Thorax*, **48**: 580–1.

Bani-Sadr, F., Gueit, I., and Humbert, G. (1996). Vasculitis related to hepatitis A vaccination. *Clin Infect Dis*, **22**: 596.

Barbaud, A., Deschildre, A., Waton, J., *et al.* (2013). Hypersensitivity and vaccines: an update. *Eur J Dermatol*, **23**: 135–41.

Baskan, E.B., Yilmaz, E., Saricaoglu, H., *et al.* (2007). Detection of parvovirus B19 DNA in the lesional skin of patients with Behcet's disease. *Clin Exp Dermatol*, **32**: 186–90.

Bassi, N., Luisetto, R., Del Prete, D., *et al.* (2012). Induction of the "ASIA" syndrome in NZB/NZWF1 mice after injection of complete Freund's adjuvant (CFA). *Lupus*, **21**: 203–9.

Bechah, Y., Capo, C., Raoult, D., and Mege, J.L. (2008). Infection of endothelial cells with virulent Rickettsia prowazekii increases the transmigration of leukocytes. *J Infect Dis*, **197**: 142–7.

Beekhuizen, H., van de Gevel, J.S., Olsson, B., *et al.* (1997). Infection of human vascular endothelial cells with *Staphylococcus aureus* induces hyperadhesiveness for human monocytes and granulocytes. *J Immunol*, **158**: 774–82.

Begier, E.M., Langford, C.A., Sneller, M.C., *et al.* (2004). *Polyarteritis nodosa* reports to the Vaccine Adverse Event Reporting System (VAERS): implications for assessment of suspected vaccine-provoked vasculitis. *J Rheumatol*, **31**: 2181–8.

Beretta, L., Caronni, M., Vanoli, M., and Scorza, R. (2001). Churg-Strauss vasculitis with brain involvement following hepatitis B vaccination. *Clin Exp Rheumatol*, **19**: 757.

Birck, R., Kaelsch, I., Schnuelle, P., *et al.* (2009). ANCA-associated vasculitis following influenza vaccination: causal association or mere coincidence? *J Clin Rheumatol*, **15**: 289–91.

Blank, M., Krause, I., Fridkin, M., *et al.* (2002). Bacterial induction of autoantibodies to beta2 glycoprotein-I accounts for the infectious etiology of antiphospholipid syndrome. *J Clin Invest*, **109**: 797–804.

Bourgeais, A.M., Dore, M.X., Croue, A., *et al.* (2003). Cutaneous polyarteritis nodosa following hepatitis B vaccination. *Ann Dermatol Venereol*, **130**: 205–7.

Bouroz-Joly, J., Hachulla, E., Vermersch, P., *et al.* (1998). Between the infection and the vaccine, one must choose. *Rev Med Interne*, **19**(Suppl. 2): S286–9.

Brightman, C.A., Scadding, G.K., Dumbreck, L.A., *et al.* (1989). Yeast-derived hepatitis B vaccine and yeast sensitivity. *Lancet*, **1**: 903.

Buonocore, S., Flamand, V., Claessen, N., *et al.* (2004). Dendritic cells overexpressing Fas-ligand induce pulmonary vasculitis in mice. *Clin Exp Immunol*, **137**: 74–80.

Castresana-Isla, C.J., Herrera-Martinez, G., and Vega-Molina, J. (1993). Erythema nodosum and Takayasu's arteritis after immunization with plasma derived hepatitis B vaccine. *J Rheumatol*, **20**: 1417–18.

Chen, R.T., Pless, R., and Destefano, F. (2001). Epidemiology of autoimmune reactions induced by vaccination. *J Autoimmun*, **16**: 309–18.

Cioc, A.M., Sedmak, D.D., Nuovo, G.J., *et al.* (2002). Parvovirus B19 associated adult Henoch Schönlein purpura. *J Cutan Pathol*, **29**: 602–7.

Corman, L.C. and Dolson, D.J. (1992). Polyarteritis nodosa and parvovirus B19 infection. *Lancet*, **339**: 491.

Damjanov, I. and Amato, J.A. (1979). Progression of renal disease in Henoch-Schönlein purpura after influenza vaccination. *JAMA*, **242**: 2555–6.

de Carvalho, J.F., Pereira, R.M., and Shoenfeld, Y. (2008). Systemic polyarteritis nodosa following hepatitis B vaccination. *Eur J Intern Med*, **19**: 575–8.

De Keyser, F., Naeyaert, J.M., Hindryckx, P., *et al.* (2000). Immune-mediated pathology following hepatitis B vaccination. Two cases of polyarteritis nodosa and one case of pityriasis rosea-like drug eruption. *Clin Exp Rheumatol*, **18**: 81–5.

Duggal, T., Segal, P., Shah, M., *et al.* (2013). Antineutrophil cytoplasmic antibody vasculitis associated with influenza vaccination. *Am J Nephrol*, **38**: 174–8.

Famularo, G., Nicotra, G.C., Minisola, G., and De Simone, C. (2006). Leukocytoclastic vasculitis after influenza vaccination. *J Clin Rheumatol*, **12**: 48–50.

Ferri, C. and Zignego, A.L. (2000). Relation between infection and autoimmunity in mixed cryoglobulinemia. *Curr Opin Rheumatol*, **12**: 53–60.

Finkel, T.H., Török, T.J., Ferguson, P.J., *et al.* (1994). Chronic parvovirus B19 infection and systemic necrotising vasculitis: opportunistic infection or aetiological agent? *Lancet*, **343**: 1255–8.

Finsterer, J., Artner, C., Kladosek, A., *et al.* (2001). Cavernous sinus syndrome due to vaccination-induced giant cell arteritis. *Arch Intern Med*, **161**: 1008–9.

Fox, B.C. and Peterson, A. (1998). Leukocytoclastic vasculitis after pneumococcal vaccination. *Am J Infect Control*, **26**: 365–6.

Gerecitano, J., Friedman-Kien, A., and Chazen, G.D. (1997). Allergic reaction to varicella vaccine. *Ann Intern Med*, **126**: 833–4.

Gherardi, R.K. and Authier, F.J. (2003). Aluminum inclusion macrophagic myofasciitis: a recently identified condition. *Immunol Allergy Clin North Am*, **23**: 699–712.

Ghose, M.K., Shensa, S., and Lerner, P.I. (1976). Arteritis of the aged (giant cell arteritis) and fever of unexplained origin. *Am J Med*, **60**: 429–36.

Gil, E., Lutalo, P., and D'Cruz, D. (2011). Systemic vasculitis: a dual diagnosis. *BMJ Case Rep*, doi: 10.1136/bcr.10.2011.4968.

Gocke, D.J., Hsu, K., Morgan, C., *et al.* (1970). Association between polyarteritis and Australia antigen. *Lancet*, **2**: 1149–53.

Goodman, M.J., Nordin, J.D., Belongia, E.A., *et al.* (2010). Henoch-Schönlein purpura and polysaccharide meningococcal vaccine. *Pediatrics*, **126**: e325–9.

Grotto, I., Mandel, Y., Ephros, M., *et al.* (1998). Major adverse reactions to yeast-derived hepatitis B vaccines – a review. *Vaccine*, **16**: 329–34.

Guillevin, L. (2004). Virus-induced systemic vasculitides: new therapeutic approaches. *Clin Dev Immunol*, **11**: 227–31.

Guillevin, L., Guittard, T., Boura, R., *et al.* (1983). [Etiology and precipitating factors of necrotizing angiitis with respiratory manifestations. 5 case reports]. *Ann Med Interne (Paris)*, **134**: 625–8.

Hanissian, A.S., Martinez, A.J., Jabbour, J.T., and Duenas, D.A. (1973). Vasculitis and myositis secondary to rubella vaccination. *Arch Neurol*, **28**: 202–4.

Huugen, D., Xiao, H., van Esch, A., *et al.* (2005). Aggravation of anti myeloperoxidase antibody-induced glomerulonephritis by bacterial lipopolysaccharide: role of tumor necrosis factor alpha. *Am J Pathol*, **167**: 47–58.

Hyla-Klekot, L., Kucharska, G., and Cieslak, W. (2005). Necrotizing glomerulonephritis in decursu vasculitis after vaccination against influenza. *Pol Merkur Lekarski*, **19**: 75–7.

Jariwala, S., Vernon, N., and Shliozberg, J. (2011). Henoch-Schönlein purpura after hepatitis A vaccination. *Ann Allergy Asthma Immunol*, **107**: 180–1.

Jennette, J.C., Falk, R.J., Bacon, P.A., *et al.* (2013). 2012 revised International Chapel Consensus Conference Nomenclature of Vasculitides. *Arthritis Rheum*, **65**(1): 1–11.

Jonville-Béra, A.P., Autret, E., Galy-Eyraud, C., and Hessel, L. (1996). Thrombocytopenic purpura after measles, mumps and rubella vaccination: a retrospective survey by the French regional pharmacovigilance centres and Pasteur-Mérieux sérums et vaccins. *Pediatr Infect Dis J*, **15**(1): 44–8.

Kasper, H.U., Dries, V., Drebber, U., *et al.* (2004). Florid ischemic cholangitis due to leukocytoclastic vasculitis. *J Gastroenterol*, **39**: 188–91.

Katzav, A., Kivity, S., Blank, M., *et al.* (2012). Adjuvant immunization induces high levels of pathogenic antiphospholipid antibodies in genetically prone mice: another facet of the ASIA syndrome. *Lupus*, **21**: 210–16.

Kelsall, J.T., Chalmers, A., Sherlock, C.H., *et al.* (1997). Microscopic polyangiitis after influenza vaccination. *J Rheumatol*, **24**: 1198–202.

Kelso, J.M., Jones, R.T., and Yunginger, J.W. (1993). Anaphylaxis to measles, mumps, and rubella vaccine mediated by IgE to gelatin. *J Allergy Clin Immunol*, **91**: 867–72.

Kerleau, J.M., Lévesque, H., Lair, G., *et al.* (1997). Is hepatitis B vaccination a new cause of necrotizing vasculitis? *Rev Med Interne*, **18**: 491–2.

Kivity, S., Agmon-Levin, N., Blank, M., and Shoenfeld, Y. (2009). Infections and autoimmunity – friends or foes? *Trends Immunol*, **30**: 409–14.

Konishi, M., Koarada, S., Yamaguchi, K., *et al.* (2011). Case of mycroscopic polyangiitis and giant cell arteritis after influenza vaccination. *Nihon Rinsho Meneki Gakkai Kaishi*, **34**: 154–61.

Le Goff, P., Fauquert, P., Youinou, P., and Hoang, S. (1988). Periarteritis nodosa following vaccination against hepatitis B. *Presse Med*, **17**: 1763.

Ledermann, J.A. and Hoffbrand, B.I. (1993). Dapsone in allergic vasculitis: its use in Henoch-Schönlein disease following vaccination. *J R Soc Med*, **76**: 613–14.

Lehmann, H.W., von Landenberg, P., and Modrow, S. (2003). Parvovirus B19 infection and autoimmune disease. *Autoimmun Rev*, **2**: 218–23.

Leventhal, J.S., Berger, E.M., Brauer, J.A., and Cohen, D.E. (2012). Hypersensitivity reactions to vaccine constituents: a case series and review of the literature. *Dermatitis*, **23**: 102–9.

Lidar, M., Lipschitz, N., Langevitz, P., and Shoenfeld, Y. (2009). The infectious etiology of vasculitis. *Autoimmunity*, **42**: 432–8.

Mader, R., Narendran, A., Lewtas, J., *et al.* (1993). Systemic vasculitis following influenza vaccination – report of three cases and literature review. *J Rheumatol*, **20**: 1429–31.

Maillefert, J.F., Sibilia, J., Toussirot, E., *et al.* (1999). Rheumatic disorders developed after hepatitis B vaccination. *Rheumatology*, **38**: 978–83.

Marks, J.S. and Halpin, T.J. (1980). Guillain-Barré syndrome in recipients of A/New Jersey influenza vaccine. *JAMA*, **243**: 2490–4.

Masse, I. and Descoffres, M.C. (1998). Hypersensitivity vasculitis after hepatitis B vaccination. *Presse Med*, **27**: 965–6.

Mastroiacovo, P. (1976). Measles vaccination and Schönlein-Henoch purpura. *Minerva Pediatr*, **28**: 1591.

Mathieson, P.W., Qasim, F.J., Esnault, V.L., and Oliveira, D.B. (1993). Animal models of systemic vasculitis. *J Autoimmun*, **6**: 251–64.

Matussek, A., Strindhall, J., Stark, L., *et al.* (2005). Infection of human endothelial cells with Staphylococcus aureus induces transcription of genes encoding an innate immunity response. *Scand J Immunol*, **61**: 536–44.

Melo Gomes, S., Glover, M., Malone, M., and Brogan, P. (2013). Vasculitis following HPV immunization. *Rheumatology (Oxford)*, **52**: 581–2.

Millikan, L.E. and Flynn, T.C. (1999). Infectious etiologies of cutaneous vasculitis. *Clin Dermatol*, **17**: 509–14.

Miron, D., Fink, D., and Hashkes, P.J. (2003). Kawasaki disease in an infant following immunisation with hepatitis B vaccine. *Clin Rheumatol*, **22**: 461–3.

Misiani, R., Bellavita, P., Fenili, D., *et al.* (1992). Hepatitis C virus infection in patients with essential mixed cryoglobulinemia. *Ann Intern Med*, **117**: 573–7.

Molina, V. and Shoenfeld, Y. (2005). Infection, vaccines and other environmental triggers of autoimmunity. *Autoimmunity*, **38**: 235–45.

Muniz, A.E. (2003). Lymphocytic vasculitis associated with the anthrax vaccine: case report and review of anthrax vaccination. *J Emerg Med*, **25**: 271–6.

Nakayama, T., Aizawa, C., and Kuno-Sakai, H. (1999). A clinical analysis of gelatin allergy and determination of its causal relationship to the previous administration of gelatin containing acellular pertussis vaccine combined with diphtheria and tetanus toxoids. *J Allergy Clin Immunol*, **103**: 321–5.

Orbach, H., Agmon-Levin, N., and Zandman-Goddard, G. (2010). Vaccines and autoimmune diseases of the adult. *Discov Med*, **9**: 90–7.

Perez, C. and Maravi, E. (2000). Polymyalgia rheumatica following influenza vaccination. *Muscle Nerve*, **23**: 824–5.

Piaggio, E., Ben Younes, A., Desbois, S., *et al.* (2005). Hepatitis B vaccination and central nervous system demyelination: an immunological approach. *J Autoimmun*, **24**: 33–7.

Pimentel, M.I., Vasconcellos Ede, C., and Cerbino-Neto, J. (2011). Henoch-Schönlein purpura following influenza A H1N1 vaccination. *Rev Soc Bras Med Trop*, **44**: 531.

Pou, M.A., Diaz-Torne, C., Vidal, S., *et al.* (2008). Development of autoimmune diseases after vaccination. *J Clin Rheumatol*, **14**: 243–4.

Rodriguez-Pla, A. and Stone, J.H. (2006). Vasculitis and systemic infections. *Curr Opin Rheumatol*, **18**: 39–47.

Saadoun, D., Cacoub, P., Mahoux, D., *et al.* (2001). Postvaccine vasculitis: a report of three cases. *Rev Med Interne*, **22**: 172–6.

Satoh, M., Kuroda, Y., Yoshida, H., *et al.* (2003). Induction of lupus autoantibodies by adjuvants. *J Autoimmun*, **21**: 1–9.

Schmöeller, D., Keiserman, M.W., Staub, H.L., *et al.* (2009). Yellow fever vaccination and Kawasaki disease. *Pediatr Infect Dis*, **28**: 1037–8.

Sedaghat, M., Zarei-Ghanavati, S., Shokoohi, S., and Ghasemi, A. (2007). Panuveitis and dermal vasculitis following MMR vaccination. *East Mediterr Health J*, **13**: 470–4.

Shaw, C.A. and Petrik, M.S. (2009). Aluminum hydroxide injections lead to motor deficits and motor neuron degeneration. *J Inorg Biochem*, **103**: 1555–62.

Shoenfeld, Y. (2009). Infections, vaccines and autoimmunity. *Lupus*, **18**: 1127–8.

Shoenfeld, Y. and Agmon-Levin, N. (2011). "ASIA"–autoimmune/inflammatory syndrome induced by adjuvants. *J Autoimmun*, **36**: 4–8.

Shoenfeld, Y., Zandman-Goddard, G., Stojanovich, L., *et al.* (2008). The mosaic of autoimmunity: hormonal and environmental factors involved in autoimmune diseases. *IMAJ*, **10**: 8–12.

Somer, T. and Finegold, S.M. (1995). Vasculitides associated with infections, immunization, and antimicrobial drugs. *Clin Infect Dis*, **20**: 1010–36.

Soriano, A., Verrecchia, E., Marinaro, A., *et al.* (2012). Giant cell arteritis and polymyalgia rheumatica after influenza vaccination: report of 10 cases and review of the literature. *Lupus*, **21**: 153–7.

Spaetgens, B., van Paassen, P., and Tervaert, J.W. (2009). Influenza vaccination in ANCA-associated vasculitis. *Nephrol Dial Transplant*, **24**: 3258.

Strassburg, C.P., Vogel, A., and Manns, M.P. (2003). Autoimmunity and hepatitis C. *Autoimmun Rev*, **2**: 322–31.

Tavadia, S., Drummond, A., Evans, C.D., and Wainwright, N.J. (2003). Leukocytoclastic vasculitis and influenza vaccination. *Clin Exp Dermatol*, **28**: 154–6.

Tishler, M. and Shoenfeld, Y. (2004). Vaccination may be associated with autoimmune diseases. *IMAJ*, **6**: 430–2.

Toplak, N. and Avcin, C. (2009). Vaccination of healthy subjects and autoantibodies: from mice through dogs to humans. *Lupus*, **18**: 1186–91.

Trepo, C. and Guillevin, L. (2001). Polyarteritis nodosa and extrahepatic manifestations of HBV infection: the case against autoimmune intervention in pathogenesis. *J Autoimmun*, **16**: 269–74.

Ulm, S., Hummel, M., Emig, M., *et al.* (2006). Leukocytoclastic vasculitis and acute renal failure after influenza vaccination in an elderly patient with myelodysplastic syndrome. *Onkologie*, **29**: 470–2.

Uji, M., Matsushita, H., and Iwata, S. (2005). Mycroscopic polyangiitis after influenza vaccination. *Intern Med*, **44**: 892–6.

Vanoli, M., Gambini, D., and Scorza, R. (1998). A case of Churg-Strauss vasculitis after hepatitis B vaccination. *Ann Rheum Dis*, **57**: 256–732.

Ventura, F., Antunes, H., Brito, C., *et al.* (2009). Cutaneous polyarteritis nodosa in a child following hepatitis B vaccination. *Eur J Dermatol*, **19**: 400–1.

Wada, M., Asai, J., Asai, A., *et al.* (2011). Giant cell arteritis with polymyalgia rheumatica associated with influenza vaccination. *J Dermatol*, **38**: 1099–101.

Walker, S.L., Swindells, K.J., and Chalmers, R.J. (2004). Leukocytoclastic vasculitis and influenza immunization. *Clin Exp Dermatol*, **29**: 95–6.

Watanabe, T. (2011). Henoch-Schönlein purpura following influenza vaccinations during the pandemic of influenza A (H1N1). *Pediatr Nephrol*, **26**: 795–8.

Weck, K.E., Dal Canto, A.J., Gould, J.D., *et al.* (1997). Murine gamma-herpesvirus 68 causes severe large-vessel arteritis in mice lacking interferon-gamma responsiveness: a new model for virus induced vascular disease. *Nat Med*, **3**: 1346–53.

Witort-Serraglini, E., Del Rosso, M., Lotti, T.M., and Matucci Cerininc, M. (1999). Endothelial injury in vasculitides. *Clin Dermatol*, **17**: 587–90.

Yanai-Berar, N., Ben-Itzhak, O., Gree, J., and Nakhoul, F. (2002). Influenza vaccination induced leukocytoclastic vasculitis and pauci-immune crescentic glomerulonephritis. *Clin Nephrol*, **58**: 220–3.

Yoshimoto, T., Matsuda, T., and Kamada, M. (1987). A case of Schönlein-Henoch purpura and subsequently occurred purpuric nephritis after influenza vaccination. *Hako-Ishi*, **11**: 77–80.

Zaas, A., Scheel, P., Venbrux, A., and Hellmann, D.B. (2001). Large artery vasculitis following recombinant hepatitis B vaccination: 2 cases. *J Rheumatol*, **28**: 1116–20.

Zhou, W., Pool, V., Iskander, J.K., *et al.* (2003). Surveillance for safety after immunization: Vaccine Adverse Events Report System (VAERS), United States 1991–2001. *MMWR*, **52**: 113.

24 Vaccinations in Rheumatoid Arthritis

Eitan Giat[1] and Merav Lidar[1,2]
[1]Rheumatology Unit, Sheba Medical Center, Tel Hashomer, Israel
[2]Sackler Faculty of Medicine, Tel Aviv University, Tel Aviv, Israel

Pathogenesis of rheumatoid arthritis

Rheumatoid arthritis (RA) is a systemic, autoimmune, inflammatory disease primarily affecting synovial joints. Both genetic and environmental factors are involved in the pathogenesis of RA. The strongest (Legrand *et al.*, 1984) genetic risk factor comes from major histocompatibility complex (MHC) class II HLA-DR4, which contains a "shared epitope" (Gregersen *et al.*, 1987), as well as from PTPN22 (Begovich *et al.*, 2004), PADI4 (Suzuki *et al.*, 2003), CTLA4 (Seidl *et al.*, 1998), and so on. Smoking is the strongest environmental risk factor for RA, but other factors, such as lifestyle factors, nutrition, alcohol consumption, and pollution, also play a role (Symmons *et al.*, 1997).

A pre-articular phase has been shown to precede the clinical phase of RA by many years (Firestein, 2003; McInnes *et al.*, 2007). This phase is characterized by an autoimmune state in which autoantibodies targeting IgG (rheumatoid factor) or cyclic citrullinated peptides and inflammatory cytokines are present. Smoking, which can induce citrullination of self-proteins, may accelerate the autoimmune process, which in turn initiates an inflammatory cascade in the joint, culminating in damage to the synovium and adjacent tissue.

The genetic association of MHC class II with RA implicates an important role for T cells in the early immune process. Genome-wide association studies have disclosed an association with additional T cell-related genes, such as PTPN22 (Chang *et al.*, 2008). This gene encodes a protein tyrosine phosphatase that is involved in signaling pathways of the immune response. The mutation is thought to alter the responsiveness of T and B cell receptors, promoting an autoimmune response. $CD4^+$ T cells, which make up more than half of the synovial cell population, are paramount in synovial inflammation, where they can induce the production of a repertoire of autoantibodies against citrullinated antigens (reviewed by Fousteri *et al.*, 2013).

The importance of B lymphocytes in RA pathogenesis is underscored by the presence of autoantibodies. Seropositive RA patients – that is, patients expressing rheumatoid factor – are prone to complications such as bone erosions and extra-articular manifestations (van Zeben *et al.*, 1992). In addition, most RA patients are positive for anticitrullinated peptide antibodies (ACPAs), which are relatively specific for RA (Avouac *et al.*, 2006) and are implicated in the pathogenesis of RA (Rantapaa-Dahlqvist *et al.*, 2003; Kuhn *et al.*, 2006). Citrullinated proteins can induce autoantibodies, such as ACPA, which can perpetuate the initial inflammatory process (Reparon-Schuijt *et al.*, 2001) by targeting antigens in the synovia. Both type II collagen (He *et al.*, 2000; Backlund *et al.*, 2002; Nissim *et al.*, 2005) and cartilage antigen glycoprotein-39 (gp39) (Cope *et al.*, 1999; Baeten *et al.*, 2004) have been implicated as putative self-antigens.

Proinflammatory cytokines are paramount in the pathogenesis of RA. GM-CSF is produced

Vaccines and Autoimmunity, First Edition. Edited by Yehuda Shoenfeld, Nancy Agmon-Levin, and Lucija Tomljenovic.

by synovial macrophages and itself activates antigen-presenting cells (APCs) and induces differentiation and proliferation of macrophages (Alvaro-Gracia *et al.*, 1989). Tumor necrosis factor (TNF) is a key player in inflammation. It induces proliferation of both T cells and B cells and upregulates the expression of endothelial adhesion molecules and the expression of proinflammatory molecules, such as collagenase, matrix metalloproteinase-3, and prostaglandins, by synovial cell inflammation (Brennan *et al.*, 1992). IL-17, which is produced by Th17 cells, upregulates cyclooxygenase-2 (COX-2) and prostaglandin E2 expression and enhances the activation of osteoclasts (Stamp *et al.*, 2004a,b) resulting in collagen destruction and bone resorption. The blocking of these cytokines has been shown to be effective in treating RA, establishing their role in the inflammatory process (van den Berg *et al.*, 2013).

The proliferation and activation of immune cells in the joints strongly depends on a supporting vascular bed. Inflammation is accompanied by the formation of new synovial blood vessels, which is critical for the development of RA. The newly formed blood vessels express adhesion molecules such as ICAM-1 (Gerritsen *et al.*, 1993), E-selectin, and P-selectin (Lally *et al.*, 2005), which promote extravasation of leukocytes to the synovium. Chemokines such as CXCL8/IL-8 and CXCL5/ENA-78 enhance the migration of immune cells to the synovial tissue (Haringman *et al.*, 2004). Synoviocytes, in turn, contribute to the inflammatory process by producing metalloproteinases, proteases, and other enzymes that mediate cartilage and bone destruction (Lipsky, 2007).

Do vaccinations induce RA?

It is estimated that RA has a heritability of 60%, leaving a substantial proportion of risk to environmental factors. Immunizations have previously been proposed as potential environmental triggers for RA. In the Norfolk Arthritis Register database, 19 of the first 588 patients reported receiving a tetanus vaccination within 6 weeks prior to the onset of arthritis. Similarly, a transient rise in RA titer was recorded in 10 out of 245 military recruits 2–3 weeks after receiving concomitant immunization against tetanus, typhoid, paratyphoid, mumps, diphtheria, polio, and smallpox. However, only two showed a persistent elevation in titer and none developed arthritis (Symmons *et al.*, 1993).

Several mechanisms have been proposed to explain the putative association between vaccination and the initiation of RA (Symmons *et al.*, 1993), the most prominent of which are molecular mimicry and nonspecific immune system activation. Infectious agents have long been suggested as a triggering agent for RA, despite lack of evidence for an infectious agent in the inflamed joint (Harris, 1990). However, as an infectious agent may ignite an autoimmune process that perpetuates long after the infectious agent has been eradicated (Wilder *et al.*, 1991), the hypothesis remained unshaken. Some have suggested that, since immunizations tend to mimic infectious agents, they are able to initiate an autoimmune process in a similar manner (Symmons *et al.*, 1993).

Alternatively, the adjuvant administered in the vaccine, which is used to boost the immune system, may be the trigger of the autoimmune cascade (Shoenfeld *et al.*, 2011). The ability of an adjuvant to induce arthritis and autoimmunity has been shown in several animal models (Cruz-Tapias *et al.*, 2013). In adjuvant-induced arthritis (AIA, also known as collagen-induced arthritis) models, susceptible rats immunized with complete Freund's adjuvant (CFA) containing a high concentration of Mycobacterium tuberculosis develop arthritis, which resembles reactive arthritis (Trentham *et al.*, 1977). The intraperitoneal injection of pristane, an alkane found in mineral oil that is commonly used to obtain ascitic fluid enriched in antibodies, can induce a lupus-like disease in some mouse strains and an RA-like disease in some rat strains (Vingsbo *et al.*, 1996). Aluminum adjuvant has been implicated in Gulf War syndrome (GWS), in which soldiers developed a variety of symptoms, including joint and muscle aches, rash, sleep disorders, and chronic fatigue, following deployment in Iraq (Israeli *et al.*, 2009). As numerous immunizations containing relatively high amounts of aluminum were administered to these soldiers prior to deployment, cause and effect has been suggested. However, GWS is an ill-defined illness and drawing conclusions about its pathogenesis is complicated (Eisen *et al.*, 2005; Institute of Medicine, 2010). Importantly, aluminum adjuvants are employed in numerous vaccinations administered all over the world (Sivakumar *et al.*, 2011), so it is likely that other factors unique to the Gulf War contributed to the development of this chronic multisymptom illness, in addition to, or instead of, aluminum. Moreover, if aluminum was the culprit in this ill-defined syndrome, it could well have been through a neurotoxic mechanism rather than an autoimmune one (Petrik *et al.*, 2007). Indeed, high levels of aluminum have been shown to cause dementia in dialysis patients.

Despite the concern that vaccines may induce autoimmunity via mechanisms of molecular mimicry or adjuvant stimulation, human studies have failed to show a clear association between vaccine administration and the development of RA. As previously mentioned, Leonard and Robertson (1956) reported both monoarthritis and polyarthritis in military recruits immunized against typhoid, tetanus, and smallpox. During the 1976 swine flu outbreak, only 100 of the 44 million Americans who were immunized later filed claims of rheumatologic complications (Symmons *et al.*, 1993): a rate that is far lower than would be expected for a population of that size had it not been vaccinated (Kurland *et al.*, 1984). The Norfolk Arthritis Register initially reported an increased incidence of tetanus immunizations within the 6 weeks prior to the onset of arthritis among 588 patients with new-onset arthritis (Symmons *et al.*, 1993). However, this observation was uncontrolled, and was not repeated in later reports from the same group.

Although arthritis is thought to be common following rubella vaccination, a randomized, placebo-controlled, double-blind study of 546 healthy women aged 18–41 indicated only a marginal increased incidence of acute joint manifestations in rubella vaccine recipients (30%) compared to placebo recipients (20%) (Tingle *et al.*, 1997). Similarly, a larger, retrospective, case-matched, computerized data-based cohort study of 971 women aged 15–59 years found no evidence of increased risk of new-onset RA following administration of rubella vaccine (Ray *et al.*, 1997).

RA has also been reported following hepatitis B virus (HBV) vaccination in small case series, and RA flares have been documented in HBV vaccine recipients. However, a controlled study failed to confirm these observations (reviewed by Elkayam *et al.*, 2002). The WHO's Global Advisory Committee on Vaccine Safety (2008) has concluded that there is no convincing evidence supporting an association between HBV and RA. Taken together, the sum of these reports indicate that, in spite of the hypothetical role immunizations may play in the initiation of RA, clinical data do not support such an association. It may be concluded that vaccines play an insignificant role in the pathogenesis of RA.

Safety and efficiency of vaccinations in RA patients

RA patients are exposed to an increased risk of infection through the pathomechanisms of their disease and the immunosuppressive therapy they receive to control it. Vaccination is by far the safest and most effective strategy for preventing serious infections. Lamentably, actual vaccination rates among RA patients are lower than in the general population (Feuchtenberger *et al.*, 2012), partly due to a misperception among caring physicians that their patients' relative immunosuppressive state is a contraindication to vaccination. In this respect, it is important to differentiate between live-attenuated vaccinations, which may indeed be hazardous to immunosuppressed patients, and vaccinations containing killed or recombinant agents. Also, when assessing the safety of vaccines in RA patients, the level of immunosuppression, including steroid dose, disease-modifying antirheumatic drug (DMARD), and biologic agent use, should be taken into account.

Although safety concerns are overemphasized, the lowered efficacy of vaccinations in RA is less commonly appreciated. For example, in a study from the Netherlands in which 400 patients with juvenile idiopathic arthritis (JIA) were compared to 2176 healthy controls aged 1–19 years, lower antibody concentrations and seroprotective titers against mumps, rubella, diphtheria, and tetanus (but not measles) were observed in the former (Heijstek *et al.*, 2012). It has been shown that the immune response mounted by different vaccinations in patients with RA varies, possibly depending on concomitant medications (Hua *et al.*, 2013). As there is a lack of studies addressing clinical end points in this patient population, and as the humoral response may not reflect clinical effectiveness (van Assen *et al.*, 2011), it is conceivable that even when immune response is impaired, vaccines confer protection in RA patients that is comparable to that of the general population. Clinical practice shows us that this may indeed be the case.

The European League Against Rheumatism (EULAR) 2011 task force recommendations (van Assen *et al.*, 2011) include the assessment of vaccination status as part of the initial investigation of patients presenting with a suspected autoimmune disease. Besides routine immunizations, initial work-up should also include vaccination history of hepatitis A, HBV, influenza, *Neisseria meningitides*, rubella (for women of childbearing age), *Streptococcus pneumoniae*, and tetanus toxoid (TTd). If any of these vaccinations are missing, it is recommended that they be promptly administered. The EULAR task force recommends vaccinating during times of disease stability, though this is based on expert opinion only, as most vaccination studies have excluded unstable patients.

Both the EULAR and the American College of Rheumatology (ACR) (Singh *et al.*, 2012) recommend avoiding all live-attenuated vaccines (MMR, varicella, live-attenuated influenza vaccine, herpes zoster vaccine (HZV), yellow fever, oral typhoid, bacillus Calmette–Guérin (BCG), and rotavirus) in immunosuppressed patients whenever possible, though the degree of immunosuppression rendering a vaccine unsafe has not been determined in RA and is still a matter of ongoing controversy. The Advisory Committee on Immunization Practices (ACIP) of the US Centers for Disease Control and Prevention (CDC) recommends avoiding live-attenuated vaccines in immunosuppressed patients, yet permits administration of live vaccines (such as HZV) to patients treated with short-term corticosteroid therapy (<14 days), moderate doses of corticosteroids (<20 mg/day prednisone or equivalent), intra-articular, bursal, or tendon corticosteroid injections, long-term alternate-day treatment with low to moderate doses (<20 mg/day prednisone or equivalent) of short-acting systemic corticosteroids, and methotrexate (MTX, <0.4 mg/kg/week), azathioprine (<3.0 mg/kg/day), or 6-mercaptopurine (<1.5 mg/kg/day) (Kroger *et al.*, 2006). Similarly, the EULAR task force recommends not administrating live vaccines to immunosuppressed patients in general, while noting that MMR, varicella, and HZV may be considered in those with mild levels of immunosuppression (van Assen *et al.*, 2011).

The following sections will discuss data concerning vaccines that have special safety or efficacy issues in RA. Additional vaccines are summarized in Table 24.1.

Measles, mumps, and rubella vaccine

The rubella vaccine that has been used for the past 30 years is the Wistar RA 27/3, which contains a live-attenuated virus, conferring lifelong protection in almost all cases. Given that reactive arthritis is elatively common following rubella vaccination, occurring in about 5% of subjects, concern over rubella vaccine-induced RA disease flare is prevalent. However, controlled studies have failed to show an increased frequency of persistent arthritis or arthralgia in rubella vaccine recipients (Elkayam *et al.*, 2002).

The current mumps vaccine is a Jeryl Lynn strain of a live-attenuated mumps virus. This vaccine was licensed in 1967 and, like Wistar RA 27/3, confers lifelong protection in most individuals. The mumps vaccine has relatively few, if any, reported adverse effects, besides transient, mild allergic reactions.

The measles vaccine available today is a live-attenuated Edmonston–Enders (Moraten) virus strain, and its most common adverse effect is transient fever.

These vaccines are not commercially available as a single-antigen vaccine but rather as a combination of measles, mumps, and rubella vaccines (MMR) or of measles, mumps, rubella, and varicella vaccines (MMRV).

As most MMR immunizations are performed in childhood, there are few data available concerning the effect of MMR administration in adults. The safety of the MMR vaccine was confirmed in large studies performed in pediatric JIA patients. While patients were found to have lower antibody concentrations and seroprotection rates against mumps and rubella compared with healthy controls, they had higher antibody concentrations and similar protection rates against measles. A booster MMR immunization had no effect on disease activity and was immunogenic in a randomized controlled study among 137 children with JIA.

As the MMR vaccine contains live viruses, both the EULAR and the ACR recommend against vaccinating immunosuppressed patients. However, the EULAR task force emphasizes that the MMR may be safe in immunosuppressed patients and permits its administration in mildly immunosuppressed patients on a case-by-case basis.

Diphtheria, tetanus, and pertussis vaccine

Since the introduction of the diphtheria toxoid, diphtheria has become rare in the industrialized world, yet it continues to occur elsewhere. The diphtheria toxoid is not available as a single antigen, but rather as a combination with either TTd (DT,Td), the acellular pertussis vaccine (DTaP, TdaP), or the HBV or polio vaccine (DTaP-HepB-IPV (Pediarix) and DTaP-IPV/Hib (Pentacel)). According to CDC guidelines, immunosuppression is not a contraindication to receiving diphtheria toxoid.

Tetanus is an acute and often fatal disease that occurs worldwide. The pathogenic organism, *C. tetani*, can be found in soil and in the intestinal tracts of animals and humans. Reported cases of tetanus infection have steadily declined since the introduction of an antitoxin and the subsequent

Table 24.1 Summary of the efficacy and safety of vaccines in RA patients, and recommendations for their use

Pathogen	Vaccine	Indications and efficacy	Safety
Measles	Wistar RA 27/3 live-attenuated virus, available only in combination with mumps and rubella (MMR)	Indicated in children and high-risk patients Scarce data in adults, but pediatric data suggest MMR to be immunogenic	Some studies suggest arthritis as a possible side effect of rubella vaccination, but larger cohorts and randomized trials show no increase in disease activity MMR is contraindicated in immunosuppressed patients
Mumps	Jeryl Lynn live-attenuated virus, available only in combination with measles and rubella (MMR)	See Measles	See Measles
Rubella	Edmonston–Enders (Moraten) live-attenuated virus, available only in combination with measles and rubella (MMR)	See Measles	See Measles
Diphtheria	Inactivated diphtheria toxin, unavailable as single vaccine	Routinely recommended in childhood and in nonimmunized persons Seems efficient in RA patients	No contraindications, including immunocompromised hosts
Tetanus	Toxoid, available alone or in combination with other vaccines (pertussis, diphtheria, *Haemophilus influenzae*, hepatitis B)	Efficacy in RA and in the immunosuppressed is unclear No specific recommendations in RA patients by ACR In case of high risk of infection within 24 weeks of rituximab, EULAR recommends passive immunization	Safe in RA and in immunosuppressed patients
Pertussis	Acellular vaccine, available only in combination with other vaccines	Efficacy is usually 80–85% in children, but effect may wane with time Few data exist regarding efficacy in RA patients No special recommendations in RA	Considered safe in RA
Streptoccocus pneumoniae	23-valent polysaccharide vaccine (23PSV) and 13-valent conjugate vaccine (13PCV)	ACIP recommends 23PSV for all adults over 65 years and for immunocompromised or hyposplenic/asplenic patients under 65 ACIP recommends 13PCV for immunocompromised hosts Efficacy is impaired in patients receiving rituximab and methotrexate, and may also be impaired in anti-TNF therapy EULAR task force recommends vaccinating with 23PSV before B cell-depleting therapy ACR recommends 23PSV and, for immunocompromised or hyposplenic/asplenic patients, a one-time 23PSV revaccination or 13PCV after 5 years	Seems to be safe in RA No contraindications in immunosuppressed patients

(continued)

Table 24.1 (*Continued*)

Pathogen	Vaccine	Indications and efficacy	Safety
Influenza	Trivalent inactivated influenza vaccine and trivalent live-activated influenza vaccine	ACIP recommends annual inactivated influenza vaccination for all persons aged 6 months and older ACR and EULAR strongly recommend vaccinating RA patients with inactivated vaccine, preferably before the initiation of DMARDs or biological treatment Inactivated vaccine seems to be efficient in RA patients, even under DMARDs or biological treatment, with the exception of rituximab	Live-attenuated vaccine is not recommended for anyone over 50 years old Live-attenuated vaccine is contraindicated in immunosuppressed patients Inactivated virus is safe in RA patients
Hepatitis B	Recombinant HBsAg: Recombivax HB (Merck), Engerix-B (GlaxoSmithKline).	Routinely recommended in childhood and in high-risk adults Increased risk of infection in RA, especially if undergoing anti-TNF therapy Reactivation reported following immunosuppressive treatment or immediately after its discontinuation Recommended by ACR only if hepatitis risk factors are present (e.g. intravenous drug abuse, multiple sex partners in the previous 6 months, health care personnel), preferably before initiating DMARD/biological therapy Recommended by EULAR only with increased risk of infection (travel to or residence in endemic countries, increased risk of exposure due to medical profession, infected family member or contact), in the absence of protective antibodies	Post-vaccine arthritis reported by some, but disproved by larger and controlled studies
Varicella zoster	Live-attenuated vaccine	Low (60–70%) efficacy in adults and very low (18%) efficacy in elderly (over 80), but decreases the severity of the disease and post-herpetic neuralgia No efficacy data in RA Recommended by ACIP for RA patients not under immunosuppressive therapy	Contraindicated in immunosuppressed patients ACR and EULAR recommend against vaccination of RA patients on biological therapy EULAR recommends vaccinating only patients seropositive for varicella zoster Recent data suggest varicella zoster vaccine to be safe even with concurrent biological therapy
Tuberculosis	BCG (live bacterial vaccine	CDC guidelines in adults: consider in high-risk health personnel	Risk of dissemination in immunosuppressed individuals Not recommended in RA by EULAR guidelines Contraindicated by ACR guidelines with concomitant anti-TNF therapy
Cholera	Inactivated vaccine (Dukoral, mORC-Vax, Shanchol)	Not routinely recommended Not recommended to all travelers Low efficacy	Inadequate data in RA No special contraindications in RA

Table 24.1 (*Continued*)

Pathogen	Vaccine	Indications and efficacy	Safety
Haemophilus influenzae	Polysaccharide–protein conjugate: PedvaxHIB (Merck) ActHIB, OmniHib (Sanofi-Pasteur)	Indicated in children and in asplenic adults Inadequate efficacy data in RA patients, but shows some efficacy in other immunocompromised patients (Meerveld-Eggink *et al.*, 2009)	Considered to be safe in RA (Davies and Woo, 2002)
Japanese encepha-litis	Inactivated mouse brain or culture-derived vaccine	Recommended for long-term high-risk travelers No efficacy data in RA Culture-derived vaccine may be associated with less hypersensitivity than brain-derived vaccine	No data exist on the use of culture-derived vaccine in immunocompromised persons or patients receiving immunosuppression, but limited data in brain-derived vaccine suggest a similar safety profile
Polio-myelitis	Inactivated poliovirus (injectable) and (oral) live-attenuated vaccine	Efficacy after three doses of either injected or oral vaccine is 100% Oral vaccine probably confers lifelong protection Antibody levels remain stable during anti-TNF therapy	Immunosuppression is associated with an increased risk of vaccine-associated paralytic polio, a very rare side effect of oral vaccine, and is therefore contraindicated in immunosuppressed RA patients Live vaccine can be transmitted from family members Inactivated vaccine is safe in RA patients
Rabies	Inactivated human diploid cell and purified chick embryo vaccines	Protective antibody titers decline in time No available data on efficacy in RA Immunosuppressed patients should avoid activities for which rabies pre-exposure prophylaxis is indicated Antibody titers should be monitored in immunosuppressed patients exposed to rabies, to ensure vaccine efficacy	No contraindication in RA or immunosuppressed patients
Nisseria meningitis-dis	(Quadrivalent) polysaccharide vaccine, capsular polysaccharide conjugate vaccine	Recommended for persons at increased risk for meningococcal disease, and for hyposplenic/asplenic RA patients Unavailable efficacy data in RA patients	No contraindications in immunosuppressed patients
Tick-borne encepha-litis	Inactivated cell culture-derived vaccines in different countries	Indicated in travelers to and residents of endemic areas No available efficacy data in RA or immunosuppressed patients	No data available, but probably safe in RA
Typhoid fever	(Oral) live-attenuated vaccine and capsular polysaccharide vaccine	Vaccines have only 50–80% efficacy No available efficacy data in RA	Live vaccine is contraindicated in immunosuppressed patients IgM rheumatic factor has been reported to be elevated after the older (and now unavailable) vaccine, but no RA has been reported after any vaccine

(continued)

Table 24.1 (*Continued*)

Pathogen	Vaccine	Indications and efficacy	Safety
Yellow fever	Live-attenuated vaccine	Indicated for travelers to endemic areas Inadequate data in RA patients. Some data suggest a lower response in patients on anti-TNF (Scheinberg *et al.*, 2010)	Risk of viscerotropic disease and mortality in elderly patients Contraindicated in immunosuppressed patients Small reports suggest more frequent moderate/severe local reactions in patients under steroidal therapy (Kerneis *et al.*, 2013) and possible increase in other DMARDs and biological (Mota *et al.*, 2009)

RA, rheumatoid arthritis; MMR, measles, mumps, and rubella; EULAR, European League Against Rheumatism; ACR, American College of Rheumatology; ACIP, Advisory Committee on Immunization Practices; TNF, tumor necrosis factor; DMARD, disease-modifying antirheumatic drug; HBsAg, hepatitis B surface antigen; BCG, bacillus Calmette–Guérin; CDC, Centers for Disease Control and Prevention; igM, immunoglobulin M

introduction of an inactive TTd vaccination. There are two available types of toxoid: an adsorbed (aluminum salt-precipitated) toxoid and a fluid toxoid. They have a similar seroconversion rate, but the adsorbed toxoid induces higher antitoxin titers and its protection lasts longer. The toxoid is available both as a single-antigen preparation and in different combinations with diphtheria toxoid, acellular pertussis vaccine, HBV vaccine, or *Haemophilus influenzae* (Hib). Although never studied in a controlled vaccine trial, the toxoid is considered to have a clinical efficacy of virtually 100%, though its effect may wane after a decade. Routine boosters are therefore recommended every 10 years. Immunosuppression is not a contraindication to receiving diphtheria toxoid.

Pertussis was a common childhood diseases and a major cause of childhood mortality prior to the vaccination era, and remains a major health problem in developing countries. The whole-cell pertussis vaccine was the first available vaccine and was widely used in the 20th century. A series of four doses was 70–90% effective in preventing serious pertussis disease, although efficacy waned with time, with little or no protection 5–10 years after the last dose. Safety issues led to the development of the purified (acellular) pertussis vaccine, which contains purified, inactivated components of *B. pertussis* cells. Acellular pertussis vaccine is not available as a single-antigen vaccine, but rather in combinations with tetanus and diphtheria toxoids and, recently, with HBV and polio vaccines. Immunosuppression is not a contraindication to receiving pertussis vaccine.

The efficacy of diphtheria, tetanus, and pertussis (DTP) vaccines in RA patients is unclear. Early small studies showed inconsistent antibody-level results post-TTd administration in RA patients compared to controls (Larson *et al.*, 1951; Whaley *et al.*, 1971). RA patients failed to show affinity maturation compared to controls, despite similar antibody levels (Devey *et al.*, 1987). As mentioned earlier, a recent large-scale study from the Netherlands showed a significantly lower prevalence of protective antibody concentrations for tetanus in JIA patients compared to healthy controls (OR 0.1; 0.05–0.30) (Heijstek *et al.*, 2012). In contrast to *in vitro* studies, CDC surveillance data do not show an increased risk for tetanus infection among RA or immunosuppressed patients. Accordingly, the ACR has no special tetanus vaccine recommendations regarding RA patients. The EULAR task force considers the efficacy for TTd vaccination in patients under immunosuppressive therapy to be comparable to that in healthy controls, including those treated with rituximab 24 weeks earlier. Due to lack of data regarding the efficacy of TTd vaccine within 24 weeks after treatment with rituximab, the EULAR recommends that in case of a serious risk of contracting tetanus (e.g. major and/or contaminated wounds), this subgroup of RA patients should also be passively immunized with tetanus immunoglobulins.

Pneumococcal vaccine

Streptococcus pneumoniae is an important cause of pneumonia, bacteremia, and meningitis. Pneumococci account for up to 36% of adult community-acquired pneumonia, 50% of hospital-acquired pneumonia, and 13–19% of all cases of bacterial meningitis in the United States. Bacterial pneumonia and other respiratory infections are an important cause of morbidity and mortality among RA patients (Naz *et al.*, 2007), underscoring the importance of preventive vaccination against pneumococcus. Pneumococcal vaccines are divided into pneumococcal polysaccharide vaccines (PSVs) and pneumococcal conjugate vaccines (PCVs). PSVs are composed of purified preparations of pneumococcal capsular polysaccharide. The currently used PSV is a 23-valent polysaccharide vaccine (PPSV23) containing antigens from 23 types of pneumococcal bacteria, which are responsible for 88% of bacteremic pneumococcal disease. PCVs are composed of purified capsular polysaccharide conjugated to highly immunogenic crossreactive material 197 (CRM_{197}), a nontoxic diphtheria toxoid protein. The first PCV included seven serotypes of *S. pneumonia,* which covered only a minority of invasive strains, but in 2010, a 13-valent PCV (PCV13) containing 6 additional serotypes was licensed in the United States, covering over 60% of pneumococcal invasive disease.

Pneumococcal vaccines seem to be safe in RA patients, though efficacy may be impaired. RA patients receiving rituximab or MTX exhibit an impaired humoral response to pneumococcal vaccination. Some studies suggest that a proportion of RA patients may not respond adequately to pneumococcal vaccination once on TNF-α blockade therapies, but others find no significant reduction in humoral response (reviewed by Hua *et al.*, 2013).

Accordingly, the EULAR task force recommends "strongly considering" vaccinating RA patients with the 23-valent polysaccharide pneumococcal vaccination prior to B cell-depleting therapy. ACIP recommends that adults with immunocompromising conditions, not previously immunized with either PCV13 or PPSV23, should receive a dose of PCV13 first, followed by a dose of PPSV23 at least 8 weeks later. A second PPSV23 dose is recommended 5 years after the first. The ACR's uniform recommendation to vaccinate before initiating therapy with a DMARD or any biologic agent includes administration of the pneumococcal vaccine.

H1N1 influenza vaccine

Influenza is a highly infectious viral illness affecting the respiratory tract that can lead to a severe respiratory illness, especially in RA patients, who have an increased mortality from pulmonary infections. Two types of influenza vaccines are available: a trivalent inactivated influenza vaccine (TIV) and, since 2003, a live-attenuated influenza vaccine (LAIV). The TIV can prevent up to 90% of influenza infections in young healthy vaccinees, but is only 30–40% effective among persons aged 65 years and older. Despite its modest efficacy in the elderly, TIV is effective in preventing complications and death in this age group, reducing hospitalizations by 50% and mortality by 80%.

The TIV has been shown to reduce hospital admissions and mortality in RA patients and its administration is recommended by the ACR and EULAR. The vaccine is beneficial in patients treated with DMARDs and biologic agents, with the exception of rituximab. Nonetheless, the ACR's recommendation is to administer the influenza vaccine before starting any immune-modulatory therapy, when possible. Although sufficiently powerful data are lacking, TIV safety among RA patients seems to be comparable to that in non-RA patients.

As in the case of other live-attenuated vaccines, both the EULAR and the ACR recommend against the administration of LAIV in immunosuppressed patients (van Assen *et al.*, 2011; Singh *et al.*, 2012), despite its proven superiority in preventing influenza among children.

Hepatitis B vaccine

HBV is a widely prevalent small DNA virus, chronically infecting more than 350 million people worldwide. RA patientsface a greater risk of HBV infection, especially those who receive biologic therapies. There are two recombinant HBV vaccines available today, differing in the composition of antigens and adjuvants, but with similar efficacies and indications. Following a three-dose course, both vaccines have 90% efficacy, which wanes in an age-dependent manner. The ACR recommends administering HBV vaccine to patients with hepatitis risk factors (e.g. intravenous drug abuse, multiple sex partners, health care personnel), preferably before initiation of DMARD or biological therapy. Despite a study from the Vaccine Adverse Event Reporting System (VAERS) comparing hepatitis B vaccine to tetanus and diphtheria vaccines (Geier *et al.*, 2002), which suggests an increased risk for arthritis following hepatitis B vaccination, a large retrospective epidemiological study disproved this association (Ray *et al.*, 2011). Accordingly, in a controlled trial, HBV vaccination was found to be efficient in 68% of RA patients, and was not associated with any clinical or laboratory deterioration (Elkayam *et al.*, 2002).

Varicella vaccine

Varicella zoster virus (VZV) is a herpesvirus that is the causative agent of chickenpox and herpes zoster (also known as shingles). Following the primary infection (chickenpox), the virus persists in sensory nerve ganglia as a latent infection. The recurrent infection (herpes zoster) appears years later, its incidence increasing with age. Varicella is usually a mild and self-limited disease in the otherwise healthy host, but immunocompromised individuals have a high risk of disseminated disease, which may be complicated by pneumonia and encephalitis. Similarly, immunosuppression is an important risk factor for herpes zoster.

The varicella vaccine contains a live-attenuated virus. It is commercially available as a low-titer varicella vaccine, in combination with the MMR vaccine, and as a high-titer HZV. Current CDC guidelines recommend its use in patients aged 13 years and older and without evidence of varicella immunity. The efficacy of HZV in adults is 60–70%, decreasing dramatically to a mere 18% in individuals over 80. Despite the decreased efficacy, disease severity is attenuated and the occurrence of post-herpetic neuralgia is lower. ACIP recommends administering a single dose of vaccine to those aged 60 years and older, regardless of prior history of herpes zoster, and to any person with a chronic medical condition unless there is an existing contraindication or precaution.

RA patients have an increased risk of developing herpes zoster (approximately 12 cases per 1000 person years, compared to 5 cases per 1000 person years in healthy controls), which is further increased with disease severity and treatment with steroids and biological and nonbiological DMARDs, with the exception of MTX (Zhang *et al.*, 2012). No studies assessing the beneficial effects of HZV have been performed in RA patients, yet the high burden of herpes zoster in this group warrants preventive measures, at least for those with a low level of immunosuppression. ACIP recommends vaccinating patients treated with low- to intermediate-dose steroid therapy (<20 mg/day of prednisone or equivalent); nonsystemic steroid therapy, such as intra-articular, bursal, or tendon injections; or low-dose therapy of MTX (<0.4 mg/kg/week), azathioprine (<3.0 mg/kg/day), or 6-mercaptopurine (<1.5 mg/kg/day). ACIP and ACR do not recommend the administration of HSV with concurrent biological treatment. While the EULAR task force does not reject these recommendations, it does emphasize that they are based on expert opinion only, suggesting that HZV might be an exception to the general recommendation of avoiding live-attenuated vaccines in immunosuppressed patients, and that it can be considered in mildly immunosuppressed RA patients on a case-by-case basis. The task force also advises the administration of HZV only to RA patients who are seropositive for varicella zoster antibodies, as a cautionary measure to prevent primary varicella infection with the vaccine strain. Despite these recommendations, in a recent, large retrospective cohort study of older patients with RA and other autoimmune inflammatory rheumatic diseases, vaccination was not associated with an increase in herpes zoster or primary varicella, even when given to patients on biologic therapy, implying its safety in all rheumatic patients (Zhang *et al.*, 2012). Moreover, the vaccine was efficacious in this population, reducing herpes zoster incidence by approximately 40% during a median of 2 years of follow-up. Further evidence is needed in order to solidify these findings and revise current guidelines.

Bacille Calmette–Guérin vaccine

Despite over 90 years of use and controversial efficacy, the BCG vaccine is still the only vaccine available for the prevention of tuberculosis (TB). Vaccination policies vary widely across the world, and different BCG substrains with variable efficacies are reported (reviewed by Zwerling *et al.*, 2011). In patients with RA, the risk of TB is increased, especially in those receiving anti-TNF monoclonal antibody treatment. This emphasizes the need for preventive measures in RA patients. Alas, the majority of TB cases in RA are reactivations of latent TB infections contracted earlier, which are not prevented by vaccination. Furthermore, the BCG vaccine contains live-attenuated mycobacteria, leading to a risk of disseminated BCG infection in immunosuppressed patients. Therefore, the EULAR does not recommend vaccinating RA patients with the BCG vaccine, and it is contraindicated by the ACR in patients undertaking anti-TNF therapy (van Assen *et al.*, 2011).

Vaccines in RA therapy

Increasing knowledge and understanding of the immune mechanisms underlying the pathogenesis of RA has been translated into improved treatment over the past 2 decades. The most notable advancement has undoubtedly been the introduction of biological immune modulators such as anti-TNF agents, anti-IL-6 agents, and others. Despite these major accomplishments, a large proportion of patients do not respond to therapy, calling for new strategies for modulation of the immune system. Vaccines could serve as a powerful tool in this regard, and, indeed, strategies employing vaccines in the therapy of RA are currently under research, both in animal models and in humans, with encouraging results.

A common reason for therapy failure while using anti-TNF monoclonal antibodies is the development of a humoral response, producing antidrug antibodies that target the monoclonal therapy (Bartelds *et al.*, 2011). To address this problem, different groups have utilized vaccines aimed at inducing high titers of endogenous neutralizing anticytokine antibodies against TNF. The goal of these vaccines is to break the natural Th tolerance to autoantigens and induce neutralizing anticytokine autoantibodies (reviewed by Delavallee *et al.*, 2010).

A common approach to inducing autoimmunity is to modify a self-protein by linking it to a foreign carrier protein, in order to activate autoreactive B cells. This complex, classically made of an inactive cytokine and a carrier non-self protein, induces Th cell proliferation, stimulating the production of autoantibodies by self-cytokine-specific B cells. The activation stimulus is enhanced by coupling repetitive antigens to the carrier protein, overcoming the natural B cell tolerance. Different carrier proteins have been used for this purpose, including virus-like particles, keyhole limpet hemocyanin (KLH), and ovalbumin, and they show encouraging results in mouse models. TNF coupled to KLH induced the production of high titers of neutralizing anti-human TNF-α (hTNF-α) antibodies in mice (Le Buanec *et al.*, 2006), and is now in phase II clinical trial.

An alternative strategy (Dalum *et al.*, 1999) for the induction of anti-TNF neutralizing antibodies is the addition of immunodominant T helper epitopes to a recombinant human TNF-α molecule. These epitopes are presented by APCs to Th cells, which in turn induce the differentiation of B cells into plasmocytes, producing antihapten antibodies. This strategy was successful in creating an active immunization against self-TNF-alpha and ameliorating arthritis symptoms in a collagen-induced arthritis mouse model. Other groups have successfully used a plasmid-based DNA vaccine expressing TNF to reduce inflammation in mouse and rat arthritis models (Wildbaum *et al.*, 2000).

Although most therapeutic studies have concentrated on TNF as the central cytokine/antigen, additional cytokines have been targeted by experimental vaccines. Vaccination with recombinant IL-17 bound to virus-like particles induced neutralizing anti-IL-17 antibodies, reducing arthritis in a murine collagen-induced arthritis model (Rohn *et al.*, 2006). Other groups have developed vaccines against IL-1, using KLH or virus-like particles as a carrier protein, inducing high titers of anti-IL-1 autoantibodies in mice. Many other anticytokine vaccines have been developed and studied, including an immunogenic IL-6 analog vaccine and DNA vaccines expressing MIF, RANTES, IL-18, MCP-1, and more (Delavallee *et al.*, 2010).

Another potential therapy in RA is the use of T cell vaccines. Here, autologous T cell lines or clones recognizing autoantigens are isolated from patients' blood, irradiated, and used as a T cell vaccine to induce a specific immune response in the host's T cells directed against the autoimmune (vaccine) T cells (Prakken *et al.*, 2007). The regulatory mechanisms induced by T cells are poorly understood. Nevertheless, this strategy has been successful in mouse models and has shown encouraging results in a small pilot study of 15 RA patients: 10 showed a clinical response, defined by ACR 50 improvement criteria (Chen *et al.*, 2007).

Altogether, the use of vaccines holds promising therapeutic potential in RA, and they carry the potential to replace anticytokine monoclonal antibodies in the future. Judging by animal models and preliminary studies, vaccines seem to be safe and effective, though maintaining the immune response may require additional boosting doses. Future studies will determine whether this approach will become common practice.

References

ACIP. (2011). General recommendations on immunization – recommendations of the Advisory Committee on Immunization Practices (ACIP). *MMWR*, **60** (2): 1–64.

Alvaro-Gracia, J.M., Zvaifler, N.J., and Firestein, G.S. (1989). Cytokines in chronic inflammatory arthritis. IV. Granulocyte/macrophage colony-stimulating factor-mediated induction of class II MHC antigen on human monocytes: a possible role in rheumatoid arthritis. *J Exp Med*, **170**(3): 865–75.

Avouac, J., Gossec, L., and Dougados, M. (2006). Diagnostic and predictive value of anti-cyclic citrullinated protein antibodies in rheumatoid arthritis: a systematic literature review. *Ann Rheum Dis*, **65**(7): 845–51.

Backlund, J., Carlsen, S., Hoger, T., *et al.* (2002). Predominant selection of T cells specific for the glycosylated collagen type II epitope (263–270) in humanized transgenic mice and in rheumatoid arthritis. *Proc Natl Acad Sci USA*, **99**(15): 9960–5.

Baeten, D., Steenbakkers, P.G., Rijnders, A.M., *et al.* (2004). Detection of major histocompatibility complex/human cartilage gp-39 complexes in rheumatoid arthritis synovitis as a specific and independent histologic marker. *Arthritis Rheum*, **50**(2): 444–51.

Bartelds, G.M., Krieckaert, C.L., Nurmohamed, M.T., *et al.* (2011). Development of antidrug antibodies against adalimumab and association with disease activity and treatment failure during long-term follow-up. *JAMA*, **305**(14): 1460–8.

Begovich, A.B., Carlton, V.E., Honigberg, L.A., *et al.* (2004). A missense single-nucleotide polymorphism in a gene encoding a protein tyrosine phosphatase (PTPN22) is associated with rheumatoid arthritis. *Am J Hum Genet*, **75**(2): 330–7.

Brennan, F.M., Maini, R.N., and Feldmann, M. (1992). TNF alpha – a pivotal role in rheumatoid arthritis? *Br J Rheumatol*, **31**(5): 293–8.

Chang, M., Rowland, C.M., Garcia, V.E., *et al.* (2008). A large-scale rheumatoid arthritis genetic study identifies association at chromosome 9q33.2. *PLoS Genet*, **4**(6): e1000107.

Chen, G., Li, N., Zang, Y.C., *et al.* (2007). Vaccination with selected synovial T cells in rheumatoid arthritis. *Arthritis Rheum*, **56**(2): 453–63.

Cope, A.P., Patel, S.D., Hall, F., *et al.* (1999). T cell responses to a human cartilage autoantigen in the context of rheumatoid arthritis-associated and nonassociated HLA-DR4 alleles. *Arthritis Rheum*, **42**(7): 1497–507.

Cruz-Tapias, P., Agmon-Levin, N., Israeli, E., *et al.* (2013). Autoimmune (auto-inflammatory) syndrome induced by adjuvants (ASIA) – animal models as a proof of concept. *Curr Med Chem*, **20**(32): 4030–6.

Dalum, I., Butler, D.M., Jensen, M.R., *et al.* (1999). Therapeutic antibodies elicited by immunization against TNF-alpha. *Nat Biotechnol*, **17**(7): 666–9.

Davies, K. and Woo, P. (2002). Immunization in rheumatic diseases of childhood: an audit of the clinical practice of British Paediatric Rheumatology Group members and a review of the evidence. *Rheumatology (Oxford)*, **41**(8): 937–41.

Delavallee, L., Duvallet, E., Semerano, L., *et al.* (2010). Anti-cytokine vaccination in autoimmune diseases. *Swiss Med Wkly*, **140**: w13108.

Devey, M.E., Bleasdale, K., and Isenberg, D.A. (1987). Antibody affinity and IgG subclass of responses to tetanus toxoid in patients with rheumatoid arthritis and systemic lupus erythematosus. *Clin Exp Immunol*, **68**(3): 562–9.

Eisen, S.A., Kang, H.K., Murphy, F.M., *et al.* (2005). Gulf War veterans' health: medical evaluation of a U.S. cohort. *Ann Intern Med*, **142**(11): 881–90.

Elkayam, O., Yaron, M., and Caspi, D. (2002). Safety and efficacy of vaccination against hepatitis B in patients with rheumatoid arthritis. *Ann Rheum Dis*, **61**(7): 623–5.

Feuchtenberger, M., Kleinert, S., Schwab, S., *et al.* (2012). Vaccination survey in patients with rheumatoid arthritis: a cross-sectional study. *Rheumatol Int*, **32**(6): 1533–9.

Firestein, G.S. (2003). Evolving concepts of rheumatoid arthritis. *Nature*, **423**(6937): 356–61.

Fousteri, G., Liossis, S.N., and Battaglia, M. (2013). Roles of the protein tyrosine phosphatase PTPN22 in immunity and autoimmunity. *Clin Immunol*, **149**(3): 556–65.

Geier, D.A. and Geier, M.R. (2002). A one year followup of chronic arthritis following rubella and hepatitis B vaccination based upon analysis of the Vaccine Adverse Events Reporting System (VAERS) database. *Clin Exp Rheumatol*, **20**(6): 767–71.

Gerritsen, M.E., Kelley, K.A., Ligon, G., *et al.* (1993). Regulation of the expression of intercellular adhesion molecule 1 in cultured human endothelial cells derived from rheumatoid synovium. *Arthritis Rheum*, **36**(5): 593–602.

Global Advisory Committee on Vaccine Safety. (2008). Global Advisory Committee on Vaccine Safety, 12–13 December 2007. *Wkly Epidemiol Rec*, **83**(4): 37–44.

Gregersen, P.K., Silver, J., and Winchester, R.J. (1987). The shared epitope hypothesis. An approach to understanding the molecular genetics of susceptibility to rheumatoid arthritis. *Arthritis Rheum*, **30**(11): 1205–13.

Haringman, J.J., Ludikhuize, J., and Tak, P.P. (2004). Chemokines in joint disease: the key to inflammation? *Ann Rheum Dis*, **63**(10): 1186–94.

Harris, E.D. Jr., (1990). Rheumatoid arthritis. Pathophysiology and implications for therapy. *N Engl J Med*, **322**(18): 1277–89.

He, X., Kang, A.H., and Stuart, J.M. (2000). Accumulation of T cells reactive to type II collagen in synovial fluid of patients with rheumatoid arthritis. *J Rheumatol*, **27**(3): 589–93.

Heijstek, M.W., van Gageldonk, P.G., Berbers, G.A., *et al.* (2012). Differences in persistence of measles, mumps, rubella, diphtheria and tetanus antibodies between children with rheumatic disease and healthy controls: a retrospective cross-sectional study. *Ann Rheum Dis*, **71**(6): 948–54.

Hua, C., Barnetche, T., Combe, B., and Morel, J. (2013). Effect of methotrexate, anti-TNFalpha and rituximab on the immune response to influenza and pneumococcal vaccines in patients with rheumatoid arthritis: a systematic review and meta-analysis. *Arthritis Care Res (Hoboken)*, **66**(7): 1016–26.

Institute of Medicine. (2010). *Gulf War and Health, Volume 8: Update of Health Effects of Serving in the Gulf War*. Washington, DC: The National Academies Press.

Israeli, E., Agmon-Levin, N., Blank, M., *et al.* (2009). Adjuvants and autoimmunity. *Lupus*, **18**(13): 1217–25.

Kerneis, S., Launay, O., Ancelle, T., *et al.* (2013). Safety and immunogenicity of yellow fever 17D vaccine in adults receiving systemic corticosteroid therapy: an observational cohort study. *Arthritis Care Res (Hoboken)*, **65**(9): 1522–8.

Kroger, A.T., Atkinson, W.L., Marcuse, E.K., *et al.* (2006). General recommendations on immunization: recommendations of the Advisory Committee on Immunization Practices (ACIP). *MMWR*, **55**(RR-15): 1–48.

Kuhn, K.A., Kulik, L., Tomooka, B., *et al.* (2006). Antibodies against citrullinated proteins enhance tissue injury in experimental autoimmune arthritis. *J Clin Invest*, **116**(4): 961–73.

Kurland, L.T., Molgaard, C.A., Kurland, E.M., *et al.* (1984). Lack of association of swine flu vaccine and rheumatoid arthritis. *Mayo Clin Proc*, **59**(12): 816–21.

Lally, F., Smith, E., Filer, A., *et al.* (2005). A novel mechanism of neutrophil recruitment in a coculture model of the rheumatoid synovium. *Arthritis Rheum*, **52**(11): 3460–9.

Larson, D.L. and Tomlinson, L.J. (1951). Quantitative antibody studies in man. I. The effect of adrenal insufficiency and of cortisone on the level of circulating antibodies. *J Clin Invest*, **30**(12:2): 1451–5.

Le Buanec, H., Delavallee, L., Bessis, N., *et al.* (2006). TNFalpha kinoid vaccination-induced neutralizing antibodies to TNFalpha protect mice from autologous TNFalpha-driven chronic and acute inflammation. *Proc Natl Acad Sci USA*, **103**(51): 19 442–7.

Legrand, L., Lathrop, G.M., Marcelli-Barge, A., *et al.* (1984). HLA-DR genotype risks in seropositive rheumatoid arthritis. *Am J Hum Genet*, **36**(3): 690–9.

Leonard, B.J. and Robertson, P.W. (1956). Some delayed complications of inoculation. *Br Med J*, **2**(5000): 1029–32.

Lipsky, P.E. (2007). Why does rheumatoid arthritis involve the joints? *N Engl J Med*, **356**(23): 2419–20.

McInnes, I.B. and Schett, G. (2007). Cytokines in the pathogenesis of rheumatoid arthritis. *Nat Rev Immunol*, **7**(6): 429–42.

Meerveld-Eggink, A., van der Velden, A.M., Ossenkoppele, G.J., *et al.* (2009). Antibody response to polysaccharide conjugate vaccines after nonmyeloablative allogeneic stem cell transplantation. *Biol Blood Marrow Transplant*, **15**(12): 1523–30.

Mota, L.M., Oliveira, A.C., Lima, R.A., *et al.* (2009). [Vaccination against yellow fever among patients on immunosuppressors with diagnoses of rheumatic diseases]. *Rev Soc Bras Med Trop*, **42**(1): 23–7.

Naz, S.M. and Symmons, D.P. (2007). Mortality in established rheumatoid arthritis. *Best Pract Res Clin Rheumatol*, **21**(5): 871–83.

Nissim, A., Winyard, P.G., Corrigall, V., *et al.* (2005). Generation of neoantigenic epitopes after posttranslational modification of type II collagen by factors present within the inflamed joint. *Arthritis Rheum*, **52**(12): 3829–38.

Petrik, M.S., Wong, M.C., Tabata, R.C., *et al.* (2007). Aluminum adjuvant linked to Gulf War illness induces motor neuron death in mice. *Neuromolecular Med*, **9**(1): 83–100.

Prakken, B. and Albani, S. (2007). Exploiting T cell crosstalk as a vaccination strategy for rheumatoid arthritis. *Arthritis Rheum*, **56**(2): 389–92.

Rantapaa-Dahlqvist, S., de Jong, B.A., Berglin, E., *et al.* (2003). Antibodies against cyclic citrullinated peptide and IgA rheumatoid factor predict the development of rheumatoid arthritis. *Arthritis Rheum*, **48**(10): 2741–9.

Ray, P., Black, S., Shinefield, H., *et al.* (2011). Risk of rheumatoid arthritis following vaccination with tetanus, influenza and hepatitis B vaccines among persons 15–59 years of age. *Vaccine*, **29**(38): 6592–7.

Ray, P., Black, S., Shinefield, H., *et al.* (1997). Risk of chronic arthropathy among women after rubella vaccination. Vaccine Safety Datalink Team. *JAMA*, **278**(7): 551–6.

Reparon-Schuijt, C.C., van Esch, W.J., van Kooten, C., *et al.* (2001). Secretion of anti-citrulline-containing peptide antibody by B lymphocytes in rheumatoid arthritis. *Arthritis Rheum*, **44**(1): 41–7.

Rohn, T.A., Jennings, G.T., Hernandez, M., *et al.* (2006). Vaccination against IL-17 suppresses autoimmune arthritis and encephalomyelitis. *Eur J Immunol*, **36**(11): 2857–67.

Scheinberg, M., Guedes-Barbosa, L.S., Mangueira, C., *et al.* (2010). Yellow fever revaccination during infliximab therapy. *Arthritis Care Res (Hoboken)*, **62**(6): 896–8.

Seidl, C., Donner, H., Fischer, B., *et al.* (1998). CTLA4 codon 17 dimorphism in patients with rheumatoid arthritis. *Tissue Antigens*, **51**(1): 62–6.

Shoenfeld, Y. and Agmon-Levin, N. (2011). "ASIA" – autoimmune/inflammatory syndrome induced by adjuvants. *J Autoimmun*, **36**(1): 4–8.

Singh, J.A., Furst, D.E., Bharat, A., *et al.* (2012). 2012 update of the 2008 American College of Rheumatology recommendations for the use of disease-modifying antirheumatic drugs and biologic agents in the treatment of rheumatoid arthritis. *Arthritis Care Res (Hoboken)*, **64**(5): 625–39.

Sivakumar, S.M., Safhi, M.M., Kannadasan, M., *et al.* (2011). Vaccine adjuvants – current status and prospects on controlled release adjuvancity. *Saudi Pharm J*, **19**(4): 197–206.

Stamp, L.K., Cleland, L.G., and James, M.J. (2004a). Upregulation of synoviocyte COX-2 through interactions with T lymphocytes: role of interleukin 17 and tumor necrosis factor-alpha. *J Rheumatol*, **31**(7): 1246–54.

Stamp, L.K., James, M.J., and Cleland, L.G. (2004b). Paracrine upregulation of monocyte cyclooxygenase-2 by mediators produced by T lymphocytes: role of interleukin 17 and interferon-gamma. *J Rheumatol*, **31**(7): 1255–64.

Suzuki, A., Yamada, R., Chang, X., *et al.* (2003). Functional haplotypes of PADI4, encoding citrullinating enzyme peptidylarginine deiminase 4, are associated with rheumatoid arthritis. *Nat Genet*, **34**(4): 395–402.

Symmons, D.P., Bankhead, C.R., Harrison, B.J., *et al.* (1997). Blood transfusion, smoking, and obesity as risk factors for the development of rheumatoid arthritis: results from a primary care-based incident case–control study in Norfolk, England. *Arthritis Rheum*, **40**(11): 1955–61.

Symmons, D.P. and Chakravarty, K. (1993). Can immunisation trigger rheumatoid arthritis? *Ann Rheum Dis*, **52**(12): 843–4.

Tingle, A.J., Mitchell, L.A., Grace, M., *et al.* (1997). Randomised double-blind placebo-controlled study on adverse effects of rubella immunisation in seronegative women. *Lancet*, **349**(9061): 1277–81.

Trentham, D.E., Townes, A.S., and Kang, A.H. (1977). Autoimmunity to type II collagen an experimental model of arthritis. *J Exp Med*, **146**(3): 857–68.

van Assen, S., Agmon-Levin, N., Elkayam, O., *et al.* (2011). EULAR recommendations for vaccination in adult patients with autoimmune inflammatory rheumatic diseases. *Ann Rheum Dis*, **70**(3): 414–22.

van den Berg, W.B. and McInnes, I.B. (2013). Th17 cells and IL-17 a – focus on immunopathogenesis and immunotherapeutics. *Semin Arthritis Rheum*, **43**(2): 158–70.

van Zeben, D., Hazes, J.M., Zwinderman, A.H., *et al.* (1992). Clinical significance of rheumatoid factors in early rheumatoid arthritis: results of a follow up study. *Ann Rheum Dis*, **51**(9): 1029–35.

Vingsbo, C., Sahlstrand, P., Brun, J.G., *et al.* (1996). Pristane-induced arthritis in rats: a new model for rheumatoid arthritis with a chronic disease course influenced by both major histocompatibility complex and non-major histocompatibility complex genes. *Am J Pathol*, **149**(5): 1675–83.

Whaley, K., Glen, A.C., MacSween, R.N., *et al.* (1971). Immunological responses in Sjögren's syndrome and rheumatoid arthritis. *Clin Exp Immunol*, **9**(6): 721–32.

Wildbaum, G., Youssef, S. and Karin, N. (2000). A targeted DNA vaccine augments the natural immune response to self TNF-alpha and suppresses ongoing adjuvant arthritis. *J Immunol*, **165**(10): 5860–6.

Wilder, R.L. and Crofford, L.J. (1991). Do infectious agents cause rheumatoid arthritis? *Clin Orthop Relat Res*, **265**: 36–41.

Zhang, J., Xie, F., Delzell, E., *et al.* (2012). Association between vaccination for herpes zoster and risk of herpes zoster infection among older patients with selected immune-mediated diseases. *JAMA*, **308**(1): 43–9.

Zhang, N., Wilkinson, S., Riaz, M., *et al.* (2012). Does methotrexate increase the risk of varicella or herpes zoster infection in patients with rheumatoid arthritis? A systematic literature review. *Clin Exp Rheumatol*, **30**(6): 962–71.

Zwerling, A., Behr, M.A., Verma, A., *et al.* (2011). The BCG World Atlas: a database of global BCG vaccination policies and practices. *PLoS Med*, **8**(3): e10010.

25 Undifferentiated Connective-Tissue Diseases

Maria Martinelli,[1,2] Carlo Perricone,[3] and Yehuda Shoenfeld[1,4]

[1]Zabludowicz Center for Autoimmune Diseases, Sheba Medical Center, Tel Hashomer, Israel
[2]Rheumatology Division, Department of Medicine, University of Brescia, Brescia, Italy
[3]Rheumatology, Department of Internal and Specialized Medicine, Sapienza University of Rome, Rome, Italy
[4]Incumbent of the Laura Schwarz-Kipp Chair for Research of Autoimmune Diseases, Sackler Faculty of Medicine, Tel Aviv University, Tel Aviv, Israel

Introduction

Undifferentiated connective-tissue diseases (UCTDs) are clinical conditions characterized by signs, symptoms, and laboratory tests suggestive of a systemic autoimmune disease but which do not fulfill the criteria for any defined connective-tissue disease (CTD), such as systemic lupus erythematosus (SLE), systemic sclerosis (SSc), primitive Sjögren syndrome (pSjS), polymyositis and dermatomyositis (PM-DM), antiphospholipid syndrome (APS), mixed connective-tissue diseases (MCTDs), or rheumatoid arthritis (RA).

The first description of these kinds of "undefined" clinical pictures dates back to 1969 (Sabo, 1969). LeRoy *et al.* (1980) first proposed the concept of undifferentiated connective-tissue syndromes (UCTSs), and these forms of systemic disease have been a matter of debate ever since. As expressed by the concept of the "mosaic of autoimmunity" (Shoenfeld *et al.*, 2008a,b), in these diseases, multiple factors interact in many complex ways to lead to the development of an autoimmune disease. Some shared mechanisms or genetic backgrounds might explain some very common clinical manifestations which transversely present CTDs, such as Raynaud's phenomenon, arthralgia/arthritis, and the presence of antinuclear antibodies (ANAs) and/or rheumatoid factor.

Classification criteria

The identification of a specific form of CTD is based on clinical experience, guided by widely accepted classification criteria. The main aims of these criteria are to improve the possibility of comparing data from different sources, to guide clinicians in the diagnostic process, and to serve teaching purposes (Fries *et al.*, 1994).

In rheumatology patients, clinical manifestations suggestive of systemic CTD but not fulfilling any existing criteria are quite common: 12–20% of patients initially asking for a rheumatologic evaluation may at least temporarily be diagnosed as affected by "undefined" or "undifferentiated" CTD. How to classify these kinds of clinical entities and which criteria to adopt when posing a diagnosis of UCTD have been and continue to be a matter of debate among physicians. The most recent review on UCTDs by Mosca *et al.* (2012) points to the necessity of using approved criteria for these diseases, since the great differences still present in the data may be mostly attributed to the different assessments used to verify or exclude a diagnosis UCTD. The lack of universally

Vaccines and Autoimmunity, First Edition. Edited by Yehuda Shoenfeld, Nancy Agmon-Levin, and Lucija Tomljenovic.

accepted criteria renders classical epidemiological data unavailable, as the existing data are hardly comparable (Mosca *et al.*, 2007). For example, the presence of ANAs is required by some but not all authors, so their positivity ranges from 58 to 100% in "UCTD" patients. When the presence of ANA is required for the definition of UCTD, patients usually present with a single antibody specificity, most frequently anti-Ro/SSA or anti-RNP antibodies. Usually, UCTD patients do not present with new autoantibody specificities even after long-term follow-up. The same authors have proposed that only persistently olygosymptomatic conditions should be considered as UCTDs, and have offered preliminary classification criteria, as follows: (i) signs and symptoms suggestive of a CTD, but not fulfilling criteria for defined CTDs; (ii) positive ANAs; and (iii) disease of at least 3 years (Mosca *et al.*, 1999, 2011) However, these criteria suffer from many limitations: first, they do not specify a necessary cut-off for ANA titer to be considered positive, running the risk of including nonrelevant low-titer positivities; moreover, they cannot distinguish stable UCTDs from an incomplete form of CTD that, even if not full-blown, shows manifestations specific enough to suggest diagnosis of defined disease. The latter would turn out to be a false-negative diagnosis if strictly evaluated according to the existing criteria for CTDs (Table 25.1). These specific manifestations could therefore be considered exclusion criteria, as Doria *et al.* (2005) have proposed (Table 25.2).

Table 25.1 Patients who do not fulfill classification criteria (Doria *et al.*, 2005)

Patients with unclassifiable CTD
False negative based on existing CTD criteria
True unclassifiable:
• Incomplete CTD
• UCTD

Table 25.2 Clinical manifestations and autoantibody reactivities that may be considered specific for a definite CTD (Doria *et al.*, 2005).

Clinical manifestations	Autoantibody
Malar rash	Anti-dsDNA
Subacute CLE	Anti-Sm
Chronic CLE	Anti P protein
Skin sclerosis	Anti-Scl-70
Heliotrope rash	Anti-centromere
Gottron's plaques	Anti-La/SSB
Erosive arthritis	Anti-Jo1
	Anti-Mi-2

CLE, cutaneous lupus erythematosus

Patients truly affected by unclassifiable CTDs sometimes run the risk of remaining undiagnosed and of being thrown together into a cauldron, where less attention is paid in determining the etiology and the precise mechanisms leading to such a difficult to validate condition (Perricone and Shoenfeld, 2013). This, in turn, may delay the discovery of proper and specific treatments. How to classify these systemic undefined clinical manifestations has animated many debates on this topic and still challenges clinicians today. Do they represent a mild spectrum of CTDs, an evolutionary phase of classic CTDs, or a "diathesis" awaiting other provoking factors (Ganczarczyk *et al.*, 1989; Greer and Panush, 1989; Venables, 1998; Al Attia *et al.*, 2006)?

In addition, it is not uncommon to observe patients presenting with a single organ involvement (such as lungs, nervous system, or liver) with positive ANAs and very unspecific clinical manifestations (acrocyanosis, alopecia, fatigue, or arthralgias) who are suspected to have a CTD. As these patients very often do not fulfill the classification criteria for existing CTDs, they may be classified as having UCTD.

As an example, interesting data are available in the literature regarding idiopathic interstitial pneumonias (IIPs). A subtype of IIP has been identified on the basis of different histological findings and a better response to treatment; this condition has been defined by some authors as "idiopathic nonspecific interstitial pneumonia" (NSIP). It has been observed that about one-third of patients with NSIP meet the criteria for UCTD. Compared to NSIP patients who do not fulfill the criteria for UCTD (Suda *et al.*, 2010), those who do have different clinical characteristics, with significantly better outcome. Therefore, some authors have suggested using the term "lung-dominant CTD" to identify conditions characterized predominantly by lung involvement, mild extrapulmonary clinical symptoms, and positive ANAs (Fischer *et al.*, 2010; Lunardi *et al.*, 2011). Collaborative studies aimed at defining these "single–organ"-dominant diseases may be of interest and may guide treatment improvement and follow-up of patients (Corte *et al.*, 2012).

Course of UCTD

Alarcon *et al.* (1991) reported that about 50% of patients with CTD of less than 1 year's duration had a UCTD which subsequently evolved to a defined CTD, resolved, or remained undifferentiated. Similarly, Mosca *et al.* (2012) summarized the three different possible outcomes of UCTDs which they and others had reported.

Evolution to defined CTD

The evolution of UCTD to defined CTD usually occurs within the first 5 years of disease. UCTDs can evolve not only into SLE but also into SSc, pSjS, MCTD, systemic vasculitis, PM-DM, and RA (Danieli *et al.*, 1998, 1999; Williams *et al.*, 1998, 1999; Bodolay *et al.*, 2003).

Mosca *et al.* (2008) followed a "historical" cohort of 91 UCTD patients for 5 years and concluded that follow-up confirmed UCTD comprises a distinct group of mild diseases and that the rate of evolution to defined CTDs is higher during the first years after onset. They also stated that patients who maintain an undifferentiated profile during follow-up seem to run a decreasing risk of developing a defined CTD. In a recent Letter to the Editor of *Clinical and Experimental Rheumatology* (Mosca *et al.*, 2013) they described the outcome of their cohort after 5 additional years of follow-up. This second part of the study included 83 patients (F: 80, M: 3), of whom 53% remained undifferentiated over a mean follow-up of 181 months, confirming the existence of "stable UCTD" as a distinct clinical entity with a simplified clinic-serological profile. The most common manifestations of stable UCTD were joint involvement, Raynaud's phenomenon, leukopenia, and thrombocytopenia. ANAs were positive in 100%, anti-Ro/SSA in 40%, anti-RNP in 19%, anti-dsDNA in 19%, ACLAIgG in 13%, anti-La/SSB in 6%, and anti-Sm in 2% of these patients. Very interestingly, the authors noted that during this extra period of observation, 36% of patients developed a defined CTD: 22 developed SLE, 3 pSjS, 3 RA, 1 SSc, and 1 MCTD. The evolution to SLE occurred earlier in the disease course (mean disease duration 62.4 months), while the development of the other CTDs occurred later (mean 165.8 months).

There is evidence of clinical and laboratory parameters predictive for the evolution of the disease. Some suggest a general tendency to further evolution, such as leukopenia (Cavazzana *et al.*, 2001), multiple-autoantibody reactivity (Mosca *et al.*, 2002), and the appearance of new autoantibody reactivities (Calvo-Alen *et al.*, 1996; Vila *et al.*, 2000; Swaak *et al.*, 2001). Others are more suggestive of an evolution to more specific CTDs (Table 25.3).

Table 25.3 Clinical manifestations and laboratory tests predictive of evolution to a specific CTD

Clinical manifestations	Laboratory analyses	Probable evolution
Malar rash Serositis Fever Oral ulcers Photosensitivity	Anti-dsDNA Anti-cardiolipin	SLE
Arthritis on onset	Rheumatoid factor ESR alterations	RA
Sclerodactily Raynaud's phenomenon Oesophageal dismotility	ANA nucleolar pattern	SSc
Sicca	Anti-Ro/SSA Increase of IgG anti-Ro/SSA during follow-up	Pss

SLE, systemic lupus erythematosus; RA, rheumatoid arthritis; SSc, systemic sclerosis; pSS, primitive Sjögren syndrome

In recent years, vitamin D, 25(OH)D_3, has attracted great attention because its deficiency has been associated with a higher incidence of autoimmune diseases (Prietl *et al.*, 2013; Sabbagh *et al.*, 2013). This can be explained by the fact that vitamin D acts at several levels in the immune system to maintain immune tolerance, and its deficiency has proved to be a plausible environmental risk factor for autoimmune disease by basic, genetic, and epidemiological studies. Vitamin D deficiency may be an important factor in autoimmune rheumatic disease, both in initiating disease development and in worsening the course once present (Gatenby *et al.*, 2013). Moreover, it has been demonstrated that UCTD patients with lower levels of vitamin D are more likely to evolve to definite forms of CTD (Cutolo, 2008; Zold *et al.*, 2008, 2010) and that that vitamin D supplementation helps restore the balance between pro- and anti-inflammatory processes in UCTD patients (Zold *et al.*, 2011).

Resolution

UCTDs can resolve with time. Both clinical and laboratory manifestations can normalize: this happens within 5 years of onset in 12% of cases and within 10 years in 25%.

Stability of clinical picture

Studies on UCTD patients show that there are diseases that persist "undefined" even after many years and suggest that the probability of evolving into a specific CTD declines with time from onset. The possibility of an evolution even after a long time of stability has recently been highlighted, however.

"Stable UCTDs" are normally characterized by mild clinical manifestations, such as arthralgia (37–80%), arthritis (14–70%), Raynaud's phenomenon (45–60%), leukopenia (11–42%), anemia, xerostomia (7–40%), and xerophtalmia (8–36%). Some 7–13% of UCTD patients are also affected by a form of autoimmune thyroid disease (Danieli *et al.*, 2000). Normally, major organs are spared (e.g. kidneys, lungs, heart, central nervous system (CNS)). Nonetheless, there are studies that show a possible lung involvement in UCTD patients, predominantly in the form of lymphocyte alveolitis (Kumanovics *et al.*, 2001), and, as previously stated, a lung-dominant presentation of the clinical picture has been described. An onset with cardiac tamponade is absolutely an exception (Hari *et al.*, 2012).

Etiopathogenesis

There is still a great lack of knowledge concerning the etiopathogenesis of UCTDs. Some data suggest a vitamin D deficit might influence T cell homeostasis, causing a Th17-Th1/Treg imbalance and leading to a predominance of proinflammatory cytokines (IL-12, INF-γ, IL-23, IL-17, IL-6). A plausible association between silicone gel implants and the development of UCTD has been reported in the literature (Rasheed *et al.*, 1995). The physical and biological properties of silicone gel-filled implants and their behavior *in vivo* are indeed compatible with the hypothesis that they may contribute to the development of CTD. They have been studied mostly in relation to scleroderma manifestations (Spiera *et al.*, 1994), even if it was sometimes not possible to clearly find a link (Park *et al.*, 1998). A very recent etiopathological hypothesis concerns a correlation between some vaccines and the development of UCTD. Bruzzese *et al.* (2013) described the onset of UCTD following hepatitis B virus (HBV) vaccination in a 43-year-old woman who started suffering from a constellation of unspecific symptoms soon after the second dose of HBV vaccine. The percentage of $CD4^+$ $CD25^+$ Tregs has proved to be higher in nonresponders to HBV vaccine, and it has been hypothesized that Tregs may be involved in negative regulation of the immune responses to HBV vaccination; this might, in turn, predispose to the development of autoimmunity (Li *et al.*, 2010).

It appears evident that more detailed classification to define the borders of UCTD and to understand its etiopathogenesis is needed. It would also be very helpful to identify biomarkers or genetic backgrounds that would allow for a better diagnosis, follow-up, and treatment of UCTD patients. The need to define clinical pictures is not just for the sake of classification, but rather is fundamental to clinical practice, as it enables physicians to compare data, to develop a common diagnostic method, and, finally, to choose appropriate treatments. Rheumatologists often encounter patients who cannot be included in any well-defined diagnostic category: the term "undifferentiated" has been used to describe all such conditions. This term is purely descriptive. It is by now widely accepted that, in order to understand diseases and therefore create effective therapies, physicians should focus on underlying etiopathogenic mechanisms more than on the clinical picture. This is the only possible way to find causal instead of symptomatic treatments and to be able to plan prevention strategies. This is especially true in the field of autoimmunity, in which pathogenic mechanisms and etiological grounds are often unclear and undefined. They almost never have a one-to-one relationship with the signs and symptoms to which they lead, because the causes of each disease are multiple and interact with one another in complicated and diverse ways, encompassing the "mosaic of autoimmunity." This renders it difficult to address the eliciting triggers of autoimmune diseases, and often leads clinicians to try to get a handle on an autoimmune process that is already long in action.

The depiction of the autoimmune/inflammatory syndrome induced by adjuvants (ASIA) has brought together patients sharing a related clinical picture. More relevantly, however, it has shifted the attention toward the common triggers: the immune adjuvants (Shoenfeld and Agmon-Levin, 2011; Perricone *et al.*, 2013). While ASIA is still achieving a more precise definition, it's increasingly clear that the best way to conceive of this syndrome is to think of it as a spectrum of different diseases that can be combined by shared manifestations and plausible causes. Many different entities falling within this spectrum have been recognized to date, including siliconosis (Caldeira and Ferreira, 2012), post-vaccine phenomena (Cerpa-Cruz *et al.*, 2013), macrophagic myofasciitis (MMF) (Gherardi and Authier, 2012), and Gulf War syndrome (GWS) (Israeli, 2012). Recently, sick building syndrome (SBS) was also proposed as a candidate for the ASIA spectrum (Israeli and Pardo, 2011). All of these diseases satisfy several criteria for fibromyalgia (Buskila and Sarzi-Puttini, 2008) and chronic fatigue syndrome (CFS). The latter comprises severely disabling conditions that have a number of prominent symptoms in common and coincide in many individuals. While little is known of their etiology, both conditions are characterized by an aberrant immune response. Recently, Shoenfeld and others suggested a role for an adjuvant mechanism in the pathogenesis of these conditions. Adjuvants (e.g. silicone, alum, pristane, infectious components) were found to induce autoimmunity by themselves in different animal models, and may possibly provoke autoimmune/autoinflammatory diseases in humans.

Like ASIA, UCTDs are still being defined, and lack precise definitive criteria. It has been underlined that classifying these conditions can be problematic and is still a matter of debate. Moreover, patients affected by these diseases often

Table 25.4 Prevalence of typical signs and symptoms in ASIA and UCTD (Perricone and Shoenfeld, 2013)

Signs and symptoms	ASIA	UCTD
Connective tissue-specific		
Arthralgias/arthritis	+	+++
Myalgias/myopathy/muscle weakness	+	+
Raynaud's phenomenon	+/−	++
Leukopenia	+/−	++
Anemia	+/−	+
Thrombocytopenia	+	+
Xerophthalmia	+	++
Xerostomia	+	++
Photosensitivity	+/−	+
Serositis	+/−	+
Malar rash	+/−	+
Oral aphthosis	+	+
Neurological/cognitive impairment	+	+
Aspecific		
Fever	+	+
Chronic fatigue/sleep disturbances	+	+
Gastrointestinal disturbances	+	+/−
Antibodies		
ANA	+	+++
Anti-dsDNA	+/−	+
Anti-SSA	+/−	++
Anti-SSB	+/−	+
Anti-Sm	+/−	+
Anti-RNP	+/−	+
Evolution to definite disease	+/+++	+/+++
Exposure to an external stimulus (infection, vaccine, silicone, adjuvant)	>99.9%	Undetermined
Linkage with HLA	+	+/− (HLA A1-B8-DR3-DQ2 haplotype)

+/− reported in less than 1% of subjects or only in case reports; +, reported in less than 30% of subjects; ++, reported in 30–60% of subjects; +++, reported in more than 60% of subjects suffer from a great delay in being diagnosed and treated. UCTDs and ASIA share many common signs and symptoms. Furthermore, very little is known about the etiopathogenesis of UCTDs, but there are hints in the literature of a possible role for adjuvants, in the form of both vaccines and silicone. This suggests UCTDs might actually be considered part of the expanding ASIA spectrum. In support of this idea, Table 25.4 shows the striking similarities between ASIA and UCTD.

Conclusions

Since the birth of medicine, physicians have been classifying patients by covering under the same umbrella those sharing similar clinical features. The term "undifferentiated" has been used to include all those conditions in which a well-defined diagnosis could not be reached. However, in recent years, science has demonstrated that, in order to create effective treatments, the underlying etiopathogenic mechanism is more important than the clinical picture. This is especially true in the field of autoimmunity. Indeed, when dealing with cancer, the aim of therapy is to destroy the specific demented cell. When dealing with infectious diseases, physicians pay less care if the patient has fever or splenomegaly, as the objective is to kill the pathogenic microorganism. However, in the field of autoimmunity, the causes of each disease are multiple, encompassing the "mosaic of autoimmunity." Most of the pathogenic mechanisms and etiological grounds are unclear, undefined, or at least hard to find.

The discovery (or, rather, the depiction) of ASIA has for the first time changed this view toward the gathering of patients who share a related picture but, more relevantly, are linked by the same pathogenic mechanism: the adjuvant. ASIA and UCTD share a common fate: both are "undifferentiated," meaning that there is no specific clinical feature characterizing each entity, and both can shift toward a definite autoimmune disease. Indeed, the overlap between the two conditions is evident, as shown in Table 25.4. Physicians should not perhaps be dwelling too much on classifying patients, but instead should aim at discovering the mechanisms underlying the pathogenesis of autoimmune diseases, as elucidating these mechanisms will be the key to finding effective preventative therapeutic strategies.

References

Al Attia, H.M. (2006). Borderline systemic lupus erythematosus (SLE): a separate entity or a forerunner to SLE? *Int J Dermatol*, **45**: 366–9.

Alarcon, G.S., Williams, G.V., Singer, J.Z., *et al.* (1991). Early undifferentiated connective tissue disease. I. Early clinical manifestation in a large cohort of patients with undifferentiated connective tissue diseases compared with cohorts of well established connective tissue disease. *J Rheumatol*, **18**: 1332–9.

Bodolay, E., Csiki, Z., Szekanecz, Z., *et al.* (2003). Five-year follow-up of 665 Hungarian patients with undifferentiated connective tissue disease (UCTD). *Clin Exp Rheumatol*, **21**: 313–20.

Bruzzese, V., Zullo, A., and Hassan, C. (2013). Connective tissue disease following hepatitis B vaccination. *J Clin Rheumatol*, **19**: 280–1.

Buskila, D. and Sarzi-Puttini, P. (2008). Fibromyalgia and autoimmune diseases: the pain behind autoimmunity. *IMAJ*, **10**: 77–8.

Caldeira, M. and Ferreira, A.C. (2012). Siliconosis: autoimmune/inflammatory syndrome induced by adjuvants (ASIA). *IMAJ*, **14**: 137–8.

Calvo-Alen, J., Alarcon, G.S., Burgard, S.L., *et al.* (1996). Systemic lupus erythematosus: predictors of its occurrence among a cohort of patients with early undifferentiated connective tissue disease: multivariate analyses and identification of risk factors. *J Rheumatol*, **23**: 469–75.

Cavazzana, I., Franceschini, F., Belfiore, N., *et al.* (2001). Undifferentiated connective tissue disease with antibodies to Ro/SSa: clinical features and follow-up of 148 patients. *Clin Exp Rheumatol*, **19**: 403–9.

Cerpa-Cruz, S., Paredes-Casillas, P., Landeros Navarro, E., *et al.* (2013). Adverse events following immunization with vaccines containing adjuvants. *Immunol Res*, **56**: 299–303.

Corte, T.J., Copley, S.J., Desai, S.R., *et al.* (2012). Significance of connective tissue disease features in idiopathic interstitial pneumonia. *Eur Res J*, **39**: 661–8.

Cutolo, M. (2008). Vitamin D or hormone D deficiency in autoimmune rheumatic diseases, including undifferentiated connective tissue disease. *Arthritis Res Ther*, **10**: 123.

Danieli, M.G., Fraticelli, P., Salvi, A., *et al.* (1998). Undifferentiated connective tissue disease: natural history and evolution into definite CTD assessed in 84 patients initially diagnosed as early UCTD. *Clin Rheumatol*, **17**: 195–201.

Danieli, M.G., Fraticelli, P., Franceschini, F., *et al.* (1999). Five-year follow-up of 165 Italian patients with undifferentiated connective tissue diseases. *Clin Exp Rheumatol*, **17**: 585–91.

Danieli, M.G., Rossetti, L., Fraticelli, P., *et al.* (2000). Autoimmune thyroid diseases in patients with undifferentiated connective tissue disease. *Clin Rheumatol*, **19**: 42–6.

Doria, A., Mosca, M., Gambari, P.F., and Bombardieri, S. (2005). Defining unclassifiable connective tissue diseases: incomplete, undifferentiated, or both? *J Rheumatol*, **32**: 213–15.

Fries, J.F., Hochberg, M.C., Medsger, T.A. Jr, *et al.* (1994). Criteria for rheumatic disease. Different types and different functions. The American College of Rheumatology Diagnostic and Therapeutic Criteria Committee. *Arthritis Rheum*, **37**: 454–62.

Fischer, A., West, S.G., Swigris, J.J., *et al.* (2010). Connective tissue disease-associated interstitial lung disease: a call for clarification. *Chest*, **138**: 251–6.

Ganczarczyk, L., Urowitz, M.B., and Gladman, D.D. (1989). Latent lupus. *J Rheumatol*, **16**: 475–8.

Gatenby, P., Lucas, R., and Swaminathan, A. (2013). Vitamin D deficiency and risk for rheumatic diseases: an update. *Curr Opin Rheumatol*, **25**: 184–91.

Gherardi, R.K. and Authier, F.J. (2012). Macrophagic myofasciitis: characterization and pathophysiology. *Lupus*, **21**: 184–9.

Greer, J.M. and Panush, R.S. (1989). Incomplete lupus erythematosus. *Arch Intern Med*, **149**: 2473–6.

Hari, P., Kondur, A., Manickam, P., and Afonso, L. (2012). An unusual case of undifferentiated connective tissue disease presenting as cardiac tamponade. *Rheumatol Int*, **32**: 231–4.

Kumanovics, G., Zibotics, H., Juhasz, E., *et al.* (2001). Subclinical pulmonary involvement assessed by bronchoalveolar lavage in patients with early undifferentiated connective tissue disease. *Clin Exp Rheumatol*, **19**: 551–9.

Israeli, E. (2012). Gulf War syndrome as a part of the autoimmune (autoinflammatory) syndrome induced by adjuvant (ASIA). *Lupus*, **21**: 190–4.

Israeli, E. and Pardo, A. (2011). The sick building syndrome as a part of the autoimmune (auto-inflammatory) syndrome induced by adjuvants. *Mod Rheumatol*, **21**(3): 235–9.

LeRoy, E.C., Maricq, H.R., and Kahaleh, M.B. (1980). Undifferentiated connective tissue syndromes. *Arthritis Rheum*, **23**: 341–3.

Li, J.C., Tan, D.M., Liu, H.B., and Li, K.C. (2010). [The roles of CD4+CD25+ T regulatory cells and Foxp3 mRNA expression in the subjects with and without responses to hepatitis B virus vaccination]. *Zhonghua Gan Zang Bing Za Zhi*, **18**: 726–30.

Lunardi, F., Balestro, E., Nordio, B., *et al.* (2011). Undifferentiated connective tissue disease presenting with prevalent interstitial lung disease: case report and review of literature. *Diagnostic Pathol*, **6**: 50.

Mosca, M., Neri, R., and Bombardieri, S. (1999). Undifferentiated connective tissue diseases (UCTD): a review of the literature and a proposal for preliminary classification criteria. *Clin Exp Rheumatol*, **17**: 615–20.

Mosca, M., Neri, R., Bencivelli, W., *et al.* (2002). Undifferentiated connective tissue disease: analysis of 83 patients with a minimum followup of 5 years. *J Rheumatol*, **29**: 2345–9.

Mosca, M., Tani, C., and Bombardieri, S. (2007). Undifferentiated connective tissue diseases (UCTD): a new frontier for rheumatology. *Best Pract Res Clin Rheumatol*, **21**: 1011–23.

Mosca, M., Tani, C., Neri, C., *et al.* (0000) [Analysis of the evolution to defined connective tissue diseases of patients with "early unidifferentiated connective tissue diseases (UCTD)"]. *Reumatismo*, **60**: 35–40.

Mosca, M., Tani, C., Talarico, R., and Bombardieri, S. (2011). Undifferentiated connective tissue diseases (UCTD): simplified systemic autoimmune diseases. *Autoimm Rev*, **10**: 256–8.

Mosca, M., Tani, C., Carli, L., and Bombardieri, S. (2012). Undifferentiated CTD: a wide spectrum of autoimmune diseases. *Best Pract Res Clin Rheumatol*, **26**: 73–7.

Mosca, M., Tani, C., Carli, L., *et al.* (2013). Analysis of the evolution of UCTD to defined CTD after a long term follow-up. *Clin Exp Rheumatol*, **31**: 471.

Park, A.J., Black, R.J., Sarhadi, N.S., *et al.* (1998). Silicone gel-filled breast implants and connective tissue diseases. *Plast Reconstruct Surg*, **101**: 261–8.

Perricone, C. and Shoenfeld, Y. (2013). Hepatitis B vaccination and undifferentiated connective tissue disease: another brick in the wall of the autoimmune/inflammatory syndrome induced by adjuvants (Asia). *J Clin Rheumatol*, **19**: 231–3.

Perricone, C., Colafrancesco, S., Mazor, R.D., *et al.* (2013). Autoimmune/inflammatory syndrome induced by adjuvants (ASIA) 2013; unveiling the pathogenic, clinical ad diagnostic aspects. *J Autoimmun*, **47**: 1–16.

Prietl, B., Treiber, G., Pieber, T.R., and Amrein, K. (2013). Vitamin D and immune function. *Nutrients*, **5**: 2502–21.

Rasheed, A., Lipstein-Kresch, E., and Kalra, J. (1995). Undifferentiated connective tissue disease after silicone-gel testicular implantation. *J Clin Rheumatol*, **1**: 310.

Sabbagh, Z., Markland, J., and Vatanparast, H. (2013). Vitamin D status is associated with disease activity among rheumatology outpatients. *Nutrients*, **5**: 2268–75.

Sabo, I. (1969). The lanthanic or undifferentiated collagen disease. *Hiroshima J Med Sci*, **18**: 259–64.

Shoenfeld, Y. and Agmon-Levin, N. (2011). "ASIA" – autoimmune/inflammatory syndrome induced by adjuvants. *J Autoimmun*, **36**: 4–8.

Shoenfeld, Y., Blank, M., Abu-Shakra, M., *et al.* (2008a). The mosaic of autoimmunity: prediction, autoantibodies, and therapy in autoimmune diseases – 2008. *IMAJ*, **10**(1): 13–19.

Shoenfeld, Y., Gilburd, B., Abu-Shakra, M., *et al.* (2008b). The mosaic of autoimmunity: genetic factors involved in autoimmune diseases – 2008. *IMAJ*, **10**(1): 3–7.

Spiera, R.F., Gibofsky, A., and Spiera, H. (1994). Silicone gel filled breast implants and connective tissue disease: an overview. *J Rheumatol*, **21**: 239–45.

Suda, T., Kono, M., Nakamura, Y., *et al.* (2010). Distinct prognosis of idiopathic nonspecific interstitial pneumonia (NSIP) fulfilling criteria for undifferentiated connective tissue disease (UCTD). *Respir Med*, **104**: 1527–34.

Swaak, A.J., van de Brink, H., Smeenk, R.J., *et al.* (2001). Incomplete lupus erythematosus: results of a multicentre study under the supervision of the EULAR Standing Committee on International Clinical Studies Including Therapeutic Trials (ESCISIT). *Rheumatology (Oxford)*, **40**: 89–94.

Venables, P.J. (1998). Undifferentiated connective tissue diseases: mixed or muddled? *Lupus*, **7**: 73–4.

Vila, L.M., Mayor, A.M., Valentin, A.H., *et al.* (2000). Clinical outcome and predictors of disease evolution in patients with incomplete lupus erythematosus. *Lupus*, **9**: 110–15.

Williams, H.J., Alarcon, G.S., Neuner, R., *et al.* (1998). Early undifferentiated connective tissue disease. V. An inception cohort 5 years later: disease remissions and changes in diagnoses in well established and undifferentiated connective tissue diseases. *J Rheumatol*, **25**: 261–8.

Williams, H.J., Alarcon, G.S., Joks, R., *et al.* (1999). Early undifferentiated connective tissue disease (CTD). VI. An inception cohort after 10 years: disease remissions and changes in diagnoses in well established and undifferentiated CTD. *J Rheumatol*, **26**: 816–25.

Zold, E., Szodoray, P., Gaal, J., *et al.* (2008). Vitamin D deficiency in undifferentiated connective tissue disease. *Arthritis Res Ther*, **10**: R123.

Zold, E., Szodoray, P., Kappelmayer, J., *et al.* (2010). Impaired regulatory T-cell homeostasis due to vitamin D deficiency in undifferentiated connective tissue disease. *Scand J Rheumatol*, **39**: 490–7.

Zold, E., Szodoray, P., Nakken, B., *et al.* (2011). Alfacalcidol treatment restores derailed immune-regulation in patients with undifferentiated connective tissue disease. *Autoimmun Rev*, **10**: 155–62.

26 Vaccines, Infections, and Alopecia Areata

Yaron Zafrir,[1,2] Sharon Baum,[1] Nancy Agmon-Levin,[2,3] and Yehuda Shoenfeld[2,4]

[1]Department of Dermatology, Sheba Medical Center, Tel Hashomer, Israel
[2]Zabludowicz Center for Autoimmune Diseases, Sheba Medical Center, Tel Hashomer, Israel
[3]Sackler Faculty of Medicine, Tel Aviv University, Tel Aviv, Israel
[4]Incumbent of the Laura Schwarz-Kipp Chair for Research of Autoimmune Diseases, Sackler Faculty of Medicine, Tel Aviv University, Tel Aviv, Israel

Introduction

Alopecia areata (AA) is an autoimmune disease characterized by one or more well demarcated oval and round noncicatricial patches of hair loss. The disease usually involves the scalp but may affect any hair-bearing parts of the body, including the eyebrows, beard, and body hair. It may include the entire scalp (alopecia totalis, AT) or the entire body (alopecia universalis, AU). Furthermore, due to its chronic relapsing nature and its profound effect on physical appearance, patients may experience a devastating loss of quality of life and self-esteem (Picardi *et al.*, 2003; Tosti *et al.*, 2006; Alkhalifah *et al.*, 2010a,b). Although it is considered to be a common autoimmune disease, its pathogenesis is barely understood, and the available therapies are rarely satisfying (Delamere *et al.*, 2008; Alkhalifah *et al.*, 2010a,b).

Epidemiology

AA is the most prevalent autoimmune skin disease, affecting more than 5 million people in the United States alone (Safavi *et al.*, 1995; McMichael *et al.*, 2007; Cooper *et al.*, 2009). Its prevalence varies with ethnicity and location: it is estimated to be between 0.1 and 0.2% in the United States and as high as 3.8% in Singapore. The lifetime risk of AA in the United States is 1.7% (Safavi *et al.*, 1992, 1995; Tan *et al.*, 2002; Bolognia *et al.*, 2012).

AA may appear at any age, affecting both children and adults, with a peak incidence during the second and third decades of life (15–29 years). Two-thirds of patients present with the first episode of AA before the age of 30 years. Interestingly, in contrast to other autoimmune diseases, no sex predilection is found in AA, although in one study by Alkhalifah *et al.* (2010a,2010b), more men were affected.

AA, like other autoimmune diseases, has an increased overall risk of concomitant disorders such as lupus erythematosus, vitiligo, and autoimmune thyroid disease (Thomas and Kadyan, 2008; Barahmani *et al.*, 2009; Kuchabal *et al.*, 2010; Kumar *et al.*, 2013). In addition, there is a high prevalence of atopic features (atopic dermatitis and allergic rhinitis) (Chu *et al.*, 2011).

Clinical manifestations

AA is commonly manifested by hair loss in a well demarcated oval area over normal-appearing skin (with no erythema and no scarring). The disease may involve any hair-bearing part of the body, but

Vaccines and Autoimmunity, First Edition. Edited by Yehuda Shoenfeld, Nancy Agmon-Levin, and Lucija Tomljenovic.

it most commonly involves the scalp and the beard area. Its onset is usually abrupt and the course is unpredictable, as it may remit spontaneously or may progress to cover the entire scalp (AT) or the entire body (AU). Another pattern of AA is a band-like peripheral hair loss along the temporal and occipital scalp, termed "ophiasis." The characteristic lesion may exhibit pathognomonic exclamation-point hairs (hairs which are broader at their distal end) (Bolognia *et al.*, 2012). Nail involvement can also be seen in AA; the most common finding is nail pitting, but other abnormalities such as trachyonychia (nail plate roughness) and leukonychia punctate have also been described (Dotz *et al.*, 1985; Vañó-Galván *et al.*, 2008).

Pathogenesis

In AA, autoreactive lymphocytes affect only the anagen hair follicles (pigmented hair follicles), leading to a premature catagen phase (dystrophy of the follicle) and shedding of hairs (Messenger *et al.*, 1986). The base of the hair follicle is the only compartment that is affected; the stem cell compartment is not involved, so the hair's regrowth capacity is retained. AA is considered an organ-specific autoimmune disease and, like other autoimmune diseases, its pathogenesis is believed to be an interaction of several different factors, including genetic predisposition and both endogenous (immune system, hormonal milieu, etc.) and exogenous (environmental) factors, also called the "exposome" (Colafrancesco *et al.*, 2013).

Genetics factors

The genetic component of AA is considered to be important but is only incompletely understood. Relatives of affected family member are at increased risk of developing AA, and 8.4% of AA patients have family members affected by the same disease (Yang *et al.*, 2004). Familial cases tend to have a more severe course of disease and a much poorer prognosis (Goh *et al.*, 2006). Furthermore a concordance rate of 55% is found among identical twins (Rodriguez *et al.*, 2010). In addition, AA is strongly associated with a personal or family history of atopy and other autoimmune diseases, further suggesting a strong genetic preponderance (Safavi *et al.*, 1992; Barahmani *et al.*, 2009; Alkhalifah *et al.*, 2010a,2010b; Huang *et al.*, 2013).

Several susceptible genes have been found to be associated with AA. In a C3H/HeJ mouse model, both the major locus on mouse chromosome 17 and the minor locus on chromosome 9 are linked with the development of AA (Silva *et al.*, 2003; Sun *et al.*, 2008). IL-1 receptor antagonist gene, which is known to be associated with autoimmune diseases, is associated with the development of AA in humans (Tazi-Ahnini *et al.*, 2002). Further investigations of HLA II genes have shown an association with AA (Akar *et al.*, 2002). More specifically, HLA gene class II alleles and various subgroups are reported to be associated in different ethnic groups, such as HLA-C in Japanese AA patients (Haida *et al.*, 2013) and the HLA-DQB1 locus in Italians. There is also a link between HLA-DRB1*11 and earlier onset and more severe course (Marques *et al.*, 2006; Alzolibani, 2011; Megiorni *et al.*, 2011).

Notably, patients suffering from Down's syndrome have a higher frequency of AA (8.8%) (Du Vivier and Munro, 1975). Two genes found on chromosome 21 may support this association: myxovirus resistance 1 (MX1) gene and the autoimmune regulator (AIRE) gene; both have been found to be associated with AA (Tazi-Ahnini *et al.*, 2000).

Environmental factors

The kaleidoscope of AA, like that of other autoimmune diseases, has been shown to be associated with various environmental factors, including emotional and/or physical stress, infections, and vaccination (Bogdanos *et al.*, 2013). These factors are all suggested to induce a change in the immune privilege area of the hair follicle, by inducing a shift in the local immune balance to a T helper 1-predominant condition. This shift in the cytokine milieu around the hair follicle is characterized by overproduction of interferon c and upregulation of major histocompatibility complex (MHC)-I expression in the proximal and outer root sheath, leading eventually to a collapse of the hair follicle immune privilege area; as a consequence, autoreactive $CD8^+$ T cells can recognize autoantigens that were concealed by the hair follicle (Israeli *et al.*, 2009)

Infectious agents have also been suggested to be associated with alopecia. One of the more well-known examples is secondary syphilis. In addition, several studies have suggested an association with other agents. For example, in one retrospective study, 12 patients developed AA following an episode of acute infectious mononucleosis confirmed by serologic tests for Epstein–Barr virus (EBV) infection in all patients

(Rodriguez *et al.*, 2008). In another report, a 4-year-old girl and an 80-year-old woman described the appearance of AA following herpes zoster (HZ) infection (Hayderi *et al.*, 2013). Swine flu infection was suggested to induce or exacerbate AA in two children, and fungal infection with *Alternaria chlamydospora* was associated with the development of AA in a 13-year-old boy (Rudnicka and Lukomska, 2012).

Vaccines and AA

The effect of vaccination on autoimmunity has been a major focus of attention in recent decades. Human vaccines contain infectious particles incorporating various adjuvants, including aluminum, the role of which is to initiate and enhance the response of the immune system to the vaccine. Immune adjuvants were considered safe in the past, but they have recently been shown to induce immune mediated and autoimmune conditions in recipients (Israeli *et al.*, 2009; Kivity *et al.*, 2009; Shoenfeld and Agmon-Levin, 2011; Lu and Hogenesch, 2013; Shaw and Tomljenovic, 2013). The correlation between vaccination and autoimmune diseases has previously been suggested in systemic sclerosis (SSc), chronic fatigue syndrome (CFS), fibromyalgia, transverse myelitis (TM) (Zafrir *et al.*, 2012), systemic lupus erythematosus (SLE) (Agmon-Levin *et al.*, 2009), antiphospholipid syndrome (APS), Guillain–Barré syndrome (GBS), Crohn's disease, macrophagic myofasciitis (MMF), vasculitis, myelitis and other diseases (Cruz-Tapias *et al.*, 2012; Dimitrijević *et al.*, 2012; Hartung *et al.*, 2012; Israeli *et al.*, 2011; Lerner, 2012; Soriano *et al.*, 2012; Stübgen, 2012). In a very large cohort, a significant correlation between influenza vaccine and rheumatoid arthritis (RA) was observed 6–12 months following immunization (Ray *et al.*, 2011). An association between hepatitis B virus (HBV) vaccine and immune-mediated neuronal damage was documented even 3 years post-vaccination (Hernán *et al.*, 2004; Mikaeloff *et al.*, 2009). Furthermore, MMF was associated up to 8 years following inoculation with HBV vaccine (Israeli *et al.*, 2011). Similarly, AA has been suggested to be induced by various vaccinations (Table 26.1). In one study, 60 patients were reported to develop alopecia following vaccination, out of which 48 had received HBV vaccination. Furthermore, 16 patients were rechallenged with reappearance of AA, confirming the plausibility of the association (Wise *et al.*, 1997).

Table 26.1 Summary of AA cases following vaccination

No. cases	Vaccination	References
60	Hepatitis B vaccinations and others	Wise *et al.* (1997)
2	Bivalent HPV vaccine	Tuccori *et al.* (2012)
1	Tetanus toxoid vaccine	Sánchez-Ramón *et al.* (2011)
1	MMR vaccine	Razeghinejad *et al.* (2009)

HBV, hepatitis B virus; HPV, human papillomavirus; TTd, tetanus toxoid; MMR, measles, mumps, and rubella

One study looking at adult-onset AA in a C3H/HeJ mouse model showed a rapid onset of clinical AA among 7-month-old mice shortly after they received HBV vaccine that was absent in the control group, which received saline (Sundberg *et al.*, 2009). However, when the study was repeated in a larger group of mice, the results were not significant, suggesting that there was no association between vaccination and AA development (Sundberg *et al.*, 2009).

Additional vaccines have occasionally been related to the development of AA. For example, tetanus toxoid (TTd) vaccine has been linked to AU development, and there are two cases of administration of bivalent human papillomavirus (HPV) vaccine being followed by telogen effluvium development (Sánchez-Ramón *et al.*, 2011; Tuccori *et al.*, 2012). In the latter case, one patient developed alopecia 1 month following the first vaccine and one developed it 3 weeks after the third vaccine. In addition, in one case development of madarosis (eyelash loss) with no scalp involvement was reported a few days after measles, mumps, and rubella (MMR) vaccination (Razeghinejad *et al.*, 2009).

Conclusions

Although AA is not a life-threatening disease, patients may experience devastating effects on their quality of life and self-esteem. A possible association between vaccination and AA has been suggested by a few case reports concerning different vaccines; the largest case series reports such a link with HBV vaccine (Wise *et al.*, 1997). Further research on the various environmental and genetic factors which may lead to the development of AA and, especially, on the link between AA and prior immunization is warranted.

References

Agmon-Levin, N., Zafrir, Y., Paz, Z., *et al.* (2009). Ten cases of systemic lupus erythematosus related to hepatitis B vaccine. *Lupus*, **18**(13): 1192–7.

Akar, A., Orkunoglu, E., Sengül, A., *et al.* (2002). HLA class II alleles in patients with alopecia areata. *Eur J Dermatol*, **12**(3): 236–9.

Alkhalifah, A., Alsantali, A., Wang, E., *et al.* (2010a). Alopecia areata update: part I. Clinical picture, histopathology, and pathogenesis. *J Am Acad Dermatol*, **62**(2): 177–88.

Alkhalifah, A., Alsantali, A., Wang, E., *et al.* (2010b). Alopecia areata update: part II. *Treatment. J Am Acad Dermatol*, **62**(2): 191–202.

Alzolibani, A.A. (2011). Epidemiologic and genetic characteristics of alopecia areata (part 1). *Acta Dermatovenerol Alp Panonica Adriat*, **20**(4): 191–8.

Barahmani, N., Schabath, M.B., Duvic, M.; National Alopecia Areata Registry. (2009). History of atopy or autoimmunity increases risk of alopecia areata. *J Am Acad Dermatol*, **61**(4): 581–91.

Bogdanos, D.P., Smyk, D.S., Invernizzi, P., *et al.* (2013). Tracing environmental markers of autoimmunity: introducing the infectome. *Immunol Res*, **56**(2–3): 220–40.

Bolognia, J.L., Jorizzo, J.L., and Schaffer, J.V. (2012). *Dermatology*. 3rd edn. Philadelphia, PA: W.B. Saunders.

Chu, S.Y., Chen, Y.J., Tseng, W.C., *et al.* (2011). Comorbidity profiles among patients with alopecia areata: the importance of onset age, a nationwide population-based study. *J Am Acad Dermatol*, **65**(5): 949–56.

Colafrancesco, S., Agmon-Levin, N., Perricone, C., and Shoenfeld, Y. (2013). Unraveling the soul of autoimmune diseases: pathogenesis, diagnosis and treatment adding dowels to the puzzle. *Immunol Res*, **56**(2–3): 200–5.

Cooper, G.S., Bynum, M.L., and Somers, E.C. (2009). Recent insights in the epidemiology of autoimmune diseases: improved prevalence estimates and understanding of clustering of diseases. *J Autoimmun*, **33**(3–4): 197–207.

Cruz-Tapias, P., Blank, M., Anaya, J.M., and Shoenfeld, Y. (2012). Infections and vaccines in the etiology of antiphospholipid syndrome. *Curr Opin Rheumatol*, **24**(4): 389–93.

Delamere, F.M., Sladden, M.M., Dobbins, H.M., and Leonardi-Bee, J. (2008). Interventions for alopecia areata. *Cochrane Database Syst Rev*, **2**: CD004413.

Dimitrijević, L., Živković, I., Stojanović, M., *et al.* (2012). Vaccine model of antiphospholipid syndrome induced by tetanus vaccine. *Lupus*, **21**(2): 195–202.

Dotz, W.I., Lieber, C.D., and Vogt, P.J. (1985). Leukonychia punctata and pitted nails in alopecia areata. *Arch Dermatol*, **121**(11): 1452–4.

Du Vivier, A. and Munro, D.D. (1975). Alopecia areata, autoimmunity, and Down's syndrome. *Br Med J*, **1**(5951): 191–2.

Goh, C., Finkel, M., Christos, P.J. and Sinha, A.A. (2006). Profile of 513 patients with alopecia areata: associations of disease subtypes with atopy, autoimmune disease and positive family history. *J Eur Acad Dermatol Venereol*, **20**(9): 1055–60.

Haida, Y., Ikeda, S., Takagi, A., *et al.* (2013). Association analysis of the HLA-C gene in Japanese alopecia areata. *Immunogenetics*, **65**(7): 553–7.

Hartung, H.P., Keller-Stanislawski, B., Hughes, R.A., and Lehmann, H.C. (2012). Guillain-Barré syndrome after exposure to influenza. *Nervenarzt*, **83**(6): 714–30.

Hayderi, L.E., Nikkels-Tassoudji, N., and Nikkels, A.F. (2013). Hair loss after varicella zoster virus infection. *Case Rep Dermatol*, **5**(1): 43–7.

Hernán, M.A., Jick, S.S., Olek, M.J., and Jick, H. (2004). Recombinant hepatitis B vaccine and the risk of multiple sclerosis: a prospective study. *Neurology*, **63**(5): 838–42.

Huang, K.P., Mullangi, S., Guo, Y., and Qureshi, A.A. (2013). Autoimmune, atopic, and mental health comorbid conditions associated with alopecia areata in the United States. *JAMA Dermatol*, **149**(7): 789–94.

Israeli, E., Agmon-Levin, N., Blank, M., and Shoenfeld, Y. (2009). Adjuvants and autoimmunity. *Lupus*, **18**(13): 1217–25.

Israeli, E., Agmon-Levin, N., Blank, M., and Shoenfeld, Y. (2011). Macrophagic myofasciitis a vaccine (alum) autoimmune-related disease. *Clin Rev Allergy Immunol*, **41**(2): 163–8.

Kivity, S., Agmon-Levin, N., Blank, M., and Shoenfeld, Y. (2009). Infections and autoimmunity –friends or foes? *Trends Immunol*, **30**(8): 409–14.

Kuchabal, S.D. and Kuchabal, D.S. (2010). Alopecia areata associated with localized vitiligo. *Case Rep Dermatol*, **2**(1): 27–31.

Kumar, S., Mittal, J., and Mahajan, B. (2013). Colocalization of vitiligo and alopecia areata: coincidence or consequence? *Int J Trichology*, **5**(1): 50–2

Lerner, A. (2012). Aluminum as an adjuvant in Crohn's disease induction. *Lupus*, **21**(2): 231–8.

Marques Da Costa, C., Dupont, E., Van der Cruys, M., *et al.* (2006). Earlier occurrence of severe alopecia areata in HLA-DRB1*11-positive patients. *Dermatology*, **213**(1): 12–14.

McMichael, A.J., Pearce, D.J., Wasserman, D., *et al.* (2007). Alopecia in the United States: outpatient utilization and common prescribing patterns. *J Am Acad Dermatol*, **57**(2 Suppl.): S49–51.

Megiorni, F., Pizzuti, A., Mora, B., *et al.* (2011). Genetic association of HLA-DQB1 and HLA-DRB1 polymorphisms with alopecia areata in the Italian population. *Br J Dermatol*, **165**(4): 823–7.

Messenger, A.G., Slater, D.N., and Bleehen, S.S. (1986). Alopecia areata: alterations in the hair growth cycle and correlation with the follicular pathology. *Br J Dermatol*, **114**(3): 337–47.

Mikaeloff, Y., Caridade, G., Suissa, S., and Tardieu, M. (2009). Hepatitis B vaccine and the risk of CNS inflammatory demyelination in childhood. *Neurology*, **72**(10): 873–80.

Lu, F. and Hogenesch, H. (2013). Kinetics of the inflammatory response following intramuscular injection of aluminum adjuvant. *Vaccine*, **31**(37): 3979–86.

Picardi, A., Pasquini, P., Cattaruzza, M.S., *et al.* (2003). Psychosomatic factors in first-onset alopecia areata. *Psychosomatics*, **44**(5): 374–81.

Ray, P., Black, S., Shinefield, H., *et al.* (2011). Risk of rheumatoid arthritis following vaccination with tetanus, influenza and hepatitis B vaccines among persons 15–59 years of age. *Vaccine*, **29**(38): 6592–7.

Razeghinejad, M.R., Nowroozzadeh, M.H., Sharifi, M., and Monabati, A. (2009). Acute-onset madarosis following MMR vaccination. *J Pediatr Ophthalmol Strabismus*, **46**(2): 123–4.

Rodriguez, T.A. and Duvic, M.; National Alopecia Areata Registry. (2008). Onset of alopecia areata after Epstein-Barr virus infectious mononucleosis. *J Am Acad Dermatol*, **59**(1): 137–9.

Rodriguez, T.A., Fernandes, K.E., Dresser, K.L., *et al.* (2010). Concordance rate of alopecia areata in identical twins supports both genetic and environmental factors. *J Am Acad Dermatol*, **62**(3): 525–7.

Rudnicka, L. and Lukomska, M. (2012). Alternaria scalp infection in a patient with alopecia areata. Coexistence or causative relationship? *J Dermatol Case Rep*, **6**(4): 120–4.

Sánchez-Ramón, S., Gil, J., Cianchetta-Sívori, M., and Fernández-Cruz, E. (2011). Alopecia universalis in an adult after routine tetanus toxoid vaccine. *Med Clin (Barc)*, **136**(7): 318.

Shaw, C.A. and Tomljenovic, L. (2013). Aluminum in the central nervous system (CNS): toxicity in humans and animals, vaccine adjuvants, and autoimmunity. *Immunol Res*, **56**(2–3): 304–16.

Safavi, K. (1992). Prevalence of alopecia areata in the First National Health and Nutrition Examination Survey. *Arch Dermatol*, **128**(5): 702.

Safavi, K.H., Muller, S.A., Suman, V.J., *et al.* (1995). Incidence of alopecia areata in Olmsted County, Minnesota, 1975 through 1989. *Mayo Clin Proc*, **70**(7): 628–33.

Serarslan, G., Savaş, N., and Yenin, J.Z. (2012). Is atopy and autoimmunity more prevalent in patients with alopecia areata? A comparative study. *J Eur Acad Dermatol Venereol*, **26**(6): 720–3.

Shoenfeld, Y. and Agmon-Levin, N. (2011). "ASIA" – autoimmune/inflammatory syndrome induced by adjuvants. *J Autoimmun*, **36**(1): 4–8.

Silva, K.A., McElwee, K.J., King, L.E., *et al.* (2003). Major locus on mouse chromosome 17 and minor locus on chromosome 9 are linked with alopecia areata in C3H/HeJ mice. *J Invest Dermatol*, **120**(5): 771–5.

Soriano, A., Verrecchia, E., Marinaro, A., *et al.* (2012). Giant cell arteritis and polymyalgia rheumatica after influenza vaccination: report of 10 cases and review of the literature. *Lupus*, **21**(2): 153–7.

Stübgen, J.P. (2012). Immune-mediated myelitis following hepatitis B vaccination. *Autoimmun Rev*, **12**(2): 144–9.

Sun, J., Silva, K.A., McElwee, K.J., *et al.* (2008). The C3H/HeJ mouse and DEBR rat models for alopecia areata: review of preclinical drug screening approaches and results. *Exp Dermatol*, **17**(10): 793–805.

Sundberg, J.P., Silva, K.A., Zhang, W., *et al.* (2009). Recombinant human hepatitis B vaccine initiating alopecia areata: testing the hypothesis using the C3H/HeJ mouse model. *Vet Dermatol*, **20**(2): 99–104.

Tan, E., Tay, Y.K., Goh, C.L., and Chin Giam, Y. (2002). The pattern and profile of alopecia areata in Singapore – a study of 219 Asians. *Int J Dermatol*, **41**(11): 748–53.

Tazi-Ahnini, R., di Giovine, F.S., McDonagh, A.J., *et al.* (2000). Structure and polymorphism of the human gene for the interferon-induced p78 protein (MX1): evidence of association with alopecia areata in the Down syndrome region. *Hum Genet*, **106**(6): 639–45.

Tazi-Ahnini, R., Cox, A., McDonagh, A.J., *et al.* (2002). Genetic analysis of the interleukin-1 receptor antagonist and its homologue IL-1L1 in alopecia areata: strong severity association and possible gene interaction. *Eur J Immunogenet*, **29**(1): 25–30.

Thomas, E.A. and Kadyan, R.S. (2008). Alopecia areata and autoimmunity: a clinical study. *Indian J Dermatol*, **53**(2): 70–4.

Tosti, A., Bellavista, S., and Iorizzo, M. (2006). Alopecia areata: a long term follow-up study of 191 patients. *J Am Acad Dermatol*, **55**(3): 438–41.

Tuccori, M., Pisani, C., Bachini, L., *et al.* (2012). Telogen effluvium following bivalent human papillomavirus vaccine administration: a report of two cases. *Dermatology*. **224**(3): 212–14.

Vañó-Galván, S., Aboín, S., Beà-Ardebol, S., and Sánchez-Mateos, J.L. (2008). Sudden hair loss associated with trachyonychia. *Cleve Clin J Med*, **75**(8): 567–8.

Wise, R.P., Katonga, P.K., and Salive, E.M. (1997). Hair loss after routine immunizations. *JAMA*, **278**(14): 1176–8.

Yang, S., Yang, J., Liu, J.B., *et al.* (2004). The genetic epidemiology of alopecia areata in China. *Br J Dermatol*, **151**(1): 16–23.

Zafrir, Y., Agmon-Levin, N., Paz, Z., *et al.* (2012). Autoimmunity following hepatitis B vaccine as part of the spectrum of "autoimmune (auto-inflammatory) syndrome induced by adjuvants" (ASIA): analysis of 93 cases. *Lupus*, **21**(2): 146–52.

27 Aluminum Particle Biopersistence, Systemic Transport, and Long-Term Safety: Macrophagic Myofasciitis and Beyond

Romain K. Gherardi,[1,2] Josette Cadusseau,[1] and François-Jérôme Authier[1,2]

[1]Faculty of Medicine, University of Paris East, Créteil, Paris, France
[2]Neuromuscular Center, H. Mondor Hospital, Créteil, Paris, France

Introduction

Over the last century, billions of humans have been vaccinated, and marked regression or eradication of several severe infectious diseases has been observed. Today, the potential applications of vaccines extend far beyond prevention of infectious diseases, and vaccination is considered to be a most promising weapon against a variety of different conditions. In general, vaccine safety has been regarded as excellent at the level of the population (Moxon and Siegriest, 2011), but adverse effects have also been reported (Agmon-Levin *et al.*, 2009). Given the considerable worldwide development of vaccination, safety signals in this field require the attention of the medical and scientific community, even if their intensity seems *a priori* to be low.

Concerns linked to the use of aluminum adjuvants (known as alum) have emerged following the recognition of their role at the origin of the so-called macrophagic myofasciitis (MMF) in 2001 (Gherardi *et al.*, 1998, 2001). MMF reveals a fundamental misconception of their adjuvant effect and points out their unexpectedly long biopersistence (Gherardi *et al.*, 2001). Recent demonstrations of their apparent capacity to migrate in lymphoid organs and to progressively accumulate in the brain (Khan *et al.*, 2013) suggest that alum adjuvant safety should be assessed in the long term, that administration of escalating doses of this compound to the population should be avoided, and that individual susceptibility factors to the development of alum adjuvant intolerance should be investigated.

Alum particles as lysosome-destabilizing adjuvants

Vaccine adjuvants have empirically been identified for their ability to enhance the adaptive immune response to a coadministered antigen. Particulate alum salts (known as alum) have been the main adjuvants approved for use in human vaccines for more than 80 years (Glenny *et al.*, 1926). They are currently used in vaccines against tetanus, hepatitis A, hepatitis B, human papillomavirus (HPV), *Haemophilus influenzae* type B (Hib), pneumococcal and meningococcal infections, and anthrax. They

Vaccines and Autoimmunity, First Edition. Edited by Yehuda Shoenfeld, Nancy Agmon-Levin, and Lucija Tomljenovic.

mainly include alum oxyhydroxyde (a crystalline compound), alum hydroxyphosphate, and amorphous alum phosphate. Alum is able to adsorb vaccine antigens on its surface. The strongest adsorption phenomenon results from ligand exchange, which involves the replacement of a surface hydroxyl on the adjuvant by a terminal phosphate group of the antigen (Lu *et al.*, 2013).

It is known that alum induces strong innate immune responses at the site of injection, as assessed by an influx of neutrophils, monocyte/macrophages, eosinophils, and MHC-II+ antigen-presenting cells (APCs), mainly dendritic cells (DCs) (Lu and Hogenesch, 2013). We observed that muscle-resident macrophages mainly located in fascias are among the first cells to sense a disturbance in muscle homeostasis (Brigitte *et al.*, 2010). Through local production of chemokines, they alert the immune system and recruit other myeloid cells, such as neutrophils, and inflammatory monocytes, which differentiate into inflammatory DCs (Brigitte *et al.*, 2010). Monocyte-derived inflammatory DCs have an immature phenotype in the muscle, specialized for antigen uptake, but upon contact with tissue debris or foreign material they migrate to the lymph node T cell paracortex, where they arrive as mature cells, expressing co-stimulatory molecules. Selective depletion studies have suggested that inflammatory DCs may be crucial for alum adjuvant activity (Kool *et al.*, 2008a,b), but eosinophils appear to play an important role too (Wang and Weller, 2008).

It has been long believed that alum ensures a long-lasting immune response through formation of a depot that slowly releases the antigen under the influence of the interstitial fluid (Flarend *et al.*, 1997; Shi *et al.*, 2001). As detailed later in the chapter, muscle biopsy findings in immunized patients have challenged the view that the injected adjuvant remains extracellular (Gherardi *et al.*, 2001) and have led to the acknowledgement that, in contrast to ancient belief, alum particles are avidly taken up by phagocytic cells (Morefield *et al.*, 2005). Due to strong antigen binding to alum particles, this has been shown to increase antigen uptake by DCs, to reduce antigen degradation, and to sustain antigen presentation *in vitro* (Ghimire *et al.*, 2012). Alum particle uptake has also been shown to promote macrophage survival (Hamilton *et al.*, 2000). *In vivo*, alum injection induces the formation of persistent alum-induced granuloma at the site of immunization (Gherardi *et al.*, 2001; Verdier *et al.*, 2005; Authier *et al.*, 2006), but local persistence of alum is not required for good immunization, because removal of the injection site as early as 2 hours after injection has no appreciable effect on antigen-specific T and B cell responses (Hutchison *et al.*, 2012).

Recent reviews have systematically pointed out that, in spite of their long usage, alum salts are mysterious compounds whose mechanisms of adjuvanticity remain uncertain and have just begun to be seriously investigated (Exley *et al.*, 2010). Alum is deficient at initiating cell-mediated immunity and skews the immune response toward a T helper type 2 (Th2) response associated with strong production of interleukin (IL)-4 and the immunoglobulin (IgG)1 antibody subtype (Ulanova *et al.*, 2001). Several pathways have been proposed to contribute to alum adjuvant effects, but proposed explanations have often been challenged by subsequent studies (McKee *et al.*, 2009). Notably, alum was shown to strongly activate the NLRP3 inflammasome (Eisenbarth *et al.*, 2008; Li *et al.*, 2008), but this finally appeared to be inessential to the adjuvant effect (Franchi and Nunez, 2008; Spreafico *et al.*, 2010). It remains true, however, that NLRP3 activation is strongly induced by alum hydroxide and other crystals, such as silica, urate sodium, and asbestos, causing IL-1b release and an activation of the downstream inflammatory cascade. Recently, alum adjuvant effects have been linked to the release of noncytokine biomolecules, including uric acid (Kool *et al.*, 2008a,b), double-stranded DNA (dsDNA) (Marichal *et al.*, 2011), and prostaglandin E2 (Kuroda *et al.*, 2011), suggesting alternate models for alum-mediated immunity. It has been suggested that the commitment of specific crystal-induced signaling pathways may explain why alum hydroxyde particles exhibit a much more irritating character than soluble alum (Shi, 2012). Alum crystals consistently bind to and aggress the plasma membrane lipid bilayer (Flach *et al.*, 2011). Alum also destabilize lysosomes (Hornung *et al.*, 2008; Lima *et al.*, 2013), which degrade endocytosed, phagocytosed, or autophagocytosed materials and play an important role in immunity. DCs possess highly controlled antigen-processing functions utilizing lysosomal proteases and pH changes that are optimal for the generation of peptides, rather than complete protein degradation (Trombetta *et al.*, 2003). Limitation of the lysosomal proteolysis of antigens is known to increase antigen presentation and immunogenicity (Delamarre *et al.*, 2006), while inhibition of lysosomal proteases enhances the stability of p:MHCII complexes, allowing their accumulation on the DC surface (Shin *et al.*, 2006). It therefore seems possible that alum-induced lysosomal blockade might play an important role in the adjuvant effect. The mechanism by which alum causes lysosomal destabilization remains unclear, but it is possible that its crystalline structure directly causes physical rupture of the membrane (Kang and Locksley, 2009). Unfortunately, the cell also uses the lysosomal and autophagy pathways to dispose of solid materials perceived as foreign or aberrant, such as pathogens or senescent organelles. The impact of particles on these pathways may, therefore, favor their biopersistence and immunotoxicity (Stern *et al.*, 2012).

MMF and alum biopersistence

In 1998, a French consortium of myopathologists described MMF as an emerging condition of unknown cause characterized by a pathognomonic lesion in muscle biopsy, mixing large macrophages with submicron- to micron-sized agglomerates of nanocrystals in their cytoplasm and lymphocytic infiltrates (Gherardi *et al.*, 1998). MMF differed from other histiocytic diseases and was always detected in the deltoid muscle in adults (Bassez *et al.*, 2003). Cytoplasmic inclusions were always found, surrounded or not by altered lysosomal membranes, and contained aluminum (Gherardi *et al.*, 2001). Their crystalline appearance appeared to be characteristic of aluminum hydroxide. Patients had normal renal function, and no exposure to alum other than that conferred by a prior immunization (100%) by vaccines containing the aluminum oxyhydroxide form of "alum" (Gherardi *et al.*, 2001). It is obvious retrospectively that the rapid emergence of MMF in France reflected the combination of (i) the replacement of the subcutaneous by the intramuscular route for vaccine injections in the early 1990s; (ii) the large-scale campaign of primovaccination of French adults against hepatitis B in the mid 1990s; and (iii) the choice of the deltoid muscle for routine muscle biopsy in France, in contrast to the preferential use of the biceps brachialis and quadriceps muscles in other countries. Notably, the subcutaneous administration of alum vaccines can induce subcutaneous lesions enclosing aluminum particle-loaded macrophages that are microscopically different from MMF, including skin pseudolymphoma in humans (Maubec *et al.*, 2005) and fibrosarcoma in cats (Madewell *et al.*, 2001).

The MMF lesion can be reproduced experimentally by intramuscular vaccination in mice, rats, and monkeys (Gherardi *et al.*, 2001; Verdier *et al.*, 2005; Authier *et al.*, 2006). The experimental lesion invariably shrinks over time (Authier *et al.*, 2006); in monkeys, it begins to disappear completely from the muscle between 6 and 12 months after a diphtheria, tetanus, and pertussis (DTP) injection corresponding to 14–21 times the human DTP-equivalent dose of alum (Verdier *et al.*, 2005).

Whether or not longstanding MMF is commonly present in a hidden form in healthy individuals cannot be determined because of the unethical character of muscle biopsy in asymptomatic individuals. This seems very unlikely, however, since a recent review of 130 consecutive deltoid muscle biopsies performed for diagnostic purposes in myalgic patients previously immunized with alum-containing vaccines revealed that most alum receivers did not have long-lasting MMF. This could be reliably assessed because age, sex ratio, number of alum-containing vaccine injections, and delay between last injection and deltoid muscle biopsy were similar in the MMF and non-MMF groups (Ragunathan-Thangarajah *et al.*, 2013). This refutes the previous, nondocumented view that every vaccinee might have longstanding MMF lesions when biopsy is performed in the deltoid muscle (Papo, 2003).

Of course, in light of experimental MMF models, it is important to systematically check the individual vaccine record in order to assess the unusually persistent character of MMF. In a recent evaluation of 583 patients collected between 1994 and 2012 (Cadusseau *et al.*, 2013), the median time between last alum administration and biopsy was 65 months. Compared to our previous reports, this was slightly increased (36 months in the initial series of 2001, collected shortly after the peak of French adult immunization; 53 months in the series of 2003: Gherardi and Authier, 2003), allowing further assessment of the highly chronic character of the lesion in affected individuals. An average number of 5.3 alum-containing shots had been administered during the 10 years prior to biopsy detecting MMF, mainly corresponding to vaccinations against hepatitis B (89.7%), tetanus (42.2%), and hepatitis A (8.8%). In practice, we consider that the MMF lesion is unusually persistent when time between last immunization and MMF detection is greater than 18 months. It is extremely important to carefully consider this point in young children, who receive multiple vaccine injections in the first year of life, thus increasing the risk of coincidental association between a constitutive muscle disease and MMF detected in the quadriceps muscle, which is used for pediatric immunizations. If the risk of such coincidental associations also potentially exists in adults, in practice it is low. For example, adult patients combining MMF and hereditary muscle disease are extremely rare, despite the intense immunization program of French patients with muscular dystrophy.

To summarize, the MMF lesion is now universally recognized as indicative of a long-lasting persistence of the alum at the site of prior intramuscular immunization. Alum-induced granulomatous lesions vary considerably in size, according to genetic background (Authier *et al.*, 2006), and the initial hypothesis made by the World Health Organization (WHO) that MMF may reflect some individual inability to clear out alum from the body remains valid (WHO, 1999). In other words, a long-lasting MMF lesion should be considered a biomarker of alum biopersistence.

MMF and myalgic encephalomyelitis/chronic fatigue syndrome

From the very beginning, we have observed that MMF is typically detected in patients with diffuse myalgias, and a strong statistical association between myalgias and MMF has been detected by general survey in different French neuromuscular centers (myalgias in 90% of patients with MMF versus 44% without MMF, $p < 0.0001$) (Gherardi *et al.*, 2001). However, alum has always been regarded as well tolerated on the basis of observations in the short term, and the exact meaning of the detection of long-standing MFM in myalgic patients remains unclear, mainly because (i) myalgias represent a poorly specific symptom that is commonly encountered in primary care practice, which may be associated with a wide spectrum of conditions, including autoimmune/inflammatory diseases, infections, endocrine or metabolic disorders, and drug intolerance, but often remains unexplained after extensive investigation, falling into the loose setting of fibromyalgia; and (ii) there is a lack of a clear biological link between the persistence of alum in macrophages at the site of immunization and the occurrence of delayed systemic clinical manifestations.

Most patients are women (70%), with a mean age at the time of biopsy of 45 years (extreme 12–83). They typically complain of (i) chronic diffuse myalgias (89%), with or without arthralgia; (ii) disabling chronic fatigue lasting more than 6 months (77%); and (iii) obvious cognitive alterations affecting memory and attention (51%) (Cadusseau *et al.*, 2013). The onset of these symptoms is typically delayed following immunization, with a median time after last injection of 7 months (range 0.5–84.0) for initial clinical symptoms, and 11 months (range 0–72) for first myalgia (Gherardi and Authier, 2003).

The onset of myalgia may follow exercise of unusual intensity. Myalgias generally begin in the lower limbs, and almost never at the site of previous vaccine injections. They gradually extend toward the top of the body to reach the paravertebral muscles and become diffuse at the time of biopsy (Gherardi and Authier, 2003). Genuine muscle weakness is rare. Myopathic electromyogram and elevation of creatine kinase (CK) are each observed in less than half of patients. Inflammatory markers are poorly contributory, but iron metabolism is frequently altered. Comparison of myalgic vaccinees with and without MMF at deltoid muscle biopsy indicates that patients with MMF rarely have fibromyalgia (the required 11 tender points of the ACR 1990 criteria for fibromyalgia are present in 16.6 versus 55.5%, $p < 0.04$) and often have delayed evoked potentials suggestive of central nervous system (CNS) demyelination (38.5 versus 5.7%, $p < 0.01$) (Ragunathan-Thangarajah *et al.*, 2013).

Chronic fatigue, often associated with sleep disturbances and headaches, is usually very disabling, with conspicuous repercussions on both the professional and personal lives of patients. A case–control study conducted under the aegis of the French regulatory agency AFSSAPS consistently showed chronic fatigue to be both significantly more frequent and more severe in patients with MMF compared to those without MFM in the deltoid muscle (AFSSAPS, 2013).

Cognitive alterations further assess CNS involvement, though they are often underestimated or not detected by routine examination. MMF patients complain of memory loss, difficulty in maintaining focus, and mood changes. Cognitive tests showed constant and distinct alterations in an initial series of 25 consecutive MFM patients without multiple sclerosis (MS), compared to 11 arthritis patients matched for severity and duration of pain, depression, and education level (Couette *et al.*, 2009). MMF patients mainly had visual memory, working memory, and dichotic listening alterations, suggestive of organic corticosubcortical impairment. Such deficits usually remain stable with time (Passeri 2011). We have now extensively evaluated the cognition of 105 MMF patients:

attention and memory complaints were reported by 102/105 (97%) and neuropsychological tests were abnormal in 93/105 (89%) of patients (Cadusseau *et al.*, 2013). We found remarkable correlations between cognitive impairment and SPECT brain imaging in MMF patients, with hyposignals predominating in the hippocampus, posterior areas, cingulum, and corpus callosum (unpublished results by Itti *et al.*). Cognitive alterations associated with MMF are reminiscent of those described after occupational alum exposure (Meyer-Baron *et al.*, 2007). Their correlation with the body burden of alum is possible, but has not yet been explored on a systematic basis (Exley *et al.*, 2009).

The combination of chronic muscle pain, chronic fatigue, and cognitive dysfunction is consistent with the so-called "myalgic encephalomyelitis/chronic fatigue syndrome" (ME/CFS), and, indeed, a majority of MMF patients meet international criteria for ME/CFS (Authier *et al.*, 2003). ME/CFS is a severe, complex, acquired illness that has been classified as a neurological disorder in the WHO International Classification of Diseases since 1969 (ICD 10 G93.3). It is distinct from fibromyalgia and psychasthenia, which are classified as musculoskeletal (M79.7) and psychiatric (F48.8) disorders, respectively. International studies have estimated the prevalence of ME/CFS at between 0.4 and 2.6% of the world population, with a total annual cost burden to society of approximately \$18.7–24.0 billion in the United States (Jason *et al.*, 2007). An ad hoc parliamentary group in the United Kingdom produced a legacy paper in 2010 which stressed that previous research into ME/CFS has produced little substantive progress but that far too much emphasis has been put on psychological research and not enough on biomedical research (All-Party Parliamentary Group on ME, 2010). Symptoms of idiopathic ME/CSF are closely similar to those of the post-infective chronic fatigue syndrome (Hickie *et al.*, 2006). Although the underlying cause of ME/CSF is currently unknown, and is likely to be heterogeneous, the illness is thought to be triggered by an abnormal immune response to an infectious or toxic agent, resulting in chronic immune activation (Landay *et al.*, 1991). ME/CFS patients consistently have increased risk of developing diffuse large B cell lymphoma and marginal-zone B cell lymphoma (Chang *et al.*, 2012). All these data support continued efforts to understand the biology of CFS.

Phagocytes and systemic diffusion of aluminum particles

As already noted, the conceptual link between long-term persistence of alum particles within macrophages at the site of previous immunization and the occurrence of adverse systemic events, in particular neurological ones, has long remained an unsolved question. On the one hand, aluminum has long been recognized as a neurotoxic metal, affecting memory, cognition, and psychomotor control, altering neurotransmission and synaptic activity, damaging the blood–brain barrier (BBB), exerting pro-oxidant effects, activating microglia and neuroinflammation, depressing the cerebral glucose metabolism and mitochondrial functions, interfering with transcriptional activity, and promoting β-amyloid and neurofilament aggregation (Tomljenovic, 2011). On the other hand, alum particles impact the immune system through their adjuvant effect and by many other means: they strongly adsorb vaccine antigens onto their surface, which protects them from proteolysis, thus forming a persistently immunogenic pseudopathogen (Rosenblum *et al.*, 2011); they may bind undesirable residual products inherent to vaccine production procedures, as shown for HPV DNA sequences (Lee, 2012) and yeast proteins (Offit and Jew, 2003), which may be potentially hazardous (Rinaldi *et al.*, 2013); and they can directly induce alum allergy (Lopez *et al.*, 1994; Bergfors *et al.*, 2005), possibly through the recently identified mechanism of metal-induced hypersensitivity (Falta *et al.*, 2013).

Of course, concerns about the biopersistence of alum largely depend upon the ability of alum particles to reach and exert toxicity in remote organs, as suggested by several studies (Wen and Wisniewski, 1985; Redhead *et al.*, 1992; Sahin *et al.*, 1994; Wang *et al.*, 2012). The reference study on alum hydroxide biodisposition was conducted using alum enriched in isotopic 26Al injected in the muscle of two rabbits; 26Al was weakly eliminated in urine (6% on day 28 endpoint) and was detected in lymph nodes, spleen, liver, and brain (Flarend *et al.*, 1997). Whether aluminum was in particulate or soluble form remained unexplored. To assess the fate of particulate material in mice, we successively performed intramuscular injections alum-containing vaccine, fluorescent latex beads, and fluorescent nanohybrids coated with precipitated alum (Khan *et al.*, 2013). These materials were quickly captured by macrophages and a large proportion left the injected muscle, mainly inside immune cells, to reach the draining lymph nodes. Particle-laden cells then left the lymphatic system and reached the blood circulation (presumably via the thoracic duct), allowing them to reach distant organs such as the spleen and liver, and, much more slowly, the brain. Recombinant chemokine injection and the use of genetically modified mice showed that systemic biodistribution of particles crucially depended on the monocyte chemoattractant MCP-1/CCL2. In the brain, particles were trapped in microglial cells. In accordance with good overall tolerance of alum, brain penetration was extremely low in normal conditions; however, brain translocation was significantly increased in the case of altered BBB or under the influence of MCP-1/CCL2 signaling (Khan *et al.*, 2013). Notably, production of this chemokine is subject to significant interindividual variations related to age, genetic, and environmental factors. We have identified the selective increase of circulating MCP-1/CCL2 as a circulating biomarker in MMF patients (Cadusseau *et al.*, 2013). The imbalance between the huge number of vaccinated individuals and the relatively low number of MMF cases suggests the crucial involvement of individual susceptibility factors in intolerance to alum. Genetically driven MCP-1/CCL2 production might represent one of these factors (Khan *et al.*, 2013).

In summary, precipitated alum and other poorly biodegradable materials taken up at the periphery by phagocytes circulate in the lymphatic and blood circulation and can enter the brain using a Trojan horse mechanism similar to that used by infectious particles (Drevets *et al.*, 2004; Eugenin *et al.*, 2006). The link between alum administration and CNS dysfunction and damage has been experimentally substantiated (Petrik *et al.*, 2007; Shaw and Petrik, 2009; Shaw *et al.*, 2013), casting doubt on alum safety (Tomljenovic and Shaw, 2011). The role of brain transport of particulate alum in alum-induced neurological and behavioral effects remains to be explored.

Beyond MMF: the ASIA concept

Neurologic conditions thought to result from gene–environment interactions, such as idiopathic ME/SFC (van Rensburg *et al.*, 2001) and MS (Exley *et al.*, 2006), have been previously associated with aluminum overload. An increased risk of developing MS in the long term following administration of alum-containing vaccine has also been reported (Hernán *et al.*, 2004; Mikaeloff *et al.*, 2009) but is not acknowledged by regulatory agencies.

Interestingly, in addition to ME/CSF, about 10% of our MMF patients had concurrent MS-like disease (Authier *et al.*, 2001), an additional 5–10% had another autoimmune disease, such as thyroiditis or diffuse inflammatory myopathies, and the remainder occasionally had low titers of various autoantibodies (Gherardi and Authier, 2003).

Yehuda Shoenfeld delineated the "autoimmune/inflammatory) syndrome induced by adjuvants" (ASIA) (Shoenfeld and Agmon-Levin, 2011). ASIA acknowledges that various combinations of (i) specific autoimmune diseases identified by well-established criteria; (ii) less specific symptoms, such as myalgia, arthralgia, chronic fatigue, and cognitive impairment (the combination of which defines ME/CFS); and (iii) circulating autoantibodies can occur after exposure to a variety of chemical or natural products with immunologic adjuvant properties. ASIA is very useful in practice, as it may alert physicians when these symptoms appear following vaccination, and helps them to put a name to such conditions.

We had previously noted that symptoms associated with MMF are strikingly similar to those described as the Gulf War syndrome (GWS), a condition strongly associated with the administration of multiple vaccinations to soldiers (Hotopf *et al.*, 2000; Cherry *et al.*, 2001), especially the anthrax vaccine, which contains alum (capable of inducing MMF; Theeler *et al.*, 2008) and possibly squalene (Asa *et al.*, 2000). On these grounds, we proposed the delineation of a vaccine adjuvant syndrome (Gherardi, 2003). Shoenfeld reasoned similarly, but added to GWS and MMF his own experience of siliconosis, a disease complex observed in patients with leaky breast silicone implants, attributed to deleterious adjuvanticity of silicone particles (Miyoshi *et al.*, 1964; Shoaib and Patten, 1996) and other post-vaccinal events. In so doing, he enlarged the causal relationship to any compound with adjuvant properties. ASIA's major and minor diagnostic criteria await international validation, and delineation of possible differences among the constitutive syndromes, possibly linked to specificities of the different causal adjuvants, deserves future studies. However, the ASIA concept has rapidly caught the attention of the international medical community, pointing to a need in the field (Vasey *et al.*, 2003). It has already extended to the field of veterinary medicine (Luján *et al.*, 2013).

Alum safety in the long term

Alum is known to be potentially neurotoxic but has been used for decades at levels considered by the industry and the regulatory agencies to constitute an acceptable compromise between its role as adjuvant and its toxic effects. The emergence of MMF in France revealed significant gaps in our knowledge of alum particles, including their exact mechanisms of action, their fate after injection, their systemic dissemination, and their safety in the long term. There has been much effort in many countries in recent years to pave the way for the delineation of novel adjuvants, but attempts to seriously examine public health questions raised by the biopersistent and neuromigrant character of alum particles have not been made. In the context of a planned expansion of immunization policies at the global level, it seems highly desirable that: (i) alum's insidious effects in the long term, notably CNS dysfunction, be investigated in order to more precisely evaluate the risk–benefit balance of ancient and novel immunization strategies; (2) escalation of alum doses be monitored at both the individual (overimmunization) and population (global alum burden) levels; (3) susceptibility factors be specifically studied, including age (with emphasis on the neonatal and aging brains, which have immature or altered BBBs) and genetic factors; and (4) plans be made for how progressive withdrawal of alum salts from human vaccines can be conducted (as has already been done for cosmetics and some veterinary vaccines). Alum should be replaced by more physiological, rapidly biodegradable, and efficient (i.e. Th1 response-inducing) adjuvants.

Many difficulties can be expected in achieving this aim. Several of the listed actions uniquely depend on appropriate public research funding, and the definition/validation/introduction of alternative adjuvants at the international level (e.g. the reintroduction of the old adjuvant calcium phosphate or the development of new-generation products) will be complex, long, and difficult. Crossing of this obstacle will represent a challenge for the industry and an efficiency test for regulators.

References

AFSSAPS (Agence Française de Sécurité sanitaire des Produits de Santé). (2003). [Exploratory epidemiological survey of macrophagic myofasciitis] Available from: http://ansm.sante.fr/var/ansm_site/storage/original/application/030593fa4e393af7cec8ff7092832215.pdf [last accessed 11 December 2014].

Agmon-Levin, N., Paz, Z., Israeli, E., Shoenfeld, Y. (2009). Vaccines and autoimmunity. *Nat Rev Rheumatol*, **5**(11): 648–52.

All-Party Parliamentary Group on ME. (2010). Inquiry into NHS service provision for ME/CFS: interim report. Available from: http://appgme.org.uk/Downloads/appg_interim_report_v2.pdf [last accessed 11 December 2014].

Asa, P.B., Cao, Y., and Garry, R.F. (2000). Antibodies to squalene in Gulf War syndrome. *Exp Mol Pathol*, **68**(1): 55–64.

Authier, F.J., Cherin, P., Creange, A., *et al.* (2001). Central nervous system disease in patients with macrophagic myofasciitis. *Brain*, **124**(5): 974–83.

Authier, F.J., Sauvat, S., Champey, J., *et al.* (2003). Chronic fatigue syndrome in patients with macrophagic myofasciitis. *Arthritis Rheum*, **48**: 569–70.

Authier, F.J., Sauvat, S., Christov, C., *et al.* (2006). AlOH3-adjuvanted vaccine-induced macrophagic myofasciitis in rats is influenced by the genetic background. *Neuromuscul Disord*, **16**: 347–52.

Bassez, G., Authier, F.J., Lechapt-Zalcman, E., *et al.* (2003). Inflammatory myopathy with abundant macrophages (IMAM): a condition distinct from macrophagic myofasciitis, sharing similarities with cytophagic histiocytic panniculitis. *J Neuropathol Exp Neurol*, **62**: 464–74.

Bergfors, E., Björkelund, C., and Trollfors, B. (2005). Nineteen cases of persistent pruritic nodules and contact allergy to aluminium after injection of commonly used aluminium-adsorbed vaccines. *Eur J Pediatry*, **164**(11): 691–7.

Brigitte, M., Schilte, C., Plonquet, A., *et al.* (2010). Muscle resident macrophages control the immune cell reaction in a mouse model of notexininduced myoinjury. *Arthritis Rheum*, **62**: 268–79.

Cadusseau, J., Ragunathan-Thangarajah, N., Surenaud, M., *et al.* (2013). Selective elevation of circulating CCL2/MCP1 levels in patients with longstanding post-vaccinal macrophagic myofasciitis and ASIA. *Curr Med Chem*, **21**(4): 511–17.

Chang, C.M., Warren, J.L., and Engels, E.A. (2012). Chronic fatigue syndrome and subsequent risk of cancer among elderly US adults. *Cancer*, **118**(23): 5929–36.

Cherry, N., Creed, F., Silman, A., *et al.* (2001). Health and exposures of United Kingdom Gulf war veterans. Part II: The relation of health to exposure. *Occup Environ Med*, **58**: 299–306.

Couette, M., Boisse, M.F., Maison, P., *et al.* (2009). *J Inorg Biochem*, **103**: 1571–8.

Delamarre, L., Couture, R., Mellman, I., and Trombetta, E.S. (2006). Enhancing immunogenicity by limiting susceptibility to lysosomal proteolysis. *J Exp Med*, **203**: 2049–55.

Drevets, D.A., Dillon, M.J., Schawang, J.S., *et al.* (2004). The Ly-6Chigh monocyte subpopulation transports Listeria monocytogenes into the brain during systemic infection of mice. *J Immunol*, **172**: 4418–24.

Eisenbarth, S.C., Colegio, O.R., O'Connor, W., *et al.* (2008). Crucial role for the Nalp3 inflammasome in the immunostimulatory properties of aluminium adjuvants. *Nature*, **453**: 1122–6.

Eugenin, E.A., Osiecki, K., Lopez, L., *et al.* (2006). CCL2/monocyte chemoattractant protein-1 mediates enhanced transmigration of human immunodeficiency virus (HIV)-infected leukocytes across the blood–brain barrier: a potential mechanism of HIV-CNS invasion and NeuroAIDS. *J Neurosci*, **26**: 1098–106.

Exley, C., Mamutse, G., Korchazhkina, O., *et al.* (2006). Elevated urinary excretion of aluminium and iron in multiple sclerosis. *Mult Scler*, **12**(5): 533–40.

Exley, C., Swarbrick, L., Gherardi, R.K., and Authier F.J. (2009). A role for the body burden of aluminium in vaccine-associated macrophagic myofasciitis and chronic fatigue syndrome. *Med Hypotheses*, **72**: 135–9.

Exley, C., Siesjo, P., and Eriksson, H. (2010). The immunobiology of aluminium adjuvants: how do they really work? *Trends Immunol*, **31**: 103–9.

Falta, M.T., Pinilla, C., Mack, D.G., *et al.* (2013). Identification of beryllium-dependent peptides recognized by CD4+ T cells in chronic beryllium disease. *J Exp Med*, **210**(7): 1403–18.

Flach, T.L., Ng, G., Hari, A., *et al.* (2011). Alum interaction with dendritic cell membrane lipids is essential for its adjuvanticity. *Nat Med*, **17**: 479–87.

Flarend, R.E., Hem, S.L, White, J.L., *et al.* (1997). In vivo absorption of aluminium-containing vaccine adjuvants using ^{26}Al. *Vaccine*, **15**: 1314–18.

Franchi, L. and Nunez, G. (2008). The Nlrp3 inflammasome is critical for aluminium hydroxide-mediated IL-1beta secretion but dispensable for adjuvant activity. *Eur J Immunol*, **38**: 2085–9.

Gherardi, R.K. (2003). Lessons from macrophagic myofasciitis: towards definition of a vaccine adjuvant-related syndrome. *Rev Neurol (Paris)*, **159**(2): 162–4.

Gherardi, R.K. and Authier, F.J. (2003). Aluminum inclusion macrophagic myofasciitis: a recently identified condition. Immunol. *Allergy Clin North Am*, **23**: 699–712.

Gherardi, R.K., Coquet, M., Chérin, P., *et al.* (1998). Macrophagic myofasciitis: an emerging entity. *Lancet*, **352**(9125): 347–52.

Gherardi, R.K., Coquet, M., Cherin, P., *et al.* (2001). Macrophagic myofasciitis lesions assess long-term persistence of vaccinederived aluminium hydroxide in muscle. *Brain*, **124**: 1821–31.

Ghimire, T.R., Benson, R.A., Garside, P., and Brewer, J.M. (2012). Alum increases antigen uptake, reduces antigen degradation and sustains antigen presentation by DCs in vitro. *Immunol Lett*, **147**(1–2): 55–62.

Glenny, A., Pope, C., Waddington, H., and Wallace, U. (1926). Immunological notes. *XVII. The antigenic value of toxoid precipitated by potassium alum. J Pathol Bacteriol*, **26**: 38–9.

Hamilton, J.A., Byrne, R., and Whitty, G. (2000). Particulate adjuvants can induce macrophage survival, DNA synthesis, and a synergistic, proliferative response to GM-CSF and CSF-1. *J Leukoc Biol*, **67**: 226–32.

Hernán, M.A., Jick, S.S., Olek, M.J., and Jick, H. (2004). Recombinant hepatitis B vaccine and the risk of multiple sclerosis: a prospective study. *Neurology*, **63**(5): 838–42.

Hickie, I., Davenport, T., Wakefield, D., *et al.* (2006). Post-infective and chronic fatigue syndromes precipitated by viral and non-viral pathogens: prospective cohort study. *BMJ*, **333**(7568): 575.

Hornung, V., Bauernfeind, F., Halle, A., *et al.* (2008). Silica crystals and aluminum salts activate the NALP3 inflammasome through phagosomal destabilization. *Nat Immunol*, **9**: 847–56.

Hotopf, M., David, A., Hull, L., *et al.* (2000). Role of vaccinations as risk factors for ill health in veterans of the Gulf war: cross sectional study. *BMJ*, **320**: 1363–7.

Hutchison, S., Benson, R.A., Gibson, V.B., *et al.* (2012). Antigen depot is not required for alum adjuvanticity. *FASEB J*, **26**: 1272–9.

Jason, A., Benton, C., Valentine, L., *et al.* (2007). The economic impact of ME/CFS: individual and societal costs. *Dynamic Medicine*, **7**: 6.

Kang, S.J. and Locksley, R.M. (2009). The inflammasome and alum-mediated adjuvanticity. *F1000 Biol Rep*, **1**: 15.

Khan, Z., Combadiere, C., Authier, F.J., *et al.* (2013). Slow CCL2-dependent translocation of biopersistent particles from muscle to brain. *BMC Med*, **11**: 99.

Kool, M., Petrilli, V., De Smedt, T., *et al.* (2008a). Cutting edge: alum adjuvant stimulates inflammatory dendritic cells through activation of the NALP3 inflammasome. *J Immunol*, **181**: 3755–9.

Kool, M., Soullié, T., van Nimwegen, M., *et al.* (2008b). Alum adjuvant boosts adaptive immunity by inducing uric acid and activating inflammatory dendritic cells. *J Exp Med*, **205**: 869–82.

Kuroda, E., Ishii, K.J., Uematsu, S., *et al.* (2011). Silica crystals and aluminum salts regulate the production of prostaglandin in macrophages via NALP3 inflammasome-independent mechanisms. *Immunity*, **34**(4): 514–26.

Landay, A.L., Jessop, C., Lennette, E.T., and Levy J.A. (1991). Chronic fatigue syndrome: clinical condition associated with immune activation. *Lancet*, **338**: 707–12.

Li, H., Willingham, S.B., Ting, J.P., and Re, F. (2008). Cutting edge: inflammasome activation by alum and alum's adjuvant effect are mediated by NLRP3. *J Immunol*, **181**: 17–21

Lima, H. Jr, Jacobson, L.S., Goldberg, M.F., *et al.* (2013). Role of lysosome rupture in controlling Nlrp3 signaling and necrotic cell death. *Cell Cycle*, **12**(12): 1868–78.

Lee, S.H. (2012). Detection of human papillomavirus (HPV) L1 gene DNA possibly bound to particulate aluminum adjuvant in the HPV vaccine Gardasil. *J Inorg Biochem*, **117**: 85–92.

Lopez, S., Pelaez, A., Navarro, L.A., *et al.* (1994). Aluminium allergy in patients hyposensitized with aluminium-precipitated antigen extracts. *Contact Dermatitis*, **31**: 37–40.

Lu, F. and Hogenesch, H. (2013). Kinetics of the inflammatory response following intramuscular injection of aluminum adjuvant. *Vaccine*, **31**(37): 3979–86.

Lu, F., Boutselis, I., Borch, R.F., and Hogenesch, H. (2013). Control of antigen-binding to aluminum adjuvants and the immune response with a novel phosphonate linker. *Vaccine*, **31**(40): 4362–7.

Luján, L., Pérez, M., Salazar, E., *et al.* (2013). Autoimmune/autoinflammatory syndrome induced by adjuvants (ASIA syndrome) in commercial sheep. *Immunol Res*, **56**(2–3): 317–24.

Madewell, B.R., Griffey, S.M., McEntee, M.C., *et al.* (2001). Feline vaccine-associated fibrosarcoma: an ultrastructural study of 20 tumors (1996–1999). *Vet Pathol*, **38**(2): 196–202.

Marichal, T., Ohata, K., Bedoret, D., *et al.* (2011). DNA released from dying host cells mediates aluminum adjuvant activity. *Nat Med*, **17**: 996–1002.

Maubec, E., Pinquier, L., Viguier, M., *et al.* (2005). Vaccination-induced cutaneous pseudolymphoma. *J Am Acad Dermatol*, **52**(4): 623–9.

McKee, A.S., Munks, M.W., MacLeod, M.K., *et al.* (2009). Alum induces innate immune responses through macrophage and mast cell sensors, but these sensors are not required for alum to act as an adjuvant for specific immunity. *J Immunol*, **183**: 4403–14.

Meyer-Baron, M., Schäper, M., Knapp, G., and van Thriel, C. (2007). Occupational aluminum exposure: Evidence in support of its neurobehavioral impact. *NeuroToxicology*, **28**: 1068–78.

Mikaeloff, Y., Caridade, G., Suissa, S., and Tardieu, M. (2009). Hepatitis B vaccine and the risk of CNS inflammatory demyelination in childhood. *Neurology*, **72**(10): 873–80.

Miyoshi, K., Miyamura, T., Kobayashi, Y., *et al.* (1964). Hypergammaglobulinemia by prolonged adjuvanticity in man disorders developed after augmentation mammoplasty. *Jpn Med J*, **2122**: 9–14.

Morefield, G.L., Sokolovska, A., Jiang, D., *et al.* (2005). Role of aluminum-containing adjuvants in antigen internalization by dendritic cells in vitro. *Vaccine*, **23**: 1588–95.

Moxon, E.R. and Siegriest, C.A. (2011). The next decade of vaccines: societal and scientific challenges. *Lancet*, **378**: 348–59.

Offit, P.A. and Jew, R.K. (2003). Addressing parents' concerns: do vaccines contain harmful preservatives, adjuvants, additives, or residuals? *Pediatrics*, **112**(6): 1394–7.

Papo, T. (2003). Macrophagic myofasciitis: focal or systemic? *Joint Bone Spine*, **70**: 242–5.

Passeri, E., Villa, C., Couette, M., *et al.* (2011). Long-term follow-up of cognitive dysfunction in patients with aluminium hydroxide-induced macrophagic myofasciitis (MMF). *J Inorg Biochem*, **105**: 1457–63.

Petrik, M.S., Wong, M.C., Tabata, R.C., *et al.* (2007). Aluminum adjuvant linked to Gulf War illness induces motor neuron death in mice. *Neuromolecular Med*, **9**(1): 83–100.

Ragunathan-Thangarajah, N., Le Beller, C., Boutouyrie, P., *et al.* (2013). Distinctive clinical features in arthromyalgic patients with and without aluminum hydroxyde-induced macrophagic myofasciitis: an exploratory study. *J Inorg Biochem*, **128**: 262–6.

Redhead, K., Quinlan, G.J., Das, R.G., and Gutteridge, J.M. (1992). Aluminium-adjuvanted vaccines transiently increase aluminium levels in murine brain tissue. *Pharmacol Toxicol*, **70**: 278–80.

Rinaldi, M., Perricone, R., Blank, M., *et al.* (2013). Anti-Saccharomyces cerevisiae autoantibodies in autoimmune diseases: from bread baking to autoimmunity. *Clin Rev Allergy Immunol*, **45**(2): 152–61.

Rosenblum, H., Shoenfeld, Y., and Amital, H. (2011). The common immunogenic etiology of chronic fatigue syndrome: from infections to vaccines via adjuvants to the ASIA syndrome. *Infect Dis Clin North Am*, **25**(4): 851–63.

Sahin, G., Varol, I., Temizer, A., *et al.* (1994). Determination of aluminum levels in the kidney, liver, and brain of mice treated with aluminium hydroxide. *Biol Trace Elem Res*, **41**: 129–35.

Shaw, C.A. and Petrik, M.S.J. (2009). Aluminum hydroxide injections lead to motor deficits and motor neuron degeneration. *J Inorg Biochem*, **103**: 1555–62.

Shaw, C.A., Li, Y., and Tomljenovic, L. (2013). Administration of aluminium to neonatal mice in vaccine-relevant amounts is associated with adverse long term neurological outcomes. *J Inorg Biochem*, **128**: 237–44.

Shi, Y. (2012). To forge a solid immune recognition. *Protein Cell*, **3**: 564–70.

Shi, Y., HogenEsch, H., and Hem, S.L. (2001). Change in the degree of adsorption of proteins by aluminum-containing adjuvants following exposure to interstitial fluid: freshly prepared and aged model vaccines. *Vaccine*, **20**: 80–5.

Shin, J.S., Ebersold, M., Pypaert, M., *et al.* (2006). Surface expression of MHC class II in dendritic cells is controlled by regulated ubiquitination. *Nature*, **444**: 115–18.

Shoaib, B.O. and Patten, B.M. (1996). Human adjuvant disease: presentation as a 414 multiple slcerosis-like syndrome. *South Med J*, **89**: 179–88.

Shoenfeld, Y. and Agmon-Levin, N. (2011). "ASIA" – autoimmune/inflammatory syndrome induced by adjuvants. *J Autoimmun*, **36**(1): 4–8.

Spreafico, R., Ricciardi-Castagnoli, P., and Mortellaro, A. (2010). The controversial relationship between NLRP3, alum, danger signals and the next-generation adjuvants. *Eur J Immunol*, **40**(3): 638–42.

Stern, S.T., Adiseshaiah, P.P., and Crist, R.M. (2012). Autophagy and lysosomal dysfunction as emerging mechanisms of nanomaterial toxicity. *Part Fibre Toxicol*, **9**: 20.

Theeler, B.J., Simper, N.B., and Ney, J.P. (2008). Polyglandular autoimmunity with macrophagic myofasciitis. *Clin Rheumatol*, **27**(5): 667–9.

Tomljenovic, L. (2011). Aluminum and Alzheimer's disease: after a century of controversy, is there a plausible link? *J Alzheimer Dis*, **23**: 567–98.

Tomljenovic, L. and Shaw, C.A. (2011). Aluminum vaccine adjuvants: are they safe? *Curr Med Chem*, **18**: 2630–7.

Trombetta, E.S., Ebersold, M., Garrett, W., *et al.* (2003). Activation of lysosomal function during dendritic cell maturation. *Science*, **299**: 1400–3.

Ulanova, M., Tarkowski, A., Hahn-Zoric, M., and Hanson, L.A. (2001).The common vaccine adjuvant aluminum hydroxide up-regulates accessory properties of human monocytes via an interleukin-4-dependent mechanism. *Infect Immun*, **69**: 1151–9.

van Rensburg, S.J., Potocnik, F.C., Kiss, T., *et al.* (2001). Serum concentrations of some metals and steroids in patients with chronic fatigue syndrome with reference to neurological and cognitive abnormalities. *Brain Res Bull*, **55**(2): 319–25.

Vasey, F.B., Zarabadi, S.A., Seleznick, M., and Ricca, L. (2003). Where there's smoke there's fire: the silicone breast implant controversy continues to flicker: a new disease that needs to be defined. *J Rheumatol*, **30**(10): 2092–4.

Verdier, F., Burnett, R., Michelet-Habchi, C., *et al.* (2005). Aluminium assay and evaluation of the local reaction at several time points after intramuscular administration of aluminium containing vaccines in the Cynomolgus monkey. *Vaccine*, **23**: 1359–67.

Wang, H.B. and Weller, P.F. (2008). Pivotal advance: eosinophils mediate early alum adjuvant-elicited B cell priming and IgM production. *J Leukoc Biol*, **83**: 817–21.

Wang, X.Y., Yao, X., Wan, Y.M., *et al.* (2012). Responses to multiple injections with alum alone compared to injections with alum adsorbed to proteins in mice. *Immunol Lett*, **149**: 88–92.

Wen, G.Y. and Wisniewski, H.M. (1985). Histochemical localization of aluminum in the rabbit CNS. *Acta Neuropathol*, **68**: 175–84.

WHO (World Health Organization). (1999). Macrophagic myofasciitis and aluminum-containing vaccines. *Wkly Epidemiol Rec*, **74**: 338–40.

28 Immune Thrombocytopenic Purpura: Between Infections and Vaccinations

Carlo Perricone,[1] Maurizio Rinaldi,[2] Roberto Perricone,[2] and Yehuda Shoenfeld[3,4]

[1]Rheumatology, Department of Internal and Specialized Medicine, Sapienza University of Rome, Rome, Italy
[2]Rheumatology, Allergology, and Clinical Immunology, Department of Internal Medicine, University of Rome Tor Vergata, Rome, Italy
[3]Zabludowicz Center for Autoimmune Diseases, Sheba Medical Center, Tel Hashomer, Israel
[4]Incumbent of the Laura Schwarz-Kipp Chair for Research of Autoimmune Diseases, Sackler Faculty of Medicine, Tel Aviv University, Tel Aviv, Israel

Definition and epidemiology

Immune thrombocytopenic purpura (ITP) is an autoimmune condition characterized by low platelet count and mucocutaneous bleeding. Clinical manifestations range from spontaneous formation of purpura and petechiae to epistaxis, bleeding of the gums, and menorrhagia. These manifestations usually occur at a platelet count lower than 2×10^4/µl (McCrae, 2011). Fatal complications include subarachnoid or intracerebral, lower gastrointestinal, or other internal bleeding, which may arise at a count $<5 \times 10^3$/µl. Thus, ITP is defined by a platelet count of less than 10^5 platelets per µl, typically without signs or symptoms of leucopenia and/or anemia, as long as overlapping disease is absent. Nonetheless, after a 6-month follow-up in 260 patients who were incidentally found to have subclinical borderline thrombocytopenia with $1.0–1.5 \times 10^5$ platelets per µl, the overall probability of developing ITP was estimated at 6.9% (95% confidence interval (CI): 4.0–12.0%), while 12% of population under consideration (85% women) was likely to experience evolution into an overlapping autoimmune disease other than ITP (95% CI: 6.9–20.8%) (Stasi *et al.*, 2006).

ITP was historically known as *morbus haemorrhagicus maculosus* or Werlhof's disease, after the physician who first described it in 1739. It has a complex immune-mediated thrombocytopenic pathogenesis. The onset can be primary or secondary following infections including, among others, *Helicobacter pylori* (Hasni, 2012), hepatitis C virus (HCV) (Pradella *et al.*, 2011), and human immunodeficiency virus (HIV) (Stasi *et al.*, 2009).

ITP onset has been reported as a rare but sometimes serious and even life-threatening adverse event following vaccine administration. It is most often observed after measles, mumps, and rubella (MMR), hepatitis A and B, diphtheria, tetanus, and acellular pertussis (DTaP), and varicella vaccinations.

The prevalence was estimated in an American cohort of 620 patients with a 2-year follow-up at 8.1 per 10^5 children under the age of 16 years (95% CI: 11.1–13.0) and 12.1 per 10^5 adults (95% CI: 11.1–13.0) (Terrell *et al.*, 2012). As many as 52 (20.2%) and 22 (8.6%) of the 257 pediatric

Vaccines and Autoimmunity, First Edition. Edited by Yehuda Shoenfeld, Nancy Agmon-Levin, and Lucija Tomljenovic.

patients showed evidence of an infection or immunization, respectively, shortly before ITP diagnosis (Yong *et al.*, 2010). ITP diagnosis is regarded as persistent if it lasts between 3 and 12 months, including in patients without a spontaneous recovery and those lacking a complete response to treatment. If the disease persists after 12 months, it is considered chronic, and it is classified as severe when bleeding occurs at onset or later, requiring treatment adjustment and additional medical care (Rodeghiero *et al.*, 2009).

ITP may be associated with several autoimmune disorders, including antiphospholipid syndrome (APS), systemic lupus erythematosus (SLE), and rheumatoid arthritis (RA) (Arkfeld and Weitz, 2009). Other disorders in which it has been found are those characterized by hypogammaglobulinemia, such as common variable immunodeficiency (CVID).

Pathogenesis

The pathogenesis of ITP is unclear. Antibody-coated platelets may undergo reticuloendothelial phagocytosis, resulting in reduced platelet survival (Imbach, 2006; McMillan, 2007). Despite evidence of autoantibodies against several platelet-surface glycoproteins, the most important being the antiglycoprotein IIb/IIIa, the initial trigger has not yet been recognized. It is likely that an uncontrolled immune response enables the development of a proinflammatory environment in genetically predisposed individuals (McMillan *et al.*, 2003; Kuwana *et al.*, 2009; Toltl *et al.*, 2011). Infectious agents may be responsible for the initiation of the autoimmune response via a molecular mimicry mechanism. In addition, a relationship between adjuvants used in vaccines and an aberrant response of the immune system has been suggested.

A wide array of aberrations may occur in the immune response during the ITP pathogenic process, affecting both innate and acquired immunity. Isolated platelets incubated with autologous lymphocytes are capable of inducing the release of interleukin (IL-) 2, whose signaling pathway then leads to a specific response from $CD4^+$ T cells activated against modified GPIIb-IIIa on activated platelets. GPIIb-IIIa and/or GPIb-IX autoantigens have been identified as the most likely ligands for antiplatelet autoantibodies. Through antigen-capture techniques, autoantibodies against GPIa-IIa and GPIV have also been reported (Provan, 2009; Provan *et al.*, 2010). These antibodies have a sensitivity of 49–66% and a specificity of 78–93%, representing the hallmark of the disease. Nevertheless, their absence does not rule out a diagnosis of ITP (McMillan *et al.*, 2003). In some antiplatelet–antibody-negative cases, a complementary T cell immune-mediated destruction mechanism or a reduction in platelet formation may explain the low platelet counts (Toltl *et al.*, 2011). Moreover, the glycoprotein antigens are continuously processed and presented after macrophage phagocytosis on the surfaces of the antigen-presenting cells (APCs), which critically contributes to autoantibody generation. This pathophysiological process has been studied by inducing the proliferation of GPIIb/GPIIIa-reactive T cell lines isolated from ITP patients and then cultured together with splenic macrophages or B cells and dendritic cells (DCs), requiring additional exogenous antigenic glycoproteins. Therefore, the reticuloendothelial system is currently regarded as a central player in maintaining antiplatelet autoantibody production (Kuwana *et al.*, 2009). $CD8^+$ and $CD4^+$ T cell-mediated responses are implicated in the pathogenesis of autoimmune thrombocytopenia. Indeed, an increase in both Th1 and Th2 cytokine release was observed when T cells from ITP patients were stimulated by platelet antigens; the Th1 cells then produced IL-2, interferons, and tumor necrosis factor (TNF), while Th2 cells secreted IL-4, IL-13, and IL-10. By contrast, $CD4^+CD25^+$ T regulatory (Treg) cells were found to be decreased in ITP patients during disease exacerbations, as well as in anti-GPIIb/IIIa-positive patients. The Treg cell count improved in patients who achieved full remission, especially after splenectomy. In addition, a refractory disease course is associated with reduced Foxp3 mRNA levels in peripheral blood mononuclear cells (PBMCs) (Sakakura *et al.*, 2007). The specific induction of Treg cells by platelet glycoprotein (GP-iTreg) and their possible *de novo* generation from nonregulatory $CD4^+CD25^-CD45RA^+$ cells in patients with ITP were also demonstrated (Zhang *et al.*, 2009).

Antibody-mediated, complement-dependent, and apoptotic mechanisms are also involved in ITP development. Platelets and megakaryocytes are bound by antiplatelet antibodies, leading to thrombocytopenia and to defective maturation/proliferation of platelet precursors (McMillan, 2000). Najaoui *et al.* (2012) showed that the major targets for complement-fixing autoantibodies are GPIIb/IIIa and GPIb/IX, and that complement fixation can occur even when autoantibody titers are low in ITP patients' sera (Peerschke *et al.*, 2010). The complement system was found to be altered in childhood ITP, with consumption of at least one component; the most affected seem to be properdin, factor H, C1q, C9, and factor B (Ohali *et al.*, 2005). In addition, B cell-activating factor (BAFF, which belongs to the TNF family), a critical cytokine for B cells, seems to be overexpressed in ITP, thus impairing self-tolerance, leading to autoimmunity (Zhou *et al.*, 2009). Similar results were reported by Emmerich *et al.* (2007), who detected a possible polymorphic site associated with susceptibility in the BAFF promoter region (-871 T/C). The -871 T/T genotype was associated with very high levels of BAFF, whereas normal levels were found in patients carrying T/C or C/C genotypes.

Infectious agents and the onset of ITP: the role of *Helicobacter pylori*

An increasing number of studies suggest an association between infectious agents and ITP onset (Stasi *et al.*, 2009). In this regard, several bacterium and virus species have been reported, including in patients with early HIV-1 infection, who may develop autoimmune thrombocytopenia with antibodies directed against an immunodominant epitope of the β3 (glycoprotein IIIa, GPIIIa) integrin, GPIIIa49-66 (Hsiao, 2000; Ma *et al.*, 2000; Ramos-Casals *et al.*, 2003; Aktepe *et al.*, 2004; Sheng Yu *et al.*, 2008; Aref *et al.*, 2009; Ursavas *et al.*, 2010; Lee *et al.*, 2011; Jayakar and Gharaie, 2012; Koh *et al.*, 2012) (Table 28.1). These antibodies seem to be able to induce thrombocytopenia via a novel complement-independent mechanism, in which platelets are fragmented by antibody-induced generation of hydrogen peroxide (H2O2) derived from the interaction of platelet nicotinamide adenine dinucleotide phosphate (NADPH) oxidase and 12-lipoxygenase (Li *et al.*, 2005). ITP has also been associated with novel influenza A infection in some case reports, although neither leucopenia nor thrombocytopenia was an uncommon hematological finding among patients with 2009 H1N1 influenza virus infection (Table 28.2) (Lee *et al.*, 2011; Koh *et al.*, 2012). There are also case reports suggesting ITP onset after rotavirus infection (Siddiqui and Chitlur, 2010).

However, one of the most interesting associations is related to infection with the Gram-negative bacterium *Helicobacter pylori* (*H. pylori*) (Franchini *et al.*, 2012; Ram *et al.*, 2012). Crossreactivity has been described between platelet-associated immunoglobulin G (PAIgG) levels and *H. pylori* cytotoxin-associated gene A (CagA) protein, suggesting that molecular mimicry with CagA protein plays a key role in the pathogenesis of a subset of ITP patients (Takahashi *et al.*, 2004). Such conclusions are also supported by data from clinical studies and a meta-analysis, in which a strict correlation between *H. pylori* eradication and a significant increase in platelet counts was observed (Franchini *et al.*, 2007; Sato *et al.*, 2011). Levels of anti-CagA antibody in platelet eluates decline after eradication therapy (Takahashi *et al.*, 2004). Furthermore, a number of studies suggest a potential benefit from eradication of *H. pylori*, mainly in children with chronic ITP (Jaing *et al.*, 2003; Kurekci *et al.*, 2004; Hayashi *et al.*, 2005; Jackson *et al.*, 2008; Rostami *et al.*, 2008; Tsumoto *et al.*, 2009). In one recent study, children from 16 centers in Italy, who were less than 18 years of age and diagnosed with chronic ITP, were screened for *H. pylori* infection (Russo *et al.*, 2011). Infection was found in 20% of 244 screened patients, 37 of whom received eradication therapy and completed a follow-up. The eradication was successful in 89% of cases. Platelet recovery was demonstrated in 39% of patients after eradication, whereas spontaneous remission was noted in only 10% of *H. pylori*-negative patients ($p < 0.005$). Among the large cohort of patients, those who underwent successful *H. pylori* eradication showed a significantly higher platelet response (Russo *et al.*, 2011). The association between *H. pylori* and chronic ITP and the effect of its eradication on thrombocytopenia were evaluated in 24 children of both sexes (mean age 8.0 ± 0.28 years, range 5.4–10.7 years) affected by ITP lasting more than 6 months. *H. pylori* was investigated by stool antigens. In 8 of the 24 patients (33.3%), *H. pylori* infection was identified and bacterial eradication was successful following 7 days of triple therapy. A follow-up of platelet counts was performed for 1 year after stool antigen detection. Six of the eight patients (75%) had a total recovery of platelet counts during the first year after bacterial eradication ($p<0.05$), while two (25%) had a partial recovery (Ferrara *et al.*, 2009). These data point to a beneficial role of *H. pylori* eradication in mainly chronic ITP, improving platelet counts in some patients. Finally, current guidelines on the management of *H. pylori* infection have extended the indications for eradication therapy to patients with ITP (Fock *et al.*, 2009; Malfertheiner *et al.*, 2012; de Korwin, 2013).

Table 28.1 Infectious agents associated with ITP onset

Microorganism	Country	Study design	Number of patients	Incidence/prevalence in patients with ITP	Clinical course	Achievement of complete remission	References
Human parvovirus B19	Turkey	Prevalence	19 adults	47%	Chronic ITP	NA	Aktepe *et al.* (2004)
EBV	Taiwan	Retrospective cohort	108 children	32.4%	Similar to patients without EBV infection	Full recovery within 26 days	Hsiao (2000)
CMV	China	Case–control	25 children with ITP infected by CMV	NA	Longer hospitalization compared to controls ($p = 0.004$)	NA	Sheng Yu *et al.* (2008)
Pulmonary TB	Turkey	Case report	46-year-old male patient with petechiae distributed over the lower extremities	NA	Platelet count $4(10^3)/\mu l$ Acid-fast bacilli observed on Ziehl–Neelsen stain of sputum	Patients treated successfully with steroids and anti-TB drugs	Ursavas *et al.* (2010)
HHV-6	China	Case–control	105 patients investigated for HHV-6, CMV, parvovirus B19	HHV-6 DNA positivity 41.0% for ITP patients	More severe symptoms in patients co-infected with HHV-6 and parvovirus B19 or CMV ($p < 0.05$)	Poorer response to treatment	Ma *et al.* (2000)
HCV	Egypt	Case–control	50 hepatic patients with HCV infection	Platelet-specific antibodies found in 86.7% of patients	Platelet count inversely correlated to levels of platelet GP-specific antibodies ($r = -0.42$, $p = 0.024$) and significantly parallel to spleen size ($p = 0.024$)	NA	Aref *et al.* (2009)
Influenza A (H1N1)	Korea	Case report	27-year-old man diagnosed with pneumonia from influenza A virus infection	-	Typical clinical signs of TTP. Patient received treatment with oseltamivir and high-dose methylprednisolone. Plasma-exchange therapy was started daily at a 1.5 dose volume of his whole blood	Full recovery was 47 days after admission	Koh *et al.* (2012)

ITP, immune thrombocytopenic purpura; NA, not available; EBV, Epstein–Barr virus, CMV, cytomegalovirus; TB, tuberculosis; HHV-6, human herpes virus 6; GP, glycoprotein; HCV, hepatitis C virus

Table 28.2 Association between influenza vaccine and ITP

Vaccine	Patient(s)	Other side effects/ complications	Note	References
H1N1	Of the 50 221 adverse reactions received in EudraVigilance for A/H1N1 vaccines (adjuvanted: 46 173; nonadjuvanted: 4048), ITP found in 28 total (23 adjuvanted, 6%; 5 nonadjuvanted, 13.2%)	314 autoimmune disorders, including T1DM, MS, GBS, and acute disseminated encephalomyelitis	No differences in the reporting of autoimmune disorders between adjuvanted and nonadjuvanted A/H1N1 vaccines	Isai *et al.* (2012)
Influenza vaccine	Healthy 3-year-old boy ITP onset 26 days after immunization	None	Full recovery with a single dose of intravenous immunoglobulin in 2 days	Mantadakis *et al.* (2010)
Influenza vaccine	Healthy 79-year-old man Onset with generalized petechiae. ITP onset 4 days after immunization Platelet count at onset: $4 \times 10^3/\mu l$	None	Refractory to high-dose immunoglobulin therapy, prednisolone, and splenectomy Partial recovery with cyclosporin A administration (platelet count: $5 \times 10^4/\mu l$)	Tsuji *et al.* (2009)
Influenza vaccine	75-year-old patient with prior autoimmune liver disease ITP onset 7 days after immunization Platelet count at onset: $5 \times 10^3/\mu l$		Full recovery with corticosteroids	Mamori *et al.* (2008)
Influenza vaccination	19-year-old patient with acute lymphoblastic leukemia Platelet count at onset: $1 \times 10^4/\mu l$ ITP onset 17 days after immunization	None	Full recovery with high-dose intravenous immunoglobulins and pulse intravenous methylprednisolone	Ikegame *et al.* (2006)
Anti-influenza vaccine	32-year-old healthy patient Onset with petechiae and ecchymoses ITP onset 15 days after immunization	Serum antiplatelet antibodies detected in high titer and bone marrow aspiration revealed an increased number of megakaryocytes	Full recovery with corticosteroid therapy within 10 days	Casoli and Tumiati (1989)
Influenza	68-year-old healthy patient ITP onset 14 days after immunization. Platelet count at onset: $3 \times 10^3/\mu l$	Gastrointestinal bleeding	Full recovery with corticosteroids and IVIg	Tishler *et al.* (2006)
Influenza	38-year-old patient with chronic obstructive pulmonary disease ITP onset 14 days after immunization Platelet count at onset: $3.2 \times 10^3/\mu l$		Full recovery after corticosteroids	Kelton (1981)
Influenza	72-year-old healthy patient ITP onset 8 days after immunization Platelet count at onset: $3 \times 10^3/\mu l$		Full recovery after corticosteroids	Granier *et al.* (2003)

ITP, immune thrombocytopenic purpura; T1DM, type 1 diabetes mellitus; MS, multiple sclerosis; GBS, Guillain–Barré syndrome

Vaccines and ITP

The role of molecular mimicry in the pathogenesis of ITP may be suggested by its onset following vaccination in some cases, especially when attenuated microorganisms are administered (O'Leary *et al.*, 2012). The vaccine–autoimmunity interplay is very similar to the established association between infections and autoimmunity. Infectious agents can cause or trigger autoimmunity through several mechanisms, including molecular mimicry, polyclonal activation, bystander activation, and the presence of super-antigens (Agmon-Levin *et al.*, 2009a). Vaccines, as well as infections, activate immune-mediated mechanisms that can induce protective immunity. The most common mechanism by which infections or vaccines induce autoimmunity is probably molecular mimicry: the epitope integrated within the infectious/vaccine antigen shares a similar structure with a self-antigen, driving toward self-reactivity. Furthermore, when polyclonal activation of B cells occurs, the increased B cell proliferation, antibody production, and formation of circulating immune complexes can result in damage to self-tissues. Notably, the increased risk of autoimmunity among recipients of a certain vaccine may stem not only from its antigen-mediated responses but also from its other constituents, including yeast proteins or extracts, adjuvants, and preservatives (Shoenfeld and Agmon-Levin, 2010; Agmon-Levin *et al.*, 2009b; Israeli *et al.*, 2009). ITP is one of the most common bleeding disorders in children, in which antiplatelet antibodies are usually detected 4–8 weeks after an infection or immunization. The development of vaccines has been and undoubtedly is still a cornerstone of medicine, although in recent decades concerns have been raised about their safety, especially regarding the autoimmune/inflammatory syndrome induced by adjuvants (ASIA), so that new immunization challenges must currently be faced (Shoenfeld and Aron-Maor, 2000; Shoenfeld and Agmon-Levin, 2010). ASIA is a new entity, entailing the association between a number of protean symptoms emerging following exposure to adjuvants. At least four conditions can be put under the ASIA umbrella: siliconosis following silicone breast implantation, Gulf War syndrome (GWS), macrophagic myofasciitis (MMF), and post-vaccination phenomena (Vera-Lastra *et al.*, 2013).

In Table 28.3, we summarize the incidence and prevalence of ITP following MMR vaccination reported in various countries, based on comparative prospective and retrospective trials and case–control studies. In one of the most recent systematic reviews, which considered 1.8 million children, 697 patients were included as potential ITP diagnoses, but this was cut to 197 cases after reviewing the charts for other hematologic disorders, sepsis or meningitis, other ITP-inducing drugs, unrecorded data, and confounding conditions (O'Leary *et al.*, 2012). The risk of development of ITP increased following MMR in the 12–19-month age group (incidence rate ratio (IRR) 5.48, 1.61–18.64, $p < 0.006$). An increased risk was also observed when other vaccines were given together with MMR or MMRV. A significantly higher risk of ITP was found following hepatitis A vaccination (IRR 23.14, 3.59–149.30, $p < 0.001$) at 7–17 years of age and following varicella and DTaP vaccination (IRR 12.14, 1.10–133.96, $p < 0.04$, and IRR 20.29, 3.12–131.83, $p < 0.002$ respectively) at 11–17 years of age (O'Leary *et al.*, 2012).

In a Canadian study, which was a part of the immunization monitoring program, 107 cases of vaccine-related ITP were recorded (77 after MMR vaccine) from 1992 to 2010 (96% symptomatic, two severe bleeding). The platelet counts improved after treatment in 93% of the pediatric patients within 3 months. Most of them received either intravenous immunoglobulin (78, 73%) or corticosteroids (21, 20%) (Sauvé *et al.*, 2010). The onset of ITP usually occurred within 6 weeks at a risk rate of 1 : 22 000–25 000 MMR vaccine doses, while the incidence of ITP following infections was 1 : 6000 for measles and 1 : 3000 for rubella (De Mattia *et al.*, 2010).

The first case of ITP in humans following antirabies vaccine intramuscular injections was recently reported in two patients: a boy and a man of 12 and 53 years of age, respectively, who received antirabies vaccines because of category II wounds (scratches) following dog bites; in both cases, they developed petechial spots, which in one case were associated with mild bleeding of the gums (Sharma *et al.*, 2012). Moreover, reactivation of ITP was reported 2 weeks after a tick-borne encephalitis vaccination in a previously treated 34-year-old woman who had achieved disease remission (Benz *et al.*, 2009).

In 2009, a 16-year-old girl developed fatigue and prolonged menorrhagia 3 months following human papillomavirus (HPV) vaccination: she was found to be anemic, with a low platelet count, and the first diagnosis of HPV vaccine-related ITP was confirmed through both bone marrow aspiration

Table 28.3 Incidence or prevalence of ITP following vaccination

Vaccine	Country	Study design	Number of patients	Number of vaccinations performed	Serious adverse effects	Incidence/prevalence of ITP	References
MMR	Cochrane review on vaccine field (Italy)	Comparative prospective or retrospective trials	14 700 000 children	1107 814 MMR vaccinations	An increased risk of ITP within 6 weeks after MMR immunization in children aged 12–23 months assessed in one case–control study (RR 6.3; 95% CI: 1.3–30.1) and in one small self-controlled case series (IRR 5.38; 95% CI: 2.72–10.62)	Increased risk of ITP within 6 weeks after MMR exposure assessed in another case–control study involving 2311 children and adolescents between 1 month and 18 years of age (OR 2.4; 95% CI: 1.2–4.7)	Demicheli *et al.* (2012)
MMR	USA	Retrospective cohort	1 036 689 children	1 107 814 MMR vaccinations	259 patients confirmed with ITP	One case of ITP per 40 000 doses	France *et al.* (2008)
MMR	Italy	Case–control study	387 cases of children with thrombocytopenia and 1924 controls	2311 vaccinations	NA	After adjusting for concurrent use of other drugs and use of antibiotics, more than twofold increased risk of developing ITP (OR 2.4; 95% CI: 1.8, 3.1)	Bertuola *et al.* (2010)
2009 H1N1 vaccines	Taiwan	Prospective cohort	NA	NA	For the interval ≥43 days after vaccination, reporting completeness was 0.1% for death, 14% for GBS, 0.1% for convulsion, <0.1% for Bell's palsy, and 0% for ITP. The estimated-to-expected ratio for Bell's palsy in the interval 0–42 days after vaccination was 1.48 (95% CI: 1.11–1.98). Reporting completeness was higher for GBS than for other adverse, events after 2009 H1N1 vaccination in adults	Chapman capture–recapture-estimated spontaneous reporting completeness within 0–42 days from vaccination was 15% for ITP	Huang *et al.* (2012)

(continued)

Table 28.3 *(Continued)*

Vaccine	Country	Study design	Number of patients	Number of vaccinations performed	Serious adverse effects	Incidence/prevalence of ITP	References
HBV vaccine, diphtheria-tetanus-acellular pertussis, MMR, and varicella vaccine	Taiwan	Retrospective cohort study	20 children with ITP	NA	Of the 12 post-vaccination cases, 5 occurred after the second dose of hepatitis B virus vaccine at 1 month of age, 4 occurred after the first dose of DTaP vaccine at 2–3 months of age, 2 occurred after the first dose of MMR vaccine at 16 months of age, and 1 occurred after the first dose of varicella vaccine at 14 months of age	NA	Hsieh and Lin (2010)
MMR	USA	Prospective cohort	1.8 million children	15 million vaccine doses during the study period	No elevated risk of ITP after any vaccine in early childhood, other than MMR in the 12–19 month age group	Significantly elevated risk of ITP after hepatitis A vaccine at 7–17 years of age, and for varicella vaccine and DTaP vaccine at 11–17 years of age. For hepatitis A, varicella, and DTaP vaccines, elevated risks were based on one to two vaccine-exposed cases	O'Leary *et al.* (2012)
Measles	China	Prospective cohort	NA	14.3 million vaccinations performed on children	Total incidence of serious adverse events after vaccination: 2.14/million doses. Incidence of anaphylactic reaction: 6.5/million for attenuated measles vaccine	1 case per 14.3 million vaccinations performed	Shu *et al.* (2011)

ITP, immune thrombocytopenia purpura; MMR, measles, mumps, and rubella; DTaP, diphtheria, tetanus, and acellular pertussis; GBS, Guillain–Barré syndrome; HBV, hepatitis B virus; IRR, *incidence rate ratio*; OR, odds ratio; CI, confidence interval; NA, not available (showing no alterations other than an increased number of megakaryocytes) and antiplatelet antibody positivity. Therefore, a cause–effect relationship was considered to be possible according to World Health Organization (WHO) criteria. Once she had recovered from 17×10^3 platelets/μl, further HPV vaccination was contraindicated for this patient. In this case, as well as in the other reports mentioned, autoimmune diseases other than ITP (or known infections associated with them, such as Epstein–Barr virus (EBV), cytomegalovirus (CMV), and human immunodeficiency virus (HIV)), were excluded (Pugnet *et al.*, 2009).

Conclusions

Vaccinations have proved to be of great advantage to the general population in preventing the spread of infectious diseases. Vaccine safety has improved in recent years, and the incidence of vaccine-induced autoimmunity is rare, but they are not yet free of risk. It is becoming apparent that it is not only the active components that could drive autoantibody production, but also the excipients, such as adjuvants (pristane, aluminum, squalene), or even the residual traces of yeast from the manufacturing process (Rinaldi *et al.*, 2013). The course of ITP can be very serious, even leading to fatal intracranial hemorrhages, although usually the platelet count improves spontaneously or normalizes after therapy. Unfortunately, in some patients, especially in adulthood and adolescence, ITP can be a chronic disease that must be continually monitored and treated.

Infections are much more likely to trigger ITP than are preventive vaccines. However, it should be borne in mind that preventive vaccines are usually administered to otherwise healthy subjects who are not yet fighting the infectious disease for which they are considered at risk. Thus, we must be careful not to cause harm to healthy individuals. Furthermore, it is critical to recognize that the induction of autoantibodies by an infectious or a vaccine-component trigger, and therefore the onset of autoimmune disease (including ITP), can occur in a period of days or years. While the short latency of post-streptococcal-induced rheumatic fever is a few weeks (Arbuckle *et al.*, 2003), the temporal relationship between vaccination and autoimmunity depends on the particular vaccine used and its associated phenomena. Finally, following from the displayed efficacy of eradication therapy in *H. pylori*-associated ITP, therapy should always be consistent with the principle of removing the pathogenic factor.

References

Agmon-Levin, N., Paz, Z., Israeli, E., and Shoenfeld, Y. (2009a). Vaccines and autoimmunity. *Nat Rev Rheumatol*, **5**: 648–52.

Agmon-Levin, N., Kivity, S., and Shoenfeld, Y. (2009b). Influenza vaccines and autoimmunity. *IMAJ*, **11**: 183–5.

Aktepe, O.C., Yetgin, S., Olcay, L., and Ozbek, N. (2004). Human parvovirus B19 associated with idiopathic thrombocytopenic purpura. *Pediatr Hematol Oncol*, **21**: 421–6.

Arbuckle, M.R., McClain, M.T., Rubertone, M.V., *et al.* (2003). Development of autoantibodies before the clinical onset of systemic lupus erythematosus. *N Engl J Med*, **349**: 1526–33.

Aref, S., Sleem, T., El Menshawy, N., *et al.* (2009). Antiplatelet antibodies contribute to thrombocytopenia associated with chronic hepatitis C virus infection. *Hematology*, **14**: 277–81.

Arkfeld, D.G. and Weitz, I.C. (2009). Immune thrombocytopenia in patients with connective tissue disorders and the antiphospholipid antibody syndrome. *Hematol Oncol Clin North Am*, **23**: 1239–49.

Benz, R., Krause, M., and Taverna, C. (2009). Immune thrombocytopenic purpura reactivation after tick-borne encephalitis vaccination. *Vaccine*, **27**: 5172–3.

Bertuola, F., Morando, C., Menniti-Ippolito, F., *et al.* (2010). Association between drug and vaccine use and acute immune thrombocytopenia in childhood: a case-control study in Italy. *Drug Saf*, **33**: 65–72.

Casoli, P. and Tumiati, B. (1989). [Acute idiopathic thrombocytopenic purpura after anti-influenza vaccination]. *Medicina (Firenze)*, **9**: 417–18.

de Korwin, J.D. (2013). New recommendations for the diagnosis and the treatment of *Helicobacter pylori infection Presse Med*, **42**(3): 309–17.

De Mattia, D., Del Vecchio, G.C., Russo, G., *et al.* (2010). Management of chronic childhood immune thrombocytopenic purpura: AIEOP consensus guidelines. *Acta Haematol*, **123**: 96–109.

Demicheli, V., Rivetti, A., Debalini, M.G., and Di Pietrantonj, C. (2012). Vaccines for measles, mumps and rubella in children. *Cochrane Database Syst Rev*, **2**: CD004407.

Emmerich, F., Bal, G., Barakat, A., *et al.* (2007). High-level serum B-cell activating factor and promoter polymorphisms in patients with idiopathic thrombocytopenic purpura. *Br J Haematol*, **136**: 309–14.

Ferrara, M., Capozzi, L., and Russo, R. (2009). Effect of Helicobacter pylori eradication on platelet count in children with chronic idiopathic thrombocytopenic purpura. *Hematology*, **14**: 282–5.

Fock, K.M., Katelaris, P., Sugano, K., *et al.* (2009). Second Asia-Pacific consensus guidelines for *Helicobacter pylori* infection. *J Gastroenterol Hepatol*, **24**: 1587–600.

France, E.K., Glanz, J., Xu, S., *et al.* (2008). Risk of immune thrombocytopenic purpura after measles-mumps-rubella immunization in children. *Pediatrics*, **121**: 687–92.

Franchini, M., Cruciani, M., Mengoli, C., *et al.* (2007). Effect of Helicobacter pylori eradication on platelet count in idiopathic thrombocytopenic purpura: a systematic review and meta-analysis. *J Antimicrob Chemother*, **60**: 237–46.

Franchini, M., Vescovi, P.P., Garofano, M., and Veneri, D. (2012). Helicobacter pylori-associated idiopathic thrombocytopenic purpura: a narrative review. *Semin Thromb Hemost*, **38**: 463–8.

Granier, H., Nicolas, X., Laborde, J.P., and Talarmin, F. (2003). Severe autoimmune thrombocytopenia following anti-influenza vaccination. *Presse Med*, **32**: 1223–4.

Hasni, S.A. (2012). Role of *Helicobacter pylori* infection in autoimmune diseases. *Curr Opin Rheumatol*, **24**: 429–34.

Hayashi, H., Okuda, M., Aoyagi, N., *et al.* (2005). *Helicobacter pylori* infection in children with chronic idiopathic thrombocytopenic purpura. *Pediatr Int*, **47**: 292–5.

Hsiao, C.C. (2000). Epstein-Barr virus associated with immune thrombocytopenic purpura in childhood: a retrospective study. *J Paediatr Child Health*, **36**: 445–8.

Hsieh, Y.L. and Lin, L.H. (2010). Thrombocytopenic purpura following vaccination in early childhood: experience of a medical center in the past 2 decades. *J Chin Med Assoc*, **73**: 634–7.

Huang, W.T., Huang, W.I., Huang, Y.W., *et al.* (2012). The reporting completeness of a passive safety surveillance system for pandemic (H1N1) 2009 vaccines: a capture-recapture analysis. *Vaccine*, **30**: 2168–72.

Kelton, J.G. (1981). Vaccination-associated relapse of immune thrombocytopenia. *JAMA*, **245**: 369–70.

Ikegame, K., Kaida, K., Fujioka, T., *et al.* (2006). Idiopathic thrombocytopenic purpura after influenza vaccination in a bone marrow transplantation recipient. *Bone Marrow Transplant*, **38**: 323–4.

Imbach, P. (2006). Development and research in idiopathic thrombocytopenic purpura: an inflammatory and autoimmune disorder. *Pediatr Blood Cancer*, **47**(5 Suppl.): 685–6.

Isai, A., Durand, J., Le Meur, S., *et al.* (2012) Autoimmune disorders after immunisation with influenza A/H1N1 vaccines with and without adjuvant: EudraVigilance data and literature review. *Vaccine*, **30**: 7123–9.

Israeli, E., Agmon-Levin, N., Blank, M., and Shoenfeld, Y. (2009). Adjuvants and autoimmunity. *Lupus*, **18**: 1217–25.

Jackson, S.C., Beck, P., Buret, A.G., *et al.* (2008). Long term platelet responses to *Helicobacter pylori* eradication in Canadian patients with immune thrombocytopenic purpura. *Int J Hematol*, **88**: 212–18.

Jaing, T.H., Yang, C.P., Hung, I.J., *et al.* (2003). Efficacy of *Helicobacter pylori* eradication on platelet recovery in children with chronic idiopathic thrombocytopenic purpura. *Acta Paediatr*, **92**: 1153–7.

Jayakar, V. and Gharaie, S. (2012). Atypical mycobacteria in a patient with HIV and ITP. *Blood*, **119**: 4585.

Koh, Y.R., Hwang, S.H., Chang, C.L., *et al.* (2012). Thrombotic thrombocytopenic purpura triggered by influenza A virus subtype H1N1 infection. *Transfus Apher Sci*, **46**: 25–8.

Kurekci, A.E., Atay, A.A., Sarici, S.U., and Ozcan, O. (2004). Complete platelet recovery after treatment of *Helicobacter pylori* infection in a child with chronic immune thrombocytopenic purpura: a case report. *Pediatr Hematol Oncol*, **21**: 593–6.

Kuwana, M., Okazaki, Y., and Ikeda, Y. (2009). Splenic macrophages maintain the anti-platelet autoimmune response via uptake of opsonised platelets with immune thrombocytopenic purpura. *J Thromb Haemost*, **7**: 322–9.

Lee, C.Y., Wu, M.C., Chen, P.Y., *et al.* (2011). Acute immune thrombocytopenic purpura in an adolescent with 2009 novel H1N1 influenza A virus infection. *J Chin Med Assoc*, **74**: 425–7.

Li, Z., Nardi, M.A., and Karpatkin, S. (2005). Role of molecular mimicry to HIV-1 peptides in HIV-1-related immunologic thrombocytopenia. *Blood*, **106**: 572–6.

Ma, X., Lu, D., Song, Y., *et al.* (2000). Relationship between human herpesvirus 6 infection and idiopathic thrombocytopenic purpura. *Zhonghua Xue Ye Xue Za Zhi*, **21**: 135–7.

Malfertheiner, P., Selgrad, M., and Bornschein, J. (2012). *Helicobacter pylori*: clinical management. *Curr Opin Gastroenterol*, **28**: 608–14.

Mamori, S., Amano, K., Kijima, H., *et al.* (2008). Thrombocytopenic purpura after the administration of an influenza vaccine in a patient with autoimmune liver disease. *Digestion*, **77**: 159–60.

Mantadakis, E., Farmaki, E., Thomaidis, S., *et al.* (2010). A case of immune thrombocytopenic purpura after influenza vaccination: consequence or coincidence? *J Pediatr Hematol Oncol*, **32**: e227–9.

McCrae, K. (2011). Immune thrombocytopenia: no longer "idiopathic." *Cleve Clin J Med*, **78**: 358–73.

McMillan, R. (2000). Autoantibodies and autoantigens in chronic immune thrombocytopenic purpura. *Semin Hematol*, **37**: 239–48.

McMillan, R. (2007). The pathogenesis of chronic immune thrombocytopenic purpura. *Semin Hematol*, **44**: S3–11.

McMillan, R., Wang, L., and Tani, P. (2003). Prospective evaluation of the immunobead assay for the diagnosis of adult chronic immune thrombocytopenic purpura. *J Thromb Haemost*, **1**: 485–91.

Najaoui, A., Bakchoul, T., Stoy, J., *et al.* (2012). Autoantibody-mediated complement activation on platelets is a common finding in patients with immune thrombocytopenic purpura. *Eur J Haematol*, **88**: 167–74.

Ohali, M., Maizlish, Y., Abramov, H., *et al.* (2005). Complement profile in childhood immune thrombocytopenic purpura: a prospective pilot study. *Ann Hematol*, **84**: 812–15.

O'Leary, S.T., Glanz, J.M., McClure, D.L., *et al.* (2012). The risk of immune thrombocytopenic purpura after vaccination in children and adolescents. *Pediatrics*, **129**: 248–55.

Peerschke, E.I., Andemariam, B., Yin, W., and Bussel, J.B. (2010). Complement activation on platelets correlates with a decrease in circulating immature platelets in patients with immune thrombocytopenic purpura. *Br J Haematol*, **148**: 638–45.

Pradella, P., Bonetto, S., Turchetto, S., *et al.* (2011). Platelet production and destruction in liver cirrhosis. *J Hepatol*, **54**: 894–900

Provan, D. (2009). Characteristics of immune thrombocytopenic purpura: a guide for clinical practice. *Eur J Haematol Suppl*, **71**: 8–12.

Provan, D., Stasi, R., Newland, A.C., *et al.* (2010). International consensus report on the investigation and management of primary immune thrombocytopenia. *Blood*, **115**: 168–86.

Pugnet, G., Ysebaert, L., Bagheri, H., *et al.* (2009). Immune thrombocytopenic purpura following human papillomavirus vaccination. *Vaccine*, **27**: 3690.

Ram, M., Barzilai, O., Shapira, Y., *et al.* (2012). Helicobacter pylori serology in autoimmune diseases – fact or fiction? *Clin Chem Lab Med*, **51**(5): 1075–82.

Ramos-Casals, M., García-Carrasco, M., López-Medrano, F., *et al.* (2003). Severe autoimmune cytopenias in treatment-naive hepatitis C virus infection: clinical description of 35 cases. *Medicine (Baltimore)*, **82**: 87–96.

Rinaldi, M., Perricone, R., Blank, M., *et al.* (2013). Anti-*Saccharomyces cerevisiae* autoantibodies in autoimmune diseases: from bread baking to autoimmunity. *Clin Rev Allergy Immunol*, **45**: 152–61.

Rodeghiero, F., Stasi, R., Gernsheimer, T., *et al.* (2009). Standardization of terminology, definitions and outcome criteria in immune thrombocytopenic purpura of adults and children: report from an international working group. *Blood*, **113**: 2386–93.

Rostami, N., Keshtkar-Jahromi, M., Rahnavardi, M., *et al.* (2008). Effect of eradication of *Helicobacter pylori* on platelet recovery in patients with chronic idiopathic thrombocytopenic purpura: a controlled trial. *Am J Hematol*, **83**: 376–81.

Russo, G., Miraglia, V., Branciforte, F., *et al.* (2011). Effect of eradication of Helicobacter pylori in children with chronic immune thrombocytopenia: a prospective, controlled, multicenter study. *Pediatr Blood Cancer*, **56**: 273–8.

Sakakura, M., Wada, H., Tawara, I., *et al.* (2007). Reduced CD4+CD25+ T cells in patients with idiopathic thrombocytopenic purpura. *Thromb Res*, **120**: 187–93.

Sato, R., Murakami, K., Okimoto, T., *et al.* (2011). Development of corpus atrophic gastritis may be associated with Helicobacter pylori-related idiopathic thrombocytopenic purpura. *J Gastroenterol*, **46**: 991–7.

Sauvé, L.J., Bettinger, J., Scheifele, D., *et al.* (2010). Canadian Immunization Monitoring Program, and Active (IMPACT). Postvaccination thrombocytopenia in Canada. *Pediatr Infect Dis J*, **29**(6): 559–61.

Sharma, S.K., Seth, T., Agrawal, N., *et al.* (2012). Immune thrombocytopenic purpura following anti-rabies vaccines. *Platelets*, **23**: 317–18.

Sheng Yu, Z., Tang, L.F., Zou, C.C., *et al.* (2008). Cytomegalovirus-associated idiopathic thrombocytopenic purpura in Chinese children. *Scand J Infect Dis*, **40**: 922–7.

Shoenfeld, Y. and Agmon-Levin, N. (2010). "ASIA" – autoimmune/inflammatory syndrome induced by adjuvants. *J Autoimmun*, **36**: 4–8.

Shoenfeld, Y. and Aron-Maor, A. (2000). Vaccination and autoimmunity – "vaccinosis": a dangerous liaison? *J Autoimmun*, **14**: 1–10.

Shu, M., Liu, Q., Wang, J., *et al.* (2011). Measles vaccine adverse events reported in the mass vaccination campaign of Sichuan province, China from 2007 to 2008. *Vaccine*, **29**: 3507–10.

Siddiqui, A.H. and Chitlur, M.B. (2010). Immune thrombocytopenic purpura in a 5-mounth-old female with rotavirus infection. *Pediatr Blood Cancer*, **54**: 633.

Stasi, R., Amadori, S., Osborn, J., *et al.* (2006). Long-term outcome of otherwise healthy individuals with incidentally discovered borderline thrombocytopenia. *Plos Med*, **3**: e24.

Stasi, R., Willis, F., Shannon, M.S., and Gordon-Smith, E.C. (2009). Infectious causes of chronic immune thrombocytopenia. *Hematol Oncol Clin North Am*, **23**: 1275–97.

Takahashi, T., Yujiri, T., Shinohara, K., *et al.* (2004). Molecular mimicry by Helicobacter pylori CagA protein may be involved in the pathogenesis of *H. pylori*-associated chronic idiopathic thrombocytopenic purpura. *Br J Haematol*, **124**: 91–6.

Terrell, D.R., Beebe, L.A., Neas, B.R., *et al.* (2012). Prevalence of primary immune thrombocytopenia in Oklahoma. *Am J Hematol*, **87**: 848–52.

Tishler, M., Levy, O., and Amit-Vazina, M. (2006). Immune thrombocytopenic purpura following influenza vaccination. *IMAJ*, **8**: 322–3.

Toltl, L.J., Nazi, I., Jafari, R., and Arnold, D.M. (2011). Piecing together the humoral and cellular mechanisms of immune thrombocytopenia. *Semin Thromb Hemost*, **37**: 631–9.

Tsuji, T., Yamasaki, H., and Tsuda, H. (2009). [Refractory idiopathic thrombocytopenic purpura following influenza vaccination]. *Rinsho Ketsueki*, **50**: 577–9.

Tsumoto, C., Tominaga, K., Okazaki, H., *et al.* (2009). Long-term efficacy of Helicobacter pylori eradication in patients with idiopathic thrombocytopenic purpura:

7-year follow-up prospective study. *Ann Hematol*, **88**: 789–93.

Ursavas, A., Ediger, D., Ali, R., *et al.* (2010). Immune thrombocytopenia associated with pulmonary tuberculosis. *J Infect Chemother*, **16**: 42–4.

Vera-Lastra, O., Medina, G., Cruz-Dominguez Mdel, P., *et al.* (2013). Autoimmune/inflammatory syndrome induced by adjuvants (Shoenfeld's syndrome): clinical and immunological spectrum. *Expert Rev Clin Immunol*, **9**: 361–73.

Yong, M., Schoonen, W.M., Li, L., *et al.* (2010). Epidemiology of paediatric immune thrombocytopenia in the General Practice Research Database. *Br J Haematol*, **149**: 855–64.

Zhang, X.L., Peng, J., Sun, J.Z., *et al.* (2009). De novo induction of platelet-specific CD4(+)CD25(+) regulatory T cells from CD4(+)CD25(−) cells in patients with idiopathic thrombocytopenic purpura. *Blood*, **113**: 2568–77.

Zhou, Z., Chen, Z., Li, H., *et al.* (2009). BAFF and BAFF-R of peripheral blood and spleen mononuclear cells in idiopathic thrombocytopenic purpura. *Autoimmunity*, **42**: 112–19.

29 Vaccinations and Type 1 Diabetes

Alessandro Antonelli, Silvia Martina Ferrari, Andrea Di Domenicantonio, Ele Ferrannini, and Poupak Fallahi
Department of Clinical and Experimental Medicine, University of Pisa, Pisa, Italy

Introduction

Autoimmune reactions to vaccinations may rarely be induced in predisposed individuals by molecular mimicry or bystander activation mechanisms. Autoimmune reactions that are considered to be reliably vaccine-associated include Guillain–Barré syndrome (GBS) and 1976 swine influenza vaccine, immune thrombocytopenic purpura and measles, mumps, and rubella (MMR) vaccine, and myopericarditis and smallpox vaccination. The suspected association between hepatitis B vaccine and multiple sclerosis (MS) has not yet been confirmed (Chen *et al.*, 2001; Tishler and Shoenfeld, 2004; Vial and Descotes, 2004).

The pathogenesis of type 1 diabetes (T1D) is complex and results from a combination of genetic, environmental, hormonal, and immunological factors. Environmental factors include viruses, pathogens, diet, toxins, and stress, as well as vaccines (Tishler and Shoenfeld, 2004). The possible association between vaccines and T1D is under debate.

Type 1 diabetes

T1D is caused by antigen-specific assaults on the insulin-producing β cells of the pancreas by T cells, leading to deficient insulin synthesis and secretion and thus dysregulation of blood glucose homeostasis. It is one of the most prevalent chronic autoimmune diseases, affecting around 36 million individuals worldwide. The global incidence of T1D is predicted to increase by 3% annually (Stanescu *et al.*, 2012).

In T1D, diabetogenic cluster of differentiation (CD)4^+ T cells (T helper, Th) are responsible for providing the cytokine microenvironment in which islet-specific $CD8^+$ T cells (cytotoxic T lymphocyte, CTL) destroy β cells. Furthermore, islet-specific Th activity may also be cytotoxic to islets at later disease stages, and CTL may not need prior costimulation before killing β cells (Zekzer *et al.*, 1998; Amrani *et al.*, 2000; Wang *et al.*, 2000).

An imbalance in the Th cell system, caused by a predominance of the Th1 response, favors the development of autoimmune diseases and has been associated with T1D (Sia, 2005). Chemokine (C-X-C motif) ligand 10 (CXCL10) is an interferon gamma (IFN-γ) inducible Th1 chemokine and reacts with its receptor, C-X-C motif receptor 3 (CXCR3), on Th1 cells. CXCL10 and CXCR3 chemokines play an important role in the pathogenesis of many autoimmune diseases (Antonelli *et al.*, 2008b).

Elevated levels of CXCL10 were detected in new-onset T1D patients and correlated with levels of glutamic acid decarboxylase (GAD)-reactive CD4T cells (Shigihara *et al.*, 2006). In addition, CXCL10 is produced by β cells and suppresses β cell proliferation (Morimoto *et al.*, 2004). Therefore, the CXCL10/CXCR3 system plays a decisive role in the pathogenesis of T1D (Antonelli *et al.*, 2008a, 2014a,2014b).

A number of autoantibodies (Abs) have also been involved in the pathogenesis of T1D: anti-GAD

Vaccines and Autoimmunity, First Edition. Edited by Yehuda Shoenfeld, Nancy Agmon-Levin, and Lucija Tomljenovic.

Ab, anti-insulin Ab, anti-islet antigen 2(IA2), and anti-CD38 Ab, among others (Pupilli *et al.*, 1999; Antonelli *et al.*, 2001a,2001b, 2002).

The etiology of T1D is unknown, though both genetic and environmental risk factors are thought to play a role in its pathogenesis (Institute for Vaccine Safety Diabetes Workshop Panel, 1999). It has been hypothesized that vaccination could trigger T1D in susceptible individuals (Classen and Classen, 1999). Although post-vaccination T1D may be biologically plausible (Chen *et al.*, 2001), cumulative evidence has not supported an increased risk of T1D following any vaccine (Institute for Vaccine Safety Diabetes Workshop Panel, 1999).

There is some evidence to suggest a role for natural infections in the pathogenesis of type 1 diabetes mellitus (T1DM) in susceptible individuals, though the mechanisms by which viral infections cause autoimmune diabetes have not been fully clarified (Antonelli *et al.*, 2008c; Coppieters *et al.*, 2012; Galleri *et al.*, 2012).

Several experiments have suggested that vaccination might exert a protecting or aggravating effect on the occurrence of diabetes, depending on the timing of vaccination (Singh, 2000).

Vaccination and diabetes in childhood

A number of studies have searched for links between diabetes and immunizations. Classen found that if the first vaccination in children is performed after 2 months of age, there is an increased risk of diabetes (Classen, 1996). In animals, he also found that certain vaccines, if given at birth, actually decrease the risk of diabetes (Classen, 1996; Classen and Classen, 1997). The latter study was based on experiments using anthrax vaccine, which is very rarely used in children or adults. Classen also compared diabetes rates with vaccination schedules in different countries, and interpreted his results as meaning that vaccination causes an increased risk of diabetes (Classen, 1996; Classen and Classen, 1997). In 2002, he suggested that vaccination of Finnish children with *Haemophilus influenzae* type b (Hib) vaccine caused clusters of diabetes 3 years later, and that his experiments in mice confirmed this association (Classen and Classen, 2002).

Classen observed that there was an epidemic of T1D and type 2 diabetes (T2D) in children. The obesity epidemic in US children has a statistically significant positive correlation with the number of vaccine doses recommended. The incidence of T2D in Japanese children decreased significantly following the discontinuation of the bacillus Calmette–Guèrin (BCG) vaccine, which is associated with an increased risk of T1D. Classen described two aberrant responses to immunization. At one extreme, immunization leads to progressive autoimmune diseases, including T1D. At the other, the body suppresses the immune system through increased cortisol activity and other countermeasures, leading to T2D and metabolic syndrome. The propensity to develop a particular response relates to race: Japanese children produce large amounts of cortisol following immunization and have a lower risk of T1D but a higher risk of T2D than do white children (Classen, 2008a,2008b,2008c).

Classen's theory was also based on the findings of an increased risk of autoimmune diabetes in diabetes-prone nonobese diabetic (NOD) mice after administration of pediatric vaccines (Classen, 1996). However, other laboratories using the same animal model were unable to reproduce these findings following similar vaccination schemes performed at 10, 12, and 14 weeks of age, and even suggested a slight reduction in the incidence of autoimmune diabetes or a moderate decrease in blood glucose levels (Ravel *et al.*, 2003).

Accumulating human data from various epidemiological studies do not support a causal association between vaccination and an increased risk of T1D. Case–control and ecological studies indicate that neither pertussis nor BCG vaccinations have a significant effect on the incidence of T1D (Dahlquist and Gothefors, 1995; Heijbel *et al.*, 1997). In a Canadian case–control study, BCG vaccination rates were similar in patients with T1D and controls, although the authors suggested a possible delayed occurrence in the onset of diabetes in birth-vaccinated compared to nonvaccinated cases (Parent *et al.*, 1997). A Swedish case–control study did not find any evidence for an increase in the risk of diabetes after BCG, smallpox, tetanus, pertussis, rubella, or mumps vaccinations, and even indicated a possible decreased risk after measles vaccination (Blom *et al.*, 1991). A large, population-based, case–control study using data from four health maintenance organizations (HMOs) in the United States examined the effects of different vaccines (DeStefano *et al.*, 2001). Children with T1DM were matched by HMO, gender, date of birth, and length of health plan enrollment to three controls each. Based on an analysis of 252 cases of T1DM and 768 controls, there was no increased risk of T1D with any of the routinely administered childhood vaccines. The risk of diabetes was not different between children vaccinated at birth with the hepatitis B vaccine and those who received their first dose at 2 months of age or later, suggesting that the timing of vaccination did not influence the likelihood of developing diabetes. A case–control study of 317 children who had a first-degree family member with T1D found no significant association between the development of β cell autoimmunity and exposure to a number of vaccines, and no effect of the timing of exposure (Graves *et al.*, 1999).

A great deal of debate surrounded a possible increase in the risk of the incidence of T1DM associated with the nationwide introduction of Hib vaccine in Finland (Classen and Classen, 1997; Karvonen *et al.*, 1999). However, a large 10-year follow-up study of over 110 000 Finnish children who participated in a clinical trial of Hib vaccine did not show any increased risk of diabetes in children first vaccinated at the age of 24 months as compared to a cohort of children born in the 24 months preceding the vaccination period (Karvonen *et al.*, 1999). Classen and Classen (2002) subsequently questioned the way in which the data had been analyzed. Their own analysis suggested an increase in the cumulative incidence of diabetes in children first vaccinated at 3 months of age who received four doses of the vaccine as compared to unvaccinated children. They also found that cases clustered at between 36 and 48 months after immunization. Another 10-year follow-up study performed in children from the United States failed to identify an increased risk of diabetes following Hib vaccination (Black *et al.*, 2002). A recent study suggests that Hib vaccine might be a risk factor in the induction of islet cell and anti-GAD Ab measured at 1 year of age (Wahlberg *et al.*, 2003). The authors propose that this vaccine produces an unspecific stimulatory polyclonal effect, which might be of clinical importance in the presence of other factors stimulating β cell autoimmunity.

Mumps vaccine, adverse effects, and diabetes

Mumps vaccine is usually given by injection to infants aged about 15 months as part of the MMR vaccine, which contains live-attenuated measles, mumps, and rubella viruses. Several attenuated strains of mumps have been developed for use in vaccines (Smorodintsev *et al.*, 1965; Gluck *et al.*, 1986; Galazka *et al.*, 1999; Plotkin and Wharton, 1999) and have been used in different locations worldwide (Galazka *et al.*, 1999; Plotkin and Wharton, 1999).

The European Commission has suggested that potential adverse reactions of mumps vaccine should be studied (Moxon *et al.*, 1996). The incidence of adverse effects varies among different strains (Galazka *et al.*, 1999; Plotkin and Wharton, 1999). Miller *et al.* (1989), Peltola *et al.* (1994), and Schwarzer *et al.* (1998) all studied the adverse effects of MMR vaccine, but they followed patients for up to 6 weeks after vaccination at most. One case of diabetes was reported 2 weeks after vaccination, which was less than the expected incidence (Peltola *et al.*, 1994). Sinaniotis *et al.* (1975) published a case report of diabetes occurring after mumps vaccination in childhood. Fescharek *et al.* (1990) analyzed spontaneous case reports of 20 cases of T1D reported in connection with the use of mumps vaccine. Cases were considered to have a relevant "temporal relationship" to mumps vaccination if they occurred within 30 days after the vaccination. There was no increased frequency of T1D following mumps vaccination.

A Finnish study (Hyöty *et al.*, 1993) of different birth cohorts noted that 0–4 year-old children born after the introduction of MMR vaccine had a higher cumulative incidence of T1D compared with nonvaccinated cohorts. The authors commented that MMR vaccination altered the epidemiology of mumps in Finland. Natural infections were eliminated and vaccination increased the number of challenged young children who otherwise would not have been infected by mumps at that stage. In very young children – targeted for MMR – the incidence of T1D continued to rise, in contrast to that in older age groups. The authors suggested there was a need for further studies to determine whether vaccine could trigger the clinical onset of T1D in young children.

A study of vaccination with mumps (Blom *et al.*, 1991) found no significant effect on the odds ratio (OR) for diabetes but concluded that this could not be interpreted unequivocally, as it might have been due to close age-matching and good adherence of the diabetic and referent (age, sex, and county-matched to each diabetic child) children to the Swedish vaccination programme. A review published in 1994 concluded that the evidence was inadequate to accept or reject a causal relation between mumps vaccine and insulin-dependent diabetes mellitus (IDDM) (Vaccine Safety Committee, Institute of Medicine, 1994). The reviewers stated, however, that it should be noted that most of the postulated mechanisms of the pathogenesis of IDDM (autoimmune, cumulative environmental effects, or persistent infection) suggested that there might be a prolonged interval between vaccination and the onset of symptoms of IDDM.

There has been considerable speculation about the role of wild mumps virus and mumps vaccine in the onset of childhood diabetes. Available data on this are incomplete and difficult to interpret, partly because several factors are thought to be involved in the development of childhood diabetes (Milne, 2001).

Hemagglutinin 1 neuraminidase 1 (H1N1)

A study was conducted to examine the risk of neurological and autoimmune disorders and T1D in people vaccinated against pandemic influenza A (H1N1) with Pandemrix (GlaxoSmithKline, Middlesex, UK) compared with unvaccinated people over 8–10 months. No change in the risk for GBS, MS, T1D, or rheumatoid arthritis (RA) was observed (Bardage *et al.*, 2011).

In another study, all suspected autoimmune disorders reported as adverse reactions to EudraVigilance from 1 October 2009 to 31 December 2010 for adjuvanted (Celtura (Novartis Vaccines and Diagnostics GmbH, Marburg, Germany), Fluval (Omninvest Ltd, Hungary), Focetria (Novartis), and Pandemrix) and nonadjuvanted (Cantgrip (Cantacuzino, Romania), Celvapan (Baxter AG, Deerfield, IL, US), and Panenza (Sanofi Pasteur S.A., Lyon, France)) pandemic influenza A/H1N1 vaccines were analyzed to determine whether adjuvanted vaccines were associated with higher reporting of autoimmune disorders than nonadjuvanted ones. The results do not suggest a difference in the reporting of autoimmune disorders between adjuvanted and nonadjuvanted pandemic influenza A/H1N1 vaccines (Isai *et al.*, 2012).

BCG and T1D

In humans, it has been hypothesized that early-age BCG vaccination could be associated with the risk of IDDM (Classen and Classen, 1997). The few studies conducted to date provide no consistent evidence of an association. However, in one study, the lower proportion of birth-vaccinated IDDM cases diagnosed at a very young age as compared with nonvaccinated cases suggests a possible temporary boost of the immune function after vaccination (Parent *et al.*, 1997). Conversely, another study reported a higher cumulative risk of IDDM in BCG-vaccinated children who were positive for islet autoantibodies (Huppmann *et al.*, 2005). In a descriptive study, BCG vaccination was associated with a lower prevalence of autoantibodies among vaccinated IDDM cases (Sanjeevi *et al.*, 2002).

Because diabetes is caused by abnormal immune mechanisms, and vaccines act by creating immunity to various diseases, some vaccines (particularly BCG) have been studied to see whether they offer protection against diabetes (Rousseau *et al.*, 2008).

The prophylactic potential of a prime-boost strategy involving BCG and the pVAXhsp65 vaccine (BCG/DNAhsp65) in diabetes induced by streptozotocin (STZ) in C57BL/6 mice and in spontaneous T1D in NOD mice was evaluated. BCG/DNAhsp65 vaccination in NOD mice induced weight gain, protection against hyperglycaemia, decreased islet inflammation, higher levels of cytokine production by the spleen, and a reduced number of regulatory T cells in the spleen with respect to nonimmunized NOD mice. In the STZ model, there was no significant difference in the clinical parameters. Even if this vaccination strategy did not protect mice in the STZ model, it was very effective in NOD mice (da Rosa *et al.*, 2013).

Although BCG does seem to be protective against diabetes in animal experiments, researchers have not been able to translate this benefit to humans. Research is still ongoing.

Vaccine and T1D in adults

In order to evaluate whether vaccination increases the risk of T1DM in active US military personnel, a retrospective cohort study was conducted among personnel aged 17–35 years (Duderstadt *et al.*, 2012). Individuals with a first-time diagnoses of T1D between 1 January 2002 and 31 December 2008 were identified using International Classification of Diseases, Ninth Revision, Clinical Modification (ICD-9-CM) codes. The study population consisted of 2 385 102 individuals followed for approximately 7 644 098 person-years of service. This included 1074 incident T1D cases. The study did not observe a significantly increased risk of T1D following vaccination with anthrax vaccine adsorbed, smallpox vaccine, typhoid vaccine, hepatitis B vaccine, MMR vaccine, or yellow fever vaccine (Duderstadt *et al.*, 2012).

Conclusions

The results of many studies do not support an association between vaccination and T1D in either young adults or children. However, available data are incomplete and difficult to interpret, partly because several factors are thought to be involved in the development of T1D. Well-designed and long-term studies into the use of vaccines and incidence of childhood diabetes are ongoing.

References

Amrani, A., Verdaguer, J., Thiessen, S., *et al.* (2000). IL-1alpha, IL-1beta, and IFN-gamma mark beta cells for Fas-dependent destruction by diabetogenic CD4(+) T lymphocytes. *J Clin Invest*, **105**: 459–68.

Antonelli, A., Baj, G., Marchetti, P., *et al.* (2001a). Human anti-CD38 autoantibodies raise intracellular calcium and stimulate insulin release in human pancreatic islets. *Diabetes*, **50**: 985–91.

Antonelli, A., Fallahi, P., Nesti, C., *et al.* (2001b). Anti-CD38 autoimmunity in patients with chronic autoimmune thyroiditis or Graves' disease. *Clin Exp Immunol*, **126**: 426–31.

Antonelli, A., Tuomi, T., Nannipieri, M., *et al.* (2002). Autoimmunity to CD38 and GAD in Type I and Type II diabetes: CD38 and HLA genotypes and clinical phenotypes. *Diabetologia*, **45**: 1298–306.

Antonelli, A., Fallahi, P., Ferrari, S.M., *et al.* (2008a). Serum Th1 (CXCL10) and Th2 (CCL2) chemokine levels in children with newly diagnosed Type 1 diabetes: a longitudinal study. *Diabet Med*, **25**: 1349–53.

Antonelli, A., Ferri, C., Fallahi, P., *et al.* (2008b). High values of CXCL10 serum levels in mixed cryoglobulinemia associated with hepatitis C infection. *Am J Gastroenterol*, **103**: 2488–94.

Antonelli, A., Ferri, C., Ferrari, S.M., *et al.* (2008c). Immunopathogenesis of HCV-related endocrine manifestations in chronic hepatitis and mixed cryoglobulinemia. *Autoimmun Rev*, **8**: 18–23.

Antonelli, A., Ferrari, S.M., Corrado, A., *et al.* (2014a). CXCR3, CXCL10 and type 1 diabetes. *Cytokine Growth Factor Rev*, **25**: 57–65.

Antonelli, A., Ferrari, S.M., Giuggioli, D., *et al.* (2014b). Chemokine (C-X-C motif) ligand (CXCL)10 in autoimmune diseases. *Autoimmun Rev*, **13**: 272–80.

Bardage, C., Persson, I., Ortqvist, A., *et al.* (2011). Neurological and autoimmune disorders after vaccination against pandemic influenza A (H1N1) with a monovalent adjuvanted vaccine: population based cohort study in Stockholm, Sweden. *BMJ*, **343**: d5956.

Black, S.B., Lewis, E., Shinefield, H.R., *et al.* (2002). Lack of association between receipt of conjugate haemophilus influenzae type B vaccine (HbOC) in infancy and risk of type 1 (juvenile onset) diabetes: long term follow-up of the HbOC efficacy trial cohort. *Pediatr Infect Dis J*, **21**: 568–9.

Blom, L., Nystrom, L., and Dahlquist, G. (1991). The Swedish childhood diabetes study. *Vaccinations and infections as risk determinants for diabetes in childhood. Diabetologia*, **34**: 176–81.

Chen, R.T., Pless, R., and Destefano, F. (2001). Epidemiology of autoimmune reactions induced by vaccination. *J Autoimmun*, **16**: 309–18.

Classen, J.B. (1996). The timing of immunization affects the development of diabetes in rodents. *Autoimmunity*, **24**: 137–45.

Classen, J.B. (2008a). Discontinuation of BCG vaccination precedes significant drop in type 2 diabetes in Japanese children. Role of inflammation and cortisol activity as a cause of type 2 diabetes. *Open Endocrinol J*, **2**: 1–4.

Classen, J.B. (2008b). Evidence childhood epidemics of type 1 and type 2 diabetes are opposite extremes of an immune spectrum disorder induced by immune stimulants. Role of race and associated cortisol activity as a major determining factor of the type of diabetes. *Diab Met Syndr: Clin Res Rev*, **3**: 67–9.

Classen, J.B. (2008c). Type 1 diabetes versus type 2 diabetes/metabolic syndrome, opposite extremes of an immune spectrum disorder induced by vaccines. *Open Endocrinol J*, **2**: 9–15.

Classen, D.C. and Classen, J.B. (1997). The timing of pediatric immunization and the risk of insulin dependent diabetes mellitus. *Infect Dis Clin Pract*, **6**: 449–54.

Classen, J.B. and Classen, D.C. (1999). Immunisation and type 1 diabetes mellitus: is there a link? *Drug Saf*, **21**: 423–5.

Classen, J.B. and Classen, D.C. (2002). Clustering of cases of insulin dependent diabetes (IDDM) occurring three years after *Haemophilus influenzae* (Hib) immunization support causal relationship between immunization and IDDM. *Autoimmunity*, **35**: 247–53.

Coppieters, K.T., Wiberg, A., and von Herrath, M.G. (2012). Viral infections and molecular mimicry in type 1 diabetes. *APMIS*, **120**: 941–9.

da Rosa, L.C., Chiuso-Minicucci, F., Zorzella-Pezavento, S.F., *et al.* (2013). Bacille Calmette–Guérin/DNAhsp65 prime-boost is protective against diabetes in non-obese diabetic mice but not in the streptozotocin model of type 1 diabetes. *Clin Exp Immunol*, **173**: 430–7.

Dahlquist, G. and Gothefors, L. (1995). The cumulative incidence of childhood diabetes mellitus in Sweden unaffected by BCG-vaccination. *Diabetologia*, **38**: 873–4.

DeStefano, F., Mullooly, J.P., Okoro, C.A., *et al.* (2001). Childhood vaccinations, vaccination timing, and risk of type 1 diabetes mellitus. *Pediatrics*, **108**: E112.

Duderstadt, S.K., Rose, C.E. Jr,, Real, T.M., *et al.* (2012). Vaccination and risk of type 1 diabetes mellitus in active component U.S. Military, 2002–2008. *Vaccine*, **30**: 813–19.

Fescharek, R., Quast, U., Maass, G., *et al.* (1990). Measles-mumps vaccination in the FRG: an empirical analysis after 14 years of use, *II. Tolerability and analysis of sponta- neously reported side effects. Vaccine*, **8**: 446–56.

Galazka, A.M., Robertson, S.E., and Kraigher, A. (1999). Mumps and mumps vaccine: a global review. *Bull World Health Organ*, **77**: 3–14.

Galleri, L., Sebastiani, G., Vendrame, F., *et al.* (2012). Viral infections and diabetes. *Adv Exp Med Biol*, **771**: 252–71.

Gluck, R., Hoskins, J.M., Wegmann, A., *et al.* (1986). Rubbini, a new live attenuated mumps vaccine virus strain for human diploid cells. *Dev Biol Stand*, **65**: 29–35.

Graves, P.M., Barriga, K.J., Norris, J.M., *et al.* (1999). Lack of association between early childhood immunizations and beta-cell autoimmunity. *Diabetes Care*, **22**: 1694–7.

Heijbel, H., Chen, R.T., and Dahlquist, G. (1997). Cumulative incidence of childhood-onset IDDM is unaffected by pertussis immunization. *Diabetes Care*, **20**: 173–5.

Huppmann, M., Baumgarten, A., Ziegler, A.G., and Bonifacio, E. (2005). Neonatal bacille Calmette–Guerin vaccination and type 1 diabetes. *Diabetes Care*, **28**: 1204–6.

Hyöty, H., Hiltunen, M., Reunanen, A., *et al.* (1993). Decline of mumps antibodies in Type 1 (insulin-dependent) diabetic children and a plateau in the rising incidence of type 1 diabetes after introduction of the mumps-measles-rubella vaccine in Finland. Chilhood diabetes in Finland Study Group. *Diabetologia*, **36**(12): 1303–8.

Institute for Vaccine Safety Diabetes Workshop Panel. (1999). Childhood immunizations and type 1 diabetes: summary of an Institute for Vaccine Safety Workshop. *Pediatr Infect Dis J*, **18**: 217–22.

Isai, A., Durand, J., Le Meur, S., *et al.* (2012). Autoimmune disorders after immunisation with influenza A/H1N1 vaccines with and without adjuvant: EudraVigilance data and literature review. *Vaccine*, **30**: 7123–9.

Karvonen, M., Cepaitis, Z., and Tuomilehto, J. (1999). Association between type 1 diabetes and *Haemophilus influenzae* type b vaccination: birth cohort study. *BMJ*, **318**: 1169–72.

Miller, C., Miller, E., Rowe, K., *et al.* (1989). Surveillance of symptoms following MMR vaccine in children. *Practitioner*, **233**: 69–73.

Milne, L.M. (2001). Difficulties in assessing the relationship, if any, between mumps vaccination and diabetes mellitus in childhood. *Vaccine*, **19**: 1018–25.

Morimoto, J., Yoneyama, H., Shimada, A., *et al.* (2004). CXC chemokine ligand 10 neutralization suppresses the occurrence of diabetes in nonobese diabetic mice through enhanced beta cell proliferation without affecting insulitis. *J Immunol*, **173**: 7017–24.

Moxon, E.R., Clemens, R., Druilhe, P., *et al.* (1996). European Commission COST/STD Initiative on European Vaccine Research. *Report of the expert panel IV. A common European network for clinical trials. Vaccine*, **14**: 624–43.

Parent, M.E., Siemiatycki, J., Menzies, R., *et al.* (1997). Bacille Calmette-Guerin vaccination and incidence of IDDM in Montreal, Canada. *Diabetes Care*, **20**: 767–72.

Peltola, H., Heinonen, O.P., Valle, M., *et al.* (1994). The elimination of indigenous measles, mumps and rubella from Finland by a 12-year, two-dose vaccination program. *N Engl J Med*, **331**: 1397–402.

Plotkin, S.A. and Wharton, M. (1999). Mumps vaccine. In: Plotkin, S.A. and Orenstein, W.A. (eds). *Vaccines*, 3rd edn. Philadelphia, PA: W.B. Saunders, pp. 267–92.

Pupilli, C., Giannini, S., Marchetti, P., *et al.* (1999). Autoantibodies to CD38 (ADP-ribosyl cyclase/cyclic ADP-ribose hydrolase) in Caucasian patients with diabetes: effects on insulin release from human islets. *Diabetes*, **48**: 2309–15.

Ravel, G., Christ, M., Liberge, P., *et al.* (2003). Effects of two pediatric vaccines on autoimmune diabetes in NOD female mice. *Toxicol Lett*, **146**: 93–100.

Rousseau, M.C., Parent, M.E., and St-Pierre, Y. (2008). Potential health effects from non-specific stimulation of the immune function in early age: the example of BCG vaccination. *Pediatr Allergy Immunol*, **19**: 438–48.

Sanjeevi, C.B., Das, A.K., and Shtauvere-Brameus, A. (2002). BCG vaccination and GAD65 and IA-2 autoantibodies in autoimmune diabetes in southern India. *Ann NY Acad Sci*, **958**: 293–6.

Schwarzer, S., Reibel, S., Lang, A.B., *et al.* (1998). Safety and characterization of the immune response engendered by two combined measles, mumps and rubella vaccines. *Vaccine*, **16**: 298–304.

Shigihara, T., Oikawa, Y., Kanazawa, Y., *et al.* (2006). Significance of serum CXCL10/IP-10 level in type 1 diabetes. *J Autoimmun*, **26**: 66–71.

Sia, C. (2005). Imbalance in Th cell polarization and its relevance in type 1 diabetes mellitus. *Rev Diabet Stud*, **2**: 182–6.

Sinaniotis, C.A., Daskalopoulou, E., Lapatsanis, P., and Doxiadis, S. (1975). Diabetes mellitus after mumps vaccination. *Arch Dis Child*, **50**: 749–50.

Singh, B. (2000). Stimulation of the developing immune system can prevent autoimmunity. *J Autoimmun*, **14**: 15–22.

Smorodintsev, A.A., Luzyanina, T.Y.A., and Mikutskaya, B.A. (1965). Data on the efficiency of live mumps vaccine from chick embryo cell cultures. *Acta Virol*, **9**: 240–7.

Stanescu, D.E., Lord, K., and Lipman, T.H. (2012). The epidemiology of type 1 diabetes in children. *Endocrinol Metab Clin North Am*, **41**: 679–94.

Tishler, M. and Shoenfeld, Y. (2004). Vaccination may be associated with autoimmune diseases. *IMAJ*, **6**: 430–2.

Vaccine Safety Committee, Institute of Medicine. (1994). Measles and mumps vaccines. In: Stratton, K.R., Howe, C.J., and Johnston, R.B. (eds). *Adverse Events Associated with Childhood Vaccines. Evidence Bearing on*

Causality. Washington, DC: National Academy Press, pp. 118–86.

Vial, T. and Descotes, J. (2004). Autoimmune diseases and vaccinations. *Eur J Dermatol*, **14**: 86–90.

Wahlberg, J., Fredriksson, J., Vaarala, O., *et al.* (2003). Vaccinations may induce diabetes-related autoantibodies in one-year-old children. *Ann NY Acad Sci*, **1005**: 404–8.

Wang, B., Maile, R., Greenwood, R., *et al.* (2000). Naive CD8+ T cells do not require costimulation for proliferation and differentiation into cytotoxic effector cells. *J Immunol*, **164**: 1216–22.

Zekzer, D., Wong, F.S., Ayalon, O., *et al.* (1998). GAD-reactive CD4+ Th1 cells induce diabetes in NOD/SCID mice. *J Clin Invest*, **101**: 68–73.

30 Narcolepsy and H1N1 vaccine

María-Teresa Arango,[1,2] Shaye Kivity,[1,3,4] Nancy Agmon-Levin,[1,5] Gili Givaty,[1,6] Joab Chapman,[1,6] and Yehuda Shoenfeld[1,7]

[1]Zabludowicz Center for Autoimmune Diseases, Sheba Medical Center, Tel Hashomer, Israel
[2]Doctoral Program in Biomedical Sciences, Del Rosario University, Bogotá, Colombia
[3]Rheumatic Disease Unit, Sheba Medical Center, Tel Hashomer, Israel
[4]The Dr Pinchas Borenstein Talpiot Medical Leadership Program 2013, Sheba Medical Center, Tel Hashomer, Israel
[5]Sackler Faculty of Medicine, Tel Aviv University, Tel Aviv, Israel
[6]Neurology Department and Sagol Neuroscience Center, Sheba Medical Center, Tel Hashomer, Israel
[7]Incumbent of the Laura Schwarz-Kipp Chair for Research of Autoimmune Diseases, Sackler Faculty of Medicine, Tel Aviv University, Tel Aviv, Israel

Introduction

Narcolepsy is a sleep disorder described as excessive sleepiness with abnormal sleep pattern characterized by uncontrollable rapid eye movement (REM) attacks, in which the preceding non-REM stage is absent. These attacks can occur at any time of the day and can be accompanied by a loss of muscle tone (cataplexy) (Fontana *et al.*, 2010; Sakurai *et al.*, 2010). The world prevalence of narcolepsy with cataplexy is between 25 and 50 per 100 000 people (Longstreth *et al.*, 2007), although it varies according to country and genetic background, with the highest prevalence reported in Japan (0.16%) (Akintomide and Rickards, 2011).

A plethora of data indicate that narcolepsy is caused by a lack of orexin (also known as hypocretin), an important neurotransmitter involved in the regulation of the sleep cycle. In humans, analysis of brain autopsy from narcolepsy patients demonstrates a loss of orexin-producing neurons in the hypothalamus (Rolls *et al.*, 2010) and low or undetectable orexin levels in the cerebrospinal fluid (CSF) (Fontana *et al.*, 2010).

Normal levels of orexin are needed for the correct function of different processes in the body, including feeding, cardiovascular regulation, emotions, and locomotion. In the sleep process, orexin causes positive feedback in the brain centers responsible for the maintenance of the wakefulness state and is indirectly involved in the repression of the centers in charge of the sleep state. If orexin neurons are removed, the centers set up a mutually inhibitory circuit, which can cause unwanted abrupt transitions and the rupture of the balance between each state. This rupture is characterized by sudden transition to REM sleep and excessive sleepiness (Ohno and Sakurai, 2008; Rolls *et al.*, 2010; Huang *et al.*, 2011).

The pathogenesis of narcolepsy has been debated for many years. It has been suggested that genetic, autoimmune, or infectious processes may be involved (recently reviewed by Mahlios *et al.*, 2013). In recent years, a considerable body of evidence has suggested an autoimmune nature to the pathogenesis of narcolepsy, perhaps involved in the loss of orexin neurons in the hypothalamus (De la Herran-Arita *et al.*, 2013; Katzav *et al.*, 2013; Mahlios *et al.*, 2013). Narcolepsy is highly associated with protective and risk polymorphisms

Vaccines and Autoimmunity, First Edition. Edited by Yehuda Shoenfeld, Nancy Agmon-Levin, and Lucija Tomljenovic.

in the human leukocyte antigen (HLA) system, especially the DQB1*06:02 risk allele (Kornum *et al.*, 2011; Mignot *et al.*, 2001), as well as with other important immune-related genes, such as T receptor and tumor necrosis factor alpha (TNF-α) genes (Fontana *et al.*, 2010). Further evidence for the pathogenesis of narcolepsy includes the discovery of novel autoantibodies against Tribbles 2 (Trib2) in narcoleptic patients in Japan (Cvetkovic-Lopes *et al.*, 2010; Kawashima *et al.*, 2010; Lim and Scammell, 2010). In addition, supporting data from concordance studies with monozygotic twins (20–35%) have revealed the importance of environment factors (Mignot, 1998). The onset of the disease is more frequent in teenagers (Silber *et al.*, 2002; Akintomide and Rickards, 2011), which may suggest possible triggers of the disease at this specific age (e.g. hormones) (Ohayon *et al.*, 2005; Nohynek *et al.*, 2012). Finally, an important aspect in the etiology of narcolepsy is some reports demonstrating a correlation between its onset and infections or H1N1 vaccination (Dauvilliers *et al.*, 2010; Kornum *et al.*, 2011).

Orexin and sleep

Two different kinds of orexins, A and B, are produced exclusively by a small group of neurons in the hypothalamic region of the brain (orexin neurons). Both proteins are highly conserved and come from the same precursor: prepro-orexin, a gene located on the long arm of chromosome 17. This precursor generates two different peptides after post-translational modifications. The orexins can bind to two types of receptors found distributed throughout the central nervous system (CNS) on neurons of the monoaminergic centers, which are responsible for the secretion of neurotransmitters, including norepinephrine, serotonin, and histamine (Ohno and Sakurai, 2008; Sakurai *et al.*, 2010). Since orexins stimulate the production of these molecules, they are closely related to different regulation processes, including feeding, emotions, and sleep balance (Ohno and Sakurai, 2008).

As for their role in maintenance of wakefulness, the secreted orexin-activating monoaminergic neurons send inhibitory signals to the ventrolateral preoptic area (VLPO), where gamma aninobutiric acid (GABA)-producing neurons are suppressed. During the sleep period, on the other hand, the neurons of the VLPO send inhibitory signals to orexin and monoaminergic neurons, thus inducing sleep. In narcolepsy, since orexin-producing neurons are not present (or are dysfunctional) in the VLPO circuit and monoaminergic neurons, inhibitory signals lead to abrupt changes in the sleep state. In fact, narcolepsy is characterized by abrupt changes in REM sleep, involving high brain activity and loss of muscle control, leading to the characteristic cataplexy. However, narcolepsy can also be found without cataplexy (Burt *et al.*, 2011; Huang *et al.*, 2011).

Genetic factors

It has been proposed that autoimmune mechanisms may contribute in the pathogenesis of narcolepsy and might be the cause of the specific loss of orexin neurons in the hypothalamus. As previously stated, narcolepsy is highly associated with HLA alleles. About 82–99% of narcolepsy patients are carriers of the DQB1*06:02 allele (Mignot *et al.*, 2001), as well as the DRB1*15:01 allele, albeit in low numbers. In contrast, just 12–38% of healthy individuals have the DQB1*06:02 allele. In addition, there are reports of protective HLA alleles in narcolepsy, such as DQB1*06:01 and DQB1*05:01, as well as in other autoimmune diseases (Kornum *et al.*, 2011).

Proteins codified by the HLA locus are involved in antigen recognition, processing, and presentation (class I and II), as well as in mechanisms of innate immunity (class III). In addition, HLA class II proteins are also involved in the recognition and attachment of the T cell receptor (TCR). This means that any alteration in the structure of either protein (HLA II or TCR) may induce changes in the affinity or specificity of this interaction (Singh *et al.*, 2013). Interestingly, a polymorphism (rs1154155) in the TCR-α chain has been associated with a higher risk in narcolepsy patients (Hallmayer *et al.*, 2009). This indicates that the combination of genetic factors may play an important role in the HLA–TCR interaction in these patients. In fact, analysis of the crystal structure of HLA II DQB1*06:02 demonstrates that an orexin derivative peptide composed of 13 amino acids may bind to the DQB1*06:02 molecule. Consequently, this peptide may be presented to T lymphocytes in carriers of this allele (Siebold *et al.*, 2004). These findings are consistent with previous reports showing that homozygotic carriers of this allele have an increased risk of developing narcolepsy (Mignot *et al.*, 2001). Therefore, it is possible to hypothesize that, in genetically susceptible individuals, there may be an immune response against orexin itself.

Recently, studies have detected high levels of antibodies against Trib2 in narcopleptic patients (Cvetkovic-Lopes *et al.*, 2010; Kawashima *et al.*, 2010; Lim and Scammell, 2010). Trib2 is a member of the Tribbles proteins family, which is related to cell cycle and signaling transduction control. Interestingly, Trib2 was first described as an autoantigen in a patient with autoimmune uveitis (Zhang *et al.*, 2005). Evidence for a possible autoimmune process related to Trib2 in narcolepsy was described using transgenic mice and sera from narcolepsy patients. The study found that anti-Trib2 antibodies from narcolepsy patient sera bound directly to mice orexin neurons. In addition, the titers of anti-Trib2 antibodies were higher in positive narcolepsy patients than in healthy controls. Remarkably, in addition to anti-orexin antibodies, titers of anti-Trib2 were also found to be higher at disease onset (Cvetkovic-Lopes *et al.*, 2010; Kawashima *et al.*, 2010).

There are some animal models of narcolepsy, such as orexin knockout mice (Chemelli *et al.*, 1999), zebrafish (Yokogawa *et al.*, 2007; Arias-Carrión, 2009), and the canine model, which occurs naturally through a mutation in the gene encoding the orexin 2 receptor (Aldrich and Reynolds, 1999; Lin *et al.*, 1999). Smith *et al.* (2004) injected total IgG from narcolepsy patients into naive mice, intravenously. They described the abrupt cessation of movements in the immunized mice after activities such as grooming. This behavior was similar to that reported in orx−/− mice (Chemelli *et al.*, 1999; Smith *et al.*, 2004). Recently, our group developed an experimental model of narcolepsy in mice by passive transfer of total IgG from narcolepsy patients into the animals' brains through intraventricular injection. Briefly, the animals were filmed during the dark phase (the awakening period) once before the injection and then weekly for a month after the procedure. The films were analyzed and abnormal sleep behavior was identified during the activity phase of those animals immunized with narcoleptic IgG patients' sera. The episodes were characterized by high mobility, especially during grooming, before and after a period of immobility (sleep attack). Notably, these episodes did not appear before the injection. Remarkably, as in human patients, the mice demonstrated hyperactivity and long-term memory deficit. Finally, immunohistochemistry analysis revealed that narcolepsy IgG-injected mice had fewer neuronal and synaptic markers and a loss of orexin-positive neurons in the lateral hypothalamus area, when compared with control IgG-injected mice (Figure 30.1). This was the first report of experimental narcolepsy induced by passive transfer to mice, supporting the autoimmune pathogenesis in narcolepsy (Katzav *et al.*, 2013).

Environmental factors

A strong relationship between the environment and the development of narcolepsy has been demonstrated. Influenza A virus and streptococcal infections have been associated with the onset of the disease (Aran *et al.*, 2009; Dauvilliers *et al.*, 2010; Viorritto *et al.*, 2012). For instance, in the United States, an analysis of sera from patients with recent-onset narcolepsy demonstrated elevated antistreptococcal antibodies, suggesting that this infection is associated with the disease (Aran *et al.*, 2009). Interestingly, fever by itself, without the diagnosis of an infectious etiology, was found to be a risk factor for narcolepsy (Picchioni *et al.*, 2007). Perhaps the most impressive evidence of an environmental role in the development of narcolepsy is related to recent vaccination with the H1N1 vaccine.

Seasonal influenza is an acute viral infection caused by one of the influenza viruses: A, B, or C. The most studied and most common infection is caused by influenza A. This virus usually infects humans and birds, but it may also affect swine. The genome of influenza viruses is divided into eight minus-strand RNA segments, which undergo recombination between segments and strains in cases of co-infection. Structurally, the virion is characterized by the presence of two major surface proteins – hemagglutinin (H) and neuroaminidase (N) – which can vary between one strain and the next. These changes are in fact the determinants of specificity by the host cell during the infection process. For this reason, the strains are named according to the subtype of each of these proteins (e.g. AH1N1 or AH3N2) (Strauss and Strauss, 2002).

Because of the seasonal behavior of the influenza infection and its seriousness, especially in children and the elderly, vaccination is recommended in most countries for susceptible individuals. Currently, a new influenza vaccine is produced each year, directed against the most prevalent strains.

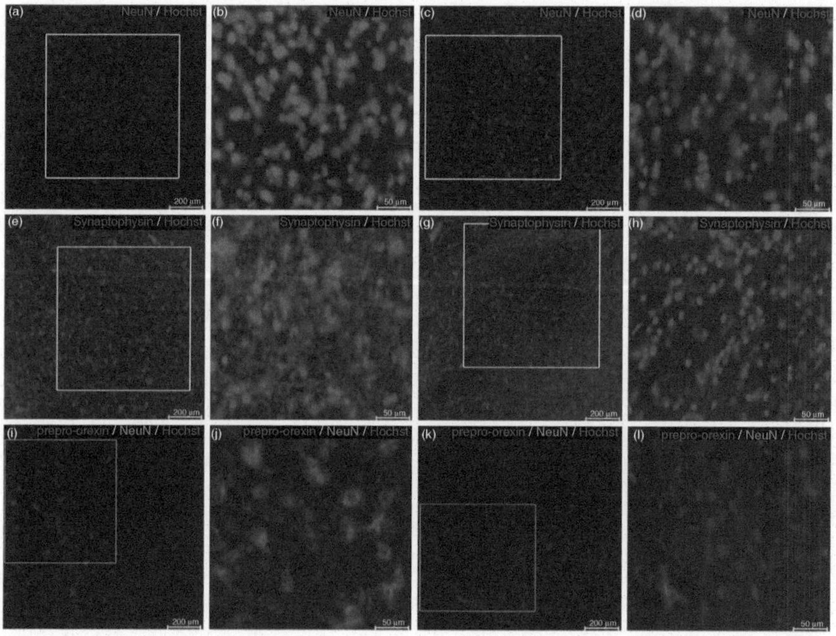

Figure 30.1 Histopathological changes induced by narcolepsy IgG. Coronal brain sections through the hypothalamus from mice injected ICV with IgG from narcolepsy patients and from healthy controls were stained for neuronal marker (NeuN) (a–d), synaptic marker (synaptophysin) (e–h), and orexin-expressing neurons (prepro-orexin) (i–l). (a,b,e,f,i,j) Representative images from control mice injected with control-IgG. (c,d,g,h,k,l) Representative images from mice injected with narcolepsy IgG. First- and third-column images are at 10× magnification and scale bar 200 μm. Second- and forth-column images are at 40× magnification and scale bar 50 μm. Reprinted from Katzav, A., Arango, M.T., Kivity, S., *et al.* (2013). Passive transfer of narcolepsy: anti-TRIB2 autoantibody positive patient IgG causes hypothalamic orexin neuron loss and sleep attacks in mice. *J Autoimmun*, **45**: 24–30. Copyright (2013), with permission from Elsevier. (For a color version of this figure, please see color plate section.)

H1N1 vaccination

Following the 2009 outbreak of pandemic influenza type AH1N1, international committees and organizations published vaccination recommendations that distinguished groups according to their morbidity and mortality risk from influenza. For example, countries in the European Union implemented different vaccination strategies, such as the vaccination of higher-risk groups (children and the elderly) in Italy and the United Kingdom and the vaccination of the entire population in Finland and Sweden (O'Flanagan *et al.*, 2011; Wijnans *et al.*, 2013). Eight different commercial vaccines were developed and authorized by the European Centre for Disease Prevention and Control, all designed from the A/California/7/2009 (H1N1) v-like strain, but with variations in other components, such us adjutants (Table 30.1). The most commonly used among European countries was Pandemrix (O'Flanagan *et al.*, 2011; Wijnans *et al.*, 2013).

Following the vaccination in Europe, an increment in the diagnosis of narcolepsy was described. In 2010, Dr Lars Palm, a child neurologist, was the first to suggest a possible association between the increased prevalence of narcolepsy and H1N1 vaccination using the ASO3-adjuvanted Pandemrix vaccine (Käll, 2013; Zhang *et al.*, 2013). Other research soon followed. For instance, in October of 2010, a task force was appointed to determine whether there was a causal relationship between the increment in narcolepsy cases in Finnish children and the vaccination campaign carried out following the 2009 H1N1 pandemic. They took as baseline the incidence of narcolepsy between 2006 and 2009 and they found a significant increase in the incidence of the disease in 2010. According to the first preliminary data to be confirmed and published, the risk of narcolepsy in the 4–19-years age group was ninefold greater among those who received the Pandemrix vaccine. These results strongly suggest an association between the onset of the disease and vaccination within a particular context (THL, 2010).

Table 30.1 Commercial H1N1 vaccines authorized for use in the 2009 pandemic by the European Centre for Disease Prevention and Control. Adapted from ECDC (2009) and GSK (2009)

Name	Producer	Components	Hemagglutinin content	Adjuvant	Adjuvant emulsion per dose
Celvapan	Baxter	Whole virion, wild-type A/California/7/2009 (H1N1)v, inactivated	7.5 µg	None	None
Pandemrix	GSK	Split-virion, reassortant, A/California/7/2009 (H1N1)v-like strain, inactivated, adjuvanted	3.75 µg	AS03	Squalene 10.69 mg α-tocopherol 11.86 mg polysorbate 80 4.86 mg
Arepanrix	GSK	Split influenza virus, A/California/7/2009 (H1N1)v-like strain (X-179A) inactivated, adjuvanted	3.75 µg	AS03	Squalene 10.69 mg α-tocopherol 11.86 mg polysorbate 80 4.86 mg
Focetria	Novartis	Surface antigens (hemagglutinin and neuraminidase), reassortant, A/California/7/2009 (H1N1)v-like strain, inactivated	7.5 µg	MF59C.1	Squalene 9.75 mg polysorbate 80 1.175 mg sorbitan trioleate 1.175 mg
Fluval P	Omnivest	Whole virion, reassortant, A/California/7/2009 (H1N1)v-like strain, inactivated, adjuvanted	6 µg	Aluminum phos-phate	Aluminum phosphate 0.33 mg Al3+
Panenza	Sanofi Pasteur	Split-virion, reassortant, A/California/7/2009 (H1N1)v-like strain, inactivated	15 µg	None	None
Celtura	Novartis	Surface antigens (hemagglutinin and neuraminidase), reassortant, A/California/7/2009 (H1N1)v-like strain, inactivated, adjuvanted	3.75 µg	MF59C.1	MF59C.1 Squalene 4.875 mg polysorbat 80 0.588 mg sorbitanoleat 0.588 mg
PanvaxH1N1	CSL	Split-virion, reassortant, A/California/7/2009 (H1N1)v-like strain, inactivated	15 µg	None	None
CANTGRIP	Cantacuzino	Split-virion, reassortant, A/California/7/2009 (H1N1)v-like strain, inactivated	15 µg	None	None

v-like, vaccine-like; GSK, GlaxoSmithKline; CSL, Commonwealth Serum Laboratories

Dauvilliers *et al.* (2010) described a series of 16 cases of diagnosed narcolepsy and cataplexy from Switzerland, the United States, the United Kingdom, and France. These patients all carried the DQB1*06:02 risk allele, had low levels of orexin in the CSF, and first devloped symptoms after receiving the ASO3-adjuvanted Pandemrix vaccine (Dauvilliers *et al.*, 2010; Zhang *et al.*, 2013). In addition, following Dr Palm's initial observation in Finland, an analysis of medical records from vaccinated individuals in this country showed a 12.7-fold increased risk of developing narcolepsy in children (Nohynek *et al.*, 2012). These results were confirmed by a large retrospective cohort study of narcolepsy diagnosis rates between 2000 and 2010, based on health care databases in Denmark, Finland, Italy, the Netherlands, Sweden, and the United Kingdom. This study also found a higher risk in the 5–19-years age group in Denmark, Finland, and Sweden when AS03-adjuvanted Pandemrix vaccine was used (THL, 2010). However, a link between H1N1 vaccination and the onset of narcolepsy was not found in Italy or the United Kingdom (Wijnans *et al.*, 2013; Zhang *et al.*, 2013). Additional reports of evidence for a higher risk of narcolepsy associated with Pandemrix came from France and England between 2012 and 2013 (Miller *et al.*, 2013; Zhang *et al.*, 2013).

Similarly, in France, one study found that immunization with the AS03 influenza vaccine raised the risk for narcolepsy by OR 6.5 (2.1–19.9) in children and OR 4.7 (1.6–13.9) in adults (age over 18). This study also demonstrated that narcolepsy patients who were vaccinated just before the onset of the disease had a quicker diagnosis and a higher number of "attacks" related to sudden REM cycle in comparison with nonvaccinated patients. In addition, neither group showed significant differences regarding infectious episodes, showing that the vaccine causes a more aggressive form of the disease (Dauvilliers *et al.*, 2013).

However, the incidence of the disease was not increased with H1N1 infection among the influenza-infected population. This is the opposite of what was found in a study in China, where a three- to fourfold increment in narcolepsy occurrence was found after the pandemic period, related not to the vaccination but to the infection itself (Han *et al.*, 2011; Nohynek *et al.*, 2012). Interestingly, after this period, the incidence of narcolepsy in the Chinese population returned to baseline (Han *et al.*, 2013). Recently, as a consequence of these contradictory findings, a new study was conducted in Finland on 45 narcoleptic patients who developed the disease following Pandemrix vaccination. The goal of this study was to identify whether a coinciding pandemic influenza infection contributed, together with the Pandemrix, to the onset of the disease. To accomplish this, serum samples were analyzed for antibodies against the NS1 viral protein, which is not included in the vaccine but is highly produced during H1N1 viral infection. No evidence of antibodies to NS1 viral protein was found in narcolepsy patients, but there were antibodies against the vaccine viral components, which indicates that these patients were not infected. Therefore, it is unlikely that an active infection by pandemic influenza was related to disease onset in this group of patients. However, these observations give additional support to the findings of an association between the AS03-adjuvanted Pandemrix vaccine and narcolepsy in a susceptible population (Melén *et al.*, 2013)

Two remarkable case reports help to further illustrate the association between H1N1 vaccination and narcolepsy. First, in Sweden, a healthy man developed narcolepsy and multiple sclerosis (MS) after vaccination against H1N1 influenza with Pandemrix. Interestingly, he was a carrier of both the DRB1*15:01 allele (associated with MS) and the DQB1*06:02 allele (associated with narcolepsy). This patient developed the first symptoms of sleep disorder and cataplexy within the first 2 months after vaccination and MS-related lesions within 5 months (Vrethem *et al.*, 2012). Second, the only case reported in South America was described in Brazil in 2010. A 19-year-old women who was also a carrier of the DQB1*06:02 allele developed narcolepsy after she was vaccinated with Arepanrix, which is the same vaccine as Pandemrix, but produced in Canada rather than the United Kingdom (Table 30.1) (Health Canada, 2009; Mendes *et al.*, 2012).

Is the adjuvant the clue?

As discussed earlier, most of the narcolepsy cases and associations mentioned in this chapter are specifically related to the AS03-adjuvanted Pandemrix vaccine. The same association has not been reported for other H1N1 adjuvanted or nonadjuvanted vaccines (Crucitti and Tsai, 2011; Waldenlind *et al.*, 2013). Further, an analysis of the results of 115 clinical trials with 79 004 subjects vaccinated with various MF59-adjuvanted and nonadjuvanted influenza vaccines did not find any risk of narcolepsy development (Tsai *et al.*, 2011), and an increase in narcolepsy cases was not found in South Korea after an analysis of the clinical records of subjects vaccinated with non-adjuvanted or MF59-adjuvanted vaccines (Choe *et al.*, 2012). All these results suggest that the differences between vaccine components can have an important effect not only on immunogenic efficiency but also on the induction of nonspecific immune responses, and perhaps autoimmunity. Of further relevance, the major difference between the AS03 and MF59 adjuvants is the presence of α-tocopherol (Table 30.1).

Notably, there are insufficient data regarding differences in the immune reaction to each adjuvant. It is known that the MF59 adjuvant does not contain α-tocopherol. Mice model studies have shown that MF59 induces the expression of cytokines and chemokines by muscle cells at the site of injection. This further results in the migration of immune cells, including monocytes and granulocytes, especially neutrophils (Calabro *et al.*, 2011). The presence of these cells amplifies the signal, causing more phagocytes to migrate to the injection site, thus increasing the chance of antigen presentation and transportation to the regional lymph nodes (Calabro *et al.*, 2011; Tsai *et al.*, 2011; O'Hagan *et al.*, 2012). In contrast, the AS03 adjuvant contains the immunomodulator α-tocopherol (a form of vitamin E), which activates the innate immune system, not only at the injection site but also at nonregional lymph nodes. This makes AS03 more potent than the squalene-only MF59. In fact, it has been demonstrated that α-tocopherol, which is present only in the AS03 adjuvant, can achieve the greatest and longest antibody response (Walker *et al.*, 2012). It is proposed that AS03 modulates the expression of cytokines, which can help increment antigen loading in monocytes and the recruitment of granulocytes in the lymph nodes, leading to enhancement of the antigen-specific adaptive immune response (Morel *et al.*, 2011; Tsai *et al.*, 2013). Moreover, it was recently demonstrated by *in vitro* analysis that α-tocopherol can increase the production of orexin, as well as proteosome activity, in a murine hypothalamic cell line. The authors suggested that the increased production of orexin fragments facilitates antigen presentation to major histocompatibility complex (MHC) class II, thus triggering an autoimmune process (Masoudi *et al.*, 2014).

Notably, these studies of the action and mechanism of the MF59 and AS03 adjuvants were conducted by comparison with other common adjuvants, such as aluminum hydroxide, and not by a direct comparison. Moreover, the protein profiles evaluated in these studies were not the same, and there are other data regarding gene expression associated with the action of MF59 (Mosca *et al.*, 2008). In summary, further studies are needed to make a proper comparison between MF59 and AS03.

It can be argued that the ability of the systemic immune response to reach the brain following vaccination is limited by the blood–brain barrier (BBB). One explanation is that the fever generated by the immune process may disrupt the BBB, thus allowing infiltration of molecules and immune cells into the CNS, perhaps causing damage to the orexin neurons in susceptible individuals (Kornum *et al.*, 2011).

Conclusions

All the evidence mentioned in this chapter suggests an important role for an immune-mediated process induced by environmental factors, especially the AS03-adjuvanted Pandemrix vaccine, in the development of narcolepsy in genetically susceptible populations. However, the precise mechanism by which the AS03-adjuvanted Pandemrix vaccine or the H1N1 infection itself might induce the onset of the disease is still not clear. There has been much speculation regarding this issue. Singh *et al.* (2013) have proposed a model which explains how H1N1 infection or vaccine can induce the loss of orexin neurons, via an interaction between infections and genetic factors (Figure 30.2). Different mechanisms may be involved in this process, including bystander activation of autoreactive B and T cells in response to the vaccine's adjuvants. Moreover, the HLA association may suggest that antigen presentation of crossreactive peptides can lead to the activation of the immune response against the orexin neurons. Another explanation is molecular mimicry between orexin neuron molecules and the vaccine, the H1N1 virus, or other infectious agents (e.g. *Streptococcus* sp.) found to be associated with the development of the disease (Kornum *et al.*, 2011; Singh *et al.*, 2013; Mahlios *et al.*, 2013). Regarding the role of the vaccine, two interesting options have been proposed. First, the AS03 adjuvant may catalyze the molecular mimicry between orexin neurons and H1N1 molecules, due both to its nature and to its method of immune system activation (Mahlios *et al.*, 2013). Second, the presentation of normal post-vaccinated events, such as fever, may favor the migration of pre-existent autoreactive cells or antibodies through the BBB, leading to the loss of orexin neurons (Kornum *et al.*, 2011).

Finally, all these data together support the relationship between the H1N1 vaccine and the development of narcolepsy under certain conditions. Therefore, these observations should raise awareness regarding the risks and benefits of H1N1 vaccination versus nonvaccination (Caplan, 2010). Perhaps in the future, the genetic and environmental background of a given individual should be taken into account before making the decision to vaccinate.

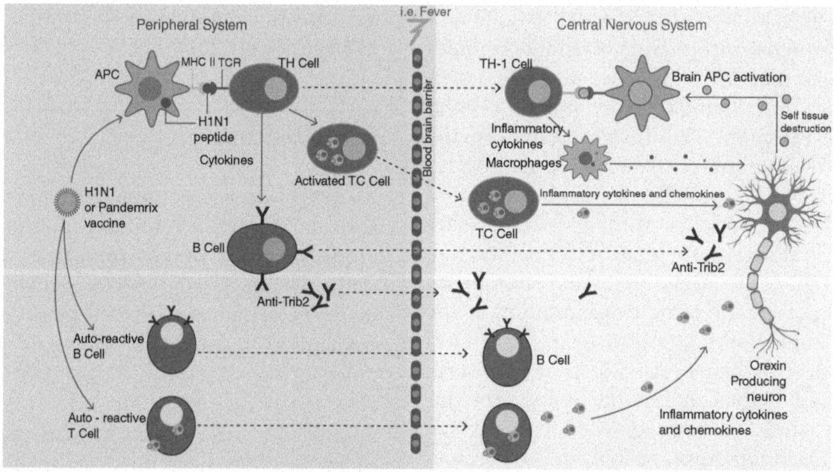

Figure 30.2 Possible pathway for H1N1 seasonal infection and Pandemrix vaccination in the onset of narcolepsy. The seasonal H1N1 influenza infection or Pandemrix vaccine could stimulate autoreactive T or B cells targeting orexin, producing neurons through the disruption of the BBB as a consequence of adverse vaccine events, such as fever, and by several other mechanisms. (i) Molecular mimicry of T cells. This describes the activation of crossreactive T cells that recognize the H1N1 epitope and then migrate to the CNS, where they recognize an antigen specific to orexin-producing neurons (crossreactivity). Activation of crossreactive T cells results in the release of cytokines and chemokines, which recruit and activate macrophages, mediating self-tissue damage. The subsequent release of orexin self-antigen and its uptake by antigen-presenting cells (APCs) perpetuates narcolepsy. (ii) Crosslink of the MHC and TCR molecules and activation of the cytotoxic T cells, which are autoreactive and specific towards orexin-producing neurons, by H1N1 antigens or Pandemrix vaccine. (iii) Molecular mimicry involving B cells and antibody-mediated disease. This could target TRIB2 as a crossreactive antigen. It would require signals from activated T cells (T cell help). (iv) Bystander activation of resting autoreactive B and T cells. This could result from general immune activation, independent of specific antigens. Current results in narcolepsy research point towards a T cell mechanism. Abbreviations: APC, antigen-presenting cell; BBB, blood–brain barrier; CNS, central nervous system; H1N1, H1N1 influenza A virus or epitopes from adjuvant vaccines; MHC, major histocompatibility complex; TCR, T cell receptor; TRIB2, Tribbles homolog 2. Reprinted and modified from Singh, A.K., Mahlios, J., and Mignot, E. (2013). Genetic association, seasonal infections and autoimmune basis of narcolepsy. *J Autoimmun*, **43**: 26–36. Copyright (2013) with permission from Elsevier. (For a color version of this figure, please see color plate section.)

References

Akintomide, G.S. and Rickards, H. (2011). Narcolepsy: a review. *Neuropsychiatr Dis Treat*, **7**: 507–18.

Aldrich, M.S. and Reynolds, P.R. (1999). Narcolepsy and the hypocretin receptor 2 gene. *Neuron*, **23**: 625–6.

Aran, A., Lin, L., Nevsimalova, S., *et al.* (2009). Elevated anti-streptococcal antibodies in patients with recent narcolepsy onset. *Sleep*, **32**: 979–83.

Arias-Carrión, O. (2009). Hypocretin system and narcolepsy. *Rev Med Chil*, **137**: 1209–16.

Burt, J., Alberto, C.O., Parsons, M.P., and Hirasawa, M. (2011). Local network regulation of orexin neurons in the lateral hypothalamus. *Am J Physiol Regul Integr Comp Physiol*, **301**: R572–80.

Calabro, S., Tortoli, M., Baudner, B.C., *et al.* (2011). Vaccine adjuvants alum and MF59 induce rapid recruitment of neutrophils and monocytes that participate in antigen transport to draining lymph nodes. *Vaccine*, **29**: 1812–23.

Caplan, A.L. (2010). Duty to warn? –the ethics of disclosing information about possible risks associated with H1N1 vaccination. *Sleep*, **33**: 1426–7.

Chemelli, R.M., Willie, J.T., Sinton, C.M., *et al.* (1999). Narcolepsy in orexin knockout mice: molecular genetics of sleep regulation. *Cell*, **98**: 437–51.

Choe, Y.J., Bae, G.R., and Lee, D.H. (2012). No association between influenza A(H1N1)pdm09 vaccination and narcolepsy in South Korea: an ecological study. *Vaccine*, **30**(52): 7439–42.

Crucitti, A. and Tsai, T.F. (2011). Explorations of clinical trials and pharmacovigilance databases of MF59®-adjuvanted influenza vaccines for associated cases of narcolepsy: a six-month update. *Scand J Infect Dis*, **43**: 993.

Cvetkovic-Lopes, V., Bayer, L., Dorsaz, S., *et al.* (2010). Elevated Tribbles homolog 2-specific antibody levels in narcolepsy patients. *J Clin Invest*, **120**: 713–19.

Dauvilliers, Y., Montplaisir, J., Cochen, V., *et al.* (2010). Post-H1N1 narcolepsy-cataplexy. *Sleep*, **33**: 1428–30.

Dauvilliers, Y., Arnulf, I., Lecendreux, M., *et al.* (2013). Increased risk of narcolepsy in children and adults after pandemic H1N1 vaccination in France. *Brain*, **136**: 2486–96.

De la Herran-Arita, A.K., Kornum, B.R., Mahlios, J., *et al.* (2013). CD4+ T cell autoimmunity to hypocretin/orexin and cross-reactivity to a 2009 H1N1 influenza A epitope in narcolepsy. *Sci Transl Med*, **5**: 216ra176.

ECDC (European Center for Disease Prevention and Control). (2009). Q&A for health professionals on vaccines and vaccination in relation to 2009 influenza A(H1N1) pandemic. Available from: http://www.ecdc.europa.eu/en/healthtopics/Documents/0911_Pandemic_Influenza_%28H1N1%29_QA_on_Vaccines_and_Vaccination_Experts.pdf [last accessed 11 December 2014].

Fontana, A., Gast, H., Reith, W., *et al.* (2010). Narcolepsy: autoimmunity, effector T cell activation due to infection, or T cell independent, major histocompatibility complex class II induced neuronal loss? *Brain*, **133**: 1300–11.

Hallmayer, J., Faraco, J., Lin, L., *et al.* (2009). Narcolepsy is strongly associated with the T-cell receptor alpha locus. *Nat Genet*, **41**: 708–11.

Han, F., Lin, L., Warby, S.C., *et al.* (2011). Narcolepsy onset is seasonal and increased following the 2009 H1N1 pandemic in China. *Ann Neurol*, **70**: 410–17.

Han, F., Lin, L., Li, J., *et al.* (2013). Decreased incidence of childhood narcolepsy 2 years after the 2009 H1N1 winter flu pandemic. *Ann Neurol*, **73**: 560.

Health Canada. (2009). Product information leaflet: Arepanrix H1N1. AS03-adjuvanted H1N1 pandemic influenza vaccine. Available from: http://www.hc-sc.gc.ca/dhp-mps/prodpharma/legislation/interimorders-arretesurgence/prodinfo-vaccin-eng.php [last accessed 11 December 2014].

Huang, W., Ramsey, K.M., Marcheva, B., and Bass, J. (2011). Circadian rhythms, sleep, and metabolism. *J Clin Invest*, **121**: 2133–41.

Käll, A. (2013). The Pandemrix-narcolepsy tragedy: how it started and what we know today. *Acta Paediatr*, **102**: 2–4.

Katzav, A., Arango, M.T., Kivity, S., *et al.* (2013). Passive transfer of narcolepsy: anti-TRIB2 autoantibody positive patient IgG causes hypothalamic orexin neuron loss and sleep attacks in mice. *J Autoimmun*, **45**: 24–30.

Kawashima, M., Lin, L., Tanaka, S., *et al.* (2010). Anti-Tribbles homolog 2 (TRIB2) autoantibodies in narcolepsy are associated with recent onset of cataplexy. *Sleep*, **33**: 869–74.

Kornum, B.R., Faraco, J., and Mignot, E. (2011). Narcolepsy with hypocretin/orexin deficiency, infections and autoimmunity of the brain. *Curr Opin Neurobiol*, **21**: 897–903.

Lim, A.S. and Scammell, T.E. (2010). The trouble with Tribbles: do antibodies against TRIB2 cause narcolepsy? *Sleep*, **33**: 857–8.

Lin, L., Faraco, J., Li, R., *et al.* (1999). The sleep disorder canine narcolepsy is caused by a mutation in the hypocretin (orexin) receptor 2 gene. *Cell*, **98**: 365–76.

Longstreth, W.T., Koepsell, T.D., Ton, T.G., *et al.* (2007). The epidemiology of narcolepsy. *Sleep*, **30**: 13–26.

Mahlios, J., De la Herran-Arita, A.K., and Mignot, E. (2013). The autoimmune basis of narcolepsy. *Curr Opin Neurobiol*, **23**: 767–73.

Masoudi, S., Ploen, D., Kunz, K., and Hildt, E. (2014). The adjuvant component alpha-tocopherol triggers via modulation of Nrf2 the expression and turnover of hypocretin *in vitro* and its implication to the development of narcolepsy. *Vaccine*, **32**: 2980–8.

Melén, K., Partinen, M., Tynell, J., *et al.* (2013). No serological evidence of influenza A H1N1pdm09 virus infection as a contributing factor in childhood narcolepsy after Pandemrix vaccination campaign in Finland. *PLoS One*, **8**: e68402.

Mendes, M.F., de Valladares Neto, C., Azevedo, R.A., and Caramelli, P. (2012). Narcolepsy after A/H1N1 vaccination. *Clinics (Sao Paulo)*, **67**: 77–8.

Mignot, E. (1998). Genetic and familial aspects of narcolepsy. *Neurology*, **50**: S16–22.

Mignot, E., Lin, L., Rogers, W., *et al.* (2001). Complex HLA-DR and -DQ interactions confer risk of narcolepsy-cataplexy in three ethnic groups. *Am J Hum Genet*, **68**: 686–99.

Miller, E., Andrews, N., Stellitano, L., *et al.* (2013). Risk of narcolepsy in children and young people receiving AS03 adjuvanted pandemic A/H1N1 2009 influenza vaccine: retrospective analysis. *BMJ*, **346**: f794.

Morel, S., Didierlaurent, A., Bourguignon, P., *et al.* (2011). Adjuvant System AS03 containing alpha-tocopherol modulates innate immune response and leads to improved adaptive immunity. *Vaccine*, **29**: 2461–73.

Mosca, F., Tritto, E., Muzzi, A., *et al.* (2008). Molecular and cellular signatures of human vaccine adjuvants. *Proc Natl Acad Sci USA*, **105**: 10 501–6.

Nohynek, H., Jokinen, J., Partinen, M., *et al.* (2012). AS03 adjuvanted AH1N1 vaccine associated with an abrupt increase in the incidence of childhood narcolepsy in Finland. *PLoS One*, **7**: e33536.

O'Flanagan, D., Cotter, S., and Mereckiene, J. (2011). Pandemic A(H1N1) 2009 influenza vaccination survey, influenza season 2009/2010. Vaccine European New Integrated Collaboration Effort (VENICE). Available from: http://venice.cineca.org/Final_Report_VENICE_Pandemic_Influenza_2009.pdf [last accessed 11 December 2014].

O'Hagan, D.T., Ott, G.S., De Gregorio, E., and Seubert, A. (2012). The mechanism of action of MF59-an innately attractive adjuvant formulation. *Vaccine*, **30**: 4341–8.

Ohayon, M.M., Ferini-Strambi, L., Plazzi, G., *et al.* (2005). How age influences the expression of narcolepsy. *J Psychosom Res*, **59**: 399–405.

Ohno, K. and Sakurai, T. (2008). Orexin neuronal circuitry: role in the regulation of sleep and wakefulness. *Front Neuroendocrinol*, **29**: 70–87.

Picchioni, D., Hope, C.R., and Harsh, J.R. (2007). A case-control study of the environmental risk factors for narcolepsy. *Neuroepidemiology*, **29**: 185–92.

Rolls, A., Schaich Borg, J., and de Lecea, L. (2010). Sleep and metabolism: role of hypothalamic neuronal circuitry. *Best Pract Res Clin Endocrinol Metab*, **24**: 817–28.

Sakurai, T., Mieda, M., and Tsujino, N. (2010). The orexin system: roles in sleep/wake regulation. *Ann NY Acad Sci*, **1200**: 149–61.

Siebold, C., Hansen, B.E., Wyer, J.R., *et al.* (2004). Crystal structure of HLA-DQ0602 that protects against type 1 diabetes and confers strong susceptibility to narcolepsy. *Proc Natl Acad Sci USA*, **101**: 1999–2004.

Silber, M.H., Krahn, L.E., Olson, E.J., and Pankratz, V.S. (2002). The epidemiology of narcolepsy in Olmsted County, Minnesota: a population-based study. *Sleep*, **25**: 197–202.

Singh, A.K., Mahlios, J., and Mignot, E. (2013). Genetic association, seasonal infections and autoimmune basis of narcolepsy. *J Autoimmun*, **43**: 26–36.

Smith, A., Jackson, M.W., Neufing, P., *et al.* (2004). A functional autoantibody in narcolepsy. *Lancet*, **364**: 2122–4.

Strauss, J. and Strauss, E. (2002). Minus-strand RNA viruses. In: Strauss, J. and Strauss, E. (eds). *Viruses and Human Disease*, 1st edn. San Diego, CA: Academic Press, pp. 147–156.

THL (National Institute for Health and Welfare). (2010). National Narcolepsy Task Force Interim Report 31 January 2011. Helsinki: National Institute for Health and Welfare.

Tsai, T., Crucitti, A., Nacci, P., *et al.* (2011). Explorations of clinical trials and pharmacovigilance databases of MF59(R)-adjuvanted influenza vaccines for associated cases of narcolepsy. *Scand J Infect Dis*, **43**: 702–6.

Tsai, T., Del Giudice, G., Crucitti, A., *et al.* (2013). Is the adjuvant solely to blame? *BMJ*, **346**: f2375.

Viorritto, E.N., Kureshi, S.A., and Owens, J.A. (2012). Narcolepsy in the pediatric population. *Curr Neurol Neurosci Rep*, **12**: 175–81.

Vrethem, M., Malmgren, K., and Lindh, J. (2012). A patient with both narcolepsy and multiple sclerosis in association with Pandemrix vaccination. *J Neurol Sci*, **321**: 89–91.

Waldenlind, L., Grundmark, B., Azarbayjani, F., *et al.* (2013). Is the risk of narcolepsy also increased with non-adjuvanted flu vaccines? *BMJ*, **346**: f2769.

Walker, W.T., de Whalley, P., Andrews, N., *et al.* (2012). H1N1 antibody persistence 1 year after immunization with an adjuvanted or whole-virion pandemic vaccine and immunogenicity and reactogenicity of subsequent seasonal influenza vaccine: a multicenter follow-on study. *Clin Infect Dis*, **54**: 661–9.

Wijnans, L., Lecomte, C., de Vries, C., *et al.* (2013). The incidence of narcolepsy in Europe: before, during, and after the influenza A(H1N1)pdm09 pandemic and vaccination campaigns. *Vaccine*, **31**: 1246–54.

Yokogawa, T., Marin, W., Faraco, J., *et al.* (2007). Characterization of sleep in zebrafish and insomnia in hypocretin receptor mutants. *PLoS Biol*, **5**: e277.

Zhang, Y., Davis, J.L., and Li, W. (2005). Identification of Tribbles homolog 2 as an autoantigen in autoimmune uveitis by phage display. *Mol Immunol*, **42**: 1275–81.

Zhang, X., Penzel, T., and Han, F. (2013). Increased incidence of narcolepsy following the 2009 H1N1 pandemic. *Somnologie*, **17**: 90–3.

31 Non-nutritional Environmental Factors Associated with Celiac Disease: Infections and Vaccinations

Aaron Lerner
Pediatric Gastroenterology and Nutrition Unit, Carmel Medical Center, B. Rappaport School of Medicine, Technion – Israel Institute of Technology, Haifa, Israel

Introduction

Gluten-induced entertopathy, gluten-sensitive enteropathy, or, more commonly, celiac disease (CD), is a lifelong autoimmune condition (Lerner *et al.*, 1996) mainly of the gastrointestinal tract, affecting the small intestine of genetically susceptible individuals. It is one of the most common chronic conditions affecting humanity. Gluten (the storage protein of wheat) and its alcohol-soluble gliadins are the offending inducers of the disease, together with structurally related molecules found in barley, rye, and, in lesser amounts, oat. As with other autoimmune conditions, additional environmental factors, such as infections and vaccinations, might play a role in the induction of CD (Reif and Lerner, 2004a,b). Tissue transglutaminase (tTG) is the autoantigen against which the abnormal immune response is directed (Reif and Lerner, 2004a,b). Two autoantibodies – anti tTG and antiendomysium – are the most abundant serological markers in CD screening (Shamir *et al.*, 2002), although two additional autoantibodies, antideaminated gliadin peptide and antineoepitope tTG, have recently been found to also be reliable (Rozenberg *et al.*, 2012). The HLA-DQ2 and HLA-DQ8 molecules are the most important predisposing genetic factors known to date, among more than 50 additional susceptible genes described by genome-wide association studies.

Much is known about the pathophysiology of the disease. The sequential chain of events leading to its induction was recently unraveled, giving hope for future therapeutic strategies (Lerner, 2010). Moreover, its epidemiology, prevalence, and clinical presentation are constantly changing, and new clinical presentations are frequently being discovered (Lerner, 1994). In fact, the classical gastrointestinal clinical picture is disappearing and extraintestinal presentations are emerging. Neurological, cutaneous, endocrine, hepatic, skeletal, hematological, oncological, gynecological, infertility, dental, behavioral, and psychiatric abnormalities are constantly being described (Branski *et al.*, 1992; Zelnik *et al.*, 2004; Lerner, 2012; Lerner *et al.*, 2012). With growing awareness among family practitioners, hematologists, dermatologists, endocrinologists, psychologists, and gastroenterologists, and now gynecologists and neurologists, the diagnosis of CD is increasingly being made throughout the lifespan. In fact, about 20% of newly diagnosed cases occur in patients who are older than 60 years. A growing domain is the association of CD with a plethora of other autoimmune diseases, many of which involve hypercoagula-

Vaccines and Autoimmunity, First Edition. Edited by Yehuda Shoenfeld, Nancy Agmon-Levin, and Lucija Tomljenovic.

bility, resulting in increased incidence of thromboembolic phenomenon (Lerner *et al.*, 2013).

Infections and CD

Recently, the concept of the infectome was introduced as a means of studying the infectious factors which contribute to the development of autoimmune diseases (Bogdanos *et al.*, 2013). It is part of the exposome, and collates all environmental factors making up the burden that leads to loss of the adaptive mechanism in the body. The infectious burden in autoimmunity includes multiple triggers (bacteria, viruses, parasites, fungi) associated with a particular condition. Recently, in a pivotal study, Fumagalli *et al.* (2011) proved that the signatures of environmental genetic adaptation pinpoint pathogens as the main selective pressure on human evolution. They found the diversity of the local environment to be the predominant driver of local adaptation in some autoimmune diseases, including CD.

The past distinction between genetic and environmental influences in the evolution of autoimmunity was somewhat artificial, and some susceptibility alleles for autoimmune diseases may be selected by pathogens. CD, being a classical autoimmune disease (Lerner *et al.*, 1996), presents multiple aspects of the infectome/autoimmunity association (Table 31.1). Most of the pathgens associated with CD presence or induction are controversial, and a clear cause–effect association is not clear. Table 31.2 summarizes these infectious agents and provides references in support of or against their association.

An exception to the infectome/CD association is the use of hookworm as a therapeutic strategy in CD. It is hypothesized that the parasite induces TH2 and IL-10 crossregulation of the TH1/TH17 inflammatory response, thus suppressing mucosal inflammation in CD (McSorley *et al.*, 2011).

A survey of the association between past infections and CD proves controversial. Neonatal infections have been associated with the risk of developing CD (Sandberg-Bennich *et al.*, 2002), although this observation was not replicated in a later study (Marild *et al.*, 2011). Repeated infectious episodes early in life have been found to increase the risk for later CD (Myleus *et al.*, 2012a,2012b), although, again, no such an association was found in a more recent study (Welander *et al.*, 2010). Since lower economical status is related to increased infectious load, one might predict higher autoimmune conditions in low-income environments. In fact, several studies suggest, on the contrary, a protective effect against CD (Kondrashova *et al.*, 2008; Plot and Amital, 2009; Plot *et al.*, 2009).

Table 31.1 Associations between the infectome and celiac disease (CD)

Association	References
Early infection and increased incidence of CD	Myleus *et al.* (2012a,b)
Antimicrobial serology incidence increased in CD	Ashorn *et al.* (2009), Papp *et al.* (2009)
Parallelism between pathogens and gluten peptides	Bethune and Khosla (2008)
Positive association between antibiotic use and subsequent CD	Marild *et al.* (2013)
Diversity and composition of microbiota in CD	Sánchez *et al.* (2011), Pozo-Rubio *et al.* (2013), Wacklin (2013)
Tissue transglutaminase antibody positivity in confirmed viral infections	Sarmiento *et al.* (2012)
Specific infectious agent related to CD	See Table 31.2

Several explanations have been suggested for the influence of pathogens on CD occurence or induction: molecular mimicry, increased mucosal permeability, bystander activation, and proinflammatory cytokine release. Most recently, the environmental factor retinoic acid was suggested to act as an adjuvant that promotes, rather than prevents, inflammatory cellular and humoral responses to fed antigen, in the presence of IL-15 (dePaolo *et al.*, 2011). This finding unveils an unexpected role for retinoic acid and IL-15 in the abrogation of tolerance to dietary antigens, as in CD. The actual pathogenic pathway by which pathgens induce autoimmunity is far from being elucidated.

ASIA

Autoimmune/inflammation syndrome induced by adjuvants (ASIA) is an entirely new syndrome that comprises a growing list of conditions in which various adjuvants induce autoimmunity/autoinflammation (Shoenfeld and Agmon-levin, 2011; Perricone *et al.*, 2013). Among these are the post-vaccination phenomena, which are linked by previous exposure to an

Table 31.2 Pathogens associated with celiac disease: pros and cons

Pathogen	Positive references	Negative references
Saccharomyces cerevisiae, pseudomonas, bacteroides caccae	Ashorn *et al.* (2009)	
Yeast (antiglycans)	Papp *et al.* (2009)	
Bacteroides species	Sánchez *et al.* (2011), Pozo-Rubio *et al.* (2013)	
Enterovirus, Epstein–Barr virus, cytomegalovirus, hepatitis C virus	Plot and Amital (2009), Plot *et al.* (2009), Sarmiento *et al.* (2012), Marconcini *et al.* (2013)	Lawler *et al.* (1994), Carlsson *et al.* (2002), Gravina *et al.* (2012)
Rotavirus	Stene *et al.* (2006), Dolcino *et al.* (2013)	
Adenovirus	Reif and Lerner (2004a,2004b), Plot and Amital (2009), Plot *et al.* (2009)	Lawler *et al.* (1994), Carlsson *et al.* (2002)
Campylobacter jejuni	Sabayan *et al.* (2007), Verdu *et al.* (2007), Riddle *et al.* (2013)	
Pneumococcus	Ludvigsson *et al.* (2008)	
Toxoplasma gondii	Rostami-Nejad *et al.* (2011a,b)	
Tuberculosis	Ludvigsson *et al.* (2011)	
Hepatitis B virus	Iglesias *et al.* (2010)	Ouakaa-Kchaou *et al.* (2010)
Helicobacter pylori		Aydogdu *et al.* (2008), Rostami-Nejad *et al.* (2011a,b)

adjuvant, resulting in similar clinical manifestations. Alum, used as a "safe" adjuvant for 90 years, is now suspected to play a role in ASIA. Several specific mechanisms relating to the adjuvanicity of aluminum compounds have recently been suggested (Lerner, 2007, 2012; Lerner *et al.*, 2012; Perricone *et al.* 2013).

Vaccination and CD

Numerous publications indicate that vaccination may trigger or worsen autoimmune diseases, as suggested by a temporal relationship between vaccination and disease induction/exacerbation (Tishler and Shoenfeld, 2004). Several mechanisms have been suggested, including release of proinflammatory cytokines during vaccine antigen introduction, induction of the Th1/Th17 proinflammatory immune pathways, molecular mimicry, modulation of autoimmune/inflammation involved genes by antipathogen antibodies, antibody shift, and so on.

Vaccine safety was checked in patients with chronic rheumatic or autoimmune diseases and no evidence of exacerbation/induction of autoimmunity was found in the majority of them (Rubinstein, 2004; Gluck and Muller-Ladner, 2008). Exceptional autoimmune manifestations included: arthritis, vasculitis, encephalitis, neuropathy, and multiple sclerosis (MS) following hepatitis B virus (HBV) vaccine; reactive arthritis following measles, mumps, and rubella (MMR) vaccine; Guillian–Barré syndrome (GBS) following influenza vaccine; and various neurological disorders following varicella vaccine. The debate over whether HBV vaccination is associated with increased risk for development of demyelinating central nervous system (CNS) disorders continues (Gluck and Muller-Ladner, 2008). It is important to mention that early vaccinations are not risk factors for CD; however, rotavirus vaccine was not included in the study (Myleus *et al.*, 2012a,b).

Rotavirus and CD

Stene *et al.* (2006) provided the first indication that a high frequency of rotavirus infections might increase the risk of CD autoimmunity in childhood in genetically predisposed individuals. In Iranian adults with positive celiac serology, however, the prevalence of active rotavirus was not significant (Rostami-Nejad *et al.*, 2010). Using a random peptide library approach, an Italian team identified a peptide (celiac peptide) in serum immunoglobulins of patients with CD (Zanoni *et al.*, 2006). This peptide shares homology with the rotavirus major neutralizing protein VP7 and with the CD autoantigen tTG. Surprisingly, antibodies directed against the celiac peptide, purified from the sera of pateints with active disease, recognized VP7. The same group, following their original observations, also demonstrated that, in active CD, a subset of anti-tTG immunoglobin A (IgA) antibodies recognize VP7, suggesting a possible involvement of rotavirus infection in CD evolution (Dolcino *et al.*, 2013). Moreover, rotavirus can induce the same mucosal injury that gluten induces in CD, and, as mentioned already, the rotavirus/CD relationship is supported by an epidemiology study (Stene *et al.*, 2006).

Encouraged by their earlier findings, Dolcino *et al.* (2013) observed that antibodies directed against VP7 predict the onset of CD and induce typical features of CD in the intestinal epithelial cell line T84. Exploring the pathogenic pathways involved using gene-array analysis, the same group showed that these antibodies modulate genes involved in apoptosis, inflammation, and epithelial barrier integrity – all typical features of CD. Taken together, these new data further support the involvement of rotavirus infection in CD pathgenesis and suggest a predictive role for antirotavirus VP7 antibodies.

An additional aspect of the interrelationship between rotavirus and CD is the effect of the former on human cellular immunity and gene expression. Comparing the patterns of gene expression in peripheral blood mononuclear cells from children affected by rotavirus diarrhea versus healthy children, the former have increased expression of genes involved in B cell differentiation, maturation, activation, and survival and a lower proportion of $CD4^+$ and $CD8^+$ T cells in the periphery, suggesting that rotavirus alters B and T cell behavior and homeostasis (Wang *et al.*, 2007). Recently, Pozo-Rubio *et al.* (2013) took a step forward by documenting the influence of rotavirus vaccine on the balance between T and B lymphocytes, finding that the B cell percentage was higher in vaccinated infants.

The involvement of rotavirus infection in the induction of autoimmunity and potential long-term effects of rotavirus vaccine have been shown in type 1 diabetes. Rotavirus infection in children genetically at risk for type 1 diabetes is associated with increased islet autoantibody levels and has been proposed to accelerate progression to diabetes (Lempainen *et al.*, 2012). Rhesus monkey rotavirus infection in adult nonobese diabetic (NOD) mice induced early diabetes onset, but this did not involve pancreatic infection (Graham *et al.*, 2008). Rather, this diabetes acceleration was associated with a Th1-biased antibody and cytokine response (Pane *et al.*, 2013). Pane *et al.* (2013) recently showed that rotavirus-stimulated splenocytes exhibited dose-dependent antigen-presenting cell (APC) and B cell activation that was independent of virus replication or strain. This activation was associated with interferon alpha (IFN-α) secretion and was prevented by rotavirus treatment with VP7-neutralizing antibody, inhibition of endosomal acidification, or TLR7 signaling and blockade of signaling through the interferon alpha/beta receptor (IFNAR). Importantly, dendritic cells (DCs) were shown to contribute to B and T lymphocyte activation following viral exposure. It was further demonstrated that rotavirus induces the activation of islet autoreactive T cells. These data provide evidence that bystander activation may be an important mechanism for lymphocyte activation during Rhesus monkey rotavirus-mediated diabetes acceleration in NOD mice (Pane *et al.*, 2014).

Conclusions

CD is an autoimmune disease induced by well-known nutritional environmental factors (nondietary factors are less studied and less well established). Several pathogens are associated with CD, but in none of them have cause–effect associations been established. Evidence is accumulating for a possible role of rotavirus in CD pathogenesis. The rotavirus VP7 shares homology with a celiac peptide and with the autoantigen tTg. Anti-VP7 antibodies are predictive for CD and modulate genes involved in CD pathophysiology. In view of the role of rotavirus in type 1 diabetes induction, the increased incidence of type 1 diabetes in CD patients, and the relationship between rotavirus, gliadin, and CD, the enigma of the rotavirus vaccine as an inducer of CD is awaiting further exploration. In fact, in a very recent publication, Parez *et al.* (2014) indicate potential safety concerns around rotavirus vaccination in Europe.

References

Aydogdu, S., Cakir, M., Yuksekkaya, H.A., *et al.* (2008). *Helicobacter pylori* infection in children with celiac disease. *Scand J Gastroenterol*, **43**(9): 1088–93.

Ashorn, S., Valineva, T., Kaukinen, K., *et al.* (2009). Serological responses to microbial antigens in celiac disease pateints during a gluten-free diet. *J Clin Immunol*, **29**(2): 190–5.

Bethune, M.T. and Khosla, C. (2008). Parallels between pathogens and gluten peptides in celiac sprue. *PLoS Pathog*, **4**(2): e34.

Bogdanos, D.P., Smyk, D.S., Invernizzi, P., *et al.* (2013). Tracing environmental markers of autoimmunity: introducing the infectome. *Immunol Res*, **56**: 220–40.

Branski, D., Ashkenazy, A., Frier, S., *et al.* (1992). Extra intestinal manifestation and associated disorders of celiac disease. In: Branski, D., Rozen, P., and Kaganoff, M.F. (ed.). *Gluten-Sensitive Enteropathy (Frontiers of Gastrointestinal Research)*. Basel: Karger, pp. 164–75.

Carlsson, A.K., Lindberg, B.A., Bredberg, A.C., *et al.* (2002). Enterovirus infection during pregnancy is not a risk factor for celiac disease in the offspring. *J Pediatr Gastroenterol Nutr*, **35**: 649–52.

dePaolo, R.W., Abadie, V., Tang, F., *et al.* (2011). Co-adjuvant effects of retinoic acid and IL-15 induce inflammatory immunity to dietary antigens. *Nature*, **471**(7337): 220–4.

Dolcino, M., Zanoni, G., Bason, C., *et al.* (2013). A subset of anti-rotavirus antibodies directed against the viral protein VP7 predicts the onset of celiac disease and induces typical features of the disease in the intestinal epithelial cell line T84. *Immunol Res*, **56**: 465–76.

Fumagalli, M., Sironi, M., Pozzoli, U., *et al.* (2011). Signatures of environmental genetic adaptation pinpoint pathogens as the main selective pressure through human evolution. *PLoS Genet*, **7**: e1002355.

Gluck, T. and Muller-Ladner, U. (2008). Vaccination in pateints with chronic rheumatic or autoimmune diseases. *Clin Infect Dis*, **46**: 1459–65.

Graham, K.L., Sanders, N., Tan, Y., *et al.* (2008). Rotavirus infection accelerates type 1 diabetes in mice with established insulitis. *J Virol*, **82**: 6139–49.

Gravina, A.G., Federico, A., Masarone, M., *et al.* (2012). Coeliac disease and C virus-related chronic hepatitis: a non-association. *BMC Res Notes*, **5**: 533–7.

Iglesias, S., Vazquer Rodrigues, S., Ulla Rocha, J.L., *et al.* (2010). Onset of celiac disease after acute hepatitis B infection. *Gastroenterol Hepatol*, **33**: 17–20.

Kondrashova, A., Mustalahti, K., Kaukinen, K., *et al.* (2008). Lower economic status and inferior environment may protect against celiac disease. *Ann Med*, **40**(3): 223–31.

Lawler, M., Humpheries, P.O., and Farrelly, C. (1994). Adenovirus 12 E1A gene detection by polymerase chain reaction in both normal and celiac duodenum. *Gut*, **35**: 1226–32.

Lempainen, J., Tauriainen, S., Vaarala, O., *et al.* (2012). Interaction of enterovirus infection and cow's milk-based formula nutrition in type 1 diabetes-associated autoimmunity. *Diabetes Metab Res Rev*, **28**: 177–85.

Lerner, A. (1994). Factors affecting the clinical presentation and time diagnosis of celiac disease: the Jerusalem and the West Bank-Gaza experience [editorial]. *Isr J Med Sci*, **30**: 294–5.

Lerner, A. (2007). Extended hypothesis: aluminum is a potential environmental factor for Crohn's disease induction. *Ann NY Acad Sci*, **1107**: 329–45.

Lerner, A. (2010). New therapeutic strategies for celiac disease. *Autoimmun Rev*, **9**: 144–7.

Lerner, A. (2012). Aluminum as an adjuvant in Crohn's disease induction. *Lupus*, **21**: 231–8.

Lerner, A., Blank, M., and Shoenfeld, Y. (1996). Celiac disease and autoimmunity. *Isr J Med Sci*, **32**: 33–6.

Lerner, A., Makhoul, B.F., and Eliakim, R. (2012). Neurological manifestations of celiac disease in children and adults. *Europ Neurolog J*, **4**: 15–20.

Lerner, A., Agmon-Levin, N., Shapira, Y., *et al.* (2013). The thrombophylic network of auto antibodies in celiac disease. *BMC Med*, **11**: 89–95.

Ludvigsson, J.F., Olen, O., Bell, M., *et al.* (2008). Coeliac disease and risk of sepsis. *Gut*, **57**(8): 1074–80.

Ludvigsson, J.F., Sanders, D.S., Maeurer, M., *et al.* (2011). Risk of tuberculosis in a large sample of patients with celiac disease – a nationwide cohort study. *Aliment Pharmacol Ther*, **33**(6): 689–96.

Marconcini, M.L., Fayad, L., Shiozawa, M.B., *et al.* (2013). Autoantibody profile in individuals with chronic hepatitis C. *Revista da Sociedade Brasilrira de Medicina*, **46**(2): 147–53.

Marild, K., Stephansson, O., Montgomery, S., *et al.* (2011). Pregnancy outcome and risk of celiac disease in offspring; a nationwide case-control study. *Gastroenterology*, **142**: 39–45.

Marild, K., Ye, W., Lebwohl, B., *et al.* (2013). Antibiotic exposure and the development of coeliac disease: a nationwide case-control study. *BMC Gastroenterol*, **13**: 109–17.

McSorley, H.J., Gaze, S., Daveson, J., *et al.* (2011). Suppression of inflammatory immune responses in celiac disease by experimental hhokworm infection. *PLoS One*, **6**(9): e24092.

Myleus, A., Stenlund, H., Hernell, O., *et al.* (2012a). Early vaccinations are not risk factors for celiac disease. *Pediatrics*, **130**: e63–70.

Myleus, A., Hernell, O., Gothefors, L., *et al.* (2012b). Early infections are associated with increased risk for celiac disease: an incident case-referent study. *BMC Pediatr*, **12**: 194–201.

Ouakaa-Kchaou, A., Gargouri, D., Kharrat, J., and Ghorbel, A. (2010). Relationship between hepatitis B virus infection and celiac disease. *Hepat Mon*, **10**: 313–14.

Pane, J.A., Webster, N.L., and Coulson, B.S. (2014). Rotavirus activates lymphocytes from non-obese diabetic mice by triggering toll-like receptor 7 signaling and interferon production in plasmacytoid dendritic cells. *PLoS Pathog*, **10**: e1003998.

Pane, J.A., Webster, N.L., Graham, K.L., *et al.* (2013). Rotavirus acceleration of murine type 1 diabetes is associated with a T helper 1-dependent specific serum antibody response and virus effects in regional lymph nodes. *Diabetologia*, **56**: 573–82.

Papp, M., Foldi, I., Altorjay, I., *et al.* (2009). Anti-microbial antibodies in celiac disease: trick or treat. *World J Gastroenterol*, **15**(31): 3891–900.

Parez, N., Giaquinto, C., Du Roure, C., *et al.* (2014). Rotavirus vaccination in Europe: drivers and barriers. *Lancet Infect Dis*, **14**: 416–25.

Perricone, C., Agmon-Levin, N., and Shoenfeld, Y. (2013). Novel pebbles in the mosaic of autoimmunity. *BMC Med*, **11**: 101–8.

Plot, L. and Amital, H. (2009). Infectious associations of celiac disease. *Autoimmun Rev*, **8**(4): 316–19.

Plot, L., Amital, H., Barzilai, O., *et al.* (2009). Infections may have a protective role in the etiopathogenesis of celiac disease. *Ann NY Acad Sci*, **1173**: 670–4.

Pozo-Rubio, T., de Palma, G., Mujico, J.R., *et al.* (2013). Influence of early environmental factors on lymphocyte subsets and gut microbiota in infants at risk of celiac disease; the PROFICEL study. *Nutr Hosp*, **28**(2): 464–73.

Reif, S. and Lerner, A. (2004a). Celiac disease and infection. In: Shoenfeld, Y. and Rose, N. (eds). *Infections and Autoimmunity*. Amsterdam: Elsevier BV, pp. 689–92.

Reif, S. and Lerner, A. (2004b). Tissue transglutaminase – the key player in celiac disease: a review. *Autoimmun Rev*, **3**: 40–5.

Riddle, M.S., Murray, J.A., Cash, B.D., *et al.* (2013). Pathogen-specific risk of celiac disease following bacterial causes of foodborne illness: a retrospective cohort study. *Dig Dis Sci*, **58**(11): 3242–5.

Rostami-Nejad, M., Rostami, K., Sanaei, M., *et al.* (2010). Rotavirus and celiac autoimmunity among adults with non-specific gastrointestinal symptoms. *Saudi Med J*, **31**: 891–4.

Rostami-Nejad, M., Rostami, K., Cheraghipour, K., *et al.* (2011a). Celiac disease increases the risk of *Toxoplasma gondii* infection in a large cohort of pregnant women. *Am J Gastroenterol*, **106**(3): 548–9.

Rostami-Nejad, M., Rostami, K., Yamaoka, Y., *et al.* (2011b). Clinical and histological presentation of *Helicobacter pylori* and gluten related gastroenteropathy. *Arch Iran Med*, **14**(2): 115–18.

Rozenberg, O., Lerner, A., Pacht, A., *et al.* (2012). A novel algorithm for childhood celiac disease serological diagnosis based upon intestinal biopsies. *Crit Rev Allerg Immunol*, **42**: 331–41.

Rubinstein, E. (2004). Vaccination and autoimmune diseases: the argument against. *IMAJ*, **6**: 433–5.

Sabayan, B., Foroughinia, F., and Imanieh, M.H. (2007). Can *Campylobacter jejuni* play a role in development of celiac disease? A hypothesis. *World J Gastroenterol*, **13**(35): 4784–5.

Sánchez, E., De Palma, G., Capilla, A., *et al.* (2011). Influence of environmental and genetic factors linked to celiac disease risk on infant gut colonization by *Bacteroides* species. *Appl Environ Microbiol*, **77**(15): 5316–23.

Sandberg-Bennich, S., Dahlquist, G., and Kallen, B. (2002). Coeliac disease is associated with intrauterine growth and neonatal infections. *Acta Paediatr*, **91**(1): 30–3.

Sarmiento, L., Galvan, J.A., Cabrera-Rode, E., *et al.* (2012). Type 1 diabetes associated and tissue transglutaminase autoantibodies in patients without type 1 diabetes and celiac disease with confirmed viral infections. *J Med Virol*, **84**: 1049–53.

Shamir, R., Eliakim, R., Lahat, N., *et al.* (2002). ELISA assay of antiendomysial antibodies in the diagnosis of celiac disease: comparison with immunofluorescence assay of anti endomysial antibodies and tissue transglutaminase antibodies. *IMAJ*, **4**: 594–6.

Shoenfeld, Y. and Agmon-levin, N. (2011). "ASIA" – autoimmune/inflammatory syndrome induced by adjuvants. *J Autoimmun*, **36**: 4–8.

Stene, L.C., Honeyman, M.C., Hoffenberg, E.J., *et al.* (2006). Rotavirus infection frequency and risk of celiac disease autoimmunity in early childhood: a longitudinal study. *Am J Gastroenerol*, **101**(10): 2333–40.

Tishler, M. and Shoenfeld, Y. (2004). Vaccination may be associated with autoimmune diseases. *IMAJ*, **6**: 430–2.

Verdu, E.F., Mauro, M., Bourgeois, J., and Armstrong, D. (2007). Clinical onset of celiac disease after an episode of *Campylobacter jejuni* enteritis. *Can J Gastroenterol*, **21**: 453–5.

Wacklin, P., Kaukinen, K., Tuovinen, E., *et al.* (2013). The duodenal microbiota composition of adult celiac disease patients is associated with the clinical manifestation of the disease. *Inflamm Bowel Dis*, **19**(5): 934–41.

Wang, Y., Dennehy, P.H., Keyserling, H.I., *et al.* (2007). Rotavirus infection alters peripheral T-cell homeostasis in children with acute diarrhea. *J Virol*, **81**: 3904–12.

Welander, A., Tjernberg, A.R., Montgomery, S.M., *et al.* (2010). Infectious disease and risk of later celiac disease in childhood. *Pediatr*, **125**: e530–6.

Zanoni, G., Navone, R., Lunardi, C., *et al.* (2006). In celiac disease, a subset of autoantibodies against transglutaminase binds toll-like receptor 4 and induces activation of monocytes. *PLoS Med*, **3**: e358.

Zelnik, N., Pacht, A., Obeid, R., and Lerner, A. (2004). Range of neurological disorders in patients with celiac disease. *Pediatrics*, **113**: 1672–6.

32 Polymyalgia Rheumatica

Alessandra Soriano[1,2] and Raffaele Manna[3]

[1]Zabludowicz Center for Autoimmune Diseases, Sheba Medical Center, Tel Hashomer, Israel

[2]Department of Clinical Medicine and Rheumatology, Campus Bio-Medico University, Rome, Italy

[3]Periodic Fevers Research Center, Department of Internal Medicine, Catholic University of the Sacred Heart, Rome, Italy

Introduction

Polymyalgia rheumatica (PMR) is an inflammatory rheumatic disorder characterized by pain and morning stiffness of the shoulders and pelvic girdles, alongside evidence of synovitis of the proximal joints and extra-articular synovial structures and laboratory findings of raised inflammatory markers. It is the most common inflammatory musculoskeletal disorder in older people, with an age-adjusted incidence of about 1 in 1000 person years (Mackie and Mallen, 2013). It may occur in "isolated" form or in association with giant cell arteritis (GCA).

The etiopathogenesis of PMR remains unknown, although significant steps have been made in our knowledge of its mechanisms of inflammation in recent years, supporting the essential role of both environmental and genetic factors (Kermani and Warrington, 2013). An association between PMR and specific polymorphisms in genes related to immune regulation has been described: genetic polymorphisms associated with disease risk or severity include intercellular adhesion molecule 1 (ICAM1), interleukin (IL-) 1 receptor antagonist, and IL-6 (González-Gay *et al.*, 1999; Boiardi *et al.*, 2006; Alvarez-Rodriguez *et al.*, 2009). Seasonal variations in incidence and differences in the geographical distribution of PMR within the same country have suggested possible environmental triggers for the disease. Nonetheless, though a close temporal relationship between epidemics of *Mycoplasma pneumoniae*, *Chlamydia pneumonia*, and *Parvovirus B19* and peaks of cases of PMR and GCA has been reported, no clear relationship between onset of PMR and viral or bacterial infections has been ascertained to date (Salvarani *et al.*, 1995; Elling *et al.*, 1996; González-Gay *et al.*, 2002).

Despite an improvement in diagnostic tools (e.g. ultrasound, magnetic resonance, fluorodeoxyglucose PET), the diagnosis of PMR remains clinical, and is determined through a combination of clinical symptoms, increased acute phase reactants, exclusion of other conditions, and response to glucocorticoids (Salvarani and Cantini, 2008). Several diagnostic and classification criteria have been formulated since 1979, based on clinical experience and collaborative initiatives (Bird *et al.*, 1979; Jones and Hazleman, 1981; Chuang *et al.*, 1982; Healey, 1984; Dasgupta *et al.*, 2012). In some cases, the diagnosis of PMR can be challenging, as several conditions may mimic the disease, including late-onset rheumatoid arthritis (RA), spondyloarthritides, vasculitides, inflammatory myopathies, and malignancies. In addition, because of the possible atypical onset of the disease, some cases may be misdiagnosed.

As previously stated, PMR is a disease of the elderly, its onset occurring mainly in subjects over the age of 70 and rarely in those under 50. Elderly people are exposed to an increased risk for severe complications from influenza infection, especially if concomitant pulmonary, cardiovascular, or metabolic diseases are present. Thus, they are recommended to undergo seasonal

Vaccines and Autoimmunity, First Edition. Edited by Yehuda Shoenfeld, Nancy Agmon-Levin, and Lucija Tomljenovic.

influenza vaccination, according to the current recommendations of the Advisory Committee on Immunization Practice (CDC, 2013).

Arthralgia, myalgia, and low-grade fever are well-known transient, mild, post-vaccinal effects that may be experienced immediately after exposure to several types of vaccines, including seasonal influenza vaccine (Inf-V). Usually, these symptoms are self-limited and well discernible from the typical findings of other musculoskeletal disorders, such as PMR. Cases of PMR following vaccine immunization are rarely described in the literature. It has to be considered that post-vaccinal PMR might be underestimated, due to the sometimes atypical clinical manifestations at onset, which may be interpreted as transient vaccine-related effects. Finally, it should be taken into account that any temporal and causal relationship between immunization and PMR onset might not be detected if an accurate clinical history – including the vaccine history – is not performed.

PMR following vaccination: evidence from the literature

A systematic literature survey of reports on PMR occurring after immunization was performed through PubMed/Medline, without any date limitation, mainly using the search term "polymyalgia rheumatica (medical subject headings (MeSH))," relating to the terms "vaccination (MeSH)" and "immunization (MeSH)." The search found 11 cases in a period of over 20 years, of which, three were associated with a concomitant GCA (see Table 32.1). Eight of the subjects were women. Patient age at time of diagnosis ranged between 61 and 91 years, with a median of 70.72. Interestingly, all described cases but one followed seasonal Inf-V. The time interval between vaccine administration and symptom onset varied between 1 day and 3 months.

The first description of a case of PMR following immunization with Inf-V dates back to 1992, when Gerth (1992) described a woman experiencing a PMR relapse some weeks after receiving Inf-V and 7 years after the diagnosis of PMR, which was already in remission. The search for a genetic predisposing background in patients experiencing PMR following immunizations was first conducted by Perez and Maravi (2000), who performed human leukocyte antigen (HLA) typing in a 61-year-old woman with a PMR onset 2 weeks following seasonal Inf-V (Inflexal Berna, 1998–99). The authors found the allele DRB1*0404, which was also found in the case described by Marti and Anton (2004).

Table 32.1 Clinical findings of cases of PMR following vaccination – isolated or in association with giant cell arteritis GCA – reported in the literature. The brands of administered vaccines and HLA typing are recorded where available. Adapted from Soriano *et al.* (2012b)

Case no.	Sex, age at time of disease onset (years)	Disease	Type of vaccination	Interval between vaccination and disease onset	HLA typing	References
1	F, 67	PMR relapse	Inf-V	2–3 weeks	NA	Gerth (1992)
2	M, 65	PMR	Inf-V (Fluvax '92)	3 weeks	NA	Brown and Bertouch (1994)
3	M, 61	PMR	Inf-V (Inflexal Berna 1998–99)	2 weeks	DRB1*0404	Perez and Maravi (2000)
4	F, 91	PMR	Inf-V (Fluarix)	1 day	NA	Liozon *et al.* (2000)
5	F, 64	PMR/GCA	Inf-V	3 days	NA	Saadoun *et al.* (2001)
6	F, 68	PMR	Tetanus	Few days	NA	Saadoun *et al.* (2001)
7	M, 79	PMR	Inf-V (Evagrip)	1 week	DRB1*0401	Marti and Anton (2004)
8	F, 80	PMR/GCA	Inf-V	3 months	NA	Soriano *et al.* (2012b)
9	F, 64	PMR	Inf-V	1 month	DR11,DR15	Soriano *et al.* (2012b)
10	F, 78	PMR/GCA	Inf-V	3 weeks	NA	Soriano *et al.* (2012b)
11	F, 61	PMR PMR relapse (2 years later)	Inf-V	2 months	NA	Soriano *et al.* (2012b)

GCA, giant cell arteritis; PMR, polymyalgia rheumatica; NA, not available

The role of the HLA–DRB1 locus in genetic patterns of PMR has been widely investigated, both in the "pure" form and in association with GCA (Dababneh *et al.*, 1998; Bartolome *et al.*, 2001). In 1998, the association between isolated PMR and HLA-DRB1*13/14 (and previously DR6) was demonstrated in a series of 86 patients; the frequency of HLA-DRB1*0401 and *0404 alleles was only marginally increased, however (Dababneh *et al.*, 1998). In general, it has been observed that the distribution of DRB1*04 is more common in GCA than in PMR (Bartolome *et al.*, 2001). Nevertheless, relapses of PMR in isolated PMR patients have been more frequently observed in those carrying HLA-DRB1*04 alleles, particularly *0401 (Gonzáles-Gay *et al.*, 1999). Moreover, both HLA-DRB1*0401 and homozygosity of the ICAM-1 codon 241 GG were significantly associated with increased risk of relapse in a series of isolated PMR patients from northwestern Spain (González-Gay *et al.*, 1999).

In specific regard to relapse cases of PMR following vaccine, and apart from the previous case described by Gerth (1992), in our literature survey, Saadoun *et al.* (2001) reported the case of a 68-year-old female patient with a previous diagnosis of PMR, clinically recovered after steroid therapy, who experienced a relapse of her disease 4 years later, when she underwent tetanus vaccination. In the case-series of GCA and PMR following Inf-V, Soriano *et al.* (2012b) described the case of a 71-year-old woman who was diagnosed with PMR following Inf-V and was in clinical remission after steroid therapy, who experienced a relapse of her disease after 2 years, following revaccination with seasonal Inf-V. As both the first onset of the disease and the PMR relapse followed vaccine administration (Inf-V), this case can be defined as a "rechallenge." Unfortunately, HLA typing was not available for this case.

Rechallenge, relapse, and exacerbation cases of autoimmune and rheumatic inflammatory disorders following immunizations with vaccines represent an intriguing issue, as they may be considered the clinical "proof of concept" of the possible role of vaccines as "triggers" of autoimmune inflammatory rheumatic disorders. As recently underlined by Shoenfeld and Agmon-Levin (2011) through the description of autoimmune/inflammatory syndrome induced by adjuvants (ASIA), any of the components of a vaccine (microbial antigen, adjuvants, preservatives) might play a role in these processes, (i) in a genetically susceptible subject and (ii) when the time interval between vaccine administration and symptom onset is plausible.

As a matter of fact, the literature survey detected post-vaccinal PMR cases encompassing a period of over 20 years; thus, PMR induced by seasonal Inf-V might not be correlated with the viral specificity of the vaccine, as the viral components included in vaccines vary from year to year, on the basis of the expected type of influenza virus, whereas a restricted number of vaccine adjuvants and preservatives have been used for decades, as they enhance the immune response to the coadministered antigens (Soriano *et al.*, 2012a). Moreover, the triggering effect of another seasonal Inf-V administered years later (as described in the rechallenge and relapse cases) or the administration of other vaccines (such as tetanus vaccination in the case described by Saadoun *et al.*, 2001) provides further support to the independence of the post-vaccination phenomena from viral strain or bacterial antigens in some cases.

Conclusions

Post-vaccinal PMR remains a very rare entity and further efforts are needed to better identify individuals at risk of developing this particular type of disorder.

An accurate clinical history, including vaccine history, is mandatory for all elderly patients fulfilling the diagnostic criteria for PMR, as the rate of many post-vaccine autoimmune and rheumatic disorders – including PMR - may be biased by underreporting. A careful risk–benefit assessment must be performed for patients already diagnosed with PMR who are in clinical remission at the time of clinical evaluation for further immunization.

Undoubtedly, further insights into the pathogenesis of post-vaccination phenomena and the identification of markers of genetic predisposition could be useful in preventing these conditions and in developing personalized and safer vaccines in the future.

References

Alvarez-Rodriguez, L., Carrasco-Marin, E., Lopez-Hoyos, M., *et al.* (2009). Interleukin-1RN gene polymorphisms in elderly patients with rheumatic inflammatory chronic conditions: association of IL-1RN*2/2 genotype with polymyalgia rheumatica. *Hum Immunol*, **70**: 49–54.

Bartolome, M.J., Martinez-Taboada, V.M., Lopez-Hoyos, M., *et al.* (2001). Familial aggregation of polymyalgia rheumatica and giant cell arteritis: genetic and T cell repertoire analysis. *Clin Exp Rheumatol*, **19**: 259–64.

Bird, H.A., Esselinckx, W., Dixon, A.S., *et al.* (1979). An evaluation of criteria for polymyalgia rheumatica. *Ann Rheum Dis*, **38**: 434–9.

Boiardi, L., Casali, B., Farnetti, E., *et al.* (2006). Relationship between interleukin 6 promoter polymorphism at position −174, IL-6 serum levels, and the risk of relapse/recurrence in polymyalgia rheumatica. *J Rheumatol*, **33**: 703–8.

Brown, M.A. and Bertouch, J.V. (1994). Rheumatic complications of influenza vaccination. *Aust NZ J Med*, **24**: 572–3.

CDC (Centers for Disease Control and Prevention). (2013). Prevention and control of seasonal influenza with vaccines. Recommendations of the Advisory Committee on Immunization Practice – United States, 2013–2014. *MMWR*, **62**(RR-07): 1–43.

Chuang, T.Y., Hunder, G.G., Ilstrup, D.M., and Kurland, L.T. (1982). Polymyalgia rheumatica: a 10-year epidemiologic and clinical study. *Ann Intern Med*, **97**: 672–80.

Dababneh, A., González-Gay, M.A., Garcia-Porrua, C., *et al.* (1998). Giant cell arteritis and polymyalgia rheumatica can be differentiated by distinct patterns of HLA class II association. *J Rheumatol*, **25**: 2140–5.

Dasgupta, B., Cimmino, M.A., Kremers, H.M., *et al.* (2012). Provisional classification criteria for polymyalgia rheumatica: a European League Against Rheumatism/American College of Rheumatology collaborative initiative. *Arthritis Rheum*, **64**: 943–54.

Elling, P., Olsson, A.T., and Elling, H. (1996). Synchronous variation of the incidence of temporal arteritis and polymyalgia rheumatic in different regions of Denmark: association with epidemics of Mycoplasma pneumonia infection. *J Rheumatol*, **23**: 112–19.

Gerth, H.J. (1992). [Polymyalgia rheumatica and influenza vaccination]. *Dtsch Med Wochenschr*, **117**(33): 1259–60.

González-Gay, M.A., García-Porrúa, C., Vázquez Caruncho, M., *et al.* (1999). The spectrum of polymyalgia rheumatica in northwestern Spain: incidence and analysis of variables associated with relapse in a ten-year study. *J Rheumatol*, **26**: 1326–32.

González-Gay, M.A., Helweg-Larsen, J., Tarp, B., *et al.* (2002). No evidence of parvovirus B19, Chlamydia pneumonia or human helper infection in temporary biopsies in patients with giant cell arteritis. *Rheumatology (Oxford)*, **41**: 445–9.

Healey, L.A. (1984). Long-term follow-up of polymyalgia rheumatica: evidence for synovitis. *Semin Arthritis Rheum*, **13**: 322–8.

Jones, J.G. and Hazleman, B.L. (1981). Prognosis and management of polymyalgia rheumatica. *Ann Rheum Dis*, **40**: 1–5.

Kermani, T.A. and Warrington, K. (2013). Polymyalgia rheumatica. *Lancet*, **381**: 63–72.

Liozon, E., Ittig, R., Vogt, N., *et al.* (2000). Polymyalgia rheumatica following influenza vaccination. *J Am Geriatr Soc*, **48**: 1533–4.

Mackie, S.L. and Mallen, C.D. (2013). Polymyalgia rheumatica. *BMJ*, **347**: f6937.

Marti, J. and Anton, E. (2004). Polymyalgia rheumatica complicating influenza vaccination. *J Am Geriatr Soc*, **52**: 1412.

Perez, C. and Maravi, E. (2000). Polymyalgia rheumatica following influenza vaccination. *Muscle Nerve*, **23**: 824–5.

Saadoun, D., Cacoub, P., Mahoux, D., *et al.* (2001). Postvaccine vasculitis: a report of three cases. *Rev Med Intern*, **22**: 172–6.

Salvarani, C. and Cantini, F. (2008). Polymyalgia rheumatica and giant cell arteritis. *Lancet*, **372**: 234–45.

Salvarani, C., Gabriel, S.E., O'Fallon, W.M., and Hunder, G.G. (1995). The incidence of giant cell arteritis in Olmsted County, Minnesota: apparent fluctuations in a cyclic pattern. *Ann Intern Med*, **123**: 192–4.

Shoenfeld, Y. and Agmon-Levin, N. (2011). "ASIA" – autoimmune/inflammatory syndrome induced by adjuvants. *J Autoimmun*, **36**: 4–8.

Soriano, A., Landolfi, R., and Manna, R. (2012a). Polymyalgia rheumatica in 2011. *Best Practice Res Clin Rheumatol*, **26**: 91–104.

Soriano, A., Verrecchia, E., Marinaro, A., *et al.* (2012b). Giant cell arteritis and polymyalgia rheumatica after influenza vaccination: report of 10 cases and review of the literature. *Lupus*, **21**: 153–7.

33 Acute Disseminated Encephalomyelitis: Idiopathic, Post-infectious, and Post-vaccination

Dimitrios Karussis and Panayiota Petrou
Department of Neurology, Multiple Sclerosis Center, and Laboratory of Neuroimmunology, The Agnes-Ginges Center for Neurogenetics, Hadassah University Hospital, Jerusalem, Ein Karem, Israel

Definition and diagnostic characteristics

Acute disseminated encephalomyelitis (ADEM) is an inflammatory demyelinating disease of the central nervous system (CNS). It is considered to be a monophasic disease, but recurrent forms are well described in 25–30% of patients (Krupp *et al.*, 2007). ADEM is defined (according to criteria proposed by the International MS Group: Krupp *et al.*, 2007) as a first-ever clinical event with presumed inflammatory or demyelinating cause and an acute or subacute onset that affects multifocal areas of the CNS, is usually polysymptomatic, and includes encephalopathy (i.e. behavioral change or altered level of consciousness). Additional criteria include: the presence of focal/multifocal lesion(s) predominantly affecting the white (but also the gray) matter, without evidence of previous destructive white-matter changes; the occurrence of clinical/radiologic improvement (though there may be residual deficits); and the absence of other etiology that could explain the event. New or fluctuating symptoms, signs, or MRI findings occurring within 3 months are considered part of the initial acute event.

In 2% of patients with ADEM (Tenembaum *et al.*, 2002), mostly following an upper respiratory infection, the disease has a hyperacute aggressive and rapidly progressive presentation (acute hemorrhagic leukoencephalitis, acute hemorrhagic encephalomyelitis, and acute necrotizing hemorrhagic leukoencephalitis), with large MRI lesions with perilesional edema and mass effect (Kuperan *et al.*, 2003; Mader *et al.*, 2004), and frequent lethal outcome. ADEM has a monophasic course in a majority of patients; if relapse or recurrence of the symptoms occurs, it usually happens within 3 months from onset and during the tapering off, or shortly after discontinuation of treatment. However, cases with relapse or recurrence of symptoms different than the original ones (involving distinct new CNS areas) have been reported; these are defined as recurrent or multiphasic acute disseminated encephalomyelitis (RADEM or MADEM) (Shoji *et al.*, 1992; Khan *et al.*, 1995; Tsai and Hung, 1996; Suwa *et al.*, 1999; Dale *et al.*, 2000; Rust, 2000; Hynson *et al.*, 2001; Tenembaum *et al.*, 2002; Mikaeloff *et al.*, 2004, 2007). The existence of such forms of ADEM remains controversial and not widely accepted. In the studies describing recurrent and multiphasic cases of ADEM, different diagnostic criteria were used for relapses, and this causes further uncertainty.

Differentiation between ADEM, multiple sclerosis (MS), and clinically isolated syndrome (CIS)

Vaccines and Autoimmunity, First Edition. Edited by Yehuda Shoenfeld, Nancy Agmon-Levin, and Lucija Tomljenovic.

(an initial-occurrence demyelinating episode that may or may not develop into MS) is still a challenge, and although diagnostic criteria exist, ADEM is still frequently considered "atypical MS" (Leake *et al.*, 2004). This is particularly true in the case of RADEM, where the border between ADEM and MS is more obscure. The lack of oligoclonal antibodies and the presence of several lymphocytes in the cerebrospinal fluid (CSF), the early involvement of CNS gray matter areas, the lack of significant dissemination in space (in the case of RADEM; usually, only expansion of previously existing lesions occurs), and the presence of fever, confusion, and headache are some of the main differentiating features between ADEM and MS (Table 33.1).

An additional issue is that of the putative transformation of ADEM into MS in some cases, and generally, the possible link between these two diseases, analogous to the link between AIDP and CIDP (acute/chronic inflammatory demyelinating polyneuropathy). Visudtibhan *et al.* (2010) reported that 3 out of 15 children with ADEM were diagnosed with MS after a mean follow-up period of 5.8 years. None of them was defined as MADEM or RADEM at any time point. Mikaeloff *et al.* (2007) showed an ADEM recurrence rate (at a different site in the CNS) of 18% after a mean follow-up period of 3.3–5.4 years. Risk factors for such recurrence included the presence of demyelination in the family, first presentation with optic neuritis, fulfillment of the MRI criteria suggested by Barkof *et al.* (1997) and McDonald *et al.* (2001) on the initial neuroimaging, and absence of sequelae after the acute episode (Mikaeloff *et al.*, 2004, 2007). Interestingly, the first three parameters are known predictive factors for further relapses in MS after a CIS, strengthening the link between RADEM onset and MS (Mikaeloff *et al.*, 2007). In another study, Pavone *et al.* (2010) reported an ADEM relapse rate of 12% after a follow-up period ranging from 4.4 to 17.1 years, but all of the patients they looked at probably did not meet criteria for RADEM or MADEM, due to the absence of encephalopathy and a lack of data about the location of the new lesions in the MRI.

In most cases, ADEM follows an infection (usually viral) or an immunization ("post-infectious or post-vaccination encephalomyelitis"). Although there is a close temporal relationship to vaccinations, ADEM occurs at low incidence following immunizations and there is no concrete evidence of a clear pathogenetic correlation. Post-vaccination ADEM accounts only for 5–10% of all cases of ADEM (Bennetto and Scolding, 2004; Huynh *et al.*, 2008).

Epidemiology

Since there are no strict diagnostic criteria for ADEM and there is usually a controversy between it and MS/CIS, its actual incidence is difficult to accurately estimate. It is considered a rather rare disease, with an incidence of 0.6–0.8 per 100 000 person years (Menge *et al.*, 2007; Huynh *et al.*, 2008; Torisu *et al.*, 2010). ADEM can occur at any age, but it is mainly a disease of children and young adults, with a mean age of onset of 5–6 years (Murthy *et al.*, 2002; Tenembaum *et al.*, 2002; Leake *et al.*, 2004). Males are more frequently affected than females (Torisu *et al.*, 2010). Two-thirds of all cases of ADEM are post-infectious (Bennetto and Scolding, 2004).

Table 33.1 Differential features of ADEM and MS

	ADEM	MS
Neurological signs	Encephalopathic signs, headache, cortical signs, altered mental status, seizures	
Other clinical signs	Fever, general malaise, myalgias, diarrheas, lemphadenopathy, rash	No prodromal signs, no non-neurological signs
CSF	OCB usually absent; frequently, pleocytosis	OCB usually present; usually, acellular
MRI	Diffuse T2 lesions, most of them gadolinium-enhancing	T1 black holes
	Gray-matter lesions	Corpus callosum involvement with discrete lesions: "Dawson's fingers"
Outcome	Relapse in up to 30% of cases, usually within 3–6 months	Relapsing course Dissemination in time or space

ADEM, acute disseminated encephalomyelitis; MS, multiple sclerosis; CSF, cerebrospinal fluid; OCB, oligoclonal band; MRI, magnetic resonance imaging

Diagnostic criteria/clinical features

ADEM is characterized by focal, or more often multifocal, demyelination of the brain or the spinal cord, resulting in variable neurological symptoms, including focal motor or sensory deficits, optic neuritis or other cranial nerve abnormalities, myelitis, and cerebellar signs. Altered mental status, confusion or other signs of encephalopathy, and cortical signs like aphasia, cortical blindness, and seizures are usually seen and are considered the clinical diagnostic hallmarks that differentiate ADEM from CIS and MS (Table 33.1).

The neurologic symptoms of ADEM often appear following a short prodromal phase of fever, malaise, headache, nausea, or vomiting. Patients subsequently develop overt neurologic symptoms subacutely, within a mean period of 4.5–7.5 days (range 1–45 days) (Tenembaum *et al.*, 2002; Pavone *et al.*, 2010). Occasionally, there is rapid progression to coma (Gupte *et al.*, 2003). The clinical presentation of ADEM is widely variable, depending on the distribution of lesions in the CNS. Although studies have reported an incidence of encephalopathy in 21–74% of patients (Dale *et al.*, 2000; Hung *et al.*, 2001; Hynson *et al.*, 2001; Murthy *et al.*, 2002; Tenembaum *et al.*, 2002; Anlar *et al.*, 2003; Gupte *et al.*, 2003; Idrissova *et al.*, 2003; Leake *et al.*, 2004; Mikaeloff *et al.*, 2004; Marchioni *et al.*, 2005; Jayakrishnan and Krishnakumar, 2010) and a meta-analysis has shown that 55% of ADEM patients have altered mental status (Pavone *et al.*, 2010), the new diagnostic criteria mandate encephalopathy for definite diagnosis. The prevalence of other neurological signs is as follows: unilateral or bilateral pyramidal signs, 60–95%; cranial nerve palsies, 22–89%; hemiparesis, 76–79%; ataxia, 18–65%; hypotonia, 34–47%; seizures, 10–47%; optic neuritis, 7–23%; and speech impairment, 5–21% (Hung *et al.*, 2001; Hynson *et al.*, 2001; Tenembaum *et al.*, 2002; Anlar *et al.*, 2003; Gupte *et al.*, 2003; Idrissova *et al.*, 2003; Leake *et al.*, 2004; Pavone *et al.*, 2010). Peripheral nervous system involvement has also been reported in adult patients (with frequencies of up to 43.6%) (Amit *et al.*, 1986; Nadkarni and Lisak, 1993; Marchioni *et al.*, 2005).

In two studies in which the new consensus definitions for ADEM diagnosis were used, the presenting signs included long tract dysfunction (79%), brainstem signs (48%), seizures (32%), and optic neuritis (6%) in one (Mikaeloff *et al.*, 2007), and ataxia (47%), hypotonia (41%), seizures (29%), thalamic syndrome (23%), hemiparesis (23%), cranial nerve palsies (18%), headache (18%), fever (12%), and ptosis (6%) in the other (Pavone *et al.*, 2010). Respiratory failure secondary to brainstem involvement or severely impaired consciousness has been reported in 11–16% of patients (Tenembaum *et al.*, 2002; Wingerchuk, 2003). One-quarter (and up to 41%) of patients with ADEM may require an ICU admission (Absoud *et al.*, 2011).

Paraclinical tests for diagnosis

Peripheral white blood counts are sometimes slightly elevated in ADEM. Lumbar puncture typically shows high protein levels and pleocytosis in the CSF, but the presence of oligoclonal antibodies in the CSF is very rare (Franciotta *et al.*, 2008). MRI is usually characterized by diffuse or multifocal white-matter lesions of increased intensity in T2 weighted imaging or FLAIR, most of them (depending on the time window of MRI testing) enhanced with gadolinium. Another characteristic finding in MRI is the absence of "old" lesions in T1 weighted images ("black holes," which could indicate a chronic disease) typically seen in MS. The distribution of the lesions in ADEM may often involve gray-matter areas (especially the thalamus) and/or cortical involvement, findings that are atypical for MS (Table 33.1).

Typical neuroimaging (MRI) findings (summarized by Marin and Callen, 2013 based on previous cohort studies) include:

1. bilateral asymmetric/symmetric involvement (rarely unilateral);
2. white matter > gray matter, but usually both affected;
3. deep/juxtacortical white matter > periventricular white matter;
4. both supratentorial and infratentorial lesions;
5. small > medium > large lesions (but often all sizes are present in the same patient); and
6. variable contrast enhancement.

Pathogenesis

The experimental animal model of MS, experimental autoimmune encephalomyelitis (EAE), bears significant similarities with ADEM, and, at least its acute form, acute monophasic EAE actually better resembles ADEM than MS. The immunopathological similarities between ADEM and EAE are particularly pronounced in post-vaccination ADEM, especially after application of primitive rubies vaccine (technically unpurified) containing fragments of myelin with antigenic properties (Huynh *et al.*, 2008), which greatly resembles the induction methods used in EAE, consisting of active immunization using myelin antigens in adjuvant (Karussis *et al.*, 1992a,b, 1999).

Particularly in post-viral or post-vaccination ADEM, the current hypothesis is that antigens of viral origin crossreact with myelin components (molecular mimicry) and, in a secondary manner, induce a hyperergic reaction, leading to the development of ADEM. Myelin proteins have shown resemblance to several viral sequences. T cells targeting human herpesvirus 6 (HHV-6) (Tejada-Simon *et al.*, 2003), coronavirus (Talbot *et al.*, 1996), influenza virus (Markovic-Plese *et al.*, 2005) and Epstein–Barr virus (EBV) (Lang *et al.*, 2002) have been shown to crossreact with myelin basic protein (MBP) antigens. Furthermore, anti-MBP T cell responses have been demonstrated in patients with post-infectious ADEM (Pohl-Koppe *et al.*, 1998; Jorens *et al.*, 2000). Demyelinating antibodies have also been shown to develop in some cases of measles (Kalimo *et al.*, 1977) and anti-MBP antibodies have been detected following vaccination with Semple rabies vaccine (Ubol *et al.*, 1990; O'Connor *et al.*, 2003).

Antimyelin antibodies, especially targeting myelin oligodendrocyte glycoprotein (MOG), have been found in children with ADEM and can be used to differentiate it from viral encephalitis (Lalive *et al.*, 2011). This finding may support the theory that ADEM is a disease that mainly targets the oligodendrocytes. Brilot *et al.* (2009) detected high serum immunoglobulin G (IgG) titers to MOG in 40% of children with ADEM or CIS but in 0% of control children affected by other neurological diseases, healthy children, or adults with inflammatory demyelinating diseases. Anti-MOG IgG titers did not differ between ADEM and CIS, and did not predict conversion from CIS to MS during a mean 2-year follow-up. However, intrathecal IgG anti-MOG antibody synthesis was only seen in CIS children and not in ADEM.

Additional antimyelin antibodies have been described in ADEM and may help in distinguishing it from MS (Van Haren *et al.*, 2013). Van Haren *et al.* (2013) found ADEM was characterized by IgG autoantibodies targeting epitopes derived from MBP, proteolipid protein, myelin-associated oligodendrocyte basic glycoprotein, and alpha-B-crystallin. In contrast, MS was characterized by IgM autoantibodies targeting MBP, proteolipid protein, myelin-associated oligodendrocyte basic glycoprotein, and oligodendrocyte-specific protein.

Another hypothesis is that the virus may activate distinct clones of antimyelin T cells in a nonspecific way and that the suppressor or regulatory cells that are supposed to control this abnormal reactivity are compromised or malfunctioning.

The possibility of a direct pathogenic involvement of known or unknown neurotropic viruses should also be considered, especially in post-infectious ADEM. In this case, the neurotropic virus may directly attack the white matter of the CNS, and the occurrence of ADEM could be the result of a reinfection or an opening of the blood–brain barrier (BBB) (after the initial infection), followed by a release of myelin protein into the peripheral blood and the lymphoid tissues, inducing an autoimmune reactivity against myelin.

Genetic predisposition to ADEM, either by a human leukocyte antigen (HLA)-associated risk or by a high susceptibility in patients with certain genetic mutations, has also been suggested. The possible mechanism in this case could be similar to that of vaccination encephalopathy, which has been found to occur at very high percentages in children with mutations in the SCN1A gene (Berkovic *et al.*, 2006; Sell and Minassian, 2006).

Iatrogenic cases of ADEM are of great importance. They are associated with the previously described immunopathogenic mechanisms; that is, the induction of immunization against myelin as a result of administration of myelin components present in "natural therapeutic" modalities or injection regiments (Goebel *et al.*, 1986). A demonstrable example of such a mechanism is the development of ADEM following experimental Alzheimer's therapy using a vaccine that contained aggregates of synthetic Aβ42 fragments of amyloid precursor protein (Orgogozo *et al.*, 2003; Rosenmann *et al.*, 2006). In experimental animals (mice), EAE may be induced with injection of Aβ42, but only when the latter is administered together with complete Freund's adjuvant (CFA). This observation points to the importance and central role of adjuvants in the induction of ADEM, and in autoimmunity in general (Agmon-Levin *et al.*, 2012; Vera-Lastra *et al.*, 2013). The pathogenic role of adjuvants in the induction of autoimmune syndromes has been highlighted by Yehuda Shoenfeld, who introduced the concept of autoimmune/inflammatory syndrome induced by adjuvants (ASIA) (Agmon-Levin *et al.*, 2012; Vera-Lastra *et al.*, 2013). In general, immunologic adjuvant is a substance that enhances antigen-specific immune responses, preferably without triggering one of its own (Israeli *et al.*, 2009). Adjuvants are commonly used in medicine to boost an immune response to treatments, such as in the case of vaccinations. The efficacy of most vaccines currently used in humans is highly dependent on the presence of an adjuvant in the immunizing inoculum (Marrack *et al.*, 2009).

Adjuvant effects are accomplished via several mechanisms that impinge on both the innate and the adaptive immune systems. Adjuvants mimic evolutionarily conserved molecules (e.g. bacterial cell walls, lipopolysaccharides (LPSs), unmethylated CpG-DNA) and bind to Toll-like receptors (TLRs). They activate dendritic cells (DCs), lymphocytes, and macrophages, which increases the release of chemokines and cytokines from T-helper and mast cells. Currently, the most widely used adjuvant in medicine is aluminum. Aluminum disrupts lysosomal functions and stimulates the production and secretion of cytokines such as interleukin (IL-) 1b, 18, and 33 (Marrack *et al.*, 2009). Adjuvants also provide physical protection to antigens, which may enable a longer exposure to the immune system and a more robust (B and T cell) response.

Adjuvants were previously thought to pose little risk, but studies in animal models and humans have demonstrated that some can inflict autoimmunity, as in the case of Tetramethylpentadecane (TMPD), known as pristane, which is capable of inducing a lupus-like disease in a murine model of systemic lupus erythematosus (SLE) (Satoh and Reeves, 1994; Reeves *et al.*, 2009).

Immunological mechanisms

As with MS and other demyelinating diseases, a genetic susceptibility to ADEM (association with the HLADQB1-0602-1501 and DRDI-1503 alleles) has been reported (Alves-Leon *et al.*, 2009).

Immunological studies have investigated the presence and secretion of cytokines and chemokines in the blood and CSF of patients with ADEM. Ichiyama *et al.* (2002) showed an increased CSF concentration of tumor necrosis factor alpha (TNF-α) and its soluble receptor 1 (sTNFR1), of IL-6, and of IL-10. The levels of sTNFR1 were associated with the intrathecal concentration of MBP, pointing to a possible link between TNF-α and the processes of demyelination. An increased serum concentration of E-selectin, as well as a trend toward increased levels of the soluble intercellular adhesion particle 1 (but not of the soluble vascular adhesion particle 1), was also observed in cases of acute ADEM (Martino *et al.*, 2005). The involvement of metalloproteinase 9 (MMP-9) and its tissue inhibitor (TIMP-1) (increased serum levels of MMP-9 and of TIMP-1) in the acute phase of ADEM has been suggested, but the evidence for this is currently unconvincing (Ichiyama *et al.*, 2006). It has been suggested that MMP-9 is more closely associated with the inflammatory process, whereas TIPM-1 acts as a modulator of MMP-1 activity.

Ishizu *et al.* (2006) showed that several cytokines are involved in the activation of microglia and macrophages, including IL-1β, -2, -4, -5, -6, -8, and -10, as well as interferon gamma (INF-γ). They also demonstrated a positive correlation between the number of cells, the protein content of the CSF, and its INF-γ, IL-6, and IL-8 concentrations. Interestingly, IL-17, which is one of the major/critical cytokines in MS pathogenesis, was not found to be increased in ADEM.

Franciotta *et al.* (2008) compared cytokine and chemokine concentrations between ADEM and MS, and showed that, in ADEM, there is a substantial increase in the concentrations of chemokines that activate neutrophils (CXCL1 and CXCL7), monocytes and T-cells (CCL3 and CCL5), Th1 cells (CXCL10), and Th3 cells (CCL1, CCL2, and CCL17). The mean concentration of CXCL7, CCLI, CCL22, and CCL17 was distinctly higher in the CSF of ADEM patients than in MS.

It is reasonable to postulate an analogy between ADEM and other neuroimmunological diseases, in which a progressive immune cascade leads to the disseminated inflammatory process and demyelination. According to this theory, the early stage of ADEM is probably initiated by a penetration of reactive peripheral T lymphocytes into the CNS, across the BBB. This extravasation is mediated by adhesion molecules, such as VCAM-1 and ICAM-1, and may also be aided by activated metalloproteinases. A number of cytokines seem to have a regulatory function. During the inflammatory process in the CNS, the macrophage and astroglial ligands are bound to the myelin sheaths, which enhances the influx of calcium protease and of other metalloproteinases endowed with a rather high affinity to the basic myelin protein.

Neuropathology

Focal, perivenous, and subependymal changes dominate the histopathological pattern of ADEM, leading to the formation of disseminated masses or conglomerates. The localization of these lesions varies between cases but there is generally a marked predisposition for the white matter at the cortical–subcortical border. Infiltrates may also be found in the cerebellum, spinal cord, and cerebral trunk. Despite the predisposition for the white matter, the cortical gray matter (and especially the thalamus) is not free of lesions, and this further differentiates ADEM from MS. The pathological process is particularly pronounced perivenously, especially around middle-sized veins, but immense infiltrates are also seen around large subcortical veins.

At the early stage, the histopathology involves mainly T cell infiltrates, accompanied by few plasma cells. At this point, the proliferation of microglia is rather insignificant. Microglial infiltrates primarily form dense syncytia, but they later undergo a transformation to macrophages of glial origin and/or rod cells. The glial infiltrates often fuse with each other to occupy greater or smaller regions of the CNS. Initially, around the veins, disintegration of myelin sheaths takes place, leading later to full demyelination, accompanied by the presence of lipid-filled macrophages. The demyelinated foci are almost never sharply delineated from the normal, unaffected nervous tissue. The myelin lesions are accompanied by rather mild axonal damage, and only sparse complete axonal destruction can be seen, mainly involving the thinner fibers. In the subsequent phases of ADEM, the glial reaction becomes more prominent.

Particularly in post-vaccination ADEM, lesions are more intense, including demyelination in the spinal cord (myelitis). The pathology in infants under the 2 years of age mainly involves a degenerative demyelination, and only very mild inflammation.

In the hyperacute form of ADEM, which is also defined as acute hemorrhagic leukoencephalitis, the dominating pathology includes brain edema with numerous small hemorrhages. The cerebral edema may cause temporal-lobe or tonsillar herniation. The histopathology of hyperacute ADEM is characterized by necrotic vascular lesions, focal demyelination, polynuclear infiltrates, and deposition of fibrin in the vessel walls and in the Virchow–Robin spaces.

In general, though ADEM and MS are both demyelinating multifocal inflammatory diseases, there are certain histopathological differences that differentiate between them (Tardieu and Mikaeloff, 2004). The inflammatory lesions in ADEM spread radially from the vessels, whereas those in MS are distinctly discrete and discontinuous. In ADEM, the macrophages are mainly concentrated around the vessels, while in MS they border the plaques. The most important differentiating feature is the sharp delineation of plaques in MS lesions, which is absent in ADEM, and astroglial reaction/fibrillary gliosis, which is seen only in MS. In the long term, only a sparse nonspecific gliosis can be detected without significant myelin loss in ADEM. Moreover, MR spectroscopy reveals an elevation of lipids and a reduction in the myoinositol/creatinine ratio, with no change in N-acetylaspartate (NAA) or choline values (Balasubramanya *et al.*, 2007; Ben Sira *et al.*, 2010); as the disease progresses, there is a reduction of NAA and an increase in choline (with corresponding reductions in NAA/creatine and NAA/choline ratios) (Bizzi *et al.*, 2001; Balasubramanya *et al.*, 2007), which is normalized as the clinical and conventional neuroimaging abnormalities resolve. This finding suggests an only transient neuroaxonal dysfunction, rather than irreversible neuroaxonal loss, in contrast to the situation in MS.

Treatment

There is no standard, evidenced-based treatment regimen for ADEM. Most data come from case reports and small series, and as yet there has been no randomized controlled trial. Most treatment approaches utilize nonspecific immunosuppressive modalities, such as corticosteroids (Dale *et al.*, 2000; Hynson *et al.*, 2001; Shahar *et al.*, 2002; Tenembaum *et al.*, 2002; Gupte *et al.*, 2003; Alexander and Murthy, 2011), intravenous immunoglobulin (IVIg) (Kleiman and Brunquell, 1995; Hahn *et al.*, 1996; Nishikawa *et al.*, 1999; Pradhan *et al.*, 1999; Revel-Vilk *et al.*, 2000; Sahlas *et al.*, 2000; Marchioni *et al.*, 2002), or plasmapheresis (Keegan *et al.*, 2002; Lin *et al.*, 2004). Beneficial responses have been reported with each of these modalities (Dale *et al.*, 2000; Hynson *et al.*, 2001; Shahar *et al.*, 2002; Tenembaum *et al.*, 2002; Gupte *et al.*, 2003; Alexander and Murthy, 2011). When there is massive cerebral edema, decompressive craniectomy has been suggested as a life-saving measure (von Stuckrad-Barre *et al.*, 2003; Refai *et al.*, 2005).

Prognosis

ADEM has a monophasic course in 70–90% of cases (Khan *et al.*, 1995; Tsai and Hung, 1996;

Suwa *et al.*, 1999; Dale *et al.*, 2000; Rust, 2000; Hynson *et al.*, 2001; Tenembaum *et al.*, 2002; Mikaeloff *et al.*, 2004, 2007; Pavone *et al.*, 2010). Typically, prognosis is excellent (Dale *et al.*, 2000; Tenembaum *et al.*, 2002). Full recovery has been noted in approximately 70–90% of patients within 6 months of onset (Dale *et al.*, 2000; Hung *et al.*, 2001; Hynson *et al.*, 2001; Murthy *et al.*, 2002; Tenembaum *et al.*, 2002; Anlar *et al.*, 2003; Gupte *et al.*, 2003; Idrissova *et al.*, 2003; Leake *et al.*, 2004; Mikaeloff *et al.*, 2004). However, subtle neurocognitive deficits (attention, executive function) may be present even years after ADEM in up to 60% of patients (Hahn *et al.*, 2003; Jacobs *et al.*, 2004; Pavone *et al.*, 2010).

Post-vaccination ADEM

General vaccines

The incidence of post-vaccination ADEM (which accounts for only 5–10% of all ADEM cases) has fluctuated over the last several decades, with a peak occurring between 1927 and 1929 and also –seemingly – during the last few years. This is probably related to the methods used for vaccine production, the amount of myelin antigen included, and –most importantly – the type of adjuvant used. The incidence of post-vaccination ADEM varies among different kinds of infectious agents. The overall incidence is estimated at 0.1–0.2 per 100 000 and a higher risk has been reported following immunization against measles. Other common causes of post-vaccination ADEM include vaccines against varicella zoster, rubella, smallpox, and influenza viruses (Huynh *et al.*, 2008). On the other hand, surprisingly, certain vaccines, such as antitetanus vaccine, have been shown to have a negative correlation with ADEM (statistically significant decreased risk) (DeStefano *et al.*, 2003).

Vaccination against hepatitis B is one of the most controversial potential causes of demyelinating disease. An increased risk for ADEM was suggested by Touzé *et al.* (2000, 2002) following a case–control study in 2002, while other, later authors (DeStefano *et al.*, 2003; Hocine *et al.*, 2007; Mikaeloff *et al.*, 2007) showed a significant risk for CIS or conversion to clinically definite MS (CDMS) following hepatitis B immunization. For the present, controlled epidemiological studies that provide a proof of causative correlation between hepatitis B vaccination and MS or other acute demyelinating diseases do not yet exist (Zipp *et al.*, 1999; Jefferson and Heijbel, 2001; Levy-Bruhl *et al.*, 2002; Hocine *et al.*, 2007; Martinez-Sernandez and Figueiras, 2013), and analysis of the existing data argues against a causal relationship between them and strongly indicates that the benefits of the vaccine clearly surpass the potential risks of CNS inflammation (Merelli and Casoni, 2000).

Post-vaccination ADEM is usually observed following primary vaccination, and much less frequently following revaccination (Huynh *et al.*, 2008). However, there are reports in which a relapse or a second neurological event was observed following repeated immunizations with the same virus (Lapphra *et al.*, 2011).

We searched PubMed for cases of post-vaccination ADEM from 1979 to 2013. In order to include all possible variants of ADEM following vaccination, we used the following terms: "post-vaccination encephalitis," "post-vaccination encephalomyelitis," "post-vaccination ADEM," "encephalomyelitis, vaccination," "optic neuritis, vaccination," "optic neuropathy vaccination." We excluded cases in which abstracts or the full text were not available in English. These criteria yielded 66 cases. We then excluded cases of MS relapse and isolated optic neuritis to leave 48 cases, presented in detail in Table 33.2. The most commonly reported vaccinations associated with ADEM/encephalomyelitis included influenza (15 cases; cases 1–15 in Table 33.2), hepatitis A or B (3 cases; cases 22–24 in Table 33.2), rabies (3 cases; cases 16–18 in Table 33.2), yellow fever (3 cases; cases 29–31 in Table 33.2), measles (2 cases; cases 25, 26 in Table 33.2), rubella (2 cases; cases 36, 37 in Table 33.2), and tetanus (2 cases; cases 34, 35 in Table 33.2). As can be seen from Table 33.2, the vast majority of post-vaccination ADEM cases are related to influenza vaccination, which could be attributed to the high percentage of the population that received the vaccine during the HIlN1 epidemic from 2009 to 2012.

In terms of the clinical presentation and the affected CNS areas, there is great diversity among reported cases of post-vaccination acute demyelinating syndromes (Table 33.2). Interestingly, in a very high proportion of patients (especially following influenza vaccination), the dominant localization of demyelination was the optic nerves and the myelon, presenting as optic neuritis and myelitis (with or without additional manifestations of ADEM) (Table 33.2). This predisposition to the spinal cord and the optic nerves is reminiscent of neuromyelitic optica (or, more generally, the NMO spectrum of diseases), which is highly associated with anti-aquaporin-4 antibodies. Indeed, a

Table 33.2 Cases of post-vaccination ADEM reported in the literature

No.	Type of vaccine	Age, gender	Delay between vaccination and symptom onset	Clinical syndrome	Response to treatment and outcome	Reference
1	Influenza	70, M	7 days	Transverse myelitis and axonal neuropathy	Partial recovery after steroids + PE	Nakamura *et al.* (2003)
2	Influenza	75, F	3 weeks	Headache, intractable hiccups, nausea and vomiting, quadriparesis, encephalopathy, brainstem involvement (left abducens palsy, dysarthria), incontinence	No response to PE and steroids; death	Shoamanesh and Traboulsee (2011)
3	Influenza	61, M	3 weeks; 3 months	Bilateral optic neuropathy and delayed ADEM	IVMP: full recovery except 50% reduction of visual acuity	Huynh *et al.* (2008)
4	Influenza	6, M	16 days	Ataxia and muscle weakness	Steroid treatment/resolved	Iyoda *et al.* (2004)
5	Influenza	83, F	8 days	Encephalopathy: GCS: 9, fever, tachycardia; quadriparesis, bilateral pyramidal syndrome, perioral fasciculations; no ON	Dramatic response to PE; died later of pneumonia	Machicado *et al.* (2013)
6	Influenza	61, M	2 weeks	Post-vaccination myelitis; suffered in the past from post-infection ADEM	i.v. steroids and IVIG; full recovery	Ravaglia *et al.* (2004)
7	Influenza	60, F	10 days	Post-vaccination myelitis; suffered in the past from post-infection ADEM	i.v. steroids and IVIG; full recovery	Ravaglia *et al.* (2004)
8	Influenza	NA	NA	Encephalomyelitis with headache, urinary retention, bilateral optic disc swelling, and a bilateral visual defect (bilateral optic neuropathy)	Recovery after i.v. steroids	Vilain *et al.* (2000)
9	Influenza	62, M	5 days	Transverse myelitis and axonal neuropathy	Partial recovery after steroids; steroids + PE in the second	Nakamura *et al.* (2003)
10	H1N1 influenza	2, M	25 days	Fever, lethargy, and recurrent seizures; no ON	Full recovery after steroid treatment	Fujii *et al.* (2012)
11	H1N1 influenza	33, F	15 days	Myelitis; hypoesthesia below the Th7 level. CSF showed increased levels of myelin basic protein and positivity for oligoclonal IgG antibodies and increased anti-MBP abs; MRI: brain and spinal lesions	Improvement with steroids	Maeda and Idehara (2012)
12	H1N1 influenza	36, M	10 days	Quadriparesis-ADEM + acute sensorimotor neuropathy and anti-GM2 abs in serum	Marked improvement with steroids	Hoshino *et al.* (2012)

13	H1N1 influenza	34 months, M	5 days	Seizure and left hemiparesis	Complete recovery with steroids	Lee *et al.* (2011)
14	H1N1 influenza	NA, F	4 days	Myelitis (sensory)	Improved without treatment	Arcondo *et al.* (2011)
15	HINI influenza	2, M	4 days; 6 days	Neurological signs after both doses: first vaccination induced gait ataxia and the second was followed by bilateral ON; MRI: diffuse white-matter lesions compatible with ADEM	Full recovery after steroids	Lapphra *et al.* (2011)
16	Rabbies	31, M	5 months	Seizures, left hemiparesis, loss of memory, and behavioral disorders; corpus callosum involvement	Paresis resolved; persisting neurological signs: seizures, behavioral changes	Gamboa *et al.* (1983)
17	Rabbies	24, M	1 week	Gait disturbance, incontinence of urine, and increasing lethargy	Died after 37 years of encephalopathy; demyelinating lesions in pathology	Iizuka *et al.* (1993)
18	Rabbies	45, M	14 days	ADEM with qyadripyramidal syndrome and ataxia; brain MRI normal, long lesion – myelitis – in cervical spine; relapse with ON 1 month after ADEM; NMO-like disease	Improvement with steroids	Kulkarni *et al.* (2004)
19	Polyvaccination	27, M	10 days	Altered consciousness, seizures, hemiplegia, meningeal involvement	Death (day 21)	Labauge *et al.* (1979)
20	Early summer encephalitis	36, M	After repeated immunization	Recurrent retrobulbar neuritis, grand mal epilepsy, focal motor clonic-tonic attack; ADEM	Partial improvement	Schattenfroh (2004)
21	Group A+C meningo-coccal vaccine	25, F	NA	ADEM	Fast disappearance of symptoms and gradual resolution of lesions in MRI after i.v. MP	Py and Andre (1997)
22	Hepatitis B	40, M	6 weeks	Recurrent ADEM previously; stable for 5 years; intranuclear ophthalmoplegia, visual loss in the left eye, and worsening of the previous cerebellar and pyramidal signs; seizures	Partially resolved after steroids	Cabrera-Gomez *et al.* (2002)

(continued)

Table 33.2 (*Continued*)

No.	Type of vaccine	Age, gender	Delay between vaccination and symptom onset	Clinical syndrome	Response to treatment and outcome	Reference
23	Recombinant hepatitis B	39, F	4 weeks after the 2nd dose	Complete right homonymous hemianopsia and severe dyslexia; large lesion in the occipital lobe extended into the splenium of corpus callosum (tumefactive ADEM); relapse 11 days after receiving the 3rd dose, 5 months after the 2nd dose: severe dyslexia, severe bilateral vision loss, moderate left pyramidal hemiparesis, and hemianopsia.	Treatment with craniotomy and dexamethasone; residual dyslexia and complete right homonymous; follow-up MRI showed almost complete resolution of the previous findings, with the exception of a proencephalic cyst	Konstantinou *et al.* (2001)
24	Hepatitis A	23, F	3–7 days	ADEM (somnolence, tetraparesis, gaze paresis, ataxia, MRI lesions in white matter and basal ganglia) + motor axonal neuropathy, high protein in CSF, anti-GM1 abs, *Campylobacter jejuni*	Improvement of CNS signs with steroids but not of axonal neuropathy	Huber *et al.* (1999)
25	Lyssa	NA	10 days	Encephalopathy	Died 3 days later; demyelinating lesions in CNS and optic nerves	Varga *et al.* (1979)
26	Measles	7, M	3 days	Fever, vomiting, convulsion, and loss of consciousness; multifocal lesions in the basal ganglia in MRI	Complete resolution of lesions within 2 weeks	Shu *et al.* (2011)
27	Measles	19, F	4 days	Pains, hypertension, flaccid quadriparesis, pyramidal signs, sensory level: PRES + myeloradiculopathy	Recovery after pulses of steroids	Hamano *et al.* (2010)
28	Diptheria, tetanus toxoid, whole-cell pertussis	6 months, M	6 days	Acute necrotizing encephalopathy: lethargy, hypotonia, and focal clonic seizures; hemorrhagic lesions, white matter, and thalamic lesions in MRI	Recovery without treatment	Aydin *et al.* (2010)
29	Yellow fever	53, M	NA	HIV-positive patient; rapidly evolving fatal myelomeningo-encephalitis following a live-attenuated yellow fever vaccination	Death	Kengsakul *et al.* (2002)
30	Yellow fever	56, M	45 days	Longitudinal myelitis (NMO-like) without encephalitis (progressive paraparesis, urinary retention); YFV IgM abs detected in CSF	Symptoms improved 5 days later	Chaves *et al.* (2009)

31	Yellow fever	NA		ADEM (no further information available)	NA	Miravalle *et al.* (2009)
32	Smallpox	19, M	12 days	PVE + ADEM	Full recovery with IVIG, steroids, and vaccina immunoglobulin	Van Dam *et al.* (2009)
	Typhoid polys		17 days	Diffuse white-matter lesions		
	Anthrax doses 1–2		17 and 5 days	Negative CSF cultures and PCR for VZV1,2; positive IgM orthopoxvirus		
33	Smallpox	30, M	10 days	Mental status changes, seizures; lesions consistent with demyelinating disease on MRI	Neurologic sequelae still present 1 year later	Sejvar *et al.* (2005)
34	Tetanus	NA	NA	ADEM with quadripyramidal syndrome and predominantly seizures; large lesions in the white matter in MRI	Recovery without steroids	Hamidon and Raymond (2003)
35	Tetanus	28, F	15 days	Spastic tetraplegia	Steroids; partial resolution	Cisse *et al.* (2012)
36	Rubella	31, F	5 days	Optic neuritis and myelitis (NMO-like ?)	Partial recovery with steroids	Kline *et al.* (1982)
37	Rubella	14, M	16 days	Tetraparesis with retention of urine and total sensory loss below Th1; nuchal rigidity and Lhermitte's sign; MRI detected bilateral high-intensity lesions in the cerebral white matter and cervical spinal cord; elevated cell count, protein, and MBP in CSF	Rapid improvement with steroids	Tsuru *et al.* (2000)
38	Japanese encephalitis	15, M	3 weeks	Bilateral optic neuritis; longidutinal myelitis; NMO-like disease; AQP4 antibodies negative	Recovery after steroids	Furukawa *et al.* (2011)
39	Diphtheria/ tetanus/ polyomyelitis	7, M	NA	ADEM after the first dose, ON after the second	Full recovery	Mancini *et al.* (1996)
40	HPV	20, F	28 days after second immunization	Headache, vomiting, followed by seizures; -OCB negative; pleocytosis and increased protein; normal VEP; diffuse white-matter lesions, swelling, and subcortical bleedings; one Gd-enchancing lesion	Relapsed twice after steroid cessation, required immunosuppression	Wildemann *et al.* (2009)
41	HPV	15, F	23 days after second immunization	Headache, nausea, fever, vertigo, diplopia, somnolence; 10 cells and normal protein in CSF; OCB negative; diffuse brain, brainstem, and cervical lesions	Neurological recovery and resolution of lesions after steroid treatment	Schaffer *et al.* (2008)

(continued)

Table 33.2 *(Continued)*

No.	Type of vaccine	Age, gender	Delay between vaccination and symptom onset	Clinical syndrome	Response to treatment and outcome	Reference
42	HPV	17, F	4 months after 3rd dose	Brown Sequard syndrome, followed by spastic paraparesis; IgG antibodies for AQP4 positive; spinal cord lesion > three segments on 1st episode, expanding to the brainstem and longer lesion with optic nerve involvement on the 2nd event; no brain lesions	Partial improvement with steroids, PE, and Rituximab; at 5 months: still paraparesis and ON	Menge *et al.* (2012)
43	HPV	14, F	4 months post 3rd dose	Optic neuritis and 4 months later myelitis; MRI: brain and spine normal at first episode and long spinal lesion at second episode; IgG antibodies for AQP4 positive	Steroids, Rituximab, mycophenolate; stopped recurrences	Menge *et al.* (2012)
44	HPV	13, F	Post 2nd immunization, time unknown	Transverse myelitis; optic nerve and spinal lesions	No response to steroids and PE; Rituximab started	Menge *et al.* (2012)
45	HPV	18, F	5 months after 2nd immunization	Clinical signs of myelitis and visual loss	NA	Menge *et al.* (2012)
46	HPV	16, F	10 days post 2nd immunization	Bilateral visual loss and mild pyramidal signs; chiasmatic lesion and tumefactive parietooccipital lesion in brain MRI; MRI of spine normal	No further demyelinating events within 18 months' follow-up; patient remained with severe visual loss	DiMario *et al.* (2010)
47	HPV	16, F	21 days post 3rd immunization	Pseudoathetosis; longidutinal spinal lesion and diffuse white-matter brain lesions	NA	Sutton *et al.* (2009)
48	HPV	25, F	16 days post 2nd immunization	Acute hemiparesis; cortical lesion with edema in brain MRI (tumefactive demyelination?); CSF: 9 cells, OCB positive	4 months later, the patient suffered from a second event and MRI showed a new lesion (diagnosed then as CDMS)	Sutton *et al.* (2009)

PE, plasma exchange; ADEM, acute disseminated encephalomyelitis; GCS, Glasgow coma scale; IVIG, intravenous gammaglobulin; CSF, cerebrospinal fluid; IgG, immunoglobin G; MBP, myelin basic protein; MRI, magnetic resonance imaging; CNS, central nervous system; NMO, neuromyelitis optica significant number of post-vaccination NMO case reports have been published recently (Furukawa *et al.*, 2011; Kitazawa *et al.*, 2012; Menge *et al.*, 2012). This raises the possibility of crossreactivity between aquaporin epitopes and certain viral proteins.

Usually, the symptoms of ADEM appear a few days following immunization (mean 14.2 days; Table 33.2), but there are cases in which the clinical presentation was delayed (more than 3 weeks, or even up to 5 months post-vaccination; this is found in approximately one-third of all reported cases in Table 33.2). The clinical course of post-vaccination ADEM varies, but it does not seem to differ significantly from that of "regular" ADEM. Notably, a substantial percentage of reported post-vaccination cases subsequently developed MS (Table 33.2). Vaccinations have also been linked to a relapse of MS or myelitis (Larner and Farmer, 2000). Such deterioration or relapse of MS has been described in association with several vaccines (as summarized by Loebermann *et al.*, 2011), including recently human papillomavirus (HPV) vaccination for protection against gynecological cancers (Merelli and Casoni, 2000; Sutton *et al.*, 2009). Optic neuritis represents a unique paradigm of such association of vaccines with demyelinating events. There are numerous reports of isolated (either unilateral or bilateral) optic neuritis following various types of vaccinations against infectious agents (van de Geijn *et al.*, 1994), including measles (Stevenson *et al.*, 1996; Arshi *et al.*, 2004; Moradian and Ahmadieh, 2008), anthrax (Kerrison *et al.*, 2002), rubella (Stevenson *et al.*, 1996; Arshi *et al.*, 2004; Moradian and Ahmadieh, 2008), hepatitis A and B (Albitar *et al.*, 1997; Erguven *et al.*, 2009; Huang *et al.*, 2009), influenza (Perry *et al.*, 1979; Ray and Dreizin, 1996; Hull and Bates, 1997; Laffon-Pioger *et al.*, 2010; Crawford *et al.*, 2012; Rubinov *et al.*, 2012), pneumonococcal (Kitazawa *et al.*, 2012), meningococcal (Laria *et al.*, 2006), rabies (Dadeya *et al.*, 2004; Gupta *et al.*, 2004), and BCG (Yen and Liu, 1991) vaccines.

The immunopathogenic mechanisms of post-vaccination ADEM are similar to those described in "regular"/idiopathic ADEM and mainly include molecular mimicry between the proteins of the viruses used for the vaccination and CNS myelin components, or else are related to the adjuvants used in the vaccines (Israeli *et al.*, 2009; Agmon-Levin *et al.*, 2012; Vera-Lastra *et al.*, 2013). Vaccination with adjuvants can induce a nonspecific activation of the immune system and can cause a break of self-tolerance and subsequent stimulation of myelin-responsive lymphocytes, which may be further accelerated by defective regulatory cells/circuits in genetically susceptible individuals (as already described). Another potentially important mechanism that might contribute to the break of self-tolerance after vaccination could be related to the route of administration: vaccines administered parenterally and not via the "natural" method of infection could carry a greater risk for induction of an autoimmune reaction, by bypassing the control mechanisms of self-tolerance.

Human papillomavirus vaccination

HPV is a sexually transmitted infection that can cause anogenital or oropharengeal cancer in males and females. Gardasil is a noninfectious quadrivalent vaccine made up of virus-like particles of the major capsid L1 protein of HPV types 6, 11, 16, and 18 that has been shown to be safe and possibly efficient for the prevention of cervical, vulval, and vaginal dysplasia (Bryan, 2007; Malagon *et al.*, 2012; Schiller *et al.*, 2012). However, its efficacy in the prevention of cervical cancer is yet to be proven, as the follow-up of vaccinated patients is still too short to allow definite conclusions to be drawn (cervical cancer usually takes 15 years or more to develop). The side effects of this vaccination are reportedly similar to those of other vaccines. Only a few cases of post-vaccination ADEM have been reported in the literature (Table 33.2). Interestingly, in most of these cases, the dominant presentation was of optic neuritis and/or myelitis with longitudinal spinal lesions (Menge *et al.*, 2012), and in most of them there was a high incidence of recurrence (second event of demyelination/neurological signs a few days to a few months following vaccination).

Conclusions

ADEM is an acute demyelinating disease of the CNS that usually affects the very young. Although well defined clinically, its pathogenesis remains not fully understood. Its resemblance to other –chronic – demyelinating diseases, and especially MS, raises the possibility of common immunopathogenetic paths. Some might claim that ADEM is to MS what AIDP is to CIDP. In support of this are reports of the "transformation" of ADEM into MS and our increasing knowledge of recurrent or relapsing types of ADEM. Others insist that ADEM is a completely different nosological entity, in terms of pathogenesis, course, and prognosis, and that cases of the "transformation" of ADEM to MS were really just MS from the beginning. There are indeed several clinical and paraclinical parameters that clearly differentiate between the two diseases. ADEM is almost always a post-infectious disease, and molecular mimicry and antibody-mediated autoimmune mechanisms seem to play a crucial role in its pathogenesis.

Of special interest is post-vaccination ADEM, which accounts for 5–10% of all ADEM cases. The widespread use of vaccinations in recent years (including new types of influenza vaccines and vaccines against HPV for the prevention of gynecological malignancies) has caused an increase in reported cases of ADEM, often with unique (NMO spectrum-like) manifestations. The central role of adjuvants in post-vaccination ADEM (and related conditions) has lately been highlighted. Treatments with steroids or antibody-targeting modalities usually have a favorable effect, and the prognosis is generally good. However, severe, hyperacute, and even lethal forms of ADEM do exist.

References

Absoud, M., Parslow, R.C., Wassmer, E. *et al.* (2011). Severe acute disseminated encephalomyelitis: a paediatric intensive care population-based study. *Mult Scler*, **17**(10): 1258–61.

Agmon-Levin, N., Hughes, G.R.V., and Shoenfeld, Y. (2012). The spectrum of ASIA: "Autoimmune (Auto-inflammatory) Syndrome induced by Adjuvants." *Lupus*, **21**(2): 118–20.

Albitar, S., Bourgeon, B., Genin, R., *et al.* (1997). Bilateral retrobulbar optic neuritis with hepatitis B vaccination. *Nephrol Dial Transplant*, **12**(10): 2169–70.

Alexander, M. and Murthy, J.M. (2011). Acute disseminated encephalomyelitis: treatment guidelines. *Ann Indian Acad Neurol*, **14**(Suppl. 1): S60–4.

Alves-Leon, S.V., Veluttini-Pimentel, M.L., Gouveia, M.E., *et al.* (2009). Acute disseminated encephalomyelitis: clinical features, HLA DRB1*1501, HLA DRB1*1503, HLA DQA1*0102, HLA DQB1*0602, and HLA DPA1*0301 allelic association study. *Arq Neuropsiquiatr*, **67**(3A): 643–51.

Amit, R., Shapira, Y., Blank, A. and Aker, M. (1986). Acute, severe, central and peripheral nervous system combined demyelination. *Pediatr Neurol* **2**(1): 47–50.

Anlar, B., Basaran, C., Kose, G., *et al.* (2003). Acute disseminated encephalomyelitis in children: outcome and prognosis. *Neuropediatrics*, **34**(4): 194–9.

Arcondo, M.F., Wachs, A. and Zylberman, M. (2011). [Transverse myelitis associated with anti-influenza A (H1N1) vaccination]. *Medicina (B Aires)*, **71**(2): 161–4.

Arshi, S., Sadeghi-Bazargani, H., Ojaghi, H., *et al.* (2004). The first rapid onset optic neuritis after measles-rubella vaccination: case report. *Vaccine*, **22**(25–26): 3240–2.

Aydin, H., Ozgul, E. and Agildere, A.M. (2010). Acute necrotizing encephalopathy secondary to diphtheria, tetanus toxoid and whole-cell pertussis vaccination: diffusion-weighted imaging and proton MR spectroscopy findings. *Pediatr Radiol*, **40**(7): 1281–4.

Balasubramanya, K.S., Kovoor, J.M.E., Jayakumar, P.N., *et al.* (2007). Diffusion-weighted imaging and proton MR spectroscopy in the characterization of acute disseminated encephalomyelitis. *Neuroradiology*, **49**(2): 177–83.

Barkhof, F., Filippi, M., Miller, D.H., *et al.* (1997). Comparison of MRI criteria at first presentation to predict conversion to clinically definite multiple sclerosis. *Brain*, **120**(Pt 11): 2059–69.

Ben Sira, L., Miller, E., Artzi, M., *et al.* (2010). 1H-MRS for the diagnosis of acute disseminated encephalomyelitis: insight into the acute-disease stage. *Pediatr Radiol*, **40**(1): 106–13.

Bennetto, L. and Scolding, N. (2004). Inflammatory/post-infectious encephalomyelitis. *J Neurol Neurosurg Psychiatry*, **75**(Suppl. 1): 22–8.

Berkovic, S.F., Harkin, L., McMahon, J.M., *et al.* (2006). De-novo mutations of the sodium channel gene SCN1A in alleged vaccine encephalopathy: a retrospective study. *Lancet Neurol*, **5**(6): 488–92.

Bizzi, A., Uluğ, A.M., Crawford, T.O., *et al.* (2001). Quantitative proton MR spectroscopic imaging in acute disseminated encephalomyelitis. *AJNR*, **22**(6): 1125–30.

Brilot, F., Dale R.C., Selter R.C., *et al.* (2009). Antibodies to native myelin oligodendrocyte glycoprotein in children with inflammatory demyelinating central nervous system disease. *Ann Neurol*, **66**(6): 833–42.

Bryan, J.T. (2007). Developing an HPV vaccine to prevent cervical cancer and genital warts. *Vaccine*, **25**(16): 3001–6.

Cabrera-Gomez, J.A., Echazabal-Santana, N., Garcia Gonzalez, L., *et al.* (2002). [A severe episode in a patient with recurrent disseminated acute encephalitis due to vaccination against hepatitis B. For or against vaccination?]. *Rev Neurol*, **34**(4): 358–63.

Chaves, M., Riccio, P., Patrucco, L., *et al.* (2009). Longitudinal myelitis associated with yellow fever vaccination. *J Neurovirol*, **15**(4): 348–50.

Cisse, F.A., Sidibe, H., Morel, Y., *et al.* (2012). [Acute disseminated encephalomyelitis after tetanus vaccination of a pregnant woman in Senegal]. *Med Sante Trop*, **22**(1): 103–5.

Crawford, C., Grazko, M.B., Raymond, W.R., *et al.* (2012). Reversible blindness in bilateral optic neruritis associated with nasal flu vaccine. *Binocul Vis Strabolog Q Simms Romano*, **27**(3): 171–3.

Dadeya, S., Guliani, B.P., Gupta, V.S., *et al.* (2004). Retrobulbar neuritis following rabies vaccination. *Trop Doct*, **34**(3): 174–5.

Dale, R.C., de Sousa, C., Chong, W.K., *et al.* (2000). Acute disseminated encephalomyelitis, multiphasic disseminated encephalomyelitis and multiple sclerosis in children. *Brain,* **123**(Pt 12): 2407–22.

DeStefano, F., Verstraeten, T., Jackson, L.A., *et al.* (2003). Vaccinations and risk of central nervous system demyelinating diseases in adults. *Arch Neurol,* **60**(4): 504–9.

DiMario, F.J., Hajjar, M., and Ciesielski, T. (2010). A 16-year-old girl with bilateral visual loss and left hemiparesis following an immunization against human papilloma virus. *J Child Neurol,* **25**(3): 321–7.

Erguven, M., Guven, S., Akyuz, U., *et al.* (2009). Optic neuritis following hepatitis B vaccination in a 9-year-old girl. *J Chin Med Assoc,* **72**(11): 594–7.

Franciotta, D., Columba-Cabezas, S., Andreoni, L., *et al.* (2008). Oligoclonal IgG band patterns in inflammatory demyelinating human and mouse diseases. *J Neuroimmunol,* **200**(1–2): 125–8.

Fujii, K., Suyama, M., Chiba, K., *et al.* (2012). Acute disseminated encephalomyelitis following 2009 H1N1 influenza vaccine. *Pediatr Int,* **54**(14): 539–41.

Furukawa, Y., Komai, K. and Yamada, M. (2011). Neuromyelitis optica after Japanese encephalitis vaccination. *Eur J Neurol,* **18**(2): e26–7.

Gamboa, E.T., Cowen, D., Eggers, A., *et al.* (1983). Delayed onset of post-rabies vaccination encephalitis. *Ann Neurol,* **13**(6): 676–8.

Goebel, H.H., Walther, G., and Meuth, M. (1986). Fresh cell therapy followed by fatal coma. *J Neurol,* **233**(4): 242–7.

Gupta, V., Bandyopadhyay, S., Bapuraj, J.R. and Gupta, A. (2004). Bilateral optic neuritis complicating rabies vaccination. *Retina,* **24**(1): 179–81.

Gupte, G., Stonehouse, M., Wassmer, E., *et al.* (2003). Acute disseminated encephalomyelitis: a review of 18 cases in childhood. *J Paediatr Child Health,* **39**(5): 336–42.

Hahn, J.S., Siegler, D.J., and Enzmann, D. (1996). Intravenous gammaglobulin therapy in recurrent acute disseminated encephalomyelitis. *Neurology,* **46**(4): 1173–4.

Hahn, C.D., Miles, B.S., MacGregor, D.L., *et al.* (2003). Neurocognitive outcome after acute disseminated encephalomyelitis. *Pediatr Neurol,* **29**(2): 117–23.

Hamano, T., Takeda, T., Morita, H., *et al.* (2010). Posterior reversible encephalopathy syndrome following measles vaccination. *J Neurol Sci,* **298**(1–2): 124–6.

Hamidon, B.B. and Raymond, A.A. (2003). Acute disseminated encephalomyelitis (ADEM) presenting with seizures secondary to anti-tetanus toxin vaccination. *Med J Malaysia,* **58**(5): 780–2.

Hocine, M.N., Farrington, C.P., Touze, E., *et al.* (2007). Hepatitis B vaccination and first central nervous system demyelinating events: reanalysis of a case-control study using the self-controlled case series method. *Vaccine,* **25**(31): 5938–43.

Hoshino, T., Uchiyama, Y., Ito, E., *et al.* (2012). Simultaneous development of acute disseminated encephalomyelitis and Guillain-Barre syndrome associated with H1N1 09 influenza vaccination. *Intern Med,* **51**(12): 1595–8.

Huang, E.H.Z., Lim, S.A., Lim, P.L., and Leo, Y.S. (2009). Retrobulbar optic neuritis after hepatitis A vaccination in a HIV-infected patient. *Eye (Lond),* **23**(12): 2267–71.

Huber, S., Kappos, L., Fuhr, P., *et al.* (1999). Combined acute disseminated encephalomyelitis and acute motor axonal neuropathy after vaccination for hepatitis A and infection with *Campylobacter jejuni*. *J Neurol,* **246**(12): 1204–6.

Hull, T.P. and Bates, J.H. (1997). Optic neuritis after influenza vaccination. *Am J Ophthalmol,* **124**(5): 703–4.

Hung, K.L., Liao, H.T., and Tsai, M.L. (2001). The spectrum of postinfectious encephalomyelitis. *Brain Dev,* **23**(1): 42–5.

Huynh, W., Cordato, D.J., Kehdi, E., *et al.* (2008). Post-vaccination encephalomyelitis: literature review and illustrative case. *J Clin Neurosci,* **15**(12): 1315–22.

Hynson, J.L., Kornberg, A.J., Coleman, L.T., *et al.* (2001). Clinical and neuroradiologic features of acute disseminated encephalomyelitis in children. *Neurology,* **56**(10): 1308–12.

Ichiyama, T., Shoji, H., Kato, M., *et al.* (2002). Cerebrospinal fluid levels of cytokines and soluble tumour necrosis factor receptor in acute disseminated encephalomyelitis. *Eur J Pediatr,* **161**(3): 133–7.

Ichiyama, T., Kajimoto, M., Suenaga, N., *et al.* (2006). Serum levels of matrix metalloproteinase-9 and its tissue inhibitor (TIMP-1) in acute disseminated encephalomyelitis. *J Neuroimmunol,* **172**(1–2): 182–6.

Idrissova, Z.R., Boldyreva, M.N., Dekonenko, E.P., *et al.* (2003). Acute disseminated encephalomyelitis in children: clinical features and HLA-DR linkage. *Eur J Neurol,* **10**(5): 537–46.

Iizuka, H., Amano, N., Izeki, E., *et al.* (1993). A case of antirabies inoculation encephalitis with a long clinical course. *Jpn J Psychiatry Neurol,* **47**(3): 603–8.

Ishizu, T., Minohara, M., Ichiyama, T., *et al.* (2006). CSF cytokine and chemokine profiles in acute disseminated encephalomyelitis. *J Neuroimmunol,* **175**(1–2): 52–8.

Israeli, E., Agmon-Levin, N., Blank, M. and Shoenfeld, Y. (2009). Adjuvants and autoimmunity. *Lupus,* **18**(13): 1217–25.

Iyoda, K., Mitsuda, N., Ogawa, K., and Okazaki, T. (2004). [The head MRI by fluid attenuated inversion recovery (FLAIR) was useful for diagnosis in the patient with post influenza vaccinal acute disseminated encephalomyelitis (ADEM)]. *No To Hattatsu,* **36**(5): 401–6.

Jacobs, R.K., Anderson, V.A., Neale, J.L., *et al.* (2004). Neuropsychological outcome after acute disseminated encephalomyelitis: impact of age at illness onset. *Pediatr Neurol,* **31**(3): 191–7.

Jayakrishnan, M.P. and Krishnakumar, P. (2010). Clinical profile of acute disseminated encephalomyelitis in children. *J Pediatr Neurosci*, **5**(2): 111–14.

Jefferson, T. and Heijbel, H. (2001). Demyelinating disease and hepatitis B vaccination: is there a link? *Drug Saf*, **24**(4): 249–54.

Jorens, P.G., VanderBorght, A., Ceulemans, B., *et al.* (2000). Encephalomyelitis-associated antimyelin autoreactivity induced by streptococcal exotoxins. *Neurology*, **54**(7): 1433–41.

Kalimo, K.O., Marttila, R.J., Ziola, B.R., *et al.* (1977). Radioimmunoassay of herpes-simplex and measles virus antibodies in serum and cerebrospinal fluid of patients without infectious or demyelinating diseases of the central nervous system. *J Med Microbiol*, **10**(4): 431–8.

Karussis, D.M., Slavin, S., Ben-Nun, A., *et al.* (1992a). Chronic-relapsing experimental autoimmune encephalomyelitis (CR-EAE): treatment and induction of tolerance, with high dose cyclophosphamide followed by syngeneic bone marrow transplantation. *J Neuroimmunol*, **39**(3): 201–10.

Karussis, D.M., Slavin, S., Lehmann, D., *et al.* (1992b). Prevention of experimental autoimmune encephalomyelitis and induction of tolerance with acute immunosuppression followed by syngeneic bone marrow transplantation. *J Immunol*, **148**(6): 1693–8.

Karussis, D., Vourka-Karussis, U., Mizrachi-Koll, R. and Abramsky, O. (1999). Acute/relapsing experimental autoimmune encephalomyelitis: induction of long lasting, antigen-specific tolerance by syngeneic bone marrow transplantation. *Mult Scler*, **5**(1): 17–21.

Keegan, M., Pineda, A. A., McClelland, R.L., *et al.* (2002). Plasma exchange for severe attacks of CNS demyelination: predictors of response. *Neurology*, **58**(1): 143–6.

Kengsakul, K., Sathirapongsasuti, K. and Punyagupta, S. (2002). Fatal myeloencephalitis following yellow fever vaccination in a case with HIV infection. *J Med Assoc Thai*, **85**(1): 131–4.

Kerrison, J.B., Lounsbury, D., Thirkill, C.E., *et al.* (2002). Optic neuritis after anthrax vaccination. *Ophthalmology*, **109**(1): 99–104.

Khan, S., Yaqub, B.A., Poser, C.M., *et al.* (1995). Multiphasic disseminated encephalomyelitis presenting as alternating hemiplegia. *J Neurol Neurosurg Psychiatry*, **58**(4): 467–70.

Kitazawa, Y., Warabi, Y., Bandoh, M., *et al.* (2012). Elderly-onset neuromyelitis optica which developed after the diagnosis of prostate adenocarcinoma and relapsed after a 23-valent pneumococcal polysaccharide vaccination. *Intern Med*, **51**(1): 103–7.

Kleiman, M. and Brunquell, P. (1995). Acute disseminated encephalomyelitis: response to intravenous immunoglobulin. *J Child Neurol*, **10**(6): 481–3.

Kline, L.B., Margulies, S.L., and Oh, S.J. (1982). Optic neuritis and myelitis following rubella vaccination. *Arch Neurol*, **39**(7): 443–4.

Konstantinou, D., Paschalis, C., Maraziotis, T., *et al.* (2001). Two episodes of leukoencephalitis associated with recombinant hepatitis B vaccination in a single patient. *Clin Infect Dis*, **33**(10): 1772–3.

Krupp, L.B., Banwell, B., and Tenembaum, S. (2007). Consensus definitions proposed for pediatric multiple sclerosis and related disorders. *Neurology*, **68**(16 Suppl. 2): S7–12.

Kulkarni, V., Nadgir, D., Tapiawala, S., *et al.* (2004). Biphasic demyelination of the nervous system following anti-rabies vaccination. *Neurol India*, **52**(1): 106–8.

Kuperan, S., Ostrow, P., Landi, M.K., and Bakshi, R. (2003). Acute hemorrhagic leukoencephalitis vs ADEM: FLAIR MRI and neuropathology findings. *Neurology*, **60**(4): 721–2.

Labauge, R., Marty-Double, C., Pages, M., *et al.* (1979). [Postvaccinal encephalitis in adult. A case with anatomo-clinical report (author's transl.)]. *Rev Neurol (Paris)*, **135**(45321): 803–13.

Laffon-Pioger, M., Rocher, F., Cohen, M., *et al.* (2010). [Bilateral optic neuropathy with loss of vision after an influenza vaccination in a patient suffering from mixed connective tissue disease]. *Rev Neurol (Paris)*, **166**(12): 1024–7.

Lalive, P.H., Hausler, M.G., Maurey, H., *et al.* (2011). Highly reactive anti-myelin oligodendrocyte glycoprotein antibodies differentiate demyelinating diseases from viral encephalitis in children. *Mult Scler*, **17**(3): 297–302.

Lang, H.L.E., Jacobsen, H., Ikemizu, S., *et al.* (2002). A functional and structural basis for TCR cross-reactivity in multiple sclerosis. *Nat Immunol*, **3**(10): 940–3.

Lapphra, K., Huh, L., and Scheifele, D.W. (2011). Adverse neurologic reactions after both doses of pandemic H1N1 influenza vaccine with optic neuritis and demyelination. *Pediatr Infect Dis J*, **30**(1): 84–6.

Laria, C., Alio, J., Rodriguez, J.L., *et al.* (2006). [Optic neuritis after meningococcal vaccination]. *Arch Soc Esp Oftalmol*, **81**(8): 479–82.

Larner, A.J. and Farmer, S.F. (2000). Myelopathy following influenza vaccination in inflammatory CNS disorder treated with chronic immunosuppression. *Eur J Neurol*, **7**(6): 731–3.

Leake, J.A., Albani, S., Kao, A.S., *et al.* (2004). Acute disseminated encephalomyelitis in childhood: epidemiologic, clinical and laboratory features. *Pediatr Infect Dis J*, **23**(8): 756–64.

Lee, S.T., Choe, Y.J., Moon, W.J., *et al.* (2011). An adverse event following 2009 H1N1 influenza vaccination: a case of acute disseminated encephalomyelitis. *Korean J Pediatr*, **54**(10): 422–4.

Levy-Bruhl, D., Desenclos, J.C., Rebiere, I., and Drucker, J. (2002). Central demyelinating disorders and hepatitis B vaccination: a risk-benefit approach for pre-adolescent vaccination in France. *Vaccine*, **20**(16): 2065–71.

Lin, C.-H., Jeng, J.-S., and Yip, P.-K. (2004). Plasmapheresis in acute disseminated encephalomyelitis. *J Clin Apher*, **19**(3): 154–9.

Loebermann, M., Winkelmann, A., Hartung, H.P., *et al.* (2011). Vaccination against infection in patients with multiple sclerosis. *Nat Rev Neurol*, **8**(3): 143–51.

Machicado, J.D., Bhagya-Rao, B., Davogustto, G. and McKelvy, B.J. (2013). Acute disseminated encephalomyelitis following seasonal influenza vaccination in an elderly patient. *Clin Vaccine Immunol*, **20**(9): 1485–6.

Mader, I., Wolff, M., Niemann, G., and Küker, W. (2004). Acute haemorrhagic encephalomyelitis (AHEM): MRI findings. *Neuropediatrics*, **35**(2): 143–6.

Maeda, K. and Idehara, R. (2012). Acute disseminated encephalomyelitis following 2009 H1N1 influenza vaccination. *Intern Med*, **51**(14): 1931–3.

Malagon, T., Drolet, M., Boily, M.C., *et al.* (2012). Cross-protective efficacy of two human papillomavirus vaccines: a systematic review and meta-analysis. *Lancet Infect Dis*, **12**(10): 781–9.

Mancini, J., Chabrol, B., Moulene, E., and Pinsard, N. (1996). Relapsing acute encephalopathy: a complication of diphtheria-tetanus-poliomyelitis immunization in a young boy. *Eur J Pediatr*, **155**(2): 136–8.

Marchioni, E., Marinou-Aktipi, K., Uggetti, C., *et al.* (2002). Effectiveness of intravenous immunoglobulin treatment in adult patients with steroid-resistant monophasic or recurrent acute disseminated encephalomyelitis. *J Neurol*, **249**(1): 100–4.

Marchioni, E., Ravaglia, S., Piccolo, G., *et al.* (2005). Postinfectious inflammatory disorders: subgroups based on prospective follow-up. *Neurology*, **65**(7): 1057–65.

Marin, S.E. and Callen, D.J.A. (2013). The magnetic resonance imaging appearance of monophasic acute disseminated encephalomyelitis: an update post application of the 2007 consensus criteria. *Neuroimaging Clin N Am*, **23**(2): 245–66.

Markovic-Plese, S., Hemmer, B., Zhao, Y., *et al.* (2005). High level of cross-reactivity in influenza virus hemagglutinin-specific CD4+ T-cell response: implications for the initiation of autoimmune response in multiple sclerosis. *J Neuroimmunol*, **169**(1–2): 31–8.

Marrack, P., McKee, A.S., and Munks, M.W. (2009). Towards an understanding of the adjuvant action of aluminium. *Nat Rev Immunol*, **9**(4): 287–93.

Martinez-Sernandez, V. and Figueiras, A. (2013). Central nervous system demyelinating diseases and recombinant hepatitis B vaccination: a critical systematic review of scientific production. *J Neurol*, **260**(8): 1951–9.

Martino, D., Branson, J.A., Church, A.J., *et al.* (2005). Soluble adhesion molecules in acute disseminated encephalomyelitis. *Pediatr Neurol*, **33**(4): 255–8.

McDonald, W.I., Compston, A., Edan, G., *et al.* (2001). Recommended diagnostic criteria for multiple sclerosis: guidelines from the International Panel on the diagnosis of multiple sclerosis. *Ann Neurol*, **50**(1): 121–7.

Menge, T., Kieseier, B.C., Nessler, S., *et al.* (2007). Acute disseminated encephalomyelitis: an acute hit against the brain. *Curr Opin Neurol*, **20**(3): 247–54.

Menge, T., Cree, B., Saleh, A., *et al.* (2012). Neuromyelitis optica following human papillomavirus vaccination. *Neurology*, **79**(3): 285–7.

Merelli, E. and Casoni, F. (2000). Prognostic factors in multiple sclerosis: role of intercurrent infections and vaccinations against influenza and hepatitis B. *Neurol Sci*, **21**(4 Suppl. 2): S853–6.

Mikaeloff, Y., Suissa, S., Vallee, L., *et al.* (2004). First episode of acute CNS inflammatory demyelination in childhood: prognostic factors for multiple sclerosis and disability. *J Pediatr*, **144**(2): 246–52.

Mikaeloff, Y., Caridade, G., Husson, B., *et al.* (2007). Acute disseminated encephalomyelitis cohort study: prognostic factors for relapse. *Eur J Paediatr Neurol*, **11**(2): 90–5.

Miravalle, A., Biller, J., Silva, E., *et al.* (2009). Acute disseminated encephalomyelitis: yellow fever vaccination. *Arq Neuropsiquiatr*, **67**(3A): 710–11.

Moradian, S. and Ahmadieh, H. (2008). Early onset optic neuritis following measles-rubella vaccination. *J Ophthalmic Vis Res*, **3**(2): 118–22.

Murthy, S.N.K., Faden, H.S., Cohen, M.E., and Bakshi, R. (2002). Acute disseminated encephalomyelitis in children. *Pediatrics*, **110**(2 Pt 1): e21.

Nadkarni, N. and Lisak, R.P. (1993). Guillain-Barre syndrome (GBS) with bilateral optic neuritis and central white matter disease. *Neurology*, **43**(3): 842–3.

Nakamura, N., Nokura, K., Zettsu, T., *et al.* (2003). Neurologic complications associated with influenza vaccination: two adult cases. *Intern Med*, **42**(2): 191–4.

Nishikawa, M., Ichiyama, T., Hayashi, T., *et al.* (1999). Intravenous immunoglobulin therapy in acute disseminated encephalomyelitis. *Pediatr Neurol*, **21**(2): 583–6.

O'Connor, K.C., Chitnis, T., Griffin, D.E., *et al.* (2003). Myelin basic protein-reactive autoantibodies in the serum and cerebrospinal fluid of multiple sclerosis patients are characterized by low-affinity interactions. *J Neuroimmunol*, **136**(1–2): 140–8.

Orgogozo, J.M., Gilman, S., Dartigues, J.F., *et al.* (2003). Subacute meningoencephalitis in a subset of patients with AD after Abeta42 immunization. *Neurology*, **61**(1): 46–54.

Pavone, P., Pettoello-Mantovano, M., Le Pira, A., *et al.* (2010). Acute disseminated encephalomyelitis: a long-term prospective study and meta-analysis. *Neuropediatrics*, **41**(6): 246–55.

Perry, H.D., Mallen, F.J., Grodin, R.W., and Cossari, A.J. (1979). Reversible blindness in optic neuritis associated with influenza vaccination. *Ann Ophthalmol*, **11**(4): 545–50.

Pohl-Koppe, A., Burchett, S.K., Thiele, E.A., and Hafler, D.A. (1998). Myelin basic protein reactive Th2 T cells are found in acute disseminated encephalomyelitis. *J Neuroimmunol*, **91**(1–2): 19–27.

Pradhan, S., Gupta, R.P., Shashank, S., and Pandey, N. (1999). Intravenous immunoglobulin therapy in acute disseminated encephalomyelitis. *J Neurol Sci*, **165**(1): 56–61.

Py, M.O. and Andre, C. (1997). [Acute disseminated encephalomyelitis and meningococcal A and C vaccine: case report]. *Arq Neuropsiquiatr*, **55**(3B): 632–5.

Ravaglia, S., Ceroni, M., Moglia, A., *et al.* (2004). Post-infectious and post-vaccinal acute disseminated encephalomyelitis occurring in the same patients. *J Neurol*, **251**(9): 1147–50.

Ray, C.L. and Dreizin, I.J. (1996). Bilateral optic neuropathy associated with influenza vaccination. *J Neuroophthalmol*, **16**(3): 182–4.

Reeves, W.H., Lee, P.Y., Weinstein, J.S., *et al.* (2009). Induction of autoimmunity by pristane and other naturally occurring hydrocarbons. *Trends Immunol*, **30**(9): 455–64.

Refai, D., Lee, M.C., Goldenberg, F.D., and Frank, J.I. (2005). Decompressive hemicraniectomy for acute disseminated encephalomyelitis: case report. *Neurosurgery*, **56**(4): e872; disc. E871.

Revel-Vilk, S., Hurvitz, H., Klar, A., *et al.* (2000). Recurrent acute disseminated encephalomyelitis associated with acute cytomegalovirus and Epstein-Barr virus infection. *J Child Neurol*, **15**(6): 421–4.

Rosenmann, H., Grigoriadis, N., Karussis, D., *et al.* (2006). Tauopathy-like abnormalities and neurologic deficits in mice immunized with neuronal tau protein. *Arch Neurol*, **63**(10): 1459–67.

Rubinov, A., Beiran, I., Krasnitz, I., and Miller, B. (2012). Bilateral optic neuritis after inactivated influenza vaccination. *IMAJ*, **14**(11): 705–7.

Rust, R.S. (2000). Multiple sclerosis, acute disseminated encephalomyelitis, and related conditions. *Semin Pediatr Neurol*, **7**(2): 66–90.

Sahlas, D.J., Miller, S.P., Guerin, M., *et al.* (2000). Treatment of acute disseminated encephalomyelitis with intravenous immunoglobulin. *Neurology*, **54**(6): 1370–2.

Satoh, M. and Reeves, W.H. (1994). Induction of lupus-associated autoantibodies in BALB/c mice by intraperitoneal injection of pristane. *J Exp Med*, **180**(6): 2341–6.

Schaffer, V., Wimmer, S., Rotaru, I., *et al.* (2008). HPV vaccine: a cornerstone of female health a possible cause of ADEM? *J Neurol*, **255**(11): 1818–20.

Schattenfroh, C. (2004). [Acute disseminated encephalomyelitis after active immunization against early summer encephalitis]. *Nervenarzt*, **75**(8): 776–9.

Schiller, J.T., Castellsague, X., and Garland, S.M. (2012). A review of clinical trials of human papillomavirus prophylactic vaccines. *Vaccine*, **30**(Suppl. 5): F123–38.

Sejvar, J.J., Labutta, R.J., Chapman, L.E., *et al.* (2005). Neurologic adverse events associated with smallpox vaccination in the United States, 2002–2004. *JAMA*, **294**(21): 2744–50.

Sell, E. and Minassian, B.A. (2006). Demystifying vaccination-associated encephalopathy. *Lancet Neurol*, **5**(6): 465–6.

Shahar, E., Andraus, J., Savitzki, D., *et al.* (2002). Outcome of severe encephalomyelitis in children: effect of high-dose methylprednisolone and immunoglobulins. *J Child Neurol*, **17**(11): 810–14.

Shoamanesh, A. and Traboulsee, A. (2011). Acute disseminated encephalomyelitis following influenza vaccination. *Vaccine*, **29**(46): 8182–5.

Shoji, H., Kusuhara, T., Honda, Y., *et al.* (1992). Relapsing acute disseminated encephalomyelitis associated with chronic Epstein-Barr virus infection: MRI findings. *Neuroradiology*, **34**(4): 340–2.

Shu, M., Liu, Q., Wang, J., *et al.* (2011). Measles vaccine adverse events reported in the mass vaccination campaign of Sichuan province, China from 2007 to 2008. *Vaccine*, **29**(18): 3507–10.

Stevenson, V.L., Acheson, J.F., Ball, J., and Plant, G.T. (1996). Optic neuritis following measles/rubella vaccination in two 13-year-old children. *Br J Ophthalmol*, **80**(12): 1110–11.

Sutton, I., Lahoria, R., Tan, I., *et al.* (2009). CNS demyelination and quadrivalent HPV vaccination. *Mult Scler*, **15**(1): 116–19.

Suwa, K., Yamagata, T., Momoi, M.Y., *et al.* (1999). Acute relapsing encephalopathy mimicking acute necrotizing encephalopathy in a 4-year-old boy. *Brain Dev*, **21**(8): 554–8.

Talbot, P.J., Paquette, J.S., Ciurli, C., *et al.* (1996). Myelin basic protein and human coronavirus 229E cross-reactive T cells in multiple sclerosis. *Ann Neurol*, **39**(2): 233–40.

Tardieu, M. and Mikaeloff, Y. (2004). What is acute disseminated encephalomyelitis (ADEM)? *Eur J Paediatr Neurol*, **8**(5): 239–42.

Tejada-Simon, M.V., Zang, Y.C.Q., Hong, J., *et al.* (2003). Cross-reactivity with myelin basic protein and human herpesvirus-6 in multiple sclerosis. *Ann Neurol*, **53**(2): 189–97.

Tenembaum, S., Chamoles, N., and Fejerman, N. (2002). Acute disseminated encephalomyelitis: a long-term follow-up study of 84 pediatric patients. *Neurology*, **59**(8): 1224–31.

Torisu, H., Kira, R., Ishizaki, Y., *et al.* (2010). Clinical study of childhood acute disseminated encephalomyelitis, multiple sclerosis, and acute transverse myelitis in Fukuoka Prefecture, Japan. *Brain Dev*, **32**(6): 454–62.

Touzé, E., Gout, O., Verdier-Taillefer, M.H., *et al.* (2000). [The first episode of central nervous system demyelinization and hepatitis B virus vaccination]. *Rev Neurol (Paris)*, **156**(3): 242–6.

Touzé, E., Fourrier, A., Rue-Fenouche, C., *et al.* (2002). Hepatitis B vaccination and first central nervous system demyelinating event: a case-control study. *Neuroepidemiology*, **21**(4): 180–6.

Tsai, M.L. and Hung, K.L. (1996). Multiphasic disseminated encephalomyelitis mimicking multiple sclerosis. *Brain Dev*, **18**(5): 412–14.

Tsuru, T., Mizuguchi, M., Ohkubo, Y., *et al.* (2000). Acute disseminated encephalomyelitis after live rubella vaccination. *Brain Dev*, **22**(4): 259–61.

Ubol, S., Hemachudha, T., Whitaker, J.N., and Griffin, D.E. (1990). Antibody to peptides of human myelin basic protein in post-rabies vaccine encephalomyelitis sera. *J Neuroimmunol*, **26**(2): 107–11.

Van Dam, C.N., Syed, S., Eron, J.J., *et al.* (2009). Severe postvaccinia encephalitis with acute disseminated encephalomyelitis: recovery with early intravenous immunoglobulin, high-dose steroids, and vaccinia immunoglobulin. *Clin Infect Dis*, **48**(4): e47–9.

van de Geijn, E.J., Tukkie, R., van Philips, L.A., and Punt, H. (1994). Bilateral optic neuritis with branch retinal artery occlusion associated with vaccination. *Doc Ophthalmol*, **86**(4): 403–8.

Van Haren, K., Tomooka, B.H., Kidd, B.A., *et al.* (2013). Serum autoantibodies to myelin peptides distinguish acute disseminated encephalomyelitis from relapsing-remitting multiple sclerosis. *Mult Scler*, **19**(13): 1726–33.

Varga, S.M., Szabo, I., and Balazs, L. (1979). [Hyperacute encephalomyelitis following lyssa vaccination]. *Morphol Igazsagugyi Orv Sz*, **19**(1): 27–31.

Vera-Lastra, O., Medina, G., Cruz-Dominguez, M.D.P., *et al.* (2013). Autoimmune/inflammatory syndrome induced by adjuvants (Shoenfeld's syndrome): clinical and immunological spectrum. *Expert Rev Clin Immunol*, **9**(4): 361–73.

Vilain, S., Waterschoot, M.P., and Mavroudakis, N. (2000). Encephalomyelitis and bilateral optic perineuritis after influenza vaccination. *Bull Soc Belge Ophtalmol*, **2000**(277): 71–3.

Visudtibhan, A., Tuntiyathorn, L., Vaewpanich, J., *et al.* (2010). Acute disseminated encephalomyelitis: a 10-year cohort study in Thai children. *Eur J Paediatr Neurol*, **14**(6): 513–18.

von Stuckrad-Barre, S., Klippel, E., Foerch, C., *et al.* (2003). Hemicraniectomy as a successful treatment of mass effect in acute disseminated encephalomyelitis. *Neurology*, **61**(3): 420–1.

Wildemann, B., Jarius, S., Hartmann, M., *et al.* (2009). Acute disseminated encephalomyelitis following vaccination against human papilloma virus. *Neurology*, **72**(24): 2132–3.

Wingerchuk, D.M. (2003). Postinfectious encephalomyelitis. *Curr Neurol Neurosci Rep*, **3**(3): 256–64.

Yen, M.Y. and Liu, J.H. (1991). Bilateral optic neuritis following bacille Calmette-Guerin (BCG) vaccination. *J Clin Neuroophthalmol*, **11**(4): 246–9.

Zipp, F., Weil, J.G., and Einhaupl, K.M. (1999). No increase in demyelinating diseases after hepatitis B vaccination. *Nat Med*, **5**(9): 964–5.

34 Fibromyalgia, Chronic Fatigue, Functional Disorders, and Vaccination: Where Do We Stand?

Jacob N. Ablin[1] and Dan Buskila[2]

[1]Departments of Rheumatology, Tel Aviv Sourasky Medical Center and Sackler Faculty of Medicine, Tel Aviv University, Tel Aviv, Israel

[2]Rheumatic Disease Unit, Department of Medicine, Soroka Medical Center, Beersheba, Israel

Introduction

Fibromyalgia syndrome (FMS) represents a unique entity within the field of rheumatology, differing from the more "classical" rheumatological disorders both in pathogenesis and in its modes of management. Focusing on the ability of the central nervous system (CNS) to process and modulate pain (Yunus, 2007), rather than on the interaction between the immune and the musculoskeletal systems, the syndrome of fibromyalgia has evolved over recent years into a novel paradigm, leading both clinicians and researchers to deal with chronic pain as a nosological entity in and of itself (Niv and Devor, 2004), rather than a mere side phenomenon associated with other disorders. Our understanding of the alterations which take place in the CNS, both at a spinal level and in pain-processing areas within the brain, has been greatly enhanced through implementation of novel functional imaging modalities, which have provided evidence that enhanced pain processing in FMS is an objective ("real") phenomenon (Gracely *et al.*, 2002). Elsewhere, progress has been made regarding the documentation and analysis of the ways in which the CNS is able to adapt and react to pain. Thus, for example, it has become evident that under normal circumstances, the CNS is able to activate descending inhibitory signals to the spinal levels in order to dampen the input of pain. This activity, known as conditioned pain modulation (CPM, previously termed diffuse noxious inhibitory control, DNIC), appears to be compromised among many FMS patients (Julien *et al.*, 2005), offering a possible pathogenetic explanation for enhanced processing of pain. At the same time, multiple lines of evidence have implicated genetic factors as contributing factors in the development of FMS (Buskila *et al.*, 2007). Thus, the interaction between genetic predisposition and various triggers that act upon the CNS during life appears to be the background against which FMS develops. That being said, however, the precise nature of these triggers remains to be elucidated, not to mention the pathways through which they might act in order to instigate central sensitization in the CNS of a genetically predisposed individual. This point is particularly poignant in view of the diverse nature of the triggers that have been associated with FMS, which would appear to defy a simple unifying explanation. Thus, physical trauma (particularly to the cervical spine) (Buskila, 2000), emotional stress (to various stressors) (Martinez-Lavin, 2007), and a variety of infections (Ablin *et al.*, 2006) have been etiologically linked with FMS. In this chapter, we shall focus our discussion on the

Vaccines and Autoimmunity, First Edition. Edited by Yehuda Shoenfeld, Nancy Agmon-Levin, and Lucija Tomljenovic.

possibility of vaccinations acting as triggers for the development of FMS.

Allen (1998) put forth the hypothesis that FMS (then termed fibrositis) and chronic fatigue might represent a reaction to a rubella immunization received 2 decades previously, noting an apparent epidemiological association between introduction of a new rubella vaccine (strain RA27/3) in 1979 and the debut of publications regarding chronic fatigue in subsequent years. Later, a causal relationship was recognized between this vaccine and a spectrum of musculoskeletal complications, including FMS, arthralgia, arthritis, and various nonspecific symptoms (Weibel and Benor, 1996).

Rubella seronegativity is routinely screened for in early pregnancy, and the rubella vaccine is administered postpartum. A randomized, placebo-controlled trial compared complications developing after this procedure with outcome after administration of placebo (Tingle *et al.*, 1997). RA27/3 rubella vaccine was associated with acute arthralgia and arthritis, while the increase in chronic arthralgia and arthritis was marginally significant. Only a small difference was found in frequency of persistent myalgia between recipients of RA27/3 vaccine (15%) and placebo (9%), while no difference in frequency of acute myalgia was observed. The risk of post-RA27/3 vaccine arthropathy may be higher among women exhibiting very low prevaccination antibody levels (Mitchell *et al.*, 2000).

Thus, while acute arthralgia and myalgia can be causally attributed to rubella vaccine, current data appear to be insufficient in order to determine a causal relationship with the development of FMS.

Lathrop *et al.* (2002) reviewed adverse events related to Lyme vaccination in the United States between December 1998 and July 2000. Arthralgia, myalgia, and pain were the most common reactions, together accounting for over 66% of adverse events. Notably, these nonspecific symptoms are to some extent overlapping with those of FMS.

Gulf War syndrome and vaccination

Gulf War syndrome (GWS) is an unusual case of a functional disorder strictly situated in a specific historical and geographical circumstance. The syndrome was described among veterans of the military conflict occurring in 1990–91 in the Persian Gulf and was eventually diagnosed among thousands of soldiers from different armies involved in the military coalition formed in that conflict. This syndrome, which developed among young individuals of presumably high prior health status, called attention to the profound capacity of specific environmental exposures to cause the development of functional symptoms in a specific setting.

Characterized by chronic fatigue, musculoskeletal symptoms, malaise, and cognitive impairment (Fukuda *et al.*, 1998; Lange *et al.*, 2001), GWS clinically overlaps with both post-traumatic stress disorder (PTSD) and FMS/chronic fatigue syndrome (CFS), as well as with other functional disorders (Ford *et al.*, 2001; Binder and Campbell, 2004). While deployment in the Gulf War evidently carried similarities to multiple other military conflicts, the specific circumstances also imparted unique exposures, which differed from many others examples. The extreme climate of the Persian Gulf is one such factor, and exposure to various chemicals (pesticides, depleted uranium) is another. The fact that most army personnel ultimately were not involved in actual combat but were exposed to prolonged waiting (and presumably to ongoing stress) is another specific factor to consider. Finally, the intense exposure of allied soldiers to vaccinations (due to high concern regarding the use of biological weapons), all administered within a short timeframe and concomitant with the other exposures, make the circumstances of GWS unique.

In a comparison performed between Gulf War veterans and veterans of the Bosnian conflict, for instance (Unwin *et al.*, 1999), the multiple vaccinations administered to soldiers in the Gulf War were identified as an exposure relatively unique to this conflict.

Hotopf *et al.* (2000) analyzed the relationship between ill health after the Gulf War and vaccine administration before and during deployment. Measures assessed included fatigue, PTSD, distress, health perception, physical function, and the presence of "multisymptom illness." Multiple vaccine administration *prior* to deployment was associated with only one of these six measures (PTSD), while administration of vaccines *during* deployment in the Gulf was associated with all five other measures. Thus, while multiple vaccinations in and of themselves do not appear harmful, the combination of vaccination administration and concurrent deployment-associated stress (and possible other factors) may increase the risk of developing functional symptoms. In a subsequent study, the severity of initial symptoms and associated psychological distress (Hotopf *et al.*, 2004) was found to be more important in long-run symptom perpetuation than multiple vaccine exposure.

The mechanism through which vaccination exposure might lead to the development of functional symptoms is not completely understood. Rook and Zumla (1997) have raised the possibility that a shift from Th1- to Th2-type reactions could be of pathogenic significance. In support of this theory, CFS patients have an increased frequency of allergic reactions, low natural killer (NK) cell activity, and low levels of interferon gamma (IFN-γ) and interleukin (IL-) 2. Administration of pertussis vaccine, a known Th2 inducer in the Gulf War, could contribute to this shift (Mu and Sewell, 1993). Deployment-related stress could also have Th2-inducing effects, mediated thorough increased cortisol levels and decreased androgen levels (e.g. DHEA) (Bernton *et al.*, 1995; Ramirez *et al.*, 1996). Increased Th2 activation has been reported in CFS unrelated to GWS (Skowera *et al.*, 2004), but the application of these trends to FMS remains to be proven.

Asa *et al.* (2000) have described the presence of antibodies to squalene among 95% of patients in a small cohort of GWS patients, as well as among 100% of patients who received predeployment vaccinations but were not deployed. In this study, none of the deployed Persian Gulf veterans who did not develop symptoms exhibited antisqualene antibodies (Asa *et al.*, 2000). While these results appear convincing, a larger study subsequently conducted among 579 US Navy veterans found no association between squalene antibody status and chronic multisymptom illness (Phillips *et al.*, 2009).

Recent research has tended to reclassify symptoms previously described under the title of "GWS" into a broader category of symptoms associated with deployment in areas of combat, the so-called "post-deployment syndrome" (PDS) (Lewis *et al.*, 2012). This syndrome is typically associated with chronic widespread pain, fatigue, and cognitive difficulties. While various toxic exposures endured during deployment, such as handling of pesticides and exposure to depleted uranium, have been linked with subsequent morbidity (Macfarlane *et al.*, 2005), a clear association between PDS and vaccinations has not been established.

An unusual case of chronic fatigue, FMS, demyelination, and autoantibodies has been described following the combination of hepatitis B vaccination and silicone breast implant rupture (Agmon-Levin and Shoenfeld, 2008). The authors in this case raised the hypothesis that an interaction between silicon, acting as an adjuvant, and exposure to the vaccine acted to augment the immune response.

Macrophagic myofasciitis (MMF) is a recently described entity, clinically characterized by diffuse myalgia, arthralgia, and muscle weakness (Gherardi *et al.*, 1998). Chronic fatigue is another major symptom (Authier *et al.*, 2003). The pathology of MMF is characterized by pathognomonic focal muscle fiber infiltration by large periodic acid–Schiff (PAS)-positive macrophages, intermingled with $CD8^+$ T cells. MMF has repeatedly been linked with vaccine-associated exposure to aluminum (Gherardi *et al.*, 2001; Exley *et al.*, 2009), although non-vaccine-related MMF has also been described (Park *et al.*, 2005). While MMF appears to be a relevant differential diagnosis for patients diagnosed with new-onset FMS and chronic fatigue, the distinct pathological and immunological findings associated with MMF (which are not typical of FMS) indicate that an etiologic link between vaccine-associated aluminum exposure and FMS cannot be assumed.

Vaccinations and chronic fatigue

FMS and CFS are conditions with considerable clinical overlap. Fatigue is an inherent symptom of FMS, along with disturbed nonrefreshing sleep, while muscular pain is a common symptom among those diagnosed as suffering from CFS. Nonetheless, chronic widespread pain, which is the hallmark of FMS, is less typical of "pure" CFS, so a clinical distinction can be made in a proportion of patients.

At the same time, fatigue is a frequent complaint among patients suffering from chronic autoimmune and inflammatory disorders, such as systemic lupus erythematosus (SLE). CFS has been associated with a broad spectrum of immunological disruptions, which, however, have often proven to be nonspecific.

Ortega-Hernandez and Shoenfeld (2009) have extensively reviewed the association between CFS and both infection and vaccination, and have discussed the etiological conundrum raised by these associations. While they found vaccinations against rubella (Morag *et al.*, 1999), Allen (1988), Q fever (Madariaga *et al.*, 2003), and hepatitis B (Delage *et al.*, 1993) to be associated with the risk of developing CFS, meningococcal vaccine (Magnus *et al.*, 2009), poliovirus (Vedhara *et al.*, 1997), and influenza vaccine (Sleigh *et al.*, 2000) were not. Intriguingly, *Staphylococcus* toxoid vaccine actually appeared to have a protective effect (Andersson *et al.*, 1998).

What about the ASIA syndrome?

Autoimmune/inflammatory syndrome induced by adjuvants (ASIA) is a recently described entity that encompasses symptoms from a number of previous clinical syndrome, including siliconosis, GWS, MMF, and post-vaccination phenomena linked with exposure to an adjuvant (Shoenfeld and Agmon-Levin, 2011). This syndrome is characterized by nonspecific and specific manifestations of autoimmune disease, associated with exposure to squalene, aluminum hydroxide, silicone, mineral oil, guaiacol, and iodine gadital (Vera-Lastra *et al.*, 2013).

The clinical symptoms of ASIA syndrome induced by hepatitis B vaccination include fatigue (42%) and neuropsychiatric (70%), musculoskeletal (59%), and gastrointestinal (50%) symptoms (Zafrir *et al.*, 2012). Elevated titers of autoantibodies were documented in 80% of sera tested. Thus, ASIA syndrome, as currently described, holds considerable clinical overlap with the FMS/CFS spectrum of disease. On the other hand, autoantibodies are not a universal finding among FMS patients, and no specific antibodies have been identified to date. The possibility that FMS represents a case of subtle inflammation within the CNS, such as microglia cell activation, is a novel and provocative concept that needs to be further explored. Activated microglia may contribute to pain conditions through production of proinflammatory cytokines, chemokines, and extracellular proteases, in addition to exhibiting a modulated cell-surface-receptor and ion-channel profile (Schomberg and Olson, 2012). Thus, the possibility that subtle manifestations of autoimmunity and inflammation within the CNS may play a role in the pathogenesis of FMS is worth further attention. As the concepts of adjuvant-associated autoimmunity and subtle neuroinflammation associated with central sensitization (and chronic pain) continue to evolve, further research is called for in order to evaluate the possibility of an etiological link between these two exciting fields.

Conclusions

Within the complex etiological scheme evolving for FMS, a variety of environmental exposures have been recognized as or suspected of being potential triggers, presumably capable of instigating central sensitization of the genetically prone individual. Within this context, a variety of vaccinations have been documented as being associated with a range of symptoms at least partially overlapping with FMS, such as widespread musculoskeletal pain and fatigue.

GWS poses a unique circumstance in which a multisymptom functional disorder developed among many thousands of healthy young individuals, following exposure to a constellation of environmental and stressful circumstances, including the administration of multiple simultaneous vaccinations within a short period.

The current evolution of the ASIA syndrome, as well as intriguing indications regarding a role for previously unrecognized CNS inflammation (e.g. microglia activation) in the pathogenesis of central sensitization and chronic pain, indicates that we may currently be standing on the brink of a new era of understanding of the enigma of chronic pain.

References

Ablin, J.N., Shoenfeld, Y., and Buskila, D. (2006). Fibromyalgia, infection and vaccination: two more parts in the etiological puzzle. *J Autoimmun*, **27**(3): 145–52.

Agmon-Levin, N. and Shoenfeld, Y. (2008). Chronic fatigue syndrome with autoantibodies- the result of an augmented adjuvant effect of hepatitis-B vaccine and silicone implant. *Autoimmun Rev*, **8**(1): 52–5.

Allen, A.D. (1988). Is RA273 rubella immunization a cause of chronic fatigue? *Med Hypotheses*, **27**(3): 217–20.

Andersson, M., Bagby, J.R., Dyrehag, L.E., and Gottfries, C.G. (1998). Effects of staphylococcus toxoid vaccine on pain and fatigue in patients with fibromyalgia/chronic fatigue syndrome. *Eur J Pain*, **2**(2): 133–42.

Asa, P.B., Cao, Y., and Garry, R.F. (2000). Antibodies to squalene in Gulf War syndrome. *Exp Mol Pathol*, **68**(1): 55–64.

Authier, F.J., Sauvat, S., Champey, J., *et al.* (2003). Chronic fatigue syndrome in patients with macrophagic myofasciitis. *Arthritis Rheum*, **48**(2): 569–70.

Bernton, E., Hoover, D., Galloway, R., and Popp, K. (1995). Adaptation to chronic stress in military trainees. *Ann NY Acad Sci*, **774**(1): 217–31.

Binder, L.M. and Campbell, K.A. (2004). Medically unexplained symptoms and neuropsychological assessment. *J Clin Exp Neuropsychol*, **26**(3): 369–92.

Buskila, D. (2000). Musculoskeletal injury as a trigger for fibromyalgia/posttraumatic fibromyalgia. *Curr Rheumatol Rep*, **2**(2): 104–8.

Buskila, D., Sarzi-Puttini, P., and Ablin, J.N. (2007). The genetics of fibromyalgia syndrome. *Pharmacogenomics*, **8**(1): 67–74.

Delage, G., Salit, I., Pennie, R., *et al.* (1993). [The possible relation between hepatitis B vaccination and chronic fatigue syndrome]. *Union Med Can,* **122**(4): 278.

Exley, C., Swarbrick, L., Gherardi, R.K., and Authier, F.J. (2009). A role for the body burden of aluminium in vaccine-associated macrophagic myofasciitis and chronic fatigue syndrome. *Med Hypotheses,* **72**(2): 135–9.

Ford, J.D., Campbell, K.A., Storzbach, D., *et al.* (2001). Posttraumatic stress symptomatology is associated with unexplained illness attributed to Persian Gulf War military service. *Psychosom Med,* **63**(5): 842–9.

Fukuda, K., Nisenbaum, R., Stewart, G., *et al.* (1998). Chronic multisymptom illness affecting Air Force veterans of the Gulf War. *JAMA,* **280**(11): 981–8.

Gherardi, R.K., Coquet, M., Cherin, P., *et al.* (1998). Macrophagic myofasciitis: an emerging entity. *Lancet,* **352**(9125): 347–52.

Gherardi, R.K., Coquet, M., Cherin, P., *et al.* (2001). Macrophagic myofasciitis lesions assess long-term persistence of vaccine-derived aluminium hydroxide in muscle. *Brain,* **124**(9): 1821–31.

Gracely, R.H., Petzke, F., Wolf, J.M., and Clauw, D.J. (2002). Functional magnetic resonance imaging evidence of augmented pain processing in fibromyalgia. *Arthritis Rheum,* **46**(5): 1333–43.

Hotopf, M., David, A., Hull, L., *et al.* (2000). Role of vaccinations as risk factors for ill health in veterans of the Gulf war: cross sectional study. *BMJ,* **320**(7246): 1363–7.

Hotopf, M., David, A., Hull, L., *et al.* (2004). Risk factors for continued illness among Gulf War veterans: a cohort study. *Psychol Med,* **34**(4): 747–54.

Julien, N., Goffaux, P., Arsenault, P., and Marchand, S. (2005). Widespread pain in fibromyalgia is related to a deficit of endogenous pain inhibition. *Pain,* **114**(1–2): 295–302.

Lange, G., Tiersky, L.A., DeLuca, J., *et al.* (2001). Cognitive functioning in Gulf War illness. *J Clin Exp Neuropsychol,* **23**(2): 240–9.

Lathrop, S.L., Ball, R., Haber, P., *et al.* (2002). Adverse event reports following vaccination for Lyme disease: December 1998–July 2000. *Vaccine,* **20**(11): 1603–8.

Lewis, J.D., Wassermann, E.M., Chao, W., *et al.* (2012). Central sensitization as a component of post-deployment syndrome. *NeuroRehabilitation,* **31**(4): 367–72.

Macfarlane, G.J., Hotopf, M., Maconochie, N., *et al.* (2005). Long-term mortality amongst Gulf War Veterans: is there a relationship with experiences during deployment and subsequent morbidity? *Int J Epidemiol,* **34**(6): 1403–8.

Madariaga, M.G., Rezai, K., Trenholme, G.M., and Weinstein, R.A. (2003). Q fever: a biological weapon in your backyard. *Lancet Infect Dis,* **3**(11): 709–21.

Magnus, P., Brubakk, O.R., Nyland, H., *et al.* (2009). Vaccination as teenagers against meningococcal disease and the risk of the chronic fatigue syndrome. *Vaccine,* **27**(1): 23–7.

Martinez-Lavin, M. (2007). Stress, the stress response system, and fibromyalgia. *Arthritis Res Ther,* **9**(4): 216.

Mitchell, L.A., Tingle, A.J., Grace, M., *et al.* (2000). Rubella virus vaccine associated arthropathy in postpartum immunized women: influence of preimmunization serologic status on development of joint manifestations. *J Rheumatol,* **27**(2): 418–23.

Morag, M., Morag, A., Reichenberg, A., *et al.* (1999). Psychological variables as predictors of rubella antibody titers and fatigue- a prospective, double blind study. *J Psychiatr Res,* **33**(5): 389–95.

Mu, H.H. and Sewell, W.A. (1993). Enhancement of interleukin-4 production by pertussis toxin. *Infect Immun,* **61**(7): 2834–40.

Niv, D. and Devor, M. (2004). Chronic pain as a disease in its own right. *Pain Pract,* **4**(3): 179–81.

Ortega-Hernandez, O.D. and Shoenfeld, Y. (2009). Infection, vaccination, and autoantibodies in chronic fatigue syndrome, cause or coincidence? *Ann NY Acad Sci,* **1173**: 600–9.

Park, J.H., Na, K.S., Park, Y.W., *et al.* (2005). Macrophagic myofasciitis unrelated to vaccination. *Scand J Rheum,* **34**(1): 65–7.

Phillips, C.J., Matyas, G.R., Hansen, C.J., *et al.* (2009). Antibodies to squalene in US Navy Persian Gulf War veterans with chronic multisymptom illness. *Vaccine,* **27**(29): 3921–6.

Ramirez, F., Fowell, D.J., Puklavec, M., *et al.* (1996). Glucocorticoids promote a TH2 cytokine response by CD4+ T cells *in vitro*. *J Immunol,* **156**(7): 2406–12.

Rook, G.A. and Zumla, A. (1997). Gulf War syndrome: is it due to a systemic shift in cytokine balance towards a Th2 profile? *Lancet,* **349**(9068): 1831–3.

Schomberg, D. and Olson, J.K. (2012). Immune responses of microglia in the spinal cord: contribution to pain states. *Exp Neurol,* **234**(2): 262–70.

Sleigh, K.M., Danforth, D.G., Hall, R.T., *et al.* (2000). Double-blind, randomized study of the effects of influenza vaccination on the specific antibody response and clinical course of patients with chronic fatigue syndrome. *Can J Infect Dis,* **11**(5): 267.

Skowera, A., Cleare, A., Blair, D., *et al.* (2004). High levels of type 2 cytokine-producing cells in chronic fatigue syndrome. *Clin Exp Immunol,* **135**(2): 294–302.

Shoenfeld, Y. and Agmon-Levin, N. (2011). "ASIA" – autoimmune/inflammatory syndrome induced by adjuvants. *J Autoimmun,* **36**(1): 4–8.

Tingle, A.J., Mitchell, L.A., Grace, M., *et al.* (1997). Randomised double-blind placebo-controlled study on adverse effects of rubella immunisation in seronegative women. *Lancet,* **349**(9061): 1277–81.

Unwin, C., Blatchley, N., Coker, W., *et al.* (1999). Health of UK servicemen who served in Persian Gulf War. *Lancet,* **353**(9148): 169–78.

Vedhara, K., Llewelyn, M.B., Fox, J.D., *et al.* (1997). Consequences of live poliovirus vaccine administration in chronic fatigue syndrome. *J Neuroimmunol*, **75**(1): 183–95.

Vera-Lastra, O., Medina, G., Cruz-Dominguez, M.D.P., *et al.* (2013). Autoimmune/inflammatory syndrome induced by adjuvants (Shoenfeld's syndrome): clinical and immunological spectrum. *Exp Rev Clin Immunol*, **9**(4): 361–73.

Weibel, R.E. and Benor, D.E. (1996). Chronic arthropathy and musculoskeletal symptoms associated with rubella vaccines. A review of 124 claims submitted to the National Vaccine Injury Compensation Program. *Arthritis Rheum*, **39**(9): 1529–34.

Yunus, M.B. (2007). Fibromyalgia and overlapping disorders: the unifying concept of central sensitivity syndromes. *Semin Arthritis Rheum*, **36**(6): 339–56.

Zafrir, Y., Agmon-Levin, N., Paz, Z., *et al.* (2012). Autoimmunity following hepatitis B vaccine as part of the spectrum of autoimmune (auto-inflammatory) syndrome induced by adjuvants (ASIA): analysis of 93 cases. *Lupus*, **21**(2): 146–52.

35 Bullous Dermatoses, Infectious Agents, and Vaccines

Yaron Zafrir,[1,2] Nancy Agmon-Levin,[2,3] and Sharon Baum[1]

[1]Department of Dermatology, Sheba Medical Center, Tel Hashomer, Israel
[2]Zabludowicz Center for Autoimmune Diseases, Sheba Medical Center, Tel Hashomer, Israel
[3]Sackler Faculty of Medicine, Tel Aviv University, Tel Aviv, Israel

Introduction

The skin, our largest organ, serves as a barrier and thus maintains our internal environment. It is composed of three layers: the epidermis, which prevents fluid loss and infection; the dermis, which is composed of connective tissue and blood vessels, giving the skin its flexibility and nutrients; and the subcutis, a layer of fat that provides thermoregulation and serves as an energy storage area. In between the epidermis and the dermis lies the basement membrane, which normally keeps the skin intact via various adhesion proteins. There are four subregions of adhesion: the basal keratinocytes cytoskeleton, hemidesmosome-anchoring filament complexes, the lamina densa region, and the sublamina densa region (Bolognia and Rapini, 2012).

Autoimmune bullous disorders encompasses a heterogeneous group of disorders characterized by the presence of blisters and autoantibodies against structural components of the skin: desmosomal proteins (in pemphigus), adhesion molecules of the dermal–epidermal junction (in pemphigoid diseases), and epidermal/tissue transglutaminase (in dermatitis herpetiformis). These autoantibodies are targeted against various components of the epidermal desmosome and hemidesmosome, inducing the loss of adhesion between the epidermal and dermal layers, which results in blister formation. The most common autoimmune bullous diseases are bullous pemphigoid (BP) and pemphigus vulgaris (PV) (Eming and Hertl, 2006; Bolognia and Rapini, 2012; Sticherling and Erfurt-Berge, 2012).

Diagnosis of autoimmune bullous disorders relies upon a combination of the clinical picture and the detection of autoantibodies in the skin and/or serum of the patient. The diagnostic gold standard for autoimmune bullous diseases is the detection of autoantibodies in the skin or mucous membranes by direct immunofluorescence (DIF) microscopy of a perilesional biopsy (Mihai and Sitaru, 2007; Schmidt and Zillikens, 2010; Bolognia and Rapini, 2012; Sticherling and Erfurt-Berge, 2012). For example: in PV, DIF detects tissuebound autoantibodies demonstrating intercellular IgG and C3 and Linear C3 and IgG along the dermo–epidermal junction in BP.

Epidemiology

BP usually appears in the elderly, showing a peak incidence in the seventh decade of life, with no difference in incidence between the genders (Bologna and Rapini, 2012).The reported incidence of BP among adults varies from 6.6 cases

Vaccines and Autoimmunity, First Edition. Edited by Yehuda Shoenfeld, Nancy Agmon-Levin, and Lucija Tomljenovic.

per million in France and Germany to 43 cases per million in the United Kingdom (Langan *et al.*, 2008). BP appears to have no gender preference and no ethnic or racial predilection. The incidence among infants and children is believed to be much lower (Bolognia and Rapini, 2012; Fuertes *et al.*, 2013). The mean age of PV onset is between 40 and 60 years, and it usually affects both genders equally. Depending on ethnicity and geographic area, the incidence varies from 0.5 to 3.2 cases per 100 000 person years. The incidence of PV is higher among those of Jewish ancestry and those of Mediterranean origin (Lombardi *et al.*, 1992; Stanley, 1999).

Clinical manifestations

The clinical course of BP is of spontaneous exacerbations and remissions. However, in about half of treated patients, the disease will "completely remit" within 2–6 years. Its clinical manifestations include tense blisters, which arise on either normal or erythematous skin and, following rapture, leave shallow erosions. In some cases, patients may present first with urticarial lesions, termed "urticarial stage BP," which precede blister formation. The blisters in infantile BP tend to occur more frequently on the palms, soles, and face, and rarely affect the genital area (Bernard *et al.*, 2009; Bolognia and Rapini, 2012). Oral cavity involvement is observed in 10–35% of patients, mainly affecting the buccal mucosa (Bernard *et al.*, 2009; Bolognia and Rapini, 2012; Sticherling and Erfurt-Berge, 2012). BP can be fatal, with a yearly mortality rate of 6–40%, especially among elderly patients and those with an extensive disease involving a large body surface area (Joly *et al.*, 2005).

In PV patients, the primary lesion involves flaccid blisters; however, as these blisters rupture easily, the first lesions that are usually encountered by the physician are erosions. PV is divided into a mucosal-dominant type, involving only the mucosa, and a mucocutaneous type, with both cutaneous and mucosal involvement. It can involve the oral mucosa only or other membranes as well, such as the pharynx or larynx, leading to dysphagia and hoarseness (Schmidt and Zillikens, 2010; Bolognia and Rapini, 2012; Sticherling and Erfurt-Berge, 2012). Classically, the lesions in the oral mucosa appear first, followed months later by the skin lesions.

Histology and immunohistochemistry features of BP

Histological features include subepidermal blister and dermal inflammatory infiltrate, composed of lymphocytes, histiocytes, and eosinophils (Bolognia and Rapini, 2012, Sticherling and Erfurt-Berge, 2012).

Immunohistological methods include DIF, which depicts deposition along the dermal–epidermal junction of IgG and/or C3 in a linear pattern. Indirect immunofluorescence (IIF) demonstrates circulating IgG, which bind to the basement membrane of normal stratified squamous epithelia. Salt split test, in which the tissue is incubated in 1 M NaCl, leading to separation of the epidermis from the dermis at the level of the lamina lucida (a component of the basement membrane), helps to differentiate between BP and other subepidermal bullous diseases, though the location of antibodies at the bottom or the roof of the blister (Beutner *et al.*, 1968; Bolognia and Rapini, 2012, Sticherling and Erfurt-Berge, 2012).

Histology and immunohistochemistry features of PV

Histological features include acantholysis (separation) at the level of the deep epidermis, forming a flaccid bulla (Bolognia and Rapini, 2012).

DIF is the gold-standard method for the diagnosis of PV, demonstrating deposition of IgG autoantibodies directed against the cell surface of keratinocytes. Antibodies of the IgM subtype and C3 may also be present, but they are less frequently seen. IIF performed on monkey esophagus demonstrates circulating IgG autoantibodies in the serum. Enzyme-linked immunosorbent assay is suitable for both the diagnosis and the monitoring of serum autoantibody (Schmidt and Zillikens, 2010; Bolognia and Rapini, 2012; Tampoia *et al.*, 2012).

Pathogenesis

The etiology of both BP and PV is poorly understood, but as in other autoimmune diseases it has been shown to be associated with various environmental factors, including emotional and/or physical stress, infections, and vaccinations (Bolognia and Rapini, 2012).

BP is characterized by two autoantibodies, BP 230 (BPAG1) and BP 180 (BPAG2), which are directed against a protein called hemidesmosome, which is part of the dermo–epidermal adhesion complex and normally promotes the adhesion between the epithelial cells (skin and mucosa). BPAG1 is a cytoplasmic protein that anchors keratin intermediate filaments to the hemidesmosome, while BPAG2 is a transmembrane protein that promotes keratinocyte adhesion to epidermal basement membrane; its main antigenic site is a small extracellular noncollagenous region called NC16A (Chimanovitch *et al.*, 2000; Sitaru *et al.*, 2002; Thoma-Uszynski *et al.*, 2006). Other autoantibodies directed against alpha 6 integrin and laminin-5 have also been identified (Bekou *et al.*, 2005; Kiss *et al.*, 2005). Autoantibodies binding to the basement membrane lead to activation of the classic complement cascade and the C3 amplification mechanism. This in turn results in migration of leukocytes, release of inflammatory mediators, degranulation of mast cells, and recruitment of eosinophils. Interleukin (IL-) 16 has been found to be responsible for recruitment of $CD4^+$ helper T cells to the skin and for the induction of IL-2 receptors (Frezzolini *et al.*, 2004).

It is believed that the inflammatory cells release proteases such as matrix metalloproteinase-2, 9, and 13, leading to the degradation of hemidesmosomal proteins and thus forming blisters (Verraes *et al.*, 2001).

The autoantibodies that take part in the pathogenesis of PV are directed against different parts of the desmosome: desmoglein 1 (Dsg1) and desmoglein 3 (Dsg3). The desmosome is an intercellular cell–cell adhesion structure acting as a transmembrane device, and its role is to anchor the intracellular intermediate filaments (Bolognia and Rapini, 2012; Wolff *et al.*, 2008). Dsg1 is expressed more intensely in the superficial layers of the epidermis, while Dsg3 is more pronounced in the lower portion of the epidermis, primarily in the basal layer. Dsg1 and Dsg3 are both expressed in patients with the mucocutaneous form of PV; those with only the mucosal form have only antidesmoglein 3 autoantibodies.

The binding of antibody to Dsg leads to loss of Dsg function, and as a result to loss of cell–cell adhesion and to blister formation within the epidermis of the skin and/or mucosal surfaces. The interaction between the antibody and the protein can occur either by a direct effect on desmosomal adherence or by induction of a cellular process that results in acantholysis (Lombardi *et al.*, 1992).

Genetic factors

Overexpression of certain human leukocyte antigen (HLA) class II alleles, such as HLA-DQB1*0301, DRB1*04, DRB1*1101, and DQB1*0302, is more prevalent in BP patients than in the general population (Delgado *et al.*, 1996; Bolognia and Rapini, 2012; Ruocco *et al.*, 2013).

PV is associated with certain HLA class II loci, such as HLA-DR4 and HLADR14 alleles (DRB1*0401 and DRB1*0402, which is prevalent in Ashkenazi Jewish, Iranian, and Sardinian patients). Other loci include DRB1*1401 (common among Japanese and Italian patients) and two DQB1 alleles (DQB1*0302 and DQB1*0503), which are strongly associated with PV. In dermatitis herpetiformis, about 80–90% of patients are associated with the genotypes HLA DR3 and HLA DQw (Bolognia and Rapini, 2012; Joly *et al.*, 2005).

Environmental factors

BP and PV have long been associated with malignancies, other autoimmune diseases, exposure to ultraviolet light, post-radiation therapy, systemic drugs, infectious agents, and vaccines. The induction of autoimmune diseases in general and of bullous dermatoses in particular can be explained by either molecular mimicry between the infectious agent antigens (also present in vaccines) and self-proteins or by a nonspecific activation of the innate immune system by any other ingredient (e.g. adjuvant included in vaccines), which promotes activation and expansion of autoreactive T cells. Notably, such links are mostly related to a temporal association, as only a few case reports and case series show evidence of the infectious agents in the lesions themselves (Fellner, 1993; Venning and Wojnarowska, 1995; Vassileva, 1998; Rzany and Weller, 2001; Sagi *et al.*, 2011; Ruocco *et al.*, 2013).

Infectious agents

The link between infectious agents and blistering diseases had already been reported by Sagi *et al.* (2011), who found that sera from BP and PV patients demonstrated a significantly higher prevalence of antibodies to hepatitis B virus (HBV), hepatitis C virus (HCV), helicobacter pylori, toxoplasma gondii, and cytomegalovirus (CMV) than healthy controls. Other studies suggested that the herpetoviridae might be a trigger for the development of PV. Tufano *et al.* (1999)

Table 35.1 Summary of BP cases following vaccination in adults

Reference	Gender	Age (years)	Vaccination	Time from vaccination to appearance of symptoms
Bodokh *et al.* (1994)	M	83	Influenza	2 days
Lear *et al.* (1996)	M	74	Influenza	10 days
Fournier *et al.* (1996)	M	84	Influenza	1 day
Downs *et al.* (1998)	M	86	Influenza	4 weeks
Downs *et al.* (1998)	M	90	Influenza	4 weeks
Downs *et al.* (1998)	M	72	Influenza	5 weeks
Downs *et al.* (1998)	F	83	Influenza	3 weeks
Venning and Wojnarowska (1995)	M	83	Tetanus booster	1 day
Fournier *et al.* (1996)	F	84	Tetanus booster	1 day
Sezin *et al.* (2013)	M	29	Tetanus booster	2 days
Walmsley and Hampton (2011)	M	81	Swine flu	2 weeks
Chacón and Sinha (2011)	M	72	Herpes zoster	2 weeks

M, male; F, female

Table 35.2 Summary of BP cases following vaccination in children

Reference	Gender	Age (years)	Vaccination	Time from vaccination to appearance of symptoms
Oranje *et al.* (1991)	M	3	dTP, polio	unknown
Cambazard *et al.* (1994)	M	4	dTP, polio	5 hrs
Baykal *et al.* (2001)	M	3.5	dTP, polio	1 day
Amos *et al.* (1998)	F	3	dTP, polio, HBV	3 days
Xiao *et al.* (2007)	M	3	dTP, polio, HBV	2 weeks
Khaled *et al.* (2010)	F	5	dTP, Polio, HBV	4 weeks
Mérida *et al.* (2005)	M	2	dTP, polio, HBV, Hib	1 week
Valdivielso-Ramos *et al.* (2011)	F	2	dTP, polio, HBV, Hib, Meningococcal, pneumococcus	3 weeks
Hafiji *et al.* (2010)	M	2	dTP, polio, Hib, pneumococcus	8 days
Cunha *et al.* (1998)	M	2.5	dTP, polio, influenza	5 days
Toyama *et al.* (2009)	F	5	dTP	3 days
Toyama *et al.* (2009)	F	5	BCG	9 days
Hiroo *et al.* (1999)	M	4	BCG	3 days
Erbagci (2002)	M	12 years	HBV	7 days

M, male; F, female; HBV, hepatitis B virus; dTP, diphtheria, pertussis, and tetanus; Hib, *Haemophilus influenzae* type B; BCG, bacillus Calmette–Guérin

demonstrated the presence of DNA sequences of both herpes simplex virus (HSV) (71%) and Epstein–Barr virus (EBV) (5%) in the skin of PV patients. Similarly, Ruocco *et al.* (1996) depicted 10 pemphigus patients associated with HSV infection. Seven of them had a positive viral culture and five had positive serology. In another work, Wang *et al.* (2005) found HHV 8 DNA sequences in 36.1% of skin lesions and 30.8% of peripheral blood mononuclear cells of pemphigus patients, in comparison with 5.6% and 7.8%, respectively, in controls.

Vaccines and BP

Many case reports propose a temporal association between the appearance of BP and immunization of adults (Table 35.1) and young children (Table 35.2). The most common vaccines related to the development of BP are influenza vaccines, diphtheria, tetanus, and pertussis, followed by HBV, BCG, polio, and herpes zoster (HZ) vaccines. Furthermore, reactivation of BP following influenza vaccination was reported in one case report (Bodokh *et al.*, 1994).

According to reports in the medical literature, it appears that gender may be of importance to the link between vaccines and BP. Notably, among adults, 83% of subjects who developed BP following immunization were males. In the adult group, mean age of patients was 76 years, mean latency period between inoculation and symptom initiation was 13 days (range 1 day to 5 weeks), and the most common vaccine related to the development of BP was influenza vaccine (Table 35.1).

Among the pediatric group, male gender again was prominent, with 64% of this group being males. Most children developed their clinical manifestations at a few months of age, except for one 12-year-old child. The mean latency period between immunization and clinical manifestations among children was 9 days (range a few hours to 4 weeks). The most common vaccines related to the development of BP among children were diphtheria, pertussis, and tetanus (dTP), polio, and HBV vaccines (Table 35.2).

Vaccines and PV

Various vaccines have been associated with the appearance of new-onset PV, including influenza vaccine (Mignogna *et al.*, 2000), HBV vaccine (Berkun *et al.*, 2005), anthrax vaccine (Muellenhoff *et al.*, 2004), typhoid booster (Bellaney and Rycroft, 1996), and rabies vaccine (Yalçin and Alli, 2007). Three of these cases were women and four were adults. Their average age was 36 years and the latency period from last vaccine was approximately 25 days. In addition, exacerbation of PV after vaccination was reported following influenza vaccine (De Simone *et al.*, 2008) and tetanus vaccine (Korang *et al.*, 2002).

Conclusions

The possible association between vaccines and bullous diseases such as BP and PV is still a matter of debate and is supported only by case reports. Importantly, this plausible association has been related to different vaccines, most notably anti-influenza vaccine, dTP, polio, and hepatitis. Further research, including long-term follow-up studies, on the various environmental factors that may lead to the development of autoimmune bullous dermatoses and especially on the association between immunization and the development of these diseases is warranted.

References

Amos, B., Deng, J.S., Flynn, K., and Suarez, S. (1998). Bullous pemphigoid in infancy: case report and literature review. *Pediatr Dermatol*, **15**(2): 108–11.

Baykal, C., Okan, G., and Sarica, R. (2001). Childhood bullous pemphigoid developed after the first vaccination. *J Am Acad Dermatol*, **44**(2): 348–450.

Bekou, V., Thoma-Uszynski, S., Wendler, O., *et al.* (2005). Detection of laminin 5-specific auto-antibodies in mucous membrane and bullous pemphigoid sera by ELISA. *J Invest Dermatol*, **124**(4): 732–40.

Bellaney, G.J. and Rycroft, R.J. (1996). Pemphigus vulgaris following a hyperimmune response to typhoid booster. *Clin Exp Dermatol*, **21**(6): 434–6.

Berkun, Y., Mimouni, D., and Shoenfeld, Y. (2005). Pemphigus following hepatitis B vaccination – coincidence or causality? *Autoimmunity*, **38**: 117–19.

Bernard, P., Reguiai, Z., Tancrède-Bohin, E., *et al.* (2009). Risk factors for relapse in patients with bullous pemphigoid in clinical remission: a multicenter, prospective, cohort study. *Arch Dermatol*, **145**: 537–42.

Beutner, E.H., Jordon, R.E., and Chorzelski, T.P. (1968). The immunopathology of pemphigus and bullous pemphigoid. *J Invest Dermatol*, **51**: 63–80.

Bodokh, I., Lacour, J.P., Bourdet, J.F., *et al.* (1994). Reactivation of bullous pemphigoid after influenza vaccination. *Therapie*, **49**(2): 154.

Bolognia, J.L., Jorizzo, J.L., Schaffer, J.V. *et al.* (2012). *Dermatology*, 3rd edn. Philadelphia, PA: W.B. Saunders.

Cambazard, F., Thivolet, J., and Mironneau, P. (1994). Bullous pemphigoid in a 4-month-old boy. *Br J Dermatol*, **131**(3): 449–51.

Chacón, G.R. and Sinha, A.A. (2011). Bullous pemphigoid after herpes zoster vaccine administration: association or coincidence? *J Drugs Dermatol*, **10**(11): 1328–30.

Chimanovitch, I., Hamm, H., Georgi, M., *et al.* (2000). Bullous pemphigoid of childhood: autoantibodies target the same epitopes within the NC16A domain of BP180 as autoantibodies in bullous pemphigoid of adulthood. *Arch Dermatol*, **136**(4): 527–32.

Cunha, P.R., Thomazeski, P.V., Hipolito, E., *et al.* (1998). Bullous pemphigoid in 2-month-old infant. *Int J Dermatol*, **37**(12): 935–93.

De Simone, C., Caldarola, G., D'Agostino, M., *et al.* (2008). Exacerbation of pemphigus after influenza vaccination. *Clin Exp Dermatol*, **33**(6): 718–20.

Delgado, J.C., Turbay, D., Yunis, E.J., *et al.* (1996). A common major histocompatibility complex class II allele HLA-DQB1* 0301 is present in clinical variants of pemphigoid. *Proc Natl Acad Sci USA*, **93**: 8569–71.

Downs, A.M., Lear, J.T., Bower, C.P., and Kennedy, C.T. (1998). Does influenza vaccination induce bullous pemphigoid? A report of four cases. *Br J Dermatol*, **138**(2): 363.

Eming, R. and Hertl, M. (2006). Autoimmune bullous disorders. *Clin Chem Lab Med*, **44**: 144–9.

Erbagci, Z. (2002). Childhood bullous pemphigoid following hepatitis B immunization. *J Dermatol*, **29**(2): 781–5.

Fellner, M.J. (1993). Drug-induced bullous pemphigoid. *Clinics Dermatol*, **11**: 515–20.

Fournier, B., Descamps, V., Bouscarat, F., *et al.* (1996). Bullous pemphigoid induced by vaccination. *Br J Dermatol*, **135**(1): 153–4.

Fuertes, I., Luelmo, J., Leal, L., *et al.* (2013). Refractory childhood pemphigoid successfully treated with rituximab. *Pediatr Dermatol*, **30**(5): e96–7.

Frezzolini, A., Cianchini, G., Ruffelli, M., *et al.* (2004). Interleukin-16 expression and release in bullous pemphigoid. *Clin Exp Immunol*, **137**(3): 595–600.

Hafiji, J., Bhogal, B., Rytina, E., and Burrows, N.P. (2010). Bullous pemphigoid in infancy developing after the first vaccination. *Clin Exp Dermatol*, **35**(8): 940–1.

Hiroo, S., Akimichi, M., and Takuo, T. (1999). [A case of juvenile bullous pemphigoid]. *Nishinihon J Dermatol*, **61**(5): 585–7.

Joly, P., Benichou, J., Lok, C., *et al.* (2005). Prediction of survival for patients with bullous pemphigoid: a prospective study. *Arch Dermatol*, **141**: 691–8.

Khaled, A., Kharfi, M., Fazaa, B., and Kamoun, M.R. (2010). Bullous eruption in a five-month-old girl. *CMAJ*, **182**(12): 1325–7.

Kiss, M., Perényi, A., Marczinovits, I., *et al.* (2005). Autoantibodies to human alpha6 integrin in patients with bullous pemphigoid. *Ann NY Acad Sci*, **1051**: 104–10.

Korang, K., Ghohestani, R., Krieg, T., *et al.* (2002). Exacerbation of pemphigus foliaceus after tetanus vaccination accompanied by synthesis of auto-antibodies against paraneoplastic pemphigus antigens. *Acta Derm Venereol*, **82**(6): 482–3.

Langan, S.M., Smeeth, L., Hubbard, R., *et al.* (2008). Bullous pemphigoid and pemphigus vulgaris – incidence and mortality in the UK: population based cohort study. *BMJ*, **337**: a180

Lear, J.T., Tan, B.B., and English, J.S. (1996). Bullous pemphigoid following influenza vaccination. *Clin Exp Dermatol*, **21**(5): 392.

Lombardi, C., Borges, P.C., Chaul, A., *et al.* (1992). Environmental risk factors in endemic pemphigus foliaceus (*Fogo selvagem*). *J Invest Dermatol Symp Proc*, **98**: 847–50.

Mérida, C., Martínez-Escribano, J.A., Frías, J.F., *et al.* (2005). Bullous pemphigoid in an infant after vaccination. *Actas Dermosifiliogr*, **96**(4): 255–7.

Mignogna, M.D., Muzio, L.L., and Ruocco, E. (2000). Pemphigus induction by influenza vaccination. *Int J Dermatol*, **39**: 800.

Mihai, S. and Sitaru, C. (2007). Immunopathology and molecular diagnosis of autoimmune bullous diseases. *J Cell Mol Med*, **11**(3): 462–81.

Muellenhoff, M., Cukrowski, T., Morgan, M., and Dorton, D. (2004). Oral pemphigus vulgaris after anthrax vaccine administration: association or coincidence? *J Am Acad Dermatol*, **50**(1): 136–9.

Oranje, A.P., Vuzevski, V.D., van Joost, T., *et al.* (1991). Bullous pemphigoid in children. Report of three cases with special emphasis on therapy. *Int J Dermatol*, **30**(5): 339–42.

Ruocco, V., Wolf, R., Ruocco, E., and Baroni, A. (1996). Viruses in pemphigus: a casual or causal relationship? *Int J Dermatol*, **35**: 782–4.

Ruocco, V., Ruocco, E., Lo Schiavo, A., *et al.* (2013). Pemphigus: etiology, pathogenesis, and inducing or triggering factors: facts and controversies. *Clin Dermatol*, **31**(4): 374–81.

Rzany, B. and Weller, N. (2001). Epidemiology of autoimmune skin disorders. In: Hertl, M. (ed.). *Autoimmune Diseases of the Skin*. Wien, New York: Springer Verlag, pp. 21–38.

Sagi, L., Baum, S., Agmon-Levin, N., *et al.* (2011). Autoimmune bullous diseases the spectrum of infectious agent antibodies and review of the literature. *Autoimmun Rev*, **10**(9): 527–35.

Schmidt, E. and Zillikens, D. (2010). Modern diagnosis of autoimmune blistering skin diseases. *Autoimmun Rev*, **10**: 84–9.

Sezin, T., Egozi, E., Hillou, W., *et al.* (2013). Anti-laminin-332 mucous membrane pemphigoid developing after a diphtheria tetanus vaccination. *JAMA Dermatol*, **149**(7): 858–62.

Sitaru, C., Schmidt, E., Petermann, S., *et al.* (2002). Autoantibodies to bullous pemphigoid antigen 180 induce dermal-epidermal separation in cryosections of human skin. *J Invest Dermatol*, **118**: 664–71.

Stanley, J.R. (1999). Pemphigus. In: Freedberg, I.M., Eisen, A.Z., Wolff, K., *et al.* (eds). *Fitzpatrick's Dermatology in General Medicine*, 5th edn. New York: McGraw-Hill, pp. 654–66.

Sticherling, M. and Erfurt-Berge, C. (2012). Autoimmune blistering diseases of the skin. *Autoimmun Rev*, **11**(3): 226–30.

Tampoia, M., Giavarina, D., Di Giorgio, C., and Bizzaro, N. (2012). Diagnostic accuracy of enzyme-linked immunosorbent assays (ELISA) to detect anti-skin autoantibodies in autoimmune blistering skin diseases: a systematic review and meta-analysis. *Autoimmun Rev*, **12**(2): 121–6.

Thoma-Uszynski, S., Uter, W., Schwietzke, S., *et al.* (2006). Autoreactive T and B cells from bullous pemphigoid (BP) patients recognize epitopes clustered in distinct regions of BP180 and BP230. *J Immunol,* **176**(3): 2015–23.

Toyama, T., Nakamura, K., Kuramochi, A., *et al.* (2009). Two cases of childhood bullous pemphigoid. *Eur J Dermatol,* **19**(4): 368–71.

Tufano, M.A., Baroni, A., Buommino, E., *et al.* (1999). Detection of herpesvirus DNA in peripheral blood mononuclear cells and skin lesions of patients with pemphigus by polymerase chain reaction. *Br J Dermatol,* **141**: 1033–9.

Valdivielso-Ramos, M., Velázquez, D., Tortoledo, A., and Hernanz, J.M. (2011). Infantile bullous pemphigoid developing after hexavalent, meningococcal and pneumococcal vaccinations. *An Pediatr (Barc),* **75**(3): 199–202.

Vassileva, S. (1998). Drug-induced pemphigoid: bullous and cicatricial. *Clin Dermatol,* **16**: 379–87.

Venning, V.A. and Wojnarowska, F. (1995). Induced bullous pemphigoid. *Br J Dermatol,* **132**(5): 831–2.

Verraes, S., Hornebeck, W., Polette, M., *et al.* (2001). Contribution of neutrophil elastase and matrix metalloproteinase 9 in the degradation of BP180 (type XVII collagen) in human bullous pemphigoid. *J Invest Dermatol,* **117**: 1091–6.

Walmsley, N. and Hampton, P. (2011). Bullous pemphigoid triggered by swine flu vaccination: case report and review of vaccine triggered pemphigoid. *J Dermatol Case Rep,* **5**(4): 74–6.

Wang, G.Q., Xu, H., Wang, Y.K., *et al.* (2005). Higher prevalence of human herpesvirus 8 DNA sequence and specific IgG antibodies in patients with pemphigus in China. *J Am Acad Dermatol,* **52**: 460–7.

Wolff, K., Goldmsith, L.A., Katz, S.I., *et al.* (2008). *Fitzpatrick's Dermatology In General Medicine,* 7th edn. Wien, New York: McGraw Hill.

Xiao, T., Li, B., Wang, Y.K., *et al.* (2007). Childhood bullous pemphigoid treated by i.v. immunoglobulin. *J Dermatol,* **34**(9): 650–3.

Yalçin, B. and Alli, N. (2007). Pemphigus vulgaris following antirabies vaccination. *J Dermatol,* **34**(10): 734–5.

36 Infections, Vaccinations, and Chronic Fatigue Syndrome

Hussein Mahagna, Naim Mahroum, and Howard Amital

Department of Medicine B, Sheba Medical Center, Tel Hashomer, Sackler Faculty of Medicine, Tel Aviv University, Tel Aviv, Israel

Introduction

Chronic fatigue syndrome (CFS) is a heterogeneous disorder affecting more than 267 people in ever 100 000 (Abbi and Natelson, 2013; Moss-Morris *et al.*, 2013). It has been estimated that in the United States, approximately 1 million people suffer from CFS symptoms (Reynolds *et al.*, 2004). Women are nearly twice as likely to be affected as men (Appel *et al.*, 2007). Similar prevalence rates have been found in different geographic locations and in diverse ethnicities (Steele *et al.*, 1998). The pathophysiology and etiology of CFS are unknown, since there are no characteristic physical signs or diagnostic laboratory abnormalities (Whistler *et al.*, 2003).

CFS patients suffer from disabling fatigue, headaches, concentration difficulties, and memory deficits (90%). Additional symptoms, such as sore throat (85%), tender lymph nodes (80%), skeletal muscle pain and feverishness (75%), sleep disruption (70%), psychiatric problems (65%), and rapid pulse (10%), are often observed (Appel *et al.*, 2007; Friedberg, 2010; Lewis *et al.*, 2013; Werker *et al.*, 2013). Due to these complaints, patients often face familial, social, and vocational crises (Bell *et al.*, 2001).

The diagnosis of CFS is complex, due to its similarity to other ill-defined disorders presenting similarly, such as fibromyalgia syndrome (FMS), Gulf War syndrome (GWS), and Sjögren syndrome (SjS) (Sirois and Natelson, 2001; Abbi and Natelson, 2013; Lewis *et al.*, 2013; Werker *et al.*, 2013).

Fukuda *et al.* (1994) described the significant overlap between CFS and FMS, considering CFS a subclass of prolonged fatigue. They proposed a method for its proper diagnosis: a patient must present with four or more symptoms contemporary for at least 6 months. Characteristics excluding patients from CFS are: active medications, past or current major depressive disorders, alcohol abuse, and severe obesity. However, some of the criteria are difficult to interpret, and opinions differ about the classification of chronic fatigue cases with a history of psychiatric illnesses (Matthews *et al.*, 1988). It is important to stratify patients with suspected CFS by the assessment of four criteria: (i) coexisting medical or neuropsychiatric condition not explaining the chronic fatigue; (ii) level of fatigue, including subjective and performance aspects; (iii) total duration of fatigue; and (iv) level of overall functional performance (Fukuda *et al.*, 1994). All of these evaluations can be performed with available instruments, Medical Outcomes Study Short Form 36, and Sickness Impact Profile (Bergner *et al.*, 1981; Ware and Sherbourne, 1992; Schwartz *et al.*, 1993; Piper *et al.*, 1998; Piper and Cella, 2010).

Vaccines and Autoimmunity, First Edition. Edited by Yehuda Shoenfeld, Nancy Agmon-Levin, and Lucija Tomljenovic.

Etiology

CFS was first described in the 1980s, when it was thought to be a consequence of a viral or bacterial infection. One of the first suspected pathogens was Epstein–Barr virus (EBV), because patients often have higher titers of IgM to the EBV viral capsid antigen (Lerner *et al.*, 2004). In addition, antibodies against cytomegalovirus (CMV) and human herpes virus 6 (HHV-6) are also detected more often in CFS patients (Ablashi *et al.*, 2000; Lerner *et al.*, 2004; Yao *et al.*, 2010), although other reports have failed to repeat these results (Soto and Straus, 2000).

Another virus family studied as a possible cause of CFS is the enteroviridae, because RNA copies were detected in the muscle biopsies of CFS patients but not in a healthy control group (Bowles *et al.*, 1993; Lane *et al.*, 2003). Other studies have failed to demonstrate positive serological tests or PCR for enteroviridae, however (Straus 1996). Parvovirus B19 is considered to be one of most probable causes of CFS, because there are some case reports of patients with a chronic course of fatigue following infection, fulfilling the criteria for CFS diagnosis. In one of these studies, the stress index was significantly associated with development of fatigue during the acute phase of parvovirus B19 infection, and also with chronic fatigue and arthritis occurring 1–3 years following the acute parvovirus B19 infection, with an odd ratio (OR) of 25.7 (95% CI: 1.7–121.9; p = 0.005) (Lane *et al.*, 2003; Kerr, 2005; Kerr and Mattey, 2008). In addition, a higher prevalence of Mycoplasma infections has been reported in CFS patients than in healthy subjects (Nijs *et al.*, 2002).

Although the pathophysiology of the disease is not yet well known, molecular mimicry and autoimmune processes have been suggested to be involved, because of reports of post-infectious onset and the presence of autoantibodies. This is also thought to be true of the role of vaccinations in the onset of the illness.

Vaccinations and CFS

It is known that vaccinations can cause self-limiting fatigue and flu-like symptoms. CFS has been reported to emerge following vaccination in several reports, including after measles, mumps, and rubella (MMR), Pneumovax, influenza, hepatitis B virus (HBV), tetanus, typhoid, and poliovirus vaccines (Devanur and Kerr, 2006). A case of CFS onset following the double effect of exposure to silicone and HBV vaccine was reported. It was suggested that the breast implants and vaccination acted as facilitators and triggers for the emergence of CFS onset in the patient (Agmon-Levin and Shoenfeld, 2008). Such reports have made researchers concerned regarding the role of vaccinations in the onset of CFS and the safety of their use in CFS patients.

HBV is one of the most controversial vaccines with regard to the potential risk of inducing CFS. Several researchers advocate for a contributory role of the vaccine in the development of CFS (House, 1992; O'Sullivan, 1992; Delage *et al.*, 1993; Agmon-Levin and Shoenfeld, 2008), while others claim the vaccine is safe, with minimal adverse effects (Zuckerman, 2004).

In Norway, a vaccine against *Neisseria meningitides* group B was administered to teenagers in 1988–89. Relative risk for CFS and myalgic encephalomyelitis was calculated in a case–control study in 2007, with 201 cases diagnosed at one of two hospitals and 389 controls. The adjusted OR for these two conditions was 1.06 (95% CI: 0.67–1.66) for subjects who received the active vaccine as opposed to those who did not (Magnus *et al.*, 2009).

Another study reported a case of CFS associated with aluminum overload, suggesting that vaccination involving aluminum-containing adjuvants could trigger a cascade of immunological events, which are associated with immune-disrupting conditions, including CFS and macrophagic myofasciitis (Exley *et al.*, 2009). Previous studies indicated that, although aluminum-based adjuvants may persist at the site of injection for years ("vaccine tattoo"), this does not reflect the existence of a diffuse inflammatory muscular disease and is not associated with a specific clinical disease (Siegrist, 2003, 2005).

Several studies have investigated the safety and efficacy of vaccines in CFS patients. One double-blind randomized study checked the safety of oral poliovirus vaccination. It was not found to be clinically contraindicated in CFS patients, but there was evidence of minimally altered immune reactivity to the live vaccine virus; the objective responses to the vaccine revealed differences between patients and controls, increased poliovirus isolation, earlier peak proliferative responses, lower T cell subsets on certain days post-vaccination, and a trend for reduced gamma-interferon in the CFS vaccine group (Vedhara *et al.*, 1997).

One study tested the effect of staphylococcal toxoid on patients with fibromyalgia and CFS. In this double-blind randomized study, treatment with staphylococcus toxoid injections over 6 months led to significant improvement in patients with FMS and CFS. The proportion of patients with a symptom reduction of ≥50% showed on an intention-to-treat analysis as 32/49 (65%) responders in the vaccinated group, compared to 9/49 (18%) in the placebo group ($p < 0.001$) (Zachrisson *et al.*, 2002).

Regarding influenza vaccination, it appears to provide protective antibody levels without worsening CFS symptoms or causing excessive adverse effects (Sleigh *et al.*, 2002). Recently, a study aimed at comparing the humoral and cellular immune responses following influenza vaccination found that CFS patients have comparable outcomes to healthy controls. Putative aberrations in immune responses in CFS patients were not evident for immunity toward influenza. Standard seasonal influenza vaccination is thus justified and, when indicated, should be recommended for patients suffering from CFS (Prinsen *et al.*, 2012).

In conclusion, except for several case reports, there are no studies that indicate vaccines might have a deleterious effect in patients with CFS. However, it is possible that various vaccines or exposures to various pathogens might take part in the induction of CFS.

References

Abbi, B. and Natelson, B.H. (2013). Is chronic fatigue syndrome the same illness as fibromyalgia: evaluating the "single syndrome" hypothesis. *QJM*, **106**(1): 3–9.

Ablashi, D.V., Eastman, H.B., Owen, C.B., *et al.* (2000). Frequent HHV-6 reactivation in multiple sclerosis (MS) and chronic fatigue syndrome (CFS) patients. *J Clin Virol*, **16**(3): 179–91.

Agmon-Levin, N. and Shoenfeld, Y. (2008). Chronic fatigue syndrome with autoantibodies – the result of an augmented adjuvant effect of hepatitis-B vaccine and silicone implant. *Autoimmun Rev*, **8**(1): 52–5.

Appel, S., Chapman, J., and Shoenfeld, Y. (2007). Infection and vaccination in chronic fatigue syndrome: myth or reality? *Autoimmunity*, **40**(1): 48–53.

Bell, D.S., Jordan, K., and Robinson, M. (2001). Thirteen-year follow-up of children and adolescents with chronic fatigue syndrome. *Pediatrics*, **107**(5): 994–8.

Bergner, M., Bobbitt, R.A., Carter, W.B., and Gilson, B.S. (1981). The Sickness Impact Profile: development and final revision of a health status measure. *Med Care*, **19**(8): 787–805.

Bowles, N.E., Bayston, T.A., Zhang, H.Y., *et al.* (1993). Persistence of enterovirus RNA in muscle biopsy samples suggests that some cases of chronic fatigue syndrome result from a previous, inflammatory viral myopathy. *J Med*, **24**(2–3): 145–60.

Delage, G., Salit, I., Pennie, R., *et al.* (1993). [The possible relation between hepatitis B vaccination and chronic fatigue syndrome]. *Union Med Can*, **122**(4): 278–9.

Devanur, L.D. and Kerr, J.R. (2006). Chronic fatigue syndrome. *J Clin Virol*, **37**(3): 139–50.

Exley, C., Swarbrick, L., Gherardi, R.K., and Authier, F.J. (2009). A role for the body burden of aluminium in vaccine-associated macrophagic myofasciitis and chronic fatigue syndrome. *Med Hypotheses*, **72**(2): 135–9.

Friedberg, F. (2010). Chronic fatigue syndrome, fibromyalgia, and related illnesses: a clinical model of assessment and intervention. *J Clin Psychol*, **66**(6): 641–65.

Fukuda, K., Straus, S.E., Hickie, I., *et al.* (1994). The chronic fatigue syndrome: a comprehensive approach to its definition and study. International Chronic Fatigue Syndrome Study Group. *Ann Intern Med*, **121**(12): 953–9.

House, A. (1992). Alleged link between hepatitis B vaccine and chronic fatigue syndrome. *CMAJ*, **146**(7): 1145.

Kerr, J.R. (2005). Pathogenesis of parvovirus B19 infection: host gene variability, and possible means and effects of virus persistence. *J Vet Med B Infect Dis Vet Public Health*, **52**(7–8): 335–9.

Kerr, J.R. and Mattey, D.L. (2008). Preexisting psychological stress predicts acute and chronic fatigue and arthritis following symptomatic parvovirus B19 infection. *Clin Infect Dis*, **46**(9): e83–7.

Lane, R.J., Soteriou, B.A., Zhang, H., and Archard, L.C. (2003). Enterovirus related metabolic myopathy: a postviral fatigue syndrome. *J Neurol Neurosurg Psychiatry*, **74**(10): 1382–6.

Lerner, A.M., Beqaj, S.H., Deeter, R.G., and Fitzgerald, J.T. (2004). IgM serum antibodies to Epstein-Barr virus are uniquely present in a subset of patients with the chronic fatigue syndrome. *In Vivo*, **18**(2): 101–6.

Lewis, I., Pairman, J., Spickett, G., and Newton, J.L. (2013). Is chronic fatigue syndrome in older patients a different disease? – a clinical cohort study. *Eur J Clin Invest*, **43**(3): 302–8.

Magnus, P., Brubakk, O., Nyland, H., *et al.* (2009). Vaccination as teenagers against meningococcal disease and the risk of the chronic fatigue syndrome. *Vaccine*, **27**(1): 23–7.

Matthews, D.A., Lane, T.J., and Manu, P. (1988). Definition of the chronic fatigue syndrome. *Ann Intern Med*, **109**(6): 511–12.

Moss-Morris, R., Deary, V., and Castell, B. (2013). Chronic fatigue syndrome. *Handb Clin Neurol*, **110**: 303–14.

Nijs, J., Nicolson, G.L., De, B.P., *et al.* (2002). High prevalence of Mycoplasma infections among European chronic fatigue syndrome patients. Examination of four Mycoplasma species in blood of chronic fatigue syndrome patients. *FEMS Immunol Med Microbiol,* **34**(3): 209–14.

O'Sullivan, S.J. (1992). Alleged link between hepatitis B vaccine and chronic fatigue syndrome. *CMAJ,* **147**(4): 399–402.

Piper, B.F. and Cella, D. (2010). Cancer-related fatigue: definitions and clinical subtypes. *J Natl Compr Canc Netw,* **8**(8): 958–66.

Piper, B.F., Dibble, S.L., Dodd, M.J., *et al.* (1998). The revised Piper Fatigue Scale: psychometric evaluation in women with breast cancer. *Oncol Nurs Forum,* **25**(4): 677–84.

Prinsen, H., de Vries, I.J., Torensma, R., *et al.* (2012). Humoral and cellular immune responses after influenza vaccination in patients with chronic fatigue syndrome. *BMC Immunol,* **13**: 71.

Reynolds, K.J., Vernon, S.D., Bouchery, E., and Reeves, W.C. (2004). The economic impact of chronic fatigue syndrome. *Cost Eff Resour Alloc,* **2**(1): 4.

Schwartz, J.E., Jandorf, L., and Krupp, L.B. (1993). The measurement of fatigue: a new instrument. *J Psychosom Res,* **37**(7): 753–62.

Siegrist, C.A. (2003). [Vaccine adjuvants and macrophagic myofasciitis]. *Bull Acad Natl Med,* **187**(8): 1511–18.

Siegrist, C.A. (2005). [Vaccine adjuvants and macro phagic myofasciitis]. *Arch Pediatr,* **12**(1): 96–101.

Sirois, D.A. and Natelson, B. (2001). Clinicopathological findings consistent with primary Sjogren's syndrome in a subset of patients diagnosed with chronic fatigue syndrome: preliminary observations. *J Rheumatol,* **28**(1): 126–31.

Sleigh, K.M., Marra, F.H., and Stiver, H.G. (2002). Influenza vaccination: is it appropriate in chronic fatigue syndrome? *Am J Respir Med,* **1**(1): 3–9.

Soto, N.E. and Straus, S.E. (2000). Chronic fatigue syndrome and herpesviruses: the fading evidence. *Herpes,* **7**(2): 46–50.

Steele, L., Dobbins, J.G., Fukuda, K., *et al.* (1998). The epidemiology of chronic fatigue in San Francisco. *Am J Med,* **105**(3A): 83S–90S.

Straus, S.E. (1996). Chronic fatigue syndrome. *BMJ,* **313**(7061): 831–2.

Vedhara, K., Llewelyn, M.B., Fox, J.D., *et al.* (1997). Consequences of live poliovirus vaccine administration in chronic fatigue syndrome. *J Neuroimmunol,* **75**(1–2): 183–95.

Ware, J.E. Jr, and Sherbourne, C.D. (1992). The MOS 36-item short-form health survey (SF-36). *I. Conceptual framework and item selection. Med Care,* **30**(6): 473–83.

Werker, C.L., Nijhof, S.L., and van de Putte, E.M. (2013). Clinical practice: chronic fatigue syndrome. *Eur J Pediatr,* **172**(10): 1293–8.

Whistler, T., Unger, E.R., Nisenbaum, R., and Vernon, S.D. (2003). Integration of gene expression, clinical, and epidemiologic data to characterize chronic fatigue syndrome. *J Transl Med,* **1**(1): 10.

Yao, K., Crawford, J.R., Komaroff, A.L., *et al.* (2010). Review part 2: human herpesvirus-6 in central nervous system diseases. *J Med Virol,* **82**(10): 1669–78.

Zachrisson, O., Regland, B., Jahreskog, M., *et al.* (2002). Treatment with staphylococcus toxoid in fibromyalgia/chronic fatigue syndrome–a randomised controlled trial. *Eur J Pain,* **6**(6): 455–66.

Zuckerman, A.J. (2004). Safety of hepatitis B vaccines. *Travel Med Infect Dis,* **2**(2): 81–4.

37 Myositis and Vaccines

Ignasi Rodriguez-Pintó,[1] and Yehuda Shoenfeld[2,3]

[1]Department of Autoimmune Disease, Hospital Clínic de Barcelona, Barcelona, Spain
[2]Zabludowicz Center for Autoimmune Diseases, Sheba Medical Center, Tel Hashomer, Israel
[3]Incumbent of the Laura Schwarz-Kipp Chair for Research of Autoimmune Diseases, Sackler Faculty of Medicine, Tel Aviv University, Tel Aviv, Israel

Introduction

The word "myositis" comes from Latin and is composed of the root "myos," meaning muscle, and the suffix "itis," meaning inflammation. Thus, myositis refers to the inflammation of the muscles. Vaccination refers to the administration of a vaccine to induce immunity, although immunity can also be acquired by natural infections or vertical passive transfer.

Inflammation of the muscle can be elicited by several processes: direct traumatic injury, direct viral or bacterial infection, or idiopathic inflammatory myopathy (IIM). Additionally, a separate disease has recently been reported following vaccination: macrophagic myofasciitis (MMF) (Gherardi *et al.*, 1998), which has been attributed to the adjuvant contained in the vaccines (Israeli *et al.*, 2011). In this chapter, we refer only to IIM; MMF is discussed in Chapter 27.

IIM is a heterogeneous group of skeletal muscles diseases in which inflammation occurs without a recognized cause (Dalakas *et al.*, 2003). Although there are no uniform accepted criteria, it is widely agreed that IIM can be classified into five major subsets: dermatomyositis (DM), polymyositis (PM), inclusion body myositis (sIBM), nonspecific myositis (NSM), and immune-mediated necrotizing myopathy (IAM) (Hoogendijk *et al.*, 2004). In this chapter, we will explore the relationship between the three main subsets – DM, PM, and sIBM – and vaccination, as no reports have been found linking NSM or IAM with vaccination.

Epidemiologic studies in IIM are hampered by a lack of uniform diagnostic criteria, but the mean incidence rate for IIM has been found to be approximately 1.2×10^{-6} to 8.7×10^{-6} (Mastaglia and Phillips, 2002). IIM prevalence is around 1.1×10^{-6} cases (Ahlström *et al.*, 1993), although one study has shown it to be as high as 142.4×10^{-6} (Tan *et al.*, 2013).

IIM has a bimodal age of distribution that peaks in childhood and again at between 45 and 55 years. DM is the most common inflammatory myopathy, while PM is the least common. Sporadic sIBM is the most common above the age of 50, reaching 97% of cases in some series (Tan *et al.*, 2013). PM commonly occurs after the second decade of life, while DM has been noted to occur both in adults and in children, in the form of juvenile DM (Tan *et al.*, 2013). All studies show a female predominance, in contrast to the male predominance reported in IBM (Nakanishi *et al.*, 2013; Tan *et al.*, 2013).

Most authors believe that IIM results from an autoimmune process triggered by environmental factors in genetically predisposed individuals (Prieto and Grau, 2010). Several viruses and vaccines have been proposed as possible environmental triggers for the development of IIM (Altman *et al.*, 2008; Prieto and Grau, 2010; Stübgen, 2014).

Vaccines and Autoimmunity, First Edition. Edited by Yehuda Shoenfeld, Nancy Agmon-Levin, and Lucija Tomljenovic.

Pathogenesis of myositis

Although these entities share many similarities, their immunopathology seems to have several important differences.

DM begins with the activation of the complement and the formation of membrane attack complexes (MACs) (Kissel *et al.*, 1986) deposited on the cells, conforming the intimal layer of the endomysial capillaries. This capillaritis leads to atrophy of the perifasciular fibers and compensatory dilatation of remaining capillaries, due to hypoperfusion and ischemia (Dalakas, 2012).

In PM and sIBM, the fundamental process is of $CD8^+$ T cell-mediated cytotoxicity. Upon activation, cytotoxic $CD8^+$ T cells induce muscular cell necrosis following recognition of major histocompatibility complex (MHC)-I–antigen complex when they are co-stimulated by B7 family molecules. Additionally, the generation of cytokines by the muscle probably creates a proinflammatory environment, which leads to disease chronicity (Dalakas, 2013).

Degenerative features have been found in sIBM, consisting of intracellular vacuoles and variable accumulation of amyloid and amyloid variable-related molecules. It has been suggested that, after a long period, the continuous stimulation of inflammatory factors may induce a cell stress response that anticipates the accumulation of β-amyloid (Muth *et al.*, 2009).

Vaccination and loss of tolerance

The autoimmune origin of IIM is supported by the presence of inflammatory infiltrates in the biopsies, the evidence of complement-mediated cytotoxicity, the upregulated expression of MHC products, its association with other autoimmune disorders, and its response to immunosuppressors.

It is unclear what breaks the tolerance and drives the immune response to induce IIM (Dalakas, 2013). Infections and vaccines have been proposed as possible triggers (Leff *et al.*, 1992; Orbach and Tanay, 2009), but no virus has been found by PCR amplification of nucleotides present in the muscle samples of IIM patients (Leff *et al.*, 1992).

Myositis associated with vaccines

All three major subsets of IIM have been associated with vaccines. The first report of IIM following vaccination was published in 1964, when Bitum *et al.* (1964) reported a series of 13 cases of children with DM, one of whom developed it following smallpox vaccination. Since then, several cases have been reported in the literature, associating different vaccines with the development of IIM (Orbach and Tanay, 2009). Although case reports are no longer published by certain medical journals (Hessel, 2013), they continue to play an important role in calling attention to rare complications associated with drugs and in generating new hypotheses to stimulate further research (Hessel, 2013).

It was for this reason that the US Centers for Disease Control and Prevention (CDC) and Food and Drug Administration (FDA) created the Vaccine Adverse Event Reporting System (VAERS) in 1991, as a vaccine safety surveillance program (Chen *et al.*, 1994). It collects reports of cases that develop adverse events after the administration of vaccines licensed in the United States. It receives 30 000 reports annually, with more than 200 000 reports added since 1990. It is aimed at detecting new, unusual, or rare vaccine adverse events, monitoring increases in known adverse events, and identifying potential patient risk factors for particular adverse events. Despite the well known intrinsic underestimation bias of this type of passive surveillance (Strom, 1989; Moride *et al.*, 1997), 119 cases of IIM had been reported to the VAERS database up to June 2013 (Stübgen, 2014). Of these, 33 were classified as PM, 85 as DM, and only 1 as sIBM.

While Kurland *et al.* (1985) did not find an increase in the incidence of PM/DM in patients vaccinated against flu in 1976, a slight increase in the frequency of PM/DM may be difficult to detect via normal prospective epidemiological studies, since IIM is a rare disease. Some studies (Winkelmann *et al.*, 1968; Koch *et al.*, 1976; Lyon *et al.*, 1989) have thus tried to show the role of vaccines in the development of IIM, but their retrospective design could lead to information bias, hampering the finding of an association.

Dermatomyositis

DM is an inflammatory myopathy that is differentiated from the other IIMs mainly by its skin involvement (Mastaglia and Phillips, 2002). The major clinical manifestations of DM are muscle weakness and skin rash.

Muscle weakness can affect all four limbs and sometimes the diaphragm, leading to quadriparesis and respiratory failure. The classical skin manifestations of DM include heliotrope rash on the upper eyelids, the "V sign," and Gotrron papules (a skin dermatitis that affects the hand joints spreading the phalanges).

DM has been reported following almost every vaccine, but only a few studies have attempted to elucidate the possible relationship between DM and vaccination. Koch *et al.* (1976) conducted a case–control study of 42 cases of childhood DM and 42 controls in order to assess whether distinctive etiology-related patterns existed among DM patients. They found that only two patients had a time-related exposure to vaccines (both developed the disease 1 month following diphtheria, pertussis, and tetanus (dTP) vaccination). However, as a retrospective case–control study, Koch *et al.* (1976) relied primarily on the memories of families in obtaining data on vaccine exposure for both cases and controls; conclusions are hence limited by this possible bias.

An increasing number of vaccine time-related cases of DM have been reported in the literature and in post-licensure surveillance systems (Table 37.1).

Table 37.1 Dermatomyositis (DM) cases related to vaccines

Vaccine	No. of cases	Reference
Anthrax	1	VAERS database
Diphtheria toxoid		Ehrengut (1978)
Diphtheria, tetanus, and pertussis (dTP)	4	Ehrengut (1978), Thieffry *et al.* (1967), VAERS database
Hepatitis A virus (HAV)	3	VAERS database
Hepatitis B virus (HBV)	17	Altman *et al.* (2008), VAERS database
Herpes zoster (HZ) vaccine	1	VAERS database
Human papillomavirus vaccine (HPV4)	8	VAERS database
Influenza trivalent (seasonal vaccine)	13	Jani *et al.* (1994), VAERS database
Influenza (H1N1) monovalent	1	Ferri *et al.* (2012), Stübgen (2014), VAERS database
Measles, mumps, and rubella (MMR)	4	VAERS database
Meningococcal vaccine	3	VAERS database
Mycobacterium bovis (BCG)	2	Ehrengut (1978), Kåss *et al.* (1978)
Poliovirus vaccine (inactivated)	2	Stübgen (2014), VAERS database
Tetanus and diphtheria toxoid	2	VAERS database
Typhoid vaccine	1	VAERS database
Varicella vaccine	4	Bitum *et al.* (1964), Wharton *et al.* (2003)
Yellow fever vaccine	1	VAERS database

Polymyositis

PM is one of the major forms of IIM, but, as a standalone clinical entity, it is an uncommon and frequent misdiagnosed disorder (Dalakas *et al.*, 2003). It is characterized by its subacute evolution over weeks to months. PM mainly affects adults and only rarely children. It presents with weakness that starts at the proximal muscles, affecting the distal ones at late states. It normally preserves facial muscles and does not involve the skin. There are no diagnostic instrumental or biological tests nor pathological features on which to base the diagnosis of PM. As an exclusion diagnosis, it has been very difficult to perform studies with a homogeneous PM sample (Mastaglia and Phillips, 2002).

Some reports have associated PM with previous immunization, and especially with with hepatitis B vaccine (Table 37.2). However, no study conducted has excluded DM patients. Thus, the association between vaccine and PM has not been properly studied and this issue requires more research.

Inclusion body myositis

sIBM is an IIM characterized by slow progression, occurrence in the elderly, male predominance, and resistance to steroid therapy. Accepted clinical diagnostic criteria affirm that sIBM affects mainly men above 45 years old, although it can occur before this age (Griggs *et al.*, 1995; Rose, 2013).

Clinical weakness affects all four limbs, predominantly the distal muscles (Rose, 2013). sIBM was recently introduced as a separate group of IIM and especially as different from PM. Thus, few studies have been carried out to assess its etiology. However, a man between 44 and 65 years old who developed sIBM after receiving HBV vaccine has been reported to VAERS (Stübgen, 2014).

Vaccines associated with myositis

Live-attenuated

Measles, mumps, and rubella vaccine

Measles, mumps, and rubella are serious viral disease that can lead to severe disabilities and even to fatal outcome. They are particular prevalent in low-income countries (CDC, 2005). However, measles is considered eliminated in the United States (Strebel *et al.*, 2004). Combined

Table 37.2 Polymiositis (PM) cases related to vaccines

Vaccine	No. of cases	References
Anthrax	1	VAERS database
Diphtheria, tetanus, and pertussis (dPT)	3	VAERS database
Hepatitis A virus (HAV)	2	VAERS database
Hepatitis B virus (HBV)	1	Ramírez-Rivera *et al.* (2003), VAERS database
Human papillomavirus vaccine (HPV4)	3	VAERS database
Influenza trivalent (seasonal)	13	VAERS database
Influenza (H1N1) monovalent		Ferri *et al.* (2012)
Lyme vaccine	1	VAERS database
Meningococcal vaccine	3	VAERS database
Mumps vaccine		
Pneumococcal vaccine	1	VAERS database
Poliovirus vaccine (inactivated)	1	VAERS database
Tetanus toxoid	3	VAERS database
Typhoid vaccine	1	VAERS database
Yellow fever vaccine	1	VAERS database

live-attenuated measles, mumps, and rubella (MMR) vaccine was introduced in the United States in 1978 (Schwarz *et al.*, 1975; van Panhuis *et al.*, 2013). It is recommended for preschool-aged children over 12 months and adults (McLean *et al.*, 2013). MMR vaccine has been shown to be useful in preventing clinical measles and mumps (Demicheli *et al.*, 2012), but no studies have been carried out to address its effectiveness in preventing rubella (Demicheli *et al.*, 2012). Although MMR vaccine is a safe vaccine, it is well known that it is linked to an increased risk of aseptic meningitis (Dourado *et al.*, 2000), febrile convulsions (Ward *et al.*, 2007), and idiopathic thrombocytopenic purpura (Black *et al.*, 2003).

Although epidemiological studies have not addressed the risk of IIM after MMR vaccination, in a retrospective study assessing the environmental risk factors for different IIM phenotypes, Rider *et al.* (2010) found that those patients who presented as a recurrent IIM disease were more likely to have a previous history of vaccination against MMR than those who had a monocyclical disease. Hanissian *et al.* (1973) reported on a 6-year-old boy who developed a generalized muscular atrophy and weakness 2 days after receiving rubella virus vaccine. The biopsy found acute inflammatory perivascular infiltrates, without vessel wall necrosis. Additionally, other cases of DM have been reported to the VAERS system since 1990.

Vaccinia (smallpox) vaccine

Vaccinia was eradicated in 1979, through the efforts of the World Health Organization (WHO), and its vaccination is no longer recommended for the general population (Wehrle, 1980). However, vaccinia vaccine is still recommended for those working with orthopoxvirus and those who might have to take care of possible terrorism smallpox cases (Wharton *et al.*, 2003). Smallpox vaccine is recognized to be related to encephalitis, myopericarditis, eczema vaccinatum, and progressive vaccinia (Cono *et al.*, 2003; Halsell *et al.*, 2003). DM was reported in a 5-year-old boy who had received the smallpox vaccine (Bitum *et al.*, 1964; Ehrengut, 1978). This patient developed fever, skin manifestations, myositis, and calcinosis. He continued handicapped 10 years later.

Inactivated/killed

Poliomyelitis (IPV)

Although only a few cases of poliomyelitis have been reported worldwide in the last few years, mainly in African countries, inactivated poliovirus vaccine is routinely recommended in childhood in the developed world (Prevots *et al.*, 2000; CDC, 2009). There are no studies addressing the incidence of IIM in poliomyelitis-vaccinated patients (Stratton *et al.*, 1994). However, up to December 2013, several cases of DM had been reported following the administration of polio vaccine (Thieffry *et al.*, 1967; Ehrengut, 1978), and a case in the VAERS database developed PM.

Toxoid

Diphtheria and tetanus (part of DTaP combined immunization)

Diphtheria component is a purified preparation of inactivated diphtheria toxin. Single-antigen diphtheria toxoid is not distributed: it is always administered in combination with tetanus toxoid (TTd) or with tetanus and pertussis vaccine. Tetanus vaccine contains an exotoxin produced by *Clostridium tetani*, while pertussis vaccine is commercially combined with diphtheria and TTd. Diphtheria, tetanus, and acellular pertussis (DTaP) vaccine was initially licensed in the United States by the FDA in 1991 (CDC, 1991). It is now recommended for children from 15 months of age up to their 7th birthday (CDC, 1997). Immunization against tetanus continues to be recommended for all residents in many countries (CDC, 2011).

Epidemiological safety studies have not reported IIM as a side effect of DT vaccine (Beytout *et al.*, 2009). However, Ehrengut (1978) reported two cases who developed a DM shortly after receiving a vaccine against diphtheria, while another case, cited earlier, followed dTP-polio vaccination (Thieffry *et al.*, 1967). Cotteril and Shapiro (1978) reported a case of an 11-year-old girl who developed DM after receiving a diphtheria–tetanus booster, tuberculosis, and cholera vaccine. She had the classical clinical features of DM, with eyelid edema and heliotrope sign. In their report, the authors proposed that the dT booster could have "primed" her immunologically, and that subsequent immunization with tuberculosis/cholera and polio booster triggered the DM.

Cold vaccine

Ehrengut (1978) reported a case of a 47-year-old patient who developed DM after taking what was called "cold vaccine." The cold vaccine was a mixture of killed bacteria (mainly *Pneumococcus* sp., *Streptococcus* sp., *Haemophilus* sp., and *Micrococcus* sp.) and was prescribed as a prophylaxis for "common cold." This vaccine is no longer used.

Subunit/conjugate

Hepatitis B

Hepatitis B vaccine consists of purified inactivate HBAg particles obtained either from yeast, through recombinant DNA technology, or from the plasma of chronic carriers. The incidence of hepatitis B declined by 78% from 1990 to 2005 (Report *et al.*, 2006). Hepatitis B vaccination has been related to multiple sclerosis (MS), Guillain–Barré syndrome (GBS), lupus, and rheumatoid arthritis (RA), among other autoimmune disease (Khamaisi *et al.*, 2004; Agmon-Levin *et al.*, 2009). Maillefert *et al.* (1999) sent questionnaires to nine rheumatology departments in order to obtain an overview of rheumatic disorders occurring within 2 months after hepatitis B vaccination. They reported 22 cases, but none included IIM.

However, Altman *et al.* (2008) reported a case of a 6-year-old child who developed juvenile DM 1 week after being vaccinated for hepatitis B virus (HBV). She presented a classic skin rash but had no Gottron's sign and no muscular weakness. Her muscle biopsy showed perimysial inflammatory changes involving the vascular structures.

Another case of IIM following hepatitis B vaccination was described by Ramírez-Rivera *et al.* (2003), and several more have been reported directly to the VAERS surveillance system.

Influenza (injection) and H1N1

Seasonal influenza vaccines are trivalent, containing a mixture of influenza A and B strains. A quadrivalent formulation is expected to replace the trivalent one in 2014, which will include a new B strain (Report, 2013). Additionally, in 2009, a monovalent vaccine was produced against H1N1 pandemic strain (FDA, 2009). Influenza vaccine is recommended for all persons above 6 months of age (Report, 2013).

Seasonal influenza vaccine is regarded as safe, although there are some reports of an increase in the risk of GBS associated with it (Tokars *et al.*, 2012). Guissa *et al.* (2012) conducted a trial to assess the efficacy and safety of influenza A H1N1 vaccine in 31 cases with juvenile DM and 81 controls, but, probably due to the lack of power of this study, they were not able to find evidence for IIM flare, whether assessed by fever, myalgia, or CK levels. Kurland *et al.* (1985) did not show an association between 1976 seasonal influenza vaccine and DM; neither did Chazan *et al.* (2002) when they looked at the incidence of DM in those taking myotoxic drugs. However, the development of IIM has been reported in an increasing number of cases following influenza virus vaccine (Jani *et al.*, 1994; Plotkin *et al.*, 2000; Ferri *et al.*, 2012). Plotkin *et al.* (2000) reported on a 68-year-old man who developed an acute rhabdomyolisis triggered by influenza vaccination while taking statins for a long period. Jani *et al.* (1994) reported a case of a 68-year-old woman who developed a myositis with a typical heliotrope hyperpigmentation over the eyelids 2 weeks after vaccination, starting at her left arm, where the vaccine was administered. Ferri *et al.* (2012) reported three patients who developed an IIM between 5 days and 1 month after receiving the pandemic A H1N1 and the seasonal trivalent influenza vaccine. Additionally, some other cases have been reported to the VAERS system.

Human papillomavirus vaccine

Although human papillomavirus (HPV) is a necessary factor for the development of cervical cancer, it is not sufficient. Indeed, most women infected with HPV will not develop cervical cancer in their lifetime, as 70% of HPV infections will resolve within a year, and as many as 90% will resolve within 2 years, without any medical intervention (Markowitz *et al.*, 2007). Of those HPV infections that do not resolve, only a small proportion will progress to cancer, over a period of several decades (Ostör, 1993; Markowitz *et al.*, 2007). Other cofactors are necessary for promotion of cervical cancer (Castellsagué and Muñoz, 2003; Castle, 2004). These appear to act through various immune-suppressing mechanisms, thus increasing the likelihood of HPV infection becoming persistent, which is a major prerequisite for cervical cancer progression (Castle, 2004).

In the United States, the quadrivalent HPV vaccine against four HPV types (6, 11, 16, and 18) was licensed in 2006. It is recommended for females aged 11 or 12 years old and for females aged 13–26 years not previously vaccinated (Markowitz *et al.*, 2007). Although this vaccine was licensed only 8 years ago, eight associated cases of DM have already been reported to the VAERS surveillance system.

Bacillus Calmette–Guérin

Bacillus Calmette–Guérin (BCG) contains living Calmatte–Guérin baccillus, an attenuated strain of *Mycobacterium bovis*. It is no longer recommended in most developed countries. Nevertheless, although its protective effect is not clear (Rodrigues *et al.*, 1993), it continues to be widely used in several countries for the prevention of disseminated tuberculosis infection (CDC, 1996).

Two cases of DM following BCG vaccine administration have been reported (Kåss *et al.*, 1978). The first was a 14-year-old boy who was vaccinated twice with BCG; 6 weeks after the second shot, he developed typical DM, with classical skin involvement and weakness. The second case, a 12-year-old boy, developed DM 2–3 weeks after receiving the BCG vaccine; his weakness spread from the limb where the vaccine was administered to the other limbs, while at the same time the typical rash appeared. Additionally, Manganelli *et al.* (2002) reported on a patient who developed focal myositis of the gastrocnemius muscles after BCG vaccination.

Other vaccines

As assessed by the VAERS, other cases of either DM or PM have been reported related to other vaccines, including meningococcal, pneumococcal, anthrax, yellow fever, typhoid fever, Lyme, and hepatitis A virus (HAV) vaccines. Whether all these cases were caused by these vaccines is not known, but all of them were time-related by the physician who administered the vaccine.

Conclusions

Several authors (Orbach and Tanay, 2009; Stübgen, 2014) have reviewed the association between vaccine administration and the development of an inflammatory myopathy, but there are few well designed studies that have directly addressed this issue. Studies performed to date lack power in some cases and have been unable to find a conclusive association between vaccination and IIM. It is not possible to exclude a relationship between vaccination and IIM, however, and vaccines probably do cause IIM in genetically predisposed individuals.

References

Agmon-Levin, N., Zafrir, Y., Paz, Z., *et al.* (2009). Ten cases of systemic lupus erythematosus related to hepatitis B vaccine. *Lupus*, **18**: 1192–7.

Ahlström, G., Gunnarsson, L.G., Leissner, P., and Sjödén, P.O. (1993). Epidemiology of neuromuscular diseases, including the postpolio sequelae, in a Swedish county. *Neuroepidemiology*, **12**: 262–9.

Altman, A., Szyper-Kravitz, M., and Shoenfeld, Y. (2008). HBV vaccine and dermatomyositis: is there an association? *Rheumatol Int*, **28**: 609–12.

Beytout, J., Launay, O., Guiso, N., *et al.* (2009). Safety of Tdap-IPV given one month after Td-IPV booster in healthy young adults: a placebo-controlled trial. *Hum Vaccin*, **5**: 315–21.

Bitum, S., Daeschner, C.W., Travis, L.B., *et al.* (1964). Dermatomyositis. *J Pediatr*, **64**: 101–31.

Black, C., Kaye, J.A., and Jick, H. (2003). MMR vaccine and idiopathic thrombocytopaenic purpura. *Br J Clin Pharmacol*, **55**: 107–11.

Castellsagué, X. and Muñoz, N. (2003). Chapter 3: Cofactors in human papillomavirus carcinogenesis–role of parity, oral contraceptives, and tobacco smoking. *J Natl Cancer Inst Monogr*, **31**, 20–28.

Castle, P.E. (2004). Beyond human papillomavirus: the cervix, exogenous secondary factors, and the development of cervical precancer and cancer. *J Low Genit Tract Dis*, **8**: 224–30.

CDC (Centers for Disease Control and Prevention). (1991). Food and Drug Administration approval of use of diphteria and tetanus toxoids and acellular pertussis vaccine. *MMWR*, **40**: 881–2.

CDC (Centers for Disease Control and Prevention). (1996). The role of BCG vaccine in the prevention and control of tuberculosis in the United States. *A joint statement by the Advisory Council for the Elimination of Tuberculosis and the Advisory Committee on Immunization Practices. MMWR*, **45**: 1–18.

CDC (Centers for Disease Control and Prevention). (1997). Pertussis vaccination: use of acellular pertussis vaccines among infants and young children. Recommendations of the Advisory Committee on Immunization Practices (ACIP). *MMWR*, **46**: 1–25.

CDC (Centers for Disease Control and Prevention). (2005). Global Measles and Rubella Laboratory Network, January 2004–June 2005. *MMWR*, **54**: 1100–4.

CDC (Centers for Disease Control and Prevention). (2009). Updated recommendations of the Advisory Committee on Immunization Practices (ACIP) regarding routine poliovirus vaccination. *MMWR*, **58**: 829–30.

CDC (Centers for Disease Control and Prevention). (2011). Updated recommendations for use of tetanus toxoid, reduced diphtheria toxoid and acellular pertussis (Tdap) vaccine from the Advisory Committee on Immunization Practices, 2010. *MMWR*, **60**: 13–15.

Chazan, B., Weiss, R., Tabenkin, H., *et al.* (2002). Influenza vaccine does not produce myopathy in patients taking statins. *J Fam Pract*, **51**: 986–8.

Chen, R.T., Rastogi, S.C., Mullen, J.R., *et al.* (1994). The Vaccine Adverse Event Reporting System (VAERS). *Vaccine*, **12**: 542–50.

Cono, J., Casey, C.G., and Bell, D.M. (2003). Smallpox vaccination and adverse reactions. *Guidance for clinicians. MMWR*, **52**: 1–28.

Cotteril, J.A. and Shapiro, H. (1978). Dermatomyositis after immunisation. *Lancet*, **2**: 1158–9.

Dalakas, M.C. (2012). Pathogenesis and therapies of immune-mediated myopathies. *Autoimmun Rev*, **11**: 203–6.

Dalakas, M.C. (2013). Pathophysiology of autoimmune polyneuropathies. *Presse Med*, **42**: e181–92.

Dalakas, M.C., Hohlfeld, R., and Marinos, C. (2003). Polymyositis and dermatomyositis. *Lancet*, **362**: 971–82.

Demicheli, V., Rivetti, A., Debalini, M.G., and Di Pietrantonj, C. (2012). Vaccines for measles, mumps and rubella in children. *Cochrane Database Syst Rev*, **2**: CD004407.

Dourado, I., Cunha, S., Teixeira, M.G., *et al.* (2000). Outbreak of aseptic meningitis associated with mass vaccination with a urabe-containing measles-mumps-rubella vaccine: implications for immunization programs. *Am J Epidemiol*, **151**: 524–30.

Ehrengut, W. (1978). Dermatomyositis and vaccination. *Lancet*, **1**: 1040–1.

FDA (Food and Drug Administration). (2009). FDA approves vaccines for 2009 H1N1 influenza virus. Available from: http://www.fda.gov/newsevents/newsroom/pressannouncements/ucm182399.htm [last accessed 11 December 2014].

Ferri, C., Colaci, M., Manzini, C.U., *et al.* (2012). Polymyositis following pandemic influenza A (H1N1) and 2009–10 seasonal trivalent vaccines. *Case Rep Rheumatol*, **2012**: 836930.

Gherardi, R., Coquet, M., Cherin, P., *et al.* (1998). Macrophagic myofasciitis: an emerging entity. *Lancet*, **352**: 347–52.

Griggs, R.C., Askanas, V., DiMauro, S., *et al.* (1995). Inclusion body myositis and myopathies. *Ann Neurol*, **38**: 705–13.

Guissa, V.R., Pereira, R.M.R., Sallum, A.M.E., *et al.* (2012). Influenza A H1N1/2009 vaccine in juvenile dermatomyositis: reduced immunogenicity in patients under immunosuppressive therapy. *Clin Exp Rheumatol*, **30**: 583–8.

Hanissian, A.S., Martinez, A.J., Jabbour, J.T., and Duenas, D.A. (1973). Vasculitis and myositis secondary to rubella vaccination. *Arch Neurol*, **28**(3): 202–4.

Halsell, J.S., Riddle, J.R., Atwood, J.E., *et al.* (2003). Myopericarditis following smallpox vaccination among vaccinia-naive US military personnel. *JAMA*, **289**: 3283–9.

Hessel, E.A. (2013). Why we should continue to publish case reports. *J Cardiothorac Vasc Anesth*, **27**: 825–7.

Hoogendijk, J.E., Amato, A.A., Lecky, B.R., *et al.* (2004). 119th ENMC international workshop: trial design in adult idiopathic inflammatory myopathies, with the exception of inclusion body myositis, 10–12 October 2003, Naarden, The Netherlands. *Neuromuscul Disord*, **14**: 337–45.

Israeli, E., Agmon-Levin, N., Blank, M., and Shoenfeld, Y. (2011). Macrophagic myofaciitis a vaccine (alum) autoimmune-related disease. *Clin Rev Allergy Immunol*, **41**: 163–8.

Jani, F.M., Gray, J.P., and Lanham, J. (1994). Influenza vaccine and dermatomyositis. *Vaccine*, **12**: 1484.

Kåss, E., Straume, S., and Munthe, E. (1978). Dermatomyositis after B.C.G. vaccination. *Lancet*, **1**: 772.

Khamaisi, M., Shoenfeld, Y., and Orbach, H. (2004). Guillain-Barré syndrome following hepatitis B vaccination. *Clin Exp Rheumatol*, **22**: 767–70.

Kissel, J.T., Mendell, J.R., and Rammohan, K.W. (1986). Microvascular deposition of complement membrane attack complex in dermatomyositis. *N Engl J Med*, **314**: 329–34.

Koch, M.J., Brody, J.A., and Gillespie, M.M. (1976). Childhood polymyositis: a case–control study. *Am J Epidemiol*, **104**: 627–31.

Kurland, E.M., Kurland, L.T., Mulder, D.W., *et al.* (1985). Lack of association of A/NJ/76 (swine flu) vaccine and polymyositis. *Neuroepidemiology*, **4**: 125–37.

Leff, R.L., Love, L.A., Miller, F.W., *et al.* (1992). Viruses in idiopathic inflammatory myopathies: absence of candidate viral genomes in muscle. *Lancet*, **339**: 1192–5.

Lyon, M.G., Bloch, D.A., Hollak, B., and Fries, J.F. (1989). Predisposing factors in polymyositis-dermatomyositis: results of a nationwide survey. *J Rheumatol*, **16**: 1218–24.

Maillefert, J.F., Sibilia, J., Toussirot, E., *et al.* (1999). Rheumatic disorders developed after hepatitis B vaccination. *Rheumatology (Oxford)*, **38**: 978–83.

Manganelli, S., De Stefano, R., Malandrini, A., *et al.* (2002). Bilateral recurrent focal myositis of gastrocnemius muscles after BCG vaccination. *Rheumatology (Oxford)*, **41**: 1074–6.

Markowitz, L.E., Dunne, E.F., Saraiya, M., *et al.* (2007). Quadrivalent human papillomavirus vaccine: recommendations of the Advisory Committee on Immunization Practices (ACIP). *MMWR*, **56**: 1–24.

Mastaglia, F.L. and Phillips, B.A. (2002). Idiopathic inflammatory myopathies: epidemiology, classification, and diagnostic criteria. *Rheum Dis Clin North Am*, **28**: 723–41.

McLean, H.Q., Fiebelkorn, A.P., Temte, J.L., and Wallace, G.S. (2013). Prevention of measles, rubella, congenital rubella syndrome, and mumps, 2013: summary recommendations of the Advisory Committee on Immunization Practices (ACIP). *MMWR*, **62**: 1–34.

Moride, Y., Haramburu, F., Requejo, A.A., and Bégaud, B. (1997). Under-reporting of adverse drug reactions in general practice. *Br J Clin Pharmacol*, **43**: 177–81.

Muth, I.E., Barthel, K., Bähr, M., *et al.* (2009). Proinflammatory cell stress in sporadic inclusion body myositis muscle: overexpression of alphaB-crystallin is associated with amyloid precursor protein and accumulation of beta-amyloid. *J Neurol Neurosurg Psychiatry*, **80**(12): 1344–9.

Nakanishi, H., Koike, H., Matsuo, K., *et al.* (2013). Demographic features of Japanese patients with sporadic inclusion body myositis: a single-center referral experience. *Intern Med*, **52**: 333–7.

Orbach, H. and Tanay, A. (2009). Vaccines as a trigger for myopathies. *Lupus*, **18**: 1213–16.

Ostör, A.G. (1993). Natural history of cervical intraepithelial neoplasia: a critical review. *Int J Gynecol Pathol*, **12**: 186–92.

Plotkin, E., Bernheim, J., Ben-Chetrit, S., *et al.* (2000). Influenza vaccine – a possible trigger of rhabdomyolysis induced acute renal failure due to the combined use of cerivastatin and bezafibrate. *Nephrol Dial Transplant*, **15**: 740–1.

Prevots, D.R., Burr, R.K., Sutter, R.W., and Murphy, T.V. (2000). Poliomyelitis prevention in the United States. Updated recommendations of the Advisory Committee on Immunization Practices (ACIP). *MMWR*, **49**: 1–22; quiz CE1–7.

Prieto, S. and Grau, J.M. (2010). The geoepidemiology of autoimmune muscle disease. *Autoimmun Rev*, **9**: A330–4.

Ramírez-Rivera, J., Vega-Cruz, AM., and Jaume-Anselmi, F. (2003). Polymyositis: rare complication of hepatitis B vaccination. An unusual cause of toxic shock syndrome. *Bol Asoc Med P R*, **95**: 13–16.

Report, M.W. (2013). Prevention and control of seasonal influenza with vaccines. Recommendations of the Advisory Committee on Immunization Practices – United States, 2013–2014. *MMWR*, **62**: 1–43.

Report, M.W., Mast, E.E., Weinbaum, C.M., *et al.* (2006). A comprehensive immunization strategy to eliminate transmission of hepatitis B virus infection in the United States: recommendations of the Advisory Committee on Immunization Practices (ACIP) Part II: immunization of adults. *MMWR*, **55**: 1–33; quiz CE1–4.

Rider, G.L, Wu, L., Mamyrova, G., *et al.* (2010). Environmental factors preceding illness onset differ in phenotypes of the juvenile idiopathic inflammatory myopathies. *Rheumatology (Oxford)*, **49**: 2381–90.

Rodrigues, L.C., Diwan, V.K., and Wheeler, J.G. (1993). Protective effect of BCG against tuberculous meningitis and miliary tuberculosis: a meta-analysis. *Int J Epidemiol*, **22**: 1154–8.

Rose, M.R. (2013). 188th ENMC International Workshop: Inclusion Body Myositis, 2–4 December 2011, Naarden, The Netherlands. *Neuromuscul Disord*, **23**(12): 1044–55.

Schwarz, A.J., Jackson, J.E., Ehrenkranz, N.J., *et al.* (1975). Clinical evaluation of a new measles-mumps-rubella trivalent vaccine. *Am J Dis Child*, **129**: 1408–12.

Stratton, K.R., Howe, C.J., and Johnston, R.B. (1994). Adverse events associated with childhood vaccines other than pertussis and rubella. *Summary of a report from the Institute of Medicine*. *JAMA*, **271**: 1602–5.

Strebel, P.M., Henao-Restrepo, A.-M., Hoekstra, E., *et al.* (2004). Global measles elimination efforts: the significance of measles elimination in the United States. *J Infect Dis*, **189**(Suppl.): S251–7.

Strom, B.L. (1989). Choosing among the available approaches to pharmacoepidemiologic studies. In: Strom, B.L. and Kimmel, S.E. (eds). *Textbook of Pharmacoepidemiology*. New York: Churchill Livingston, pp. 245–56.

Stübgen, J. (2014). A review on the association between inflammatory myopathies and vaccination. *Autoimmun Rev*, **13**: 31–9.

Tan, J.A., Roberts-Thomson, P.J., Blumbergs, P., *et al.* (2013). Incidence and prevalence of idiopathic inflammatory myopathies in South Australia: a 30-year epidemiologic study of histology-proven cases. *Int J Rheum Dis*, **16**: 331–8.

Thieffry, S., Arthuis, M., Martin, C., *et al.* (1967). [Infantile dermatomyositis. Study of 8 personal cases]. *Ann Pediatr (Paris)*, **14**: 554–78.

Tokars, J.I., Lewis, P., DeStefano, F., *et al.* (2012). The risk of Guillain-Barré syndrome associated with influenza A (H1N1) 2009 monovalent vaccine and 2009–2010 seasonal influenza vaccines: results from self-controlled analyses. *Pharmacoepidemiol Drug Saf*, **21**: 546–52.

van Panhuis, W.G., Grefenstette, J., Jung, S.Y., *et al.* (2013). Contagious diseases in the United States from 1888 to the present. *N Engl J Med*, **369**: 2152–8.

Ward, K.N., Bryant, N.J., Andrews, N.J., *et al.* (2007). Risk of serious neurologic disease after immunization of young children in Britain and Ireland. *Pediatrics*, **120**: 314–21.

Wehrle, P.F. (1980). A reality in our time–certification of the global eradication of smallpox. *J Infect Dis*, **142**: 636–8.

Wharton, M., Strikas, R.A., Harpaz, R., *et al.* (2003). Recommendations for using smallpox vaccine in a pre-event vaccination program. Supplemental recommendations of the Advisory Committee on Immunization Practices (ACIP) and the Healthcare Infection Control Practices Advisory Committee (HICPAC). *MMWR*, **52**: 1–16.

Winkelmann, R.K., Mulder, D.W., Lambert, E.H., *et al.* (1968). Course of dermatomyositis-polymyositis: comparison of untreated and cortisone-treated patients. *Mayo Clin Proc*, **43**: 545–56.

Index

Figures are indicated by italic page numbers, Tables by bold numbers.

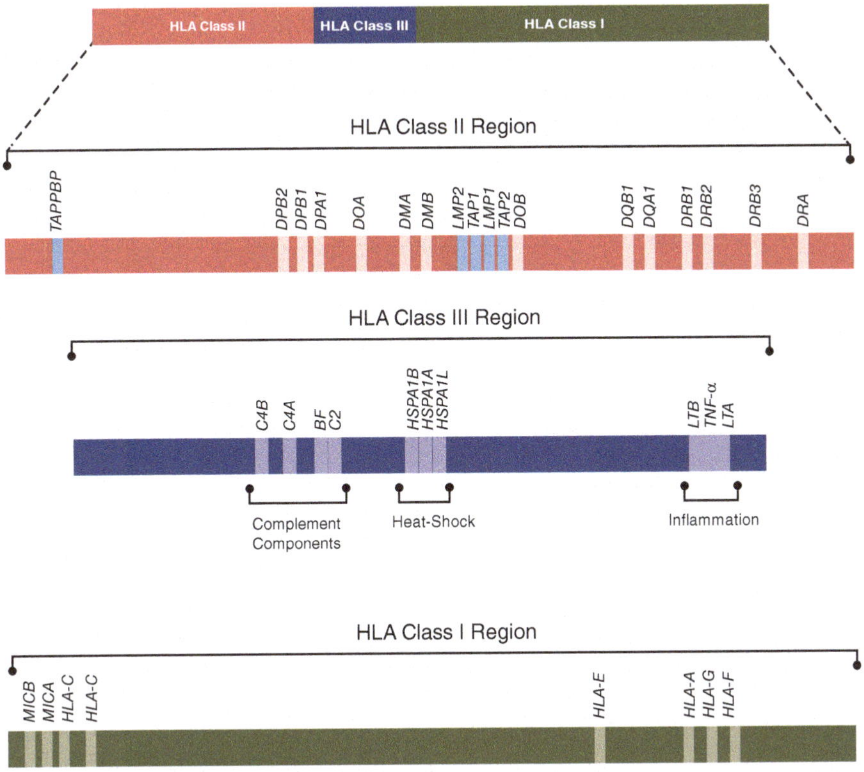

Figure 6.1 Map of the human HLA. The region is conventionally divided into three subregions: the class I, II, and III regions. Each contains numerous genes – only a few of the most relevant are shown here. Abbreviations: *TAPBP*, Tapasin; *LMP1* and *LMP2*, large multifunctional proteases 1 and 2; *TAP1* and *TAP2*, transporter associated with antigen processing 1 and 2; *C2*, *C4A*, and *C4B*, complement components 2, 4A, and 4B; *BF*, complement factor B; *HSPA1A* and *HSPA1B*, heat-shock protein 1A A-type and B-type; *HSPA1L*, heat-shock protein 1A-like; *LTA* and *LTB*, lymphotoxins A and B; *TNFA*, tumor necrosis factor α; and *MICA* and *MICB*, major histocompatibility complex class I chain genes A and B.

Vaccines and Autoimmunity, First Edition. Edited by Yehuda Shoenfeld, Nancy Agmon-Levin, and Lucija Tomljenovic.

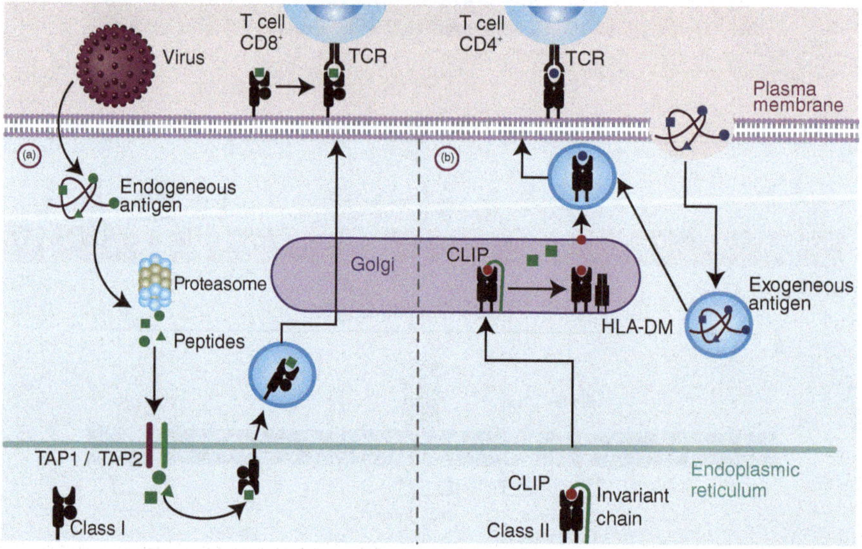

Figure 6.2 Antigen processing by HLA class I and II molecules. (a) Class I antigen processing and presentation occurs when proteins in the cytosol are degraded by the proteosome into small peptides and then are transported by transporter associated with antigen processing (TAP) into the endoplasmic reticulum (ER) lumen. HLA class I molecules are synthesized, translocated, and assembled into the lumen of the ER, where they load the peptide; HLA class I–peptide complexes then leave the ER and move through the Golgi apparatus to the plasma membrane, where they present the joined peptide to the T cell receptor (TCR) of CD8+ T cells. (b) Class II presentation occurs when extracellular proteins are phagocytized and then degraded into small peptides. These peptides are then sorted into vesicles, where they interact with the HLA class II molecules. HLA class II α and β chains, class II-associated invariant peptide (CLIP), and the invariant chain (Ii) molecules are located and assembled in the lumen of the ER, where they cannot bind peptides because the complex occupies the peptide-binding site. Heterotrimers leave the ER and pass through the Golgi apparatus to fuse with vesicles. The Ii is degraded and, with the help of HLA-DM and HLA-DO, a peptide can be joined. Complexes of HLA class II and peptide are relocated to the plasma membrane, where they can be recognized by CD4+ T cells.

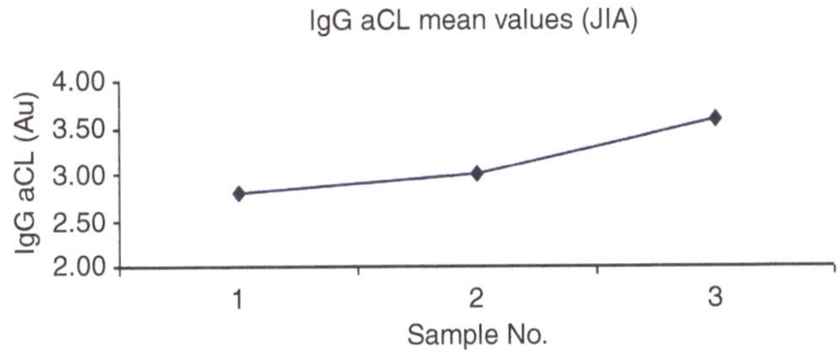

Figure 9.1 Mean values of IgG aCL before and 1 and 6 months after influenza vaccination in patients with juvenile idiopathic arthritis (JIA). The difference between the mean values of IgG aCL before and 6 months after the vaccination was not statistically significant ($p = 0.05$) (Author's unpublished data). aCL, anticardiolipin antibodies; AU, arbitrary units; 1, sample before vaccination; 2, sample 1 months after vaccination; 3, sample 6 months after vaccination.

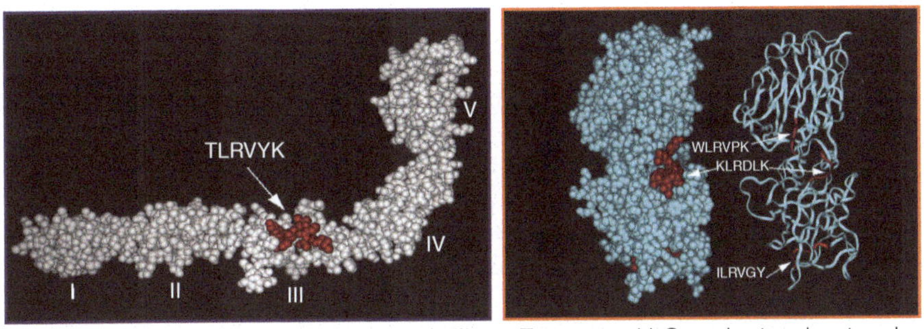

▸ β2GP1: Peptide TLRVYK is located on domain III

▸ Tetanus toxoid: One mimotope is external and 2 mimotopes are cryptic

Figure 15.1 Homology between β2GP1-related peptide and TTd.

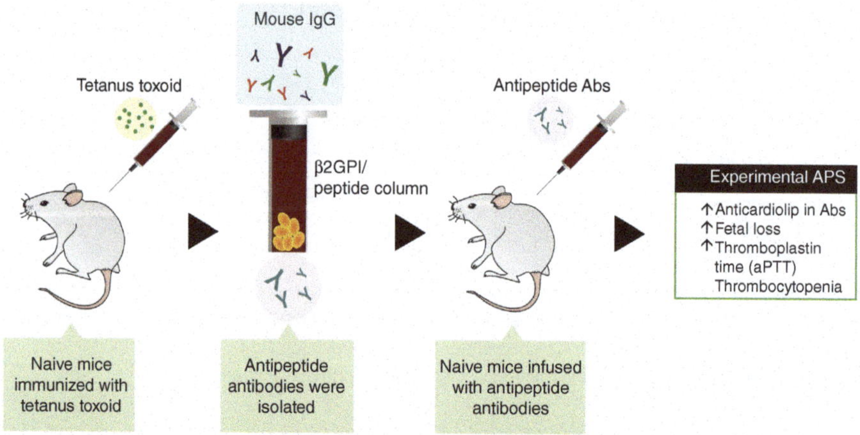

Figure 15.2 Active immunization of mice with TTd induces anti-β2GPI antibodies. The anti-β2GPI antibodies, when passively infused into another set of naive mice, induce experimental APS.

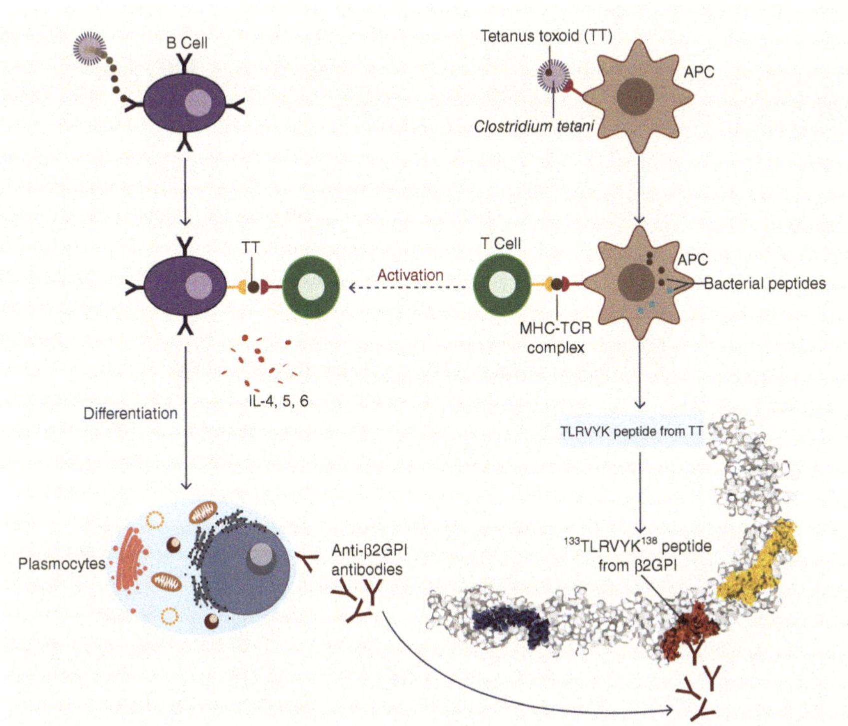

Figure 15.3 Molecular mimicry between β2GPI and TTd.

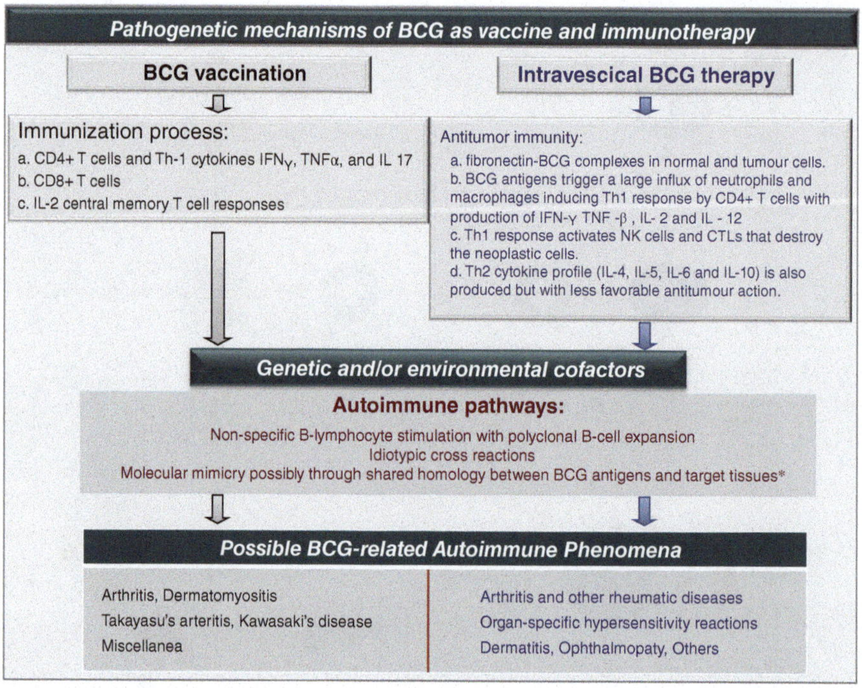

Figure 21.1 Summary of the main pathogenetic mechanisms correlated with BCG vaccination or intravescical BCG immunotherapy for bladder cancer, as well as the possible BCG-driven autoimmune disorders. *Crossreaction between the BCG heat-shock protein HSP65 and the cartilage proteoglycan link protein represents one example of a possible molecular mimicry pathogenetic mechanism.

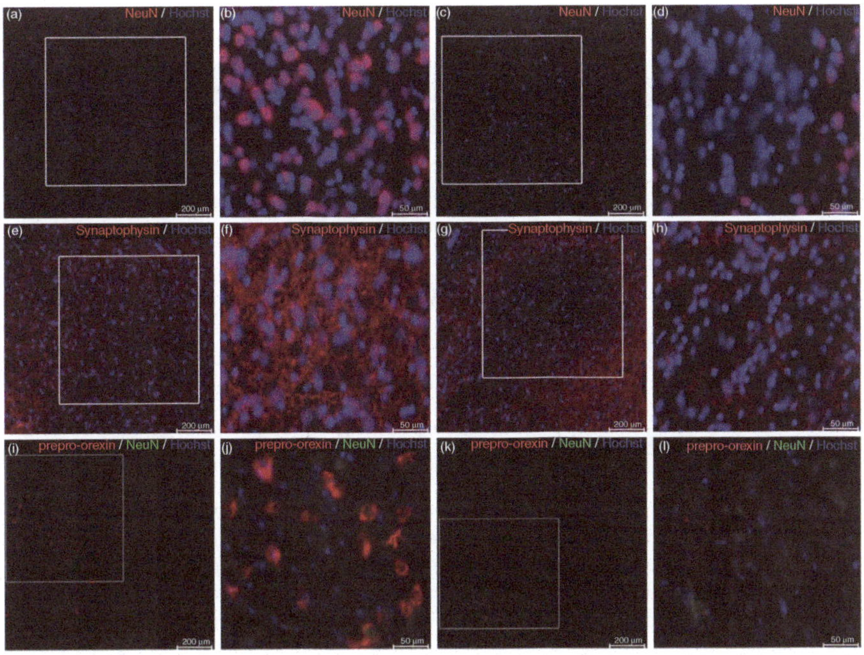

Figure 30.1 Histopathological changes induced by narcolepsy IgG. Coronal brain sections through the hypothalamus from mice injected ICV with IgG from narcolepsy patients and from healthy controls were stained for neuronal marker (NeuN) (a–d), synaptic marker (synaptophysin) (e–h), and orexin-expressing neurons (prepro-orexin) (i–l). (a,b,e,f,i,j) Representative images from control mice injected with control-IgG. (c,d,g,h,k,l) Representative images from mice injected with narcolepsy IgG. First- and third-column images are at 10× magnification and scale bar 200 µm. Second- and forth-column images are at 40× magnification and scale bar 50 µm. Reprinted from Katzav, A., Arango, M.T., Kivity, S., *et al.* (2013). Passive transfer of narcolepsy: anti-TRIB2 autoantibody positive patient IgG causes hypothalamic orexin neuron loss and sleep attacks in mice. *J Autoimmun*, **45**: 24–30. Copyright (2013), with permission from Elsevier.

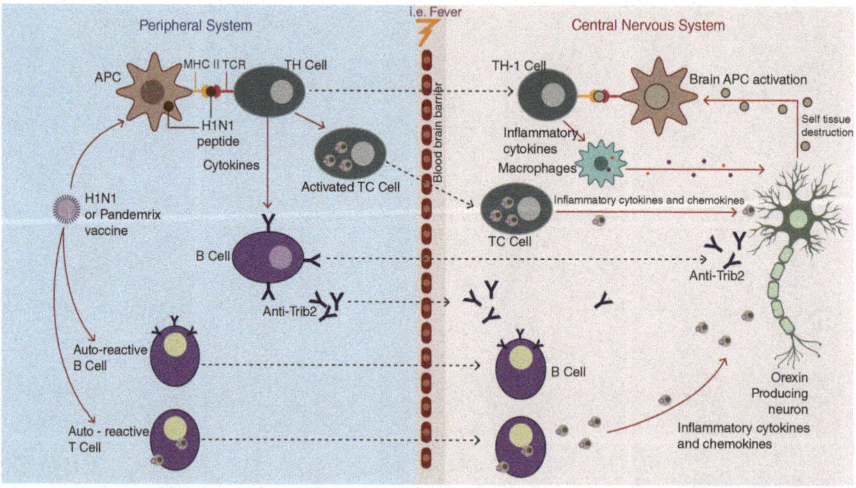

Figure 30.2 Possible pathway for H1N1 seasonal infection and Pandemrix vaccination in the onset of narcolepsy. The seasonal H1N1 influenza infection or Pandemrix vaccine could stimulate autoreactive T or B cells targeting orexin, producing neurons through the disruption of the BBB as a consequence of adverse vaccine events, such as fever, and by several other mechanisms. (i) Molecular mimicry of T cells. This describes the activation of crossreactive T cells that recognize the H1N1 epitope and then migrate to the CNS, where they recognize an antigen specific to orexin-producing neurons (crossreactivity). Activation of crossreactive T cells results in the release of cytokines and chemokines, which recruit and activate macrophages, mediating self-tissue damage. The subsequent release of orexin self-antigen and its uptake by antigen-presenting cells (APCs) perpetuates narcolepsy. (ii) Crosslink of the MHC and TCR molecules and activation of the cytotoxic T cells, which are autoreactive and specific towards orexin-producing neurons, by H1N1 antigens or Pandemrix vaccine. (iii) Molecular mimicry involving B cells and antibody-mediated disease. This could target TRIB2 as a crossreactive antigen. It would require signals from activated T cells (T cell help). (iv) Bystander activation of resting autoreactive B and T cells. This could result from general immune activation, independent of specific antigens. Current results in narcolepsy research point towards a T cell mechanism. Abbreviations: APC, antigen-presenting cell; BBB, blood–brain barrier; CNS, central nervous system; H1N1, H1N1 influenza A virus or epitopes from adjuvant vaccines; MHC, major histocompatibility complex; TCR, T cell receptor; TRIB2, Tribbles homolog 2. Reprinted and modified from Singh, A.K., Mahlios, J., and Mignot, E. (2013). Genetic association, seasonal infections and autoimmune basis of narcolepsy. *J Autoimmun*, **43**: 26–36. Copyright (2013) with permission from Elsevier.